International Conference on Design and Application of Engineering Materials (IC-DAEM 2018)

Emerging the Knowledge of Materials Design to Achieve Better Materials Integrity

IOP Conference Series: Materials Science and Engineering
Volume 547

Bandung, Indonesia
6 – 7 September 2018

ISBN: 978-1-5108-9382-5
ISSN: 1757-8981

Printed from e-media with permission by:

Curran Associates, Inc.
57 Morehouse Lane
Red Hook, NY 12571

Some format issues inherent in the e-media version may also appear in this print version.

Printed by Curran Associates, Inc. (2019)

For permission requests, please contact the Institute of Physics at the address below.

Institute of Physics
Dirac House, Temple Back
Bristol BS1 6BE UK

Phone: 44 1 17 929 7481
Fax: 44 1 17 920 0979

techtracking@iop.org

Additional copies of this publication are available from:

Curran Associates, Inc.
57 Morehouse Lane
Red Hook, NY 12571 USA
Phone: 845-758-0400
Fax: 845-758-2633
Email: curran@proceedings.com
Web: www.proceedings.com

Table of contents

Volume 547

International Conference on Design and Application of Engineering Materials (IC-DAEM)

6–7 September 2018, Bandung, Indonesia

Accepted papers received: 10 May 2019
Published online: 5 September 2019

Preface

Papers

Hydrogen Embrittlement

Biomedic

Welding

Surface Treatment

Corrosion

Refractory Material

Fabrication

Metal Casting

Characterization

Composite

Metal Solidification

Biomedic

Welding

Failure Analysis

Heat Treatment

Finite Element Analysis

Surface Treatment

Composite

Corrosion

Energy

Composite

Heat Treatment

Renewable material

Materials Degradation

Composite

Failure Analysis

Metal Alloy

Metal Casting

Surface Treatment

Composite

Energy

Metal Alloy

Composite

Corrosion

Materials Testing

Biomedic

Materials Testing

Composite

Renewable material

Heat Treatment

Composite

Charactarization

Heat Treatment

Failure Analysis

Editorial Preface on the Proceeding of International Conference on Design and Application of Engineering Materials (IC-DAEM).

International Conference on Design and Application of Engineering Materials 2018 (IC-DAEM) was the first international conference initiated by Materials Science and Engineering Research Group, Faculty of Mechanical and Aerospace Engineering, Institut Teknologi Bandung, Indonesia. The conference was held in Bandung Indonesia on 6th – 7th September 2018 in conjunction with annual national meeting entitled Seminar Nasional Metalurgi dan Material XI 2018 (SENAMM XI). The theme of the conference was *"Emerging the knowledge of materials design to achieve better materials integrity"* and was held as an honour for the retirement of Prof. Mardjono Siswosuwarno, which is considered as one of the pioneers in the field of failure analysis in Indonesia.

The topics of interest of IC-DAEM 2018 included but were not limited to: materials design, development of synthesis and characterization method, experimental and computational work related to this topic, application aspect of engineering materials including materials degradation and failure analysis. Four keynote lectures were delivered on the topics of *Production of Rare Earth Ortho-Ferrites from end-of-life Nd Magnets, Characterization of Calcium Phosphate Ceramics Synthesized by Heterogeneous Nucleation for Artificial Bone Substitutes, Development of Steel Matrix Composites Used for Metallic Die, and Effective Steps in Performing Failure Analysis*.

Papers presented in IC-DAEM 2018 are included in this proceeding after going through peer reviewing process. This proceeding contains papers with broad topics on design and application of engineering materials, such as (1) Inspection, Maintenance, Techniques and Managements, (2) Method and Case of Engineering Failure Analysis, (3) Time Dependent Properties and Degradation of Engineering Materials, (4) Computational Modelling and Simulation, (5) Development of Synthesis and Characterization Methods, (6) Laboratory and Full Scale-Scale Experiments, (7) Materials and Alloy Design, and (8) Materials Property Improvement. We hope that these papers will immensely benefit the research community working in Materials Science and Engineering, both in the academic and industrial fields.

IOP Conf. Series: Materials Science and Engineering **547** (2019) 011002 doi:10.1088/1757-899X/547/1/011002

Peer review statement

All papers published in this volume of *IOP Conference Series: Materials Science and Engineering* have been peer reviewed through processes administered by the proceedings Editors. Reviews were conducted by expert referees to the professional and scientific standards expected of a proceedings journal published by IOP Publishing.

Hydrogen Induced Cracking of API X52 and X60 Sour Service Steels Subjected to Pre-Strain under Prolonged H_2S Exposure

A.A. Korda*, T. Taufiq
Department of Metallurgical Engineering, Faculty of Mining and Petroleum Engineering, Bandung Institute of Technology

*E-mail: akhmad@mining.itb.ac.id

Abstract. The effect of plastic deformation on API X52 and X60 sour service steels on hydrogen induced cracking was studied. Plastic deformation was performed with a tensile pre-strain with the percentage variation of strain. The HIC test was carried out with a duration of 168 hours of hydrogen charging to investigate the cracking behavior. After hydrogen charging, the cross sections of both steels were observed to find out hydrogen cracks. Ultrasonic testing was carried out on the specimen surface to see the distribution of laminated cracks. The hydrogen crack surface was observed using SEM. The results showed that the values of CSR, CLR and CTR increased with increasing percentage of pre-strain. Ultrasonic testing showed that the area influenced by HIC increased with increasing percentage of pre-strain. The susceptibility of API X52SS and X60SS to HIC increases with increasing percent of pre-strain. API Steel X52SS has better HIC resistance than X60SS on high pre-strain.

Keywords: hydrogen induced cracking, sour service, pre-strain, X52, X60

1. Introduction

Hydrogen Induced Cracking (HIC) is one of the most frequent problems in oil and gas pipelines when exposed to sour service environments containing hydrogen sulphide (H_2S) [1,2]. Steels exposed to sour environments absorb hydrogen atoms on their surface through H_2S corrosion reaction. The mechanism of the HIC begins with the diffusion of hydrogen atoms into the metal lattice. The hydrogen atoms then combine again to form hydrogen molecules. When reacting with water, H_2S can cause blistering failure, hydrogen embrittlement, hydrogen-induced cracking, sulfide stress cracking and so on. Such failure may decrease the lifetime of the pipeline.

In general, HIC is caused by three factors i.e. corrosive environment, stress components, and material susceptibility. Material susceptibility is influenced by many factors, including microstructure, presence of inclusion, and chemical composition [3-6]. Corrosive environments containing water and H_2S on oil and gas production are defined as sour services. The stress can be a residual stress that can come from strain during pipe making. The thickness of the plate and the radius of the bending affect the magnitude of the strain received by the pipe during pipe manufacturing. In the smaller pipe diameter, the strain received will be greater [7-9].

There have been several studies that have been done to study the phenomenon of HIC on different types of alloys. For example Wan Keun Kim et al. [10] discusses the effect of metallurgical factors on HIC. Hitoshi Asahi et al. [11] describes the effect of the integration elements on HIC resistance. Mohtadi-Bonab et al. [12-16] has conducted extensive studies of HIC on steels. Several other studies were more concentrated on modification of microstructures and certain phases to improve HIC resistance. Until now, there are few references to discuss the effect of plastic deformation on HIC. Research on the effect of plastic strain on steel on HIC resistance is important because in pipeline manufacture, the steel plate undergoes plastic deformation. The strain generated by the plastic deformation can cause residual stress and may decrease its HIC resistance. The purpose of this study was to investigate the effect of plastic deformation on API X52 and X60 steels on HIC cracking under prolonged H_2S exposure environment.

2. Materials and Methods

The steel used in this study was steel plate according to API 5L X52 and X60 specifications. The X52 steel used is specifically designed for sour-service conditions. Unlike the X52, X60 steel is not intended for the sour-service environment even though its chemical composition is still within the range for the sour-service design. X52 and X65 steels used respectively have a thickness of 6.3 and 14 mm. Table 1 shows the chemical composition of the two steels used in this study. The test specimens were taken from the HRC plate at ½ wide position. The position is selected to be a test specimen because there is more segregation and more susceptible to hydrogen-induced cracking (HIC) so it can be observed for the purposes of this study. The plate is then prepared into a tensile test specimen to be stretched using a tensile test machine. Strain applied is 4%; 8%; and 12%. Fig. 1 shows the specimen dimension for tensile pre-strain. The pre-strain gauge length section was then taken for HIC testing.

Table 1. Chemical composition of the steels (wt%)

	C	Mn	P	S	N	Cu	Mo	Si	Ni
X52	0.039	0.673	0.0057	0.0017	0.037	0.024	0.003	0.241	0.167
X60	0.056	1.008	0.008	0.002	0.026	0.019	0.006	0.222	0.219
	Cr	V	Ti	Nb	Al	B	Co	As	Fe
X52	0.01	0.001	0.002	0.032	0.044	0.001	0.025	0.025	Bal.
X60	0.114	0.001	0.003	0.046	0.035	0.005	0.044	0.022	Bal.

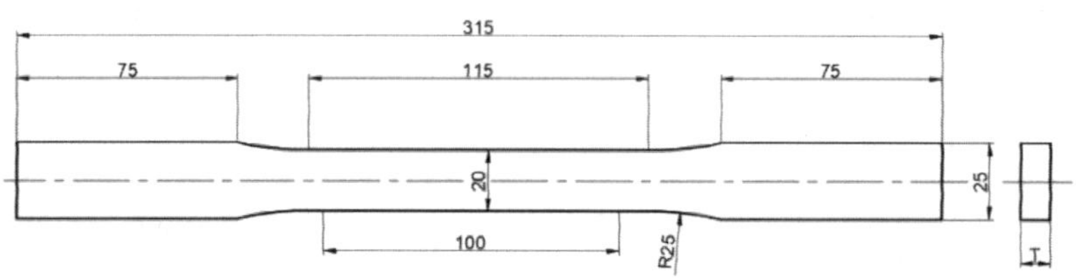

Fig. 1. Specimen dimension for tensile pre-strain (in mm). The pre-strain gauge length section was then taken for HIC testing.

IOP Conf. Series: Materials Science and Engineering **547** (2019) 012001 doi:10.1088/1757-899X/547/1/012001

Samples without strain are also prepared as a comparison. The specimens that have been pre-strained were observed for their microstructure. Hardness and tensile testing are also performed on pre-strained specimens. Meanwhile, other pre-strained samples were also prepared for hydrogen induced cracking testing. Tensile tests were carried out following the ASTM E-8 standard [17]. Hardness testing was performed using Vickers microhardness with a load of 300 grams with a 15-second indentation period. The steel plate hardness was tested for as-received material, after pre-strained, and after HIC testing. Measurements of hardness values were performed in the longitudinal direction and on the thickness of the plate.

Observations using optical microscope were also performed to see the presence of inclusions before HIC testing and cracks that emerged after HIC testing. The specimens were transversally cut. The observations were performed on the short-transverse side to see the distribution of cracks and cross sections of the cracks. The length and thickness of the crack is then measured using a scale on an optical microscope. The observed cracks were then calculated to the values of the crack ratio.

Ultrasonic testing (UT) is performed before and after HIC testing. The purpose of UT testing prior to HIC is to identify the cleanliness of the steel. If a laminate or other defect is found, the plates in that section will not be used for HIC testing. After HIC testing, UT is reestablished to determine the crack distribution area inside the steel plate. The crack may be a laminate inside the test specimen. The crack area was then reconstructed to show the severity of the hydrogen damage to the steel plate.

Fig. 2 shows schematic diagram of HIC test. The test generally follows the NACE TM0284-2003 standard [18]. However, the standard followed was for manual preparation of test solutions, preparation of test specimens, and computation of crack comparisons. The duration of the test is set to exceed the standard duration of 96 hours. This study used a duration of 168 hours (7 days) so that cracks could be identified significantly for research purposes. Testing is done twice for each variation.

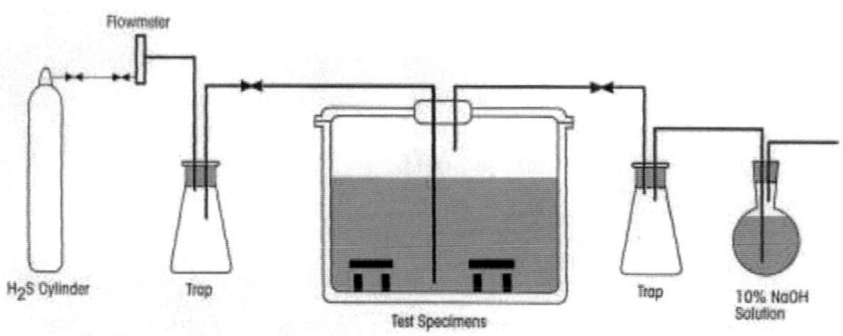

Fig. 2. Schematic diagram of HIC test assembly [18]

According to NACE standards, HIC test is done on pipes or vessels. However, in the present test, the specimen used is a stretched steel plate as a simulation of the actual strain experienced by the plate during the piping and installation process. The dimensions of the test specimens follow a predetermined standard with a size of 100 ± 1 mm and a width of 20 ± 1 mm. The specimens were then grinded using abrasive paper with grit 320, rinsed with ethanol to remove the oil, dried, and then stored in the desiccators. The specimen should be used in less than 24 hours to avoid further contamination before it is inserted into the HIC test container.

IOP Conf. Series: Materials Science and Engineering **547** (2019) 012001 doi:10.1088/1757-899X/547/1/012001

The test solution used was solution A according to NACE TM0284-2003 standard [18]. The solution should contain 5% by weight of NaCl and 0.5% by weight of CH_3COOH dissolved in distilled water. Initial pH should be 2.7 ± 0.1. All the reagents added in the preparation of the test solution have a threshold of approximately 1 percent of each set value. The test is carried out at a temperature of 25 ± 3 ° C. This solution is then purified (purging) using nitrogen gas for a minimum of 1 (one) hour at a flow rate of 100 mL per minute per liter of test solution. After purging has finished, the solution is saturated using H_2S gas at a flow rate of 200 mL per minute per liter of test solution for the first 60 minutes. Furthermore, H_2S is sufficiently bubbled at a lower flow rate (100 mL per minute) to maintain the acidity value of the solution in the pH range of 2.7 - 4.0. The pH measurements were repeated after the solution was finished saturated using H_2S. The specimen cross-section which has been tested by HIC is observed with an optical microscope to show hydrogen cracking. Fig. 3a shows the test specimens and crack dimensions to be used in calculating CLR, CTR, and CSR at the cross section. While Fig. 3b shows the location of UT testing on the surface of the specimen. The marked areas show examples of cracked areas identified by UT. The value of crack sensitivity ratio (CSR), crack length ratio (CLR), and crack thickness ratio (CTR) are calculated according to Eqs. (1) - (3) as follows [18]:

$$CSR = \frac{\sum(a \times b)}{(W \times T)} \times 100\% \qquad (1)$$

$$CLR = \frac{\sum a}{W} \times 100\% \qquad (2)$$

$$CTR = \frac{\sum b}{T} \times 100\% \qquad (3)$$

where a is the cracking length, b represents the crack thickness, W is the specimen width, T is the specimen thickness.

The crack surface due to HIC was observed using Scanning Electron Microscope (SEM). Crack surfaces were observed by opening an existing crack after being exposed to liquid nitrogen. The observation was conducted on X60 specimen subjected to a plastic strain of 12% and has been exposed to the HIC test solution for 168 hours.

3. Results and Discussion

3.1. Effect of Plastic Strain on Microstructure and Mechanical Properties

Pre-strain causes the grains to be elongated in the direction of tensile loading. Fig. 4a to 4d respectively show the observed microstructure for 0% pre-strained X52, 12% pre-strained X52, 0% pre-strained X60 and 12% pre-strained X60. While Fig.5 shows the relationship between the change in the mean length of the grain measured in the direction of the tensile load by the variation of the pre-strain percentage for the studied steels.

Fig. 6 shows the relationship between strain and strength values for API X52 and X60, which have subjected to pre-strain. Meanwhile, Fig. 7 shows the relationship between strain with hardness for both steels after pre-strain. Figs. 6 and 7 show that for API X52 and X60 steels, the greater the percentage of pre-strain applied, the more strength and hardness of the two steels.

A hardness increase of approximately 5% was observed for each steel having a 4% increase in strain. The increase in strength and hardness due to the application of plastic deformation is called strain hardening, which is commonly observed in metal alloys.

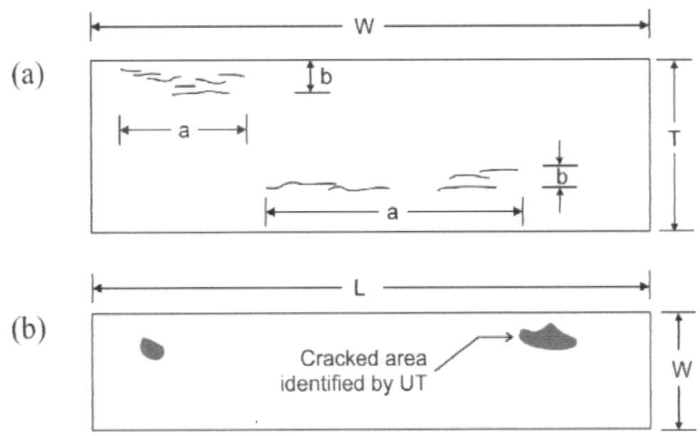

Fig. 3. (a) Test specimens and crack dimensions to be used in calculating CLR, CTR, and CSR at the cross-section [18], (b) testing the UT on the surface of the specimen. The marked areas show examples of cracked areas identified by UT

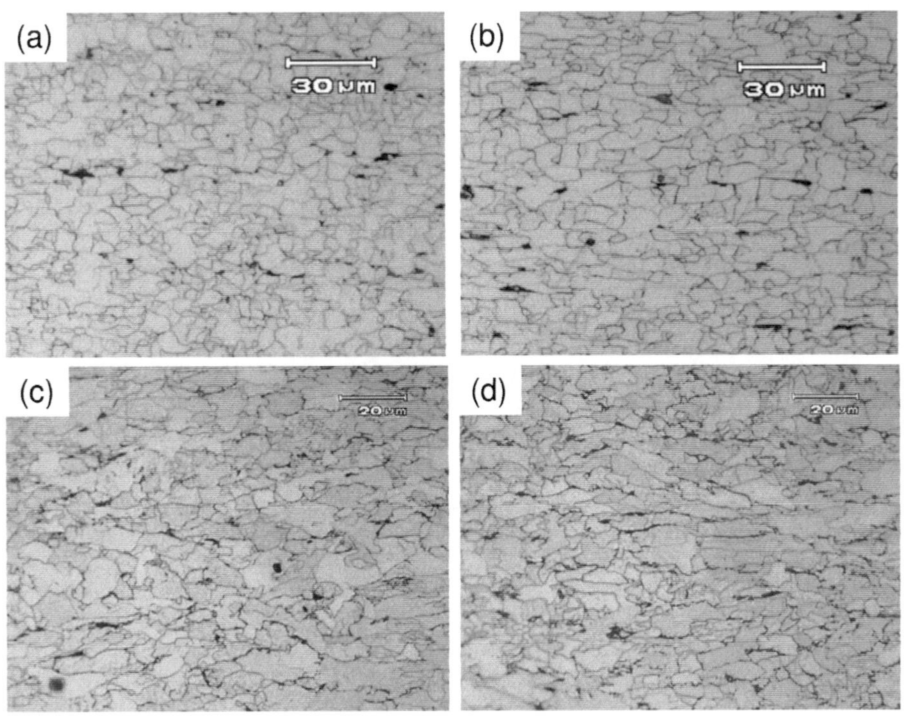

Fig. 4. Microstructure observation of the investigated steels (a) 0% pre-strained X52, (b) 12% pre-strained X52, (c) 0% pre-strained X60 and (d) 12% pre-strained X60.

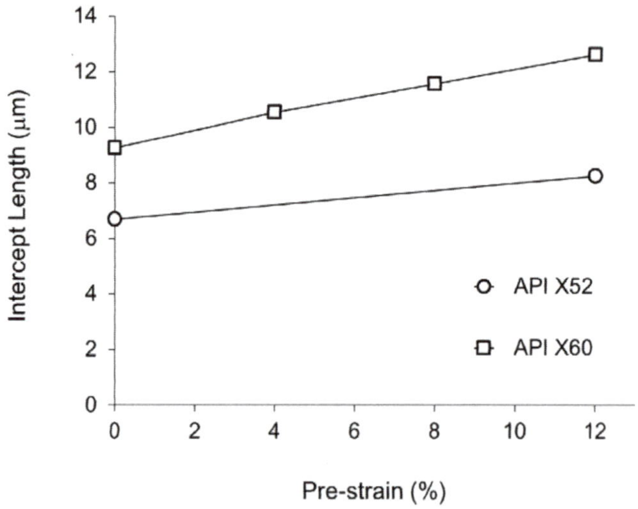

Fig. 5. Relationship between the change in the mean length of the grain measured in the direction of the tensile load by the variation of the pre-strain percentage for the studied steels

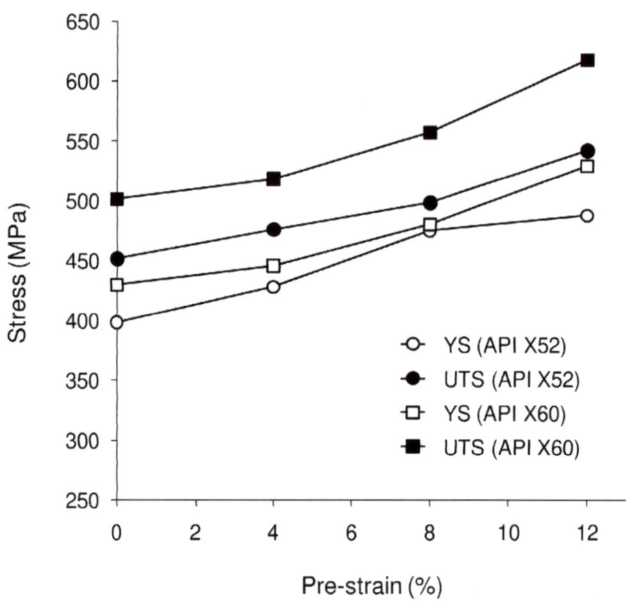

Fig.6. Relationship between pre-strain and strength for X52 and X60 steels

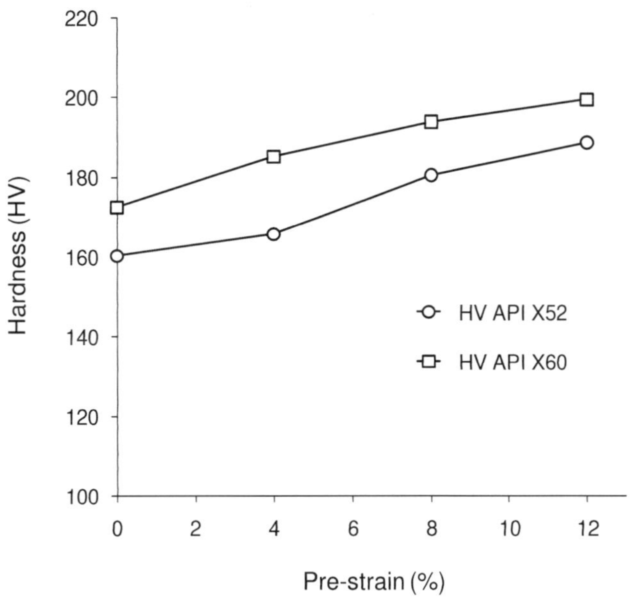

Fig. 7. Relationship between strain and hardness for X52 and X60 after pre-strain

3.2. Observation of Cracks and Fracture Surface

Figs. 8 and 9 show a crack observation on the cross-section of steel plate X52 and X60. In both images it is seen that both of the steel have cracks that occur due to exposure to hydrogen. In the figure also seen that the crack lengthwise in the direction of rolling. Steel that experienced a larger percentage of pre-tension strain exhibited a longer crack size. At the same OM observation scale, X52 steels with low yield strength exhibit a wider crack gap and clearer visible cracks compared to X62 steels. Fig. 9 shows that for X60 steel the HIC cracks propagated by stepwise manner.

Prior to the HIC test, a metallographic examination of the X52 and X60 steel plates was performed. The inclusions were found in X60 steel as shown in Fig. 10a. Through EDS testing it is known that the inclusion is manganese sulphide. Meanwhile, inclusion is not found in X52 steel. Considering its chemical composition, X60 steels are more likely to form more inclusions than X52 steels. All HIC cracks are known to initiate from inclusion and precipitate, and propagate in the direction of rolling [12,13,19,20].

In X52 steel the step-wise crack was not observed. This is because X52 steel has a better level of steel cleanliness. It should be noted that the occurrence of step-wise cracking is also influenced by the non-metallic inclusion distribution in alloys [21]. If inclusion is not distributed or even too small in the alloy, step-wise crack is not formed.

Cracks in X52 steel after the HIC test were found in the half-thickness of the plates. The area is where the occurrence of segregation. The segregation region is the most preferred area for hydrogen in the steel. Fig. 10b shows the location of segregation in X52 steel. The hard phase accumulates in the center of the steel plate during hot rolled steel making. Due to melting point differences, some elements forming hard phases are rejected into the center of the cross section of the steel plate. HIC crack propagates through the hard phase in these steels that are mostly located in the central part of the pipe plate [13,15,22-24]. In the area, the accumulation of inclusions causes a decrease in fracture toughness and makes the area very susceptible to HIC. The empty space between the inclusions and the metal matrix provides an atmosphere suitable for the concentration of hydrogen atoms thus making nucleation of the HIC crack [14].

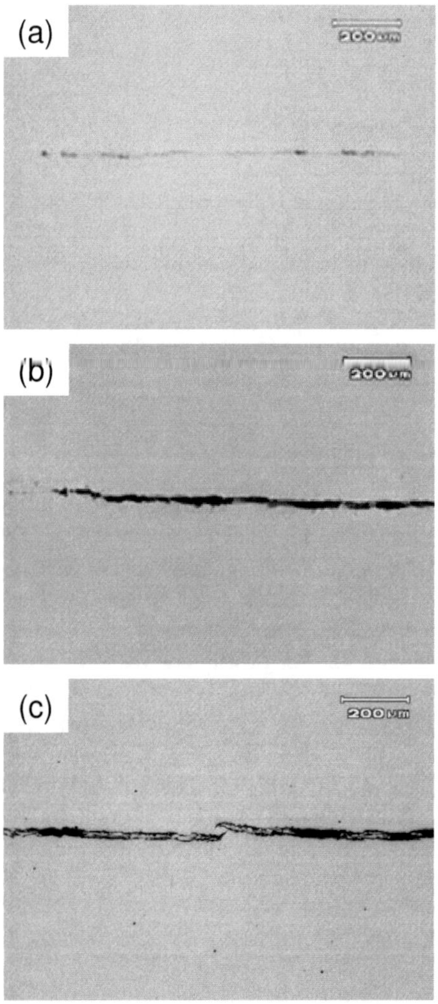

Fig. 8. Observation of cracks in cross-section of steel plate X52 after HIC test for (a) 4% pre-strain, (b) 8% pre-strain, (c) 12% pre-strain

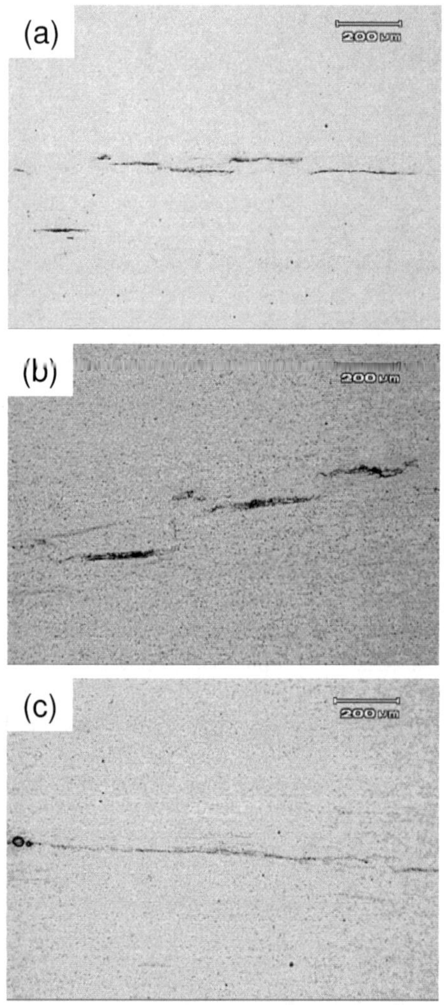

Fig. 9. Observation of cracks in cross-section of steel plate X60 after HIC test for (a) 4% pre-strain, (b) 8% pre-strain, (c) 12% pre-strain

Fig. 11 shows the fracture surfaces observed for X60 specimens after having 12% pre strain and exposure on H2S for 168 hours. Planar cracks were opened for observation purposes. It appears that the fracture mode is transgranular brittle. HIC cracks propagate dominantly in transgranular manner [14]. In the picture also appears the presence of secondary cracks.

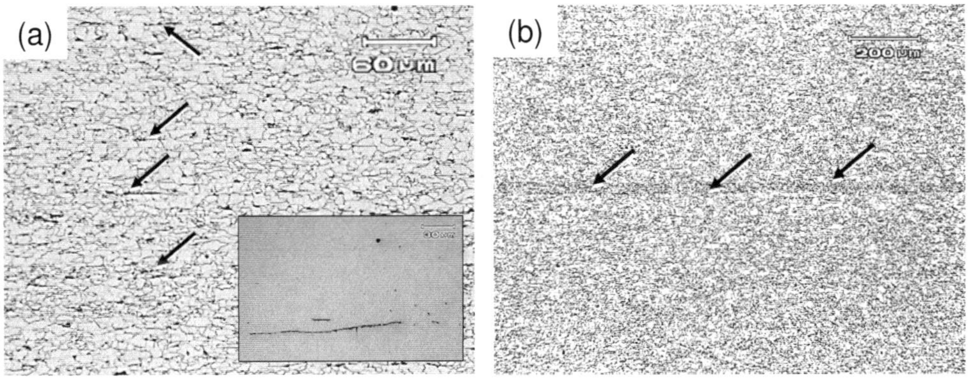

Fig. 10. The microstructure was observed on ½ thick of (a) steel X60 and (b) steel X52. The arrows indicate the presence of inclusions and segregation in steel X60 and X52, respectively.

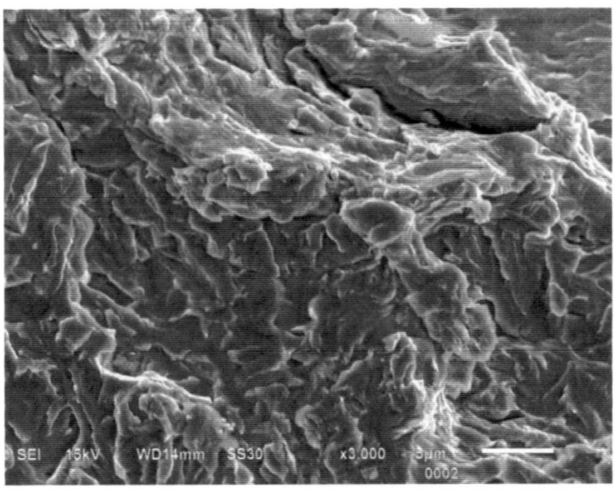

Fig. 11. Fracture surface observation for X60 specimen after subjected to 12% pre-stain and HIC testing for 168 hours

3.3. Parameters and Distribution of HIC Cracks

All HIC parameters including CLR, CTR and CSR for both X52 and X60 steels that have undergone plastic strain were observed. Observation of distribution of cracked areas formed was also conducted to provide further information HIC susceptibility of both the steels. Fig. 12 shows the relationship between pre-strain and crack ratio (CLR and CTR) for X52 and X60 steels. From the figure it is seen that the value of CLR for steel X60 is higher than X52 steel for all variations of pre-strain. Nevertheless, the CTR values of the two steels are relatively similar, and do not experience significant changes with the increased application of pre-strain variations. This suggests that cracks tend to elongate with increasing pre-strain. Meanwhile, as shown in Fig. 13, CSR for both steels has the same tendency as CLR. Steel X60 has a higher susceptibility to HIC than X52. The CSR of both steels increases with increasing pre-strain. Susceptibility to HIC is believed to be strongly associated with increased strength and hardness, especially for X60 steels that have experienced significant strength and hardness after pre-strain.

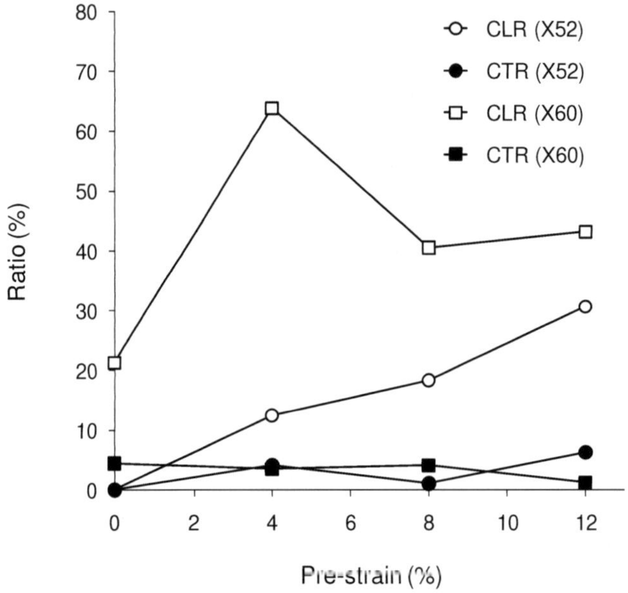

Fig. 12. Relationship between pre-strain and crack ratio (CLR and CTR) for X52 and X60

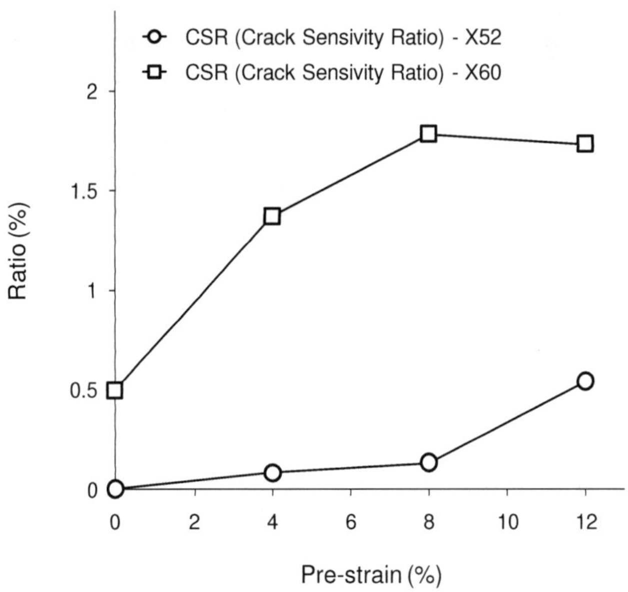

Fig. 13. Relationship between pre-strain and CSR for X52 and X60

Mohtadi-Monab et al. [16] comparing the HIC parameters on the X70 and X60 found that the X70 is more susceptible to HIC. Cracks propagate across deformed grains, as observed using EBSD mapping [14]. Grains with high stored energy levels are more susceptible to HIC cracking. In steels with higher pre-strain levels, stored energy is also higher and causes cracks to form easily and propagate in the plastically deformed steel.

The distribution of cracked HIC areas examined using the UT from the surface of the specimen also exhibits a similar trend. Figs. 14 and 15 show the results of the reconstruction of cracked areas resulting from observations using the UT method. Areas containing cracks are represented by shaded areas. Two specimens were tested for each pre-strain variation. In the figures it is seen that for non-pre-strained (0% pre-strain) steels, the crack areas in both X52 and X60 are identified. It is seen that the crack area in X60 steel is wider than X52 steel. It should be noted again that this test was carried out with H_2S exposure times that passed the general test standard, which is 168 hours.

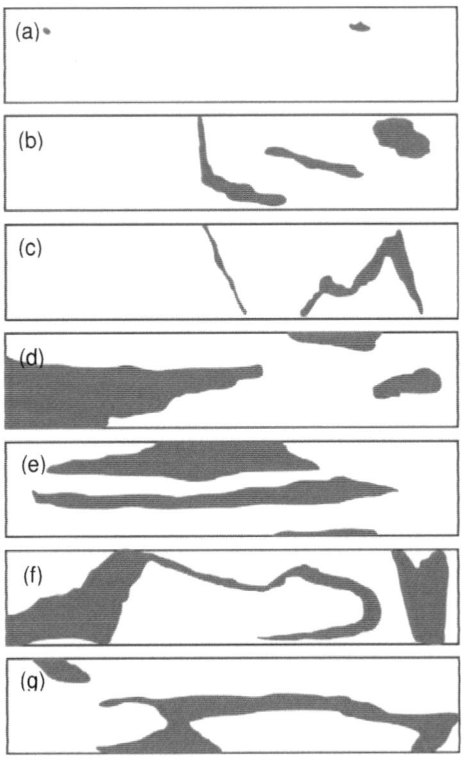

Fig. 14. The observed laminate crack area from the upper side of X52 plate after HIC test for (a) 0% pre-strain, (b & c) 4% strain, (d & e) 8% strain and (f & g) 12% strain

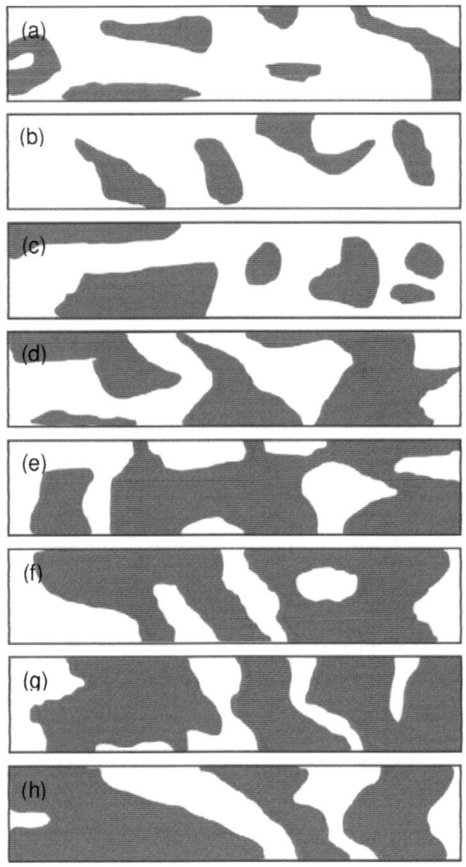

Fig. 15. The observed laminate crack area from the upper side of X60 plate after HIC test for (a & b) 0% pre-strain, (c & d) 4% strain, (e & f) 8% strain and (g & h) 12% strain

The relationship between pre-strain and crack distribution area for X52 and X60 steels is shown in Fig. 16. In the figure it is clear that the increase in crack distribution area for X60 steels increases sharply compared to X52 steel. The presence of inclusions and increased hardness for X60 steels contributes and combines in increasing susceptibility to HIC cracking.

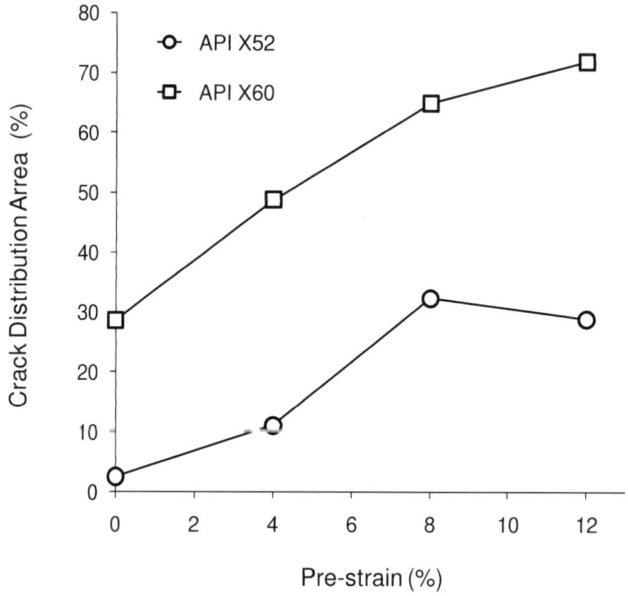

Fig. 16. Relationship between pre-strain and crack distribution area for X52 and X60 steels

4. Conclusions

Based on results from the investigation of hydrogen induced cracking of API X52 and X60 sour service steels subjected to pre-strain under prolonged H_2S exposure, the following conclusion were drawn. Yield strength, tensile strength and hardness of the investigated steels increased with increasing tensile pre-strain. The observed HIC crack propagates in the direction of rolling. The segregation region is the most preferred area for HIC cracking in the X52 steel, while in X60 steel where inclusions were found, the HIC cracks initiate on the inclusions. HIC cracks propagate in transgranular manner. The value of CLR for steel X60 is higher than X52 steel for all variations of pre-strain. The CTR values of the two steels are relatively similar, and do not change significantly with pre-strain variations. Cracks tend to elongate with increasing pre-strain. Steel X60 has a higher susceptibility to HIC than X52. The CSR and crack distribution area of both steels increases with increasing pre-strain. Susceptibility to HIC is associated with an increase in strength and hardness after tensile pre-strain.

5. Acknowledgements

The authors are grateful to PT Krakatau Steel for supporting this work.

6. References

IOP Conf. Series: Materials Science and Engineering **547** (2019) 012001 doi:10.1088/1757-899X/547/1/012001

[1] Zhao, M.C., Tang, B., Shan, Y.Y. and Yang, K., Metallurgical and Materials Transactions A, 34A, (2003), 1089–1096.
[2] Lopez, H.F., Bharadwaj, R., Albarran, J.L. and Martinez, L., Metallurgical and Materials Transactions A, 30A, (1999), 2419–2428.
[3] Iino, M., Metallurgical Transactions 10A, (1979), 1691.
[4] Beidokhti, B., Dolati, A. and Koukabi, A.H., Materials Science and Engineering A507, (2009), 167–173.
[5] Carneiro, R.A., Ratnapuli, R.C. and Lins, V.F.C., Materials Science & Engineering A357 (2003), 104–110.
[6] Domizzi, G., Anteri, G. and Ovejero-Garcia, J., Corrosion Science 43, (2001), 325–339.
[7] Weng, C.C. and White, R.N., Journal of Structural Engineering, 116, (1990), 24–39.
[8] Law, M., Prask, H., Luzin, V. and Gnaeupel-Herold, T., Materials Science and Engineering A, 437 (2006), 60–63.
[9] Li, S.J., Zhou, C.Y., Li, J., Pan, X.M. and He, X.H., International Journal of Mechanical Sciences 131–132, (2017), 572–585.
[10] Kim, W.K., Koh, S.U., Yang, B.Y. and Kim, K.Y., Corrosion Science and Technology, Vol. 6, No. 3, (2007), 96–102.
[11] Asahi, H., Sogo, Y. and Ueno, M., Higashiyama, H., Metallurgical Transactions A, Vol. 19, Issue 9, (1988), 2171–2177.
[12] Mohtadi-Bonab, M.A., Eskandari, M. and Szpunar, J.A., Materials Science and Engineering A, 620, (2015), 97–106.
[13] Mohtadi-Bonab, M.A., Szpunar, J.A., Basu, R. and Eskandari, M., International Journal of Hydrogen Energy, 40, (2015), 1096–1107.
[14] Mohtadi-Bonab, M.A., Eskandari, M., Karimdadashi, R. and Szpunar, J.A., Metals and Materials International, Vol. 23, No. 4, (2017), 726–735.
[15] Mohtadi-Bonab, M. A., Szpunar, J.A. and Razavi-tousi, S.S., International Journal of Hydrogen Energy, 38, (2013), 13831–13841.
[16] Mohtadi-Bonab, M.A., Szpunar, J.A. and Razavi-Tousi, S.S., Engineering Failure Analysis, 33, (2013), 163–175.
[17] ASTM E8. Standard Test Methods for Tension Testing of Metallic Materials, in Annual Book or ASTM Standards, ASTM International, 2013.
[18] NACE Standard TM0284-2003. Evaluation of Pipeline and Pressure Vessel Steels for Resistance to Hydrogen-Induced Cracking, NACE International, 2003.
[19] Venegas, V., Caleyo, F., Baudin, T., Espina-Hernández, J.H. and Hallen, J.M., Corrosion Science, 53, (2011), 4204–4212.
[20] Tamehiro, H., Takeda, T., Matsuda, S., Yamamoto, K. and Okumura, N., Transactions of Iron Steel Institute Japan, 25, (1985), 982–988.
[21] Gonzalez, J.L., Ramirez, R., Hallen, J.M. and Guzman, R.A., Corrosion, 53, 12, (1997), 935–943.
[22] Moon, J., Park, C. and Kim, S.J., Metals and Materials International, 18, (2012), 613–617.
[23] Moon, J., Kim, S. J. and Lee, C., Metals and Materials International, 19, (2013), 45–48.
[24] Jin, T. Y., Liu, Z. Y. and Cheng, Y.F., International Journal of Hydrogen Energy, 35, (2010), 8014–8021.

The Antimicrobial Activity and Characterization of the Cast Titanium Copper Alloys with Variations of Copper Content

Akhmad A. Korda*, S. Munawaroh, Eddy A. Basuki
Department of Metallurgical Engineering, Faculty of Mining and Petroleum Engineering, Bandung Institute of Technology

E-mail: akhmad@metallurgy.itb.ac.id

Abstract. The effect of Cu addition on antimicrobial activity of Cast Ti-Cu Alloys against *Staphylococcus aureus* were studied. The Ti-Cu alloys were prepared with a variation of 0.5, 1, 2, 3, 4, 8 and 10 wt% of copper. Pure Ti metal was also prepared and tested as a comparison. The antibacterial was performed by killing activity test of colony forming unit (CFU) method with variation of contact time. Optical microscope observation, XRD, and the Vickers hardness test were carried out to characterize the investigated alloys. The results showed that Ti-Cu alloys were able to decrease the amount of bacteria by time. However, the activity of killing of bacteria in the varied range of Cu content did not show significantly different result. The highest bacterial kill ability in the alloy was observed in addition of 3 wt% Cu.

Keywords: Antimicrobial alloy, Ti-Cu alloys, Copper content, *Staphylococcus aureus*

1. Introduction

Titanium has been used as a biomaterial applied to dental, implant and cardiac valve retaining structures [1]. Titanium which is very light and has good mechano-chemical is more tolerant to the human body than stainless steel and cobalt alloys. It can also form a layer of titanium oxide (TiO_2) quickly and spontaneously in surface called as a passive layer, so it is not dissolved in interstitial fluid and prevents the release metal ions which can react with tissue [2].

However, titanium has no antimicrobial properties so serious infections often occur during implant surgery. One of the bacteria which causes infection in the implant is the bacteria of *Staphylococcus aureus (S. aureus)* and *Staphylococcus epidermis (S. epidermis)*. The infection which occurs in the implant has reached 2–30% of cases found in various literature. Mahan [3] reported that 75% of infections occurred in the pin tract region. *S. aureus* and *S. epidermis* bacteria form biofilm layers which are difficult to clean clinically because they are protected by phagocytosis and antibiotics [3]. Direct microbial contamination of opened wounds (such as postoperative wounds) or post-traumatic infections (such as chronic osteomyelitis after opened fracture) is the cause of nosocomial infections [4,5].

Antimicrobial properties can be obtained with the addition of copper elements. Copper has an antimicrobial nature which is proven to rapidly kill bacteria and in the future copper alloys will be introduced as surfaces and objects that humans frequently touch [6]. The copper toxin is caused by Cu (I) ions in the bacterial cell. Cu (I) ions can damage cell membranes and kill bacteria. In a bacterial cell, there is an equilibrium of copper content or a homoeostatic condition, there is an excess copper exhaust mechanism by reducing copper to cuprous or Cu (I) ions. This Cu ion (I) is toxic for bacteria [7].

The amount of Cu released from the copper surface can increase significantly when contaminated with bacteria. This can occur due to the difference in the concentration of copper between the metal surface and the solution [8]. In addition, titanium combined with copper is reported [9] that it has good biocompatible properties, high corrosion resistance, low melting temperature, and it is easy when casting. Since 1950, many studies have focused on Ti-Cu alloys for industrial applications. The results of investigation show that Ti-Cu alloys have high mechanical strength and ability to be easily formed.

In the last 20 years, titanium dental foundry technology has made great progress. Kikuchi [10] showed that the Ti-Cu alloy as cast has a high yield strength reaching 650 MPa, a value greater than pure Ti and Co-Cr alloy dental casting. Yao et al. [11] showed that heat treatment could strengthen Ti-2.5% Cu alloy by forming Ti_2Cu intermetallic particles after the β phase was decomposed into αTi and Ti_2Cu.

Titanium applied as an implant combined copper become antimicrobial implant prevents the emersion of infection. Investigation of the antibacterial activity of Ti-Cu alloys has been carried out [12,13]. However, studies covering a larger percentage of Cu content are still very limited. Nevertheless, studies for low grade copper have not been performed on the casting process. The results of these studies will be compared with the other studies to determine the properties of antimicrobial in low grade Cu.

2. Materials and Method

The Ti-Cu samples used in this investigation included alloys with seven Cu variants 0.5, 1, 2, 3, 4, 8, and 10 wt%. Samples were produced by mini DC EAF furnace using a non-consumable tungsten electrode and water-cooled copper crucible in argon atmosphere with current flowed 70 A for 3 minutes. Five samples were cut in size of 100×50×20 mm, then homogenized at 800°C for 48 hours using muffle and furnace cooled to be prepared in the observation of microstructures, mechanical behavior, X-ray diffraction (XRD), and antimicrobial tests. Microstructural characterization was performed by optical microscopy and X-ray diffraction. Metallographic preparation included sandpapering with SiC paper in 60, 150, 240, 320, 400, 600, 800, 1000, 1500, and fineness of 2000 grit. Polishing used diamond and ethyl alcohol as a lubricant, and etching used "Kroll Etchant" with a solution of 5 vol.%HF, 10 vol.%HNO_3, and 85 vol.% H_2O, for 10-minutes dipping time.

Antimicrobial test began with the preparation of medium, physiological solution, bacterial culture, and bacterial colony-killing test. Nutrient broth (NB) is a liquid medium used to grow the inoculant of *S. aureus* bacteria in solution form. It was made by adding 13 g of nutrient broth powder to one litre of distilled water. After that, mixed, heated and sterilized. Nutrient agar (NA) is medium to observe the growth of bacterial colonies, made of 28 g of nutrient agar powder in one litre of distilled water and stir using magnetic stirred. Furthermore, nutrient agar was sterilized and poured in every sterile petri dish each 10–15 ml. Physiological solutions were used to inhibit bacterial growth, created by 0.85%(w/v) NaCl and dissolved in aquades. These solutions were inserted in every falcon tube for 9 ml, then sterilized. The entire mediums, solutions, and equipment for antimicrobial test should be sterilized in the autoclave for 20 minutes to prevent contamination. Fig. 1 shows schematic dilution of physiological solution to incubation in petri dishes at 24 hours.

The first step for liquid bacterial cultures was the bacterial culture stock of *S. aureus* in the tube taken with oce wire (sterilized by burning on fire bunsen) swiped at agar surface and transferred in sterile nutrient broth up to 2–3 oce. Medium nutrient broth which had been inoculated *S. aureus* bacteria then incubated in orbital shaker at room temperature for 18–24 hours before used. The bacterial colony-killing test was begun by cleaning laminar using 70% alcohol and exposed by UV light for 5–10 minutes. *S. aureus* culture was taken 1 ml with a micropipette and inserted in a falcon containing 9 ml physiological solution, then shaken by vortex for 15–20 seconds.

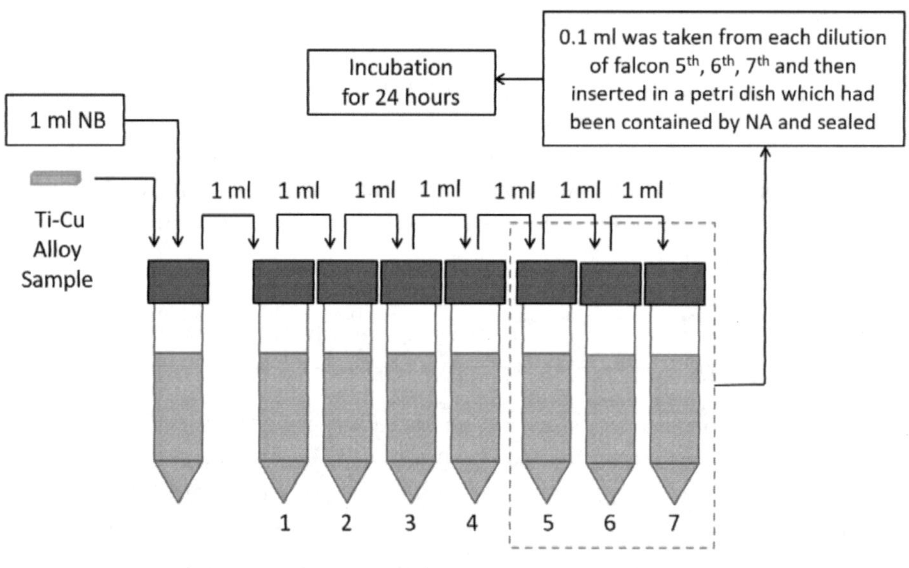

Fig. 1. Schematic dilution of physiological solution to incubation in petri dishes at 24 hours

For the second step, the alloy samples taken with tweezers in an alcohol bath was dried near fire, and incorporated in physiological solution content 1 ml of bacterial culture (dilution 0), shaken by vortex for 15–20 seconds. The physiological solution in the falcon was the sample stock. The solution was taken at dilution 0 and fed into 9 ml of physiological solution up to the 7th dilution. Each physiological solution at 5th, 6th, 7th dilution were taken using 0.1 ml micropipette and included in a petri dish containing sterile nutrient agar, thereupon flattened it in petri dish used L stick until dry. The second step were repeated with the duration of sample immersed in 1, 2, 3 hours in sequence. Petri dishes which been dried in the seal were incubated for 18–24 hours, then counted the number of the bacterial colonies formed on each petri dish.

3. Result and Discussion

3.1. Microstrutural characterization

Microstructure of Ti-Cu alloy with contents of 0.5, 1, 2, 3, 4, 8, 10 wt% Cu was examined. Fig. 2 shows the Ti-Cu alloy binary phase diagram [14]. In this phase diagram, the maximum solubility of Cu in Ti is 2.1 wt% Cu at 790°C, and the eutectoid point is in the composition of 7.1 wt% Cu. Above eutectoid temperature (790°C) Ti-Cu alloy containing 0.5–2 wt% Cu is in the alpha-Ti region. Alloy Ti-Cu with content of 3 and 4 wt% Cu is in the region between α-Ti and β-Ti. While alloys with 8 and 10 wt% content are in the beta-Ti and Ti_2Cu regions.

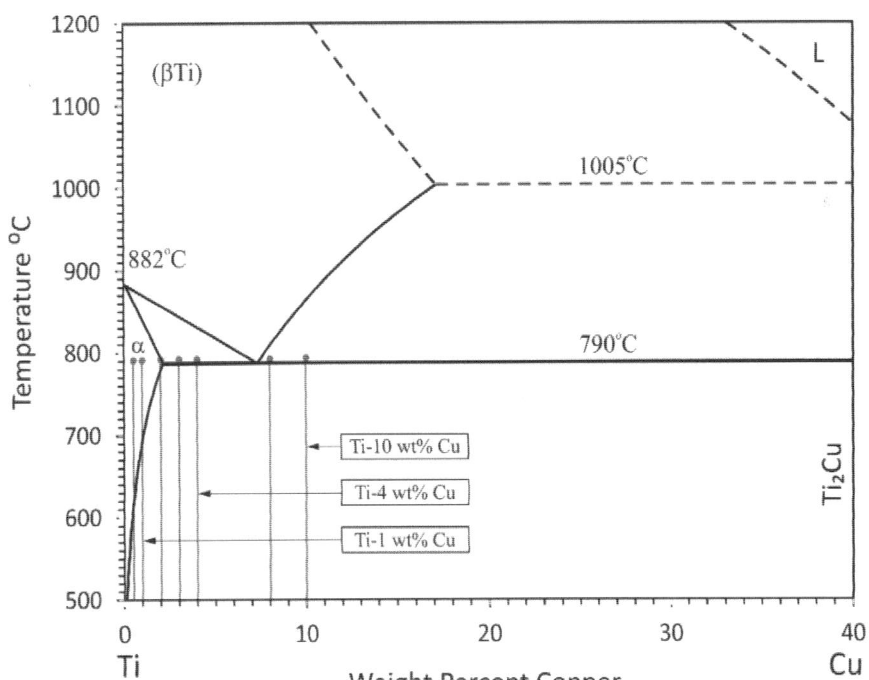

Fig. 2. Ti-Cu binary phase diagram [14]

Homogenization at 800°C for 48 hours followed by furnace cooling will provide a phase and microstructure that correspond to the Ti-Cu phase diagram. Below eutectoid temperature, the following reactions occurred. In the Ti-Cu alloys with composition in 0.5–2 wt% Cu intervals, Ti_2Cu particles undergo precipitation from the saturated α phase due to the decreasing solubility of Cu at alpha phase with decreasing temperature from homogenization temperature (800°C) to room temperature:

$$\alpha\ (\text{supersaturated}) \rightarrow \alpha + Ti_2Cu \quad (1)$$

For Ti-3 wt% Cu and Ti-4 wt% Cu alloys, below eutectoid temperature, α (eutectoid) phase is formed from β phase with Ti_2Cu particles precipitated. The α (proeutectoid) phase is formed from saturated α phases, as follows:

$$\alpha\ (\text{proeutec.}) + \beta \rightarrow \alpha\ (\text{proeutec.}) + \alpha\ (\text{eutec.}) + Ti_2Cu\ (\text{eutec.}) \quad (2)$$

For Ti-8 wt% Cu and Ti-10 wt% Cu alloys, below eutectoid temperature, α (eutectoid) and Ti_2Cu phases are formed from β phase, as follows:

$$\beta + Ti_2Cu\ (\text{proeutec.}) \rightarrow \alpha\ (\text{eutec.}) + Ti_2Cu\ (\text{eutec.}) + Ti_2Cu\ (\text{proeutec.}) \quad (3)$$

Fig. 3 shows the microstructure of Ti-Cu alloys with variations of 0.5, 1, 2, 3, 4, 8 and 10 wt% Cu. In Figs. 3a-c which shows the Ti alloy with a content of 0.5–2 wt% Cu, the visible microstructure is α-Ti and Ti_2Cu intermetallic. α-Ti is indicated by light phase while Ti_2Cu is shown by a dark phase in the form of fine elongated particles. The Ti_2Cu intermetallic particles became more pronounced by increasing wt% Cu. For Ti-3% Cu, the elongated precipitates begin to appear denser. At 4 and 8 wt% Cu (Figs. 3e & 3f), the eutectoid lamellar (dark colour) structure appears to be clearly observed. Meanwhile in Fig. 3g, hypereutectoid structure appear for 10 wt% Cu where lamellar structures are present with Ti_2Cu intermetallic phases. In this figure the lamellar structure is shown by a dark color while the Ti_2Cu (proeutectoid) precipitate is indicated by the bright phase. This lamellar structure has also been previously observed by Hayama [15] who studied Ti-Cu alloys for hypoeutectoid, eutectoid and hypereutectoid compositions.

Fig. 4 shows the relationship between Cu content in Ti-Cu alloys and Cu content in α or Ti_2Cu phases. The Cu content of the phases is calculated simply by using Ti-Cu phase diagrams for variations of Cu content in Ti-Cu alloys. In the figure, it can be seen that the Cu content in Ti_2Cu continues to increase with increasing Cu content in the alloy, while the Cu content in the α phase is relatively unchanged. This relationship needs to be made to consider the presence of Cu elements in solid solutions or intermetallic compounds in their role in bacterial killing activities.

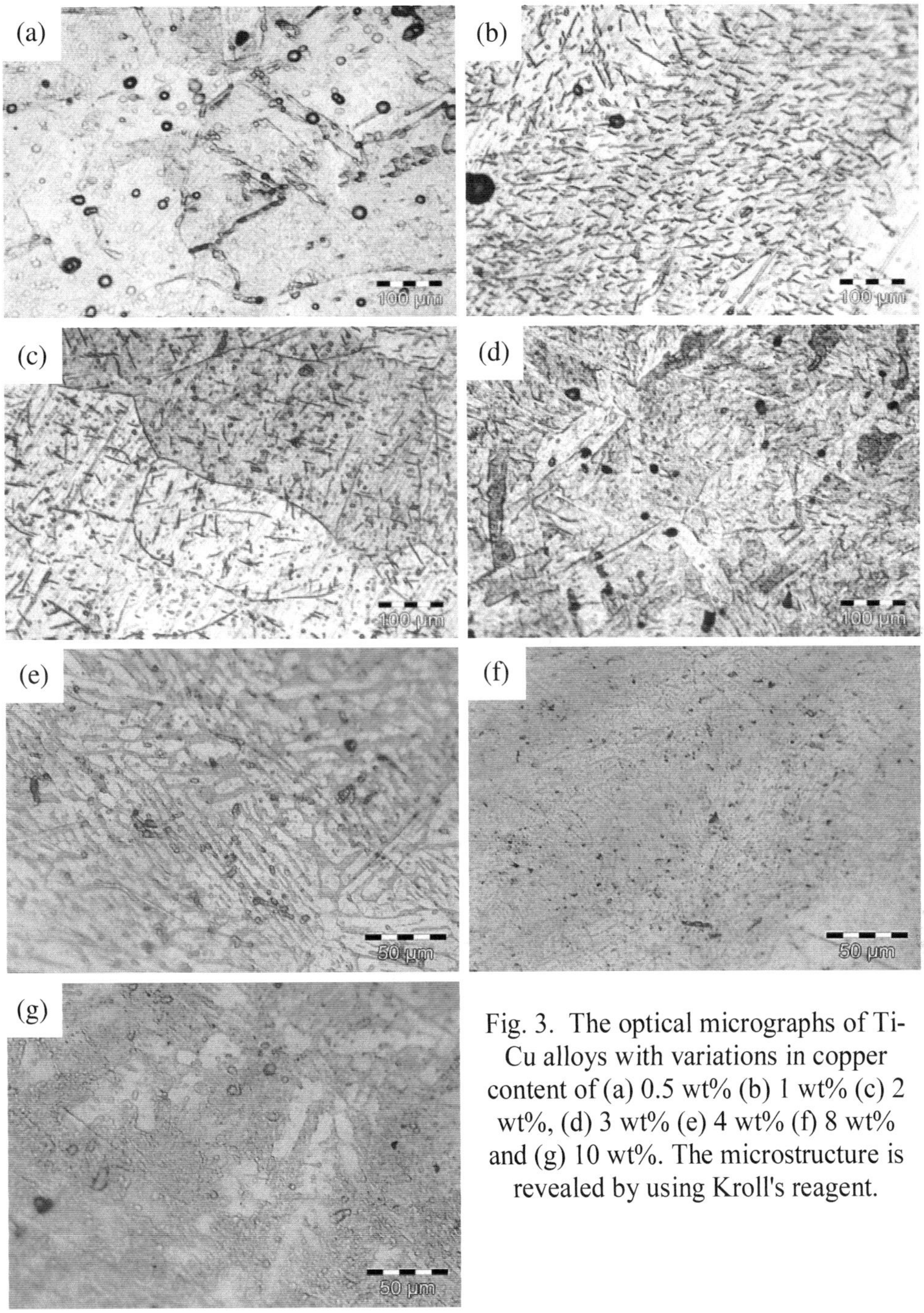

Fig. 3. The optical micrographs of Ti-Cu alloys with variations in copper content of (a) 0.5 wt% (b) 1 wt% (c) 2 wt%, (d) 3 wt% (e) 4 wt% (f) 8 wt% and (g) 10 wt%. The microstructure is revealed by using Kroll's reagent.

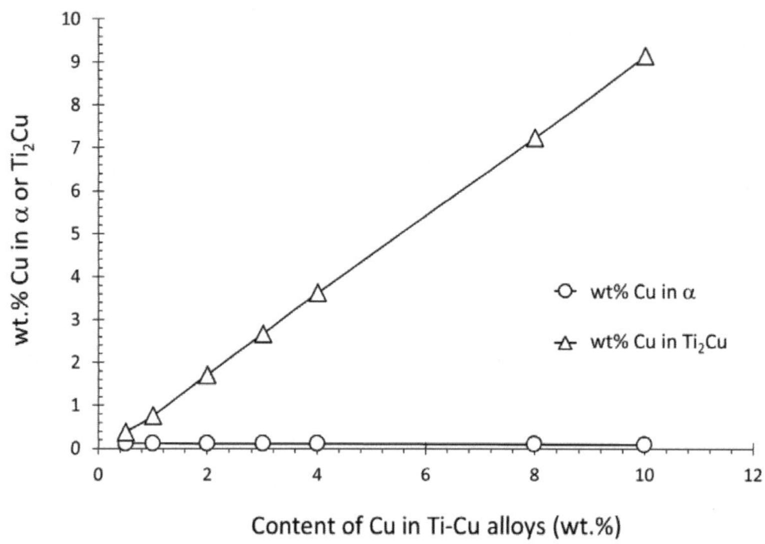

Fig. 4. Relationship between Cu content in Ti-Cu alloys and Cu content in alpha or Ti_2Cu phases

The presence of phases in the Ti-Cu alloy was also examined using XRD. Fig. 5 presents the XRD pattern of the studied Ti-Cu alloy. The Ti-Cu alloy tested was Ti-Cu alloy containing Cu of 1 and 10 wt%. In the figure, diffraction peaks can be found for the alpha and Ti_2Cu phases. Furthermore, it was seen that in the addition of 10 wt% Cu, Ti_2Cu phase diffraction peaks were apparent and increased.

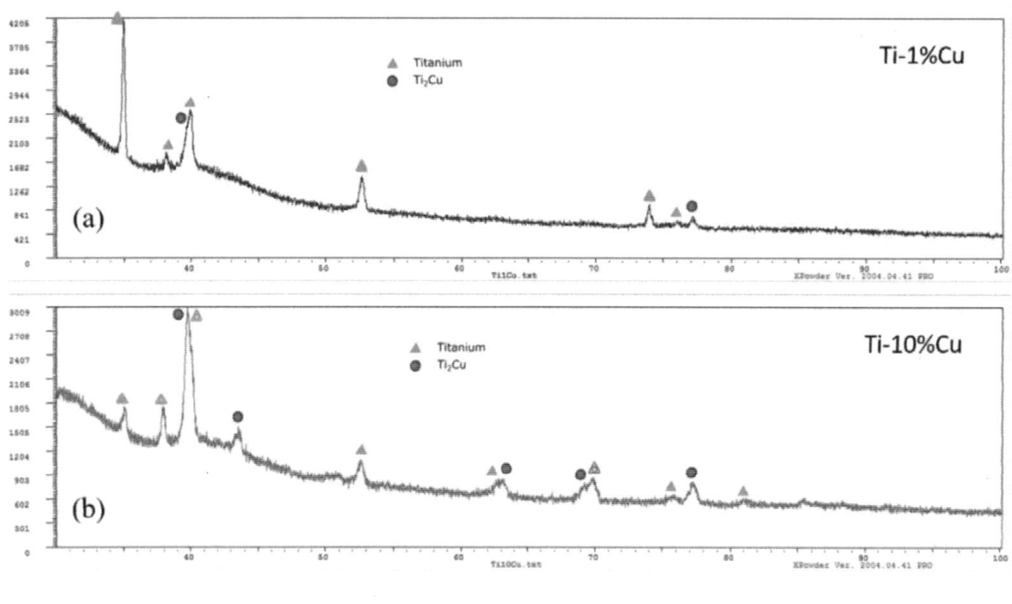

Fig. 5. XRD patterns for (a) Ti-1%Cu and (b) Ti-10%Cu alloys

3.2. Mechanical behavior

Fig. 6 shows the relationship between hardness and variation of Cu content in Ti-Cu alloys. Pure titanium is also shown as a comparison. The hardness of pure titanium was 181 HV, while the Ti-Cu alloys had hardness between 301–647 HV which was higher than pure titanium. In the figure, it can be seen that the addition of Cu to titanium increases the hardness of the alloy. In general, the higher the copper content, the higher the hardness of Ti-Cu alloys. At a content of 10 wt% Cu, the hardness reached the highest compared to other wt% Cu. The increase in hardness at the addition of 10 wt% Cu on Ti-Cu alloy is mainly due to the increase in the hard phase of Ti_2Cu intermetallic compounds. Increased hardness occurs due to the strengthening mechanism of solid solution hardening of α phase and precipitation strengthening with the presence of Ti_2Cu. This mechanism inhibits the movement of dislocations thereby increasing the hardness of the Ti-Cu alloy. High hardness will provide better wear resistance for implants. High hardness itself tends to provide high strength. Increased hardness with increasing Cu content is in accordance with the results obtained by Kikuchi [16]. However Hayama [15] reported different results in which an increase in Cu content did not increase hardness, and he even found a slight decrease in hardness with increasing Cu content in Ti-Cu alloys.

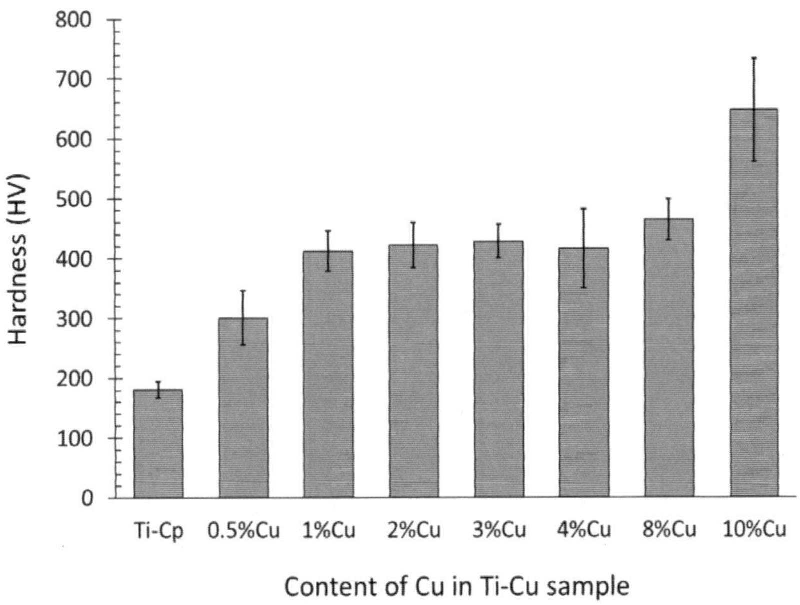

Fig. 6. Vickers hardness as a function of Cu content in Cu-Ti samples

3.3. The effect of Cu composition Ti-Cu alloys to antimicrobial properties

Fig. 7 shows colonies of *S. aureus* bacteria incubated after 24 hours at 6th dilution. Figs.7a and 7b are colony images for Ti-1% Cu alloys exposed for 0 and 3 hours. Figs. 7c and 7d are images of colonies for Ti-2% Cu alloys which are exposed for 0 and 3 hours. Figs. 7e and 7f are images of colonies for Ti-3% Cu alloys which are exposed for 0 and 3 hours. In the left figures (7a, 7c

IOP Conf. Series: Materials Science and Engineering **547** (2019) 012002 doi:10.1088/1757-899X/547/1/012002

and 7e), bacterial colonies were clearly observed, with nearly the same amount at the 6th dilution at 0-hour exposure. But after a 3-hours exposure, the bacterial colonies seemed to shrink and decrease. It is seen that Ti-Cu alloy is able to kill and reduce the bacterial colonies over time.

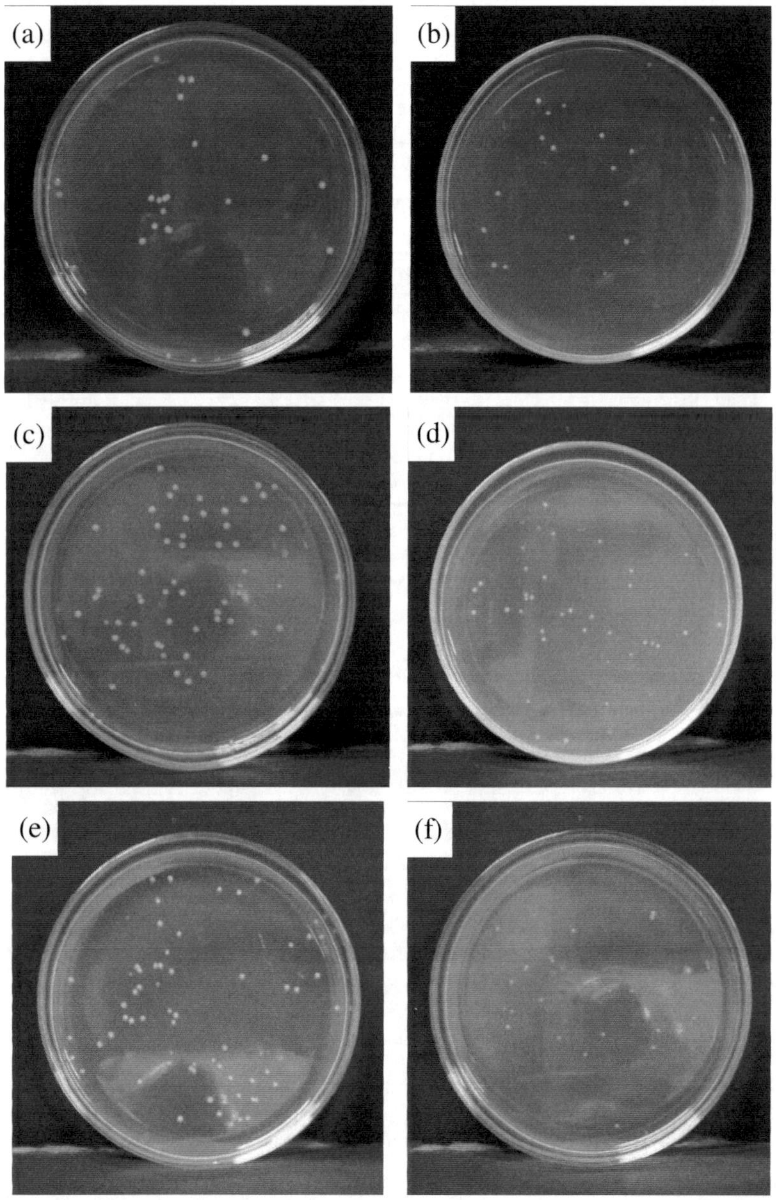

Fig. 7. Photographs of bacterial colonies after incubation at the 6th dilution for alloys: (a) Ti-1% Cu at 0 hour, (b) Ti-1% Cu at 3 hour, (c) Ti-2% Cu at 0 hour, (d) Ti-2% Cu at 3 hour, (e) Ti-3% Cu at 0 hour and (f) Ti-3% Cu at 3 hour

Fig.8 shows a graph of the ability to kill bacteria by Ti-Cu alloys. In the figure, it can be seen the relationship between the time of exposure of Ti-Cu alloys to the number of bacterial colonies. The exposure period is carried out with a time variation of 0, 60, 120 and 180 minutes. The number of bacterial colonies is seen decrease with increasing the exposure time by Ti-Cu alloys. Pure titanium was also tested as a control variable in this experiment.

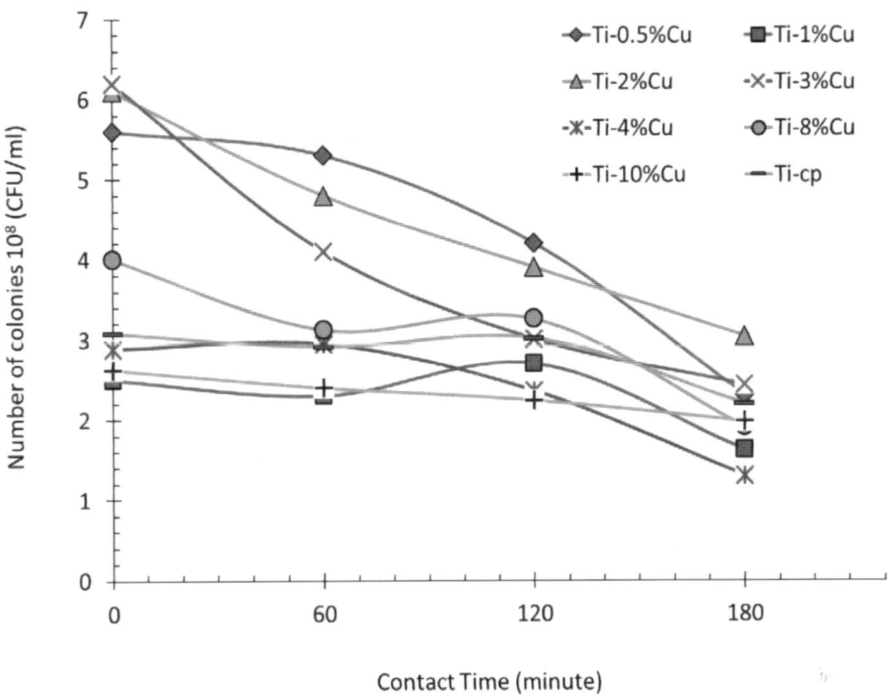

Fig. 8. The killing power of bacterial colonies in Ti-Cu alloys

From Fig. 8 it can be seen that pure titanium appears to reduce bacterial colonies. However, pure titanium does not have the ability to kill bacteria except in the form of TiO_2 compounds. In this case, titanium is oxidized due to cutting of the sample during preparation which produces heat and forms a TiO_2 layer on the surface area. Sanding and sterilization using alcohol is not enough to remove this layer, so TiO_2 also kills bacteria and decreases the number of bacterial colonies. TiO_2 has photocatalyst properties that can absorb light and produce various chemical reactions. UV rays with H_2O and O_2 react with TiO_2 layers to form hydroxy (OH •) radicals on the surface and kill bacteria [17]. However, the killing power of TiO_2 is insignificant compared to Cu ions.

Increasing copper content in the Ti-Cu alloy is expected to increase the number of dead bacterial colonies. However, the results obtained from the test did not prove the statement. Fig. 9 shows a bar graph of the number of bacterial colonies that died during 3-hour exposure by Ti-Cu alloys. In the figure there is no tendency to decrease the number of bacterial colonies with variations in Cu content. Alloy with a content of 0.5 wt% Cu even has the ability to kill distant bacteria compared to 10 wt% Cu. Together with Ti-1% Cu and Ti-4% Cu, Ti-10% Cu showed the ability to kill low bacteria in a 3-hour exposure interval. Meanwhile the Ti-Cu alloy with Cu content of 2, 0.5 and 3 wt% Cu showed the highest ability to kill bacteria, respectively.

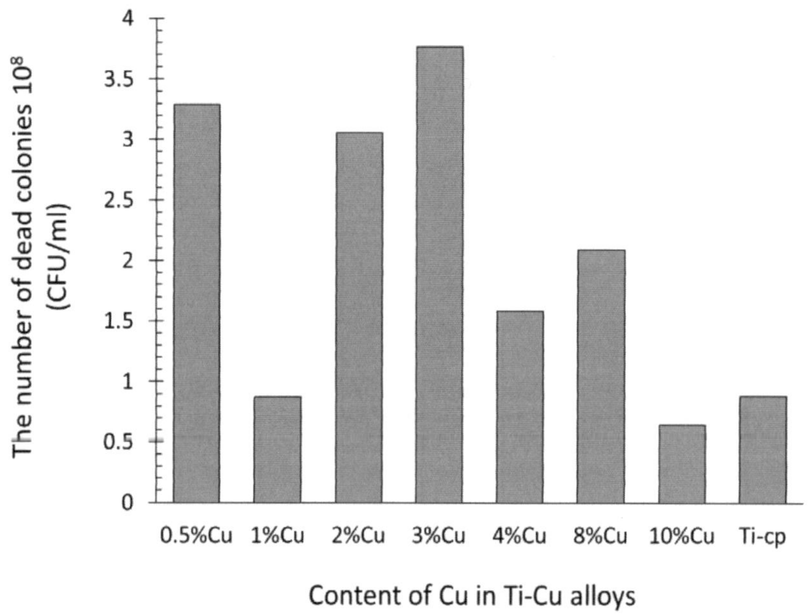

Fig. 9. Number of dead bacterial colonies after being exposed to Ti-Cu alloys for 3 hours

Based on the theory, copper can kill bacteria by releasing Cu ions on the surface and dissolving into the surface of bacteria which then causes the cell membrane to break so that the components inside are lost. Stainless steel combined with copper has been designed and fabricated. In this study, it was explained that copper can release Cu ions from steel so that the corrosion formed in the steel is minimal in the biological environment [18].

Recommended daily intake of Cu for an adult per day by WHO is 2-3 mg [19]. Ma et al. [20] investigated the cytotoxicity of Cu on the human body stating that the daily average release Cu for a typical 50×15×2 mm bone fixation plate made of Ti-6Al-4V-5Cu Alloy is about 0.0001 μg, which is far below the recommended number. Therefore, the use of Ti-Cu alloy with 10 wt% Cu can be considered still in a safe range.

Antibacterial properties that did not show an increase in the addition of Cu to Ti-Cu alloys studied and even without a tendency were thought to be closely related to the state of the Cu element in the alloy, namely whether it is in a solute state in the α phase or as Ti_2Cu precipitates. Ma et al. [20] stated that the Ti-6Al-4V-5Cu alloy was easier to release interstitial Cu atoms than in the Ti2Cu stable phase. Previously, in Fig. 2 we have shown a relationship between Cu content in Ti-Cu alloys and Cu content in α or Ti_2Cu phases. In the figure it is shown that the number of Cu atoms in the α phase is relatively unchanged, while the number of atoms in Ti_2Cu increases with increasing Cu content in the Ti-Cu alloys. An increase in the amount of Ti_2Cu

does not give a role in antibacterial ability. Thus, it is the Cu atoms in the α phase which are easier to release from the alloy and play a role in killing bacterial colonies.

4. Conclusions

Ti-Cu alloys with content of 0.5–3 wt% Cu shows α-Ti and Ti_2Cu intermetallic elongated particles. On the addition of 4 wt% Cu, the alloy showed the presence of eutectoid lamellar structure. Meanwhile, the hypereutectoid structures (lamellar structures and Ti_2Cu phases) appear for the addition of 8 and 10 wt% Cu. The addition of Cu to titanium increases the hardness of the Ti-Cu alloy. The higher the copper content, the higher the hardness of Ti-Cu alloys. At a content of 10 wt% Cu, the hardness of the alloy reaches the highest value compared to hardness at the addition of other wt% Cu. There was no strong correlation between antimicrobial ability and variation of Cu content in Ti-Cu alloys. The highest bacterial kill rate is found in addition 3 wt% Cu. The Ti_2Cu phase increases with increasing Cu content in the alloy, but does not provide a significant role in antibacterial ability. The Cu atom in the alpha phase which is more easily released from the alloy, plays a role in killing bacterial colonies.

5. Acknowledgements

Authors gratefully acknowledge Mycology Laboratory and Department of Metallurgical Engineering, Bandung Institute of Technology for supporting the research.

6. References

[1] Williams, D.F., Titanium for Medical Applications. In Brunette, D.M., Tengvall, P., Textor, M. and Thomsen, P. (Eds.), Titanium in Medicine, 2001 ed., Berlin, 13-24.
[2] Oldani, C., Dominguez, A., Titanium as a biomaterial for implants. In: Fokter, S. (Ed.), Recent Advances in Arthroplasty. InTech, Rijeka, Croatia; 2012. ISBN: 978-953-307-990-5. Available from: http://www.intechopen.com/books/recent-advances-in-arthroplasty/titanium-as-a-biomaterial-for-implants.
[3] Shirai, T., Tsuchiya, H., Shimizu, T., Ohtani, K., Zen, Y., Tomita, K., Journal of Biomedical Materials Research - Part B Applied Biomaterials, 91(1), 2009, 373-380
[4] Erayman, I., Erdi, F., Kalkan, E., Karatas, Y., Kaya, B., Keskin, F., Izci, E.K., Uyar, M., International Journal of Clinical and Experimental Medicine, 9(4), 2016, 7334-7338.
[5] Peng, J., Ren, Y., He, W., Li, Z., Yang, J., Liu, Y., Zheng, Z., Kates, S.L., Schwarz, E.M., Xie, C., Xu, Y., Journal of Bone and Joint Infection, 2(3), 2017, 149-153.
[6] O'Gorman, J., Humphrey, H., Journal of Hospital Infection, 81(4), 2012, 217-223
[7] Grass, G., Rensing , C., Solioz, M., Metallic Copper as an Antimicrobial Surface, Journal of American Society for Microbiology, 77(5), 2011, 1541-1547.
[8] Borkow, G., Use of Biocidal Surfaces for Reduction of Healthcare Acquired Infections, Springer International, 2014, pp. 85-101
[9] Osório, W.R., Cremasco, A., Andrade, P.N., Garcia, A., Caram, R., Electrochimica Acta, 55, 2010, 759-770.

[10] Kikuchi, M., Takahashi, M., Okuno, O., Elastic Moduli of cast Ti-Au, Ti-Ag, and Ti-Cu alloys. Journal of Dental Materials, 22, 2006, 641-646.
[11] Yao, X., Sun, Q.Y., Xiao, L., Sun, J., Journal of Alloys and Compounds, 484, 2009, 196-202.
[12] Zhang, E., Li, F., Wang, H., Liu, J., Wang, C., Li, M., Yang, K., Materials Science and Engineering C, 33, 2013, 4280–4287
[13] Ren, L., Ma, Z., Li, M., Zhang, Y., Liu, W., Liao, Z., Yang, K., Journal of Materials Science and Technology, 30(7), 2014, 699-705.
[14] Baker, H., Okamoto, H., Henry, S.D., ASM Handbook, Vol. 3, Alloy Phase Diagrams, ASM International Materials Park, Ohio, 1992, ISBN-10: 0871703815, ISBN-13: 978-0871703811.
[15] Hayama, A.O.F., Andrade, P.N., Cremasco, A., Contieri, R.J., Afonso, C.R.M., Caram, R., Journal of Materials and Design, 55, 2014, 1006-1013.
[16] Kikuchi, M., Takada, Y., Kiyosue, S., Yoda, M., Woldu, M., Cai, Z., Okuno, O., Okabe, T., Dental Materials, 19, 2003, 174–181.
[17] Epling, G.A., Lin, C., Chemosphere, 46, 2002, 561–570.
[18] Xia, J., Yang, C., Xu, D., Sun, D., Nan, L., Sun, Z., Li, Q., Gu, T., Yang, K., Biofouling, 31(6), 2015, 481-492.
[19] IPCS, Copper: Environmental Health Criteria 200, International Programme on Chemical Safety, World Health Organization, Geneva, 1998.
[20] Ma, Z., Ren, L., Liu, R., Yang, K., Zhang, Y., Liao, Z., Liu, W., Qi, M., Misra, R.D.K., Journal of Materials Science & Technology, 31, 2015, 723-732.

Effect of Shielding Gas and Filler Metal to Microstructure of Dissimilar Welded Joint Between Austenitic Stainless Steel and Low Carbon Steel

Dewin Purnama[1,a], Herry Oktadinata[2b]

[1] Department of Mechanical Engineering, Politeknik Negeri Jakarta. Depok, Indonesia
[2] Department of Mechanical Engineering, Universitas Jenderal Achmad Yani, Bandung, Indonesia

[a)] Corresponding author: dewinto@gmail.com
[b)] Corresponding author: herry.oktadinata@yahoo.com

Abstract: Welding of dissimilar metals involves different types of metals which have different chemical composition, mechanical and thermal properties. This paper presents a study on dissimilar metal weld between type 304 stainless steel and SS400 carbon steel. The steel plates were welded by using different filler metals (solid wire, flux-cored wire) and different shielding gases (100%CO_2, 90%Ar+10%CO_2). Then the weldments were observed by optical microscope, scanning electron microscope and EDX analysis. It seems that the weldments which were welded by using 100%CO_2 shielding gas have significant amount of inclusions in weld metal compared to 90%Ar+10%CO_2 shielding gas mixture. The EDX analysis shown the inclusions are oxides which consist of some elements (C, Si, Cr, Mn) with O. The weldment which was welded by using 90%Ar+10%CO_2 shielding gas mixture and solid wire has much less inclusions in weld metal.

Keywords: dissimilar metal weld (DMW), 304 stainless steel, SS400 carbon steel, microstructure, inclusions

1. Introduction

Dissimilar metal weld (DMW) which applied for joining of low carbon steels to austenitic stainless steels is frequently applied in the automobile, nuclear plants, petroleum and petrochemical industries. A serious concern is noticeable in recent years where DMW joints indicated a small-localized zones near fusion boundary which show high hardness region [1]. The difference in chemistry, mechanical and thermal properties of metals influence complexity of DMW, so DMW is challenging and need more attention compared to similar metal weld [2, 3].

Fusion welding is generally applied for joining of similar and dissimilar metals. Gas–tungsten arc welding (GTAW) is preferred for similar or dissimilar metals weld consider good quality result and good appearance of workpiece. However, to achieve good quality joints, DMW requires proper selection of welding parameters, filler metal, shielding gas, welding method, etc. Austenitic stainless steel has an excellent combination of corrosion resistance and mechanical properties so that often used in petrochemical, power plant, automotive, oil and gas industries [2].

DMWs are made from two metals which have significant different in chemical composition and mechanical properties. When dissimilar metals are joined by an arc welding process, mixing of parent metal and filler metal in the fusion zone produce different mechanical properties and this phenomenon influence the distribution of residual stress after welding [4]. Gas metal arc welding (GMAW) method is commonly applied in steel fabrication for both similar and dissimilar metal welding, on the other hand GTAW method usually is selected to achieve high quality welds in a various metals, but it does not usually used in thick metals due to its low deposition rate so that poor productivity. In fusion welding operation, welding parameters such as voltage, current, welding speed, polarity, filler metal, shielding gas and gas flow rate have all a significantly influence on weld quality [5].

Interaction factor of filler metal, welding speed, and current have affected the mechanical properties of DMW between stainless steel (AISI 304) and carbon steel. The results of ultimate tensile strength and elongation for DMW between AISI 304 stainless steel and carbon steel showed there was significant different due to interaction factor of filler metal, welding speed, and current [3].

Specification of filler metal has significant affect to weldment. From the study of the various experiments for DMW between 316LN austenitic stainless steel and alloy 800, selection of the nickel-based consumables show better tensile strength and improved thermal stability compared to 316 austenitic steel welding wire [6].

Research of weldability of AISI 4140 and AISI 316 by GTAW method with and without filler metal (autogeneous) have been reported by Madduru et al. [7]. Carbides was found at the heat affected zone (HAZ) of AISI 316 side when GTAW method without filler metal (autogeneous) was performed. It would be probably occurred due to carbon element diffused from AISI 4140 low alloy steel to AISI 316 stainless steel. Hardness distribution of GTAW steel weldments with and without filler metal show clearly that the highest hardness was measured at the HAZ of AISI 4140 in autogeneous GTA weldments [7].

M.T. Liao et al. [8] reported that the microstructure evolution of stainless steel weldment are influenced by the shielding gas which used to protect the weld pool from atmospheric gas such as nitrogen and oxygen. The shielding gas type not only influence microstructure of the weldment but also influence the shape and weld penetration. During welding, the shielding gas also interacts with the filler metal to produce microstructure evolutions which influence mechanical properties and corrosion resistance of particular weld deposits. The spatter rates increase and ferrite number decreases with increasing the CO_2 percentage of the Ar+CO_2 shielding gas mixture from 2 to 20%. The increase of CO_2 percentage will raise carbon content in the weld deposits, therefore the ferrite number will decrease and nickel equivalent become larger. Also, the increase of CO_2 content will speed up the consumption of Cr and Si elements due to oxidation effect and make the Cr equivalent smaller [8].

D. Katherasan et al. [9] studied a 316L stainless steel which was welded by flux-cored arc welding (FCAW) method with various shielding gases. Impact toughness and ferrite amount of the weld metals tend to decreased by increasing the CO_2 in shielding gas mixtures, when impact test was performed at temperatures of 25 ^{0}C, -100 ^{0}C and -196 ^{0}C. The average values of impact toughness decreased with decreasing test temperatures and increasing the CO_2 percentage in shielding gas mixtures [9]. On the other hand, the CO_2 percentage in the shielding gas mixtures has significantly influenced on the size of undercut weld defects. The width, length and volume of undercut weld defects decreased with increasing CO_2 percentage. When the CO_2 percentage was more than 20%, welding spatters deteriorated the weld bead appearance [10].

During welding operation, the weld pool is protected by the shielding gas from direct contact to the atmospheric gas. Proper selection of the shielding gas influence welding quality. Argon is widely used for shielding gas, argon often mixtured with some other gases that can be inert such as helium, or active gas such as CO_2, O_2 or H_2 [5].

Gasem et al. [1] suggested that the fusion boundary on the carbon steel side is the problem area of DMWs joint, because the hard zones present along the fusion boundary. These hard zones are very thin and noncontinuous layers (typically 0.025 mm wide) with a hard phase. Microscopic observation by using Energy Dispersive X-Ray Spectrometry (EDS) on some different DMWs revealed that the hard zones containing 3-5% chromium and 2-3% nickel [1]. Microstructure analysis and EDS observation indicated that these hard zones containing chromium carbides in a martensite matrix. Filler metals which containing chromium element will produce a similar hard zone in the fusion boundary, it explains why E309 (austenitic filler metal with 24 wt% Cr) resulting welds with high amount of hard zone, and hence susceptible to sulfide stress cracking compared to Inco 182 (Ni-base filler metal with 15 wt %Cr) [1].

In this study, the effects of various filler metal and shielding gases on the microstructure of weldments were investigated. The specimens were observed with optical microscope and scanning electron microscope.

2. Materials and Methods

In these experiments, 304 austenitic stainless steel and SS400 low carbon steel plates were used as the base metals. The chemical composition of both 304 and SS400 steel plates are provided in Table 1 and 2.

Eight samples of 304 austenitic stainless steel plates were cut with the dimension of 90 x 50 x 3 mm. And another eight samples of SS400 low carbon steel plates were cut with the dimension of 120 x 80 x 6 mm. Those plates 304 austenitic stainless steel and SS400 low carbon steel were fillet welded as shown in Figure 1.

TABLE 1. Chemical analysis of the present 304 steel (wt%)

C	Cr	Ni	Mn	Si	P	S	Fe
0.025	18.2	8.36	1.18	0.338	0.019	0.005	bal.

TABLE 2. Chemical analysis of the present SS400 steel (wt%)

C	Mn	Si	P	S	Cr	Al	Cu	Fe
0.14	0.705	0.255	0.009	0.007	0.015	0.048	0.006	bal.

The welding processes in this experiment were gas metal arc welding (GMAW) and flux-cored arc welding (FCAW). Welding was performed with the flat position (1F). The various filler metal and shielding gas which used on each specimens is shown in Table 3.

The metallographic specimens were prepared to perform microstructural observation of the weld joint (Figure 2). The specimens were mirror polished and etched with 2% nital solution to show microstructure of low carbon steel, and etched by aqua regia solution (HNO_3 and HCl in a ratio of 1:1) to find the microstructure of weld metal with optical microscope and scanning electron microscope.

TABLE 3. The welded specimens with various filler metal and shielding gas

Specimen	Welding method	Filler metals	Diameter of filler metal (mm)	Shielding gas
S1	GMAW	ER309LSi	0.9	100% CO_2
S2	GMAW	ER309LSi	0.9	90%Ar+10%CO_2
S3	FCAW	E316L	1.2	100% CO_2
S4	FCAW	E316L	1.2	90%Ar+10%CO_2
S5	FCAW	E308L	1.2	100% CO_2
S6	FCAW	E308L	1.2	90%Ar+10%CO_2
S7	FCAW	E309L	1.2	100% CO_2
S8	FCAW	E309L	1.2	90%Ar+10%CO_2

IOP Conf. Series: Materials Science and Engineering **547** (2019) 012003 doi:10.1088/1757-899X/547/1/012003

FIGURE 1. Welded samples

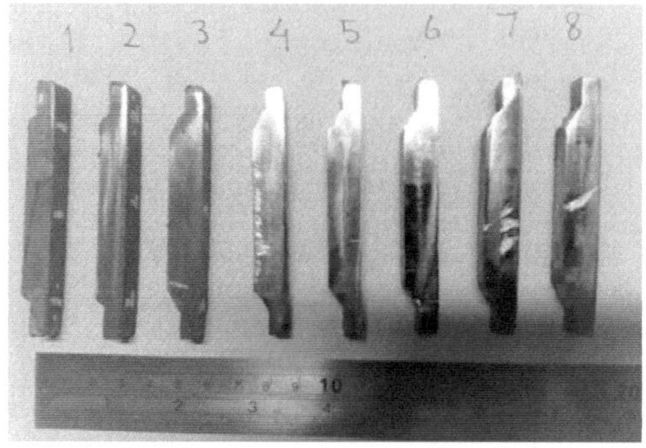

FIGURE 2. The metallographic specimens were cut from each welded samples

3. Results and Discussions

The microstructures were observed on each specimens using optical microscope. The observation regions including BM, HAZ, and WM.

(a) (b)

FIGURE 3. Optical microscope images of specimen S4 : (a) HAZ of SS400 side, 50X magnification. (b) Fusion line between SS400 base metal and weld metal, 500X magnification.

Microstructure observation of the BM revealed that SS400 carbon steel was composed of ferrite (light region) and pearlite (dark region), and 304 stainless steel was composed of austenitic grains as shown in Figure 4.

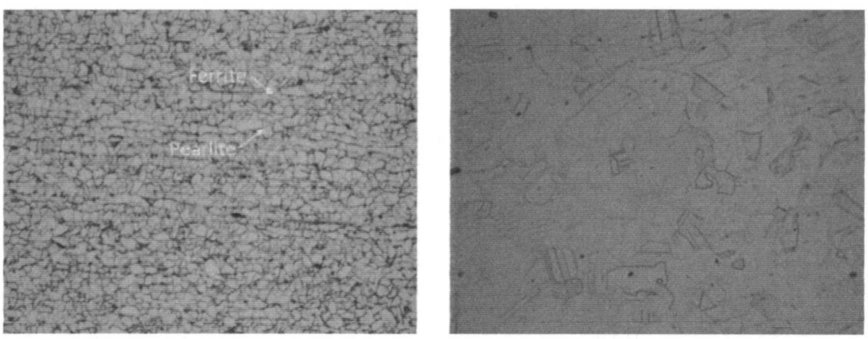

(a) (b)

FIGURE 4. Optical microscope images of parent metal: SS400 carbon steel (a), 304 stainless steel (b)

The HAZ of SS 400 side consist of martensite and coarse grain heat affected zone (CGHAZ) as shown in Figure 5. The microstructure in the fusion boundary have a partially mixed zone, that is a narrow region where the composition gradually changes from the WM to the BM.

IOP Conf. Series: Materials Science and Engineering **547** (2019) 012003 doi:10.1088/1757-899X/547/1/012003

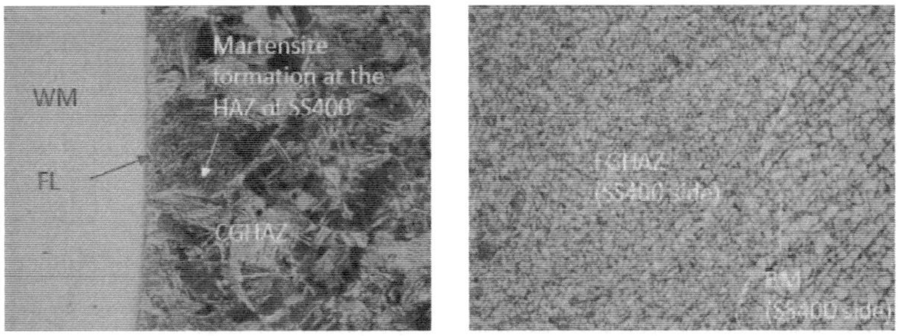

FIGURE 5. Optical microscope images on specimen S1 at the HAZ of SS400 side

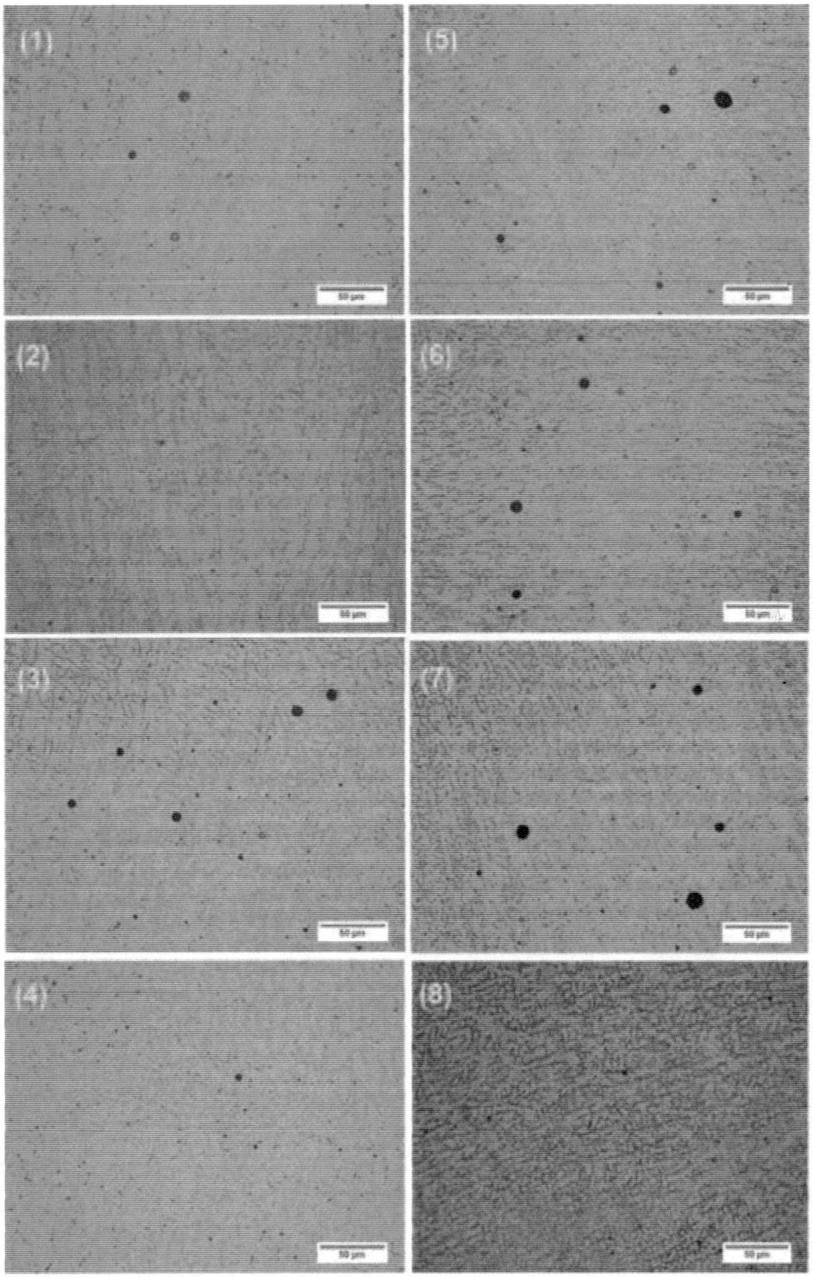

FIGURE 6. Optical microscope images on specimens S1 to S8 at weld metal (WM)

Each specimens were observed using optical microscope and scanning electron microscope. The microstructure of WM was observed with optical microscope using 500X magnification for all specimens S1 to S8 and the results are shown in Figure 6. Specimens which were welded using 100% CO_2 as shielding gas (S1, S3, S5, S7) show more inclusions than specimens which were welded using mixture gas 90%Ar+10%CO_2 (S2, S4, S6, S8). It seems the usage of mixture gas 90%Ar+10%CO_2 as shielding provided better protection from the atmospheric gas compared to CO_2 shielding gas. The inclusions were found in the weld metal may came from the CO_2 of shielding gas which was decomposed at high temperature during welding.

The area fractions that quantitatively measured by image analyzer show specimen S2 (0.17%), S4 (0.42%), S6 (0.34%) and S8 (0.24%) which welded with 90%Ar+10%CO_2 mixture gas have less inclusions than other specimens using 100% CO_2 (0.59 to 0.75%).

Specimen S2 which welded by combination of GMAW method and mixture gas 90%Ar+10%CO_2 has much less inclusions than other specimens. It would be probably the shielding gas used more effectively protect the weld pool from contact with the atmospheric gas during welding operation. Solid wire that using in this GMAW may contributed to reduce inclusions as well.

Figure 6 show that size and volume of inclusions tend to increase with increasing CO_2 content of the shielding gas. These are probably oxide inclusions such as silicon oxides or manganese oxides. The EDX analysis confirmed the inclusions likely combination of (C, Si, Cr, Mn) with O. The oxides present in WM may caused decomposition of CO_2 to carbon and oxygen in the WM, and oxygen interact with filler metal at high temperature during welding operation.

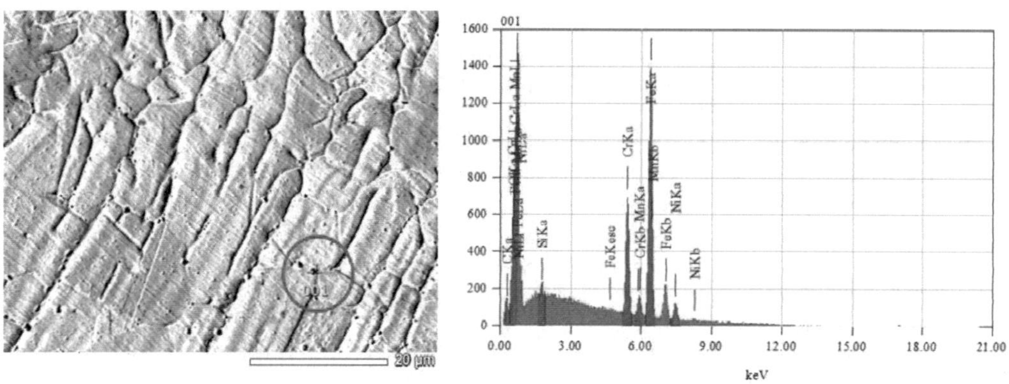

ZAF Method Standardless Quantitative Analysis
Fitting Coefficient : 0.3226

Element	(keV)	Mass%	Error%	Atom%	Compound	Mass%	Cation	K
C K*	0.277	3.80	0.21	14.28				0.8877
O K*	0.525	3.28	0.16	9.26				4.2501
Si K*	1.739	0.51	0.20	0.82				0.3072
Cr K	5.411	18.50	0.30	16.08				21.1098
Mn K*	5.894	1.04	0.41	0.86				1.0516
Fe K	6.398	65.74	0.43	53.20				65.7361
Ni K	7.471	7.14	0.72	5.50				6.6576
Total		100.00		100.00				

FIGURE 7. EDX analysis of specimen S5 showing inclusion elements in the weld metal

From 316L WM to SS400 BM side, Cr & Ni tend to decreased as shown in Figure 8. It could be probably due to Cr & Ni diffuse towards SS400 BM from 316L WM.

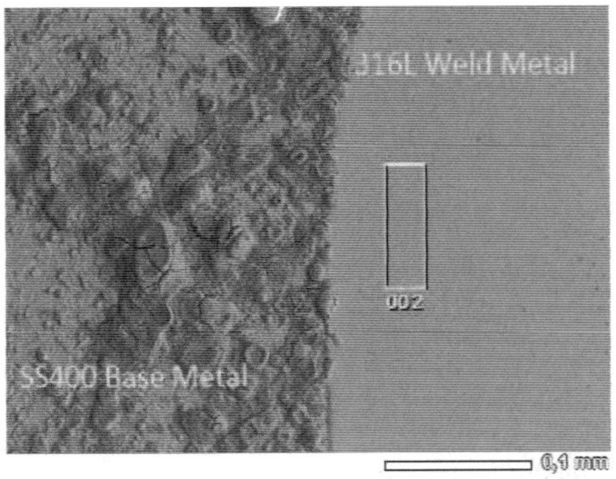

ZAF Method Standardless Quantitative Analysis
Fitting Coefficient : 0.2472

Element	(keV)	Mass%	Error%	Atom%
F K	0.677	3.38	0.15	9.14
Si K	1.739	0.87	0.15	1.59
S K	2.307	0.74	0.10	1.18
Cr K	5.411	16.40	0.22	16.19
Mn K	5.894	1.04	0.30	0.97
Fe K	6.398	69.29	0.31	63.68
Ni K	7.471	8.29	0.53	7.25
Total		100.00		100.00

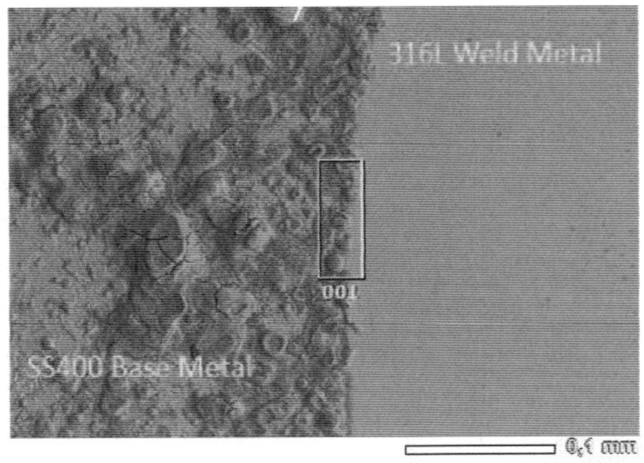

ZAF Method Standardless Quantitative Analysis
Fitting Coefficient : 0.3348

Element	(keV)	Mass%	Error%	Atom%
O K	0.525	19.04	0.21	45.03
Cr K	5.411	3.89	0.37	2.83
Fe K	6.398	74.98	0.55	50.80
Ni K	7.471	2.08	0.93	1.34
Total		100.00		100.00

FIGURE 8. SEM-EDS analysis of specimen S4 (SS400 side, near fusion zone).

From 309L WM to SS400 BM, Cr & Ni tend to decreased as shown in Figure 9. It could be probably due to Cr & Ni diffuse towards SS400 BM from 309L WM.

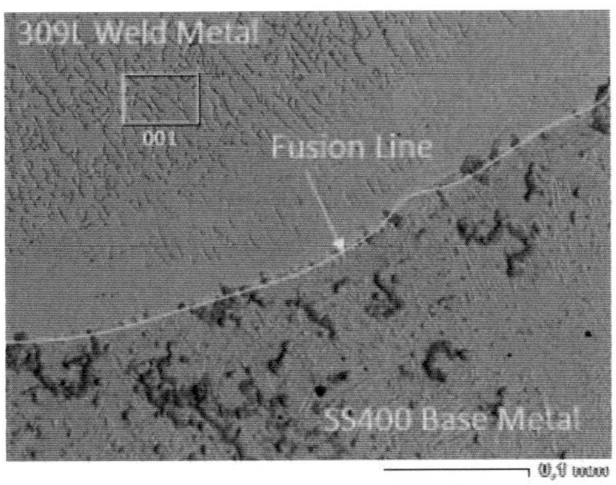

ZAF Method Standardless Quantitative Analysis
Fitting Coefficient : 0.2617

Element	(keV)	Mass%	Error%	Atom%
Cr K	5.411	20.14	0.23	21.42
Mn K	5.894	1.55	0.31	1.56
Fe K	6.398	66.80	0.33	66.16
Ni K	7.471	11.52	0.56	10.85
Total		100.00		100.00

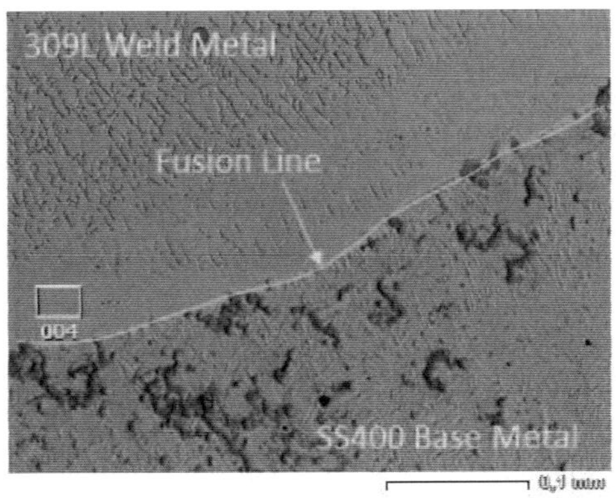

ZAF Method Standardless Quantitative Analysis
Fitting Coefficient : 0.2733

Element	(keV)	Mass%	Error%	Atom%
Cr K	5.411	12.43	0.27	13.27
Fe K	6.398	80.96	0.40	80.47
Ni K	7.471	6.61	0.68	6.25
Total		100.00		100.00

FIGURE 9. SEM-EDS analysis of specimen S8 (SS400 side, near fusion zone).

4. Conclusions

The conclusions of this DMW study can be summarized as follows.

1. Effect of different types of filler metal and shielding gas has significantly influenced the microstructure of WM for DMW of 304 stainless steel and SS400 low carbon steel.
2. DMWs that use 100% CO_2 shielding gas produced more inclusions than 90%Ar + 10%CO_2 mixture gas. The inclusions were found oxides such as silicon oxides or manganese oxides. It may caused decomposition of CO_2 to carbon and oxygen in the WM, and oxygen interact with filler metal at high temperature during welding operation.
3. Chromium and nickel content in the WM tend to decreased towards the SS400 BM. It could be probably due to Cr & Ni diffuse towards Carbon Steel BM from Stainless Steel WM.

IOP Conf. Series: Materials Science and Engineering **547** (2019) 012003 doi:10.1088/1757-899X/547/1/012003

5. References

[1] Gasem M. Fallatah, Anwar K. Sheikh, Zafarullah Khan, John K. Boah, *Reliability of Dissimilar Metal Welds Subjected to Sulfide Stress Cracking, The 6th Saudi Engineering Conference*, 2002.

[2] Anirban, Mechanical *Properties and Metallurgical Characterization of Dissimilar Welded Joints between AISI 316 and AISI 4340, Trans Indian Inst Met*, 2016.

[3] S. Kaewkuekool, B. Amornsin, *A Study of Parameters Affecting to Mechanical Property of Dissimilar Welding Between Stainless Steel (AISI 304) and Low Carbon Steel, Proceedings of the 1st WSEAS International Conference on Material Science*, 2008.

[4] E. Ranjbarnodeh, S. Seradjadeh, A.H. Kokabi, A. Fischer, *Effect of welding parameters on residual stresses in dissimilar joint of stainless steel to carbon steel, J Mater Sci 46:3225–3232*, 2011.

[5] G. Costanza, A. Sili, M.E. Tata, *Weldability of austenitic stainless steel by metal arc welding with different shielding gas, 21st European Conference on Fracture, ECF21, 20-24 June 2016, Catania, Italy, Procedia Structural Integrity 2 3508–3514*, 2016.

[6] M. Sireesha, V. Shankar, S.K. Albert, S. Sundaresan, *Microstructural Features of Dissimilar Welds Between 316LN Austenitic Stainless Steel and Alloy 800, Materials Science and Engineering A292 74-82*, 2000.

[7] Madduru, A. Aldrin, M.Mohan, *Assessment of Mechanical Properties of AISI 4140 and AISI 316 Dissimilar Weldments, 7th International Conference on Materials for Advanced Technologies, Procedia Engineering 75 29–33*, 2014.

[8] M.T Liao, W.J. Chen, *The effect of shielding-gas compositions on the microstructure and mechanical properties of stainless steel weldments, Materials Chemisrry and Physics 55 145-151*, 1998.

[9] D. Katherasan, P. Sathiya, A. Raja, *Shielding Gas Effects on Flux Cored Arc Welding of AISI 316L (N) Austenitic Stainless Steel Joints, Material and Design 45 43-51*, 2013.

[10] R. Zong, J. Chen, C. Wu, G.K Padhy, *Influence of shielding gas on undercutting formation in gas metal arc welding, Journal of Materials Processing Technology 234 169-176*, 2016.

Effect of Applied Voltage and Coating Time on Nano Hydroxyapatite Coating on Titanium Alloy Ti6Al4V Using Electrophoretic Deposition for Orthopaedic Implant Application

Nuzul Ficky Nuswantoro[1] Ikhwal Budiman[2] Andre Septiawarman[2] Djong Hon Tjong[3] Menkher Manjas[1] Gunawarman[2]*

[1]Biomedical Science Department, Faculty of Medicine, Andalas University, Padang, Indonesia

[2]Mechanical Engineering Department, Engineering Faculty, Andalas University, Padang, Indonesia

[3]Biological Department, Mathematic and Natural Science Faculty, Andalas University, Padang, Indonesia

*Corresponding author (E-mail): gunawarman@eng.unand.ac.id

Abstract. Nanoparticles of Hydroxyapatite (HA) were coated onto biocompatible titanium alloy, Ti6Al4V ELI using Electrophoretic Deposition (EPD) in order to improve its bioactivity. Electrophoretic Deposition (EPD) has been selected as the coating method because of the simplicity of the instrument, inexpensive cost, and ability to coat complicated products. This study, therefore, aims to investigate the effect of voltage and coating time of EPD process on increasing of implant mass as a parameter of deposition rate, coating thickness, and surface coverage of the HA on implant screw prototype products. Voltages were controlled in the range of 3, 5, and 7 volt and coating times were in the range of 3, 5, and 7 minutes. Surface morphology was examined using scanning electron microscopy (SEM). Coating thickness was measured by coating thickness gauges. While surface coverage was determined using ImageJ software. Based on the result, applied voltage and coating time affects the mass growth of samples and HA coating thickness in positive correlation. However, on the surface coverage, applied voltage and coating time reach the optimum value at 5 volt and 5 minutes. The best HA coating in which fulfilling the standard for orthopaedic implants was obtained at 5 volts for 5 minutes with mass growth is 0.00107 g, coating thickness are 79.13 µm, and surface coverage is 97.89%. HA coating thickness that produced in this research has fulfilled the desired coating thickness for orthopaedic implant application (50-100µm). SEM micrographs show that nano-HA is coated the alloy surface uniformly at these parameters. It can be concluded that these parameters can be applied to coat titanium Ti6Al4V ELI with HA for improving bioactivity in the orthopaedic application.

Keywords: Hydroxyapatite, Electrophoretic Deposition, Ti6AL4V ELI, Orthopedic Implant

1. Introduction

Nanoparticles of Hydroxyapatite (HA) are coated onto biocompatible titanium alloy, Ti6Al4V ELI using Electrophoretic Deposition (EPD) in order to improve its bioactivity in orthopaedic implant application[1]–[3]. Hydroxyapatite (HA) had been used worldwide for coating material on a metallic implant for orthopaedic application in order to accelerate bone fracture healing[4]–[8]. The increasing cases of metallic implant failure made the hydroxyapatite coating become one of the good solutions for the problems such as aseptic loosening, metal ion release that can be the trigger metal allergy, and chronic inflammation[9]–[11]. Electrophoretic Deposition (EPD) has chosen as coating method because of the simplicity of the instrument and its making, inexpensive cost, and ability to coat things with the complicated design[12]–[16]. EPD used electric current to move the HA particle through electrode in the suspension of ethanol and HA[14], [17], [18]. Desired HA coating quality can be adjusted by optimizing the voltage and coating time[19]–[21].

Ti6Al4V ELI is the titanium alloy that designed to use for orthopaedic implant application. This material has good mechanical properties such as desired strength for load bearing and dynamic load. However, the inertness of titanium made the Ti6Al4V ELI still have no bioactivity, so that, it needs to be coat with HA for gaining bioactivity properties[22]–[26]. This research aimed to analyze the effect of voltage and coating time of EPD process toward HA coating that produced on the surface of titanium alloy, Ti6Al4V.

2. Materials and Methods

2.1 Sample Preparation

Ti6Al4V ELI Titanium bar in length of 400 mm and 4 mm in diameter was cut and lathe into screw shape with type M3 x 0.5. Screw length and diameter were 5 mm and 3 mm, respectively. Then, the screws were sandblasted to refine the screw surface and clean the remaining waste from the lathe process. After that, screws were cleaned using ultrasonic cleaner in 15 minutes.

2.2 Coating Process using Electrophoretic Deposition

Pretreatment was conducted before the coating process. Screws were submerged in ethanol, acetone, and HNO3 in 15 minutes, respectively. After that, screws were remaining in NaOH for about one hour. HA suspension was made of 1.8g HA powder and 50 ml ethanol, and pH was set about 4. Nano HA powder that used was the commercial Nano HA. The anode was graphite and cathode was Ti6Al4V titanium. Voltages were in the range of 3, 5, and 7 volts and coating time were in the range of 3, 5, and 7 minutes. After the coating process screws were sintered in a vacuum furnace at 700oC with holding time about one hour. This method was modified from the research that has been conducted before[14], [16], [18].

2.3 Hydroxyapatite (HA) coating Characterization

Characterization of HA coating was conducted by an optical microscope that directly connected with a digital camera to obtain the microstructure images of HA coating and continued to assess surface coverage value using software ImageJ. In addition to, screws mass growth was also assessed to confirm the deposition of HA on the screws surface. After that, HA coating thickness was assessed using coating thickness gauges.

3. Results and Discussion

3.1 Surface Morphology of HA coating on the screw surface

Figure 1 shows the HA coating morphology that captured by an optical microscope for each voltages treatment. Based on the image it is shown that there are significant differences on the surface morphology of each sample. The enhancement of voltage will produce HA coating that spread evenly but also has optimum value. Sample from 5-volt treatment shows the best surface coverage of HA coating especially at the thread of screw region. Thread of screw is the important part of the screw that directly contacted with the bone tissue after the implantation, so it is necessary to sure that this region has been coated very well with no empty space that possible the corrosion after long time application.

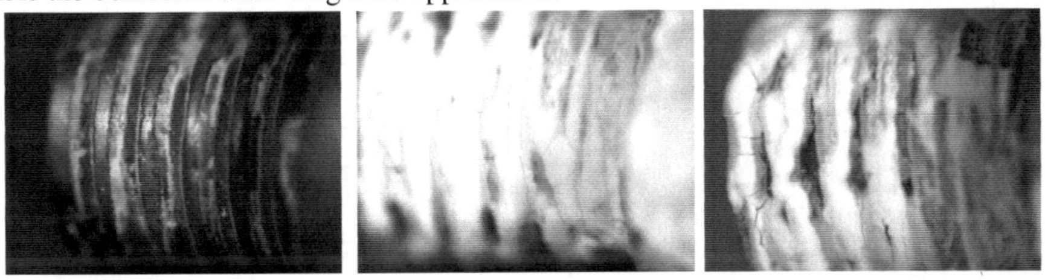

3 volts 5 volts 7 volts

Figure 1. Optical Microscope imaging of surface morphology of HA coating on the screw surface at different voltage

However, the sample from 7-volt treatment shows cracks on the surface of the HA coating. It can be assumed that HA coating that produced by 7-volt treatment has a high thickness that made the bonding between HA particle and titanium surface become weak and lead to cracking. It is also mean that voltage applied higher than 5 volts would produce HA coating with low quality, in other words, there is an optimum value of the applied voltage that should be considered to produce HA coating on Ti6Al4V ELI using EPD.

A research that conducted previously also revealed the same result with the current research. Increasing of the applied voltage from 5V to 20V will produce HA coating that spread more evenly and thicker. However, applying high voltage on EPD process will produce crack on the coating layer that indicated the high voltage is not suitable to produce good HA coating layer on the Ti6Al4V ELI implant surface[17]. Using high voltage up to 200 volts seems to be not recommended to produce desirable HA coating for orthopaedic implant application because the difficulties to produce the crack-free coating that would be impacted the performance of the implants[2].

3 volts 5 volts 7 volts

Figure 2. SEM imaging of surface morphology of HA coating on screw surface at different voltage

SEM imaging (fig.2) also shows the same result with optical microscope images. Sample from 5-volt treatment shows the best surface coverage and there are no cracks can be found. It can be assumed that sample from 5-volt treatment can be sample candidate to produce orthopaedic implant using EPD process. Optical Microscope imaging of surface morphology of HA on the screw surface at different coating time (Fig. 3) showing a similar result with the voltage treatment (Fig. 1). Sample from 5 minutes shows the best result because the HA powder spread more evenly and cracks are not found especially at the thread of screws region. The SEM imaging of the samples also shows the same result (Fig. 4). Another research also revealed the same result with this current research. The deposition of HA particle will increase in compliance with the increasing of coating time. However, it seems to be there is an optimum point of coating time that will produce a better HA coating layer[12], [16], [27]. Based on this result it can be assumed that using coating time up to 7 minutes will produce HA coating that will have the crack on its surface.

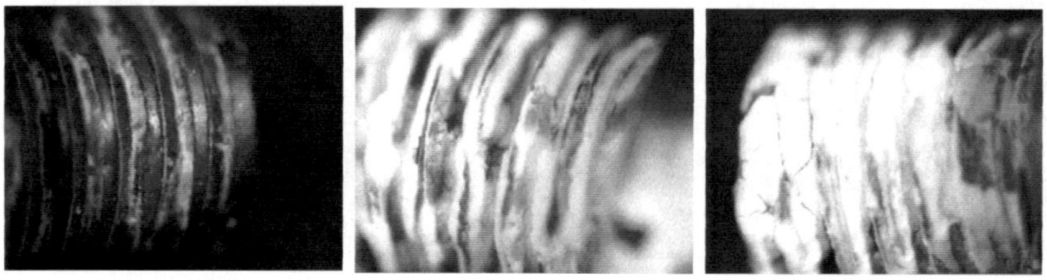

3 minutes 5 minutes 7 minutes

Figure 3. Optical Microscope imaging of surface morphology of HA coating on the screw surface at different coating time

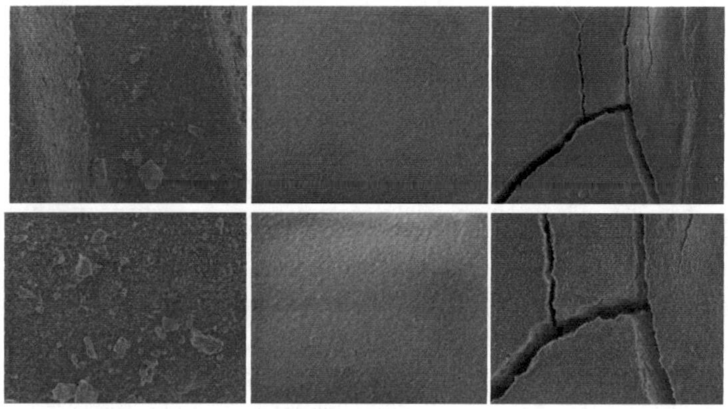

3 minutes 5 minutes 7 minutes

Figure 4. SEM imaging of surface morphology of HA coating on screw surface at different coating time

3.2 Mass Growth of Specimen

Table 1. Mass Growth of Specimen at Different Voltage

No.	Voltage	M_1 (g)	M_2 (g)	ΔM (g)	Average Δm (g)
1	3	0.1848	0.1852	0.0004	0.000333
2		0.1965	0.1968	0.0003	
3		0.2120	0.2123	0.0003	
1	5	0.2372	0.2380	0.0008	0.001067
2		0.2182	0.2192	0.0010	
3		0.2100	0.2114	0.0014	
1	7	0.2212	0.2223	0.0011	0.001567
2		0.2139	0.2153	0.0014	
3		0.2307	0.2329	0.0022	

Table 1 shows the mass growth of specimen after EPD process at a different voltage. This result shows the same pattern with the imaging result from optical microscope and SEM. Based on this result it can be seen that the enhancement of applied voltage will increase the mass of the screws. This data are collected to confirm the existence of HA particle that adheres on the surface of Ti6Al4V screws. Enhancement of screws mass will show the deposition of HA particle on the screws surface. Based on the data, it can be seen that the specimen from 7-volt treatment has the highest mass growth. It means that the enhancement of voltage will increase the HA particle deposition. It can be caused by the enhancement of potential differences on the cathode and anode that produce high energy to throws HA particle onto screws surface that causes higher HA particle deposition. Figure 5 shows the graphic of mass growth of screws at a different voltage in average. Another research also revealed the same result with this current research, increasing the applied voltage will increase the deposition of HA particles significantly[16], [27].

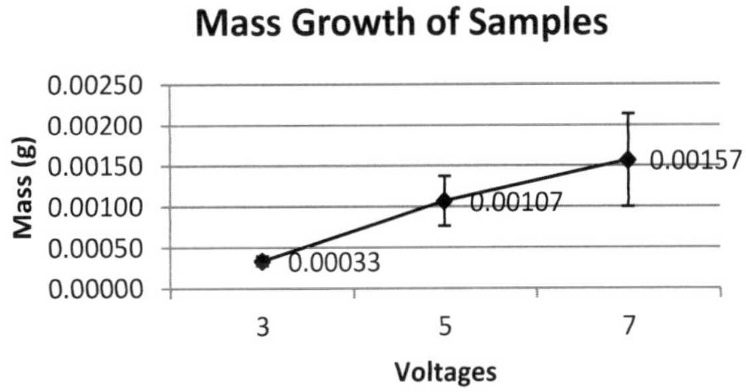

Figure 5. Average of screw's mass growth at different voltage

In compliance with the mass growth of specimen from voltage treatment, the mass growth from the specimen for coating time treatment also revealed a significant result. Table 2 shows the mass growth of specimen after EPD process at different coating time. Based on the data, it can be seen that the enhancement of coating time will increase the mass of the screws. The highest mass growth is found at the specimen from 7 minutes treatment. It can be concluded that enhancement of coating time will increase the deposition of HA particle on the screws surface. Another research also revealed the same result with this current research, increasing the applied voltage will increase the deposition of HA particles significantly[12], [16], [27]. Figure 6 shows the graphic of mass growth of screws at different coating time in Figure 6 shows the graphic of mass growth of screws at different coating time on average. However, this data has not shown the optimum value yet. So that, it is necessary to continue the investigation to understanding the effect of voltage and coating time of the EPD process on the Ti6Al4V implant that could produce optimum HA coating for better orthopaedic implants.

Table 2. Mass Growth of Specimen at Different Coating Time

No.	Time (minutes)	M_1(g)	M_2(g)	ΔM(g)	Average Δm (g)
1	3	0.2002	0.02005	0.0003	0.00043
2		0.2226	0.2231	0.0005	
3		0.2519	0.2524	0.0005	
4	5	0.2372	0.238	0.0008	0.00093
5		0.21	0.2114	0.0014	
6		0.2182	0.2188	0.0006	
7	7	0.1991	0.2009	0.0018	0.0014
8		0.2376	0.2394	0.0018	
9		0.2236	0.2244	0.0008	

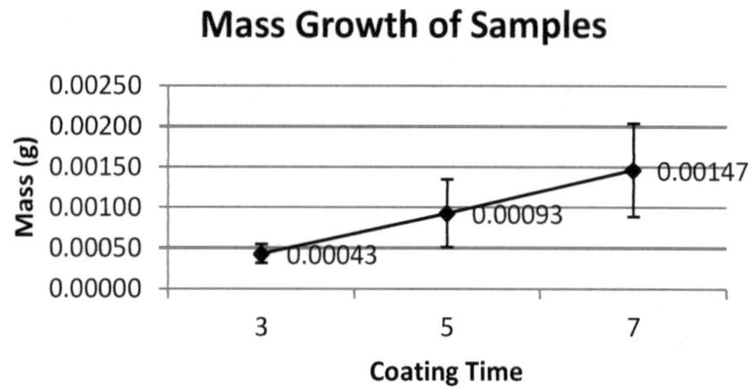

Figure 6. Average of screw's mass growth at different coating time

3.3 Coating Thickness

Table 3. HA Coating Thickness that produce at different voltage

No.	Voltage	Thickness (μm)	Average (μm)
1	3	57.6	54.23
2		54.4	
3		50.7	
1	5	81.2	79.13
2		79.1	
3		77.1	
1	7	90.9	92.13
2		93.1	
3		92.4	

Table 3 shows the data of HA coating thickness that produce on the surface of Ti6Al4V ELI screws implant. Based on the data, it can be seen that the enhancement of applied voltage on the EPD process will increase the HA coating thickness that produces on the surface of the screws. The highest coating thickness is 93.1 μm found at the 7-volt specimen and lowest coating thickness is 50.7 μm found at the 3-volt specimen. However, all the thickness data have fulfilled the characteristic for the orthopaedic implant application. The optimums HA coating thickness that suitable for orthopaedic implant application are between 50-100 μm[28]–[30]. But, the coating thickness is not the only parameter that should be considered in evaluating the ability of an orthopaedic implant since the surface coverage, chemical composition on the implant surface, hydrophobic feature, etc must be considered[28].

In addition to, the effect of HA coating thickness on the titanium implant surface must be concerned for the next research. Because the HA coating thickness on the titanium implant will affect the ability of an implant to keep the biocompatibility and bioactivity features. Except, HA coating thickness also affect the corrosion resistant and shear strength of the metal implant as the result of the reaction that influenced by the biological environment of the human body[31], [32]. Figure 7 shows the HA coating thickness that produces at a different voltage in average. Based on the graphic it can be concluded that enhancement of applied voltage will increase the coating thickness.

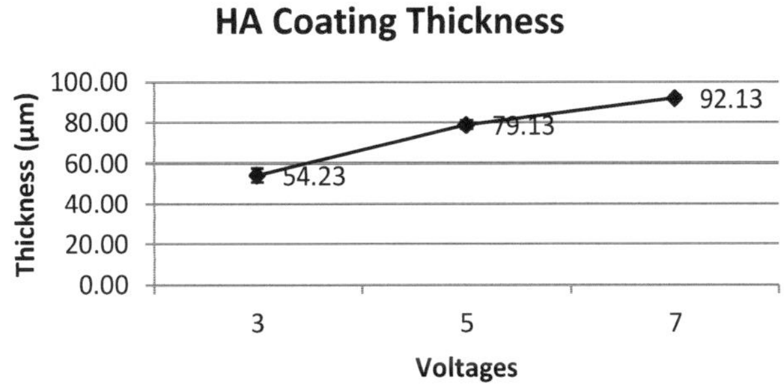

Figure 7. Average of HA coating thickness that produce at different voltage

HA coating thickness that produces on the surface of Ti6Al4V ELI screws implant at different coating time also showing the same pattern with the specimen from voltage treatment. Enhancement of coating time will increase the HA coating thickness on the Ti6Al4V ELI screws implant surface. The highest coating thickness is 94.9 µm found at 7 minutes specimen and the lowest coating thickness is 58.62 µm found at 5 minutes specimen (Table 4 and Fig. 8.).

Table 4. HA coating thickness that produce at different coating time

No.	Time (minutes)	Thickness (µm)	Average (µm)
1	3	66.66	63.94
2		66.52	
3		58.62	
1	5	77.56	78.16
2		79.8	
3		77.14	
1	7	94.9	90.38
2		87.8	
3		88.44	

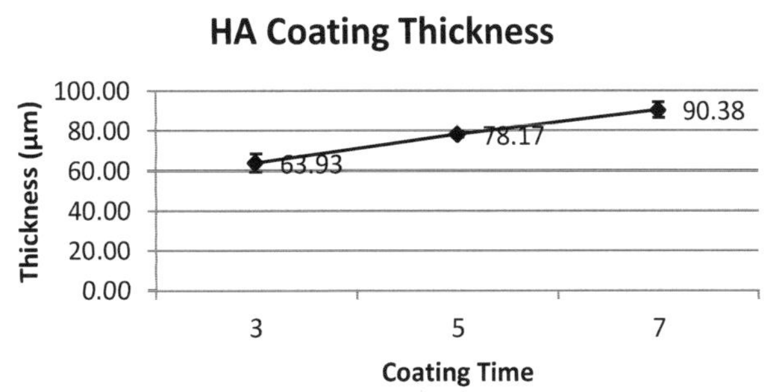

Figure 8. Average of HA coating thickness that produce at different coating time

3.4 Surface Coverage

Figure 9 shows the surface coverage data of HA coating on the surface of Ti6Al4V ELI screws implant. Based on the data it can be seen that the highest surface coverage is about 97% found at the 5-volt specimen and the lowest surface coverage is 73% found at the 3-volt specimen. However, at the 7-volt specimen the surface coverage value is decreased. It can be concluded that there is an optimum value of the applied voltage that can produce better HA coating on the Ti6Al4V ELI screws implant. Surface coverage is an important parameter for evaluating the quality of HA coating on the implant. Data from coating time treatment also shows the same pattern with the voltage treatment. The best surface coverage is about 82% found at 5 minutes specimen (Fig. 10.). Another research also revealed that HA coating that prepared at a constant low voltage consisted of fine HA particle and was dense. While the HA coating that prepared at constant high voltage consisted of big HA particle and was porous[15].

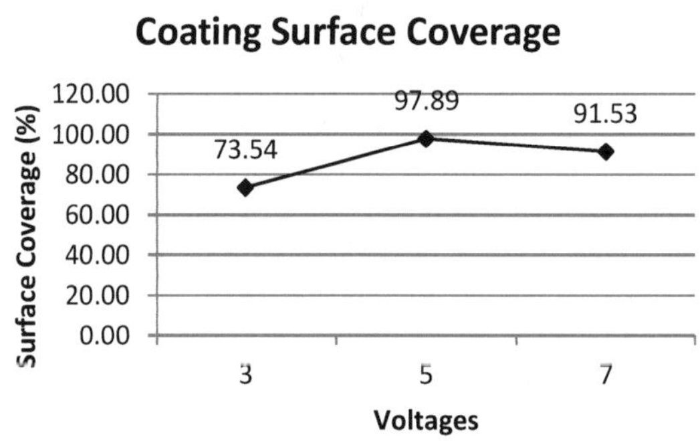

Figure 9. Surface Coverage of HA coating at different voltage

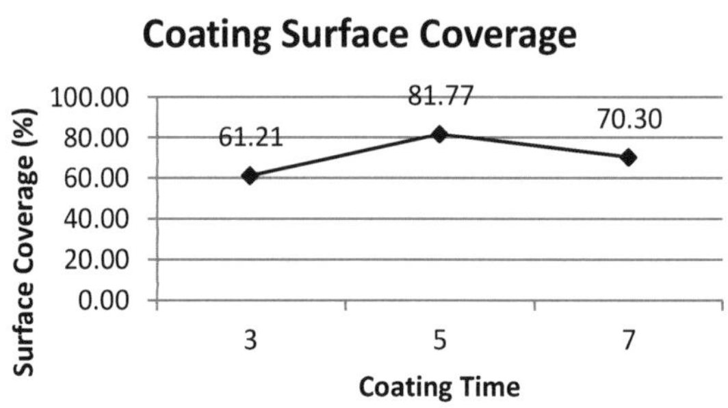

Figure 10. Surface Coverage of HA coating at different coating time

4. Conclusions

1. Applied voltage and coating time affects the mass growth of samples and HA coating thickness in positive correlation.
2. However, on the surface coverage, applied voltage and coating time reach the optimum value at 5 volt and 5 minutes.
3. The best HA coating in which fulfilling the standard for orthopaedic implants was obtained at 5 volts for 5 minutes with mass growth is 0.00107 g, coating thickness are 79.13 μm, and surface coverage is 97.89%.
4. HA coating thickness that produced in this research has fulfilled the desired coating thickness for orthopaedic implant application (50-100μm).

5. Acknowledgements

The corresponding author thanks Ministry of Research and Education, Republic of Indonesia, for providing financial support for Pendidikan Magister Menuju Doktor Untuk Mahasiswa Unggul (PMDSU) research program on the contract year 2018. Thank you to Mr T.O. and his crew from Wira Usaha Workshop for the great help in making the Ti6Al4V screws. Also, thank you to the anonymous reviewer for the advice to make this paper journal better.

6. References

[1] C. T. Kwok, P. K. Wong, F. T. Cheng, and H. C. Man, "Applied Surface Science Characterization and corrosion behavior of hydroxyapatite coatings on Ti6Al4V fabricated by electrophoretic deposition," *Appl. Surf. Sci.*, vol. 255, pp. 6736–6744, 2009.

[2] V. O. Kollath *et al.*, "AC vs . DC Electrophoretic Deposition of Hydroxyapatite on Titanium," *J Eur Ceram Soc*, pp. 1–12, 2013.

[3] D. Juliadmi, V. R. Fauzi, Gunawarman, H. Nur, and M. H. Idris, "Hydroxyapatite Coating on Titanium Alloy Ti-6Al-4V with Electrophoretic Deposition (EPD) for Dental Root Application," *Int. J. Adv. Sci. Eng. Informational Technol.*, vol. 7, no. 6, pp. 2152–2158, 2017.

[4] F. Schwarz *et al.*, "Effects of surface hydrophilicity and microtopography on early stages of soft and hard tissue integration at non-submerged titanium implants: an immunohistochemical study in dogs.," *J. Periodontol.*, vol. 78, no. 11, pp. 2171–2184, 2007.

[5] J. D. Voigt and M. Mosier, "Hydroxyapatite (HA) coating appears to be of benefit for implant durability of tibial components in primary total knee arthroplasty.," *Acta Orthop.*, vol. 82, no. 4, pp. 448–459, 2011.

[6] N. Safuan, I. Sukmana, M. R. A. Kadir, and D. Noviana, "The Evaluation of Hydroxyapatite (HA) Coated and Uncoated Porous Tantalum for Biomedical Material Applications," *ScieTech. J. Phys. Conf. Ser.*, vol. 495, 2014.

[7] S. D. Cook, K. a Thomas, J. F. Kay, and M. Jarcho, "Hydroxyapatite-Coated Titanium for Orthopedic Implant Applications," *Clin Orthop Relat Res*, vol. 232, pp. 225–243, 1988.

[8] N. F. Nuswantoro, Gunawarman, M. R. Saputra, I. P. Nanda, M. H. Idris, and A. Arafat, "Microstructure Analysis of Hydroxyapatite Coating on Stainless Steel 316L Using Investment Casting Technique for Implant Application," *Int. J. Adv. Sci. Eng. Informational Technol.*, vol. 8, no. 5, pp. 2168–2174, 2018.

[9] D. J. Langton, S. S. Jameson, T. J. Joyce, N. J. Hallab, S. Natu, and A. V. F. Nargol, "Early failure of metal-on-metal bearings in hip resurfacing and large-diameter total hip replacement A CONSEQUENCE OF EXCESS WEAR," *J Bone Jt. Surg [Br]*, vol. 92–B, pp. 38–46, 2010.

[10] D. J. Langton *et al.*, "Adverse reaction to metal debris following hip resurfacing," *J Bone Jt. Surg [Br]*, vol. 93–B, pp. 164–171, 2011.

[11] D. J. Langton *et al.*, "Blood metal ion concentrations after hip resurfacing arthroplasty A COMPARATIVE STUDY OF ARTICULAR SURFACE REPLACEMENT," *J Bone Jt. Surg [Br]*, vol. 91–B, no. 10, pp. 1287–1295, 2009.

[12] R. Anand, "Electrophoretic Deposition of Hydroxyapatite on Ti6Al4V," 2012.

[13] Z. Feng and Q. Su, "Electrophoretic Deposition of Hydroxyapatite Coating," *J Mater Sci Technol*, vol. 19, no. 1, pp. 30–32, 2003.

[14] A. R. Boccaccini, S. Keim, R. Ma, Y. Li, and I. Zhitomirsky, "Electrophoretic deposition of biomaterials.," *J. R. Soc. Interface*, vol. 7, pp. S581–S613, 2010.

[15] X. Meng, T. Kwon, Y. Yang, J. L. Ong, and K. Kim, "Effects of Applied Voltages on

Hydroxyapatite Coating of Titanium by Electrophoretic Deposition," *J. Biomed. Mater. Res. Part B Appl. Biomater.*, vol. 78B, pp. 373–377, 2006.
[16] N. F. Nuswantoro, I. Maulana, H. T. Djong, M. Manjas, and Gunawarman, "Hydroxyapatite Coating on New Type Titanium , TNTZ , Using Electrophoretic Deposition," *J. Ocean. Mech. Aerosp.*, vol. 56, no. 1, pp. 1–4, 2018.
[17] R. Drevet, N. Ben Jaber, J. Fauré, A. Tara, A. B. C. Larbi, and H. Benhayoune, "Electrophoretic deposition (EPD) of nano-hydroxyapatite coatings with improved mechanical properties on prosthetic Ti6Al4V substrates," *Surf. Coat. Technol.*, pp. 1–6, 2015.
[18] I. Zhitomirsky and L. Gal-Or, "Electrophoretic deposition of hydroxyapatite," *J. Mater. Sci. Mater. Med.*, vol. 8, pp. 213–219, 1997.
[19] S. Seuss, M. Lehmann, and A. R. Boccaccini, "Alternating Current Electrophoretic Deposition of Antibacterial Bioactive Glass-Chitosan Composite Coatings," pp. 12231–12242, 2014.
[20] A. Molaei, M. Yari, and M. R. Afshar, "Modification of electrophoretic deposition of chitosan–bioactive glass–hydroxyapatite nanocomposite coatings for orthopedic applications by changing voltage and deposition time," *Ceram. Int.*, vol. 41, no. 10, pp. 14537–14544, 2015.
[21] T. A. Rad, M. Solati-Hashjin, N. A. A. Osman, and S. Faghihi, "Improved bio-physical performance of hydroxyapatite coatings obtained by electrophoretic deposition at dynamic voltage," *Ceram. Int.*, vol. 40, no. 8, pp. 12681–12691, 2014.
[22] M. Shahrezaei, E. Mirtaheri, H. Miryousefi, and M. Saremi, "Effects of Substrate and Electrodeposition Parameters on the Microstructure of Hydroxyapatite Coating," *J Arch Mil Med*, vol. 4, no. 2, pp. 2–5, 2016.
[23] X. F. Xiao and R. F. Liu, "Effect of suspension stability on electrophoretic deposition of hydroxyapatite coatings," *Mater. Lett.*, vol. 60, pp. 2627–2632, 2006.
[24] A. Araghi and M. J. Hadianfard, "Fabrication and characterization of functionally graded hydroxyapatite/TiO2 multilayer coating on Ti-6Al-4V titanium alloy for biomedical applications," *Ceram. Int.*, 2015.
[25] B. Gunawarman, M. Niinomi, and T. Akahori, "Mechanical properties of Ti – 4 . 5Al – 3V – 2Mo – 2Fe and possibility for healthcare applications," *Mater. Sci. Eng. C*, vol. 25, pp. 296–303, 2005.
[26] Gunawarman, M. Niinomi, D. Eylon, S. Fujishiro, and C. Ouchi, "Effect of b Phase Stability at Room Temperature on Mechanical Properties in b -Rich a ⁄-b Type Ti – 4 . 5Al – 3V – 2Mo – 2Fe Alloy," *ISIJ Int.*, vol. 42, no. 2, pp. 191–199, 2002.
[27] A. A. Abdeltawab, M. A. Shoeib, and S. G. Mohamed, "Electrophoretic deposition of hydroxyapatite coatings on titanium from dimethylformamide suspensions," *Surf. Coat. Technol.*, vol. 206, no. 1, pp. 43–50, 2011.
[28] H. Daugaard, B. Elmengaard, J. E. Bechtold, T. Jensen, and K. Soballe, "The effect on bone growth enhancement of implant coatings with hydroxyapatite and collagen deposited electrochemically and by plasma spray," *J Biomed Mater Res A.*, vol. 92, no. 3, pp. 913–921, 2010.
[29] A. K. Lynn and D. L. DuQuesnay, "Hydroxyapatite-coated Ti-6Al-4V part 1 the effect of coating thickness on mechanical fatigue behaviour." pp. 1937–1946, 2002.
[30] H. Li *et al.*, "Effect of Thickness of HA-Coating on Microporous Silk Scaffolds Using Alternate Soaking Technology," *Biomed Res. Int.*, 2014.
[31] D. Tang, "Effect of substrate preheating temperature and coating thickness on residual stress in plasma sprayed hydroxyapatite coating," *Glob. Conf. Polym. Compos. Mater. (PCM). IOP Conf. Ser. Mater. Sci. Eng.*, vol. 87, 2015.
[32] M. Svehla, P. Morberg, W. Bruce, B. Zicat, and W. R. Walsh, "The Effect of Substrate

IOP Conf. Series: Materials Science and Engineering **547** (2019) 012004 doi:10.1088/1757-899X/547/1/012004

Roughness and Hydroxyapatite Coating Thickness on Implant Shear Strength," *J. Arthroplasty*, vol. 17, no. 3, pp. 304–311, 2002.

The Effect of Sintering Temperature on Bilayers Hydroxyapatite Coating of Titanium (Ti-6Al-4V) ELI by Electrophoretic Deposition for Improving Osseointegration

Dian Juliadmi[1], Harlendri[2], Djong Hon Tjong[3], Menkher Manjas[1], Gunawarman[2]*

[1]Biomedical Science, Medical Faculty, Andalas University, Padang, INDONESIA
[2]Mechanical Engineering, Engineering Faculty, Andalas University, Padang, INDONESIA
[3]Biology Department, Mathematics and Science Faculty, Andalas University, Padang, INDONESIA
*Corresponding author: gunawarman@eng.unand.ac.id

Abstract. Bilayers hydroxyapatite (HA) coating using electrophoretic deposition (EPD) method has been applied into a biocompatible metal, Ti-6Al-4V ELI, in order to improve bioactivity of the alloy and then to accelerate bone remodelling and osseointegration during implantation. Two steps EPD process wasconducted for obtaining based layer consist of HA nanoparticles using voltage 3 volt for 3 minutes. It was then followed by second layer formation consist of HA micro particles using voltage 5 volt for 5 minutes. The coated alloy was then sintered at temperatures of 700°C, 800°C and 900°C for each 10 minutes within vacuum tube furnace at the sintering rate 20°C/s. Surface morphologies were examined using optical microscope and scanning electron microscope (SEM). The result of this study showed that bilayers hydroxyapatite coating onto Ti-6Al-4V surface was produced through EPD with those parameters. Sintering treatment at temperature 800°C gives thin, dense, even and crack-free coatings. The Ti-6Al-4V ELI surface is almost fully covered by bilayers hydroxyapatite (with surface coverage 98.6%) with adequate bonding (without lost weight of coating layers). An improper coating layer is obtained for lower (700°C) and higher (900°C) temperature. The sintered samples at temperature 900°C have micro cracks, void and debonding layers. It is expected such bilayers hydroxyapatite with sintering at 800°C can initiate mineralization and then make a quick osseointegration.

Keywords: bilayers hydroxyapatite, Ti-6Al-4V ELI, electrophoretic deposition, temperature sintering, osseointegration

1. Introduction

Biomaterial implantation for fracture treatment gives an effect directly to body metabolism. Biological responses determined by biomaterial characteristics will influence bone remodelingand osseointegration as the aim of successful implantation. Osseointegration can be adjusted using materials with good mechanical properties combined high biocompatibility characteristic. This properties may be found on orthopedics implant that is titanium alloy, Ti-6Al-4V ELI (*extra low interstitial*) with compositions 80% titanium, 6% aluminum, 4% aluminum and other constituents. In structural, Ti-6Al-4V have phases α/β will transform from HCP-α structure to BCC- β structure at temperature 980°C [1]. Titanium has corrosion resistance cause of spontaneous passivation when contacts oxygen to form surface oxide layers

[2]. This layer will protect in releasing of hazardous material constituents to body environment for long term using. Therefore, surface modification through the coatings was needed in order to prevent corrossion and to improve bioactivity of materials with hydroxyapatite for achieveing osseointegration.

Hydroxyapatite are ceramic biomaterials in the molecule form $Ca_{10}(PO_4)_6(OH)_2$ having hexagonal unit [3] having characteristics for coating of biomaterial implant that are biocompatibility, bioactive [4], and osteoconductivity [5]. Particularly, hydroxyapatite nanoparticles, with low porosity, are involved in biological responses within increases density of osteoblast [6] and molecular interaction [7] for osseointegration [8]. Coating using hydroxyapatite micro particles will increase interfacial bonding through apatite mineralization. The combination of hydroxyapatite nanoparticles and micro particles as bilayers coating was aimed to improve bioactivity material to achieve osseointegration.

The coating of bilayers hydroxyapatite using electrophoretic deposition (EPD) need heat treatment to obtain better mechanical property than the other method such as plasma sprayed [9]. The advantages of this method encompass high versality and controlled product, using low temperature and simplicity procedure. The appropriate sintering temperature will effect for optimum quality of hydroxyapatite coating. Assessment of sintering temperature to bilayers hydroxyapatite (composed by nanoparticles and micro particles) deposited on material Ti-6Al-4V is needed to adjust osseointegration in orthopedic implantation.

2. Materials and Methods

Titanium alloy, Ti-6Al-4V ELI screw materials were used. Materials were abraded with mesh 800, 1000 and 1500 for homogenization of surface roughness. The cleaning was conducted using ethanol, acetone and nitric acid 25% solutions each for 15 minutes with ultra-sonication. Materials were dissolved in NaOH solution for 1 hour and air-dried using hot plate. This procedure based on work Bai et al and Erakovic et al with modification.

Hydroxyapatite solution for EPD process was prepared with 1 gram (Sigma Aldrich, USA) dissolved in 100 mL ethanol and homogenized using hot plate stirring for 1 hour prior to coating. The coating was carried out for two step that is to form nanoparticle hydroxyapatite as the first layer (3V for 3 minutes) and after air dried for 24 hours and the second layers was formed of micro particle hydroxyapatite (5V for 5 minutes). The coated materials were air dried for 24 hours and followed by sintering process at 700°C, 800°C, and 900°C temperature (holding 10 minutes and sintering rate 20°C/min) using vacuum furnace (High Temperature Vacuum Tube furnace GSL-1100). Weight measurement of coating layers was determined by weight differences between uncoated and coated material after coating process used balance. Analysis of morphology characteristics was conducted using optical microscope (Olympus LG-PS2) and scanning electron microscope (SEM Hitachi S-3400N at 10.0 kV). Imaging was used to measure coverage surface with computer analysis and the areas were covered up by hydroxyapatite particle calculated as surface coverage regions.

3. Results and Discussion

The surface of Ti-6Al-4V ELI materials covered up by bilayers hydroxyapatite obtained from EPD process (Figure 1). Principally, the nanoparticle coatings have thin, homogeny, and cover up the entire of material surface. The amorphous hydroxyapatite micro particles just coat the certain part of material surface and cohesive bind with hydroxyapatite nanoparticle. Hydroxyapatite micro particle layers are going to be agglomeration and have variance thickness that may be effected by voltage and time parameters of EPD process [10]. Deposited coatings on material surface at moment after the coating process have lower adhesion strength [11]. The following process, sintering, was needed to obtain the increasing of adhesion strength among

IOP Conf. Series: Materials Science and Engineering **547** (2019) 012005 doi:10.1088/1757-899X/547/1/012005

the coating layers onto material surface, because deposition and densification was not occurred simultaneously. The sintered layers have more dense structure due to densification that shrinkage and cracking may be occurred [12] and particularly founded in more than the 20μm thickness [11]. The crack part may increase osseointegration because it role as mechanical interlocking through nucleation apatite mechanism on crack boundary site [13].

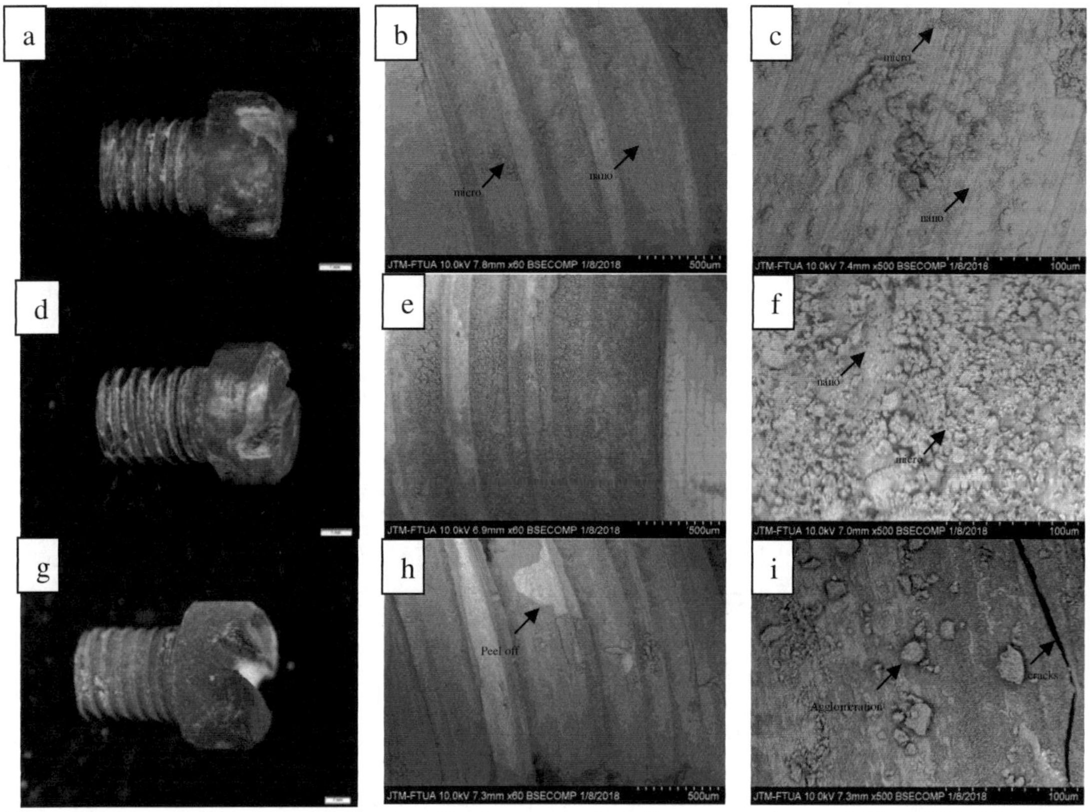

Figure 1. Morphology characteristic of hydroxyapatite bilayers coating on Ti6Al4V ELI surface effected by different sintering temperature; (a), (b) and (c) 700 ^{0}C; (d), (e) and (f) 800 ^{0}C; (g), (h) and (i) 900 ^{0}C

Sintering affects the density of bilayers hydroxyapatite coatings. Although the high temperature treatment at 900°C improves densification and reduces porosity, the coatings crack followed by peeling of the certain layers. High porosity, one of the EPD process flaws, promotes corrosion and releases titanium material ions induced by body environment. Porosity reduction by high temperature promote cracks might be caused by the difference of material thermal coefficient [14]. That will effect expansion and shrinkage difference between coating and material at heating and cooling of sintering process thus resulting thermal stress [17,18]. The Ti-6Al-4V materials have thermal coefficient ~10.3μm/mK lower than hydroxyapatites (~14 μK/mK) that effects resulting of residual tensile stresses promoted the cracking [17]. It proved that crack propagation more associate to used sintering rather than EPD process [15].

Nanoparticle and micro particle hydroxyapatite well bonds to Ti-6Al-4V ELI material surface when sintering was given at 800°C temperature.The dense, even and non-cracks of nanoparticle hydroxyapatite role as the protecting toward possibility of material corrosion and acting in biology activity at implantation. The loosening micro particle hydroxyapatite may be dissolute acting directly for bone remodeling through the forming of interfacial coatings. Using nanoparticle hydroxyapatite as intermediate between micro particle hydroxyapatite and material surfaces have suitability of thermal coefficient with titanium material cause of a number of atoms at grain boundaries increasing particle mobility [18]. It caused sintering with 800°C temperature for 10 minutes no promote cracks propagation. The previous study reports that the 800°C temperature for 1 hour is optimum condition for hydroxyapatite sintering in vacuum condition because not causes structural hydroxyapatite changes and cracks [13,21] with adhesion strength 47 ± 0.5 MPa [19]. Araghi et al [20] stated that optimum temperature effects the stable chemical characteristic of hydroxyapatite coatings because have good adhesion to material surface. Hydroxyapatite coating implanted by in vivo may occurs dissolution in the form of calcium and phosphate ions so promote nucleation and apatite crystal formation [23,24]. Densification founded relatively slow when sintered at 700°C temperature because there is still releasing of hydroxyapatite micro particles while hydroxyapatite nanoparticles bond well with material surface. The low sintering temperature is still possible to form the weak bonding with the material surface [18,25].

Table 1. Surface coverage and coating mass value of hydroxyapatite bilayers on Ti6Al4V ELI treated by different sintering temperature

Sintering Temperature (°C)	Surface coverage (%)	Coating mass (μg)
700	88.92 ± 4.24	- 3.10^{-2} ± 0.00
800	98.59 ± 0.49	0
900	26.99 ± 18.03	13.10^{-2} ± 0.00

The alteration of hydroxyapatite particle weight attached on material surface (Figure 2) shows that sintering may associate with surface coverage values (Table 1). The lowest value of surface coverage happened at high sintering temperature 900°C, but it was found the increasing of coating weight and may be caused by too high temperature. High temperature will promote oxidation of materials that may cause forming of oxide layers on material surface so there is more coating mass. Materials Ti-6Al-4V ELI is reactive toward high temperature making oxidation thus sintering should be conducted at optimum temperature in intense vacuum condition [16]. Contrary, high temperature beyond 900°C effects the loosing of hydroxyapatite coating weight in small amount [11].The low temperature of sintering is still unsufficient to increase bonding between hydroxyapatite particle and materials. However, high temperature around 900°C will effect densification, peeled off, and uncovered coating layers on material surface. For different materials and method, stainless steel coated by hydroxyapatite using investment casting, Nuswantoro et al [24] found the better sintering temperature is around 800°C-900°C which resulting high cristallinity, surface coverage, purity and hardness of hydroxyapatite coatings.

The 700°C temperature revealed that the bonding of bilayers hydroxyapatite with material surface is low. It may caused by founding particle peeling of is around 11.08% of coated surface area with weight loss of hydroxyapatite particles is around 3.10^{-2}μg.The better bonding of coating was discovered at 800°C temperature with material was covered up by hydroxyapatite approximately 98.59% with no peeling off the particles. It is needed optimum temperature is approximately 800°C in resulting the better bilayers hydroxyapatite coating on titanium materials. Chen et al [25] stated that sintering temperature at 800°C have not effect to purity of nanoparticle hydroxyapatite compounds and increasing hardness of coating layers. This coating layers also have good biocompatibility. At that sintering temperature, there is no more degradation of titanium and no metal-induced decomposition of hydroxyapatite [25]. For bilayers hydroxyapatite coating on titanium in screw form, temperature sintering at 800°C is effective in resulting good surface coverage, dense and homogeneous coating.

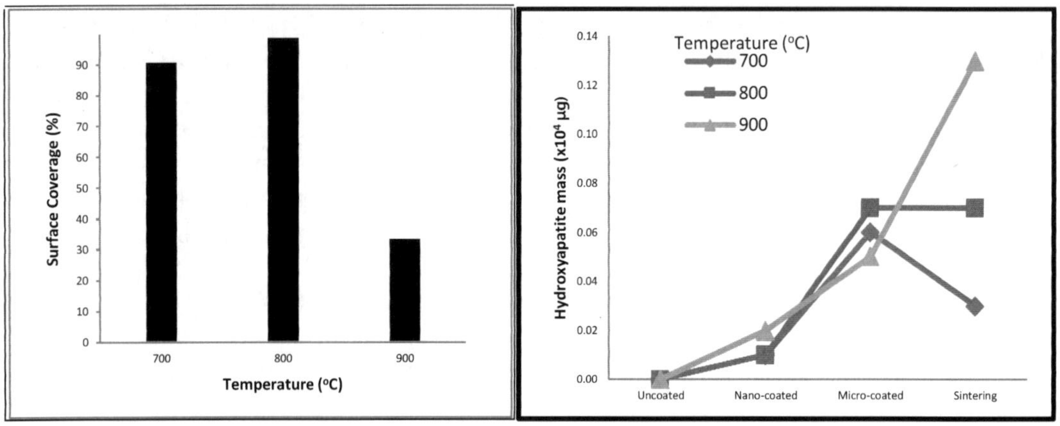

Figure 2. Surface coverage average of hydroxyapatite bilayers coating on Ti6Al4V ELI and Hydroxyapatite mass was coated on Ti6Al4V ELI surface by electrophoretic deposition and effected by sintering

The two step coating using hydroxyapatite nanoparticle and microparticle followed by sintering results good mechanical strength in which the first layers (undercoat) make a diffusion barrier to decomposition of substrat materials that results no crack, dense and no phase transformation coatings [16]. Forming first layer as interfacial can be applied in order to prevent decomposition of materials. Beside of protecting for harmful ion releasing from material, the interfacial layers act as reduction of decomposition due to sintering. Using of electrophoretic deposition is potential method [25] in resulting of bilayers hydroxyapatite suitable for biomedical application particularly for orthopedic due to increase osseointegration.

4. Conclusions

Sintering treatment at temperature 800°C gives thin, dense, even and crack-free coatings. The Ti-6Al-4V ELI surface is almost fully covered by bilayers hydroxyapatite (with surface coverage 98.6%) with adequate bonding (without lost weight of coating layers). An improper coating layer is obtained for lower (700°C) and higher (900°C) temperature. The sintered samples at temperature 900°C have micro cracks, void and debonding layers. It is expected such bilayers hydroxyapatite with sintering at 800°C can initiate mineralization and then make a quick osseointegration.

5. Acknowledgements

Corresponding author thanks to Ministry of Research and Education, Indonesia (KEMENRISTEKDIKTI) below the program of Program MenujuDoktoralSarjanaUnggulan (PMDSU) with contract number is 20/UN.16.17./PP.PMDSU/LPPM/2017.

6. References

[1] Boyer, R., Welsch, G., and E.W. Collings. 1994. *Materials Properties Handbook: Titanium Alloys*. ASM International. United States of America.
[2] Niespodziana, K., K. Jurczyk, J.Jakubowicz, and M.Jurczyk. 2010. Fabrication and properties of titanium–hydroxyapatite nanocomposites. *MaterialsChemistry and Physics,*123(1):160–5.

[3] Arifin, A., A.B. Sulong, N. Muhamad, J. Syarif, and M.I. Ramli. 2014. Material processing of hydroxyapatite and titanium alloy (HA/Ti) composite as implant materials using powder metallurgy: A review. *Materials and Design*, 55: 165-175.

[4] Kalita, S.J., A. Bhardwaj, and H.A. Bhatt. 2007.Nanocrystalline calcium phosphate ceramics in biomedical engineering.*Materials Science and Engineering C*, 27(3): 441–9.

[5] Cai, Y., X.Wang, C.K. Poh, H.C. Tan, M.T. Soe, S. Zhang and W. Wang. 2014. Accelerated bone growth in vitro by the conjugation of BMP2 peptide with hydroxyapatite on titanium alloy. *Colloids and Surfaces B: Biointerfaces*, 116: 681–6.

[6] Guo, X., J. Gough, P. Xiao, J. Liu, and Z. Shen. 2007. Initial In VItro Interaction Of Human Osteoblasts With Nanostructured Hydroxyapatite (NHa). *Advances in Bioceramics and Biocomposite II,*1-13.

[7] Dalby, M.J., M.O. Riehle, H.J. Johnstone, S. Affrossman, andA.S. Curtis. 2002. Polymer-demixednanotopography: control of fibroblast spreading and proliferation. *Tissue Engineering,* 8(6):1099-1108.

[8] Ellingsen, J.E.,C.B. Johansson, A.Wennerberg, and A. Holmen. 2004. Improved retention and bone-to-implant contact with fluoride-modified titanium implants. *Oral Maxillofacial Implants,* 19(5):659-666.

[9] Mohseni, E., E. Zalnezhad,and A.R. Bushroa. 2014. Comparative investigation on the adhesion of hydroxyapatite coating on Ti-6Al-4V implant: A Review paper. *Adhesion and Adhesives*, 48, 238-257

[10] Juliadmi, D., V.R. Fauzi, Gunawarman, H. Nur, M.H. Idris. 2018. Hydroxyapatite Coating on Titanium Alloy Ti-6Al-4V with Electrophoretic Deposition (EPD) for Dental Root Application. *Advanced Science Engineering Information Technology*, 7(6): 2152-2159.

[11] Sridhar, T.M., U.K.Mudali, and M.Subbaiyan. 2003. Sintering atmosphere and temperature effects on hydroxyapatite coated type 316L stainless steel. *CorrosionScience,*45(10): 2337–2359.

[12] Ducheyne, P., S. Radin, M.Heughebaert, and J.C.Heughebaert.1990. Calcium phosphate ceramic coatings on porous titanium: effect of structure and composition on electrophoretic deposition, vacuum sintering and in vitro dissolution. *Biomaterials*, 11(4):244–254

[13] Montenero, A., G. Ganapi, F. Ferrari, M. Cesari, E. Salvioli, L., Mattogno, S. Kaciulis, M. Fini. 2000. Sol-gel derived hydroxyapatite coatings on titanium substrate.*Materials Science*, 35(11): 2791-7.

[14] Abbasi, S., F. Golestani-Fard, H.R. Rezaie, and S.M.M. Mirhosseini. 2012. MAO-derived hydroxyapatite/TiO2 nanostructured multi-layer coatings on titanium substrate. *Applied Surface Science*, 261, 37-42.

[15] Kollath, V.O., Q.Chen, R. Closset, J. Luyten, K. Traina, S. Mullens, A.R. Boccaccini, and R. Cloots. 2013. AC vs DC Electrophoretic deposition of hydroxyapatite on titanium. *European Ceramic Society*, 33(13-14): 2715-21.

[16] Wei, M., A.J. Ruys, M.V. Swain, B.K. Milthorpe, and C.C. Sorrell. 2005. Hydroxyapatite-coated metals; Interfacial reactions during sintering. *Materials Science: Materials in Medicine,* 16(2): 101-6.

[17] Wei, M., M.V. Swain, A.J. Ruys, B.K. Milthorpeand C.C. Sorrell. 1999. Solution ripening of hydroxyapatite nanoparticles: Effects on electrophoretic deposition. *Biomedical Materials Research*, 45(1): 11-19.

[18] Yamashita, K., E. Yonehara, X. Ding, M. Nagai, T.Umegaki, and M. Matsuda. 1998.Electrophoretic coating of multilayered apatite composite on alumina ceramics.*Biomedical Materials Research*, 43(1):46–53.

[19] Chen, F., W.M. Lam, C.J. Lin, G.X. Qiu, Z.H. Wu, K.D.K. Luk, and W.W. Lu. 2007. Biocompatibility of Electrophoretical Deposition of Nanostructured Hydroxyapatite

Coating on Roughen Titanium Surface: In Vitro Evaluation Using Mesenchymal Stem Cells. *Biomedical Materials Research Part B: Applied Biomaterials*, 82(1): 183-91.
[20] Araghi, A and M.J. Hadianfard. 2015. Fabrication and characterization of functionally graded hydroxyapatite/TiO2 multilayer coating on Ti-6Al-4V titanium alloy for biomedical application. *Ceramics Int*, 41(10).
[21] Zhang, Q., J.Chen, J.Feng, Y.Cao, C.Deng, and X.Zhang. 2003.Dissolutionand mineralization behaviors of HA coatings. *Biomaterials.* 24(26):4741–8.
[22] Cheng, K., G.Shen, W.Weng, G.Han, J.M.F.Ferreira, and J.Yang. 2001.Synthesis of hydroxyapatite/fluoroapatite solid solution by a sol–gel method. *Materials Letters*, 51(1): 37–41.
[23] Ruys, A.J., M.Wei, C.C.Sorrell, M.R.Dickson, A.Brandwood, and B. K. Milthrope. 1995. Sintering effects on the strength of hydroxyapatite. *Biomaterials,* 16(5): 409–15.
[24] Nuswantoro, N.F., Gunawarman, M. R. Saputra, I. P. Nanda, M. H. Idris and A. Arafat. 2018. Microstructure Analysis of Hydroxyapatite Coating on Stainless Steel 316L Using Investment Casting Technique for Implant Application. *Advanced Science Engineering Information Technology*, 8(5): 2168-2174.
[25] Chen, F., W. M. Lam, C. J. Lin, G. X. Qiu, Z. H. Wu, K. D. K. Luk, W. W. Lu. 2007. Biocompatibility of Electrophoretical Deposition of Nanostructured Hydroxyapatite Coating on Roughen Titanium Surface: In Vitro Evaluation Using Mesenchymal Stem Cells. *Journal of Biomedical Materials Research Part B: Applied Biomaterials*, 183 191.
[26] Boccaccini, A.R., S. Keim, R. Ma, Y. Li, and I. Zhitomirsky. 2010. Electrophoretic Deposition of biomaterials, Review. *R.Society Interface*, 7: S581-613.

Effect of Addition Tahongai Leaf Extract (Kleinhovia hospita Linn.) As Organic Inhibitor on 1040 AISI Steel

Triana Yunita[1]*, Asih Wentika Putri Kusuma[2], Sari Eli Novita[3], Sulistijono[1]

Department of Materials and Metallurgical Engineering, Institut Teknologi Kalimantan
Kampus ITK Karang Joang, Balikpapan 76127, Indonesia

Email: nita@itk.ac.id

Abstract
This research will study about corrosion inhibition by using tahongai leaf extract *(Kleinhovia hospita Linn.)* in a 3.5% NaCl environment. The inhibitor is extracted using maceration process from Tahongai leaf (*Kleinhovia hospita Linn.*). Testing of antioxidant activity using DPPH method gives result that antioxidant activity of tahongai leaf extract *(Kleinhovia hospita Linn.)* is strong antioxidant (96%) compared to vitamin C (98%), functional group test by FTIR method indicates that tahongai leaf extract (*Kleinhovia hospita Linn.)* contain flavonoid shown by O-H bond (3357.41 cm^{-1}), double bond C=C (1643.36 cm^{-1}), C-H bond (2922.24 cm^{-1}, 1460.69 cm^{-1}, 1375.57 cm^{-1}), Efficiency of inhibition reach 99.9825% for addition of inhibitor with concentration 200 ppm and duration of immersion 20 days. This inhibition behavior is also supported by polarization measurements where the lowest corrosion rate of $8.3x10^{-4}$ mm/year is obtained at the same concentration and immersion time.

Keywords: Antioxidant, Organic inhibitor, Tahongai leaf (*Kleinhovia hospita Linn.*)

1. Introduction

A corrosion inhibitor is a compound that protects metal from corrosion in various ways. Therefore, it is necessary to have a good analysis and calculation on the practice of its use in order to obtain effective results [1]. Corrosion inhibitors are divided into two types: organic and inorganic inhibitors. Some inorganic compounds containing clusters having electron-free pairs such as phosphate, chromate, borate, nitrite, urea, phenylalin and other amine compounds are often used as corrosion inhibitors. However, these synthetic chemicals are dangerous chemicals because they are toxic, expensive, and environmentally unfriendly. For that, start looking for another alternative that is more economical and environmentally friendly [2]. One alternative that can be applied is by utilization of organic material as an environmentally friendly corrosion inhibitor. The requirement of organic material that can be used as a corrosion inhibitor is an organic material containing antioxidant compounds. Antioxidants in corrosive environments are able to inhibit the rate of material corrosion.

Natural antioxidants can be found in vegetables, fruits, and woody plants. Secondary metabolites in plants derived from alkaloids, flavonoids, saponins, quinones, tannins, steroids / triterpenoids [3]. One of the typical plants of East Kalimantan that grow freely in nature is tahongai plant or known by the name *Kleinhovia hospita Linn.*

Tahongai (*Kleinhovia hospita Linn.*) is a shrub with soft green leaves that grows naturally in East Kalimantan. Its habitat mostly located in the riverside. For a long time, tahongai plant has been used as medicine by Dayak people in Kalimantan. They believe the plant has various health benefits. Among others are able to lower blood sugar levels, lower cholesterol and help restore and strengthen the liver. *Kleinhovia hospita Linn.* which has various regional names such as tahongai (Dayak), paliasa (Makassar), and katimahar (Ambon) are used as traditional medicine in parts of Malaya, Indonesia and Papua New Guinea to treat scabies. The leaves and stem bark are used as hairwash for ticks, while leaf juice is used as an eye washer [4]. Leaves and stem bark of *Kleinhovia hospita Linn.* contains cyanogen compounds that are assumed to help kill ectoparasites such as lice. A number of fatty acids with cyclopropenylic rings such as scopoletin, and flavonoids such as kaempferol, quercetin and routine have been isolated from the leaves [5].

Research conducted by [6] suggests that tahongai (*Kleinhovia hospita Linn.*) leaf has anti-diabetic effect with the mechanism of inhibition of active transport of glucose. Research conducted in isolating organic compounds contained in tahongai (*Kleinhovia hospita Linn.*) leaf based on several levels of solvent polarity, showed that there were at least 96 single isolates and instrumentation indicated the presence of terpenoid and phenolic group compounds [7]. Then the results of research conducted by [8] successfully identified the class of flavonoid and alkaloids. Then on the fraction of methylene chloride from the active fraction was isolated a coumarin derived compound suspected as scopoletin [9]. The presence of these antioxidant compounds supports the utilization of the tahongai plant *(Kleinhovia hospita Linn.)* in East Kalimantan as an organic inhibitor on the application of AISI 1040 steel pipes.

2 Research Methodology

2.1 Tools and Materials

The equipment used in this study include the plastic container size of 8,000 mL, nylon yarn, wood cantilever, measuring cups capacity of 1,000 mL, measuring cups capacity of 10 mL, coarse and fine filter paper, digital scales, paper pads grade 80 to 2,000, spatula, dropper, ruler, scissors, potentiostat device, evaporator device and optical microscope. While the materials used in this study such as steel AISI 1040, technical salt solution, distilled water, and fresh tahongai *(Kleinhovia hospita Linn.)* leaves (Fig. 1) were obtained from Lempake, Samarinda, East Kalimantan.

2.2 AISI 1040 Steel Specimen Preparation

The specimen used was AISI 1040 steel with a composition (wt%): 98.6 to 99% Fe, 0.60 to 0.90% Mn, 0.370 to 0.440% C, ≤0,050% S, and ≤0,040% P. AISI 1040 steel was prepared into 2 specimen shapes. The first specimen to test weight loss cut rectangular with a size of 2 cm x 2 cm x 0.3 cm and perforated using a drill as the linking thread, while the test specimen to cut potentiostat shaped beam with a size of 2 cm x 0.5 cm x 0.4 cm for polarization testing. Coat layer is removed by grinding and polishing.

2.3 Organic Inhibitor Preparation

The process of leaf drying starts from the leaves of tahongai *(Kleinhovia hospita Linn.)* to further separated from the stalks and then weighed the wet weight. The leaves are then cleaned of dirt and then bonded per 20 leaves for further hanging and aerated in the open air without exposure to rain and sun for 30 days. Manufacture inhibitors done by maceration method by introducing 775 grams of tahongai leaves *(Kleinhovia hospita Linn.)* dry with 8,000 mL of 99.9% ethanol into a container with a capacity of 8.000 mL with immersion done for 11 x 24 hours. Then, the extraction results between the filtrate and the residue are separated using a coarse and fine paper filter. The filtrate is used as tahongai leaf extract *(Kleinhovia hospita Linn.)* which becomes an organic inhibitor. The concentration of inhibitor used in this study was 0 ppm, 100 ppm, 200 ppm, 300 ppm, 400 ppm and 500 ppm as well as variations of 10 days, 20 days, and 30 days immersion.

Fig.1 Fresh tahongai *(Kleinhovia hospita Linn.)* leaves 3-4 months old

Fig.2 (A) Fresh tahongai (*Kleinhovia hospita Linn.*) leaf ,(b) Dry tahongai (*Kleinhovia hospita Linn.*) leaf, (c) Immersion process with maseration method, (d) Filtering process with filter paper, (e) Evaporation process, (F) Taking extracts with acetone, (g) Extracted into small bottle before dried, (h) Drying process extract

2.4 Cluster Function Analysis

Functional group analysis using FTIR (Fourier Transform Infra Red) method was used to determine the bioactive compounds contained in tahongai leaf extract (*Kleinhovia hospita Linn.*). The analysis was conducted at the Material Characterization Laboratory, Department of Materials and Metallurgical Engineering, Sepuluh Nopember Institute of Technology. The test was performed using FTIR spectrophotometer (Shimadzu) with KBr pellet method.

2.5 Preparation of Electrolyte Solution

NaCl 3.5% corrosion media is made by mixing the NaCl technical salt and aquades. Calculation of inhibitor concentration: ppm (part per million) of units in mg / L, which means that in every 1 L of corrosion media there are several mg inhibitors. For example to get 100 ppm means in 1 L aquades added 100 mg inhibitor.

2.6 Weight Loss Corrosion Testing

This test is intended to look at corrosion products and to know the weight loss produced by the sample when immersed in a 3.5% corrosive NaCl medium. Testing is done by varying the concentration of inhibitor (0 ppm, 100 ppm, 200 ppm, 300 ppm, 400 ppm and 500 ppm) and soaking time (10 days, 20 days and 30 days). Corrosion testing with weight loss method is done by repeating three times. Percentage inhibition efficiency (IE%) is calculated using the following equation:

$$IE\% = (W - W_i) / W \times 100 \quad (1)$$

Where :

W = Weight loss without inhibitor (0 ppm)

W_i = Weight loss with inhibitor (100, 200, 300, 400, 500 ppm)

2.7 Potential Testing

Potentiostat tests were performed using the PGSTAT101 Autolab device at the Materials and Metallurgy Laboratory, the Materials Engineering and Metallurgy Study Program, the Kalimantan Institute of Technology. OCP (Open Circuit Potential) testing is performed to obtain data about corrosion rate. In the method of potentiodynamic polarization there are 3 electrodes, the comparative electrode Ag / AgCl, platinum auxiliary electrode, and AISI 1040 steel working electrode. This test is based on the concentration of inhibitor (0 ppm, 100 ppm, 200 ppm, 300 ppm, 400 ppm, and 500 ppm) and immersion time of 10, 20, and 30 days.

2.8 Macrostructural Observations

The observations on macrostructure were performed using Carl-Zeiss microscope in the Materials and Metallurgy Laboratory, the Material and Metallurgical Engineering Study Program, Kalimantan Institute of Technology to compare macrostructural changes and the protective layer formed on the surface of AISI 1040 steel after weight loss corrosion test with or without the addition of inhibitor on various variations of immersion and concentration.

3. Results and Discussion

3.1 Analysis of Functional Groups

Kleinhovia hospita Linn. leaf methanol extract showed strong antioxidant activity (96%) compared with vitamin C (98%) with DPPH and activity was measured based on extracts from the methanol extract of 48.9% for n-hexane, 74.0% for diethyl ether, 84.3% for ethyl acetate, and 77.1% for residuals. MTT assay showed HepG2 cells in 14%, 76%, and 80% at each concentration of 50 g/mL, 87.5 mg/mL, and 125 mg/mL (White, 2009). From the results of FTIR testing on corrosion inhibitor materials, there are various contained links shown in the peak form (Fig. 4).

Based on the results of the test analysis FTIR (Fourier Transform Infra Red) against tahongai (*Kleinhovia hospita Linn.*) leaf extract obtained bond O-H at wave number 3357.41 cm^{-1}, the double bond carbon-carbon C=C at wave number 1643.36 cm^{-1}, a single bond carbon-hydrogen C-H at wave number 2922.24 cm^{-1}, at wave number 1460.69 cm^{-1}, and the wave number 1375.57 cm^{-1}. In addition, the presence of electronegative atoms (ie, N and O) and the double bonds of the alkene and alkaline groups gives the potential of the extract as a corrosion inhibitor (Kurniawan, 2015). Indicated that flavonoid compounds contained in tahongai (*Kleinhovia hospita Linn.*) leaves extract saw bond contained in flavonoids include several functional groups, including the double bond carbon-carbon C=C, double bond carbon - oxygen C=O, single carbon-oxygen bond C-O, single carbon-hydrogen bond C-H, and single-oxygen-hydrogen O-H (Robinson, 1995).

IOP Conf. Series: Materials Science and Engineering **547** (2019) 012006 doi:10.1088/1757-899X/547/1/012006

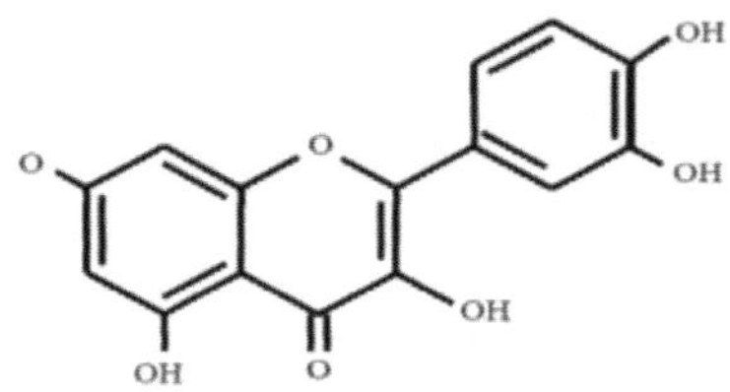

Fig. 3 Flavonoid

Raflizar [10] in his research confirmed the presence of several organic compounds such as saponins, antrakinon, cardenolin and bufadienol contained within *Kleinhovia hospita Linn* leaf extract with 70% alcohol through chromatography test.

Table.1 Differences of Organic Compounds Tahongai Extract *(Kleinhovia hospita Linn.)*

Compound	Description	Structure
Saponin	In saponin compounds there is no O-H structure and aromatic ring as obtained in FTIR test results. This means that the compound obtained from the FTIR test *Kleinhovia hospita Linn.* leaves not a saponin compound.	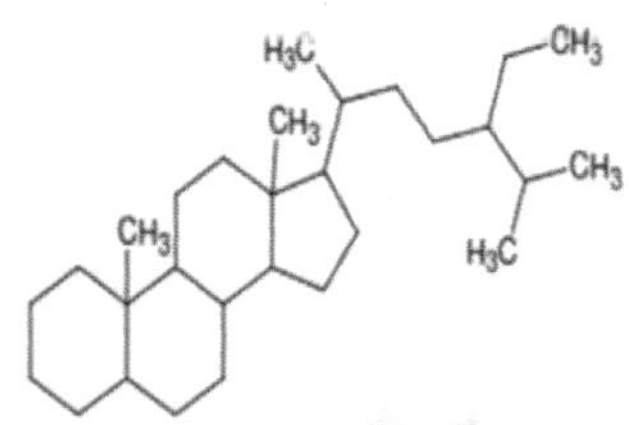
Antrakinon	In anthrakinon compound there is a functional group C = O, whereas based on FTIR test results there is no known functional group C = O. This means that the compound obtained from the FTIR test *Kleinhovia hospita Linn.* leaves not an anthracin compound.	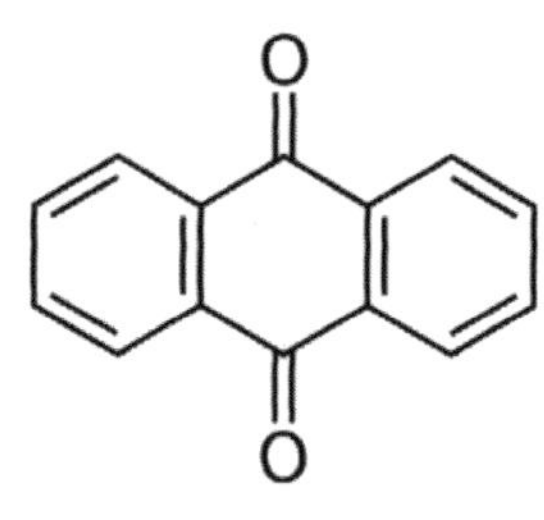
Flavonoid	The flavonoid functional groups include the C = C double bond, the C = O double bond, the C-O single bond, the single C-H bond, and the O-H single bond. From the obtained FTIR results, the widest area of the O-H hydroxyl group function. In addition to O-H there are also functional groups C-H and C = C.	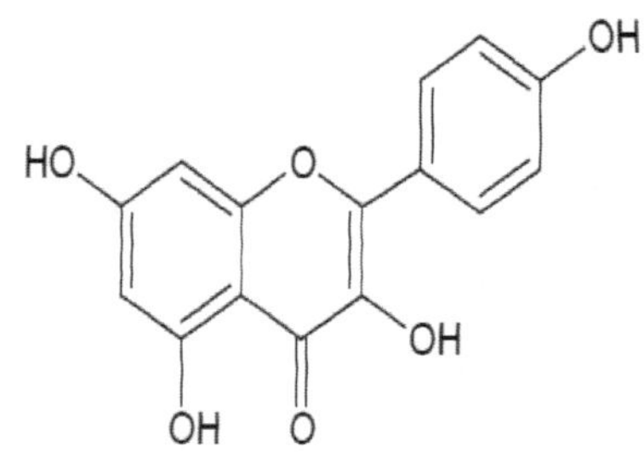

IOP Conf. Series: Materials Science and Engineering **547** (2019) 012006 doi:10.1088/1757-899X/547/1/012006

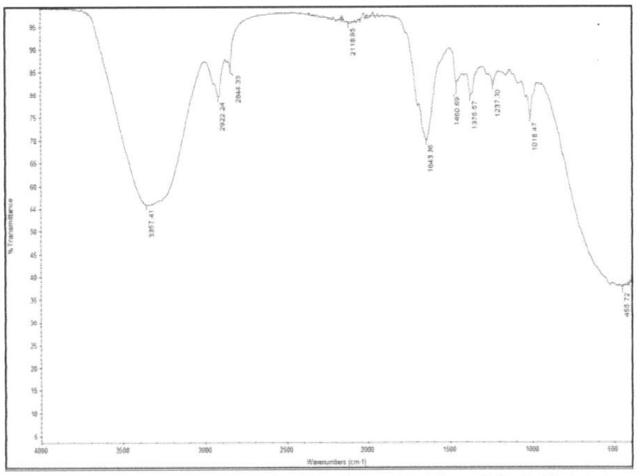

Fig. 4 FTIR results in Tahongai leaf inhibitors

Table.2 FTIR Analysis Results in Tahongai Leaf Inhibitors

Wave Number (cm^{-1})	Frequency Range (cm^{-1})	Bond	Functional Group
3357.41	3200 – 3600	O – H	Hydroxyl
2922.24	2850 – 2970	C – H	Alkane
2118.95	2100 – 2260	C ≡ C	Alkyne
1643.36	1610 – 1680	C = C	Alkene
1460.69	1340 – 1470	C – H	Alkane
1375.57	1340 – 1470	C – H	Alkane
1237.70	1180 – 1360	C – N	Amine, Amide

3.2 Calculation Result of Corrosion Efficiency by Weight Loss Method

Corrosion testing with weight loss method was performed with and without addition of inhibitor. The variable used is the level of inhibitor concentration and the duration of immersion. The test is done with three repetitions and then taken on average for the difference in weight.

From the calculation of weight loss can be seen that the efficiency of inhibition of tahongai (*Kleinhovia hospita Linn.*) leaves Highest in the solution of NaCl 3.5% is at a concentration of 200 ppm and at 20 days of soaking time with an efficiency value of 99.9825%. While the lowest efficiency is at a concentration of 300 ppm and on a 20-day immersion period with an efficiency value of 59.1916%.

The decrease in inhibitory efficiency occurring in the indication occurs due to the passive layer of the extract not being able to cover the entire surface of the steel, so that in the uncovered Fe part can be ionized and corrosion occured. The highest efficiency was achieved in the addition of 200 ppm long-period immersion inhibitors due to the formation of a strong protective layer to the maximum capable of covering the entire metal surface.

Table.3 Eficiency Calculation Data

Inhibitor Concentration (ppm)	Immersion Time (days)	Start Weight (grams)	End Weight (grams)	Weight Difference (grams)	IE %
0	10	11.6833	9.9735	1.7148	-
	20	12.9417	9.4037	3.5830	-
	30	12.3133	9.2992	3.5830	-
100	10	11.7631	11.2582	0.5049	83.2487
	20	12.5765	12.2957	0.2808	92.0633
	30	13.2082	12.7924	0.4158	88.2476
200	10	14.4097	14.0278	0.3819	89.2058
	20	10.3001	10.2998	0.0003	99.9825
	30	7.7774	7.2648	0.5126	82.9933
300	10	12.5099	12.1384	0.3715	87.6746
	20	10.7576	9.3138	1.4438	59.1916
	30	10.1478	9.8856	0.2622	91.3009
400	10	13.0628	13.0519	0.0109	99.6919
	20	12.1307	11.8302	0.3005	82.4761
	30	8.5465	8.4445	0.1020	94.0518
500	10	14.8427	14.8037	0.0390	97.7257
	20	12.5765	12.2957	0.2808	92.0633
	30	11.2216	11.0452	0.1764	94.1475

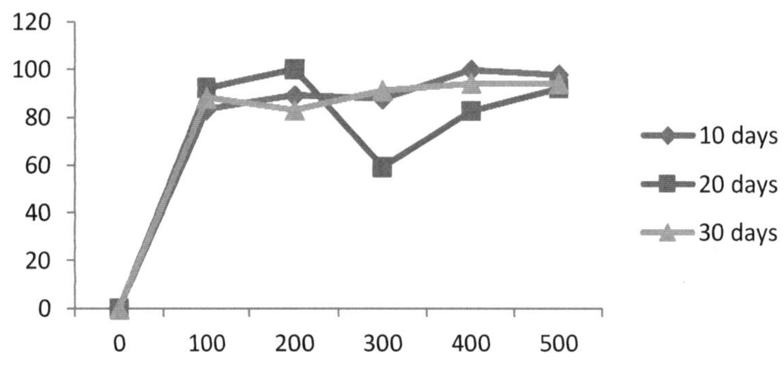

Fig. 5 Graph of Efficiency Comparison of 10, 20, and 30 days

3.3 OCP Polarization Testing Results (Open Circuit Potential) Based on Concentration

The measurement of corrosion rate of OCP (Open Circuit Potential) method using AISI 1040 steel and with variation of inhibitor 0, 100, 200, 300, 400, and 500 ppm and 10, 20 and 30 day immersion period.

From OCP results at various concentrations of inhibitors (0, 100, 200, 300, 400, and 500 ppm) at the time of immersion 0 days obtained the highest corrosion rate at 0 ppm concentration of 8.2×10^{-3} and the lowest corrosion rate at 500 ppm with a value of 2.3x10-3.

From the OCP results at various concentrations of inhibitors (0, 100, 200, 300, 400, and 500 ppm) at the 10-day immersion period obtained the highest corrosion rate at 0 ppm concentration of 1.4×10^{-2} and the lowest corrosion rate at 400 ppm with a value of 2.3×10^{-3}.

From OCP results at various concentrations of inhibitors (0, 100, 200, 300, 400, and 500 ppm) at 20 days of submersion, the highest corrosion rate was obtained at a concentration of 300 ppm at 1×10^{-2} and the lowest corrosion rate at 400 ppm of 8.4×10^{-4}.

From OCP results at various concentrations of inhibitors (0, 100, 200, 300, 400, and 500 ppm) at 30 days of maximum immersion, the highest corrosion rate at 0 ppm concentration was 7.3×10^{-3} and the lowest corrosion rate at 500 ppm with a value of 3.4×10^{-3}.

The values E_{corr}, J_{corr}, I_{corr} and CR (Corrosion Rate) respectively show the potential difference values of corrosion between metals soaked in electrolytes provided with standard reference electrodes, corrosion current densities or corrosivity rates, corrosion currents and corrosion rate values.

Table.4 Comparison of corrosion rate at 0 days

Inhibitor Concentration (ppm)	E_{corr} (mV)	J_{corr} (μA/cm²)	I_{corr} (μA)	CR (mm/year)
0	-589.320	711.790	2.59090	0.0082709
100	-466.420	339.780	1.23680	0.0039482
200	-513.920	361.400	1.31550	0.0041995
300	-529.490	265.210	965.380	0.0030818
400	-516.890	226.760	825.420	0.0026350
500	-542.440	200.290	729.050	0.0023273

Table.5 Comparison of corrosion rate at 10 days

Inhibitor Concentration (ppm)	E_{corr} (mV)	J_{corr} (μA/cm²)	I_{corr} (μA)	CR (mm/year)
0	-530.030	1.24940	4.54780	0.0145180
100	-524.930	460.800	1.67730	0.0053544
200	-402.780	376.130	1.36910	0.00437060
300	-478.170	420.410	1.53030	0.0048852
400	-484.470	198.240	721.580	0.0023035
500	-517.280	213.800	778.220	0.00248430

Table.6 Comparison of corrosion rate at 20 days

Inhibitor Concentration (ppm)	E_{corr} (mV)	J_{corr} (μA/cm²)	I_{corr} (μA)	CR (mm/year)
0	-554.220	893.600	3.25270	0.01038400
100	-540.740	337.640	1.22900	0.0039233
200	-531.890	71.8710	261.610	0.00083514
300	-531.500	945.080	3.44010	0.010982
400	-442.650	72.7250	264.720	0.00084506
500	-539.060	630.600	2.29540	0.0073276

Table.7 Comparison of corrosion rate at 30 days

Inhibitor Concentration (ppm)	E_{corr} (mV)	J_{corr} (μA/cm²)	I_{corr} (μA)	CR (mm/year)
0	-397.700	629.530	2.29150	0.0073151
100	-414.990	404.920	1.47390	0.0047051
200	-509.630	547.640	1.99340	0.0063635
300	-386.120	349.040	1.27050	0.0040558
400	-395.910	328.100	1.19430	0.0038126
500	-500.390	291.260	1.06020	0.0033845

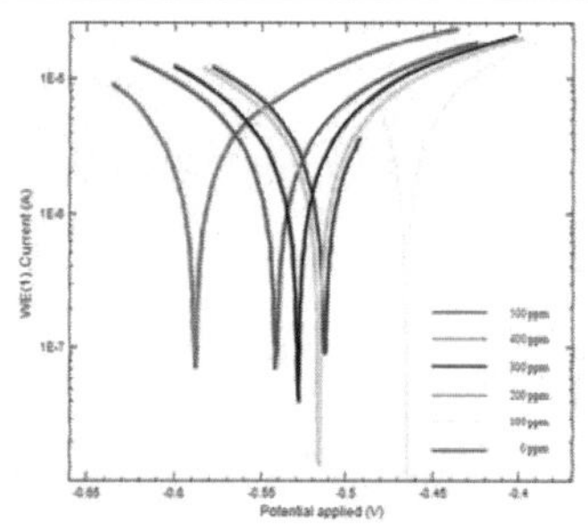

(a)

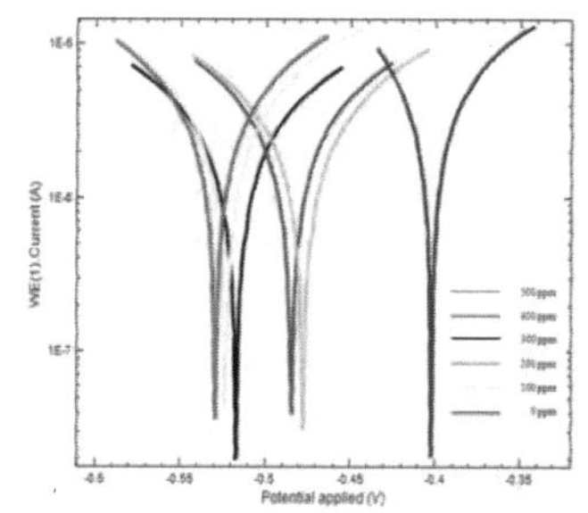

(b)

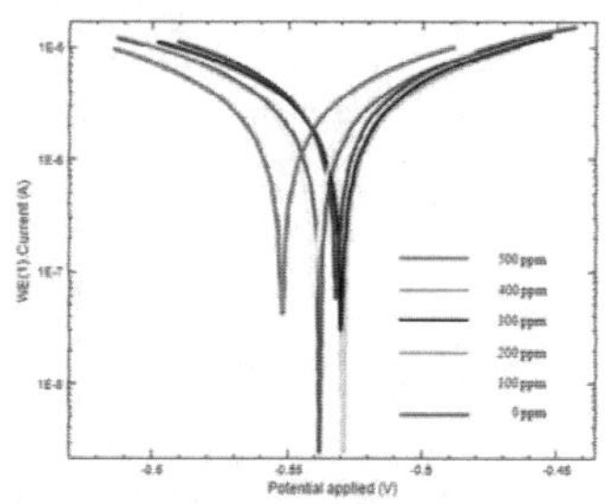

(c)

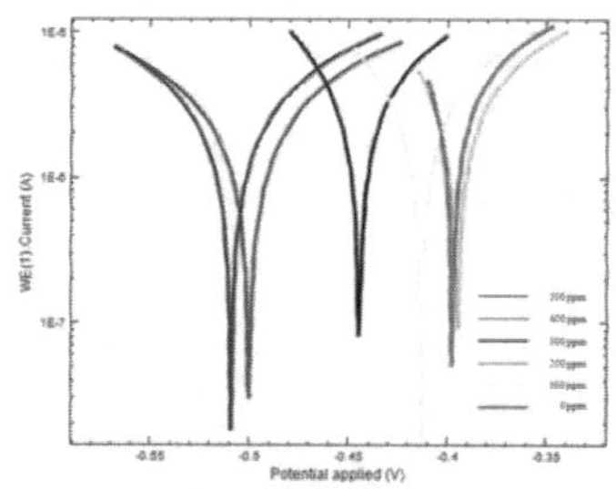

(d)

Fig. 6 Tafel curve of tahongai leaf extract inhibitor on (a) 0 days, (b) 10 days, (c) 20 days, (d) 30 days

From the result of the obtained polarization test, the highest corrosion rate was found at 0 ppm concentration with 20 days of soaking time of $1x10^{-2}$ mm/year, while the lowest corrosion rate was at 200 ppm with 20 days of soaking period of $8.4x10^{-4}$ mm/year. The overall polarization measurement results were in accordance with the weight loss corrosion test results where the lowest corrosion rate and highest efficiency was achieved in the addition of 200 ppm concentration inhibitor with 20 days of soaking time.

3.4 Mechanism of Corrosion Inhibitors

Identifying the mechanism of organic inhibitors is certainly not an easy thing. This is due to the large number of organic compounds contained in the plant. Each of these compounds will cause different reactions to each other. In this study, the focused organic compounds observed were flavonoid compounds found in tahongai (*Kleinhovia hospita Linn.*) leaf. Based on the previous polarization curve it is known that there has been a cathodic reaction which is seen from the value of E*corr* and I*corr* in the cathodic polarization region. It shows the inhibition of the reaction at the cathode in polarization. Here's what happens: Reduced oxygen levels in 3.5% NaCl solution, causing a redox reaction in solution. The decrease in oxygen levels will further decrease the formation of Fe^{2+} corrosion product. This leads to inhibition of corrosion products and reduced corrosion rate. This mechanism is called the scavenger oxygen mechanism. Judging from the various factors that may affect the reaction, the mechanism is not the only possible mechanism for the use of tahongai leaf inhibitors (*Kleinhovia hospita Linn.*) in this study.

4 Conclusion

Based on the research that has been done, it can be concluded that the lowest corrosion rate at 20 days 200 ppm with the value of $8.3x10^{-4}$ mm/year and the efficiency value of 99.9825%, the highest corrosion rate at 20 days 400 ppm with a value of $6.9x10^{-3}$ mm/year and efficiency value of 59.1916%, and medium highest corrosion rate without inhibitor at 20 days and concentration 0 ppm with corrosion rate value of $1x10^{-2}$ mm/year. The results of macro photographs obtained the best results on specimens with a 20-day immersion period of 200 ppm and the worst at 20 days 300 ppm. The mechanism of corrosion inhibition is indicated as a scavenger oxygen mechanism.

5. Bibliography

[1] Indra. (2004). "Kimia dari Inhibitor Korosi. e-USU Repository ". 8.
[2] Aziz, Abdul A.(2013). "*Solusi Mewujudkan Produksi Pangan yang Aman dan Ramah Lingkungan serta Meningkatkan Pendapatan Petani*". Yogyakarta. Jurusan Tanah,Fakultas Pertanian UGM
[3] Gordon I. (1994). "Functional *Food, Food Design, Pharmafood*". New York: Champman dan Hall.
[4] Kochummen KM.(1973)."*Hypericaceae. Tree Flora of Malaya*".2: 248–252.
[5] "*Kleinhovia hospita L. In: Faridah Hanum I, van der Maesen LJG (eds) Plant resources of South-East Asia no. 11: auxiliary plants*". Backhuys Publisher, Leiden, The Netherlands, pp 166–16

[6] "Pengaruh *Infus Daun Paliasa (Kleinhovia hospita Linn.) Terhadap Traspor Aktif Glukosa pada Usush Halus Marmut"*. Skripsi tidak diterbitkan. Makassar: Jurusan Farmasi Fakultas MIPA Universitas Hasanuddin.
[7] Noor, A. dan Kumanireng A.S., (2004).*"Isolasi dan Identifikasi Konstituen Organik Tanaman Daun Paliasa, Kleinhovia hospita Linn. pada Kelarutan Berdasarkan Kelompok Polaritasnya; Suatu Laporan Reserch"*. Jurusan Kimia FMIPA Universitas Hasanuddin. Makassar.
[8] Taebe, B., (2004). *"Standarisasi Ekstrak Daun Paliasa (Kleinhovia hospita Liin.) Sebagai Bahan Baku Sedian Fitofarmaka"*. Makalah disajikan dalam seminar hasil penelitian pada Pascasarjana Universitas Hasanuddin. Makassar 23 Pebruari 2004
[9] Wiwi.(2005). *"Eksplorasi Senyawa Kimia Daun Paliasa (Kleinhovia hospita Linn.) Pada Fraksi Aktif Metilenklorida"*. Skripsi tidak diterbitkan. Makassar: Jurusan Kimia, Fakultas MIPA Universitas Hasanuddin.
[10] Raflizar. (2000). *"Dekok Daun Paliasa (Kleinhovia hospita Linn.) sebagai Obat Raang Hati Akut"*. Badan Litbang Kesehatan Jakarta. (Online). (http://digilib.litbang.depkes.go.id/go.php?id=jkpkbppk-gdl-res-2001-raflizar-272-paliasa, diakses 5 November 2016).

Corrosion Behaviour of Collagen Coated and Uncoated Biomedical Titanium Alloy (TNTZ) Within Human Synovial Fluid

Hidayatul Fajri[1], Windy Ariani[2], Gunawarman[2*], Djong Hon Tjong[3], Menkher Manjas[4]
[1]Biomedical Science, Faculty of Medicine, Andalas University, Padang, Indonesia
[2]Mechanical Engineering, Faculty of Engineering, Andalas University, Padang, Indonesia
[3]Department of Biology, Faculty of Science, Andalas University, Padang, Indonesia
[4]Department of Surgery, Faculty of Medicine, Andalas University, Padang, Indonesia

*Corresponding author (E-mail): gunawarman@eng.unand.ac.id

Abstract. The corrosion behavior of uncoated Ti-29Nb-13Ta-4.6Zr (TNTZ) and collagen-coated TNTZ within human synovial fluid + NaCl 0.9% solution was investigated using immersion test in order to know the effectivity of collagen to increase bioactivity and also reducing corrosion rate as a coating material. Collagen was selected as coating material because of the ability of adhesion cell and increasing bone healing process. The immersion test has been used as corrosion method due to its simplicity and inexpensive. The immersion time in this study was for 2, 4, and 6 weeks. Surface morphology of material was examined by scanning electron microscope (SEM). Corrosion rate was calculated by measuring degradation of coating mass during the corrosion process. While mechanical property (hardness) of material was measured by microhardness testing. The result of this study showed that collagen coating has reduced the corrosion rate of TNTZ in the human synovial fluid where the corrosion rate of the coated TNTZ is 0.00125 mmpy and the corrosion rate of uncoated one is 0.00262 mmpy after 6 weeks immersion. On the other hand, the hardness of coated TNTZ is higher than the uncoated one. It can be concluded that the collagen coating increases both corrosion resistant and mechanical properties of TNTZ.

Keywords: Corrosion, Synovial fluid, TNTZ, collagen, immersion

1. Introduction

Metal alloys as biomaterials have to possess some characteristics such as corrosion and biocompatibility that really important especially for human utilization. Some kind of titanium alloys has been used in orthopedic like Ti6Al4Vwhrere is known to have high modulus

Young (100 GPa) than human cortical bone (30 GPa). Incompatibility of this modulus Young causes stress shielding that may cause bone resorption, moreover are loosening and premature failure implant [1]. In other hands, releasing Ti6Al4V compound (Al and V) will become toxic in the human body [2]. Based on this situation, type β titaniun has all requirements that necessary that will be needed and they have β stable elements, such as molybdenum, silicon, zirconium, and tantalum.

Some of the titaniun alloys have been reported closer to modulus elasticity of cortical bone, such as Ti29Nb13Ta4.6 Zr (TNTZ) [3], Ti25Ta25Nb [4], dan Ti35Nb4Sn [5] are 65 GPa, 55 GPa, dan 40 GPa respectively. Event another titanium alloy like Ti13Nv13Zr has been reported having significant value in cellular attachment than Ti4Al4V [6]. As the material in this research, TNTZ has some elements showed good performance. Other research found that Niobium element could increase the corrosion resistance of a material. Tantalum element [7] could be better in corrosion resistance because its oxidation stability highly [8,9] and its superior corrosion resistance also the biocompatibility of tantalum have been extensively evaluated [10]. Zirconium element (Zr) showed acceptable mechanical strength [11] and an in vivo study reports that zirconium implant exhibits good osteointegration [12,13].

Implant surface coating with a biomaterial that has bioactivity is really important to increase biocompatibility between implant and host. Coating with calcium phosphate (CaP) has been known to increase extracellular matrix (ECM) component and confirmed to upgrade osteoconductivity properties [14,15]. Though deposited CaP is one of natural composition of bone, it is different because the natural bone is not only consisted of the inorganic compound like CaP but also organic matrices like collagen and non-collagenous proteins. The organic part of bone ECM consists of collagen type 1 fibrils embedded in an amorphous substance, which consist of glycosaminoglycans and various bone proteins. ECM components participate actively in the regulation of cellular processes and responses. Therefore, implant surface modifications with components of bone ECM appears attractive to modulate specific intrinsic osteogenesis directly at the bone–implant interface [16,17]. The ECM works as a scaffold for bone forming cells and influences migration, adhesion, and differentiation of these cells [18,19]. Briefly, collagen type 1, major structural protein in ECM has been used as an organic implant coating material. Recent studies have referred to the role of a collagen coating in stimulating cellular responses, increasing bone growth and improving bone-to-implant contact [20,21,22].

Corrosion behavior classically focuses to material testing on simulated body fluid like phosphate buffer saline, sodium chloride solution, Hank solution, Ringer solution, and *bovine calf serum* and showed the essential purposed at gaining information on the involved corrosion mechanisms and analyzing the significant factors affecting the surface reactivity of biomedical alloys. Synovial fluid is a viscous fluid that is found in the synovial joints which are known to be composed mainly by inorganic salts like NaCl (Natrium Chloride) and organic molecules such as hyaluronic acid and lubricin, proteinase, and collagenase also cells [23] with alkali pH (7,4 – 10) [24]. These cells have been viewed as having a responsibility to release reactive chemical species that can change surface oxidation compositions of metal alloy [25]. Simulated body fluid disables to show the complexity of the in vivo electrolyte and the corrosion situation may determine through the contact between artificial joint and synovial fluid. Therefore, the aim of this study is to comprehend the corrosion resistance of one of new β type titanium alloy, TNTZ coated collagen in synovial fluid as attracted material.

2. Material and Methods

2.1 Material and Solution

Titanium type β Ti-29Nb-13Ta-4.6Zr (TNTZ) samples were provided in disk form, the thickness, 2-3 mm and diameters 6-7mm. All samples were polished and cleaned with nitric acid, acetone, and ethanol for 10 minutes for each solution, respectively using an ultrasonic cleaner for surface treatment. Corrosion testing was conducted within the synovial fluid (archived biological materials) that was found in human synovial joint and were added NaCl 0,9%. The temperature of the solution was kept at 37^0C.

2.2 Collagen type I coating

After polishing and cleaning process, all samples were rinsed with distilled water, air dried and autoclaved process. For the coating process, 1 mg calf skin collagen type I/mL was added into 10 m*M* acetic acid and incubated overnight in 4^0C. PBS 60 mM, pH 7.0 was added into solution (collagen + acetic acid) with ratio 1:1 and then the solution was incubated for 15 minutes in 37^0C. The solution was centrifuged at 2000 rpm for 30 minutes in 27^0C. The resulting gel was homogenized and the supernatant was removed. Next, adding PBS solution 1 mg/ml into solution and centrifuged for 15 minutes at 2000 rpm in room temperature (27^0C). The supernatant was removed and the pellet was resuspended in 60 mM phosphate buffer, pH 7.0 to a concentration about 2 mg collagen/mL [18]. TNTZ disks were incubated in the suspension at 27°C for 15 min, washed with distilled water, and air dried.

2.3 Corrosion test

Corrosion method was immersion test with all samples were soaked into corrosion solution (synovial solution + 0,9 NaCl). samples consist of collagen uncoated dan coated TNTZ. The corrosion test was done for 2,4, and 6 weeks. All samples have measured the weight before and after corrosion test. The corrosion test was done in incubator 37^0C with the stable variable. After the immersion process, the corrosion rate was measured with weight loss method based on ASTM G1 standard with the formula:

$$CR = (K \times W) / (D \times A \times T)$$

Where:
CR = corrosioon rate (mmpy or milimetre per year)
W = lost of weight (gram)
K = constant factor of rate corrosion
= $8,76 \times 10^4$
D = sample density (g/cm^3)
A = surface area of sample (cm^2)
T = time for immersion (hours).

2.4 Observation of microstructure, hardness, and thickness of samples.

The microstructure of samples was measured before and after corrosion process with a scanning electron microscope (SEM) (Hitachi High-Technologies Corporation, S-3400N Type II) with voltage was 230 Volt. The hardness value of samples was measured using Vicker hardness tester (Shimadzu HMV-2). The load was HV0.01 – HV2 (98.07N-19.614N) with pressuring time was 15 minutes and the voltage was 240V, 300VA 50/60Hz. The thickness was measured with Sanfix thickness gauge series type GF-280.

3. Results and discussion

3.1 Morphology of sample surface area

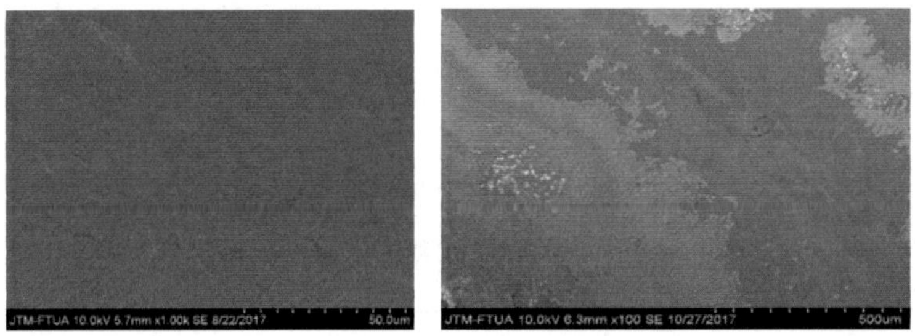

(a) (b)

Figure 1. SEM photograph of samples before the coating process (magnification about 100x). (a) before the coating process, (b) after the coating process.

The sample surfaces have been observed in Figure 1(a) before coating process. The surface is average with a little scratch after the sanding process. Some white spot was assumed as tissue that has been recorded while cleaning process of surface area. After coating process (Figure 1 (b)), the collagen almost covers all of the surfaces but the spread of collagen in surface area is uneven and has different thickness. It is related to the position of a sample while the coating process and the collagen is fibrils that effect to how it covers the surface. Some areas look like a collection of dense dots and other areas like the fiber cluster. The collagen has well attached to the surface area and will reduce the corrosion process of material that has been showed in the result of the corrosion process. Corrosion areas are extensive in samples without coating process comparing to samples with coating process (Figure 2) for each week.

Corrosion process has increased along with the duration of the immersion process. Corrosion area of 6 weeks immersion period is wider than the area of 2 weeks immersion process for uncoated and coated process based on the SEM photograph. Collagen has degraded along with corrosion test slightly as time goes by corrosion process. Collagen has biodegradable features that will disappear by the time based on the corrosion process. The corrosion level of the solution also has an effect to collagen detached duration until it gets involved to the solution. Collagen coating reduces the duration of the corrosion process, so uncoated will be corrosive quickly than collagen coated. Based on Figure 2, the corrosion form is general, crevice and fretting corrosion where these corrosion types could operate in combination [26]. General corrosion is experienced by metallic biomaterial, include titanium alloy (TNTZ) because of releasing metal ion after exposuring time in physiological fluid.

Metal ions release because dissolution of passive oxide film dan ion transport passes the film. Another research found that exposure Ti-6Al-4V in human serum formed Ti complexes with serum protein [27]. This phenomenon has been experienced by coated and uncoated TNTZ when they soaked into synovial fluid.

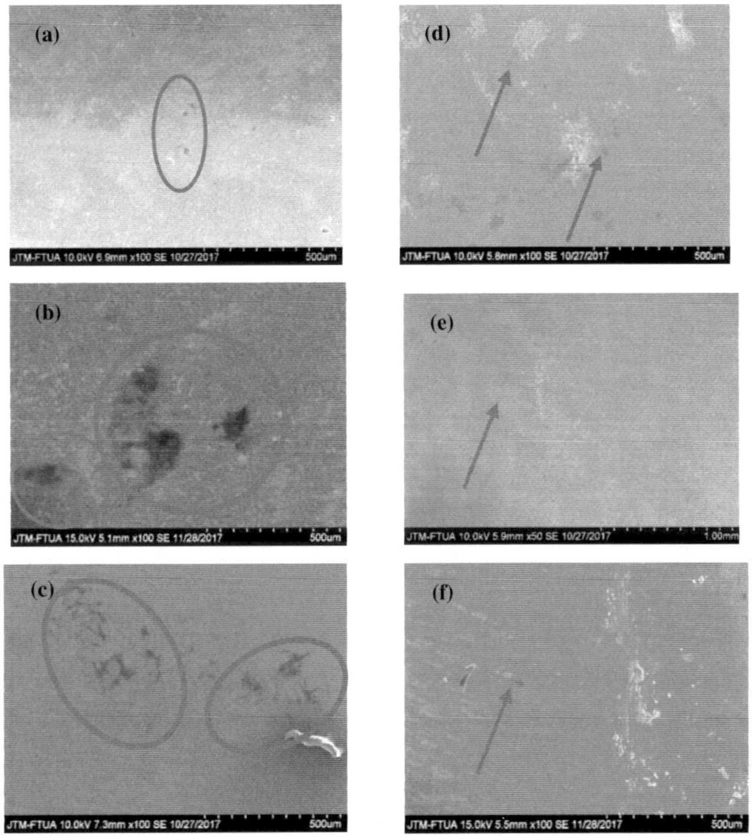

Figure 2. Morphology after corrosion testing for TNTZ uncoated, (a) 2 weeks, (b) 4 weeks, and (c) 6 weeks and TNTZ coated collagen (d) 2 weeks (e) 4 weeks (f) 6 weeks (magnification about 100x).

Pitting corrosion has been made due to surface heterogeneity generally like dislocation, inclusions, second-phase particles or mechanical strength and it all caused sites for pit initiation [28]. Uncoated samples showed more pitting corrosion than coated samples. Crevice corrosion has been found in some case of orthopedic devices [29]. It is created also at the boundary of markers fitted into the stent [26]

3.2 Corrosion rate

In Figure 3, the graphic shows that the corrosion test causes reducing the mass of TNTZ samples (Figure 3). Mass reductions tend to increase along with immersion time, respectively. The highest value of mass reduction belongs to uncoated TNTZ for 6 weeks immersion time that is 0.003 gram (Table 1). Based on all data from the graphic, we may assume that the coating process has an effect to reduce the corrosion process of samples and the value of mass reduction is directly proportional to the duration of corrosion time. Reducing of mass has effect to corrosion rate where increasing of reducing mass makes the corrosion rate becomes higher (Figure 4). The corrosion rate of the uncoated sampel is higher than the coated one in synovial fluid. Overall, corrosion behavior of titanium TNTZ is better compared to other titanium alloys (Ti6Al4V) or pure titanium.

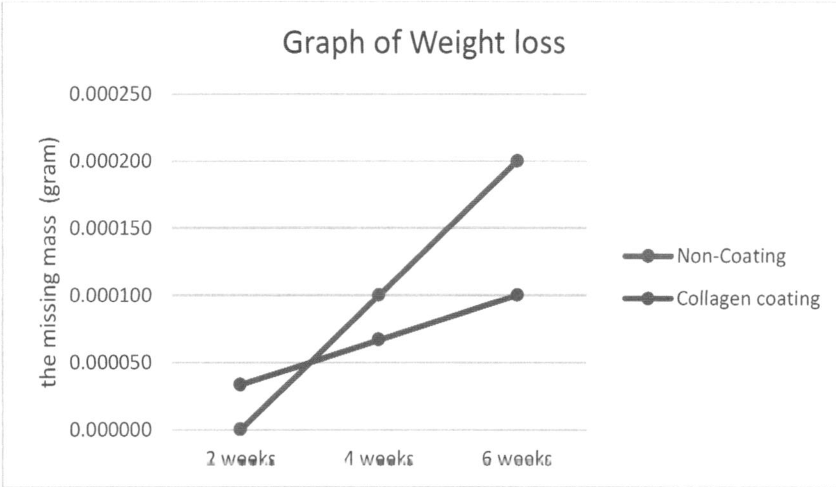

Figure 3. Graph of weight loss of TNTZ samples (coated and uncoated with collagen)

Table 1. The corrosion rate of coated and uncoated samples for each immersion time

Coating process	Weeks	Corrosion rate (mmpy)	Average of corrosion rate (mmpy)
Collagen uncoated	2	0.00165602	0.00205441
	4	0.00188331	
	6	0.00262391	
Collagen coated	2	0.00120125	0.00123278
	4	0.00124186	
	6	0.00125522	

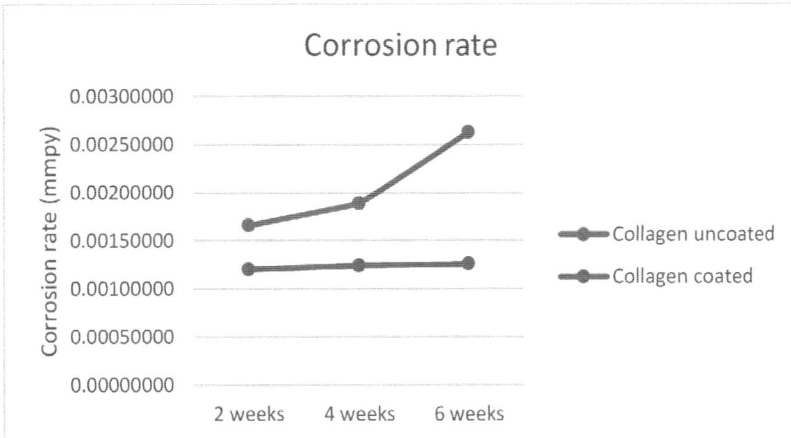

Figure 4. The corrosion rate of coated and uncoated samples for each immersion time

3.3 Hardness of samples

Hardness testing has a function to know the effect of corrosion test of TNTZ samples (coated and uncoated) to mechanical properties. Hardness values of coated and uncoated samples have been shown in Figure 5. The hardness value of the coated material is higher than uncoated samples except for time variation in 4 weeks. The hardness value of uncoated is higher than coated samples even the score is excessive to early hardness value samples before the corrosion test. It is because of the error process on taking a point of hardness test and focuses on the area with high hardness value but overall, decreasing of hardness value happened in each time based on the minimum distance of hardness value for coated and uncoated samples. The data shows that synovial fluid decreases the hardness value of all samples. It is because of degradation proses in the surface of samples and releasing of ions that cause damage in the surface of TNTZ based on SEM photograph (Figure 2).

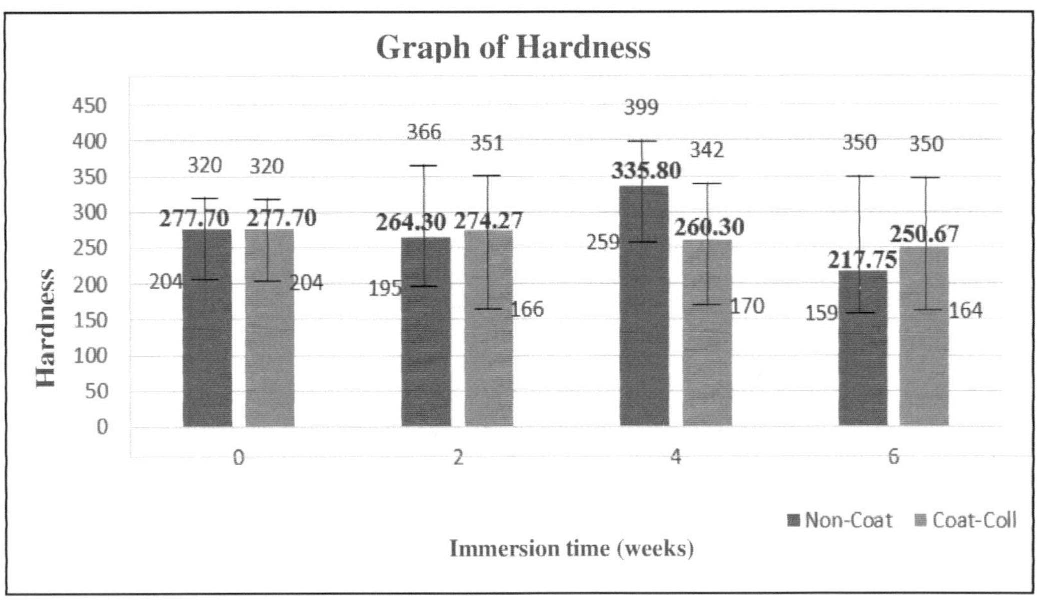

Figure 5. The hardness of each sample after corrosion process by the time

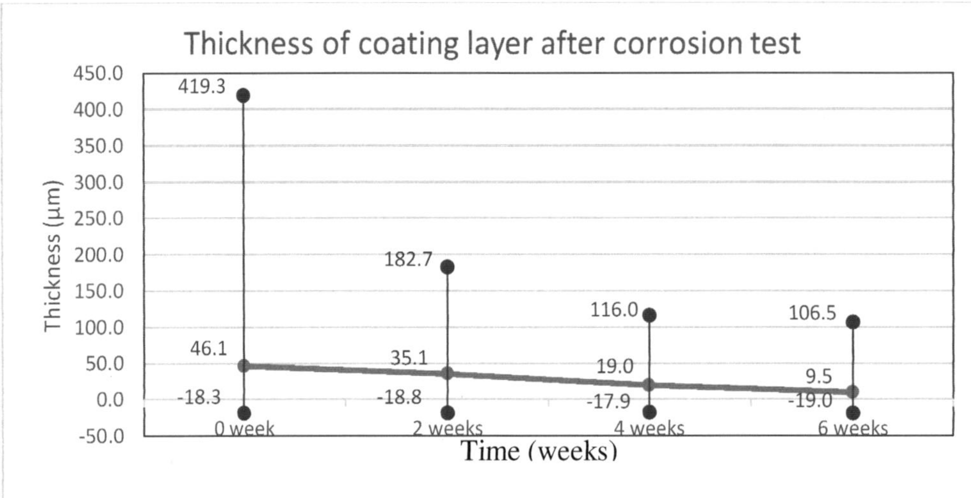

Figure 6. Graph of the degradation process of a coating layer after the corrosion test

Thickness testing of coating layer after immersion time has a function to calculate the residue of the coating layer. The data in the graph (Figure 6) have been calibrated and the result showed that the coating layer has been reduced along with the addition of corrosion time. In 6 weeks of treatment, the collagen layer still exists for about 9.5 µm. It proves that collagen coating as tissue layer does not drop off easily if it uses as a coating layer. It is assumed that collagen coating on TNTZ as joint prostheses is excellent because collagen induces cell adhesion after implantation.

4. Conclusions

The following conclusions are reached:

1. The corrosion rate of titanium TNTZ coated collagen is lower than the uncoated one. The corrosion rate of TNTZ coated collagen is $1{,}233 \times 10^{-3}$ mmpy and the uncoated one is $2{,}054 \times 10^{-3}$ mmpy.
2. Immersion time within synovial fluid has an effect to hardness value for coated and uncoated process where the hardness value becomes lower. In other hands, the duration of immersion treatment has an effect reducing the thickness of the collagen coating.

5. Acknowledgment

The authors thank Ministry of Research, Technology, and Higher Education of the Republic of Indonesia for PMDSU scholarship and PMDSU research grant with the contract number 25/UN.16.17.PP.PMDSU/LPPM/2017.

References

[1] Huiskes R, Weinans H, Rietbergen B. 1992. The relationship between stress shielding and bone resorption around total hip stems and the effects of flexible materials [J]. *Clinical Orthopaedics and Related Research*. **274** 124-134.
[2] Hallab N J, Vermes C, Messina C, Roebuck K. A, Glant T. T, Jacobs J J. 2002. Concentration- and composition- dependent effects of metal ions on human MG-63 osteoblasts. *Journal of Biomedical Materials Research*. 60(**3**) 420-433.
[3] Niinomi M. 2003.Fatigue performance and cytotoxicity of low rigidity titanium alloy, Ti-29Nb-13Ta-4.6Zr. *Biomaterials*. 24(**16**) 2673-2683.
[4] Bertrand E, Gloriant T, Gordin DM, Vasilescu E, Drob P, Vasilescu C, Drob S I. 2010. Synthesis and characterization of a new superelastic Ti-25Ta-25Nb biomedical alloy. Journal of the Mechanical Behavior of Biomedical Materials. 3(**8**) 559-564.
[5] Jung T K M H, Abumiya T, Masahashi N, Kim MS, Hanada S. 2010. Mechanical properties-graded Ti alloy implants for orthopedic applications. *Materials Science Forum*. 205-210.
[6] Park CH, Lee C S, Kim YJ, Jang J H, Suh JY, Park JW. 2011. Improved pre-osteoblast response and mechanical compatibility of ultrafine-grained Ti-13Nb-13Zr alloy. *Clinical Oral Implants Research*. 22(**7**) 735-742.
[7] Hamdy A S, El-Shenawy E, El-Bitar T. 2006. Electrochemical impedance spectroscopy study of the corrosion behavior of some niobium bearing stainless steels in 3.5% NaCl. *International Journal of Electrochemical Science*.**1** 171-180.
[8] Robin A, Rosa J L. 2000. Corrosion behavior of niobium, tantalum and their alloys in hot hydrochloric and phosphoric acid solutions. *International Journal of Refractory Metals and Hard Materials*. 18(**1**) 13-21.
[9] Sherepo K M, Red'ko I A. 2004. The use of zirconium for implants in traumatology and orthopedics. *Med Tekh*. **2** 22-24.
[10] Matsuno H, Yokoyama A, Watari F, UO M, Kawasaki T. 2001. Biocompatibility and osteogenesis of refractory metal implants, titanium, hafnium, niobium, tantalum, and rhenium. *Biomaterials*. 22(**11**) 1253-1262.
[11] Yang H P, Yang S H, Cai Y N, Hou G F, Xia J Y, Tang M T. 2011. Electrochemical behaviors of tantalum in anhydrous ethanol containing hydrogen sulfate ions. *Transactions of Nonferrous Metals Society of China*. 21(**1**) 179-184.
[12] Cabrini R L, Guglielmotti M B, Almagro J C. 1993. Histomorphometry of initial bone healing around zirconium implants in rats. *Implant Dentistry*. 2(**4**) 264-267.
[13] Kulakov O B, Doktorov A A, D'iakova S V, Denisov-Nikol'skii IU I, Grotz K A. 2005. Experimental study of osseointegration of zirconium and titanium dental implants. *Morfologiia*. 127(**1**) 52-55.
[14] Siebers, M.C., Walboomers, X.F., Leeuwenburgh, S. C., Wolke, J.G. & Jansen, J.A. 2004. Electrostatic spray deposition (ESD) of calcium phosphate coatings, an in vitro study with osteoblast-like cells. *Biomaterials*. **25** 2019–2027.
[15] Siebers, M.C., Wolke, J.G., Walboomers, X.F., Leeuwenburgh, S.C. & Jansen, J.A. 2007 In vivo evaluation of the trabecular bone behavior to porous electrostatic spray deposition-derived calcium phosphate coatings. *Clinical Oral Implants Research* **18** 354–361.
[16] Stadlinger, B., Pilling, E., Huhle, M., Mai, R., Bierbaum, S., Bernhardt, R., Scharnweber, D., Ku- hlisch, E., Hempel, U. & Eckelt, U. 2007. Influence of

extracellular matrix coatings on implant stability and osseointegration: An animal study. *Journal of Biomedical Materials Research Part B: Applied Biomaterials.* **83** 222–231

[17] de Jonge, L.T., Leeuwenburgh, S.C., Wolke, J.G. & Jansen, J.A. 2008. Organic-inorganic surface modifications for titanium implant surfaces. *Pharmaceutical Research.* **25** 2357–2369.

[18] Geissler, U., Hempel, U., Wolf, C., Scharnweber, D., Worch, H. & Wenzel, K. 2000 Collagen type 1-coating of Ti6Al4V promotes adhesion of osteoblasts. *Journal of Biomedical Materials Research.* **51** 752–760.

[19] de Jonge, L.T., Leeuwenburgh, S.C., van den Beucken, J.J., te Riet, J., Daamen, W.F., Wolke, J.G., Scharnweber, D. & Jansen, J.A. 2010. The osteogenic effect of electrosprayed nanoscale collagen/ calcium phosphate coatings on titanium. *Biomaterials* **31** 2461–2469.

[20] Rammelt, S., Illert, T., Bierbaum, S., Scharnweber, D., Zwipp, H. & Schneiders, W. 2006 Coating of titanium implants with collagen, RGD peptide, and chondroitin sulfate. *Biomaterials.* **27**: 5561– 5571.

[21] Stadlinger, B., Pilling, E., Mai, R., Bierbaum, S., Bernardt, R., Scharnweber, D. & Eckelt, U. 2008. Effect of biological implant surface coatings on bone formation, applying collagen, proteoglycans, glycosaminoglycans and growth factors. *Journal of Materials Science: Materials in Medicine.* **19**: 1043–1049.

[22] Schliephake, H., Aref, A., Scharnweber, D., Bierbaum, S. & Sewing, A. 2009. Effect of modifications of dual acid-etched implant surfaces on peri-implant bone formation. Part 1: organic coatings. *Clinical Oral Implants Research.* **20**:31–37.

[23] Valero Vidal C, Igual Muñoz A, Olsson C-OA, Mischler S. J. 2014. Adsorption of BSA on Passivated CoCrMo PVD Alloy: An EQCM and XPS Investigation. Journal of *Electrochemical Society*;161(**6**): C294–301.

[24] Munoz, A.I., Schwiesau, J., Jolles, B.M., Mischler, S. 2015. In vivo electrochemical corrosion study of a CoCrMo biomedical alloy in human synovial fluids. *Acta Biomaterialia*: 1-9

[25] Martin EJ, Pourzal R, Mathew MT, Shull KR.2013. Dominant Role of Molybdenum in the Electrochemical Deposition of Biological Macromolecules on Metallic Surfaces. *Langmuir.* **29**:4813–22.

[26] Pound, BG. 2014. Corrosion behavior of metallic materials in biomedical applications. I. Ti and its alloys. Biomedical corrosion. Part I. *Corrosion rev.* 32 (**1-2**) 1-20

[27] Hallab NJ, Mikecz K, Vermes C, Skipor A, Jacobs JJ. Differential lymphocyte reactivity to serum-derived metal-protein complexes produced from cobalt-based and titanium-based implant alloy degradation. J Biomed Mater Res 2001; **56** 427–436.

[28] Szklarska-Smialowska Z. 1986. Pitting Corrosion of Metals. Houston, TX: NACE International.

[29] Goldberg JR, Gilbert JL, Jacobs JJ, Bauer TW, Paprosky W, Leurgans S. 2002. A multicenter retrieval study of the taper interfaces of modular hip prostheses. *Clinical Orthopedic.* **401** 149–161.

Effect of Aluminum and Silicon Antioxidants Addition on Porosity, Cold Crushing Strength, Wear Resistance and Hydration Effect of Alumina Spinel Refractories

E.D. Fernando, A.A. Korda[1]
Department of Metallurgical Engineering, Faculty of Mining and Petroleum Engineering, Bandung Institute of Technology

*E-mail: akhmad@mining.itb.ac.id

Abstract. Due to their high resistance and wear resistance, alumina spinel ($MgAl_2O_4$) are widely used for slide gate on ladle. The characteristics strongly depend on the formation of ceramic phases in the matrix. Addition of antioxidants is used to prevent the graphite oxidation, possibly leading to porosity increase. Currently, other problem appears as the consequences of climate change. The object of this research is to study the effect of antioxidants addition in temperature variations. The influence of Al and Si addition is investigated by means of XRD, SEM, TGA-DTA, apparent porosity, CCS, wear resistance and hydration resistance tests. It was found that lowest porosity, highest CCS and best wear resistance were reached by sample containing Al 5% in 1400°C, while best hydration resistance was reached by sample containing Si 5%.

Keywords: spinel, antioxidant, porosity, CCS, wear resistance, hydration

1. Introduction

The increasing of steels and iron production affects refractories demand generally. In 2016, the usage of refractories for steels and iron industries reached 60% of all refractories demand in the world [1]. Refractories are widely used for slide gate in ladle. Magnesium alumina ($MgAl_2O_4$) refractory was applied due to its low thermal expansion, high melting point, stability in chemical environment and good corrosion resistance [2-5]. In advanced research, graphite was added to improve its wear resistance and cold crushing strength. However, graphite was easily oxidized on low temperature that caused the porosity increase and strength decrease. The antioxidants (Al, Si, Mg, C) were added to prevent the graphite oxidation.

Climate change is another problem in refractory industry. According to Indonesian Agency for Meteorology, Climatology and Geophysics (BMKG), Indonesia had its first wet-dry season in 2016 that the rain intensity in summer was higher than usual [6]. The condition would increase vapor and caused refractories disintegration during handling and inventory process.

Corresponding author: akhmad@mining.itb.ac.id

In this research, the effect of antioxidants addition on refractories properties with variation of temperature was studied. The aim of aluminum addition is to improve the strength and wear resistance, while the silicon addition is to improve the hydration resistance [7]. Several experimental tests were carried out for a better understanding of the antioxidants performance and the characterization of the refractory properties.

2. Materials and Methods

Commercial AR90 alumina spinel with different size aggregate was mixed with tubular alumina, phenolic resin, graphite and variations of antioxidants until homogenized. Antioxidants used were aluminum and silicon metal powder. Antioxidants were added with variations of 5 wt.% Al, 3 wt.% Al + 2 wt.% Si, 2 wt.% Al + 3 wt.% Si and 5 wt.% Si. The mechanical mixing was done twice; dry mixing and resin mixing. After mixing, aging was conducted for 24 hours in room temperature to form resin polymerization. Then, pressing was conducted using Uniaxial Pressing Machine with 50 kN. The dies set of steel has 1.5 cm diameter and 1.7 cm height. After that, drying was conducted using oven in temperature 200°C for 24 hours. Sintering was conducted right after that with different temperature of 1000, 1200, 1300 and 1400°C for 6 hours.

Porosity is a comparison between the open porous weight and total weight of the sample. Apparent porosity test refers to ASTM C-20 [8]. The value of porosity was calculated with:

$$Apparent\ Porosity = \frac{V_0}{V} \times 100\% = \frac{W - D}{W - S} \times 100\% = \frac{W - D}{V} \times 100\% \quad (1)$$

Cold crushing strength (CCS) test was conducted by giving one direction force to sample until crushed, using Uniaxial Pressure Machine. The value of CCS shows the ability of material to withstand on pressure in room temperature which refers to ASTM C133 [8]. The value of CCS was calculated with:

$$CCS = \frac{F}{A} \quad (2)$$

Wear resistance was conducted in abrading wheel with velocity of 90 rpm for 5 minutes. The value of wear resistance was the weight loss of each sample. Hydration resistance test was conducted for samples in sintering temperature of 1000°C. Samples were located in room with temperature of 25–30°C and relative humidity 68-82% [9]. The weight gain was calculated for 23 days.

Thermogravimetry Analysis/Differential Thermal Analysis (TGA/DTA) test was conducted for samples containing 2 wt.% Al + 3 wt.% Si and 5 wt.% Si antioxidants. The test was performed on the Setaram Setsys 1750 machine which operated up to 1300°C with a temperature rate of 10°C/min. X-Ray Diffraction (XRD) patterns were recorded using XRD Rigaku SmartLab machine with Cu Kα radiation. The samples tested were samples with the addition of 5 wt.% Al and 5 wt.% Si antioxidants at a sintering temperature of 1400°C. This test was performed to see the compounds that were formed. The microstructural evaluation of the sintered sample was performed by scanning electron microscopy (SEM) JEOL 6510. Samples containing 5 wt.% Al, 2 wt.% Al + 3 wt.% Si and 5 wt.% Si antioxidants were used for both tests in sintering temperature of 1400°C.

3. Results and Discussion

3.1. Thermal Characterization

In order to observe and identify the existence of organic phase, graphite and antioxidants containing in alumina spinel, thermal characterization was studied using TGA and DTA curves. It was predicted what process occurred during heating and identified the critical temperature of new phase formation. Fig. 1 shows TGA-DTA curves for samples containing 2 wt.% Al + 3 wt.% Si antioxidants. In temperature 100–400°C, water vaporization, phenolic resin decomposition, hydroxide and small compounds (e.g. CO, CO_2, CH_4) decomposition occurred [10-12], as weight loss shown in TGA curve. Fig. 2 for sample containing 5 wt.% Si shows the similar results as Fig. 1.

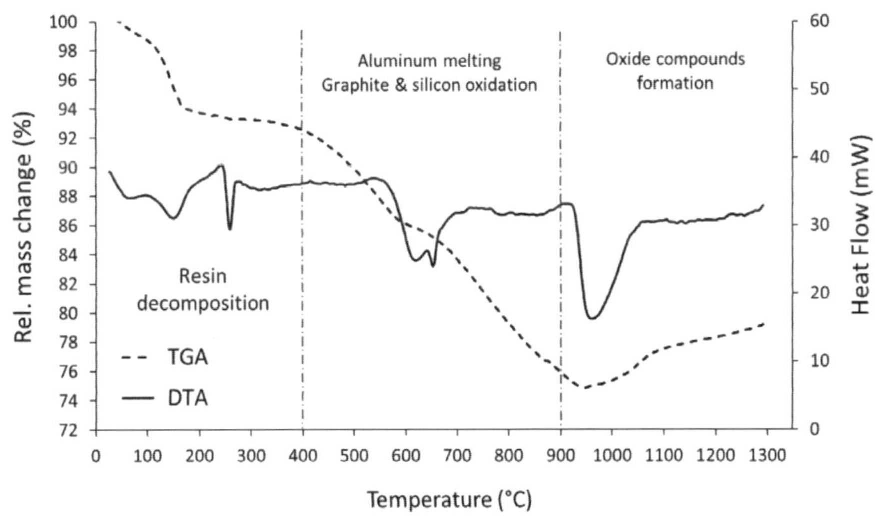

Fig. 1. TGA-DTA curve for the refractory with 2 wt. % Al + 3 wt. % Si

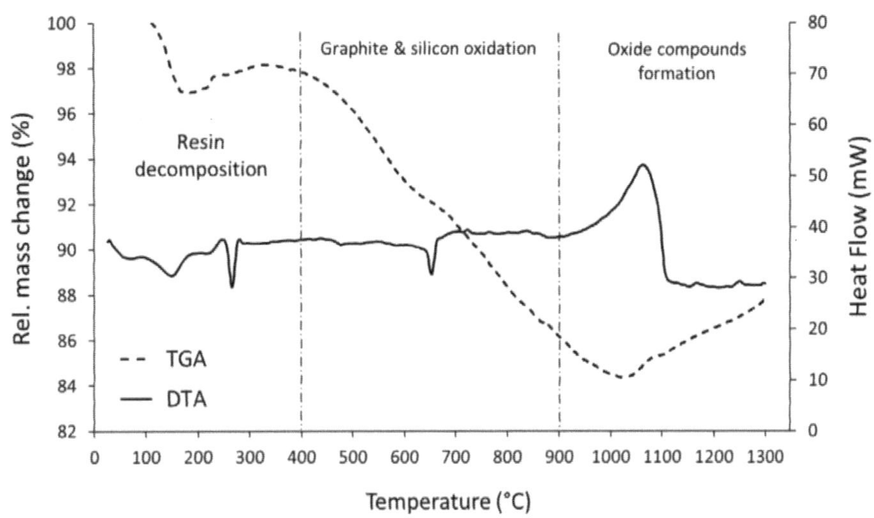

Fig. 2. TGA-DTA curve for the refractory with 5 wt. % Si

Fig. 1 shows graphite oxidation and Al oxidation into Al^2O^3 in temperature 500 800 C. In temperature 500–850°C, Al was melted and formed Al_4C_3 with graphite [11-13]. In temperature 1000°C, Si was oxidized into SiO_2, and graphite was still oxidizing to form more stable compound of CO. Carbide and oxide compounds were formed in temperature above 1000°C [10-17]. Fig. 2 shows the similar same results as Fig. 1. However, carbide and oxide compounds formation just occurred for Si elements. In temperature of 950°C, mullite was formed, as shown in XRD test as well.

3.2. Porosity, Mechanical and Hydration Properties

Antioxidants addition decreases the porosity in temperature more than 750°C [10]. Fig. 3 shows apparent porosity decreased along with sintering temperature increase as neck formation between solid particles occurred better. The event was supported by TGA-DTA curves showing weight gain occurred in temperature more than 750°C. According to TGA-DTA curves, weight loss in temperature of 200–850°C occurred because of binder decomposition and aluminum melting, and caused porosity formation in refractory. In temperature above 950°C, weight gain occurred because of oxide formation filling the porosity [9]. Sample containing 5 wt.% Al in 1400°C has the smallest value of porosity, 18.72%, caused by aluminum affinity to react with oxygen and graphite is higher than silicon. Therefore, oxide and carbide were formed better in samples containing higher aluminum addition.

Oxide and carbide compounds formation increased the cold crushing strength. Fig. 4 shows CCS increased along with sintering temperature increase as mass transfer in each particles that formed ceramics bonding occurred. Besides, neck formation between particles created bigger diameter [10]. Sample containing 5 wt.% Al in 1400°C has higher CCS than others, 39.33 MPa, caused by aluminum has lower melting point and ease to form Al_4C_3. The carbide will react with CO to form Al_2O_3 improving refectory strength [13]. The mechanism occurred as well in samples containing Si, yet required higher temperature.

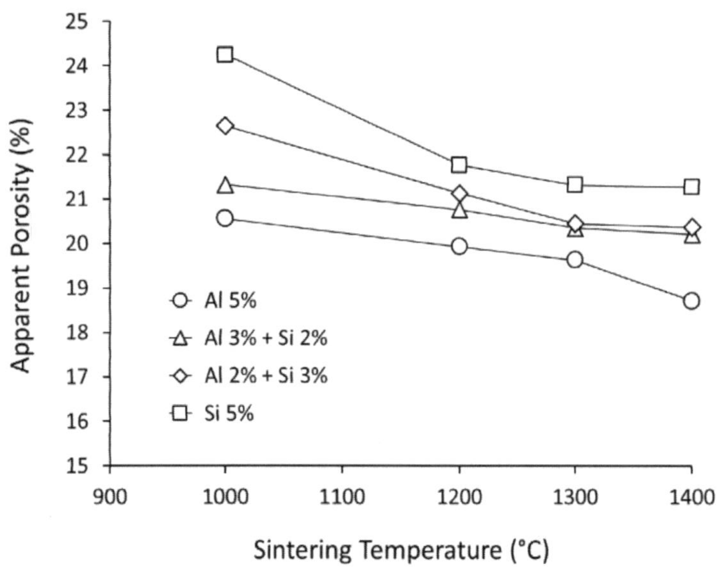

Fig. 3. Apparent porosity vs. sintering temperature for the alumina spinel refractories with aluminum and silicon antioxidants addition

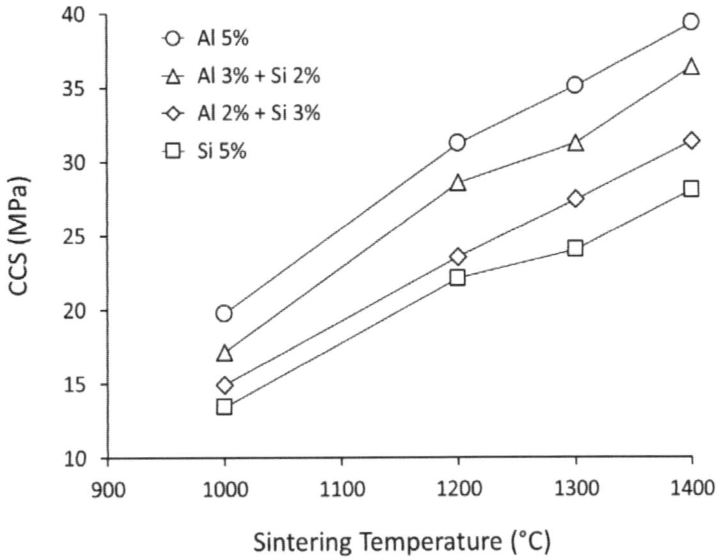

Fig. 4. Variation in cold crushed strength with sintering temperature for the alumina spinel refractories with aluminum and silicon antioxidants addition

Wear resistance shows formed ceramics bonding strength to withstand on high speed of wear. Fig. 5 shows wear resistance, shown in weight loss, increased along with wintering temperature increase as sintering process occurred better in high temperature. In temperature of 1100°C, weight loss declined steeply because ceramics bonding formed stronger. Sample containing 5 wt.% Al in 1400°C shows the best result with weight loss of 0.0028 gr. The density plays an important role to increase the wear resistance. Therefore, aluminum addition increased wear resistance, better than silicon.

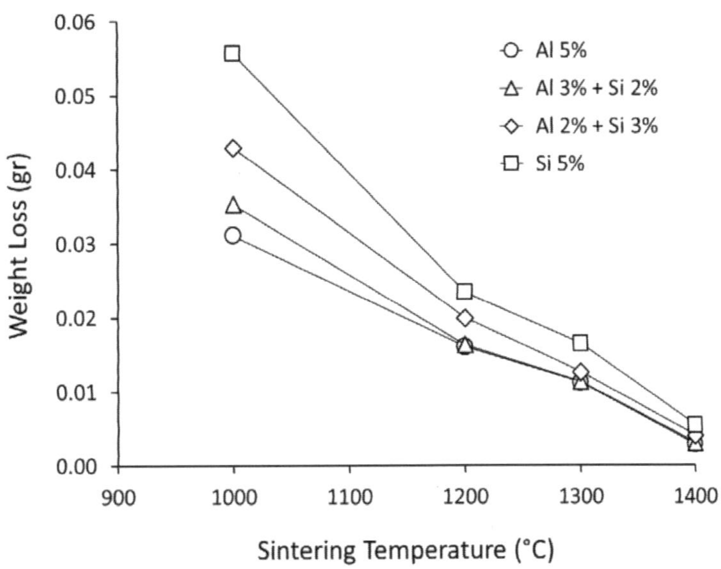

Fig. 5. Variation in wear resistance with sintering temperature for the alumina spinel refractories with aluminum and silicon antioxidants addition

Fig. 6 shows all samples adsorbed vapor as shown in weight gain every day. Sample containing 5 wt.% Al has the lowest resistance to hydration with %-weight gain 0.168%, while sample containing 5 wt.% Si has the best resistance with %-weight gain 0.071%. Al_4C_3 and Al_2O_3 in sample containing 5 wt.% Al reacted with vapor to form $Al(OH)_3$. The hydrate expanded and caused refractories disintegration [9]. On the other hand, silicon formed passive layer of gel M-S-H on refractory surface and prevented from vapor adsorption [18]. In the first three day, sample containing 5 wt.% Si had higher %-weight gain as the formation of passive layer occurred. After that, the weight gain did not occur significantly.

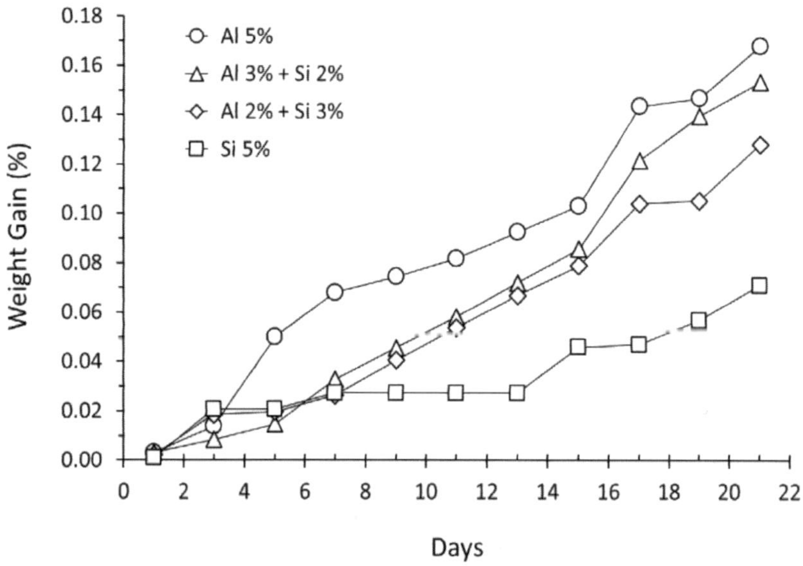

Fig. 6. Weight gain as a function of adsorption days the alumina spinel refractories with aluminum and silicon antioxidants addition

3.3. Phase Composition and Microstructure

XRD test was conducted for samples containing 5 wt.% Al and 5 wt.% Si in sintering temperature of 1400°C. Fig. 7 shows the spinel and alumina dominated the compounds formed. Alumina was a result of oxidation process during sintering, indicating aluminum powder succeeded as the antioxidant. Fig. 8 shows some compounds beside alumina and spinel were formed, such as silica, mullite and silicon. The presence of Si element shows that silicon did not work well as antioxidant.

SEM/EDS test was conducted for samples containing 5 wt.% Al, 5 wt.% Si and 2 wt.% Al + 3 wt.% Si. Fig. 9 shows the morphologies are granules in some area and flakes in another area. However, in Fig. 10, the morphologies are granules in most of the area and melted shape in different area. Fig. 9 shows EDS result for sample containing 5 wt.% Al where alumina dominated the compounds formed. The presence of graphite and magnesium were very low for both results. Fig. 11 shows several compounds were formed. The existence of alumina explains that aluminum powder worked well as antioxidant. Mullite compound was formed as well. However, silicon oxidation was not complete shown with Si existence.

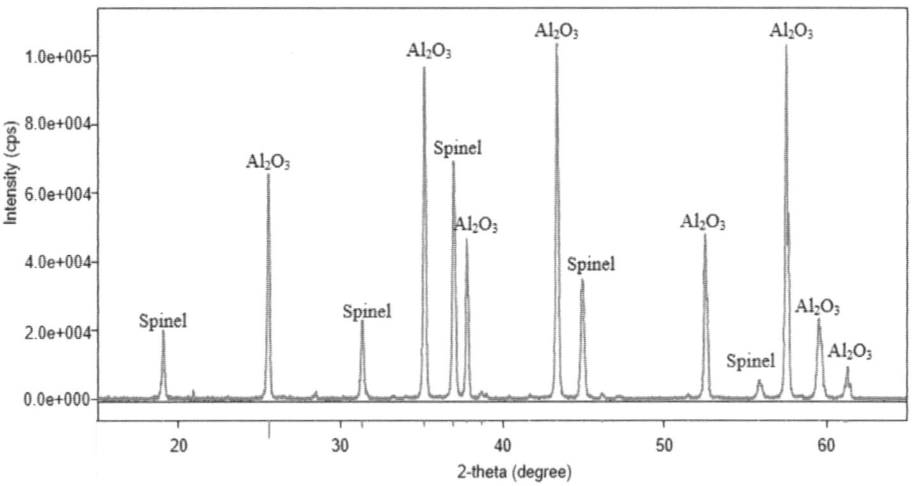

Fig. 7. XRD pattern for alumina spinel refractory with addition of 5 wt.% Al

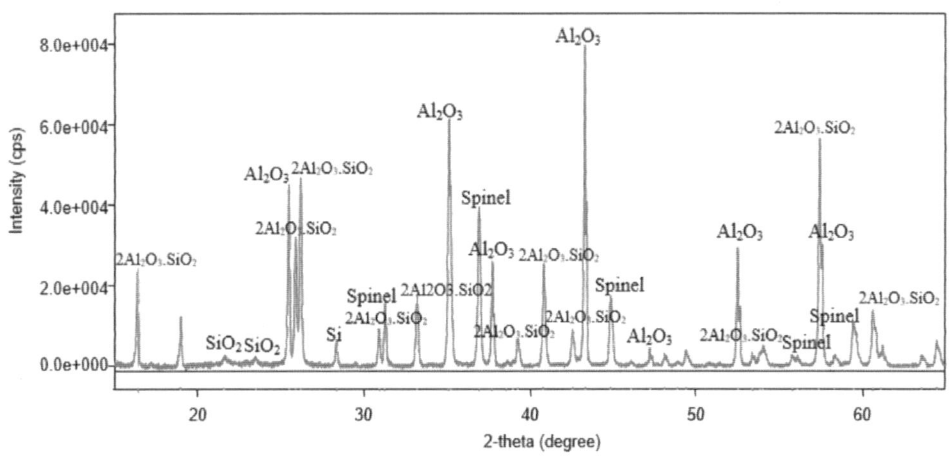

Fig. 8. XRD pattern for alumina spinel refractory with addition of 5 wt.% Si

Fig. 9. EDS analysis of sintered aluminum spinel refractory with addition of 5 wt.% Al

Fig. 10. EDS analysis of sintered aluminum spinel refractory with addition of 5 wt.% Si

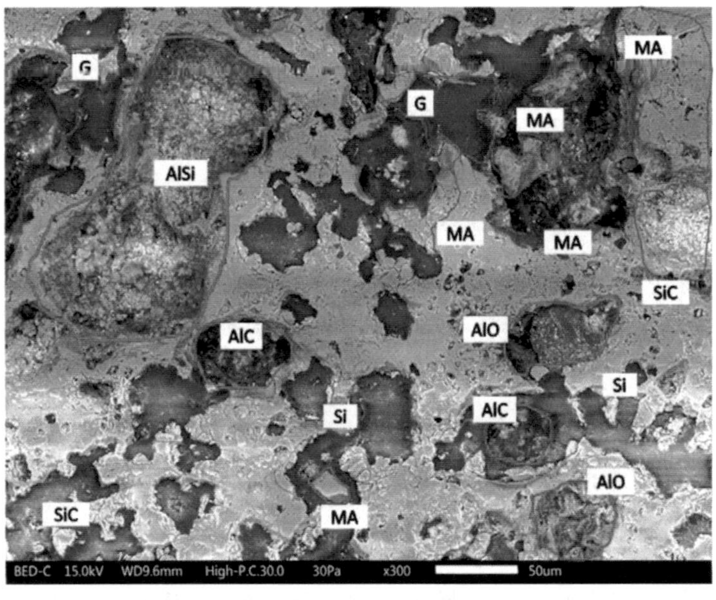

Fig. 11. EDS analysis of sintered and polished aluminum spinel refractory with addition of 2 wt.% Al + 3 wt.% Si. The marked area indicates the dominant area of each of the compounds or elements that make up the refractory

The elemental mapping characterization shown in Fig. 12 and 13 for samples with the addition of 5 wt.% Al and 5 wt.% Si antioxidants, respectively. The most dominant observed element is aluminum. This is because the refractory raw materials used were commercial AR 90 with alumina composition of more than 87%. In Fig. 12, it shows that many oxide compounds were formed in the sample. Most regions are an association of elements of Al, Mg and O showing the spinel phase. In Fig. 13 there is an area which is an association of Si and O elements showing the silica phase. There is also an area which is an association of all elements indicating the presence of oxide and carbide phases. Areas that are only associated with C without additional elements indicate the presence of non-oxidized graphite.

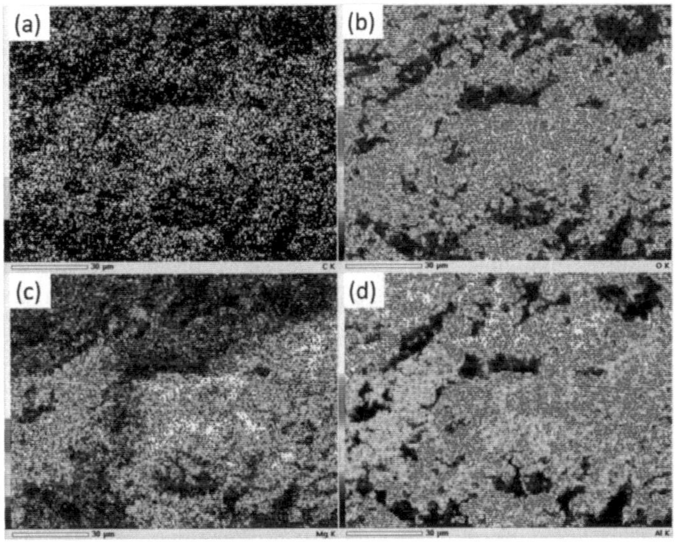

Fig. 12. Elemental mapping of sintered aluminum spinel refractory with addition of 5 wt.% Al (a) C, (b) O, (c) Mg and (d) Al

Fig. 13. Elemental mapping of sintered aluminum spinel refractory with addition of 5 wt.% Si (a) C, (b) O, (c) Mg (d) Al and (e) Si

4. Conclusions

The influence of Al and Si antioxidants on the mechanical properties, phase composition and microstructure was studied in this research. The main conclusions are as follows:

a. Alumina and spinel are the major compounds for sample containing Al while silica and mullite are the major compounds for sample containing Si.
b. Porosity decreased along with sintering temperature increase. Wear resistance and CCS increased along with sintering temperature increase. Sample containing 5 wt.% Al shows the best result for the tests.
c. Sample containing 5 wt.% Si shows the best result for hydration resistance test.

5. Acknowledgements

The authors are grateful to PT Sigma Mitra Sejati for supporting this work.

6. References

[1] http://ceramics.org/ceramic-tech-today/freedonia-world-refractory-demand-to-grow-annually-34-percent-through-2016 (accessed on December 11 2016)

[2] Banerjee, G., et. al., Slide gate refractories-recent trends, *Tehran International Conference on Refractories*, 4-6 May 2004

[3] Chaudhuri, J., et. al., New generation ladle slide gate system for performance improvement, *MPT International*, **Vol. 30**, 2007, 38-42

[4] Sainz, M.A., et. al., Thermochemical Stability of Spinel ($MgO.Al_2O_3$) Under Strong Reducing Conditions, *Material Chemistry and Physics*, **Vol. 86**, 2004, 399-408

[5] Li, J., et. al., Investigations on phase constitution, mechanical properties and hydration kinetics of aluminous cements containing magnesium aluminate spinel, *Ceramics International* **39**, 2013, 8393-8400

[6] http://www.bmkg.go.id/iklim/prakiraan-musim.bmkg (accessed on January 16 2017)

[7] Gokce, A.S., et. al., The effect of antioxidant on the oxidation behavior of magnesia-carbon refractory bricks, *Ceramic International* **34**, 2008, 323-330

[8] 2003, *Annual book of ASTM standards*, Refractories: Activated Carbon, Advance Ceramics, 15.01 pp 19

[9] Nandy, S.K., et. al., Hydration of Coked MgO-C-Al Refractories, *Ceramics International* **32**, 2006, 163-172

[10] Efendy, H., Study of antioxidants on MgO-C monolithic refractories with tar-resin binder. *Dissertation*. Faculty of Mathematics and Natural Sciences, Bandung Institute of Technology, 2008

[11] Musante, L., et. al., High temperature mechanical behavior of Al_2O_3-MgO-C refractories for steelmaking use, *Ceramics International* **37**, 2011, 1473-1483

[12] Munoz, V., Martinez, A.G., Thermal evolution of Al_2O_3-MgO-C refractories, *11th International Congress on Metallurgy & Materials SAM/CONAMET* 2011

[13] Munoz, V., Physical, chemical and thermal characterization of alumina-magnesia-carbon refractories, *Ceramics International* **40**, 2014, 9133-9149

[14] Sadrnezhaad, S.K., et. al., Effect of Si Antioxidant on the Rate of Oxidation of Carbon in MgO-C Refractory, Center of Excellence for Production of Advance Materials, Department of Materials Science and Engineering, Sharif University of Technology, Tehran, Iran, 2011

[15] Zhang, S., et. al., Thermochemistry and microstructures of MgO-C refractories containing various antioxidants, *Journal of the European Ceramic Society* **21**, 2001, 1037-1047

[16] Naghizadeh, R., et. al., Effect of Antioxidants and Atmosphere on the Phase Formation Microstructure of in situ MgO-Al_2O_3 Spinels, Department of Metallurgical and Materials Engineering, Iran University of Science and Technology, 2008

[17] Zhu, T.B., The Influence of Al and Si Additives on the Microstructure and Mechanical Properties of Low-Carbon MgO-C Refractories, *Journal of Ceramic Science and Technology* **07 [01]**, 2016, 127-134

[18] Souza, T.M., et. al., Phosphate chemical binder as an anti-hydration additive for Al_2O_3-MgO refractory castables, *Ceramics International* **40**, 2014, 1500-1512

Fabrication of Micro Dimples Pattern Using Ball End Mill

Muhamad Izdhar Bin Zaki[1], **Kushendarsyah Saptaji**[1]*, **Raden Dadan Ramdan**[2], **Tedi Kurniawan**[3]
[1]1Faculty of Manufacturing Engineering, University Malaysia Pahang, 26600 Pekan, Pahang Darul Makmur, Malaysia
[2]Faculty of Mechanical and Aerospace Engineering, Institute Technology of Bandung, Indonesia
[3]Faculty of Mechanical Engineering, Universiti Malaysia Pahang, 26600 Pekan, Pahang Darul Makmur, Malaysia

*Corresponding author (E-mail): kushendarsyah@ump.edu.my

Abstract. In this paper, the effect of workpiece tilt angle and feed direction in producing micro dimple pattern using ball end mill tools were studied. Experiments were conducted using carbide ball end mill tools, aluminium Al 6061-T6 workpiece and using CNC machine. Three workpiece tilt angles are selected 0^0, 30^0 and 45^0; and two feed directions z-axis and x-axis. Visual observations using Scanning Electron Microscope (SEM) and geometry measurements using laser scanning confocal microscope were conducted on the fabricated micro dimple patterns. The results show that the workpiece tilt angle produces less burr at the edge of the dimples compared to without tilt (0^0). In addition, the workpiece tilt angle can avoid the formation of undercut at the center of the dimple. The workpiece tilt angle of 30^0 and z-axis feed direction produced better result compared to 0^0 and 45^0 tilt angles with the value of ratio diameter/depth closer to the theoretical value.

Keywords: micro-dimple pattern, ball end mill, workpiece tilted angle, feed direction

1. Introduction

The surface profile modifications have been applied to reduce the friction in between two metal surfaces and to improve the oil lubrications especially in automotive engines. In automotive engines, most friction occurs between two slides surface inside the engine and directly affected energy consumptions and fuel savings [1,2]. One of the surface profiles that can improve the lubrication is micro dimple pattern. The micro dimple pattern can increase the performance of working surface and quality of surface [3]. Micro dimple pattern on the metal surface can act as water repellent or self-cleaning surface similar to the bio-inspired lotus leaf [1]. The micro dimple pattern creates small hydrodynamic pressure and produces less friction between the sliding surfaces [2,4-6]. At the end, the less friction sliding surfaces resulted in less energy consumption. Therefore, micro dimple pattern can be applied on the metal surfaces to increase the performance and mechanical properties of engineering components such as bearing [5]. In addition, the micro dimple pattern can also be applied in implants, bearings, turbine blades, aerofoil and also, micro lens.

Micro dimple structure can be defined as a geometric that have contour like square, circular, or elliptical with diameter of 20 µm – 4 mm and ratio of 7 to 40% and also depth of cut between 200 nm – 100 µm [7]. Micro dimple pattern can be classified into two arrangement patterns; linear / square arrangement and zigzag arrangement (Figure 1). The distance between dimples is important to ensure the surface effectiveness.

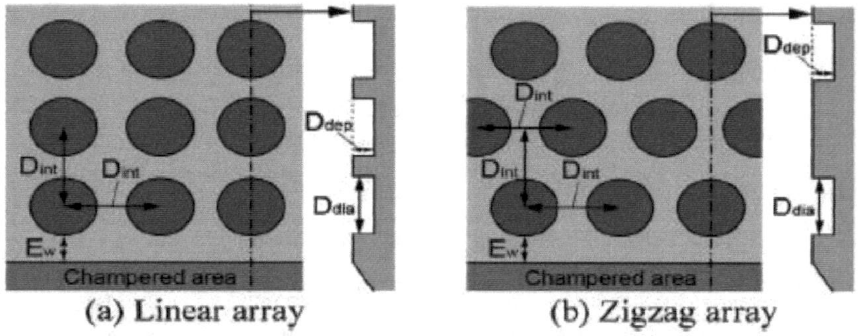

Figure 1. Two micro dimple patterns, linear / square (left) and zigzag (right) [8].

The micro dimple pattern on the metal surface can be fabricated using several methods. Kurniawan et. al. [9] fabricated micro dimple using conventional texturing (CT) by sinusoidal motion tip of tool on surface. Dai et. al [10] used Masked Laser Ablation (MLA) method with pulse Nd:YAG laser to produce micro dimple pattern. The electrochemical machining (ECM) using a micro second pulse on-time toward the surface of material was used by Byun et. al. [6] in producing micro dimple pattern. Graham et. al. [2] used the ball end mill with inclined technique to produce effective and efficient fabrication of micro dimple pattern. Ball end mill method is simple and less expensive machining method that can be used to produce micro-dimple pattern. Ball end mill is also known for making three dimensional contour shape that can be found on moulds and dies products. Ball end mill usually operates on 3-axis machine. The ball end mill or the ball nose cutter is similar to the slot bit, but the end of the mill is hemispherical.

In this paper, the ball end mill tool will be used to fabricate the micro dimple pattern. The objective of this paper is to study the feasibility of fabricating micro dimple pattern using ball end mill tool in CNC milling machine and to determine the effect of workpiece tilt angle on the quality of the micro dimple pattern. The micro dimple pattern will be produced using three different workpiece tilt angles. The idea to tilt the tool is to find which angle can produce the best results in producing micro dimple pattern. The drawing software is used to sketch the pattern of micro dimple and to simulate the ball end mill process. This method becomes a quick and efficient way to create micro dimple pattern surfaces due to execution of G-code by the software. Finally, the micro dimple pattern is fabricated using CNC machine with the optimum machining parameters such as the spindle speed, feed rate and depth of cut.

2. Experimental Setup

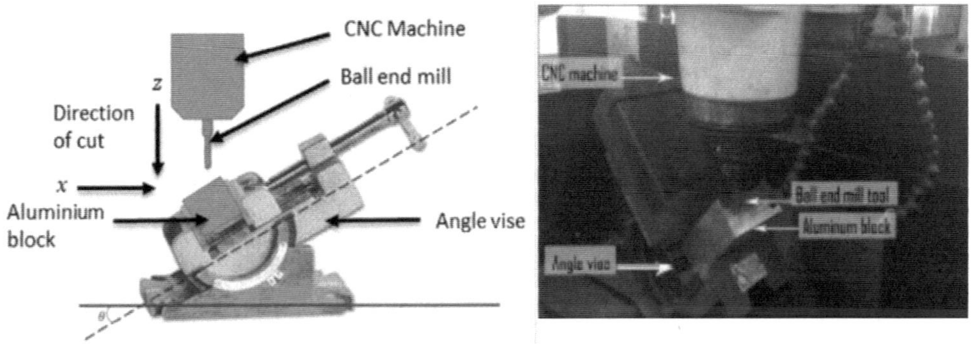

Figure 2. Schematic view (left) and experimental setup (right) for dimple pattern fabrication using workpiece tilt angle method

The cutting parameters used are depth of dimple 0.1 mm, feed rate 20 mm/min, spindle speed 4,000 rpm. These are the optimum parameters selected based on the preliminary results and machine capability. The carbide ball end mill tool having 1 mm diameter and 0.5 mm nose cutter radius is used in this experiment. The workpiece is aluminum alloys Al6061-T6. Prior to the fabrication of micro dimple patterns, the workpiece is polished to have a smooth surface with surface roughness of 0.476 μm. Experiments are conducted using Makino Ke55 CNC milling machine. The workpiece is clamped in vise to be able to tilt the workpiece (Figure 2). Three tilt angles of the workpiece are tested 0^0, 30^0 and 45^0. Two feed directions are used for the 30^0 and 45^0 which are x-axis and z-axis (Figure 3). Two dimple patterns are selected, square and zigzag. The distance between the dimples (pitch) for the square pattern is a = b = 1 mm and for zigzag pattern a = b = 1 mm and c = 0.5 mm (Figure 4).

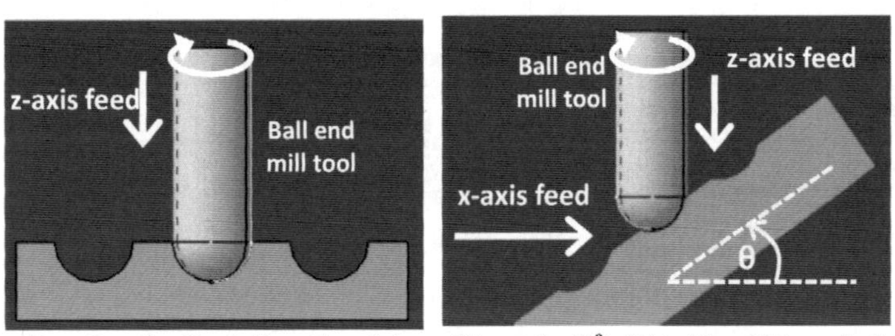

Figure 3. Illustrations for the feed direction (a) no tilt angle (0^0) (b) with a tilt angle θ (30^0 and 45^0).

The parameters used in the experiments are shown in Table 1. The CATIA software is used to design the dimple pattern and to generate the machining strategy to be used in Computer Numerical Control machine (CNC) for the machining of micro dimple pattern. Micro dimple patterns were fabricated with the size of 5 mm x 5 mm for each of combination of pattern, workpiece tilt angle and feed directions. The fabricated dimple patterns are observed visually using Scanning Electron Microscope (SEM) and analyzed using 3D Laser Scanning Confocal Microscope (Keyence).

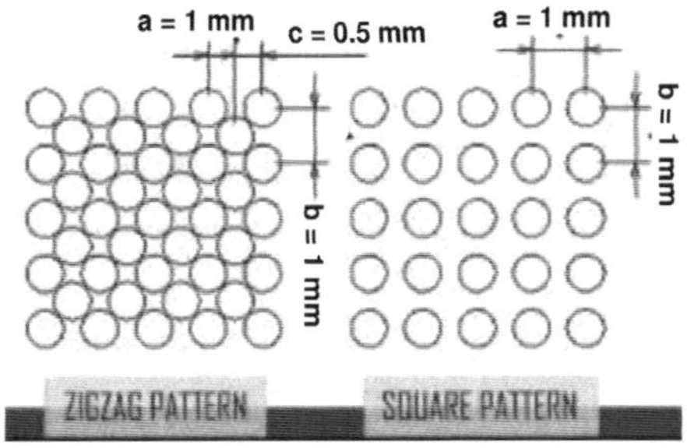

Figure 4. Distance between dimples used in this experiment for zigzag (left) and square (right) patterns

Table 1. Machining parameters used in the experiment

Tool	Carbide ball end mill tool with 1 mm diameter and ball nose radius 0.5 mm
Workpiece	Aluminum Al6061-T6
Spindle speed	4,000 rpm
Feed rate	20 mm/min
Depth of dimple	100 μm
Type of patterns	– Square (pitch = 1 mm) – Zigzag (distance between dimples a = 1 mm, b = 1 mm, c = 0.5 mm)
Workpiece tilt angle (θ)	0^0, 30^0, 45^0
Feed directions	z- axis and x-axis

3. Results and Discussion

In total 10 different micro-dimple patterns were fabricated (three tilt angles, two feed directions and two patterns). Surface quality observations are conducted using SEM and optical microscope. Dimple depth, circular diameter, and pitch are measured on a selected dimple for each of the combination of tilt angle, feed direction and pattern.

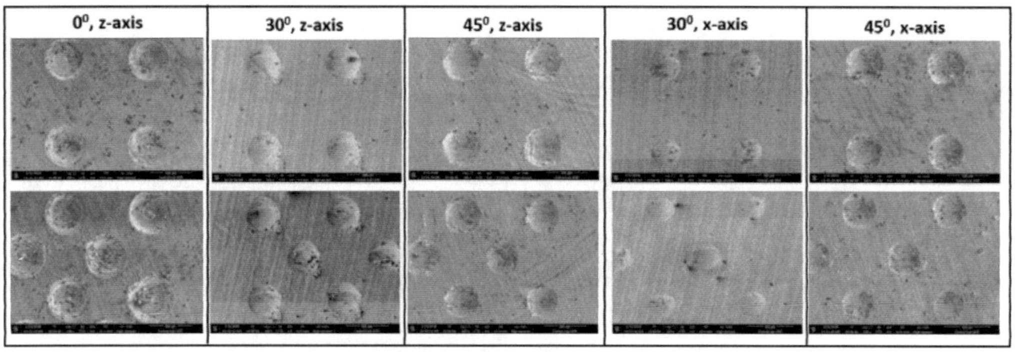

Figure 5. SEM images of the micro dimple pattern for various workpiece tilt angles and feed directions for square pattern (top) and zigzag pattern (bottom).

3.1. Visual Observations

Visual observations on the micro dimple patterns are mainly conducted on the morphology of the micro dimple pattern, edge quality and the dimple surface quality based on the workpiece tilt angle and feed direction. Figure 5 shows SEM images of the 10 fabricated micro dimple patterns for various workpiece tilt angle, feed directions, and patterns. Optical microscope images taken by 3D Laser Scanning Confocal Microscope (Keyence) on a selected dimple show some burrs existed on the edges for all patterns (Figure 6). It can be seen that the 0^0 tilt angle is seen to have more burrs compared to other tilt angle and the 45^0 tilt angle shows the shape is not perfectly circular. This can be occurred due to the material removal mechanism by the ball end mill when the workpiece is tilted at 45^0. The 30^0 workpiece tilt angle is seen to have less burrs and more precise circular shape compared with the other two tilt angles.

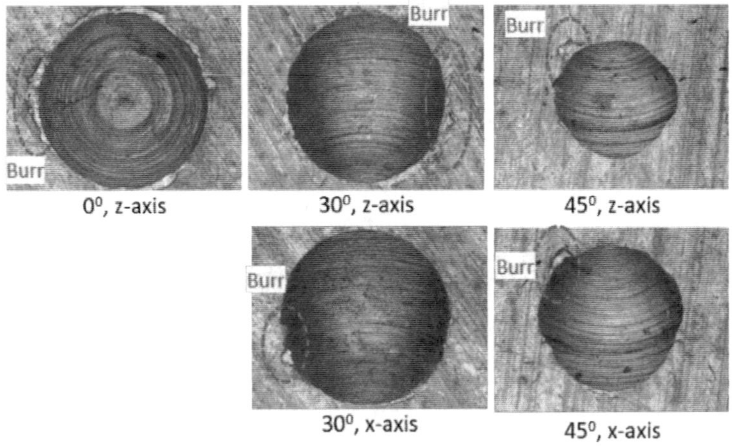

Figure 6. Optical Microscope Images for one of dimple for various workpiece tilt angle and feed direction.

3.2. *Geometrical Analysis*

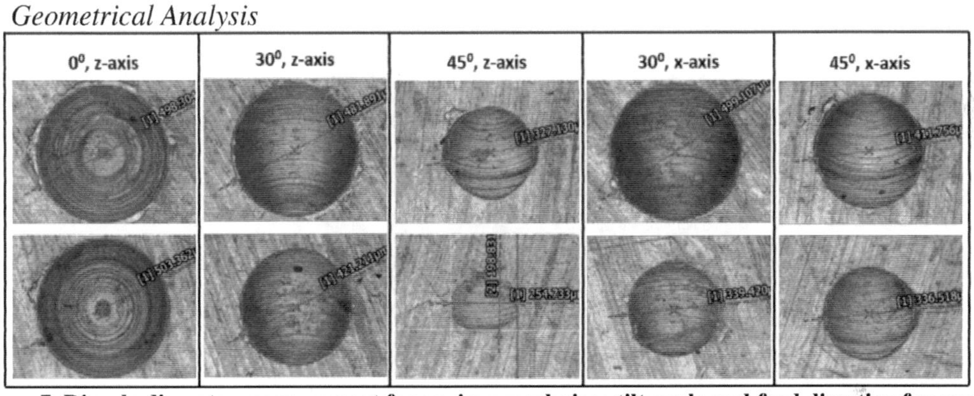

Figure 7. Dimple diameter measurement for various workpiece tilt angle and feed direction for square pattern (top) and zigzag pattern (bottom).

Table 2. Measured dimple diameter and depth for square and zigzag patterns

	Theoretical Dimple diameter (μm)	Measured Dimple diameter (μm)	Theoretical dimple depth (μm)	Measured Dimple depth (μm)	Aspect ratio (diameter/depth)
Square pattern					
0^0, z-axis	600	498.30	100	68.49	7.27
30^0, z-axis	600	481.89	100	65.06	7.41
45^0, z-axis	600	327.13	100	26.81	12.20
30^0, x-axis	600	499.11	100	67.53	7.39
45^0, x-axis	600	411.76	100	38.79	10.62
Zigzag pattern					
0^0, z-axis	600	503.36	100	73.22	6.87
30^0, z-axis	600	421.21	100	53.33	7.90
45^0, z-axis	600	254.73	100	13.34	19.10
30^0, x-axis	600	339.42	100	26.07	13.02
45^0, x-axis	600	336.52	100	29.49	11.41

Dimple diameter measured using Laser Scanning Confocal Microscope (Keyence) is shown in Figure 7. Table 2 summaries the measured diameter, depth and aspect ratio (diameter/depth) of the selected dimples for square and zigzag patterns respectively. The theoretical diameter is 600 μm based on the theoretical depth of the dimple which is 100 μm. Hence the expected aspect ratio is 6. It can be seen that dimple depth for all the combinations are less than 100 μm resulted the diameter to be less than 600 μm. This can be occurred due to the limitation of the CNC machine in achieving precision at micro meter scale. However, the optimum results are achieved for 0^0 and 30^0 workpiece tilt angle with z-axis feed direction with aspect ratio in between 7 – 8.

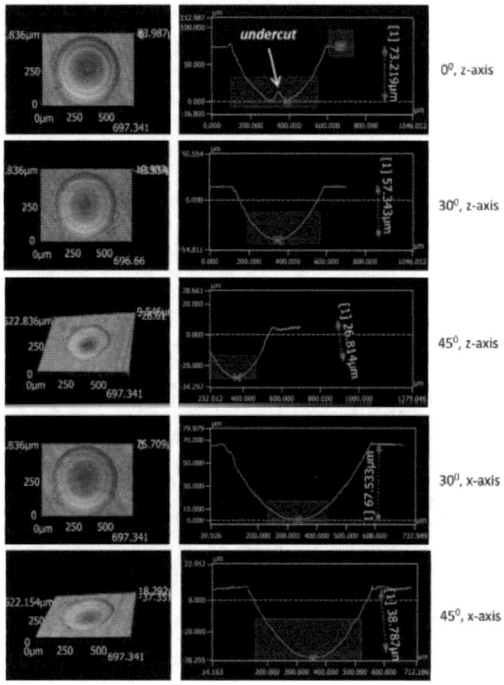

Figure 8. Depth measurement result for various workpiece tilt angle and feed direction.

Figure 8 shows the 3D profile of the dimple and 2D cross section profile of the dimple depth measured using confocal laser microscope (Keyence). It can be seen that there is a small bulge at the center of the dimple for 0^0, z-axis as a result of undercut. The other tilt angle and feed direction show no undercut at the center of dimple.

4. Conclusions

In this initial experiment, micro dimples pattern with the depth of dimple of 100 μm can be fabricated by using ball end mill, however better machining center is needed in order to have better precision and accuracy. The workpiece tilt angle produces less burr at the edge of the dimples compared to without tilt (0^0) and it can avoid the formation of undercut at the center of the dimple. The workpiece tilt angle of 30^0 and z-axis feed direction produced better result compared to 0^0 and 45^0 tilt angles with the diameter/depth ratio closer to the theoretical value. Follow-up work is currently on-going in order to improve the accuracy and precision of the fabrication of dimple pattern using ball end mill.

5. Acknowledgements

This research is supported by research grant RDU 180324 from Universiti Malaysia Pahang, Malaysia

6. References

[1] A. Malshe, K. Rajurkar, A. Samant, H. N. Hansen, S. Bapat, and W. Jiang, "Bio-inspired functional surfaces for advanced applications," *CIRP Annals*, vol. 62, no. 2, pp. 607–628, 2013.

[2] E. Graham, C. I. Park, and S. S. Park, "Fabrication of micro-dimpled surfaces through micro ball end milling," *International Journal of Precision Engineering and Manufacturing*, vol. 14, no. 9, pp. 1637–1646, 2013.

[3] T. Pratap and K. Patra, "Micro ball-end milling—an emerging manufacturing technology for micro-feature patterns," *The International Journal of Advanced Manufacturing Technology*, vol. 94, no. 5, pp. 2821–2845, 2018.

[4] T. P. and K. Patra, "Micromilling of Ti-6Al-4V Titanium Alloy Using Ball-end Tool," *IOP Conference Series: Materials Science and Engineering*, vol. 229, no. 1, p. 12011, 2017.

[5] T. Ibatan, M. S. Uddin, and M. A. K. Chowdhury, "Recent development on surface texturing in enhancing tribological performance of bearing sliders," *Surface and Coatings Technology*, vol. 272, pp. 102–120, 2015.

[6] J. W. Byun, H. S. Shin, M. H. Kwon, B. H. Kim, and C. N. Chu, "Surface texturing by micro ECM for friction reduction," *International Journal of Precision Engineering and Manufacturing*, vol. 11, no. 5, pp. 747–753, 2010.

[7] S. G. and D. C. and T. R. and A. B. M. and H. H. M. and B. Pingguan-Murphy, "Tribological investigation of diamond-like carbon coated micro-dimpled surface under bovine serum and osteoarthritis oriented synovial fluid," *Science and Technology of Advanced Materials*, vol. 16, no. 3, p. 35002, 2015.

[8] T. Sugihara and T. Enomoto, "Performance of cutting tools with dimple textured surfaces: A comparative study of different texture patterns," *Precision Engineering*, vol. 49, pp. 52–60, 2017.

[9] R. Kurniawan, G. Kiswanto, and T. J. Ko, "Micro-dimple pattern process and orthogonal cutting force analysis of elliptical vibration texturing," *International Journal of Machine Tools and Manufacture*, vol. 106, pp. 127–140, 2016.

[10] F. Z. Dai, D. P. Wen, Y. K. Zhang, J. Z. Lu, X. D. Ren, and J. Z. Zhou, "Micro-dimple array fabricated on surface of Ti6Al4V with a masked laser ablation method in air and water," *Materials & Design*, vol. 84, no. Complete, pp. 178–184, 2015.

IOP Conf. Series: Materials Science and Engineering **547** (2019) 012010 doi:10.1088/1757-899X/547/1/012010

Effect of Composition and Pouring Temperature of Cu-Sn Alloys on The Fluidity and Microstructure by Investment Casting

Sugeng Slamet[1], Suyitno[1], Indraswari Kusumaningtyas[1]
[1] Department of Mechanical and Industrial Engineering, Universitas Gadjah Mada, Yogyakarta, Indonesia

Corresponding author (E-mail): sugeng.slamet@umk.ac.id

Abstract. The alloy composition and the pouring temperature are main parameters to avoid cast defects in the metal casting process. The aim of this study was to investigate the effect of tin composition and pouring temperature on the fluidity and microstructure of Cu-Sn alloys to minimize product failure. The material used in this study was Cu with the addition of tin composition (20-25)% Sn by melting Cu = 99.9% and Sn = 99.9%. The pouring temperatures were varied T1=1000°C and T2=1100°C. Mold for fluidity test was used the wax pattern with investment casting method. The wax pattern for mold of fluidity test shaped bars with a length of 400 mm, the width of 10 mm and cavity thickness varies from 1.5-5 mm. The mold material was used natural ceramic such as clay (SiO_2). Increased of the tin composition decreases the length of fluidity, while the increased pouring temperature and the cavity thickness of the mold increase the length of fluidity. Increasing tin composition and pouring temperature increases the phase microstructure α and triggers the formation of columnar dendrites. The high rate of cast metal fluidity and the thickness of the mold cavity greatly affect the microstructure of the alloy.

1. Introduction

The tin bronze is an important alloy that the industry needs. These alloys are widely used to manufacture various engine components such as valves, bearings, pump impellers, piston rings, weapon components, gears and other [1]. The bronze alloys are more easily poured and shaped, compared to pure copper and brass [2]. The low tin bronze alloys <17wt.% Sn is used to make various machine components while high tin bronze alloys >17wt.% Sn is used to make various musical instruments. The tin bronze alloys are used to make church bells since the 11th century [3,4], musical instruments of trumpet, percussion, and gamelan [5–7]. The tin bronze alloys can be poured in various molds, forgedable, corrosion resistant and have good acoustic properties.

The casting is one of metal forming techniques that has long been used to produce various metal products. One of the most widely used metal casting techniques is Investment casting. It is also known as lost wax process or precision casting [8]. The advantage of investment casting is able to produce thin-walled cast products, having a slope and curvature with small radius variations, smooth surface, accurate shape and dimensions [8,9]. The investment casting produces a cast product approaching the end product, lowering production time and without machining process [10]. The resulting investment casting product does not require final work on its surface and high product tolerance [11]. The disadvantage of investment casting is it requires long process stages, relatively expensive cost, limited to small cast objects and difficult to add core.

The Fluidity is one of the main parameters for the success of casting. Poor fluidity makes it difficult for cast metals to flow and fill the mold cavity and can cause cast defects [12]. Fluidity is the ability of the molten metal to flow and fill every part of the mold [13]. The fluidity length is measured from the pour point to the solidity point and is the maximum distance of liquid metal flows in the mold [14]. Fluidity is influenced by several parameters such as composition, pouring temperature, viscosity, flow rate and surface tension between a liquid metal with the mold wall.

The increased composition can reduce the fluidity rate to a certain extent from the alloy composition. The increases of composition to the limit of 25wt.% Sn increases the viscosity Cu-Sn alloy [15,16]. Increased viscosity of Cu-Sn alloys occurs in the 20-30wt.% Sn composition in the β and γ intermetallic phases [17].

Increasing pouring temperatures significantly increase the length of fluidity for all types of mold material [18]. The high pouring temperature gives a long time on the molten metal until it is finally solid. The cooling rate at investment casting is slower compared sand casting [19]. High temperature gradients between mold walls and cast metals cause the solidification rate to run fast. The low solidification rates produce larger dendrite α, decrease the mechanical properties of hardness and tensile strength. The increase of the pouring temperatures 1000°C to 1200°C decreases the density and mechanical properties of 20wt.%Sn alloy [20].

The microstructure growth during the solidification process is strongly influenced by several factors, such as temperature gradients between liquids (GL), growth rate (VL) and alloy composition (CO) [21]. The growth of microstructure of dendrite arm spacing (DAS) tends to be greater in investment casting [22]. The addition of the tin composition and the increase of pouring temperature results in the change of equiaxed microstructure to dendrite columnar, secondary dendrite arm spacing growth and increased $\alpha + \delta$ phase in Cu-Sn alloys. The fluidity length and pouring temperature can reduce product defects where cast metal is able to flow and fill the entire mold cavity. The length of fluidity in cast metal is very necessary as a reference for casting, especially in thin-walled molds. The aim of this research is to investigate the influence of the addition of tin composition in the Cu(20-25)wt.%Sn and pouring temperatures of T1=1000°C and T2=1100°C to fluidity length and microstructure by investment casting method.

2. Materials and Methods

The materials used in the study were tin bronze Cu(20-25)wt.%Sn. The alloy compositions were prepared using Cu = 99,9% and Sn = 99,9% in weight ratio respectively. The alloy was melted in a crucible furnace with uses charcoal wood fuel. The composition of Cu-Sn alloys was shown in Table 1.

Table 1. Compositions of the Cu-Sn alloys

Alloys	Unsur (%)						
	Cu	Sn	Zn	Pb	Fe	Ni	Al
Cu-20wt.% Sn	79.77	20.06	0.08	0.02	0.04	0.00	<0.001
Cu-22wt.% Sn	78.16	21.39	0.03	0.14	0.13	0.10	<0.001
Cu-24wt.% Sn	75.13	23.86	0.14	0.20	0.48	0.14	<0.001
Cu-25wt.% Sn	70.71	24.90	0.05	3.51	0.59	0.02	0.041

The method used was investment casting. The pattern for fluidity test mold was made using wax. Furthermore, the wax pattern was coated with a natural ceramic material in the form of a mixture of clay, sand and water. The wax pattern for the fluidity test mold was gradually coated with clay slurry: 200 mesh for the inner surface followed by 100 mesh for the outer surface. The dewaxing and sintering treatment of the mold was carried out until it reaches the temperature of 600°C. The wax pattern for the mold of fluidity test shaped bars with a length of 400 mm, the width of 10 mm and cavity thickness varies from 1.5, 2, 3, 4, and 5 mm. The fluidity test mold and fluidity product as shown in Figure 1.

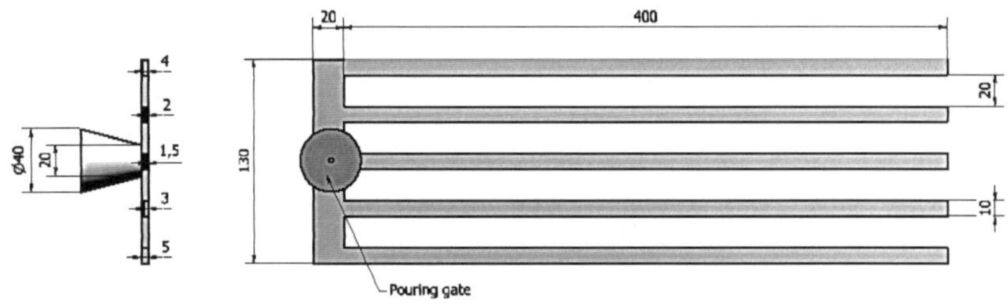

Figure 1. Wax pattern design (unit: mm)

The natural ceramic material in the form of clay to manufacture molds has been tested using SEM-EDS. The composition of a natural ceramic material was contained SiO_2 compound, as shown in Figure 2.

Figure 2. Composition of natural ceramic material by SEM-EDS

The stages of investment casting were started from pattern preparation, coating with clay slurry, dewaxing process and fluidity product as shown in Figure 3.

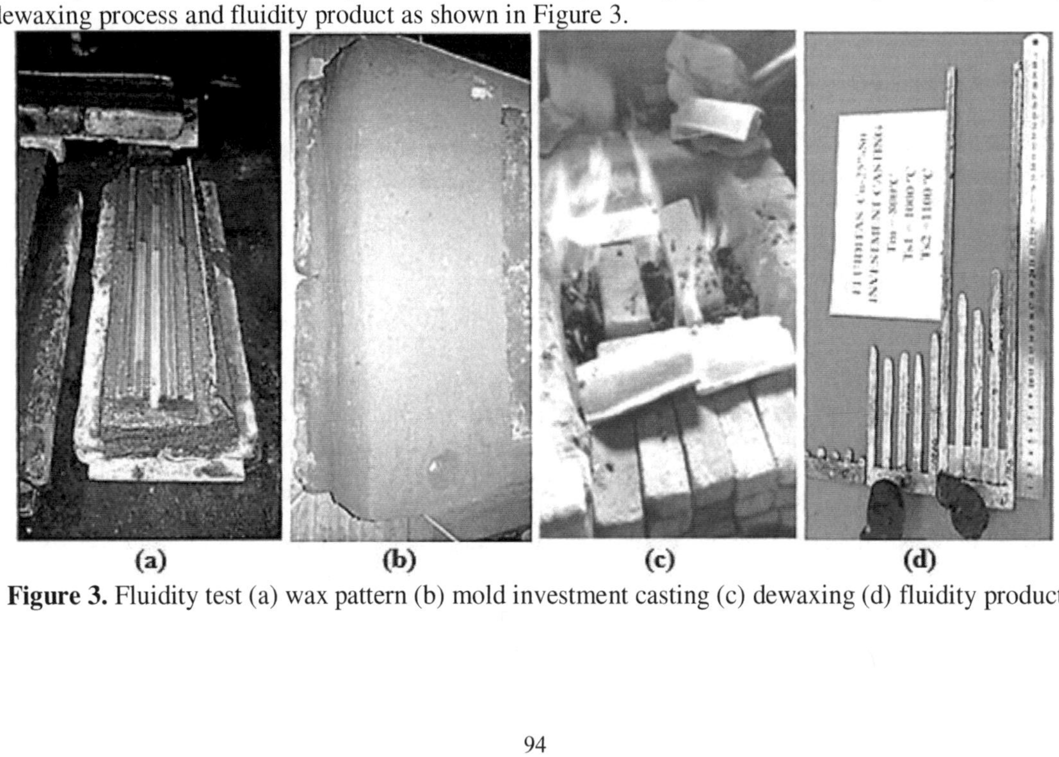

(a) (b) (c) (d)

Figure 3. Fluidity test (a) wax pattern (b) mold investment casting (c) dewaxing (d) fluidity product

The temperature of the casting is varied at superheat temperatures with the initials T1 and T2. It was the highest temperatures above the Cu alloy liquid point (20-25)wt.%Sn, where T1 =1000°C and T2 =1100°C. The length of fluidity is measured from the base of the inlet to the end of the cast metal. The microstructure observation using an optical microscope with 100X magnification. The etching was done with a mixed solution between HNO_3 and H_2O (50%:50%) ratio dipped for 5 seconds.

3. Results and Discussion

The addition of tin composition to Cu (20-25) wt.% Sn decreases the length of fluidity. The decrease in fluidity length is caused by increasing solidification rate and viscosity of the cast metal. The Cu-Sn phase diagram shows that the composition Cu(22-25)wt.%Sn is the transition phase, where the transition limit is at the composition of 22wt.%Sn. The composition (13.5-22)wt.%Sn is in the phase of α + L hypoperitectic and composition (22-25.5)wt.%Sn is in phase β + L hyperperitectic. The results of the measurement of fluidity length with the investment casting method on the tin bronze of Cu-20wt.%Sn, Cu-22wt.%Sn, Cu-24wt.%Sn and Cu-25wt.%Sn are shown in Figure 4.

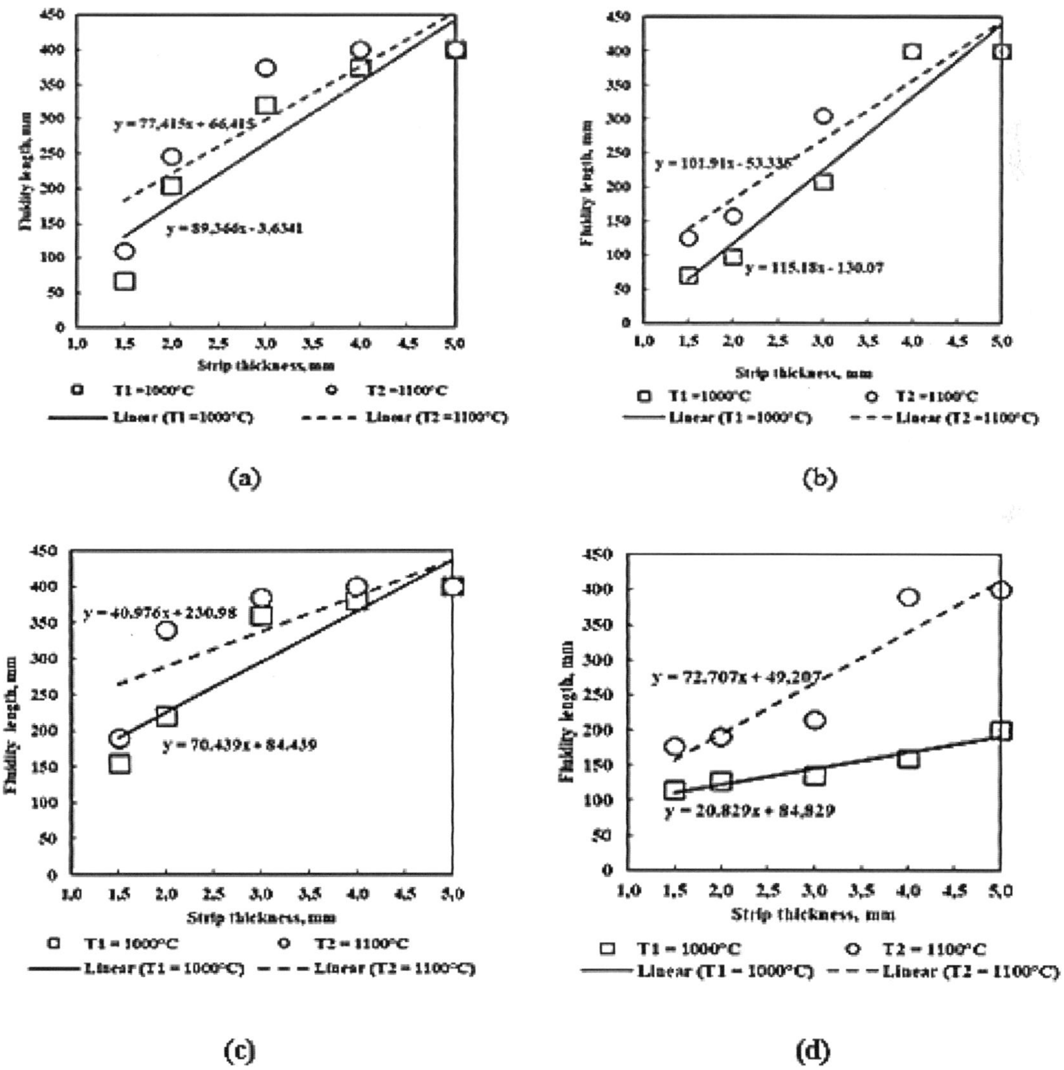

Figure 4. Fluidity of Cu-Sn alloys (a) 20wt.%Sn (b) 22wt.%Sn (c) 24wt.%Sn (d) 25wt.%Sn

The increased pouring temperature and thickness of the mold cavity increase the length of fluidity for all compositions. A high pouring temperature in the superheat area can extend the distance between the melting point and solidus point. This will increase nucleation growth in cast metals during the solidification process. The superheat temperature increases the fluidity length of all alloy compositions [23]. The investment casting method has a relatively low solidification rate.

One of the factors that influence the high solidification rate is the temperature gradient between cast metal and mold wall. The alloy composition and pouring temperature are the main factors that determine success in the casting process in molds that have thin cavities. Increased the pouring temperature and mold cavity can improve the length of fluidity. The process of dewaxing and sintering besides removing wax material from the mold also serves to reduce water content (H_2O) and increase the temperature of the mold.

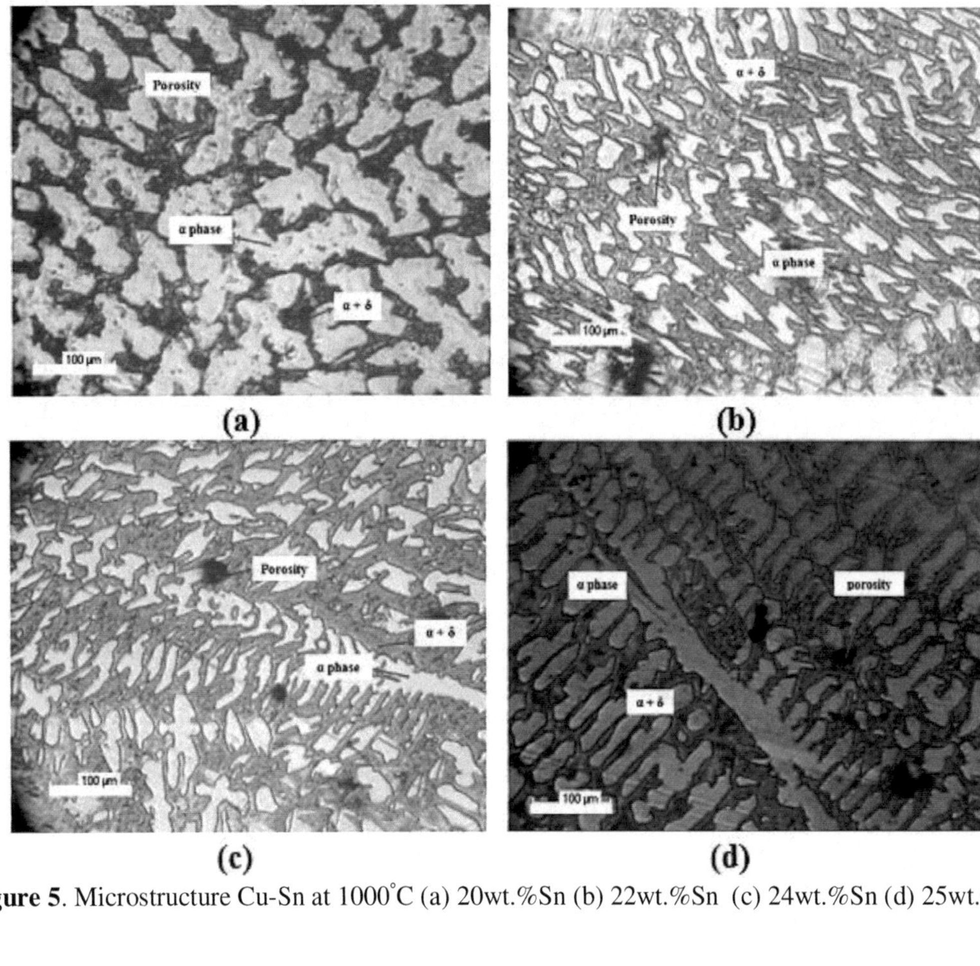

Figure 5. Microstructure Cu-Sn at 1000°C (a) 20wt.%Sn (b) 22wt.%Sn (c) 24wt.%Sn (d) 25wt.%Sn

IOP Conf. Series: Materials Science and Engineering **547** (2019) 012010 doi:10.1088/1757-899X/547/1/012010

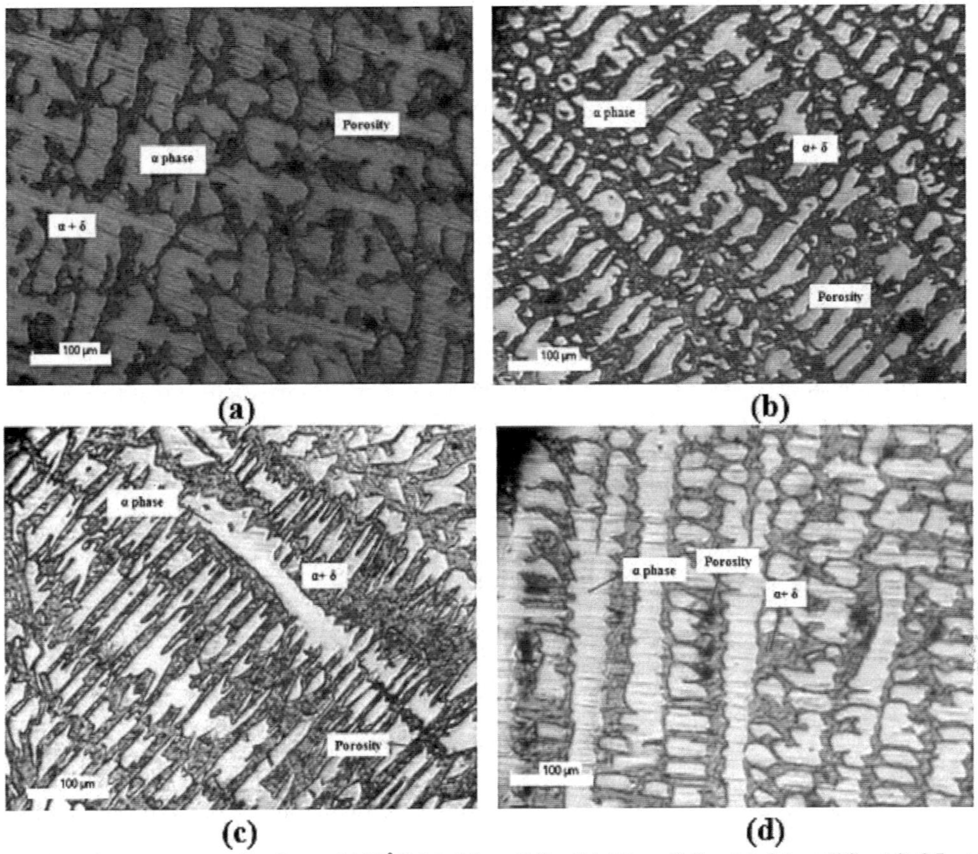

Figure 6. Microstructure Cu-Sn at 1100°C (a) 20wt.%Sn (b) 22wt.%Sn (c) 24wt.%Sn (d) 25wt.%Sn

Increasing of the tin composition and pouring temperature affect the microstructure of Cu-Sn alloys. The results of microstructural observation show that the growth of the α phase decreases with equiaxed grains morphology to dendritic columnar grains. The increase of the tin composition and the pouring temperatures increased the morphology of columnar dendritic grains and dendrite arm spacing (DAS). The increase of the tin composition and pouring temperatures increases the formation of α + δ eutectoid phase [22]. The microstructure of Cu-20wt.%Sn, Cu-22wt.%Sn, Cu-24wt.%Sn and Cu-25wt.%Sn at pouring temperature 1000°C is shown in Figure 5. Increased of the tin composition in the Cu(20-25)wt.% Sn triggers a change in equiaxed microstructure into columnar dendrites in the matrix [24]. Figure 6. shows the microstructures at pouring temperature at 1100°C.

4. Conclusions

This research can be concluded that : The increase of the pouring temperatures and cavity thickness of the mold will increase the length of fluidity Cu-Sn alloys. Increased of tin composition in the Cu(20-25)wt.%Sn decreases the fluidity length of Cu-Sn alloys. The increase of tin composition and pouring temperatures causes the change of fine equiaxed to the coarse columnar dendrite structure of the matrix and increase the α + δ eutectoid phase.

5. Acknowledgements

The research was carried out under dissertation grant by the Indonesian Ministry of Finance through Lembaga Pengelola Dana Pendidikan (LPDP). The Authors also like to thank KUBE Sampurna Tunas Muda, Godean, Sleman, Yogyakarta of the supporting this research.

References

[1] Jixin H O U, Jianjun S U N, Chengwei Z, Xuelei T and Xichen C, 2007, The structural change of Cu-Sn melt, *Physics, Mech. Astron.* **50**, 414–420.
[2] Sugita I K G and Miasa I M, 2013, Feasibility Study On The Use Of Silicon-Bronze Alloys As An Alternative Material For Balinese Musical Instruments, 1–5.
[3] Debut V, Carvalho M, Figueiredo E, Antunes J and Silva R, 2016, The sound of bronze: Virtual resurrection of a broken medieval bell, *J. Cult. Herit.***19**, 544–554.
[4] Audy J and Audy K, 2008, Analysis of bell materials: Tin bronzes, *China Foundry.* **5**, 199–204.
[5] Fletcher N, 2012, Materials And Musical Instruments, *Acousitics Aust.* **40**, 130–134.
[6] Sumarsam, 2002, Introduction to Javanese Gamelan (Javanese Gamelan-Beginners), *Syllabus.* 1–28.
[7] Goodway M, 1992, Metals of Music, *Mater. Charact.* **29**, 177–184.
[8] Pattnaik S, Karunakar D B and Jha P K, 2012, Developments in investment casting process - A review, *J. Mater. Process. Technol.* **212**, 2332–2348.
[9] Singh J, Singh R and Singh H, 2017, Dimensional accuracy and surface finish of biomedical implant fabricated as rapid investment casting for small to medium quantity production, *J. Manuf. Process.* **25**, 201–211.
[10] Chua C M C C K and Feng C W L C, 2005, Rapid prototyping and tooling techniques : a review of applications for rapid, *Int.J.Adv. Manuf Technol.* **25**, 308–320.
[11] Lee K, Blackburn K and Welch S T, 2017, A more representative mechanical testing of green state investment casting shell, *Ceram. Int.* **43**, 268–274.
[12] Caliari D, Timelli G, Bonollo F, Amalberto P and Giordano P, 2015, Fluidity of aluminium foundry alloys: Development of a testing procedure, *Metall. Ital.* **107**, 11–18.
[13] Campbell J and Harding R A, 1994, The Fluidity of Molten Metals 3205 The Fluidity of Molten Metals, TALAT Lect. 3205. 1–19.
[14] Siavashi K, 2011, The Effect of Casting Parameters on the Fluidity and Porosity of Aluminium Alloys in the Lost Foam Casting Process, thesis, University of Birmingham, United Kingdom.
[15] Tan M, Xiufang B, Xianying X, Yanning Z, Jing G and Baoan S, 2007, Correlation between viscosity of molten Cu-Sn alloys and phase diagram, *Phys. B Condens. Matter.* **387**, 1–5.
[16] Hou J, Guo H, Zhan C, Tian X and Chen X, 2006, Viscous and magnetic properties of liquid Cu – 25 wt .% Sn alloy, **60**, 2038–2041.
[17] Mudry S, Korolyshyn A, Vus V and Yakymovych A, 2013, Viscosity and structure of liquid Cu – In alloys, *J. Mol. Liq.* **179**, 94–97.
[18] Rzychoń T, Kiełbus A and Serba M, 2010, The influence of pouring temperature on the microstructure and fluidity of elektron 21 and WE54 magnesium alloys, *Arch. Metall. Mater.* **55**, 7–13.
[19] Sulaiman S and Hamouda A M S, 2001, Modeling of the thermal history of the sand casting process, *J. Mater. Process. Tech.* **113**, 245–250.
[20] Bartocha D and Baron C, 2016, Influence of Tin Bronze Melting and Pouring Parameters on Its Properties and Bells ' Tone, *Arch. Foundry Eng.* **16**, 17–22.
[21] De Souza E N, Cheung N and Garcia A, 2005, The correlation between thermal variables and secondary dendrite arm spacing during solidification of horizontal cylinders of Sn-Pb alloys, *J. Alloys Compd.* **399**, 110–117.
[22] Zeynep Taslicukur E T, Gözde S. Altug, Şeyda Polat, Ş and Hakan Atapek, 2012, A Microstructural Study on CuSn10 Bronze Produced by Sand and Investment Casting Techniques, *Met. 2012.* **5**, 23–25.
[23] Di Sabatino M, 2005, Fluidity of aluminium foundry alloys, thesis, Norwegian University of Science and Technology (NTNU), Norwegia.
[24] Nadolski M, 2017, The Evaluation of Mechanical Properties of High-tin Bronzes, *Arch. Foundry Eng.* **17**, 127–130.

Characterization of The Wear Resistance of 3D Printed Polylactic-Acid (PLA) in Water and Bovine Serum

B. Arifvianto*, A.T. Putra, B.T. Prayoga, M. Mahardika, Suyitno
Center for Innovation of Medical Equipments and Devices (CIMEDs)
Department of Mechanical and Industrial Engineering, Faculty of Engineering, Universitas Gadjah Mada
Jl. Grafika 2 Yogyakarta 55281, Indonesia

*Corresponding author (E-mail): b.arifvianto@gmail.com

Abstract. Three-dimensional (3D) printing has become an emerging technology for the fabrication of geometrically complex, patient-specific biomedical implants. However, several properties of the 3D printed material have not so far been fully understood. Among these properties, wear resistance is considered as the critical one for the 3D printed material that should be applied as a sliding component in an artificial joint prosthesis. In this research, the wear resistance of a 3D printed polylactic-acid (PLA) pin that slid over the surface of titanium plate was characterized in water and bovine serum by using a pin-on-plate tribometer. The result showed that the use of bovine serum could reduce the wear of the 3D printed PLA better than the use of the water as the lubricant. Based on the analysis of weight losses and worn surface morphology, polymer film transfer was proposed as the possible wear mechanism that occurred during the sliding of the 3D printed PLA pin over the surface of the Ti plate.

Keywords: 3D printing; polylactic-acid; titanium; wear; water; bovine serum

1. Introduction

Recently, three-dimensional (3D) printing has become potential technology that can be used to aid the fabrication of geometrically complex, patient-specific biomedical implants [1-3]. With this technology, a solid 3D material can be built by depositing multiple layers of fully or partially melted material. In order to meet the requirements of any specific applications, many studies have currently been carried out to investigate the properties of such 3D printed materials, such as their mechanical properties [4, 5], bioactivity [6] and wear resistance [7, 8]. Among these properties, wear resistance is considered of the critical one, particularly for the materials that should be able to withstand shear loads in their applications.

Up to now, the number of studies for investigating the tribological properties of 3D printed polymeric materials is still limited. The work of Sood et al. [7] briefly explained the mechanisms involved in the wearing process of a 3D printed material. They reported the following steps of the wear mechanism of the 3D printed polymeric material, i.e., (i) adhesion of the surface asperities, (ii) plastic shearing, and (iii) detachment of the tips of the softer material. Meanwhile, the work of Borges et al. [8] demonstrated the benefit of the porous structure of the 3D printed material in lowering friction coefficient and wear of this material in a lubricated sliding system. Further studies are however still needed to unravel many other issues related to the tribological behaviors of the 3D printed polymeric materials.

In this research, a preliminary study was conducted to characterize the wear resistance of a 3D printed polylactic-acid (PLA) pin in a reciprocating pin-on-plate tribometer. Two types of lubricants were used in this study, i.e., distilled water and bovine blood serum. Water has long been utilized as a lubricant for many tribological tests owing to its simple preparation. However, the use of water as a lubricant for biotribology could result a different wear mechanism from that in a periprosthetic joint [10]. Therefore, several bio-lubricants were utilized [11, 12], including bovine blood serum that had been reported to be able to mimic the properties of synovial fluid in the human body [11]. To determine possible wear mechanism in this research, characterizations of the weight losses and the worn surface morphologies and roughness of both the PLA pin and Ti plate were carried out.

2. Materials and Methods

In this research, polylactic-acid (PLA) and grade 2 titanium (Ti) were used as the materials for the cylindrical pin and the plate, respectively, in a unidirectional reciprocating pin-on-plate tribometer. The PLA pin was prepared by using 3D printing technique to obtain a cylindrical solid material with 8.1 mm and 20 mm in diameter and length, respectively. Meanwhile, the Ti plate was prepared with a dimension of 55 mm × 18 mm × 3 mm. Surface roughness of the sliding surfaces of both the PLA pin and the Ti plate were determined by using a contact stylus profilometer (Surfcom 120A, Advanced Metrology System, UK), with which the average R_a values of three different locations on the material surface were obtained. Scanning of the stylus over the surface of the material was conducted in a direction perpendicular to the raster orientation and sliding direction of the PLA pin over the surface of Ti plate. All the samples were cleaned up by rinsing in ethanol and distilled water and then dried in air prior to and after the sliding test in tribometer.

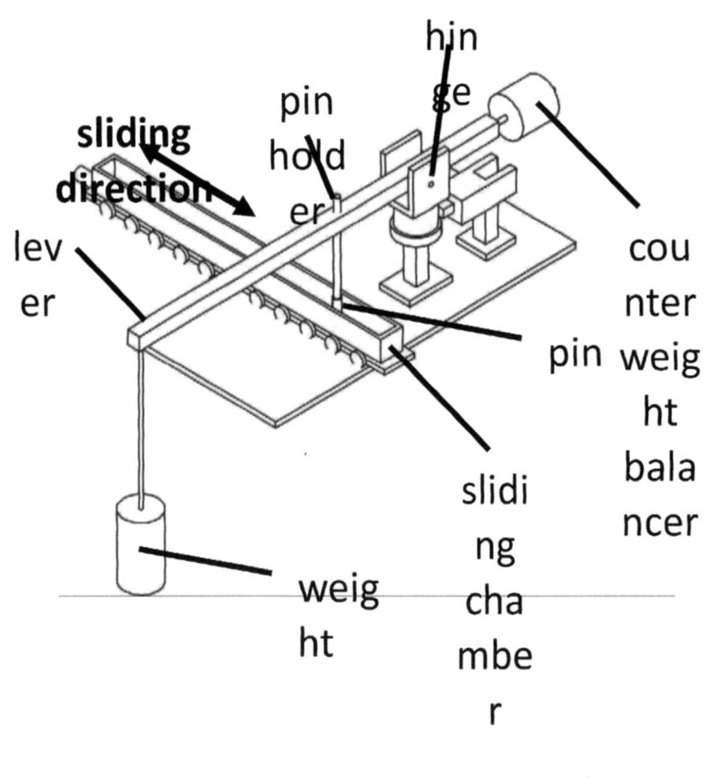

Fig. 1. Unidirectional reciprocating pin-on-plate tribometer used in this research.

The sliding test of the PLA pin was performed in a six-station unidirectional reciprocating pin-on-plate tribometer at room temperature, such as shown in Fig. 1. The PLA pin was mounted on its holder while the Ti plate was positioned inside the sliding chamber. During the test, the sliding surfaces of both these materials were immersed into the lubricant. A solid weight was added so that a normal load of 120 N could be applied on each of the PLA pin. The sliding test of the PLA pin against the titanium plate was run with a sliding distance of 25 mm and velocity of 50 mm s^{-1}, respectively, and with a sliding direction parallel to the raster orientation of the printed PLA pin. Two types of lubricants were used in this experiment, i.e., distilled water and bovine blood serum solution. The latter was prepared with a composition of 25% bovine serum and 75% of distilled water. In order to prevent the lubricant from bacterial contamination, sodium azide with volume fraction of 0.1% was added into the bovine serum lubricant. Detailed information about the experimental procedures of this research was described in the work of Prayoga et al. [9].

The weight losses of the PLA pin and Ti plate were determined in microbalance (Ohaus, Germany). Prior to weighing, all the PLA pins were cleaned up by using distilled water in an ultrasonic bath for 15 minutes and then dried at room temperature. The weight losses (w_L) of the PLA pin and the Ti plate were then calculated by using Eq. (1):

$$w_L(\%) = \frac{w_0 - w_t}{w_0} \times 100\,\% \qquad (1)$$

where w_0 and w_t are the initial weight and the weight of material after being rubbed for certain sliding distance against its countersurface in tribometer, respectively. Surface morphology and the PLA pins and Ti plates were characterized by using an electron microscope (JSM-6510LV, JEOL Ltd., Japan).

3. Results and Discussion

In this research, a preliminary work had been carried out to investigate the wear resistance of a 3D printed PLA pin that slid over the surface of Ti plate in a reciprocating pin-on-plate tribometer. Fig. 2 shows the physical appearance of all the 3D printed polylctic-acid (PLA) pins after the sliding test in the tribometer. As seen in Fig. 2a, the sliding surface of the PLA pin melted after the sliding without any lubricants for a distance of approximately 1×10^3 m over the surface of Ti plate. This finding might indicate that the local temperature at the sliding surface of the pin increased significantly during the test and exceeding the melting point of PLA material. The use of lubricant, either distilled water or bovine serum, seemed to be able to reduce local temperature rise at the contact and sliding surface between the PLA pin and the Ti plate. Therefore, all the PLA pins remained intact after the sliding test for 4×10^3 m.

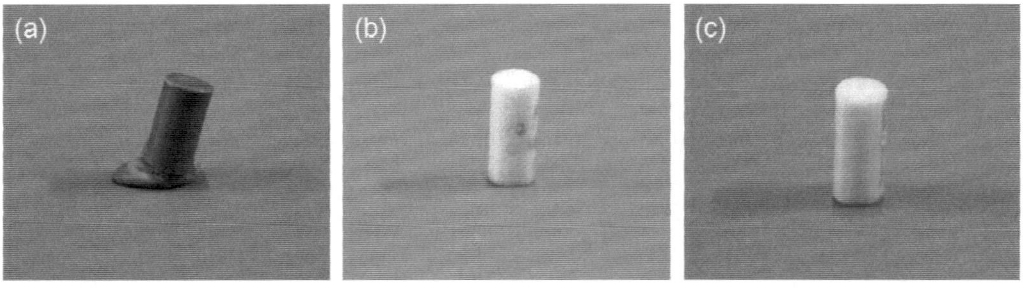

IOP Conf. Series: Materials Science and Engineering **547** (2019) 012011 doi:10.1088/1757-899X/547/1/012011

Fig. 2. The 3D printed PLA pin after sliding test (a) without lubricant, (b) with distilled water and (c) with bovine serum

Fig. 3a shows the changes in the percentage of weight losses of PLA pin due to wear that occurred during the sliding test. In general, the weight loss of the PLA pins increased with the sliding distance travelled by the pin. Interestingly, the rate of the weight loss of this material was apparently higher at a distance of less than 3×10^3 m, as indicated from the steeper slopes of the weight loss plots within the range of sliding distance from 0 to 3×10^3 m. This phenomenon occurred in both the sliding tests with water and bovine serum. Meanwhile, as seen in Fig. 3b, the plots of the weight losses of the Ti plate were also obviously influenced by the lubricant used in the test. Once tested with bovine serum lubricant, the plot of the weight losses of the Ti plate increased with the sliding distance. However, the weight of the Ti plate sliding in distilled water remained constant after a distance of 0.5×10^3 m had been reached. All these findings confirmed that the weight losses of the 3D printed PLA pin and its Ti countersurface due to wear were obviously dependent on the lubricant used in the test.

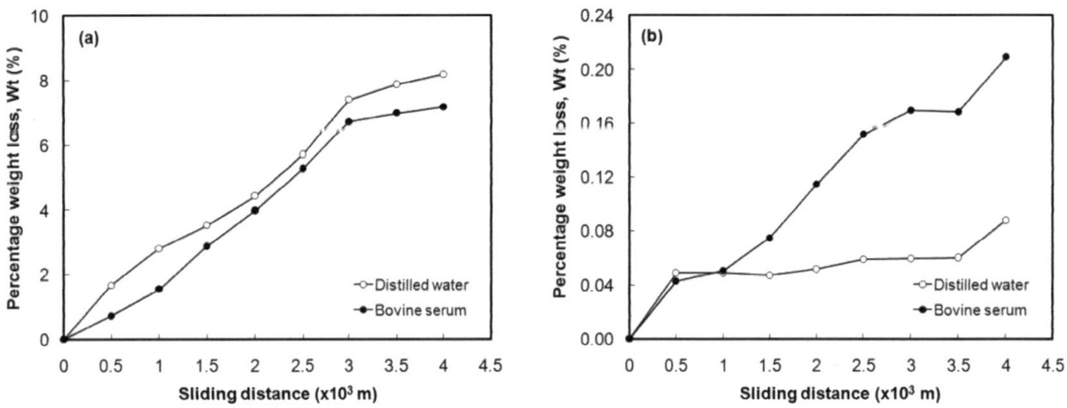

Fig. 3. Weight losses of (a) the 3D printed PLA pin and (b) Ti plate due to sliding wear

Fig. 4 shows the changes in surface roughness of the sliding surface of the 3D printed PLA pin after the test in tribometer with a sliding distance of 4×10^3 m. As demonstrated in this figure, the wear decreased the R_a value of the sliding surface of the PLA pin. As also seen in Fig. 4a, the type of lubricant used in the sliding test apparently influenced final roughness of the sliding surface of the PLA pin. In this case, after sliding for 4×10^3 m in distilled water, the worn surface of the PLA pin was slightly smoother than that tested in bovine serum. In contrary, as seen in Fig. 4b, the sliding over the PLA pin surface resulted in the increased roughness of the Ti plate.

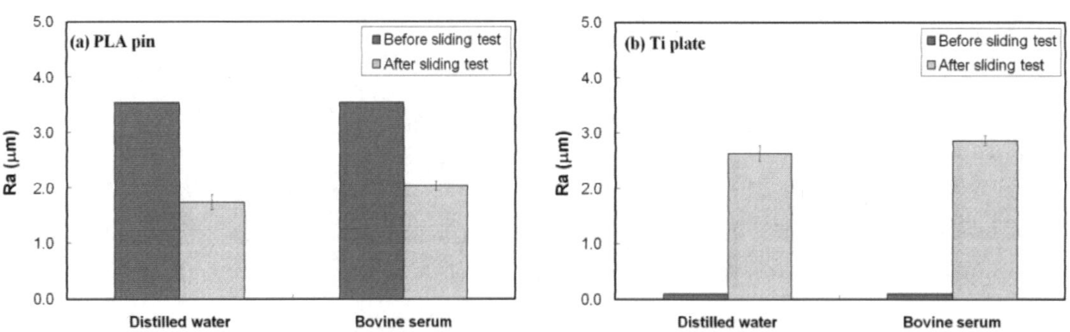

Fig. 4. The roughness of the sliding surface of (a) the 3D printed PLA pin and (b) Ti plate

Fig. 5 shows a series of micrographs displaying the morphologies of the sliding surface of the 3D printed PLA pin and Ti plate prior to and after the sliding tests. In Figs. 5a and b, the initial surface morphology of the PLA pin and Ti plate could be recognized from their relatively flat and smooth surface. A linear defect was clearly seen in Fig 5a, indicating the boundary between two adjacent rasters of the PLA material that were formed with the 3D printing technique. Meanwhile, as shown in Fig. 5b, thin scratches could be seen at the surface of the Ti plate, as a result of mechanical grinding and polishing that were carried out in the preparation of the plate prior to the sliding test.

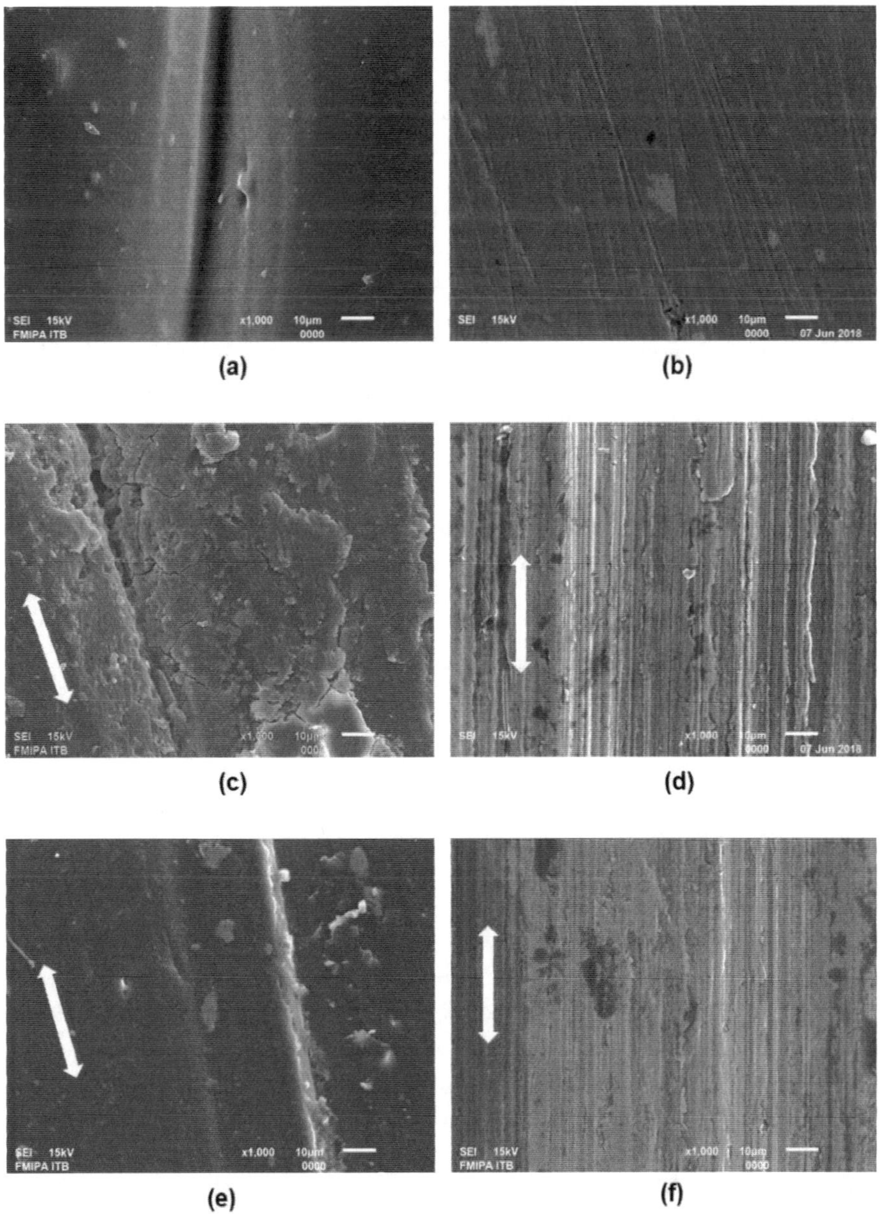

Fig. 5. Micrographs of the sliding surface of (a) the PLA pin and (b) Ti plate prior to the sliding test, (c) the PLA pin and (d) Ti plate after the sliding test in distilled water, and (e) the PLA pin and (f) Ti plate after the sliding test in bovine blood serum

Figs. 5c - f displayed typical worn surfaces of the 3D printed PLA pin and Ti plate after the test in tribometer with a sliding distance of 4×10^3 m. A white arrow was added into Figs. 5c – f to indicate the sliding direction of the PLA pin over the surface of the Ti plate. As seen in Figs. 5c and e, the sliding of the PLA pin over the surface of Ti plate generally caused scratches and crevices at the sliding surface of the pin. In Fig. 5c, it is also clearly seen that cracks were formed at the crevice and then propagated to the surrounding of the defect at the sliding surface of PLA pin that was tested in distilled water. However, as seen in Fig. 5e, this phenomenon could not be clearly seen at the worn surface of the PLA after sliding in bovine serum. In addition, fragments of the PLA material were visible in both Figs. 5c and e. Meanwhile, Figs. 5d and f shows typical worn surface of the Ti plate after a sliding test for 4×10^3 against the 3D printed PLA pin. In general, both Figs. 5d and f demonstrated similar pattern of scratches. However, the tips of the scratches at the Ti surface that was tested in distilled water were apparently sharper than those at the Ti surface sliding in bovine serum lubricant. Furthermore, some dark spots could be recognized in Figs. 5d and f, which might correspond to the presence of carbon as a trace of the wear debris of the PLA material.

In a previous work, the influence of lubricant type on the wear resistance of polymeric material that slid against the surface of metallic material had been highlighted. In their review, Brown and Clarke [10] mentioned that the use of water as the lubricant in tribological testing of an implant material resulted in massive flake-like particles with several micrometers in sizes. Meanwhile, the particle wear debris retrieved from periprosthetic tissue in total hip replacement was round and elongated in shape and having submicron sizes. Accordingly, water was then considered of being inappropriate lubricant for the tribological testing of biomedical implant material [10]. Meanwhile, Wang et al. [11] reported that submicron size particle wear debris with round an elongated shape could be obtained once bovine serum was used as the lubricant in a tribological test of ultra-high molecular weight polyethylene (UHMWPE) in hip simulator. This might demonstrate that bovine serum could mimic the properties of synovial fluid that lubricates the surface of articular joint tissue.

Based on the results presented in Figs. 3 – 5, some important findings of this study could be summarized as follows:

(1) the worn surface of the PLA pin could be recognized consisting of some grooves that stretched over the pin surface and parallel to the sliding direction against Ti plate;
(2) the weight losses of the 3D printed PLA pin due to sliding wear against Ti plate could be reduced once bovine serum was preferred as the lubricant instead of distilled water;
(3) the weight losses of Ti plate was higher once bovine serum was preferred as the lubricant rather than distilled water;
(4) after tribological test with a sliding distance of 4×10^3 m, the surface of the Ti plate lubricated with distilled water was slightly smoother than that lubricated with bovine serum;
(5) dark spots together with some deep scratches were observed at the sliding surface of the Ti plate and this might indicate the presence of carbon as a trace of the PLA material.

Considering all these findings, the wear mechanism of the 3D printed PLA pin that slid over the surface of Ti plate was possibly attributed by abrasion and polymer film transfer phenomenon. According to this mechanism, the wear debris of the polymeric pin were firstly sheared by the sharp metal asperities and detached from the sliding surface of the pin. The wear debris that trapped in the clearance between these two sliding materials then adhered, formed many small clumps and agglomerated at the surface of the Ti plate [13]. As a consequence, the presence of this transferred polymer film could add the weight of the Ti plate.

In this research, it is confirmed that the use of bovine serum could slightly reduce wear of the 3D printed PLA pin. As also inferred in Fig. 4a, the more asperity peaks of the PLA pin were removed when water was used as the lubricant, resulting in a smoother PLA pin surface. It is also reported by Cooper et al. [14] that protein containing lubricant such as bovine serum could reduce the amount of the polymer film transfer into the metallic countersurface. The plots in Fig. 3b might confirm the work of Cooper et al. [13], in which the weight loss of the Ti plate in distilled water was lower than that in bovine serum due to the more polymer film attached at the surface of this plate.

Although the current work might be able to explain possible wear mechanism of a 3D printed PLA pin that slid over the surface of Ti plate, further studies are still needed to ensure the presence of polymer traces at the worn surface of the Ti plate. Further analysis to determine friction coefficient and surface chemical composition of both the PLA pin and the Ti plate should also be carried out.

4. Conclusions

In this research, a preliminary study was carried out to characterize wear characteristics of 3D printed polylactic-acid (PLA) pin that sliding over the surface of Ti plate. Based on the results of this work, it can be concluded that the presence of lubricant could prevent the sliding surface of the PLA pin from excessive heat generation and in the end maintain the shape of the pin during the sliding test. Meanwhile, the use of bovine serum reduced the wear of the 3D printed PLA better than the use of water as the lubricant. Considering all the findings obtained in this research, the wear mechanism of the 3D printed PLA pin that slid over the Ti plate was possibly attributed by abrasion and polymer film transfer phenomenon.

5. Acknowledgment

This work was financially funded by the Department of Mechanical and Industrial Engineering, Faculty of Engineering, Universitas Gadjah Mada, with Research Grant No. 823/H1.17/TMI/LK/2018. The authors are grateful to Dr. Herianto for the fruitful discussions concerning the 3D printing technology.

References

[1] Ventola CL 2014 *P&T* **39** 704
[2] Mertz L *IEEE Pulse* 2013 **4** 15
[3] Thomas DJ *J. Orthop.* 2017 **14** 182
[4] Tymrak BM, Kreiger M and Pearce JM 2014 *Mater. Des.* **58** 242
[5] Farzadi A, Solati-Hashjin M, Asadi-Eydivand M and Osman NAA 2014 *Plos One* **9** 1
[6] Fielding GA, Bandyopadhyay A and Bose S 2012 *Dental Mater.* **28** 113
[7] Sood AK, Equbal A, Toppo V, Ohdar RK and Mahapatra SS 2012 *CIRP J. Manuf. Sci. Technol.* **5** 48
[8] Borges RA, Choudury D, and Zou M 2018 Tribol. Int. **122** 1
[9] Prayoga BT, Suyitno S and Dharmastiti R 2016 *Tribol. Industry* **38** 543
[10] Brown S and Clarke IC 2006 *Tribol. Trans.* **49** 72
[11] Wang A, Essner A, Stark C and Dumbleton JH 1996 *Biomaterials* **17** 865
[12] Guezmil M, Bensalah W and Mezlini S 2016 *J Mech. Behav. Biomed. Mater.* **63** 375
[13] Bahadur S 2000 *Wear* **245** 92
[14] Cooper JR, Dowson D and Fisher J 1993 *Clin. Mater.* **14** 295

Effect of Phenolic Resin on Density, Porosity, Hardness, Thermal Stability, and Friction Performance as A Binder in Non-Asbestos Organic Brake Pad

P Nawangsari[1*], Jamasri[1], H S B Rochardjo[1]
[1] Departement of Mechanincal and Industrial Engineering, Faculty of Engineering, Universitas Gadjah Mada, Jl. Grafika No.2 Yogyakarta 55281, Indonesia

*Corresponding author (E-mail): putri.nawangsari@mail.ugm.ac.id

Abstract. Phenolic resin is the most common resin binder used in brake pad materials. Selection of the relative amount of phenolic resin is an important to maintain structurally intact with the other ingredients of brake pad composite. In this experimental study, the effects of phenolic resin as a binder in non-asbestos organic brake pad was investigated. The brake pad composite was made using a powder metallurgy technique. Four formulations with different volume fractions 10, 15, 20, and 25 were selected. Parameters applied were 47 MPa compaction pressure, 150 ℃ molding temperature, and 130 ℃ post curing temperature with 6 hours holding time. The properties of brake pad composite were examined density, porosity, hardness, thermal stability, and friction performance (refers to SNI 09-0143-1987). The results show the increase of phenolic resin content can decrease the density, porosity, thermal stability, friction coefficient, and volume wear rate by 1%, 16.79%, 2.39%, 42.5%, 24.16%, respectively. However, the hardness increases by 19.29 %. This investigation also recommends not to select 10 and 15 volume fraction of phenolic resin due to lower resistance to mechanical stresses although the thermal stability is better than the other studied composites.

Keyword : phenolic resin, density, porosity, hardness, thermal stability, friction performance

1. Introduction

Phenolic resin have some properties such as excellent thermal behavior, high strength level, long thermal and mechanical stability, and excellent cost performance characteristic. They are widely used in various application related to production of friction composite, brake pads, and clutch lining for the automotive industry and for brake elements in rail vehicles and machine [1]. Even today, phenolic resin still represent the most important binders in commercial brake pad composite because they offer resistance to temperature of 250℃, good wetting capability, low cost, and good combination of thermal, mechanical, and tribology properties [2]. However, phenolic resin has limitations, such as high hardness which made the brake pad brittle, sensitive to humidity, and very low impact resistance [3]. Selection of the relative amount of phenolic resin in brake pad formulation as a binder is an important issue to maintain structurally intact with the other ingredients at various temperature, thus the brake pad composite is able to withstand under mechanical and thermal stresses during braking operation.

Furthermore, phenolic resins are also contribute to the friction performance. Jang *et al* [4] reported that the large of quantity phenolic resin was exhibited unstable friction coefficient as a result of the poor thermal stability phenolic resin. Kim *et al* [5] stated that increasing amount of phenolic resin with potassium titanate and CNSL were increase the hardness and the coefficient of friction at moderate temperature of about 100°C. While, increasing in amount of CNSL modified resin were decrease the friction coefficient, however the mechanical properties and wear were improve [6].

Recently, a lot papers reported the exploration of resin modified as a binder to improved mechanical properties and friction performance in fade and recovery condition [3,6–8]. However, a few modified resin are used in the commercial brake pad because of the cost factor that is not competitive [8], thus the commercial brake pad still used straight phenolic resin as a binder. Therefore, selection of the right amount of phenolic resin content is required. Phenolic resin will be effect to thermal stability of brake pad, ability to maintain mechanical properties, and ability to bind ingredients under braking condition from crumbling apart with other ingredients. In this study, the brake pad composites with a different of the volume fraction of phenolic resin were prepared. The effect of phenolic resin on density, porosity, hardness, thermal stability, and friction performance as a binder in non-asbestos organic brake pad were investigated.

2. Materials and Methods

The composition used in this study consisted of 10 ingredients, of which eight ingredients were kept constant (59%) and two ingredients, namely phenolic resin and $BaSO_4$ (41%) were varied in the volume fraction. All ingredients of the brake pad composite in the form of powder. Phenolic resin with the particle size of +270 mesh were used as a binder, and $BaSO_4$ with the particle size of 230 mesh used as an inert filler. The ingredients investigated in the formulation of brake pad composite as shown in Table 1.

Table 1. Ingredients of brake pad composites was used in this study

Sample	Phenolic resin (%)	$BaSO_4$ (%)	Friction Dust (%)	Al_2O_3 (%)	SiO_2 (%)	MgO (%)	Copper (%)	MoS_2 (%)	Graphite (%)	hBN (%)
PR-10	10	31	15	2	4	3	8	8	13	6
PR-15	15	26	15	2	4	3	8	8	13	6
PR-20	20	21	15	2	4	3	8	8	13	6
PR-25	25	16	15	2	4	3	8	8	13	6

Four different formulations, namely PR-10, PR-15, PR-20, and PR-25 were manufactured with powder metallurgy technique. The powder metallurgy stage consist of the following processes, namely, weighing, dry mixing, hot pressing, post-curing, and finishing. An analytical balance (Sartorius L 1201S) was used to weigh the ingredients sample. These ingredients were mixed in Rotation Drive STR4 for 60 minutes at 10 rpm in room temperature. Dry mixing was done to ensure the macroscopic homogeneity. Then, the powder mixture were hot pressed 47 MPa at a constant temperature of 150 °C for 15 minutes to obtained green body with the dimension of 25 mm x 25 mm x 7 mm. In the beginning of hot pressing process, the intermittent pressure was released 3-4 times to release the gas. Green body was post-cured in a furnace at a constant temperature of 130°C for 6 hours. The surface of the samples were then polished with the sand paper to obtain the smooth surface and desired thickness accordance with the dimensional of standard testing. The properties of the composite samples were examined density, porosity, hardness, thermal stability, and friction performance. The schematic of manufacturing brake pad composite was shown in Figure 1.

Characterization of the brake pad composite

Density test of composite sample was performed after post-curing process. A density of composite sample was measured using Archimedes' principle in accordance with the ASTM C-20, where the difference between the sample weight in air (w_{air}) and its weight in a fluid (w_{fluid}), divided by the density of the fluid (ρ_{fluid}). The volume of the liquid displaced, which was identical to the volume of sample [9]. The density of composite sample can be calculated using the formula (1).

$$\rho = \frac{w_{air}}{w_{air} - w_{fluida}} \, x \, \rho_{fluida} \qquad (1)$$

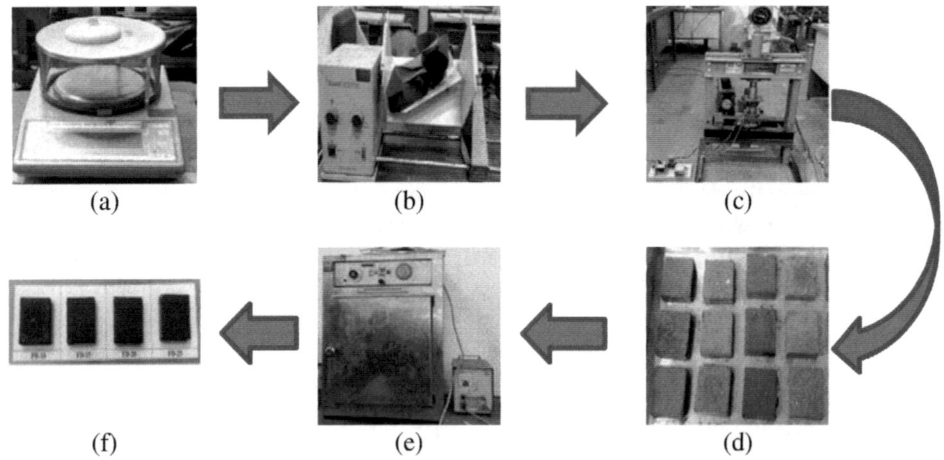

Figure 1. Stages process of making brake pad composite: (a) an analytical balance (Sartorius LC 1201S); (b) mixer (rotation drive STR4); (c) hand press hydraulic; (d) green bodies; (e) furnace (memmert 854 schwabach); (f) brake pad composites

The porosity measurement of composite sample was done based on Japanese Standard JIS D 4418: 1996. Composite samples with dimension of 25 mm x 25 mm x 5 mm ± 0.5 mm were left in the desiccator for 24 hours at room temperature. All composite samples were weighted on the analytical balance (Sartorius LC 1201S) before the samples immersed in the oil container with the constant temperature of 90±10°C for 8 hours. SAE 90 grade oil was used as the fluid of the test. Composite samples were left in container for 12 hours until the oil cools to the room temperature. Finally, all composite samples were withdrawn from the oil container and wiped with dry cloth for 5 times to remove oil from the sample surface. Composite sample was weighed again on the analytical balance (Sartorius LC 1201S). The porosity of composite sample was calculated using the following formula

$$\% \, porosity = \frac{m_2 - m_1}{\rho V} \, x \, 100 \qquad (2)$$

Where m_1 is mass of composite sample before immersed in the oil container (mg), m_2 is mass of composite sample after absorbing oil (mg), ρ is density of oil (mg/cm^3), and V is composite sample volume (cm^3).

The hardness of composite sample was measured using a Shore D hardness tester according to the ASTM D2240. At least five replications of hardness test for each sample were made and the average value was reported.

Thermal stability is very important to determine the fade resistance on braking application. TG/DTA test was used to check change in weight and energy of the sample as a function of temperature [10]. The thermal stability of composite sample was measured using a Diamond TG/DTA analyzer PerkinElmer Instruments. Composite samples were heated at temperature of 30°C to 1000°Cwith a heating rate of 10°C/minute in air condition. Alumina pan was used for this testing. TG/DTA test results are the graph of the temperature to weight loss and heat flow.

Friction and wear of composite samples were tested according to Indonesian National Standard (SNI 09-0143-1987) using a Friction Testing Machine JTM-1699 A (JTM Technology Co. Ltd). Two testing samples with dimension of 25 mm x 25 mm x 5 ± 0.2 mm were used as a stator and disc made of cast iron was used as rotor. Two samples were pressed against the disc at a contact pressure of 0.981 MPa. The disc temperature was maintained at 100, 150, 200, and 250 ∘C. The disc rotation and the number of rotation are 500 rpm and 1500 respectively at each temperature. The thickness of composite samples were noted before and after testing as much as 5 points to calculate the volume wear rate. The volume wear rate V (mm^3/ (N m) was calculated using formulation (3).

$$V = \frac{1}{2\pi R}\frac{A}{n}\frac{d_1 - d_2}{f_m} \quad (3)$$

Where R is the distance between the center of sample and the center of the rotating disc (150 mm), n is number of rotation of disc (1500), A is the area of the sample (25 mm x 25 mm), d_1 is the average thickness of the sample before it is tested (mm), d_2 is the average thickness of sample after it is tested (mm), and f_m is the average force of sliding friction (N). Drawing scheme of the friction performance test was shown in Figure 2.

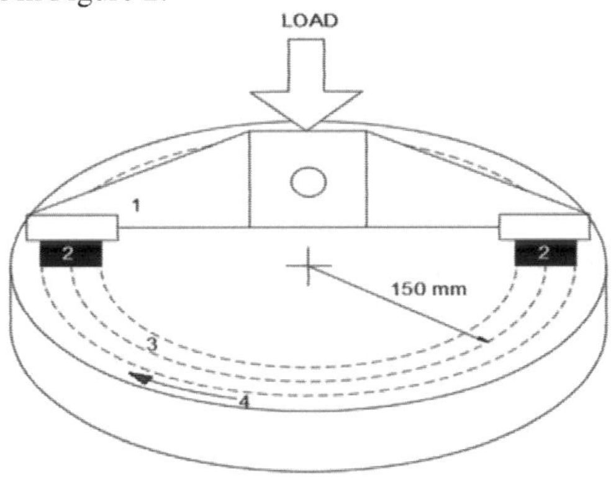

1. Stator
2. Sample
3. Sliding track
4. Rotor (Counter face disc)

Figure 2. Drawing scheme of the friction performance test

3. Results and Discussion

3.1 Density, hardness, and porosity of brake pad composite

Figure 3 shows the result test of density, porosity, and hardness with on the varying of volume fraction of phenolic resin. Density sample depends upon the ingredients of brake pad composite. The result of density test exhibits that the density of composite samples decrease with increasing volume fraction of phenolic resin. Decreasing in composite density is attributed to replacement of low density phenolic resin (ρ = 1.2 gr/cm^3) by high density of $BaSO_4$ (ρ = 4.3 gr/cm^3) in the formulation of composite samples. In addition, density results show that the composite sample density complies with "the rule of mixture" of composite.

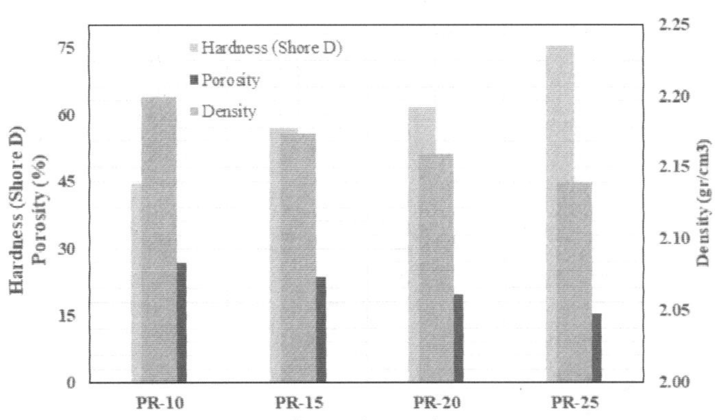

Figure 3. Density, porosity, and hardness of brake pad composites

The hardness result test of composite samples increase with increasing volume fraction of phenolic resin. Increasing volume fraction of phenolic resin will be increase amount of crosslinking during curing process thus the hardness improve. Kim *et al* [5] also reported that the hardness of brake pad increased with increasing phenolic resin. It was attributed to the high hardness of the thermosetting binder resin after curing. Meanwhile, porosity test results exhibit decrease with increasing in volume fraction of phenolic resin. Composite sample of PR-10 (filled 10 %vol phenolic resin) has the highest porosity. While, composite sample PR-25 (filled 25 %vol phenolic resin) has the lowest of porosity. The result of that porosity are caused by particle size of phenolic resin (about +270 mesh) fills the void space of particles during hot pressing process thus the contact area between particles increase and lead to decreasing in porosity. Brake pad porosity plays a role in the effectiveness of the brake system, because it absorbs heat and reduce brake noise. It is generally that the commercial products of brake pad need to maintain porosity about 10% to 14% for resulting in a good performing pads and the wear low [11].

3.2 The thermal stability of brake pad composite

The thermal analysis of four formulations, namely PR-10, PR-15, PR-20, and PR-25 were tested using Thermogravimetric and Derivative thermogravimetric Analysis (TG/DTA).

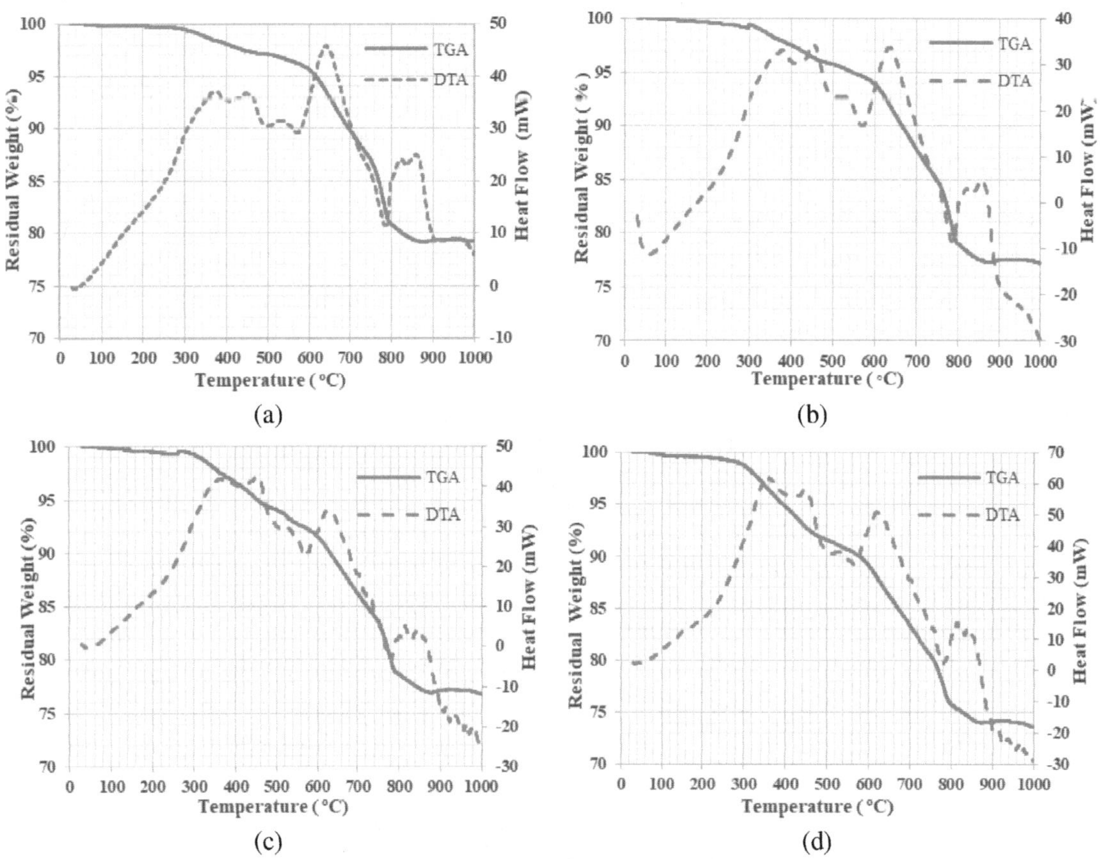

Figure 4. TG/DTA analysis of brake pad composites: (a) PR-10; (b) PR-15; (c) PR-20; (d) PR-25

Figures 4a-d show the TG/DTA curves for four composite samples with variation in volume fraction of phenolic resin. TG/DTA curves indicate the same weight losing trend, they have three weight loss steps, while the thermal decompositions occur in two stages. The initial weight loss (□1.7%) occurs between the temperatures of 100℃ to 200℃. It is attributed to the evaporation of water due to the reaction of chemical degradation phenolic resin. The degradation of four formulations are started at temperature of above 275°C. DTA curves show that the maximum thermal degradation are at 631°C for PR-10, 624 °C for PR-15, 625°C for PR-10, and 613 °C for PR-25. From the result of the maximum thermal degradation exhibits that composite sample of PR-10 has high degree of crosslinking thus it need more energy to decompose and improve the thermal resistance. In Figures 4a-d, there are a difference of slope in the TGA curves for four formulations, where the thermal degradation for PR-10 sample is more stable than the other studied composites. Thermal degradation (weight loss %) attributed to thermal stability, if the weight loss lesser meant thermal stability is better [12]. Table 2 shows that the weight loss (%) is follows: PR-10 < PR-15 < PR-20 < PR-25.

Table 2. Weight sample of brake pad composite in TG/DTA test

No	Sample Code	Initial Weight (mg)	Residue at 1000 °C (mg)	Residue at 1000 °C (%)	Maximum decomposition temperature (°C)
1	PR-10	20.165	15.964	79.17	631
2	PR-15	22.264	17.199	77.25	624
3	PR-20	19.654	15.091	76.80	625
4	PR-25	23.720	17.448	73.56	613

TG/DTA results (Table 2) shows that the thermal stability of composite samples (PR-10, PR-15, PR-20, and PR-25) are decrease with increasing volume fraction of phenolic resin. Therefore, the composite sample of PR-10 has the best thermal stability than the other studied composites.

3.3 The friction and wear performance of brake pad composite

The friction coefficient and wear of brake pad composites are presented in Figure 6a-b. That graphs show that the results of friction test only performed on composite samples of PR-20 and PR-25. Where, the composite samples of PR-10 and PR-15 are damaged during the test run, as seen in Figure 5. The damage of composite samples PR-10 and PR-15 occur at temperature of 100°C with the test duration of about 400 second for PR-10 and 700 second for PR-15. It is caused by phenolic resin is not sufficient binding all of the ingredients of composite, therefore composite samples have unable to against mechanical stresses during testing operation. It is also supported from the results of hardness value of composite samples PR-10 and PR-15. They have lower hardness value and have higher porosity than other studied composites (as seen in Fig 3).

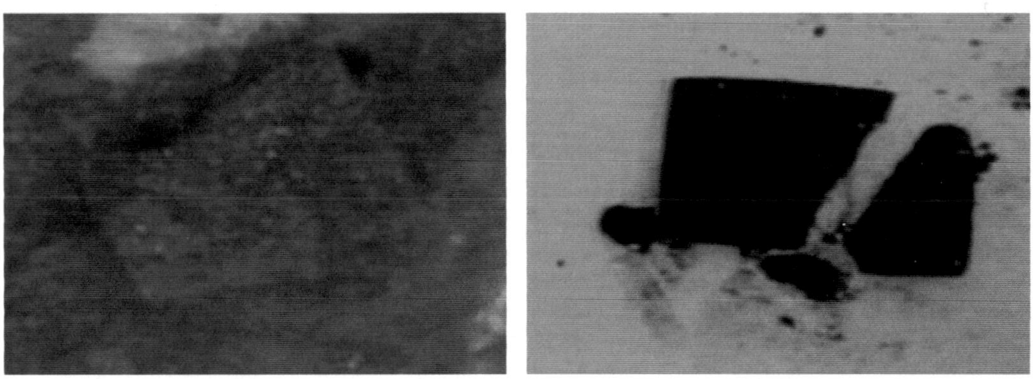

(a) (b)

Figure 5. Brake pad composites damaged during friction test: (a) PR-10; (b) PR-15

The results of friction coefficient and volume wear rate of composite samples PR-20 and PR-25 are shown in Figures 6a-b. Figure 6a shows that friction coefficient of composite sample PR-25 at temperatures of 100°C to 250°C is lower than composite sample of PR-20. It is attributed to composite sample of PR-25 has lower thermal stability than composite sample of PR-20. Cai *et al* [2] reported that the sample with poor thermal stability, causing a low magnitude of friction coefficient. In addition, composite sample of PR-20 shows the friction coefficient decrease with increasing in temperature and more stable. However, the friction coefficient of composite sample PR-25 are fluctuated against temperature changes. These fluctuations are due to non-homogeneous heat loads on the friction surfaces changing during test period. Therefore, friction coefficient will be changes continuously [13].

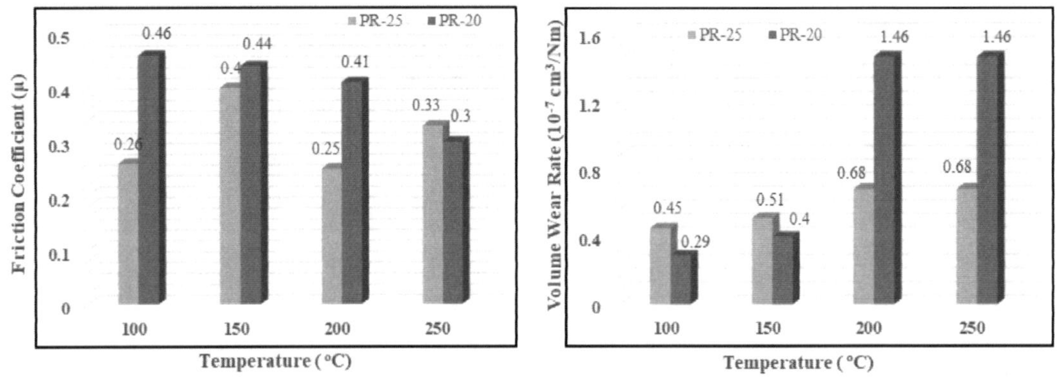

(a) (b)

Figure 6. Friction performance of brake pad composites (PR-20 and PR-25) (a) friction coefficient versus temperature; (b) Volume wear rate versus temperature

Figure 6b shows the volume wear rate of composite samples PR-20 and PR-25. The graph shows that increase in temperature lead to increasing in the volume wear rate. The composite sample of PR-25 has higher volume wear rate than composite sample of PR-20 up to temperature of 150°C. On the contrary, the volume wear rate of composite sample PR-25 is lower than composite sample PR-20 at temperature of 200 °C to 250°C. Increasing in temperature allows phenolic resin to decompose thus they are not sufficient to bind all of ingredients composite. This conditions causes increasing in the volume wear rate. In addition, the thermal stability of composite sample PR-20 is better than composite sample PR-25 up to temperature of 150°C.

4. Conclusions

The density, porosity, hardness, thermal stability, and friction performance of non-asbestos organic brake pad with a different of the volume fraction of phenolic resin were investigated. Based on the results and discussion of this study the following conclusion can be made:

1. The density decreased with increasing of the volume fraction of phenolic resin. This is because of the replacement of low density phenolic resin by high density of $BaSO_4$ in the formulation.
2. The porosity decreased with increasing of the volume fraction of phenolic resin, while the hardness increased. This is attributed to the contact area between particles increased as a result of the phenolic resin particle fills the void space during hot press process. Thus, bonding strength and hardness of composite sample increase and the porosity of composite sample decrease.
3. The thermal stability of composite samples decrease with increasing of the volume fraction of phenolic resin. The composite sample of PR-10 exhibits better thermal stability than the other studied composites due to it has the lowest weight loss as compared to the other studied composites.
4. Increasing in the volume fraction of phenolic resin will be decreased in friction coefficient and increase in volume wear rate. The composite samples of PR-10 and PR-15 are not recommended to be selected in this formulation because they have lower resistance to mechanical stresses during friction performance test although the thermal stability is better than the other studied composites.

5. Acknowledgement
This research is supported by DIKTI (Ministry of Higher Education) and LPDP (Indonesia Endowment Fund for Education).

6. References
[1] Gardziella A, Pilato A and Knop A 2000 Phenolic Resins 2nd Edition (Verlag Berlin : Springer Heidelberg) p 353
[2] Cai P, Wang Y and Wang Q 2015 Effect of resins on thermal, mechanical and tribological properties of friction materials *Tribology Int.* **87** 1–10
[3] Öztürk B and Öztürk S 2011 Effects of resin type and fiber length on the mechanical and tribological properties of brake friction materials *Tribol Lett.* **42** 339–350
[4] Jang H, Lee J S and Fash J W 2001 Compositional effects of the brake friction material on creep groan phenomena *Wear.* **250** 1477–1483
[5] Cheol Y, Hyung M, Jin S and Jang H 2008 The effect of phenolic resin , potassium titanate , and CNSL on the tribological properties of brake friction materials *Wear.* **264** (2008) 204–210
[6] Nidhi, Bijwe J and Mazumdar N 2006 Influence of amount and modification of resin on fade and recovery behavior of non-asbestos organic (NAO) friction materials *Tribol Lett.* **23** 215–222
[7] Sai Balaji M A and Kalaichelvan K 2012 Effect of Alkyl Benzene Modified Resin Matrix in relation to the friction stability in a Non Asbestos Disc brake pad *IEEE.*150–154.
[8] Gurunath P V and Bijwe J 2009 Potential exploration of novel green resins as binders for NAO friction composites in severe operating conditions *Wear.* **267** 789–796
[9] Barsoum M W 2003 Fundamentals Of Ceramics (New York : Taylor and Francis) p 83-84
[10] Bahrom M Z, Berhan M N and Mohamed R 2009 Thermal Characteristics of Light Rail Transit *J. Mekanikal.* **29** 35–51.
[11] Sundarkrishnaa K L 2013 Friction Material Composites (New York: Sringer Heidelberg) p 32
[12] Thiyagarajan V, Kalaichelvan K, Vijay R and Lenin Singaravelu D 2016 Influence of thermal conductivity and thermal stability on the fade and recovery characteristics of non-asbestos semi-metallic disc brake pad *J Brazilian Soc Mech Sci Eng.* **38** 1207–1219
[13] Anderson A E 1992 Friction, Lubrication and Wear Technology ASM Handbook (United States of Amerika : ASM International) p 569-577

Study of Strontium Effect on ADC12/Nano-Al_2O_3 Composite Characteristics with Al-Ti-B Grain Refiner Addition by Stir Casting Method

A Z Syahrial[1*], and M Al-Wafiy[1]
[1]Metallurgical and Materials Engineering Department, Universitas Indonesia, Depok, Indonesia

E-mail: anne@metal.ui.ac.id

Abstract. Cast iron are the common material used as brake-shoe component of the train, but this material has a high density of 7,25 g/cm^3, which tends to be heavy and causes more energy consumption. Aluminum composites can be used as alternative materials with less density. For improving the mechanical characteristic of the aluminum composite, grain modification can be done using strontium (Sr) addition. This research is intended to know the effect of variation of addition of Sr as a modifier on microstructure and mechanical properties of ADC12 composite material with reinforcement of 0,03 vf% nano-Al_2O_3 particles. Composite fabrication is done by stir casting method. The addition of Mg as wetting agent was 5 wt%, Al-5Ti-B as a grain refiner was 0.04 wt%, and Sr as modified grain with varying addition as much as 0.018; 0.021; 0.026; 0.028; and 0.037 wt%. To characterize the composite product, several testing is done, which is chemical composition characterization, metallographic observation, Scanning Electron Microscopy – Energy Dispersive Spectroscopy characterization, X-Ray Diffraction characterization, hardness testing, wear testing, tensile testing, and impact testing. The results of mentioned testing generally indicate an increase in mechanical properties of the material, due to the role of Sr in the morphological modification of silicon and quasi-binary Mg_2Si, secondary dendritic arm spacing size reduction, and wetting agents in the composite system. However, excess addition of Sr will increase the %porosity of the material.

Keywords: Aluminum composite, ADC12, nano-Al_2O_3 reinforcement particle, Strontium modifier

1. Introduction

In the development of the world of transportation, one that interesting is the fast-train industry. There are challenges faced by the fast-train industry, one of it is on the braking system. The braking system consists of many components, one of it is a brake shoe that works by direct contact with the fast train wheel. This component functions is to slow or stop the wheel rotation of the train when moving at a certain speed.

The basic materials used should have stable friction and wear resistance in variations of load, speed, temperature, and environment conditions, with high endurance. Material for fast-train brake shoe must also have good thermal stability to avoid distortion or cracking caused by

thermal stress due to friction. In addition, the weight of components and manufacturing processes is also an important factor to be considered in designing fast train brake shoe.
The material commonly used as a brake shoe is cast iron. However, cast iron has a high density of 7.25 g/cm^3, which tends to be heavy and causes more energy consumption. Resilience of this type of material is also low, resulting in high cost for routine maintenance, especially for brake shoe replacement [1].
Currently, the development of aluminum-based components in world of transportations is being done significantly, which can also be adapted by the fast-train industry. Aluminum alloys can be used as a base matrix with the addition of a ceramic reinforcement forming a composite system, which demonstrates promising capabilities in the needs of this type of application (brake shoe) [2]. This material has a lower density (2.7 g/cm^3) and better thermal conductivity compared to conventional materials. Bayuseno and Nugroho [3], revealed that the brake shoe with composite materials will have a lifetime that can be increased up to three times compared to conventional materials. The weight reduction in the braking system can reach 50% - 60%. Moreover, this type of material demonstrates better performance potential under heavier application conditions [4].
Alloying elements can also provide mechanical properties improvement, which are used in this study is Strontium (Sr). Sr addition was carried out on the casting of Al-Si alloys with the aim of changing the crystalline morphology of silicon from acircular flakes to fibrous rod-like form, which giving the positive effect of mechanical properties of the aluminum product [11].
The aim of this study is, to investigate experimentally the effect of variations of addition of Sr as alloying element on microstructure and mechanical properties of aluminum composite material with reinforcement of ceramic particles.

2. Materials and Methods

2.1. Materials

In this work, Aluminum ADC12, which obtained from PT Ogindo Prakarsatama, was used as the matrix, with reinforcement of 0,03 vf% nano-Al_2O_3 particles, which obtained from US Research Nanomaterials, Inc. Some addition was also made, there were 5 wt% of Magnesium as wetting agent, 0.04 wt% of Al-5Ti-B as a grain refiner, and Strontium as grain modifier with varying addition as much as 0.018; 0.021; 0.026; 0.028; and 0.037 wt%. These alloying substances were obtained from PT. Makmur Meta Graha Dinamika.

Table 1. Chemical Composition of Aluminum ADC12

Si (%)	Cu (%)	Mg (%)	Mn (%)	Fe (%)	Ni (%)	Zn (%)	Sn (%)	Others (%)	Al (%)
9,5 – 11,5	2,0 – 3,0	max 0,1	max 0,5	max 1,3	max 0,3	max 3,0	max 0,15	max 0,5	bal.

2.2. Experimental Methods

ADC12 which already weighed inserted into the tilting furnace, then raised to 700°C incremently until completely melted. In parallel, nano-Al_2O_3 powder were vibrated as the preparation before it was added to the molten ADC12. After that, insert Mg, Al-Sr, and Al-5Ti-B into the melt of ADC12, then stir for 2 minutes to get homogenous mixture. Degassing should be done to remove trapped gases inside the molten metal. Insert Al_2O_3 powder that has been preheated, slowly while stirring for 2 minutes. Then, pour melt into the mold which has been preheated before. Open the mold after the composite solidified for later removal from the mold.

2.3. Characterization

To characterize the composite product, several testing was done, which was chemical composition characterization using Optical Emission Spectroscopy, metallographic observation based on ASTM E3-11 using Zeiss Primotech Microscope, Scanning Electron Microscopy – Energy Dispersive Spectroscopy (SEM-EDS) characterization, X-Ray Diffraction (XRD) characterization which analyzed using PANanalytical HighScore Plus, hardness testing based on ASTM E18-11 (Rockwell B method), wear testing using Ogoshi Machine, tensile testing based on ASTM B557M-02A using Gotech AL-7000 LA 10, and impact testing based on ASTM E23-01 using GoTech Testing Machine. There was also a feasibility study of composite material as substitution of cast iron material in its application as component of brake shoe of train.

3. Results and Discussion

3.1. Chemical Composition

The ADC12 material used identified that comply with the compositional standards of aluminum alloys ADC12, so it is appropriate to use as a matrix in composite manufacturing processes. We observed a significant increase in magnesium value from 0.221 wt% to more than 5 wt%. This is due to the addition of magnesium of 5 wt% as a non-volatile wetting agent, resulting in a major solidification reaction no longer Al-Si-Cu but Al-Si-Mg.

The difference in strontium content obtained is at a significant value, but the excess strontium element content in the composite is still tolerable, which is below the 0.12 wt%, so it does not form an intermetallic compound which is not desired [5].

Table 2. Chemical compotition of ADC12 as a matrix and Composite Materials

Elements	Content (wt%)					
	ADC12	**Composite 1**	**Composite 2**	**Composite 3**	**Composite 4**	**Composite 5**
Al	84,80	79,700	79,800	79,200	78,633	79,500
Si	10,50	11,130	10,587	10,453	11,007	10,547
Cu	2,330	1,800	1,800	1,970	1,980	1,800
Mg	0,221	5,320	5,700	6,320	6,317	6,267
Fe	0,864	0,616	0,622	0,534	0,557	0,551
Sr		0,018	0,021	0,026	0,028	0,037
Ti	0,046	0,092	0,094	0,081	0,095	0,093
B		0,003	0,003	0,003	0,003	0,003
Cr	0,039	0,050	0,048	0,037	0,043	0,032
Mn	0,223	0,122	0,115	0,115	0,116	0,111
Bi		0,006	0,007	0,006	0,005	0,007
Ni	0,076	0,039	0,036	0,027	0,032	0,028
Zn	0,637	0,599	0,619	0,555	0,613	0,568
Sn	0,080	0,008	0,013	0,005	0,006	0,007
others	0,144	0,496	0,536	0,668	0,671	0,450

Copper contained in the ADC12 material is at a value of 1.80 - 2.00 wt%. In such content, $CuAl_2$ compound formed, which will increase the strength of the material [6]. Fe content derived from ADC12 base material is categorized high enough and can form an intermetallic phase of β-Al_5FeSi with a needle-shape morphology that can decrease the mechanical properties of the material [7].

3.2. SEM-EDS and XRD Examination

Figure 1 (a) SEM and (b) XRD observation result of composite material

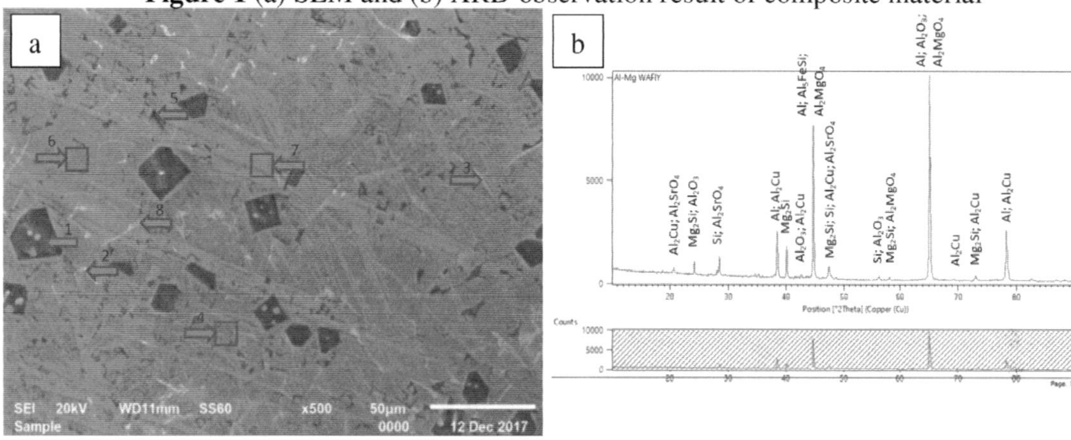

Table 3. Chemical composition of phase based on SEM-EDS examination

Number	Content (wt%)								Possible Phase
	C	O	Mg	Al	Si	Ti	Fe	Cu	
1	11,66	22,21	41,79	0,21	23,98	0,02	0,05	0,08	Mg_2Si
2	25,39	0,57	2,73	57,34	2,84	0,13	0,16	10,84	Al_2Cu
3	28,58	1,28	0,51	57,99	7,07	0,03	2,69	1,85	β-Al_5FeSi
4	8,47	2,94	4,87	80,15	2,92	0,00	0,00	0,64	Al+Mg_2Si
5	19,60	2,50	12,12	60,59	4,61	0,00	0,00	0,57	Al+Mg_2Si
6	7,97	0,70	0,73	79,65	10,22	0,01	0,09	0,64	Al+Si
7	8,89	0,81	1,36	80,11	7,73	0,14	0,29	0,67	Al+Si
8	21,39	0,84	0,16	73,71	0,91	0,02	0,08	2,90	Al_2Cu

Based on SEM-EDS observation, and also supported by the result of XRD examination, it can be seen that the presence of Mg_2Si primary phase in the form of blocky and black, which shown by arrow number 1. Mg_2Si is an intermetallic phase with low density, high melting point, good hardness characteristic and elastic modulus, also low coefficient thermal expansion. Therefore, it is clear that Mg_2Si provides excellent mechanical characteristics in composite materials [8]. Quasi-binary phase Al + Mg_2Si is indicated by arrows number 4 and 5. Al + Mg_2Si can be observed with in chinese-script morphology.

Fe content that is high enough in this study formed the intermetallic phase β-(Al_5FeSi) which has a needle-like morphology and gray as shown in arrow numbered 3. The presence of Al_2Cu intermetallic phase can also be known from the arrow number 2. Al_2Cu intermetallic phase will increase the value of UTS of the composite material [6].

In point 8 it appears that Al Cu is in a needle-like phase, based on research conducted by S. G. Irizalp and N. Saklakoglu [9], this is an Al_2Cu phase attached at the β-Al_5FeSi intermetallic phase which has needle-like form.
Aluminum as the main matrix can be observed by arrows number 6 and 7. The presence of silicon at high enough value in these points indicates that there is the existence of silicon eutectic crystals.
The presence of other elements at each SEM-EDS identification point was also observed, among them were oxygen (O) and carbon (C). The oxygen in this study is contained from the addition of Al_2O_3 reinforcing particles. While the high carbon content was identified as a result of refractory wear used in the composite fabrication process, the refractories used were SiC types.

3.3. Microstructure

Figure 2 Microstructure of Composite Materials; (1) α-Aluminum, (2) Silicon, (3) primary Mg_2Si, (4) quasi-binary Mg_2Si, and (5) β-Al_5FeSi

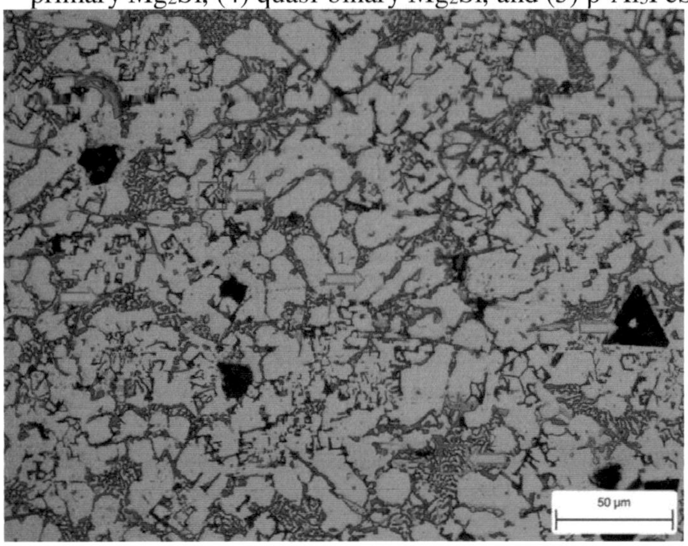

Figure 3. Microstructures of (a) ADC12 and Composite Materials with Additions of Sr (b) 0.018 wt%, (c) 0.021 wt%, (d) 0.026 wt%, (e) 0.028 wt%, (f) 0.037 wt%

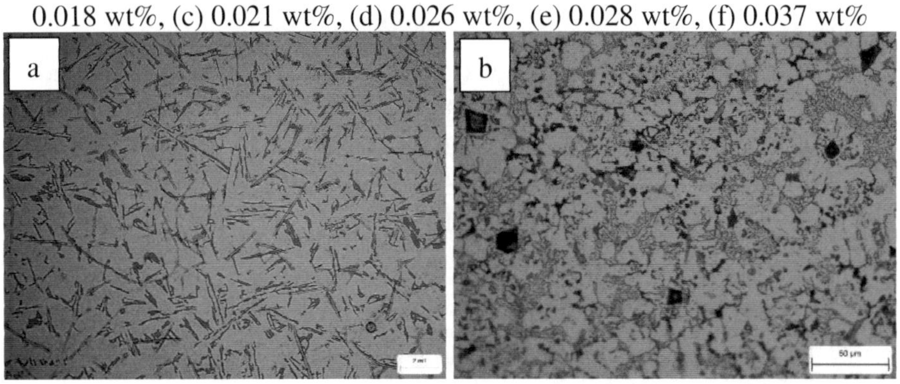

IOP Conf. Series: Materials Science and Engineering **547** (2019) 012013 doi:10.1088/1757-899X/547/1/012013

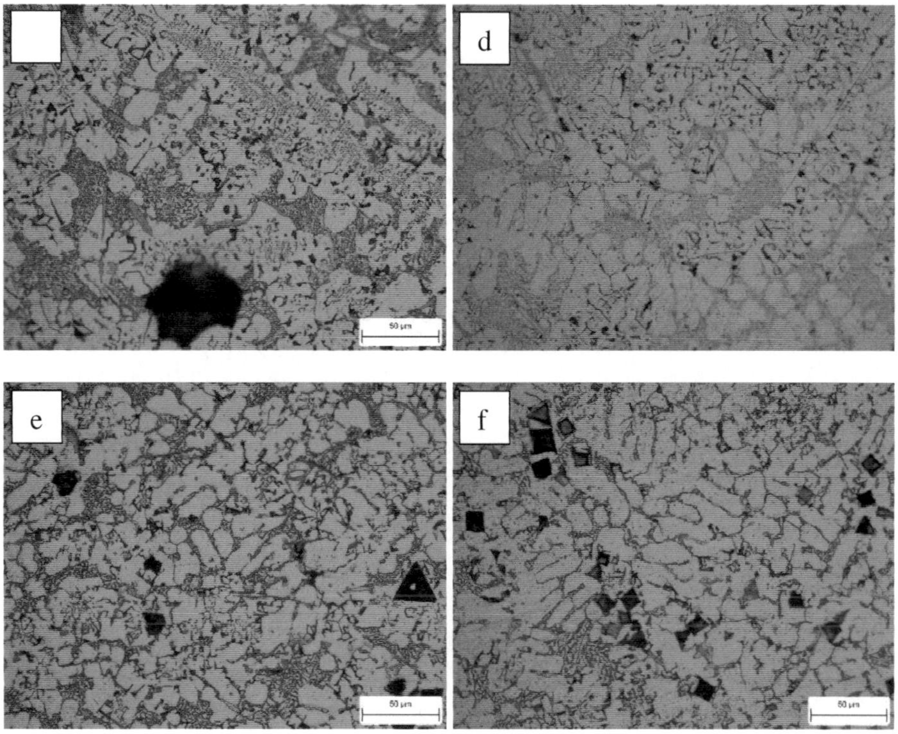

The effect of Sr addition is observed on the morphology of silicon crystals from initially acircular flakes shape modified into fibrous rod-like forms. The distance between the crystals of silicon will also be closer.

Strontium can also modify the quasi-binary Mg_2Si phase chinese-script morphology. As is known, chinese-script morphology can reduce the mechanical strength of composite materials due to the easy crack nucleation along the interface between quasi-binary Mg2Si phase with α-Aluminum. With the addition of Strontium, the Mg_2Si morphology becomes very smooth, and the chinese-script form changed into more rounded or irregular polygons.

Strontium also affects secondary dendritic arm spacing (SDAS) values of α-Aluminum grains. This is due to the addition of strontium made α-Aluminum to be more columnar, due to the properties of strontium that can initiate the growth of columnar dendrites in Al-Si alloys, thereby reducing the size of SDAS [10].

3.4. Mechanical Properties

Figure 4. Effects of SR addition on the (a) hardness, (b) wear rate, (c) ultimate tensile strength, (d) elongation, and (e) impact toughness characteristics of composite materials

IOP Conf. Series: Materials Science and Engineering **547** (2019) 012013 doi:10.1088/1757-899X/547/1/012013

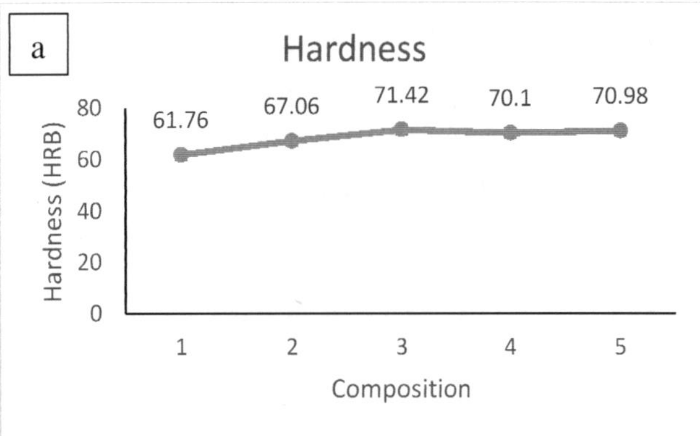

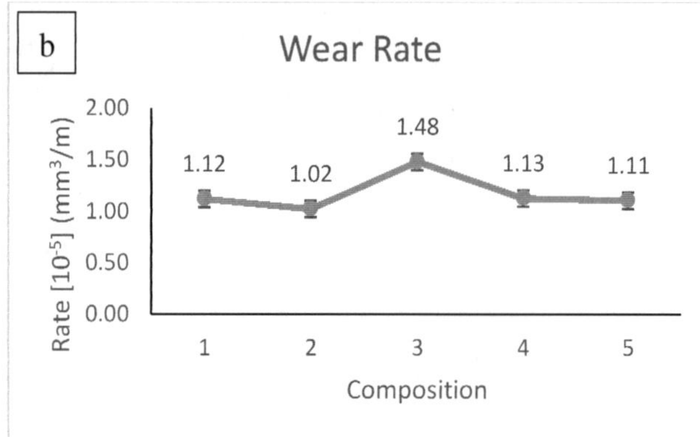

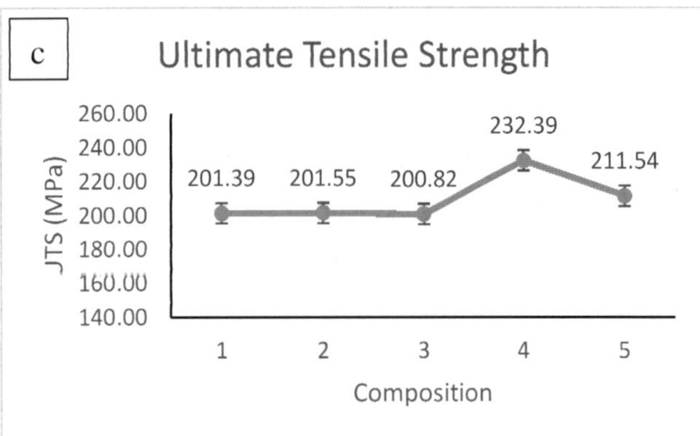

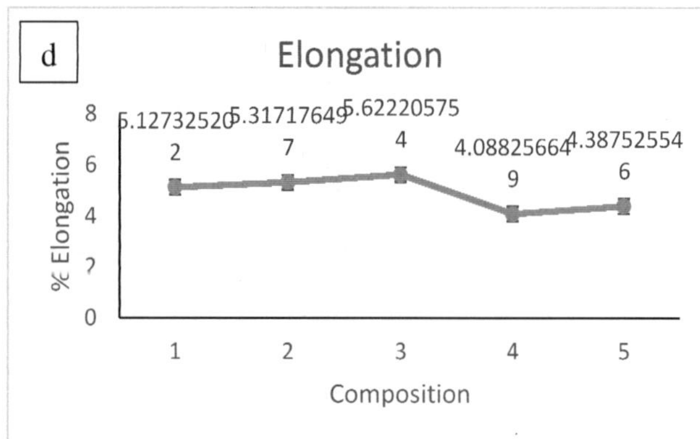

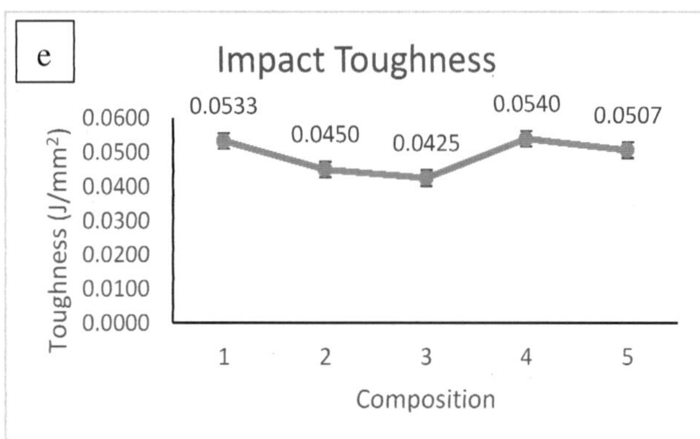

Hardness of composite material generally showed better characteristic with increasing amount of strontium added, with optimum value of 71,42 HRB in addition of Sr as much as 0,026 wt%, due to effect of modification of silicon crystals morphology and chinese-script Mg_2Si, and reduction of size SDAS [11].

The wear resistance of the composite material generally shows good characteristic with the addition of strontium, with the smallest wear rate value being 1.0223 x 10-5 mm2 in addition of strontium of 0.021 wt%, due to modification of the silicon crystals as the load-bearing phase during wear loading [12].

Tensile strength, elongation, and impact properties of composite material showed better characteristic with strontium addition, due to the effect of modification of silicon crystals morphology and chinese-script Mg_2Si [13] and the role of strontium as a wetting agent [14].

Looking at the mechanical characteristics of the composite, this material is very potential and suitable for use as a substitution material of cast iron in its application as a train brake shoe in the future [15].

4. Conclusion

- Strontium as grain modifier can modify silicon crystal morphology from acircular flakes to fibrous rod-like form, quasi-binary Mg_2Si quasi-binary chinese-script morphology becomes more refined and uniformly distributed, and forms α-aluminum to be more columnar.
- In general, there is an increase in the mechanical characteristics of the material as the amount of Strontium is added.
- The ADC12 / nano-Al_2O_3 composite material is considered feasible as a subtitution material for cast iron in the future application as a component of the train brake shoe, due to it's generally superior mechanical characteristics.

5. Acknowledgements

Authors would like to express their gratitude for financial support from the Directorate of Research and Community Services (DRPM), Universitas Indonesia, through International Publications Index for Final Year Students Projects Grant (PITTA) 2018 with contract number: 2378/UN2/R3.1/HKP.05.00/2018

6. References

[1] Bayuseno, A. P., & Nugroho, S. (2011). Pengaruh Variasi Kecepatan Putar Dalam Metode Stir Casting terhadap Densitas dan Porositas Al-SiC untuk Aplikasi Blok Rem Kereta Api. 31–35.

[2] Zhang, S., & Wang, F. (2007). Comparison of Friction and Wear Performances of Brake Material Dry Sliding Against Two Aluminum Matrix Composites Reinforced with Different SiC Particles. *Journal of Materials Processing Technology*, 182, 122–127.

[3] Bayuseno, A.P., & Chamdani, N. A. (2011). ADC12 Sebagai Material Sepatu Rem Menggunakan Pengecoran High Pressure Die Casting Dengan Variasi Temperatur Penuangan. *Jurnal Teknik Mesin Universitas Diponegoro*, 17–23.

[4] Stephenson, T. F., Warner, A. E. M., Wilson, S., Alpas, A. T., & Rohatgi, P. K. (1996). Aluminum Hybrid Composites Containing Nickel-Coated Graphite Particulate. *Processing, Properties and Applications of Cast Metal Matrix Composites*. 337–351.

[5] Khrisnan, G. R., & Nair, E. M. S. (2013). Study of Strontium Modification in Aluminium Alloy. *International Journal of Emerging Technology and Advanced Engineering*, *3*(10), 753–760.

[6] Shabestari, S. G., & Moemeni, H. (2004). Effect of copper and solidification conditions on the microstructure and mechanical properties of Al – Si – Mg alloys. *Journal of Materials Processing Technology*, 153-154, 193–198.

[7] Samuel, E., Samuel, A. M., Doty, H. W., Valtierra, S. & Samuel., F. H. (2014). Intermetallic Phases in Al–Si Based Cast Alloys : New Perspective. International Journal Cast Met. Res., 27(2), 107–114.

IOP Conf. Series: Materials Science and Engineering **547** (2019) 012013 doi:10.1088/1757-899X/547/1/012013

[8] Huang, Z. L., Wang, K., Zhang, Z. M., Li, B., Xue, H. S., & Yang, D. Z. (2015). Effects of Mg content on primary Mg2Si phase in hypereutectic Al-Si alloys. *Transactions of Nonferrous Metals Society of China (English Edition)*, *25*(10), 3197–3203.

[9] Gencalp Irizalp, S., & Saklakoglu, N. (2014). Effect of Fe-rich intermetallics on the microstructure and mechanical properties of thixoformed A380 aluminum alloy. *Engineering Science and Technology, an International Journal*, *17*(2), 58–62.

[10] Liao, H., Sun, Y., & Sun, G. (2002). Correlation between mechanical properties and amount of dendritic α-Al phase in as-cast near-eutectic Al-11.6% Si alloys modified with strontium. *Materials Science and Engineering A*.

[11] Hegde S., & Prabhu, K. N. (2008). Modification of Eutectic Silicon in Al-Si Alloys. *Journal of Materials Science*, 43, 3009–3027.

[12] V, Vijeesh. & Narayan Prabhu, K (2014). Review of Microstructure Evolution in Hypereutectic Al-Si Alloy and Its Effect on Wear Properties. Transaction of Indian Institute Met, 67, 1.

[13] Razaghian, A., Emamy, M., Najimi, A. A., & Ebrahimi, S. H. S. (2009). Sr effect on the microstructure and tensile properties of A357 aluminum alloy and Al 2 O 3 / SiC-A357 cast composites. *Materials Characterization*, *60*(11), 1361–1369.

[14] Garcia-hinojosa, J. A., Gonzalez, C. R., Juárez, J. I., & Surappa, M. K. (2004). Effect of Sr addition on an Al – 7Si – 10 vol .% SiCp cast composites, *382*, 315–320.

[15] Senen and A. P. Bayuseno, "Karakteristik Blok Rem Kereta Api Berbahan Besi Cor dan Al-SiC Berdasarkan Kekuatan Uji Tarik dan Harga Impak," Teknik, vol. 33, no. 1, pp. 42–46, 2012.

IOP Conf. Series: Materials Science and Engineering **547** (2019) 012014 doi:10.1088/1757-899X/547/1/012014

The Influence of Mold Material on Cooling Curve, Solidification Parameters, and Micro-hardness of Al–6wt.%Si in Unidirectional Solidification

D Masnur[1, 2, *], Suyitno[1], V Malau[1]
[1]Mechanical and Industrial Engineering Department, Universitas Gadjah Mada, Jl. Grafika No. 2, Yogyakarta 55281, Indonesia
[2]Mechanical Engineering Department, University of Riau, Kampus Bina Widya km 12.5 Simpang Baru, Pekanbaru 28293, Indonesia

*E-mail: dedymasnur@eng.unri.ac.id

Abstract. The influence of mold material on cooling curve, solidification parameters, and micro-hardness were investigated. Al–6wt.% Si in clay and stainless steel mold was directionally solidified using vertical Bridgman type furnace. The samples were heated up to 700°C at 1.9°C/s heat rate, held for 60 minutes, cooled, and withdrawn at 40 μm/s. The results show that the increasing of thermal conductivity by 72% affects cooling curve shape, solidification parameters, and micro-hardness. Furthermore, comparative evaluation of the exponent value of the function of micro-hardness to distance and micro-hardness to growth rate also have been made to proposed the range of exponent value for each mold material.

Keywords: unidirectional solidification, clay mold, cooling curve, micro-hardness

1. Introduction

Cooling curve is a plot of temperature dynamics and heat evolution during solidification processes related to phase transformation in the liquid. Temperature gradient, growth rate, and cooling rate are derived from the cooling curve. These parameters indicate the mechanism of microstructure evolution. The shape of cooling curve is affected by alloy type, compositions, and mold material.

Ceramics and metals were two types of mold materials, which were generally used in unidirectional solidification in a last decade. In metal type, there are carbon steel and stainless steel. Carbon steel mold was used by Araújo et. al and Barros et. al to investigate aluminum copper alloys [1,2]. Stainless steel was brought into practiced by Brito et. al and Verissimo et. al to study Al-Mg, Al-Mg-Si [3], and Mg-Zn [4] alloys. In the ceramic type, graphite [5–14] and alumina [15–20] were more common as mold material. Yet, there are no specific restrictions in mold materials type selection. Clay as mold material has been never been as an object of study. Clay has been widely applied in investment casting and it plays an important role as it can create a shell from a complicated pattern and therefore, providing an extensive information of clay as mold material will be a valuable thing.

Mold material influences the final structure of casting product by its ability to transfer heat. For instances, ceramic with 1.07 W/m.K heat transfer coefficient and stainless steel with 21.5 W/m.K heat transfer coefficient [21]. Stainless steel releases the amount of heat twenty one times bigger than the amount heat released by clay therefore this change affects the solidification behavior. However, there is still limited explanation on how much the change of heat transfer coefficient. By knowing this, ones can estimate the changes on the shape of cooling curve and solidification parameters.

Comparisons in several findings are made and dissimilarity of growth rate exponent value and microhardness values are found. Çadirli showed the relationship eutectic spacing and growth rate of Al-3wt.%Cu in ceramic mold and found the growth rate exponent value equals to 0.46 [11]. This is two-third of the growth rate exponent value obtained by Araujo et al. [1] in similar alloy composition with metal mold. Moreover, Araújo et al. compared their work to Kaya et al. and found the micro hardness values of Al–3wt.%Si in metal mold [1] are a higher than the microhardness values of Al-3wt.%Si in ceramic mold [22]. It is suspected these differences might be affected by mold material heat transfer ability.

Mold materials with its thermal conductivity is mentioned as one of major factors in determining microstructure however there have been limited explicit established explanations attributed to its effects on cooling curve and mechanical properties. Present work investigates experimentally the influences of clay and stainless steel as mold material on cooling curve, solidification parameters, microstructures, and hardness in unidirectional solidification.

2. Materials and Methods

Al– 6wt.%Si samples were prepared by weighed amount of pure Al and Al–12wt.%Si in electrical furnace. Molten alloy was casted into a metal die (50 mm in length and 8 mm in diameter). The chemical compositions are presented in Table 1.

Table 1 Specimen chemical composition

Element	Si	Cu	Fe	Mg	Mn	Cr	Ni	Zn	Ti	Ca	P	Pb	Sb	Sn	Al
%	5.995	1.385	0.527	0.430	0.093	0.045	0.030	0.49	0.023	0.000	0.000	0.033	0.001	0.012	90.93

The cast products were machined by lathe to reduce its diameter to 6 mm. Clay and stainless steel were mold materials in this experiment. The sample was inserted in 10 mm of outer diameter, 6 mm inside diameter, and 50 mm length cylinder molds. There are two types of mold, clay with 12.75 W/m.K thermal conductivity (laboratory tested) and stainless steel with 22 W/m.K respectively [23] Four thermocouples were attached in the mold and their location were 10 (T1), 20 (T2), 30 mm (T3) from base/the cooling pad, and another (T4) was at the top surface of the sample. Each position had 120° angle differences respect to another (Figure 1a).

These thermocouples are connected to a data logger interface with computer. The lower part of the mold was put on a stainless steel withdrawal bar and upper part was held by stainless steel holder in which was connected to a 25 mm of diameter ceramic pipe. Figure 1b illustrated schematically the apparatus arrangement and the initial position of specimen. Each sample was heated up to 700°C at 1.9°C/s heat rate in a vertical Bridgman type furnace and was kept for 60 minutes. Then, it was withdrawn downward at 0.04 mm/s and was cooled at the bottom with water-cooled system inside the stainless steel withdrawal bar simultaneously. The furnace temperature was maintained at 700°C. A data logger recorded the temperatures data and stroke positions during the process. The recording was stopped when the position of a cylinder was out of the furnace.

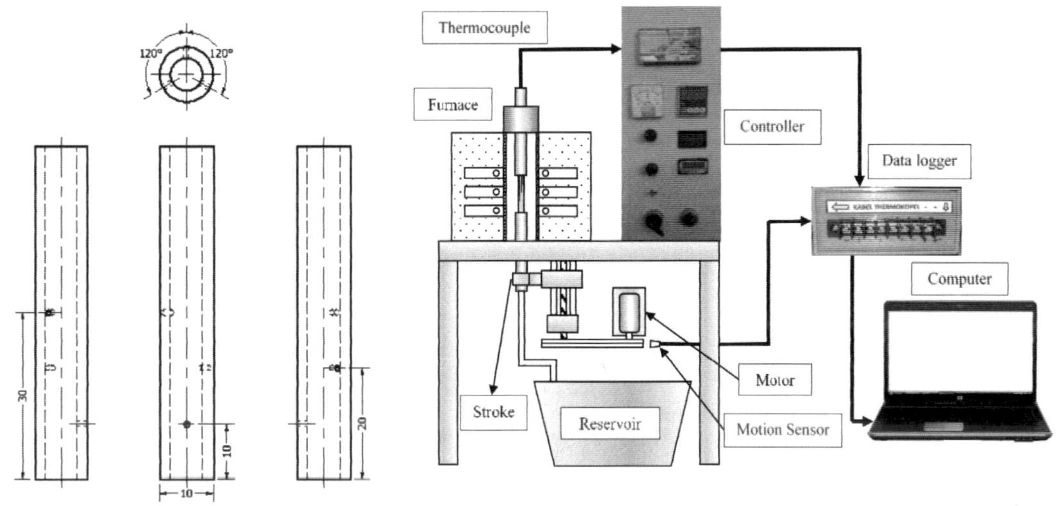

Figure 1 a. Thermocouple positions in a mold, b. Scheme of equipment.

The temperatures data were plotted as a function of time along with the stroke position. The cooling rates for each thermocouple position were determined by calculating the gradient of cooling curve line between liquidus and solidus temperature [24]. The temperature gradient for each thermocouple was calculated using the measured value of ΔT and the known value of ΔX. The growth rate of each sample was determined by dividing the distance of two thermocouples with the time taken by solid liquid interface to reach the second thermocouple. The calculating method of temperature gradient and growth rate refer to Kaya et al [11] and Çadirli [6].

Each sample was sectioned along its longitudinal direction then it was cold mounted with epoxy-resin. The longitudinal section was flattened with 800, 1000, 1200, and 2000 grit abrasive paper and subsequently polished. The samples were etched with NaOH 5 pct. for microstructural analysis. Microstructures and dendrite arms spacing observation was conducted using metallurgy microscope. Image J software was used to find the percentage of eutectic silicon. Micro-hardness values were evaluated with Boehler micro-hardness using 50g load and 10s dwelling time [11] [25]. The micro-hardness value was measured at 10, 20, 30, and 40 mm from heat extraction and 3 indentations was taken in each position.

3. Results and Discussion

3.1 Cooling curve, Solidification Parameters, and Microstructure

The liquidus and solidus temperature of Al–6wt.%Si is determined by drawing a vertical line at selected composition until it intersects with liquidus and solidus temperature line at Al-Si phase diagram [26]. The liquidus and solidus temperature are 628°C and 577°C accordingly.

Experimental cooling curve of Al–6wt.%Si in clay mold shows there are two types of slopes (Figure 2a). Steep slope at the beginning of solidification indicates a rapid solidification (high cooling rate) where a great amount of heat is released while shallow slope indicates slow cooling (low cooling rate). The cooling rate is maximum at the beginning of solidification and gradually decreases as the distance from heat extraction increases. T1 curve has the highest cooling rate by 14.569°C/s in Al–6wt.%Si cooling curve then it falls to 2.509 (T2), 2.256 (T3), and 2.287 (T4) °C/s accordingly (Figure 2b). The initial solidification time is at 9.969s as liquidus temperature (T_L) was at T1 and finished as the solidus temperature (T_S) is at 69.746s (T4).

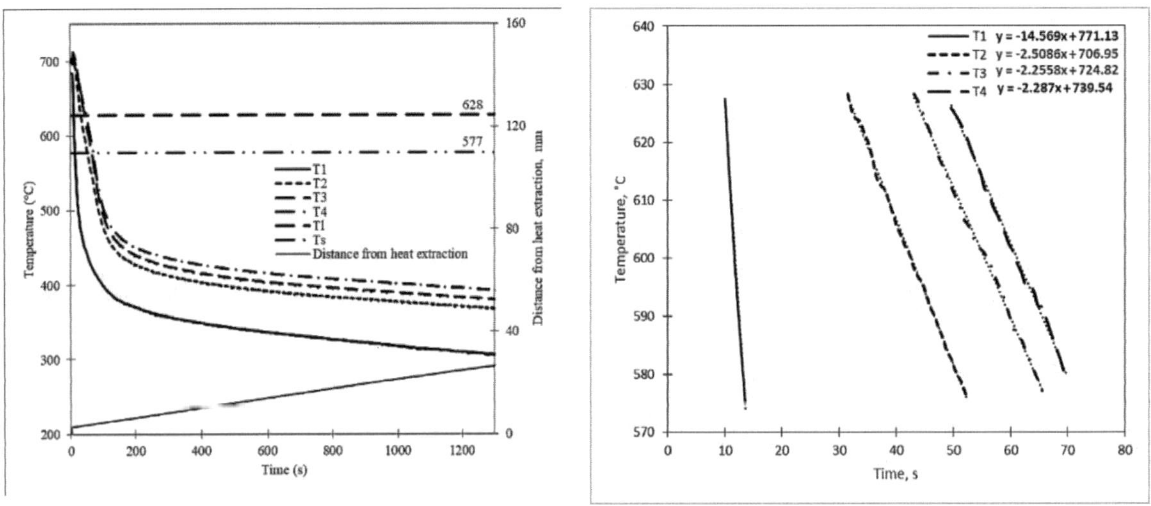

a. b.

Figure 2 a. Al– 6wt.% Si in clay mold cooling curve and distance from heat extraction, b. Cooling rate of Al–6wt.%Si in clay mold at various thermocouple positions

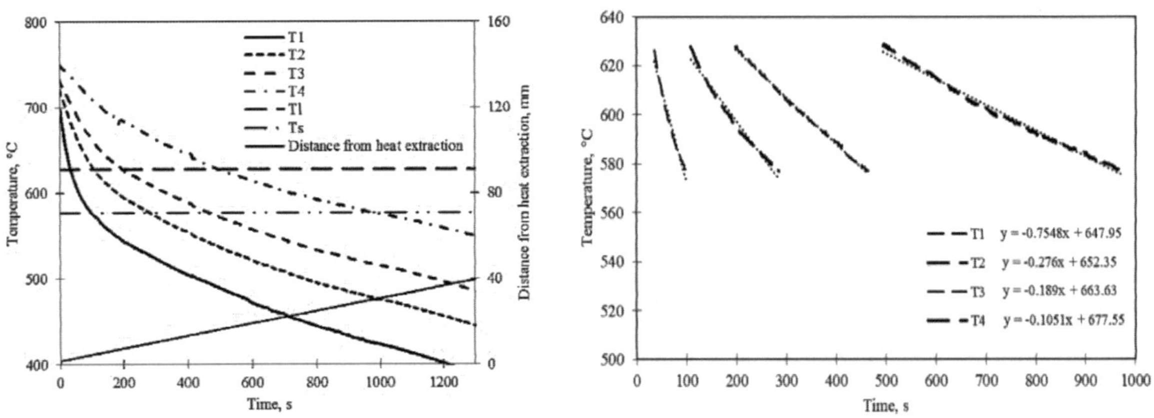

a. b.

Figure 3 a. Cooling curve of Al– 6wt.%Si with stainless steel mold, b. Cooling rates of Al– 6wt. %Si in stainless steel mold at various thermocouple positions

The slope of cooling curve of Al–6wt.%Si in stainless steel is 16 times higher in average than the slope of cooling curve of Al–6wt.%Si in clay mold. The solidification of Al–6wt.%Si in stainless steel mold occurs from 33.82s to 975.96s (Figure 3a). The major dropped of temperature is at T1 and the drop is slightly decrease as the thermocouple position away from heat extraction. The highest cooling rate is at T1 by 0.755°C/s and the lowest is at T4 by 0.105°C/s (Figure 3b).

The cooling rates are varies between each curve and it contributes to the different grain size and microstructure formation. High cooling rate at T1 in both cooling curve contribute to rapid solidification. Sudden drop of temperature does not provide allowable time for microstructures to grow and as a result fine microstructures are formed (Figure 4a, d). Coarse microstructures are found at lower cooling rate (T2, T3, and T4) (Figure 4 b, c, e, f). Clay mold isolates the heat from furnace to the samples thus the heat is extracted only at the bottom and less time is needed to solidify all the liquid. While samples in stainless steel mold obtain additional heat from the furnace, therefore, the heat is maintained and slow cooling is occurred. Increasing to nearly twice-thermal conductivity of mold material decreases solidification time about 15 times, reduces about 16 times the cooling curve slope and moves the initial position of cooling curve by 23.85s to the right.

As cooling rate increases on the one hand, it increases solidification rate and triggers the formation of fine microstructure and non-equilibrium eutectics, on the other hand it decreases solidification time and dendrite arms spacing. When solidification rate exceed the interfacial diffusion coefficient over atomic distance value, solute trapping occurs [27]. However as the cooling rate decreased, more segregation occurred, it is indicated by coarse eutectic silicon in stainless steel sample (Figure 5d, e, f).

The gaps between cooling curves are related to the temperature gradient. The highest gradient in both curve is occurred adjacent to the heat extraction and it is lessening as the distance from heat extraction increases (Figure 4a). The highest gradient is at 10 mm from heat extraction point, 7.68°C/mm for clay mold and 5.6°C/mm for stainless steel mold curve. For clay mold, it gradually decreases and hits the lowest point at 20 mm and raises to 3.9°C/mm at 40 mm (end of solidification) and for stainless steel mold, it drops to 3.08 at 20 mm and finally fells to a minimum value 2.78°C/s at 40 mm. Although the details are different both curve is showing the same trends, the temperature gradient gradually is decreases as the distance is increases.

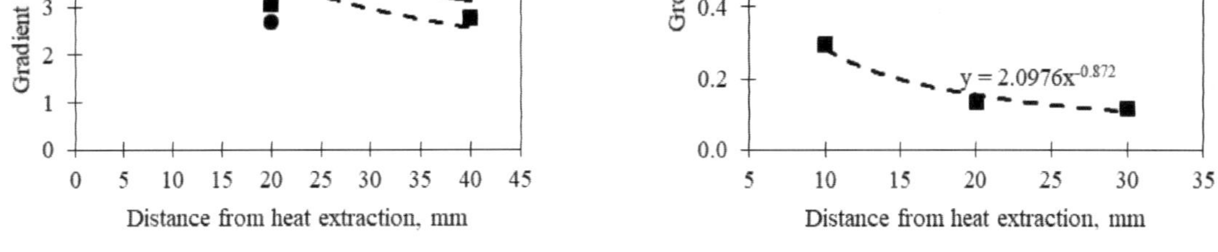

a. b.

Figure 4 a. Gradient temperature and b. growth rate of Al– 6wt. %Si at 10, 20, 40 mm in two different mold materials

The temperature gradient depends on the heat released at the bottom and the side mold. Liquid metal starts to solidify at the bottom of cylinder and a great number of heat is released.

IOP Conf. Series: Materials Science and Engineering **547** (2019) 012014 doi:10.1088/1757-899X/547/1/012014

When the clay mold wall resists the additional heat, the existing heat flow is mainly on vertical direction. This causes a significant difference between two measured points in vertical direction. Unlike the condition in the clay mold, stainless steel mold wall transfers heat from the furnace to liquid thus less temperature gradient is gained. Temperature gradient data parallels with growth rate data where the growth rate of Al–6wt. %Si in clay is higher about 0.958 mm/°C in average than the growth rate of Al–6wt. %Si in stainless steel (Figure 4b). The growth rate of clay and stainless steel are as follows 1.0 to 0.868 mm/°C and 0.296 to 0.117 mm/°C. High growth rate is confirmed by the formation of microstructures, fine columnar microstructures are found in clay mold samples. This microstructure is found at 10 to 40 mm and it getting thicker as the distance increased.

Thermal conductivity of mold material affects the shape cooling curves and solidification parameters. Increasing thermal conductivity from 12.75 W/m.K to 22 W/m.K increases cooling curve time range by increasing the total solidification time about 15 times and reduces the curve slopes about 1/16 times. To solidification parameters, it increases the cooling rate about 16 times however it reduces the gradient temperature and the growth rate about 0.958 mm/°C and about 0.595 mm/s respectively.

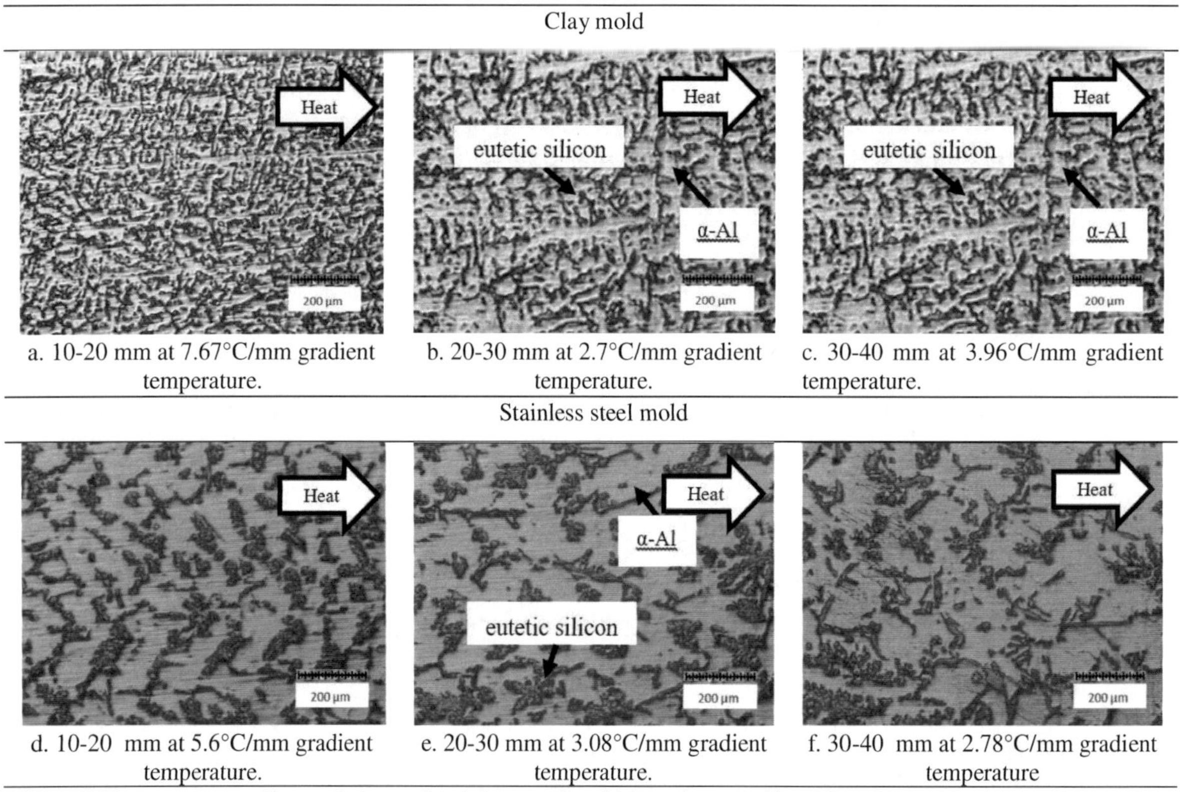

a. 10-20 mm at 7.67°C/mm gradient temperature.

b. 20-30 mm at 2.7°C/mm gradient temperature.

c. 30-40 mm at 3.96°C/mm gradient temperature.

d. 10-20 mm at 5.6°C/mm gradient temperature.

e. 20-30 mm at 3.08°C/mm gradient temperature.

f. 30-40 mm at 2.78°C/mm gradient temperature

Figure 5 Temperature gradients of clay and stainless steel mold in various positions from heat extraction and its microstructure.

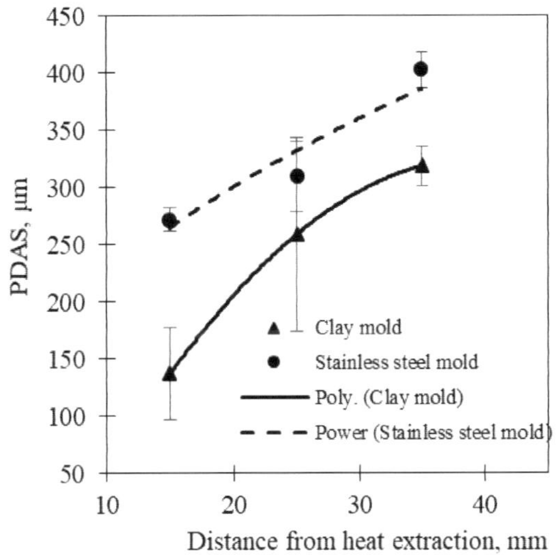

Figure 6 Primary Dendrite Arms Spacing of Al–6wt.%Si in clay and stainless steel mold

Primary Dendrite Arms Spacing (PDAS) data in both sample confirms well with solidification parameters data (Figure 6). The PDAS are reducing as the distance from heat extraction increases. It also can be observed that the PDAS of the sample in clay mold has less length than the sample in stainless steel mold, this result later will affect the hardness value.

3.2 Micro-hardness

Micro-hardness values of Al– 6wt.%Si in clay and stainless steel mold as a function of distance from heat extraction shows similar micro-hardness trend. The micro-hardness values decrease as the distance increase. Maximum values of micro-hardness are obtained from the position near to heat extraction point. This relationship agrees with the cooling rate, gradient temperature, and PDAS data. High cooling rate contributes to finer microstructures. As the position is further, finer branches disappear and thicker branches exist. However, the micro-hardness values of Al– 6wt.%Si in clay mold are lower than those in stainless steel mold. The micro-hardness values of Al– 6wt.%Si in clay mold and stainless steel are from 62.2 to 46.2 kgf/mm^2 and from 75.8 to 57.78 kg/mm^2 respectively. Complete segregation is occurred in stainless steel samples due to slow cooling and triggers the formation coarse silicon, while in the clay sample the alloying elements is trapped in α-Al. These coarse silicon microstructures have higher hardness value than α-Al hardness value and it is relatively in large amount in stainless steel samples. By applying power law method, the exponent values of clay and stainless steel mold are obtained, 0.113 and 0.037 respectively. Compared to the exponent values of similar mold material, stainless steel exponents value are lower than 0.069, the exponents value obtained by Araujo et al. [1].

Figure 8 shows the variation of micro-hardness as a function of growth rate for Al–6wt.%Si in clay and stainless mold. Both curves have similar trend, the values of micro-hardness increase as the growth rate increase. Micro-hardness and growth rate relationship are expressed in form of $HV = kV^b$. The exponent values related to growth rate of Al–6wt.%Si in clay and stainless steel were 0.2478 and 0.1334 respectively (Figure 8). The exponent value of Al–6wt.% Si in clay are higher than 0.07, the exponent value by Engin et al. [28] for Al–Ni–Fe alloys with graphite mold moreover the value are slightly under the given ranged 0.12–0.14 by Kaya et al. [22] for different binary alloy system in similar solidification condition. The exponent value of Al–6wt.%Si in stainless steel mold is smaller than 0.71, the exponent value of Al–6wt.%Si in carbon steel mold by Vasconcelos et al. [29].

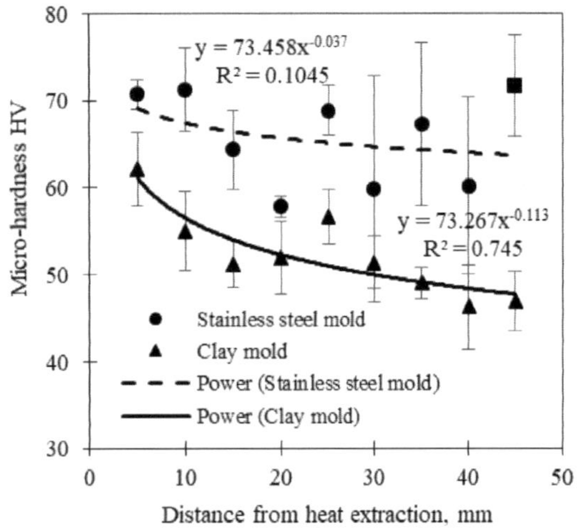

Figure 7 Microhardness of Al– 6wt.% Si as a function of position

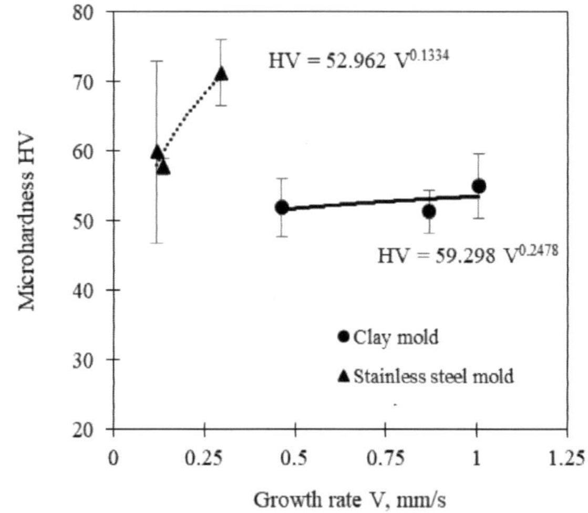

Figure 8 Microhardness of Al– 6wt.% Si as a function of growth rate

4. Conclusions

Experimental investigation of the influences of clay and stainless steel as mold material on cooling curve, solidification parameters, microstructures, and hardness in unidirectional is performed. As thermal conductivity increases by 72%:

1. The time range of the cooling curve increases about 15 times and the curve slope reduces about 1/16 times.
2. The cooling rate increases about 16 times and the gradient temperature and the growth rate reduce about 0.958 mm/°C and 0.595 mm/s respectively.
3. The PDAS reduces about 98.7 to 19.6%.
4. The micro-hardness value increases about 21.8-53.4%, the function of micro-hardness to distance exponent value reduces about 67.3%.
5. The exponent value of the micro-hardness to growth rate decreases about 46.2%.

5. Acknowledgements

This work is supported by Ministry of Research, Technology and Higher Education of the Republic of Indonesia and Indonesia Endowment Fund for Education (LPDP).

6. References

[1] Araújo E C, Barros A S, Kikuchi R H, Silva A P, Gonçalves F A, Moreira A L and Rocha O L 2017 The Role of Si and Cu Alloying Elements on the Dendritic Growth and Microhardness in Horizontally Solidified Binary and Multicomponent Aluminum-Based Alloys *Metall. Mater. Trans. A* **48** 1163–75

[2] Barros A S, Magno I A, Souza F A, Mota C A, Moreira A L, Silva M A and Rocha O L 2015 Measurements of microhardness during transient horizontal directional solidification of Al-Rich Al-Cu alloys: Effect of thermal parameters, primary dendrite arm spacing and Al_2Cu intermetallic phase *Met. Mater. Int.* **21** 429–39

[3] Brito C, Costa T A, Vida T A, Bertelli F, Cheung N, Spinelli J E and Garcia A 2015 Characterization of Dendritic Microstructure, Intermetallic Phases, and Hardness of Directionally Solidified Al-Mg and Al-Mg-Si Alloys *Metall. Mater. Trans. A* **46** 3342–55

[4] Verissimo N C, Brito C, Santos W L R, Cheung N, Spinelli J E and Garcia A 2016 Interconnection of Zn content, macrosegregation, dendritic growth, nature of intermetallics and hardness in directionally solidified Mg-Zn alloys *J. Alloys Compd.* **662** 1–10

[5] Çadırlı E, Büyük U, Engin S and Kaya H 2017 Effect of silicon content on microstructure, mechanical and electrical properties of the directionally solidified Al–based quaternary alloys *J. Alloys Compd.* **694** 471–9

[6] Çadirli E 2013 Effect of solidification parameters on mechanical properties of directionally solidified Al-Rich Al-Cu alloys *Met. Mater. Int.* **19** 411–22

[7] Çadirli E, Böyük U, Engin S, Kaya H, Maraşli N, Keşlioğlu K and Ülgen A 2010 Investigation of the effect of solidification processing parameters on the rod spacings and variation of microhardness with the rod spacing in the Sn-Cu hypereutectic alloy *J. Mater. Sci. Mater. Electron.* **21** 608–18

[8] Böyük U, Engin S and Maraşli N 2011 Microstructural characterization of unidirectional solidified eutectic Al-Si-Ni alloy *Mater. Charact.* **62** 844–51

[9] Engin S, Böyük U, Kaya H and Maraşli N 2011 Directional solidification and physical properties measurements of the zinc-aluminum eutectic alloy *Int. J. Miner. Metall. Mater.* **18** 659–64

[10] Engin S, Büyük U and Maraşli N 2016 The effects of microstructure and growth rate on microhardness, tensile strength, and electrical resistivity for directionally solidified Al-Ni-Fe alloys *J. Alloys Compd.* **660** 23–31

[11] Kaya H, Böyük U, Çadirli E and Maraşli N 2013 Influence of growth rate on microstructure, microhardness, and electrical resistivity of directionally solidified Al-7 wt% Ni hypo-eutectic alloy *Met. Mater. Int.* **19** 39–44

[12] Kaya H, Böyük U, Çadirli E and Maraşli N 2012 Measurements of the microhardness, electrical and thermal properties of the Al-Ni eutectic alloy *Mater. Des.* **34** 707–12

[13] Okayasu M and Takeuchi S 2015 Crystallization characteristics of cast aluminum alloys during a unidirectional solidification process *Mater. Sci. Eng. A* **633** 112–20

[14] Yilmaz E, Çadirli E, Acer E and Gündüz M 2016 Microstructural Evolution and Mechanical Properties in Directionally Solidified Sn-10.2 Sb Peritectic Alloy at a Constant Temperature Gradient *Mater. Res.* **19** 370–8

[15] Acer E, Çadırlı E, Erol H, Kırındı T and Gündüz M 2016 Effect of heat treatment on the microstructures and mechanical properties of Al-5.5Zn-2.5Mg alloy *Mater. Sci. Eng. A* 144–56

[16] Hazari N, Chatterjee D, Tigga D P and Das N 2015 Effect of Alloy Chemistry and Growth Rate on the Unidirectional Solidification behaviour of Single Crystal Superalloys *Trans. Indian Inst. Met.* **68** 1239–44

[17] Liu D, Li X, Su Y, Rettenmayr M, Guo J and Fu H 2014 Local melting/solidification during peritectic solidification in a steep temperature gradient: Analysis of a directionally solidified Al-25at%Ni *Appl. Phys. A Mater. Sci. Process.* **116** 1821–31

[18] Wang L, Shen J, Shang Z, Zhang J, Chen J and Fu H 2014 Effect of Dy on the microstructures of directionally solidified NiAl-Cr(Mo) hypereutectic alloy at different withdrawal rates *Intermetallics* **44** 44–54

IOP Conf. Series: Materials Science and Engineering **547** (2019) 012014 doi:10.1088/1757-899X/547/1/012014

[19] Yu J, Ren J, Li H, Fu J, Zhai Q, Luo Z and Zheng H 2014 A new approach to grow the Heusler Ni-Mn-Sn unidirectional crystal *J. Cryst. Growth* **402** 147–50
[20] Zhongwei C, Wang E and Hao X 2016 Microstructure and orientation evolution in unidirectional solidified Al-Zn alloys *Mater. Sci. Eng. A* **667** 1–8
[21] Holman J P 2005 Heat Transfer *Steam its Gener. use* 1–34
[22] Kaya H, Çadirli E, Böyük U and Maraşli N 2008 Variation of microindentation hardness with solidification and microstructure parameters in the Al based alloys *Appl. Surf. Sci.* **255** 3071–8
[23] Holman J P 1986 Heat Transfer *McGraw-Hill* 702
[24] Desrosin W, Boycho L, Scheiber V, Méndez C M, Schvezov C E and Ares A E 2015 Evolution of Metallographic Parameters during Horizontal Unidirectional Solidification of Zn-Sn Alloys *Procedia Mater. Sci.* **8** 968–77
[25] ASTM Int. 2012 ASTM E384: Standard Test Method for Knoop and Vickers Hardness of Materials *ASTM Stand.* 1–43
[26] Zolotorevsky V S, Belov N A and Glazoff M V. 2007 *Casting Aluminum Alloys - Knovel* (Elsevier)
[27] Stefanescu D M 2015 *Science and engineering of casting solidification, second edition* (Tuscaloosa: Springer Internastional Publishing)
[28] Engin S, Büyük U and Maraşli N 2016 The effects of microstructure and growth rate on microhardness, tensile strength, and electrical resistivity for directionally solidified Al-Ni-Fe alloys *J. Alloys Compd.* **660** 23–31
[29] Vasconcelos A de J, Silva C V A da, Moreira A L S, Silva M A P de S da and Rocha O F L da 2014 Influence of thermal parameters on the dendritic arm spacing and the microhardness of Al-5.5wt.%Sn alloy directionally solidified *Rem Rev. Esc. Minas* **67** 173–9

Wear Prediction of UHMWPE Cup Against Commercially Pure Titanium Hip Implant With The Nonlinear Load and Contact Area Wear Equation

Handoko[1]* Suyitno[1] Rini Dharmastiti[1] Rahadyan Magetsari[2]
[1]Department of Mechanical and Industrial Engineering, Faculty of Engineering, Universitas Gadjah Mada, Indonesia
[2]Department of Orthopedics and Traumatology, Sardjito General Hospital and Faculty of Medicine, Universitas Gadjah Mada, Indonesia

*Corresponding author (E-mail): handoko.dtm@ugm.ac.id

Abstract. Hip implants are prostheses designed for severe hip arthritis patients. One of its components, acetabular cup is mostly made from ultra-high molecular weight polyethylene (UHMWPE). This polymer has a wear problem. Its debris would cause osteolysis followed by implant failures. Numerical studies of this problem were commonly used a linear contact pressure wear equation. Recently it was found not suitable to model the polymer wear in hip implants. This study used a nonlinear equation to predict the UHMWPE volumetric wear. It states a nonlinear relation between wear, load and contact area. The computation model was a biaxial rocking motion hip simulator assembly with UHMWPE acetabular cup sliding on commercially pure titanium femoral head. Multidirectional pin on disc experiments conducted to obtain wear factor and coefficient of friction data for the computations. Commercial finite element software calculated the other parameters. Predicted volumetric wear has a good agreement with the experimental data. Lowest numerical error was found 0.53 %.

Keywords: wear prediction, hip implant, UHMWPE, cp Ti.

1. Introduction

Hip implants are lower limb prostheses useful to help patients regain their productivities. The products replace the hip joints of patients with severe arthritis or injuries. The materials of the implants must meet several critical requirements to function properly. It must be biocompatible, have moderate strengths to avoid stress shielding at the femur and wear resistant. One of the most popular biomaterials used is the metal on polymer (MOP) [1]. It consists of artificial metallic femoral head and polymer acetabular cup. Many products designed and used with sophisticated reliability for decades. This is now challenged by a trend of the increased number of younger patients undergo hip replacement surgeries. The products must be able to be in service in much longer time than before. Wear problem especially of the polymers arises. Studies shown a high volumetric wear rate of the mostly used polymer, ultra high molecular weight polyethylene (UHMWPE) leads to osteolysis and bone resorption [2, 3]. These in vivo mechanisms caused implant failures [4].

Attempts to improve wear resistant need data obtained from the tribological tests. Experimental setups to study the biomaterial wear behavior must be able to mimic the in vivo conditions. The obstacles are the cost and time duration [5]. It is costly to build simulators able to wear test biomaterials in tens or even hundreds of stations simultaneously. The cyclic frequency of the simulators is one Hertz [6] and it takes millions of cycles to obtain adequate wear data. There is another method available to shorten wear test time duration. It is the computer modeling and simulation. This process use a wear model describes the wear behavior of biomaterial pairs. One of the models mostly used is the Archard equation. The use of this equation needs a wear factor obtained experimentally and contact mechanic parameters computed numerically.

The purpose of this research is to predict volumetric wear of UHMWPE acetabular cup paired with cp Ti femoral head in the hip simulator. The reasons to choose cp Ti are the excellent biocompatibility, low modulus of elasticity compared to other metals and its high fracture toughness [7]. Unlike other previous works [8, 9, 10], this prediction used the nonlinear load and contact area wear model from [11]. This model adapts the Ratner-Lancaster relation. It states that wear rate is not linearly proportional with contact pressure as assumed in Archard equation. The nonlinear model predicts that wear rate is independent from the sliding distance. This model is unique and specially developed for the untreated UHMWPE materials multidirectional sliding in the hip simulator. Inputs for the model was the calculated contact mechanic parameters obtained with Abaqus commercial finite element software. Experimental verification from literature [12] conducted to compare the precision of both models.

2. Materials and Methods

Realistic wear simulations need wear factors for specific tribological conditions. The process to determine the factor usually requires various experimental wear tests. This study uses the same process for cp Ti - UHMWPE pairs designed as hip implant biomaterials. The wear tests were in a multi directional sliding mode as experienced by hip implants [11]. The tests used a three stations multidirectional pin on disc tribotester. UHMWPE pins have dimensions of 9 mm diameter and 17 mm length. The counter faces are 50 mm thick cast cp Ti discs. The lubricant is a simulated body fluid (SBF) solution made from 25 % v/v bovine serum, distilled water and 0.03 % wt sodium azide [13, 14, 15, 16]. Continuous tests ran with 2 MPa and 3 MPa contact pressures and a 1 Hz cyclic sliding for one million cycles. A temperature controller managed the heaters to maintain the lubricant from 37 to 39 centigrade. The purpose is to mimic the in vivo human body fluid. Wear tests were in the incremental steps of 250,000 cycles. The intention was to avoid unrealistic wear rate due to protein adsorption at the surface of the discs [17]. Specimens were dismounted, cleaned and weighted at each incremental step. Wear factor calculations for the nonlinear contact model used the equation below [11]:

$$K_1 = \frac{\Delta V}{W^{2/3}\, A^{1/3}\, N} \qquad (1)$$

with ΔV, W, A and N are the volumetric loss, applied load, contact area and number of cycles respectively. Equation (1) is different compared with the Archard equation which wear factor is:

$$K_2 = \frac{\Delta V}{W\, L\, N} \qquad (2)$$

with L is the sliding distance. The next obtained experimental data was the coefficient of friction. It is important for the main input of the explicit dynamic contact mechanic modelling and simulations.

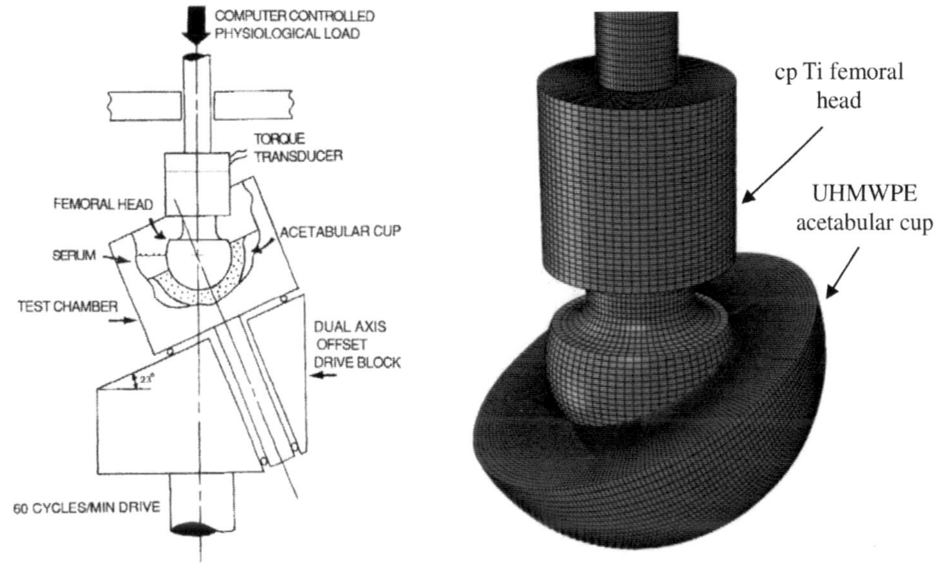

(a) Experiment schematic [12] (b) CAE model

Figure 1. The bi-axial rocking motion hip simulator model with 32 mm femoral head diameter.

The computation processes adopted the UHMWPE and cp Ti models with material properties obtained from literatures. Modulus of elasticity between the two materials differs much, hence the setup were rigid body for cp Ti and deformable for the softer UHMWPE. Other parameters such as time steps, loads and boundary conditions were carefully set similar to the experimental hip simulator setup of [12]. Computations were then conducted with Abaqus to simulate the dynamics of a biaxial rocking motion hip simulator (Fig 1). A dynamic explicit method were used to take inertial effects into account. The workstation was a four thread 3.9 Ghz CPU and 12 GB random access memory. The numerical results were used as the inputs to compute volumetric wear loss. A custom Python script developed to handle those calculations. On the linear contact pressure wear calculations (Equation 2), linear wear depths must be determined. It is useful for the UHMWPE cup geometry modifications. Geometry update at every 250,000 cycles performed for contact nodes coordinates at the worn area [18]. This repeated process performed until the desired number of cycles achieved. Predicted UHMWPE wear volume is an accumulation of calculated wear at every nodes and time steps. This volumetric wear is usually has a linear trend as experimentally shown by [14] [19, 20, 21]. In the nonlinear contact load model, wear volume was predicted only with the wear parameters at the peak load. There are two reasons for this approach. First, contact area cannot be accumulated from every nodes and every time steps just like the sliding distance on the linear model. Initial attempts in this research to calculate wear volume based on those accumulated contact area found errors up to several hundred percent compared to the experimental data. Second, the nonlinear model was developed by [11] with statistic regression which based on the maximum contact load (W_{max}). Hence, the nonlinear model offers a simple method to predict wear volume without repetitive tasks. There is a limitation to use this method. It is unable to predict the wear depth incrementally.

3. Results and Discussion

Wang et al [11] proposed a new model to calculate the UHMWPE volumetric wear rate per cycle sliding against a hard smooth counter face. The foundation of this model is a postulate that wear is proportional to the frictional work. Other findings by [22] [23] supported the development of the model as described by equation (1). Experimental verification conducted with wear tests of UHMWPE acetabular cup in a BRM hip joint simulator. The simulator provides a cross shear motion as occurred in the human hip joints. Unlike the commonly used Archard model, the data from [11] proved that the wear rate per cycle of UHMWPE is independent of sliding distance. Hence, this model is an update or at least as an alternative for the hip implant biomaterials wear prediction and computation.

Initial work of this study was the model mesh sensitivity analysis. The femoral head model used a 1 mm element size hexahedral mesh. As a rigid body model, it was not subject to mesh variations. Acetabular cup model as the object of wear analysis was tested by element type and element size variations. The tested element types are linear hexahedral (HL) and quadratic tetrahedral (TQ). Element size varied from a 1.5 mm coarse mesh up to 0.3 mm fine mesh. Single bias was applied at all radial direction from the corner close to the contact surfaces. Based on two contact mechanics parameters i.e. maximum contact pressure and maximum von Mises stress, the numerical convergence limit for both hexahedral and tetrahedral elements cannot be found (Fig. 2). The exponential trend of TQ element shows that element size less than 0.5 mm would cause severe computational costs.

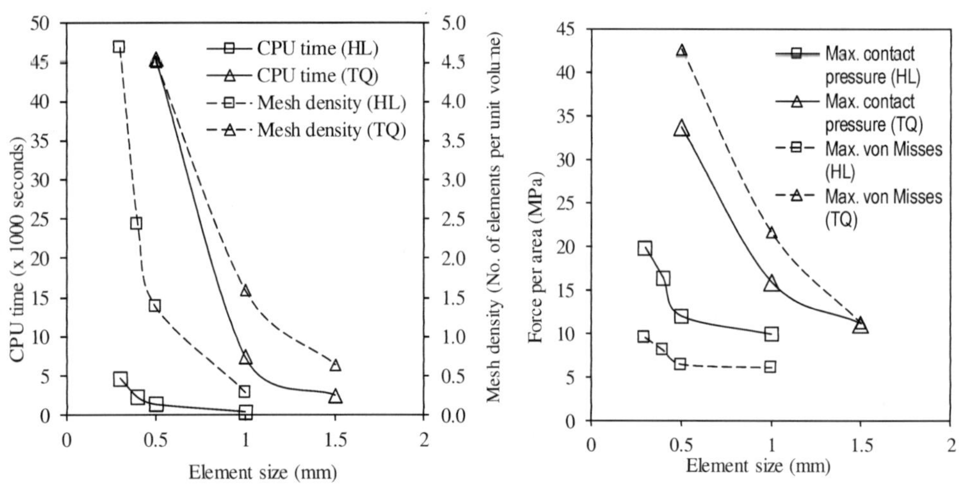

Figure 2. CPU time and mesh density (left), maximum contact pressure and maximum von Mises stress (right) as the function of element size.

Evaluation of the results taken from various mesh variations was continued with the gait load cycle (Fig. 3). Three models with 0.5 mm and 0.3 mm hexahedral and 0.5 mm tetrahedral elements were selected to be compared with applied load (Fig. 4). All element types show a similar trend, meshes must be smaller to obtain calculated load closer to the applied load. Hexahedral element perform better with curves closer to the Paul physiological load than the tetrahedral. The drawback of hexahedral element is the need to use finer element to approach the 1.96 kN maximum load. In this case is 0.5 mm or lower (Fig. 4(a)) while the tetrahedral can do the same with the 1 mm element size (Fig. 4(b)). Because equation (1) depends on both maximum load and contact area, this calculated load assessments cannot yet concluded the suitable mesh to accurately predict UHMWPE acetabular cup wear volume.

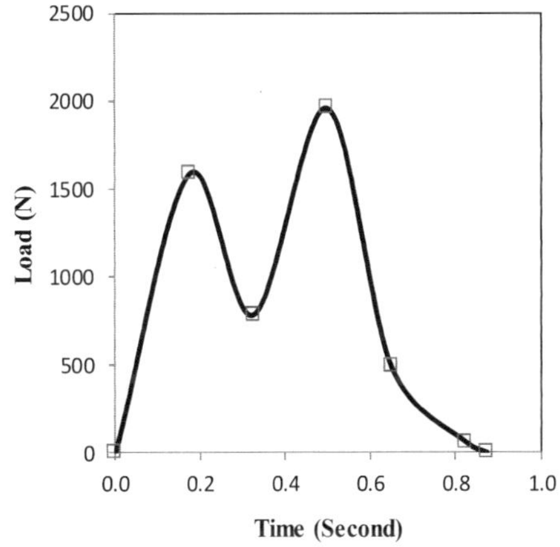

Figure 3. Load for the CAE model based on the Paul physiological load [12].

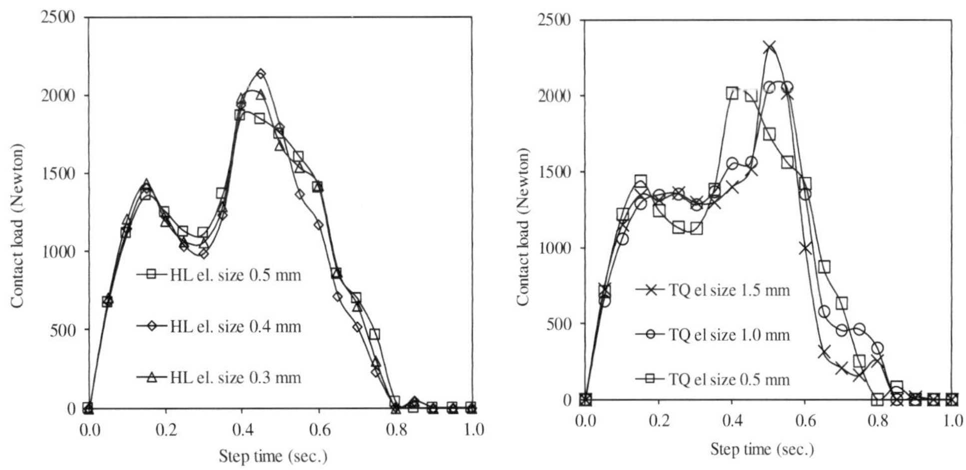

(a) Contact load by HL elements. (b) Contact load by TQ elements.

Figure 4. Calculated load at the surface of UHMWPE acetabular cup.

The next evaluation is to check whether the calculated contact pressures are appropriate or not (Fig. 5 (a)). The Paul physiological load used by [12] has a maximum pressure approximately 15 MPa applied to the hip implant specimens. Due to the dynamic of the system which rotate at one Hz and the friction between the two materials, calculated pressure magnitudes could differ from the load. Finer meshes with smaller element size tend have higher calculated contact pressures. Data from the 0.5 mm tetrahedral element has very high unrealistic contact pressures compared to the contact load. The tetrahedral element has another drawback, its computational cost is higher than the hexahedral. This is why the tests of tetrahedral elements were stopped at 0.5 mm element size (Fig. 2). The closer curves to the load were obtained from the 0.3 mm and 0.5 mm hexahedral elements. The next numerical outputs related to contact pressure is the calculated von Mises stress (Fig. 5(b)). Similar trends with contact pressure were found. Tetrahedral element predicted too high stress magnitudes, more than 40 MPa. This result is unrealistic. Wear of the polymer could be catastrophic. The fatigue failure limit of typical UHMWPE is 32 MPa. From these checks, it can be concluded that the tetrahedral element is not suitable for this wear modelling and computation.

IOP Conf. Series: Materials Science and Engineering **547** (2019) 012015 doi:10.1088/1757-899X/547/1/012015

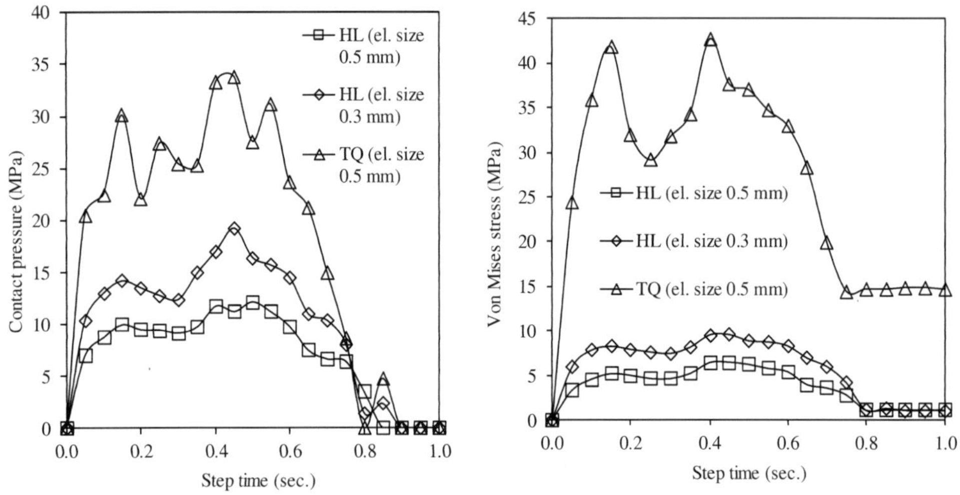

(a) Calculated contact pressures. (b) Computed von Mises stress.

Figure 5. Predicted UHMWPE cup contact pressures and von Mises stresses.

The von Mises curves of all fine hexahedral meshes show low stress values below the elastic plastic limit of UHMWPE. These are indications that the polymer material is safe and strong enough to withstand the physiological load.

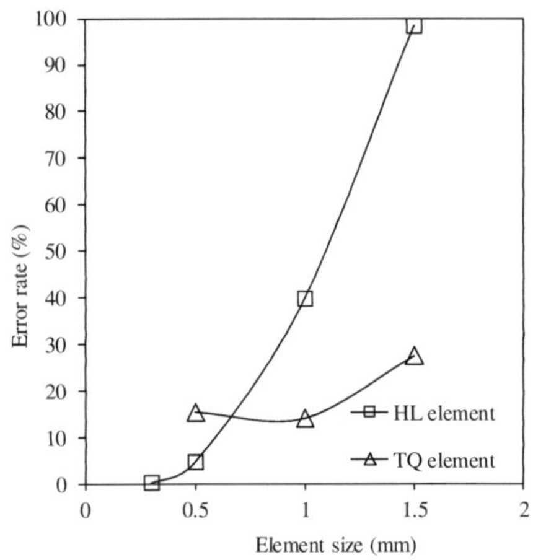

Figure 6. Errors at various element type and size.

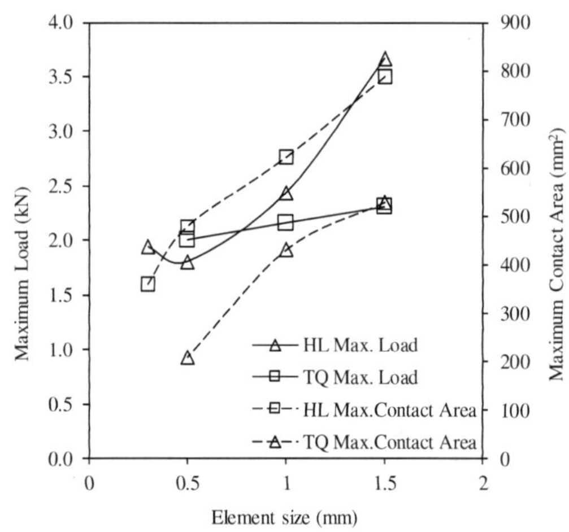

Figure 7. Maximum load and contact area at various element types.

Wear volume calculation based on equation (1) conducted on the UHMWPE acetabular cup model. The dynamic modelling needed two parameters, coefficient of friction and wear factor. Those two have to be obtained experimentally. The pin on disc wear tests delivered a coefficient of friction data equal to 0.2. Wear factors were also obtained from the tests after a million cycles. The values are $K_I = 4.02 \times 10^{-8}$ mm$^{(5/3)}$ N$^{-(2/3)}$ for 2 MPa and $K_I = 8.02 \times 10^{-8}$ mm$^{(5/3)}$ N$^{-(2/3)}$ for 3 MPa contact pressures. Wear calculation with the K_I (2 MPa) constant values were not adequate. Experimental validation with volumetric wear data from [12] at 90.7 mm^3 per one million cycles found errors up to 51% (hexahedral) and 58.45% (tetrahedral). The next was the wear calculations with K_I (3 MPa) as shown in Fig. 6. Initially the error is high on coarse meshes

IOP Conf. Series: Materials Science and Engineering **547** (2019) 012015 doi:10.1088/1757-899X/547/1/012015

(1.5 and 1 mm) but then drop drastically. At the 0.5 mm element size, acceptable error was found 4.31%. The lowest value is 0.53% obtained from the 0.3 mm element size. Meanwhile, the tetrahedral offers lower error rate with coarse element size but it cannot drop to the acceptable lower rate.

This result demonstrated that polymer wear volume can be predicted with a nonlinear contact load and contact area equation [11] and a constant wear factor (K_1). This method is relatively fast and easier but recently limited only to the wear volume. The use of Archard linear contact pressure equation is much complicated. It must be optimized with a series of wear factor data at various contact pressure ($K_2=f(P)$) as discussed by [8]. In the new model, two main calculated contact mechanics parameters needed, the maximum load and contact area. These parameters are dependent to the element size (Fig 7). Various size should be tested until the maximum load is close to the experimental, in this case equal to 1.96 kN according to [12]. Plotted curves between the maximum loads and contacts area could reveal the contact mechanic situations (Fig 8). Based on those contact areas and the acetabular cup radius, displacements (h) can be geometrically calculated. The results were plotted as a function of load (W) as shown in Fig 9. The power relations $y = ax^b$ have b exponents higher than the parabolic frictionless Hertzian exponent of 1.5 [24]. These metal on polymer contacts were less stiff than Hertzian contact.

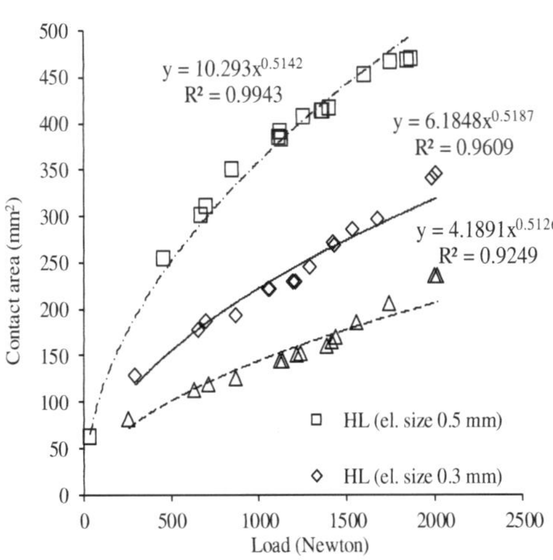

Figure 8. Load and contact area relation.

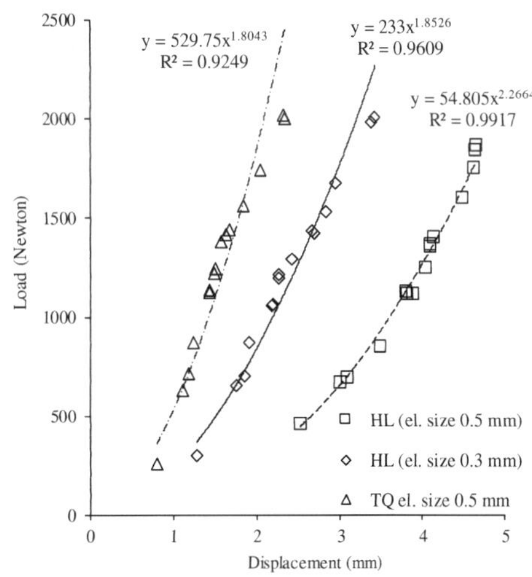

Figure 9. Load and displacement relation.

4. Conclusions

Wear volume prediction for hip implants made from a metal on polymer biomaterial pair of cp Ti and UHMWPE can utilize a nonlinear contact pressure model. The process is fast and simple. Tests conducted on the 32 mm hip implant models show that the hexahedral element is suitable for the contact mechanics calculations. Minimal error was found 0.53% with 0.3 mm element size validated with experimental data.

5. Acknowledgements

This research was supported by Kementerian Riset, Teknologi, dan Pendidikan Tinggi, Republic of Indonesia. Authors would like to thank to the Faculty of Veterinary and Faculty of Pharmacy, Universitas Gadjah Mada, for the processing of simulated body fluid and the provided precision analytical balance.

IOP Conf. Series: Materials Science and Engineering **547** (2019) 012015 doi:10.1088/1757-899X/547/1/012015

6. References

[1] Bozic KJ, Lau EC, Ong KL, Vail TP, Rubash HE, Berry D 2012 Comparative effectiveness of metal-on-metal and metal-on-polyethylene bearings in medicare total hip arthroplasty patients *The Journal of Arthroplasty* **27**(8) p 37-40

[2] Kurtz SM 2009 *UHMWPE Biomaterials Handbook* 2nd ed Academic Press

[3] Shanbhag A, Rubash HE, Jacobs JJ 2006 *Joint Replacement and Bone Resorption* New York Taylor & Francis

[4] Green G, Khan M, Haddad FS 2014 Why do total hip replacements fail ? *Orthopaedics and Trauma* **29** (2) p 79-85

[5] Kang KT, Son J, Kim HJ, Baek C, Kwon OR, Koh YG 2017 Wear predictions for UHMWPE material with various surface properties used on the femoral component in total knee arthroplasty *J Materials Science Materials in Medicine* **26** (7) p 105

[6] Cheng G, Shan X 2012 Dynamics analysis of a parallel hip joint simulator with four degree of freedoms *Nonlinear Dynamics* p 2475-2486

[7] Joshi VA 2006 *Titanium alloys An atlas of structures and fracture features* Taylor & Francis

[8] Mattei L, Puccio FD, Ciulli E 2013 A comparative study of wear laws for soft-on-hard hip implants using a mathematical wear model *Tribology International* **63** p 66-77

[9] Fialho JC, Fernandes PR, Eca L, Folgado J 2007 Computational hip joint simulator for wear and heat generation *J Biomechanics* **40** p 2358–2366

[10] Teoh S, Chan W, Thampuran R 2002 An elasto-plastic finite element model for polyethylene wear in total hip arthroplasty *J Biomechanics* **35** p 323-330

[11] Wang A, Lee R, Herrera L, Korduba L 2013 Modeling and verification of ultra-high molecular weight polyethylene wear in multi-directional sliding *Wear* **301** (1-2) p 162-167

[12] Röstlund T, Albrektsson B, Albrektsson T, McKellop H 1989 Wear of ion-implanted pure titanium against UHMWPE biomaterials *Biomaterials* **10** p 176-80

[13] Xiong D, Yang Y, Deng Y 2013 Bio-tribological properties of UHMWPE against surface modified titanium alloy *Surface and Coatings Technology* **228** (Supplement 1) p S442-S445

[14] Abdelgaied A, Brockett CL, Liu F, Jennings LM, Fisher J, Jin Z 2012 Quantification of the effect of crossshear and applied nominal contact pressure on the wear of moderately cross-linked polyethylene *In Proc IMechE Part H J Engineering in Medicine* 227 (1) p 18-26

[15] Gee S, Kang X, Zhao Y 2011 One-year biodegradation study of UHMWPE as artificial joint materials: Variation of chemical structure and effect on friction and wear behavior *Wear* **271** (9-10) p 2354-2363

[16] Leslie LJ, Williams S, Isaac G, Ingham E, Fisher J 2009 High cup angle and microseparation increase the wear of hip surface replacements *Clinical Orthopaedics and Related Research* **467** p 2259–2265

[17] Wang A, Essner A, Polineni V, Stark C, Dumbleton J 1998 Lubrication and wear of ultra-high molecular weight polyethylene in total joint replacements *Tribology International* **31** p 17-33

[18] Liu F, Galvin A, Jin Z, Fisher J 2011 A new formulation for the prediction of polyethylene wear in artificial hip joints *Proc IMechE 225 Part H*

[19] González-Mora VA, Hoffmann M, Stroosnijder R, Gil FJ 2009 Wear tests in a hip joint simulator of different CoCrMo counterfaces on UHMWPE *Materials Science and Engineering C* **29** (1) p 153-158

[20] Dressler MR, Strickland MA, Taylor M, Render TD, Ernsberger CN 2011 Predicting wear of UHMWPE decreasing wear rate following a change in direction *Wear* **271** (11-12) p 2879-2883
[21] Trommer RM, Maru MM, Filho WLO, Nykanen VPS, 2015 Multi-scale evaluation of wear in UHMWPE-metal hip implants tested in a hip joint simulator *Biotribology* **4** p 1-11
[22] Wang A, Essner A, Klein R 2001 Effect of contact stress on friction and wear of ultra-high molecular weight polyethylene in total hip replacement *In Proceedings of the Institute of Mechanical Engineers, Part H Journal of Engineering in Medicine* 215 p 133
[23] Wang A, Lee R, Herrera L, Korduba L 2013 Wear of ultra-high molecular weight polyethylene moving along a circular path in a hip simulator *Wear* **301** (1-2) p 157-161
[24] Flores, P., Lankarani, HM 2016 *Contact Force Models for Multibody Dynamics* Springer.

Influence of Tack Weld on Physical and Mechanical Properties of MIG Welding of AA5083H116

Mudjijana, M.N. Ilman, P.T. Iswanto
Department of Mechanical and Industrial Engineering
Gadjah Mada University

E-mail: mudjijana@ugm.ac.id

Abstract. The research objective is to determine the effect of tack weld on physical and mechanical properties of tandem-MIG (T-MIG) welding of AA5083H116 materials. A Tenjima 200S MIG machine, sample dimension of 300mm × 75mm × 3mm, and 0.8mm diameter ER5356 electrode were used. The distance between electrodes of T-MIG was varied of 18, 27, 36mm. Welding speed of 12.5mm/s, and welding process was shield using argon gas. First, T-MIG welding were employed for untacked and tacked welding. The leading current was 125A and trailing current was 120A, but current wave of each T-MIG welding machine was not taken into account. The welding voltage was set the same. Second, parameters resulting in better physical and mechanical properties were then employed for single-MIG (S-MIG) tack weld using 8, 10, 12mm/s welding speeds. The result shows that T-MIG tacked weld produced in bowing distortion and T-MIG untack weld in downward bending distortion. Higher peak temperature resulted in larger distortion. T-MIG tacked weld results in welding efficiency 76% being similar with that of bending strength (75%) in comparison with that of as received material. T-MIG tack weld results in better penetration and 37% shorter welding time in comparison with S-MIG tack weld.

Key words: T-MIG welding, tacked weld, AA5083H116, ER5356, physical properties, mechanical properties

1. Introduction

Aluminum alloys have widely been used in marine transportation. Ship-building industries currently concern with reducing the manufacturing time by employing tandem-MIG welding in manufacturing process. AA 5083H116 mainly containing Al - 4,5% Mg-0,15% Cr-0,7% Mn possessing density of 2,66 gr/cm^3 [1], lighter but with mechanical properties comparable to A36 steel (density of 7,85 gr/cm^3), and after being H116 treated being corrosion resistant and ease of welding has been extensively used [2]. Considering these, AA5083H116 has been used in ship building, especially for small size ships replacing A36 steel. Welding is the main process in ship manufacturing. Tandem-MIG welding [3] of Q690 steel can increase the welding speed, deposition rate, and welding quality. In tandem welding (two feeding units and two power sources), the wires may have different potentials, and the welding parameters can be set freely for each wire [4]. Numerical and experimental studies of double-wire gas metal arc welding (DW-GMAW) in plasma environment [5] show that welding defects such as undercut and humping is less in double wire-GMAW than those in single wire-GMAW. In comparison with single-MIG, tandem-MIG can refined microstructure in both weld and heat affected zone, narrowing HAZ and improving joint strength. In addition, energy efficiency and welding cost can also be saved [6]. Partially melted zone (PMZ) is an important zone of aluminum alloy welds, because PMZ in these materials is weak link in the weldments and significantly affected by welding parameters. Microstructure changes in PMZ are related not only to welding input, but also depend on the thermal history of alloy [7]. The tensile strength of a sound welded joint with proper preheating reached 90% of that of the base metal. The efficiency of MIG welding using aluminum electrode can reach up to 0.73 - 0.91 [8]. A study on the influence of clamping on welding distortion [9] of 1 mm thick DP600 overlap joint and 6 mm thick S355 T-joint

IOP Conf. Series: Materials Science and Engineering **547** (2019) 012016 doi:10.1088/1757-899X/547/1/012016

showed that the final residual stresses and distortions depend strongly on the clamping conditions. Based on the tack welding modeling of single welding processes like in real fabrication conditions results in a final distortion about 23% higher than the case of neglecting tack welding [10]. Distortion as weld defect due to extensive heat input causes the weld metal experiences sudden shrinking due to uneven thermal expansion. MIG welding of aluminum alloy plates results in relatively larger distortion [11]. In order to overcome this problem, welding of plates are commonly clamped to restrain the workpiece, increase welding speed, use smaller current, and applying tack weld. Using clamping device can only avoid distortion during welding but not for weld result. Increasing weld speed should be combined with the increase of weld current in order to reach melting point of the weld metal, such that it required more energy and frequently result in uncomplete weld pool due to over speed and the weld metal has not reach root section. Effects of phase difference on the behavior of arc and weld pool in tandem pulsed GMAW of mild steel has been investigated [12]. By increasing the phase difference, the height of bulged molten metal between the two wires decreases, the weld appearance becomes worse with decreased weld width and increased weld penetration and reinforcement. Untack and tack T-MIG weld of AA5083H116 materials have not been studied. The objective of this research is to determine the effect of tack-weld on the physical and mechanical properties of T-MIG welding of AA5083H116 materials.

2. Materials and Methods

2.1 Welding Conditions

In order to obtain consistent result, semi-automatic Tandem-MIG welding using ER 5356 electrode and AA5083H116 material containing composition a presented in Table 1 were used. Two pieces of AA5083H116 plate of 300 mm × 75 mm × 3 mm was welded using butt joint along 300 mm transverse to the rolling direction. Two Tenjima MIG 200S machines and argon were used. During the welding, the distortion at the outer side of the plate was measured using dial indicator and recorded using a video camera. The detail of the welding parameters of the T-MIG untack weld and tack weld, single MIG tack weld have been presented in Table 2, and the welding process illustrated in Fig. 1. The current wave of each welding machine was not taken into account in this welding process, and the voltage was set the same, and leading current of MIG_1 was 125 A and the trailing current MIG_2 was 120 A [13].

Table 1. Chemical compositions

Elements (wt%)	Al	Mg	Mn	Fe	Si	Cr	Ti	Cu	Zn	Ni
AA5083H116[1)]	94.31	4.50	0.57	0.29	0.16	0.088	0.027	0.022	0.021	0.008
ER5356[2)]	93.05	5.50	0.20	0.40	0.25	0.20	0.20	0.10	0.10	-

[1)] Alcoa certificate, [2)] ASME (2001)

Table 2. Welding parameter

Parameter	MIG_1	MIG_2	Single-MIG tack weld
	Tandem-MIG Untack weld and tack weld		
Electrode distance (d)	18, 27, 36 mm		
Electrode distance to specimen (h)	10 mm	10 mm	10 mm
Welding speed (S)	12.5 mm/s	12.5 mm/s	8, 10, 12 mm/s
Welding angle (θ)	80°	80°	80°
Mean welding voltage (V)	19 V	19 V	19V
Mean welding current (I)	125 A	120 A	120 A
Filler rate	27 mm/s	27 mm/s	27 mm/s
Filler diameter	0,8 mm	0,8 mm	0.8 mm

Argon *flow*	15 liter/minute	15 liter/ minute	15 liter/minute

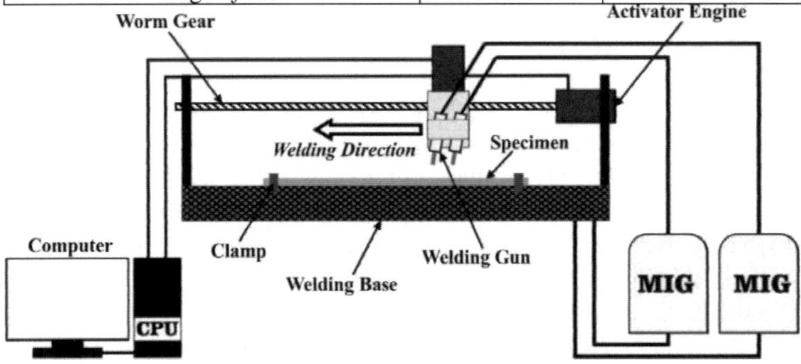

Fig. 1 Tandem-MIG welding process

2.2 Thermal cycles

Thermal cycle measurement was carried out by installing four type-K wire thermocouples as illustrated in Fig. 2. The data being obtained were then analyzed using an NI USB-9162 hardware from *National Instrument* and the result was displayed using the *Signal Express* 2010 software and further calculated using the *Microsoft Excel* software to produce thermal cycle curves.

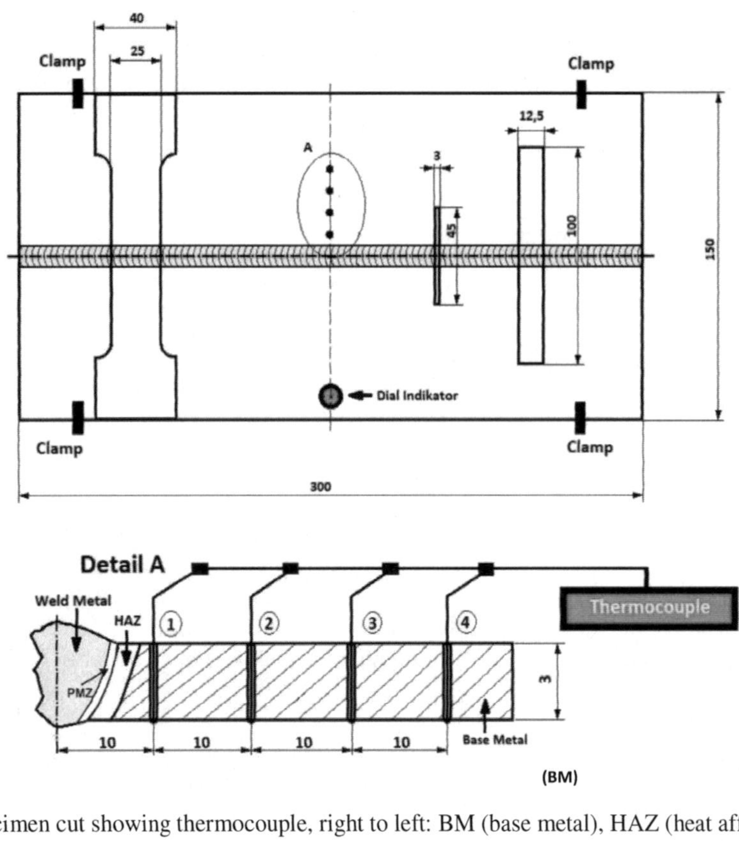

Fig 2. Specimen cut showing thermocouple, right to left: BM (base metal), HAZ (heat affected zone), PMZ (partially melted zone), weld metal (WM), and dial indicator location at welding specimen.

2.3 Distortion Patterns

The distortion of the welding results were measured along x-axis, *i.e.* along the welding line, on a milling machine table in z-axis direction, *i.e.* perpendicular to weld surface, using a dial indicator possessing an accuracy of 0.001 mm attached to a *magnetic base stand.* The measurement was done every 10 mm long of las deposit. Along y-axis the measurement were done every 12.5 mm as presented in Fig. 3. The data being obtained were then calculated using Microsoft excel software for 3D surface plotting.

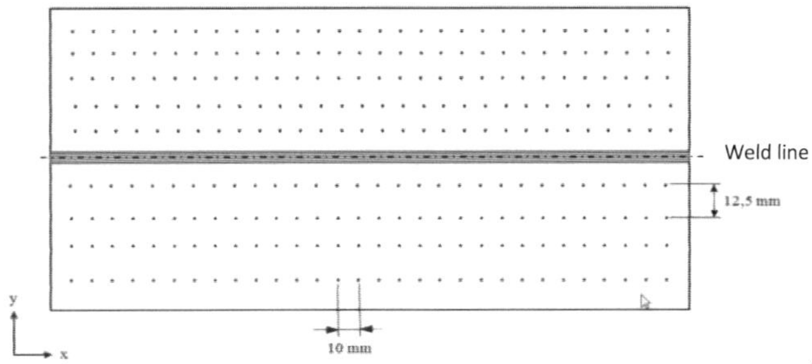

Fig. 3 Distortion points measurement

2.4 Macro and microstructure, Vickers hardness, Tensile, and bending test

Tensile and bending test were carried out using a 20-T capacity Servopulser machine and Torse universal testing machine. Observation of macro and microstructures were carried out by means of an optical microscope (OM), and a scanning electron microscope (SEM). The points being observed were the weld metal (WM), partially melted zone (PMZ), and heat affected zone (HAZ) as shown in Fig. 2. Micro Vickers hardness test was done using a 100-g load after the AA5083H116 specimen were welded using untack and tack T-MIG weld, and S-MIG tack welding process using ER5356 electrode.

3. Results and Discussion

3.1 Temperature distributions

The best weld result was obtained when at electrode distance being 18 mm of the untack weld and tack weld T-MIG, and at 8 mm/s welding speed of the Single-MIG as has been illustrated in Fig. 4. The distribution of peak temperature has also been illustrated in Fig. 5. The highest peak temperature was obtained at 328.8^{0}C for the T-MIG tack weld, and 282.8^{0}C for the T-MIG untack weld, where 46^{0}C difference were noticed where the heat input was from both leading and trailing torches. The amount of heat input can be calculated using $q = \eta\frac{VI}{v}$ where η is welding efficiency, *i.e.* tensile strength of welded divided with that of the as received material, V is voltage in Volt, I is the current in A, and v is welding speed in mm/s. The welding parameters were set the same for both tack weld and untack weld T-MIG, but resulted in 46^{0}C peak temperature difference. This difference may be caused by the current setting was being manual and current fluctuation of the MIG welding cannot be avoided due to the melting process at the electrode tips then the drop of melted weld deposit continuously happened during the welding process.

IOP Conf. Series: Materials Science and Engineering **547** (2019) 012016 doi:10.1088/1757-899X/547/1/012016

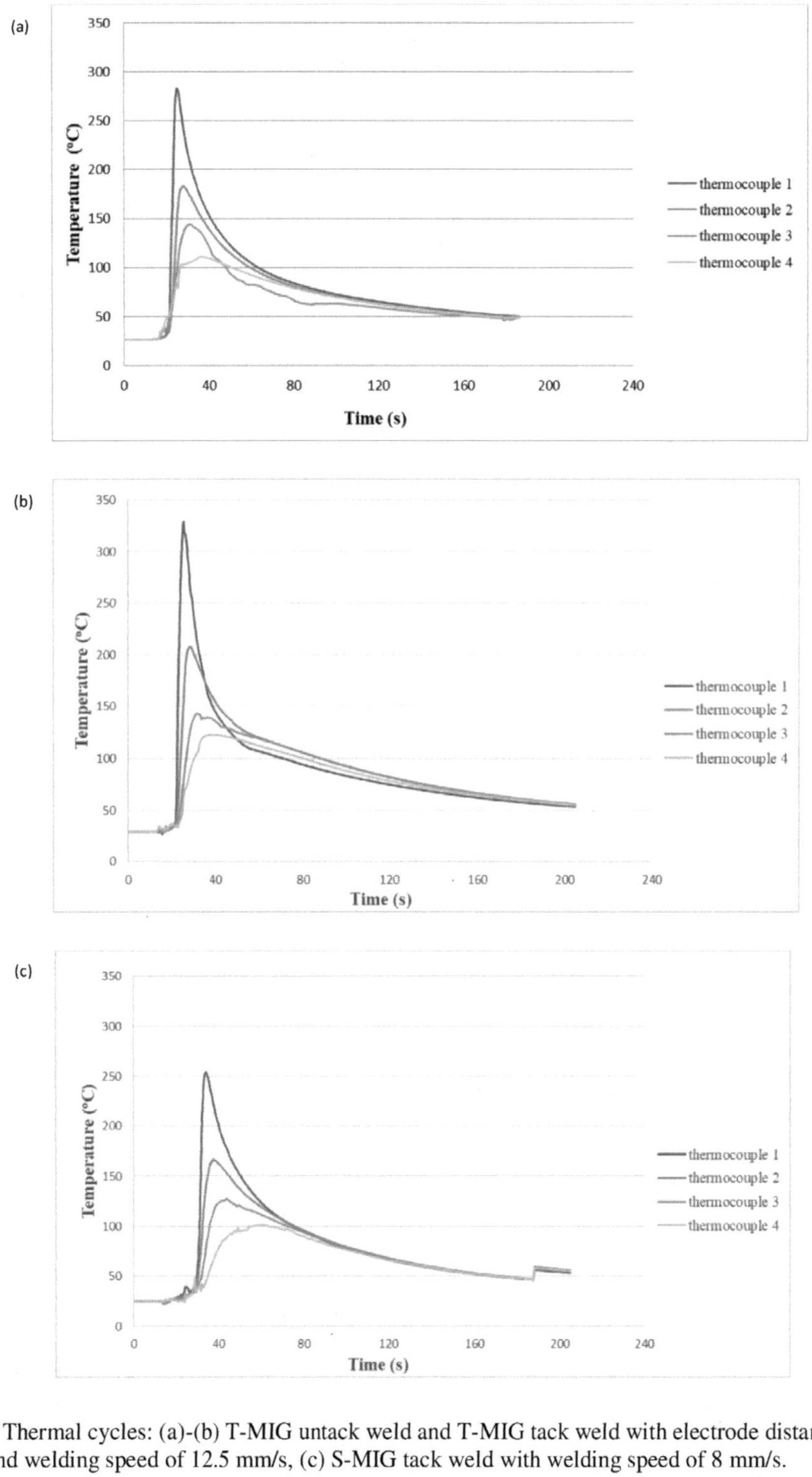

Fig. 4 Thermal cycles: (a)-(b) T-MIG untack weld and T-MIG tack weld with electrode distance of 18 mm and welding speed of 12.5 mm/s, (c) S-MIG tack weld with welding speed of 8 mm/s.

Table 2. Peak temperatures

Welding methods	Thermocouple temperatures (^{0}C)			
	1	2	3	4
T-MIG untack weld	282.8	183.1	143.8	111.0
T-MIG tack weld	328.8	207.9	143.3	122.4
S-MIG tack weld	254.0	166.6	128.0	101.8

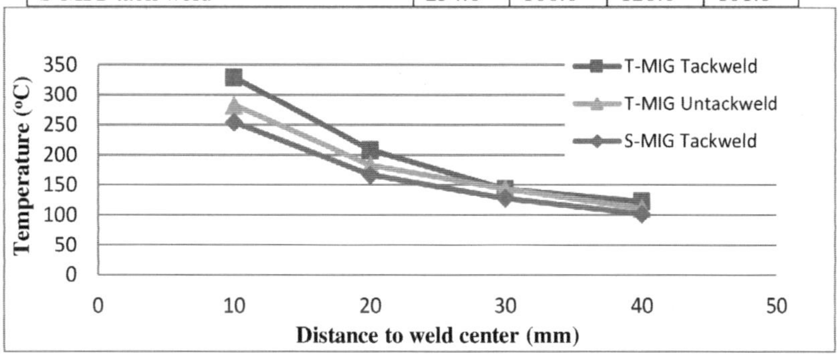

Fig. 5. Peak temperature distributions (^{0}C) vs distance from weld center (mm) of T-MIG untack weld and T-MIG tack weld with d=18 mm, and S-MIG tack weld from Fig.4 (a-c).

3.2 On line bending distortion

Fig. 6 shows displacement pattern during untack and tack T-MIG welding, and S-MIG tack welding. The untack T-MIG welding exhibited positive displacement, the magnitude of displacement increased with the increase of time showing that the welded materials had gained more heat than they released though there was no change in heat input. For the T-MIG tack welding, the displacement was observed to negative direction and smaller magnitude. For the S-MIG tack weld, the displacement was in positive-negative-positive direction forming a sine curves with 0.22 mm minimum peak. There was fluctuation in heat input due to instability of the power suply where the magnitude of the current was fluctuated. Fig. 6 shows that the welding time for 300 mm long deposit for the T-MIG tack weld is 24 seconds and that for the S-MIG tack weld is 38 seconds, thus, there is a 37% shorter time.

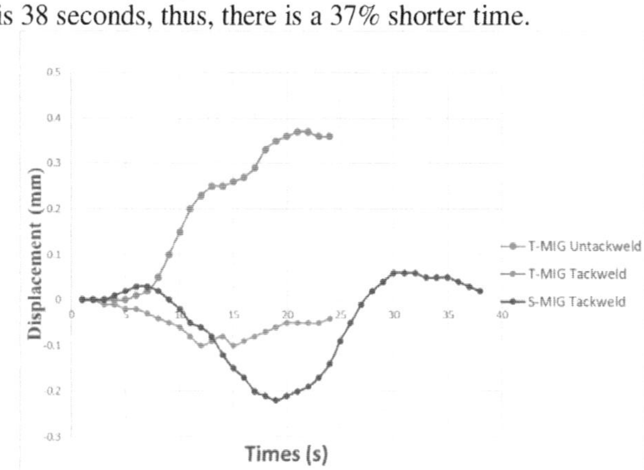

Fig. 6. Experimental development of displacement at dial indicator point, u_z: T-MIG untack weld and T-MIG tack weld, and S-MIG tack weld

IOP Conf. Series: Materials Science and Engineering **547** (2019) 012016 doi:10.1088/1757-899X/547/1/012016

3.3 Distortion Patterns

Fig. 7 shows the 3D distortion for (a) T-MIG untack weld, (b) T-MIG tack weld, and S-MIG tack weld. The distortion of the T-MIG untack weld was observed downward, while that of the T-MIG tack weld was upward or bowing with slightly larger distortion for the T-MIG tack weld. This may be caused by peak temperature of the T-MIG tack weld being 46^{0}C higher than that of the T-MIG untack weld, different from that of the S-MIG tack weld which distorted downward with smaller magnitude. Such phenomena may be caused by the peak temperature of the S-MIG tack weld being 75^{0}C lower than that of the T-MIG tack weld, such that the amount of heat input being smaller leading to smaller distortion. T-MIG tack welding resulted in bowing (upward distortion) with peak temperature being ~328.8^{0}C at 10 mm distance from welding line, while that of the T-MIG untack weld the distortion was downward with peak temperature being ~282.8^{0}C. Therefore, tack weld of the specimens affected the magnitude of the on line displacement and the magnitude of the distortion after welding.

(a)

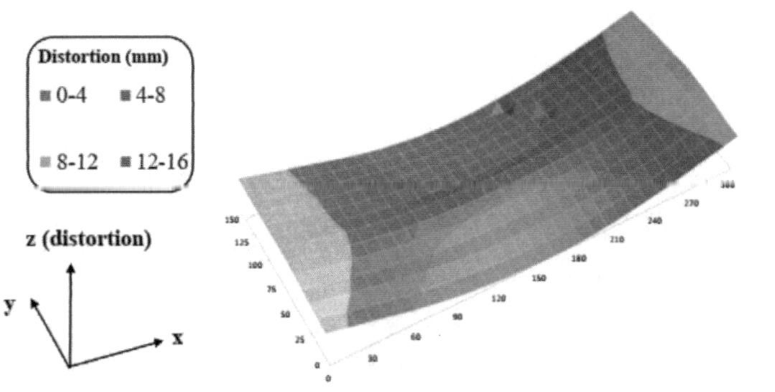

(b)

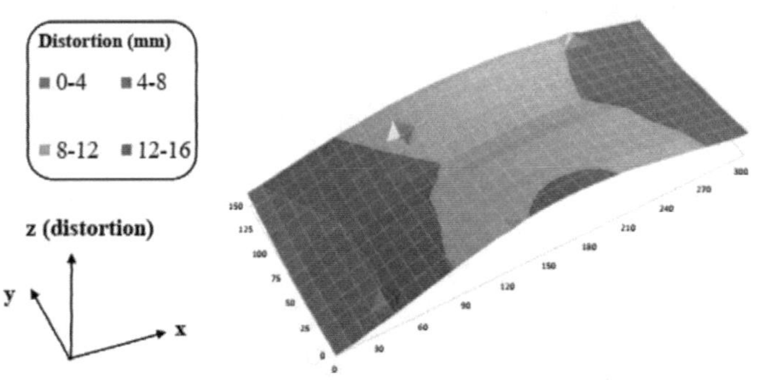

(c)

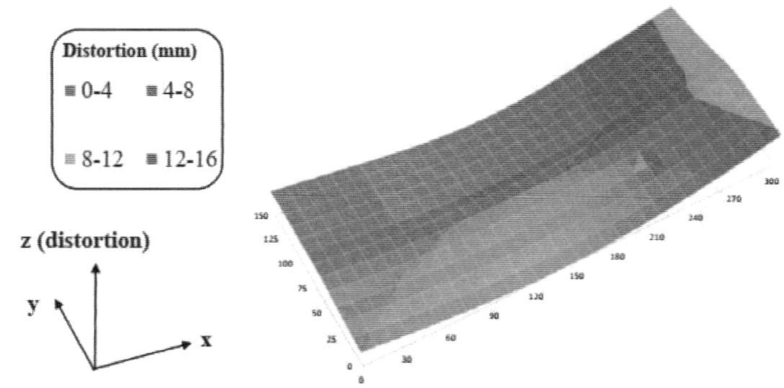

Fig. 7 Distortion (a) T-MIG untack weld, (b) T-MIG tack weld, and (c) MIG tack weld dengan v = 8 mm/s.

3.4 Macrostructures and microstructures

Macrostructures of the T-MIG untack weld, T-MIG tack weld, and S- MIG have been presented in Fig. 8 (a), (b), (c). Fig. 8 (a) shows a lack of penetration on lower left side due to movement farther from the weld center during the welding. This result is consistent with that showed in Fig. 6 for the T-MIG untack weld where u_z displacement has been developed until the completion of the welding. Fig. 8(b), the T-MIG tack weld u_z displacement being the smallest in comparison with that of T-MIG untack weld, and better fusion can be observed, but some porosity can be found that may be caused by the lack of argon gas flow rate causing intrusion of oxygen from its surrounding. This has been confirmed by the SEM and EDS as presented in Fig. 9, where 'EDS spot' (red color) analysis shows 24,13 wt% oxygen (O) content and part of it reacts with the base metal to produce Al_2O_3 and the other part forms porosity. At this spot, Al and Mg produce Al_3Mg_2 precipitate. SEM image shows porosity, micro cracks due to different cooling rate between larger and finer grains, as well as columnar structure at WM. Macrostructure of Fig. 8(c) where there was a misalignment between the two plates being welded, this was confirmed by Fig. 6 for S-MIG tack weld where u_z sines wave displacement can be observed.

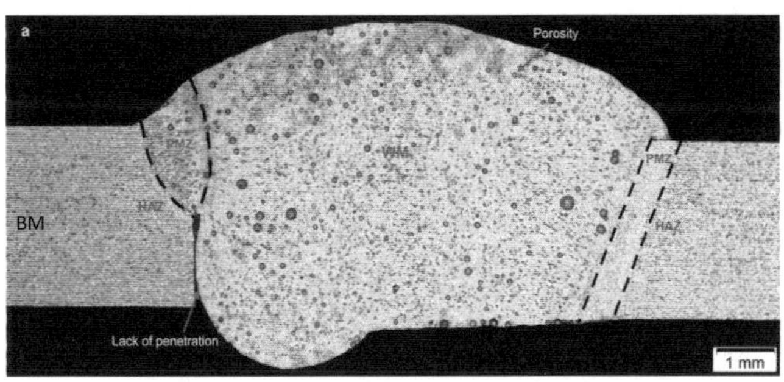

IOP Conf. Series: Materials Science and Engineering **547** (2019) 012016 doi:10.1088/1757-899X/547/1/012016

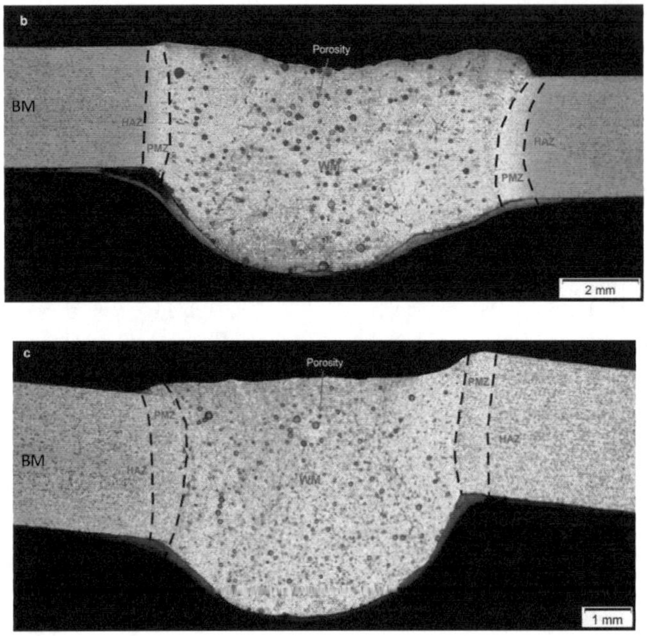

Fig. 8 Macrostructures of T-MIG untackweld (a), T-MIG tackweld (b), and S- MIG (c).

In correlation to macrostructure (Fig. 8), Vickers hardness can be divided into weld metal (WM), partially melted zone (PMZ), heat affected zone (HAZ), and base metal (BM). Weld metal was indicated by dark spots that may be porosity due to the technical argon gas being used may contain oxygen as has been explained earlier. The PMZ obtained from this process was the same as previously reported [7].

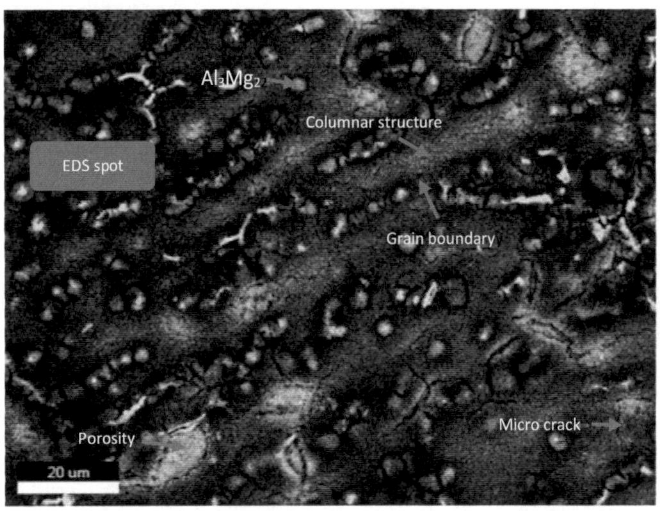

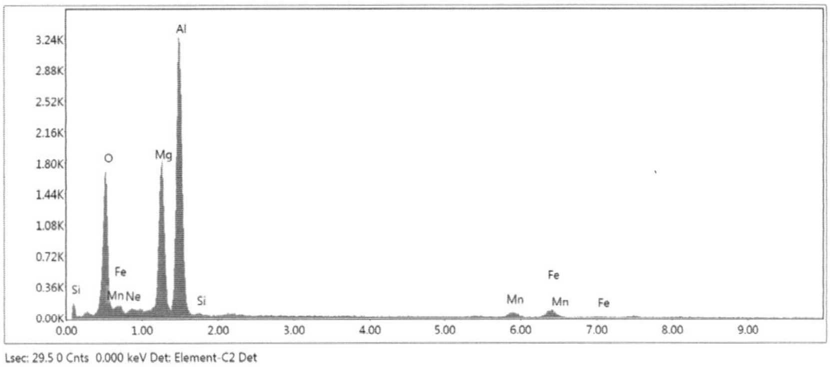

eZAF Smart Quant Results

Element	Weight %	Atomic %	Net Int.	Error %	Kratio	Z	R	A	F
O K	24.13	36.22	428.34	7.41	0.1272	1.1057	0.9592	0.4769	1.0000
NeK	1.18	1.40	25.87	15.01	0.0050	1.0537	0.9777	0.4026	1.0000
MgK	19.58	19.34	512.01	5.31	0.1404	1.0162	0.9936	0.6991	1.0097
AlK	41.51	36.95	933.80	5.35	0.2686	0.9775	1.0008	0.6605	1.0022
SiK	0.45	0.38	8.18	30.64	0.0025	0.9980	1.0076	0.5621	1.0034
MnK	5.01	2.19	28.91	14.65	0.0439	0.8160	1.0525	0.9989	1.0749
FeK	8.15	3.51	38.55	9.46	0.0715	0.8267	1.0527	1.0011	1.0598

Fig. 9 SEM and EDS results at the WM of T-MIG tack weld with electrode distance of 18 mm and welding speed of 12.5 mm/s.

Microstructures of WM, PMZ, and HAZ regions have been depicted in Fig. 10. PMZ of untack T-MIG weld and tack weld, as well as S-MIG show different microstructures in comparison with those of WM. The microstructure of PMZ tends to be equiaxed, whole that of WM are columnar and equiaxed structure. In correlation with Vickers microhardness (Fig.11), PMZ and WM possess considerably similar hardness, and transition can be observed in HAZ that may be caused by partial recrystallization of the 5083H116 base metal due to heating during the welding.

(a)

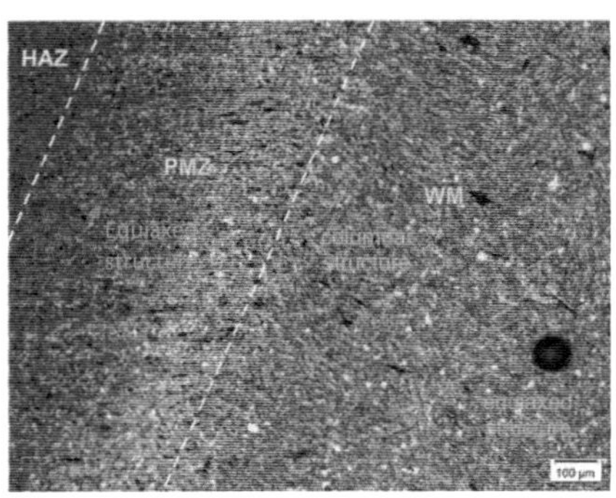

(b)

IOP Conf. Series: Materials Science and Engineering **547** (2019) 012016 doi:10.1088/1757-899X/547/1/012016

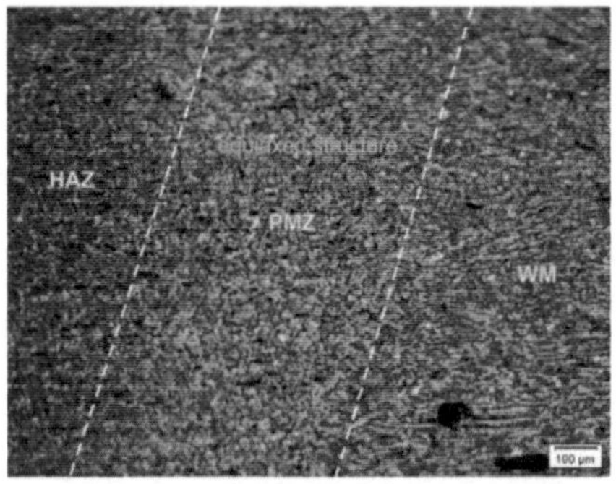

(c)

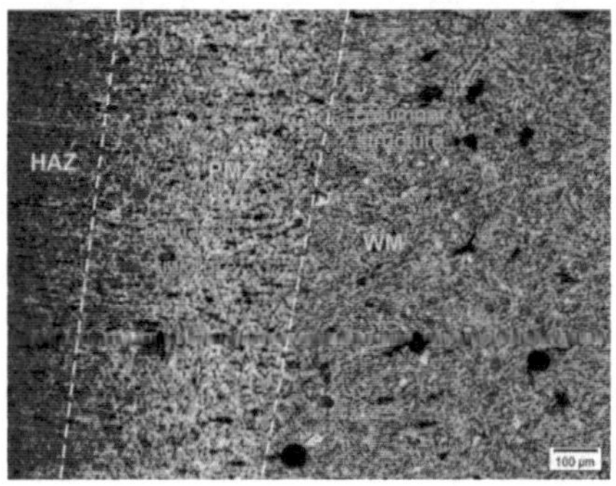

Fig. 10 Microstructures at HAZ, PMZ, and WM zone: (a) T-MIG untack weld, (b) T-MIG tack weld, and (c) S-MIG tack weld

3.5 Vickers hardness test

The result of Vickers micro-hardness test has been depicted in Fig. 11. The hardness of the T-MIG untack weld, T-MIG tack weld, and S-MIG tack weld tend to possess similar values for BM, HAZ, PMZ and WM. For the T-MIG untack weld, the hardness was found being fluctuated that can be caused by inhomogeneous grain size being among indentation points. According to Hall and Petch equation, finer grain will result in higher hardness. Average widths of the untack T-MIG weld are ~7.0 mm, ~1.0 mm, ~4.5 mm for the WM, PMZ and HAZ, respectively. Those of the tack T-MIG weld are ~9.0 mm, ~1.0 mm, and ~7.5 mm for the WM, PMZ and HAZ, respectively. While those of the tack S-MIG weld are ~6.0 mm, ~7.0 mm and ~4mm for the WM, PMZ, and HAZ, respectively. The hardness of the three welding procedures was found being noticeably proportional with that of tensile and flexural test result as presented in Fig. 12 and Fig. 13.

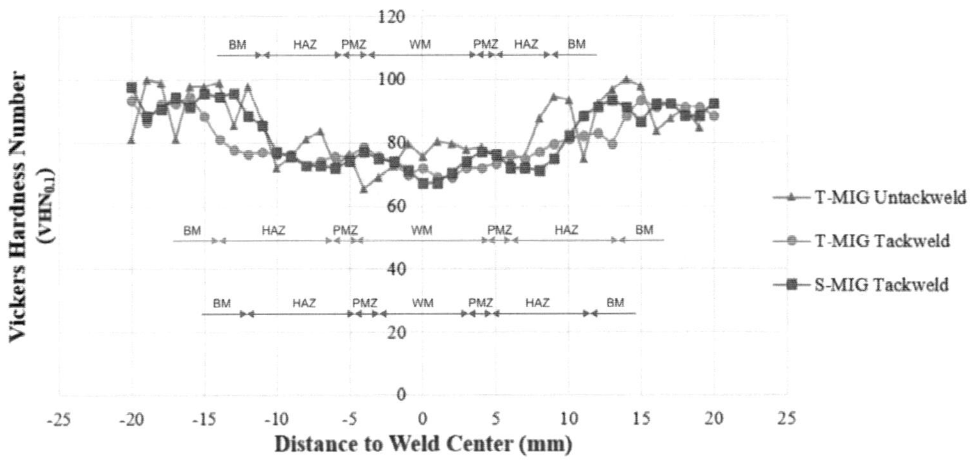

Fig. 11. Vickers hardness number ($VHN_{0.1}$) vs distance to weld center (mm) of T-MIG untackweld (blue color) and tackweld (green color), S-MIG tackweld (red color), and BM, HAZ, PMZ, WM zones.

3.6 Tensile tests

Tensile test result has been presented in Fig. 12. The T-MIG tack weld resulted in 16% higher tensile strength compared to that of the T-MIG untack weld, but 6% lower than that of the S-MIG tack weld, and 24% lower than that of the as received material. Thus, the efficiency of the T-MIG tack weld, 76%, is still applicable for being used in welding of AA5083H116 plates of ship structures.

3.7 Bending test

The result of flexural test has been presented in Fig. 13. The strength of the face and root bending for the T-MIG tack weld is similar with that of the T-MIG untack weld and S-MIG tack weld, and 75% of the bending strength of as received materials. During the bending, the outer face suffers tension and the inner face suffers compression, so that they are in good agreement with those of tensile test of the T-MIG tack weld for 76% weld efficiency, and the test are valid and reliable.

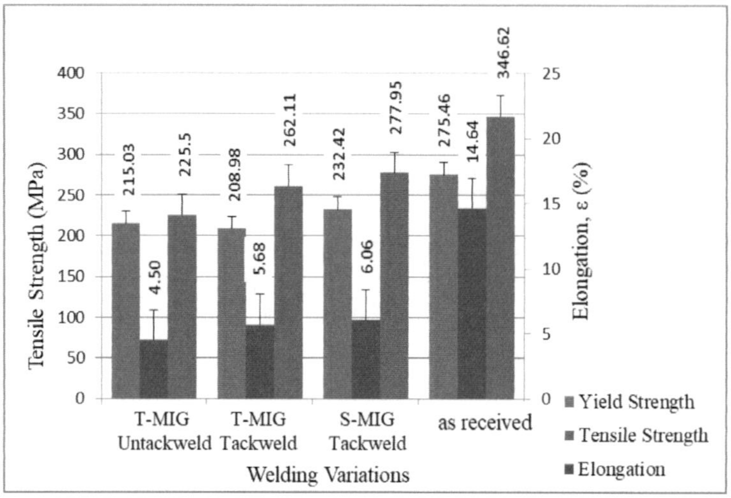

Fig. 12 Tensile strength

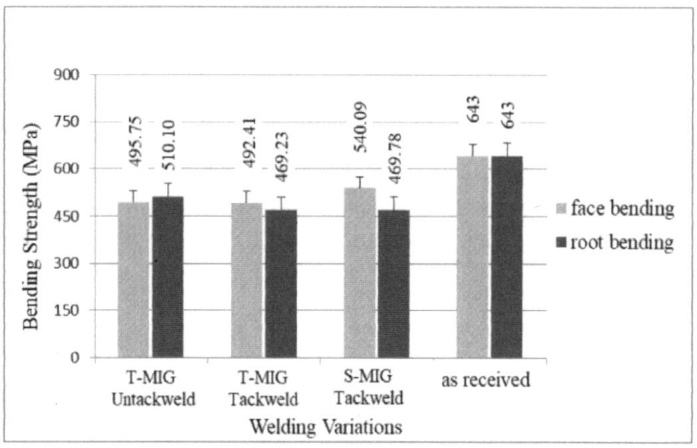

Fig. 13 Bending strength

4. Conclusions

Based on the test and analyses, the following conclusions can be drawn.

1. Whilst the distortion of T-MIG tack weld distorts upward (bowing), that of the T-MIG untack weld distorts downward.
2. Tensile strength of T-MIG tack weld were found being 16% higher than that of T-MIG untack weld where the tensile weld efficiency being 76% and consistent with bending weld efficiency being 75% with respect to those of respective as received material.
3. T-MIG tack welding can save welding time up to 37% based on the distortion measurement time during the welding.

5. Acknowledgements

The authors would like to sincerely thank to Kemenristekdikti which has provided funding of research through UGM PDD: No. 1769/UN1/DITLIT/DIT-LIT/LT/2018 and No: 789/H1.17/TMI/ LK/2018, and the Department of Mechanical and Industrial Engineering who have given the opportunity to do research, and Ekak Novianto, Sunhaji, Arif, Yasin, Chandra, Isnan, and Henry who have helped conduct the research.

6. References

[1] Ashby M F and Jones D R H 2013 *Engineering Material 2*: An Introduction Microstructures and Processing 4thed Butterworth-Heinemann p 190
[2] Lee S, Kim S, 2011 *Trans. Nonferrous Met. Soc. China* **21** 1703-709
[3] Chen-fu F, Xiao-hui M, Qing-xian Hu, Feng-jiang W, He R, Hai-song W, Yu G, Ming M, 2012 *Journal of iron and steel research, International* **19**(5) 79-85
[4] Goeke S, Kaufmann H, Hedegard J, Lundin M, Goeke S F, 2015, https://www.researchgate.net/publication/266214551, accessed date: January 17, 2018
[5] Xueping D, Huan L, Huiliang W, Jiquan, L, 2016 *Vacum* **124** 46-54
[6] Sproesser G, Pittner A, Rethmeiner M, 2016 *Procedia CIRP* **40** 642-647
[7] Rao K S, Reddy G M, Rao K P, 2005, *Materials Science and Engineering A* **403** 69-76
[8] Haelsig A, Kusch M, Mayer P, 2012 *Welding in the World* **56** 98-104
[9] Schenk T, Richardson IM, Kraska M, Ohnimus S, 2009 *Computational Materials Science* **45** 999-1005
[10] Heinze C, Schwenk C, Rethmeier M, 2012 *Journal of Materials Processing Technology* **212** 308-314
[11] Mudjijana, Ilman M N, Iswanto P T, 2017, *International Annual Engineering Seminar (InAES)*, Yogyakarta, Indonesia 1-7
[12] Chen D, Chen M, Wu C, 2015 *Journal of Materials Processing Technology* **225** 45-55
[13] Ping Y, Jiaxiang X, Kang Z, Xiaojun W, Qiang Z, 2016 *Journal of Materials Processing Technology* **229** 111-120
[14] ASME 2001 *Materials Part C – Specifications for Welding Rods, Electrodes, and Filler Metals* New York p 230

Failure Analysis of the Leaf Spring of Truck Colt Diesel Using Finite Element Method

Husaini[1], **Rizki Riantoni, Nurdin Ali and Teuku Edisah Putra**
Laboratory of Computational Mechanics
Department of Mechanical Engineering – UniversitasSyiah Kuala
Darussalam – Banda Aceh 23111, Indonesia

[1]E-mail: husainiftm@unsyiah.ac.id

Abstract. Damage is often experienced by leaf springs of diesel trucks, known as premature failure and is need to be analyzed. Thus, this study aims to analyze the causes of failure of leaf springs through analytical and numerical analyses. Hardness testing was performed using the Rockwell method. Microstructure was observed using scanning electron microscopy to identify the fracture surface. Analytical and numerical analyses were used to obtain the load distribution of each spring. The analysis results of load distribution on each spring were 26.95 kN. The chemical composition of the specimen shows that the material was tested according to AISI 5150 standard. The hardness value of leaf springs tested was HR_B 106.9. The Result of numerical and analytical shows that the strain, stress and deformation of leaf springs with maximum values occurred at the near of hole of spring. The K_I value $\approx K_{IC}$, therefore this may cause crack propagation. The failure of leaf springs was caused by fatigue during the operation marked by a beach mark and the normal stress is far below the specimen fatigue strength.

Keywords: Leaf springs, Crack initiation, Finite element method, Fatigue fracture, Stress intensity factor

1. Introduction

A leaf spring is one of the suspension system commonly used for the suspension in wheeled vehicles, function as vibrationdamper, as well as to support the vehicle frame [1].The leaf springs used in cargo vehicles such as trucks, originally deferred to as a semi-elliptical spring. Accordingly, the truck experienced load on the center point towards the rear of its body, resulted larger deflection compare with forepart [2].

The leaf springs always receive a high dynamic load and often experience fatigue failure that occurs after the component is used for a certain period in its operation [2,3]. Failure due to fatigue is a fracture mechanism that can be identified through 3 stages, the first stage is initiation of crack, then crack propagationand the final stage is fracture [4,5].

Fatigue failure is the most common failure in automotive components, particularly those involvingspring on trucks. This is due to automotive components often experience excessive load and shocks that occur due to the unevenness of the road through the wheel. The load occurs suddenly so that the wheels up and down with irregular frequencies. Leaf springs are more affected by the fatigue load, because the springs hold the mass of the entire car [6].

The spring is produced by means of hot forming such as hot rolling depend on the expected specification of spring function manufactured [2]. Leaf spring is an automotive components produced from high carbon steel as its raw material (C > 0.5%) [3].

Colt diesel truck is a small capasitytruck with the container tub on rear of its part. The trucks often carry over maximum capacity of load and operated on the unevenness of the road. The truck was operated sincetwo years ago with the speedo meter numbers of 572 km, but the leaf spring number 8 experience the fracture. It is known that thelife time of the spring still far under design life time. Thus, the failure of the leaf spring needs to be assessed experimentally and analyzed numerically.

The main objective of this research is to analyze the causes of spatter spring failure used by colt diesel trucks experimentally and numerically. Known causes of failure, so that anticipated the occurrence of the same failure in the future and can reduce serious accidents on the highway.

2. Material and Method

2.1. Material and visual inspection

The material used in this study is a fracture leaf springs of colt diesel truck. The springs experience fracture at the usage of 572 km operating as shown in Fig. 1. This Materialwas supplyed by a mechanical workshop, then it was identified the vehicle types and the loads capacity allowable using thename plate. The arrows in Fig. 1 shows the fracture part of the leaf springs near the bolt hole.

Figure 1. Material for research object

The fracture surface was visually evaluated asshown in Fig. 2. The figure shows the spring number eight experienced the fracture on the bolt hole.

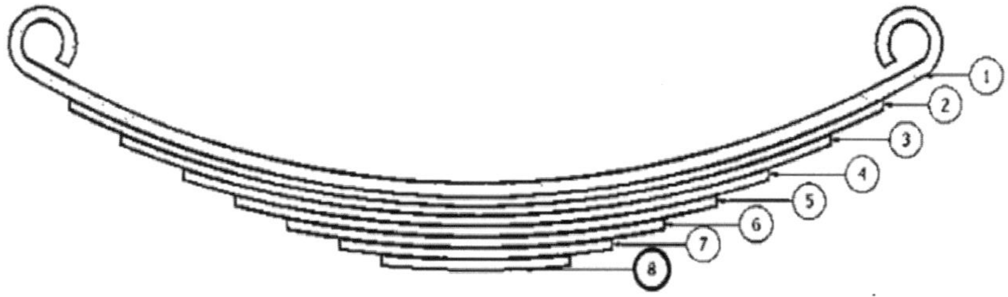

Figure 2. Failures leaf spring occurred at number eight

Visual evaluation of the leaf spring failurewas focusedin the area of bolt hole. Image of the failure surface was taken for showing the historical failure data of failure. Theimage of surface marked initial crack, beach mark and final crack as shown in Fig. 3.

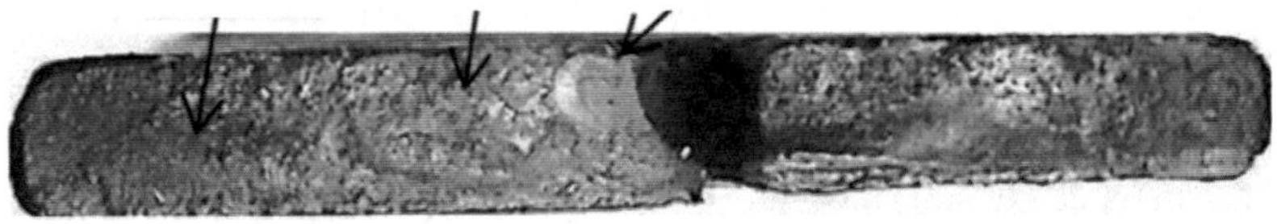

Figure 3. Image of fracture surface of leaf spring

The surface of the fracture spring identified as a fatigue fracture, characterized by a crack initiation area, which subsequently occurs crack propagation in line with fluctuating loading and final fracture [7,8]. The loads leaded by springs on the vehicle are known as fatigue loads because of their dynamic nature. The leaf springs fracture surface was evaluated using scanning electron microscopy (SEM) image. Throught SEM image, failure region can be evaluated whether there is a preliminary defect in the leaf springs as shown in Fig. 3.

Hardness test was performed using Rockwell Type Zwick/Roell ZHR to measure the surface hardness of leaf spring specimens. The indentor used was a steel ball of 1/16-in in diameter and a 100 kg of loading. The size and form of hardness sample is shown in Fig. 4 Configuration of the specimen and the points for hardness test.

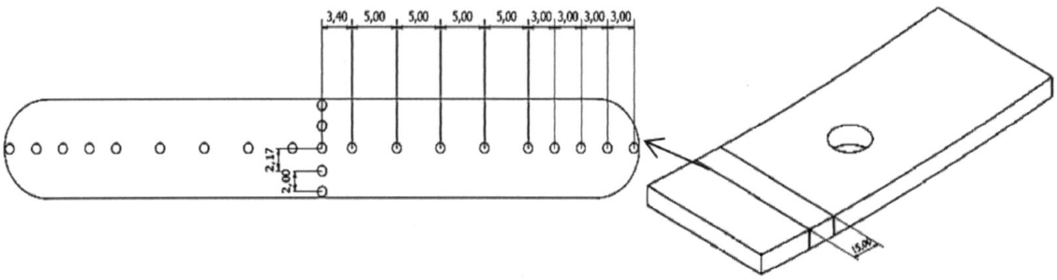

Figure 4. Configuration of the specimen and the points for hardness test

2.2. Chemical Composition Evaluation

Evaluation of chemical composition was performed on test specimens of failure leaf spring. A small part of spring sample was taken and tested. The elements of spring material are shown in Table 1 comparing to the chemical composition of the ASTM standard AISI 5150. Table 2 shows the mechanical properties of AISI 5150 used for finite element analysis.

Table 1. Chemical composition of the material (% wt)

Element	Test Result	AISI 5150 [9]
C	0.529	0.480 – 0.530
Si	0.293	0.150 – 0.300
Mn	0.762	0.700 – 0.900
P	0.0021	≤ 0,0350
S	0.02	≤ 0.04

Table 2. Mechanical properties of AISI 5150 steel

Mechanical Properties of Materials	Value	Unit
Modulus Young (E)	140	GPa
Poison's ration (μ)	0,28	-
Density (ρ)	7850	kg/m^3
Yield Strength (σ_y)	360	MPa
Tensile Strength(σ_B)	675	MPa
Shear Modulus (S)	80	GPa
Fracture Toughness, K_{IC} [10]	23	MPa.m$^{1/2}$

2.3. Analytical Analysis

Analytical analysis is performed to obtain the load distribution on each wheel, either the maximum load or minimum load during the operation of the leaf springs. As it is known that the fine crack found in the material that leaded a bending load and resulted the stress intensity factor (K_I) around the crack tip. When the value of $K_I > K_{IC}$ (fracture toughness), the crack propagation will occur, but when $K_I < K_{IC}$, crack does not spread. Based on this understanding, the stress intensity factor was calculated using available data of the failure life spring [11-13]. Further analytical solution is explained in detail in sub sections 3.3.

2.4. Numerical Analysis

The simulation of collected data was done on the bolt hole area on leaf springs number eight. The usage of hole area due to the fracture failure is around it. From the result of simulation, the stress distribution known, as well as stress intensity factor and strain also obtained [14,15].

The steps and simulation process for numerical analysis were done using FEM software by giving fixed support on the spring eye. The loading type used was centralized load given on the bottom of the spring. The truck loading can be calculated by analytical analysis as shown in Fig. 5.

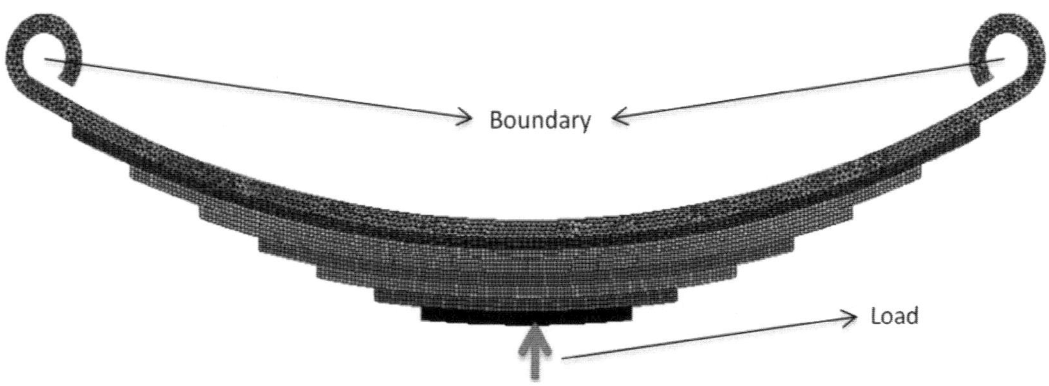

Figure 5. Determination of the boundary condition and loading

The meshing steps used the obtained nodes number of 633438 and the number of elements of 5280. The meshing result of finite element analysis for all spring is shown Fig. 6.

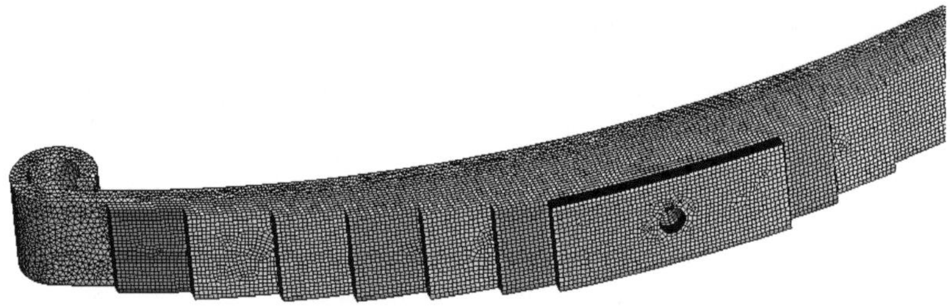

Figure 6. Meshing results on leaf springs for FEM analysis

3. Results and Discussion

3.1. Hardness of Failure Leaf Spring

The results of hardness test areshown in Fig. 7 horizontal axis from point A to B and vertical axis from C to D in the sum of 22 points. The distance between point approximately 2 to 5 mm (ref. to Figure 7.a). The hardness in the area close to the bolts hole or in middle of the spring is lower than of the ends of leaf spring. It shows that, the materials toughness on the center of the leaf spring higher than of the spring ends. This result is in agreed well with the result reported by previous researchers [2].

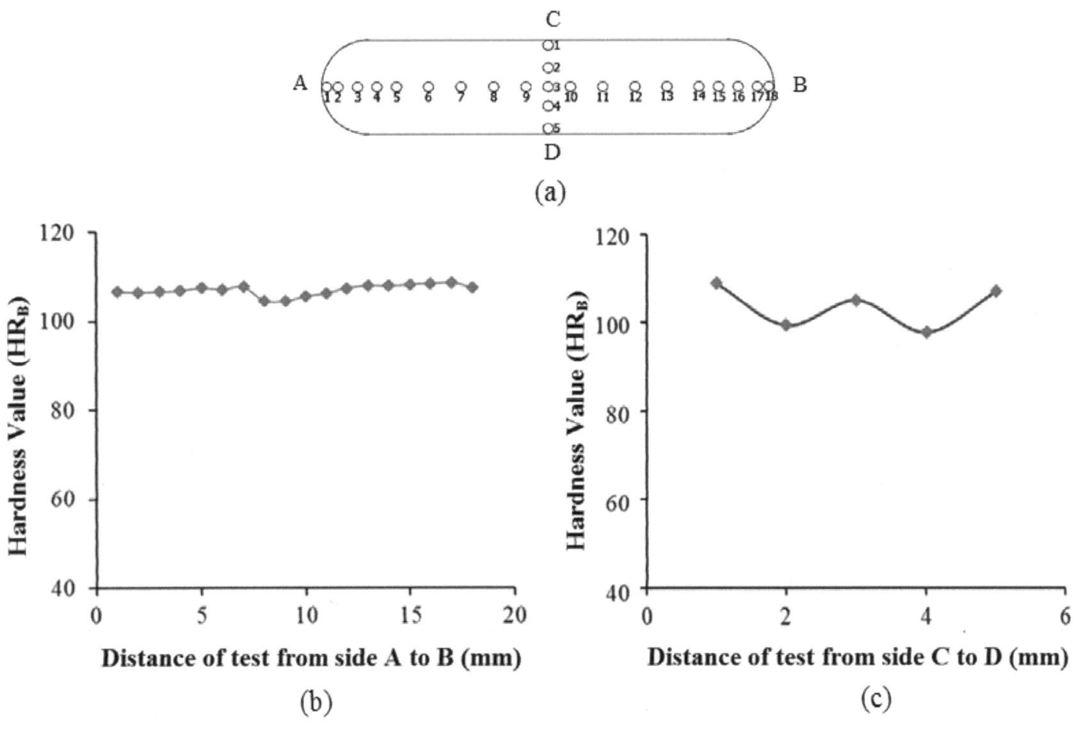

Figure 7. Material hardness testing. (a) Hardness test point at the spring. (b) Inner surface hardness graph from A to B side (see Figure 10.a). (c) Surface hardness graph from side C to D (see Figure 10.a)

The sample of leaf spring surface hardness yields different values between the center and the edges. Hardness test results show that, in the horizontal directions of A to B are 104.6 HR$_B$

minimum and 108.5 HR_B maximum. The difference in hardness value is insignificant only 3.6 %. The same results are encountered in directions of C to D are 97.8 HR_B minimum and 107.1 HR_B maximum remain under 10 % namely 8.7 %. The mean hardness was 106.5 HR_B in the A to B direction and 102.5 HR_B for the C side to D direction. The mean of the overall hardness value is 104.5 HR_B. It can be concluded that no hardening of the damaged leaf springs as seen from hardness is not much different between the points around the fracture and outside the point.

After testing the hardness depth reaches 22 mm from the outer surface of leaf springs. Hardness of the AISI 5150 material according ASTM standard is 92 HRB, while the hardness value of leaf springs of the results is 106.9 HRB. This hardness value indicates that the spring had undergone heat treatment and shot peening as surface treatment for residual stress regenerating. The high hardness difference between the value of the standard and the test results can decrease the ductility or tinkering occurs during the spring operation. This may cause the spring elasticity function to tend to decrease and when the maximum spring loading will be easy to crack [16,17].

3.2. Leaf Spring Fracture Evaluation

Microscopic observations with magnitudes of 6000 on the fracture surface of the spring identified that the crack occur around the bolt hole. The crack starts on the bolt portion then propagate to the cross section of the spring. Figure 8 shows the image of crack area.

Figure 8. Fracture surface spring shows crack area marked with iniation crack

3.3. Analytical and Finite Element Analysis

Analytical analysis aims to obtain the value of stress intensity factor on the elements with crack. For this propose, it is needed the weight of unloaded truck of 3000 kg, the vehicle weight during operation is 11000 kg and maximum allowable vehicle load standard is 7500 kg. The load distributed on each wheels is calculated using Eq. (1) [2,18].

$$F = \frac{m \times g}{4} \tag{1}$$

Where m is a vehicle weight of 11000 (kg), g is the gravity of 9, 8 (m/s^2). The maximum load leaded by each wheels is 26.950 kN.

Figure 9 shows a single-edge-crack-at-hole specimen. Obtaining a value intensity factor (K_I) is calculated using Eq. (2) in the basis of data shown in Fig. 9 [19].

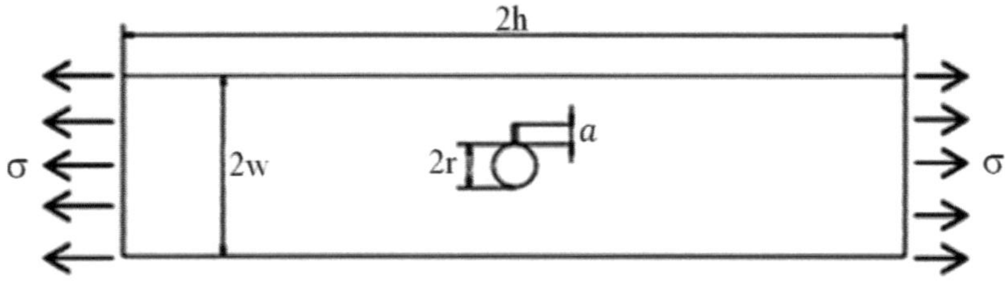

Figure 9. Crack configurations single-edge-crack-at-hole specimen for the calculation of K_I (radius, r = 5mm, wide, w = 25mm)

$$K_I = \sigma\sqrt{(\pi a)}F_{h3} \tag{2}$$

Where a is the crack length. The crack length is obtained from the visual observation with the value is 4.9 mm, σ *is* the stress that occurs in the spring (specimen). Stress occurring at the spring is calculated using finite element analysis. From finite element analysis as shown in Fig. 10 with a given load of 26950 N, the maximum value of 132.84 (MPa) is obtained. F_{hs} is a limiting correction factor calculated by referencing to previous research methods [19]. The results of boundary correction factor analysis obtained value of 2.49.

After calculation of stress intensity factor (K_I) on the spring that experienced an initial crack of 4.9 mm, then obtained K_I value of 22.53 MPa$\sqrt{m}$.

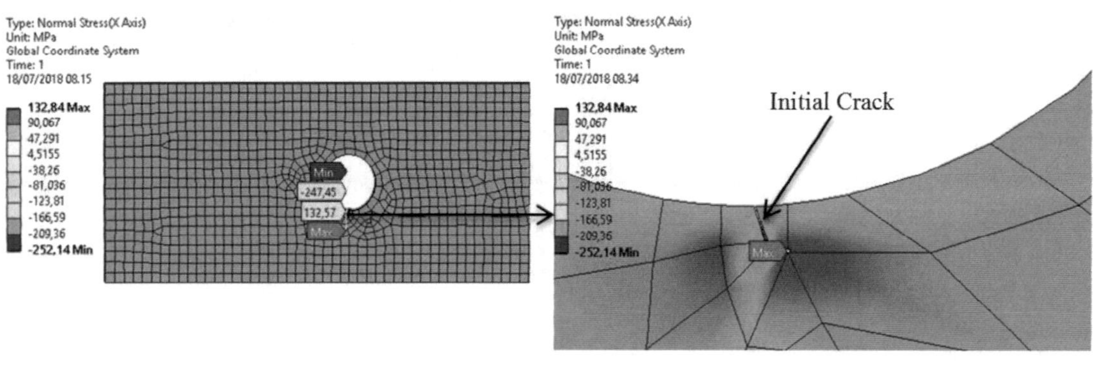

Figure 10. The normal stress of leaf spring

The maximum stress value received by spring is smaller than the yield strength of AISI 5150 material equal to 218 MPa, so it does not cause the spring number eight deformed and the mean stress on the spring number eight is well distributed.

The result of strain occurring on leaf spring number eighth with load of 26950 N shown in Fig. 11. The value of maximum strain resulted by simulation was 8.67x10^{-4} at bolt hole area and minimum was -6.27x10^{-4}. The largest shape change was on the eight spring bolt holes.

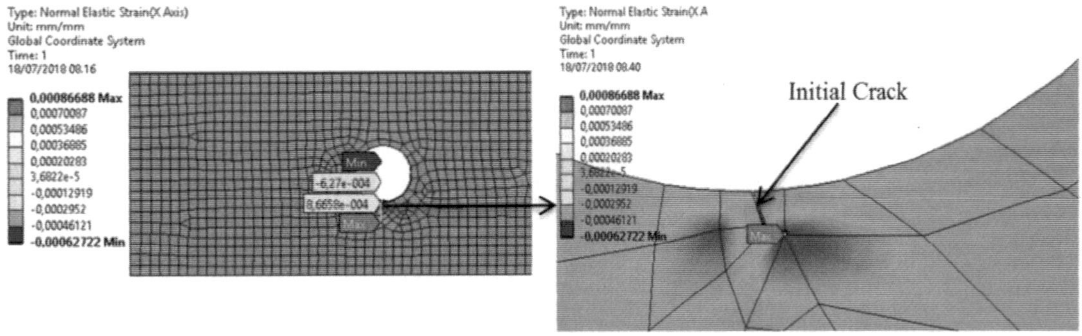

Figure 11. The normal strain of leaf spring

The results of finite element method analysis for obtaining the stress intensity factor (K_I) on leaf springs by modelling the crack length of 4.9 mm in the bolt edge crack are shown in Fig. 12.

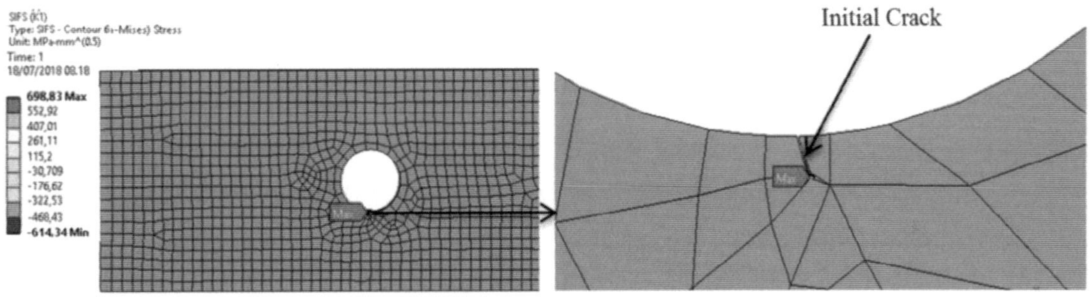

Figure 12. Stress Intensity Factor (K_I)

The stress intensity factor analysis yielded a value of 22.09 MPa$\sqrt{m}$ with a load of 26950 N. The analysis results show that the value intensity factor is close to the fracture toughness ($K_I \approx K_{IC}$). Therefore, this can cause crack propagation due to dynamic loads [9,10]. This fact states that this is one of the causes of leaf spring failure.

4. Conclusions

The failure analysis of the leaf spring through microstructure observation, analytical and numerical analysis yielded the following conclusions:

1. The truck load is approximately 11000 kg that is exceeding from 7500 kg of standard limit, above is approximately 17% of the lift power truck.
2. The maximum stress occurs on eighth sequence of leaf spring is 132.84 MPa. This stress lower than yield stress of AISI 5150 of 360 MPa.

3. The result shows that the Stress Intensity Factor has a value of 22.53 MPa, the value intensity factor is close to the fracture toughness ($K_I \approx K_{IC}$), $22.53 \approx 23$ MPa$\sqrt{m}$. The crack propagation occurred due to dynamic loading.
4. Failure was caused by fatigue fracture characterized by existent beach mark on the fracture surface.

Acknowledgements

The authors would like to express their gratitude to Syiah Kuala University for financial support for this research through the grant No. 2347/UN11.2/TU 2018.

Reference

[1] Peng W, Zhang J, Yang X, Zhu Z & Liu S 2010 Failure analysis on the collapse of leaf spring steels during colt-punching, *Engineering Failure Analysis* ;**17**:971-978

[2] Sepfitrahand Rizal Y 2013 Analisa peristiwa kegagalan pada "leaft spring" truck colt diesel pengangkut pasir, *JURNAL APTEK* ;**5**:2 (in Indonesian)

[3] Thamrin I 2009 Analisa tegangan-regangan struktur pegas daun akibat modifikasi penekanan, *JURNAL REKAYASA MESIN* ;**9**:1 (in Indonesian)

[4] Todinov MT 1999 Maximum principal tensile stress and fatigue crack for compression springs. *International Journal of Mechanical Sciences* ;**41**:357-70

[5] Husaini, Ali N, Bakhtiar A 2016 *Analysis on the surface of broken screw spring front car suspensio,* Proseding Seminar Nasional Teknologi Rekayasa (SNTR) III :111-117 (in Indonesian)

[6] Aher V K andSonawane P M 2012 Static and fatigue analysis of multi leaf spring used in the suspension system of LCV, *Internasional Journal of Engineering Research and Applications (IJERA)* ;**2**:1786-1791

[7] Qian C, Shi W, Chen Z, Yang, Song 2017 Fatigue reliability design of composite leaf springs based on ply scheme optimization, *Engineering Failure Analysis* ;**168**:40-46

[8] Ali N, Fulazzaky M A, Mustapa M S, Imran M G, Ridha M, Sujitno T 2014 Assessment of fatigue and corrosion fatigue behaviours of the nitrogen ion implanted CpTi, *International Journal of Fatigue* ;**61**:184-190

[9] Azom 2012. *AISI 5160 alloy steel (UNS G51600)*

[10] Samoila A 2011 *Liftning for lifting of the locomotive with bogies,* Karlstads University Press

[11] Husaini 2015 *Dasar-dasar Mekanika Retakan*, Syiah Kuala University Press

[12] Husaini and Zuhaimi 2016 Mixed mode fracture behavior of an aluminum alloy A6061 investigated by using compact tension shear specimens, *International Journal of Technology* ;**3**:456-462

[13] Husaini, Kishimoto K, Hanji M, Notomi M 2016 *Investigations of the mixed mode crack growth behavior of an aluminum alloy, ARPN Journal of Engineering and Applied Sciences* ; **11**; 885-890

[14] Wahyu F, Atmawan S, Muthoriq E, M K Herman 2015 *Analisa kekuatan suspense pegas daun truck dengan metode finite element,* POLITEKNOLOGI;**14**:3 (in Indonesian)

[15] Dwivedi Dev Dutt, Jain V K 2016 Design and analysis of automobile leaf spring using ansys, *Internasional Journal of Current Engineering And Scientific Research (IJCESR)* ;**3**:2394-8374
[16] Abdillah F 2010 *Analysis Kegagalan Komponen Pegas Ulir Luar K5 Pada Bogie Kereta Api*. IKIP Vetran, Semarang
[17] Akhyar H and Husaini 2016 Study on Cooling Curve Behavior During Solidification and Investigation of Impact Strength and Hardness of Recycled AL-Zn Aluminum Alloy, *Internasional Journal of Metalcasting* ;**10**:452-456
[18] Sudarsono and Gunawan Y 2012 Analisis kekuatan pegas daun (leaf spring) pada Suzuki carry future 1.5 mega cargo, *Dinamika Jurnal Ilmiah Teknik Mesin;***3**: 251-259 (in Indonesian)
[19] Newman Jr J C and Daniewicz S R 2014 Predicting crack growth in specimens with overloads and colt-worked holes with residual stresses, *Engineering Fracture Mechanics* ;**127**:252-266

Observation on Cetyl Trimethyl Ammonium Bromide Addition as Cationic Surfactant on Water-based Carbon Microfluid Performance for Quench Medium in Heat Treatment Process

C.A Ramadhani[1*], W.N. Putra[1], D. Rakhman[1], L. Oktavio[1], S. Harjanto[1]**
[1]Material and Metallurgy Department, Faculty of Engineering, Universitas Indonesia, Depok, Indonesia

*caesaria.ayu@ui.ac.id
**wahyuaji@metal.ui.ac.id

Abstract- Quenching takes an important part in the heat treatment process that controls the microstructure, thus enhance its mechanical properties. The heat treatment process starts with heating at an elevated temperature, holding time then rapid cooling to room temperature. It requires a medium with a good thermal conductivity that can be achieved by the addition of nanoparticles to the quench medium, referred to as nanofluids. In this research, carbon particles were prepared by the top-down method, where the reduction of carbon particle was done by planetary ball-mill for 15 hours at 500 rpm. Cetyl Trimethyl Ammonium Bromide is utilized as a cationic surfactant in order to reduce agglomeration at suspended particles thus increase quenching efficiency. Field-Emission Scanning Electron Microscope (FE-SEM), and Energy Dispersive X-Ray Spectroscopy (EDX) were used to observe the composition of material, particle size and particle morphology, and the change of the surface. Initial characterization by FE-SEM showed that the particle size after milling was averaged roughly at 15 μm, therefore, it was still not in the nanometer range. However, EDS result confirmed that the powder used in this research were 99% carbon. Carbon microparticles were added as the particle to distilled water as the microfluid base. Water-based carbon microfluid with a volume of 100 ml was produced by the two-step method, by mixing carbon microparticles at 0.1 wt%, and 0.5 wt% in various concentration of cationic surfactant of 1 wt%, 3 wt%, and 5wt % respectively. Samples of AISI 1045 or JIS S45C carbon steels were heat treated by austenizing at 1000°C in a heating furnace, followed by rapid quenching in microfluid as the medium quench resulting on cooling rate diagram. Mechanical properties and microstructures of the quenched samples will be observed by conducting hardness examination and metallography observation to analyze the effect of various carbon and surfactant concentration used in the water-based carbon microfluid quench medium.

Keywords: Nanofluid, Microfluid, Surfactant, Carbon, Quenching

1. Introduction

Mechanical properties of material play an important and vital role in various kind of industry. It is often achieved by heat-treating due to its effectiveness and economic factor. The principle of heat-treatment in order to enhance properties is to trap martensitic structure in the steel matrix by rapid cooling from heating at above austenitic temperature for it has a needle-like structure to hinder dislocation movements thus higher hardness value can be achieved [1]. Quench medium determine the cooling rate of heat treatment and becomes a critical factor for the material's properties and failure [2].

With around 100nm size, nanoparticles can be dispersed to quench medium to increase heat transfer efficiency and capability called nanofluid. In previous researches, nanofluid has been proved to increase the thermal conductivity of the quench medium [3-4]. Titanium dioxide nanoparticles have put an impact when added to water-based nanofluid for enhancing thermal conductivity by the addition of particle concentration [5]. Carbon as nanoparticle has hydrophobic properties and affect directly when dispersed to liquid such as water or oil, to cause agglomeration thus the performance regressed [6]. Stability and good dispersion are desired and it can be acquired by the use of surfactant. Recent studies have proved the use of surfactant to graphene nanofluid have greater stability to graphene nanofluid without surfactant due to the reduction of the contact angle between the nanoparticle and the base fluid, surface tension also increasing the wettability of nanoparticles thus interfacial tension is reduced and heat transfer rate and cooling rate can be improved[7-9]. Recent studies have proved that the greater amount of surfactant concentration compared to nanoparticle concentration used in nanofluid will improve dispersion of nanoparticle in the fluid base [10]. The use of different type of surfactant can affect different leverage to nanofluid, it is classified by the composition of the polarity of the head group which is nonionic, anionic, cationic and amphoteric [11-12]. The adsorption of amphoteric surfactant is better than anionic surfactant due to electrostatic association affecting on the better controlling ability of lubrication and wetting of nanofluid, non-ionic surfactant play a positive effect on carbon nanotubes nanofluid and cationic surfactant has the ability to reduce the viscosity of crude oil improving the clean flow of the oil [13-18].

In this study, we investigate the influence of a variety of cationic surfactant composition, Cetyl Trimethyl Ammonium Bromide (CTAB) on carbon microfluid by observation of the microstructures of S45C steel and a hardness value of the material before and after heat-treated by quenching.

2. Materials and Methods

2.1. Material and Heat Treatment Experiments

The material used for this study was S45C steel rod with carbon composition ranging from 0.42 - 0.48 wt.% in 15mm x 10mm x 10mm dimension cut by hand saw. Heat treatment was conducted in a furnace from room temperature around 25°C, then pre-heated to 600°C with 10°C/minute rate, followed by holding for 15 minutes. Pre-heat was done to avoid cracking of the material [19]. S45C steel then carried out to heat at 1000°C for one hour at an austenizing temperature to acquire austenite structure. Samples of steel then quench and immersed in carbon water-based microfluid in pure distilled water to achieve martensite structure. Quench medium used were pure distilled water to observe microstructures and hardness without microparticles, 0.1 wt% and 0.5 wt% carbon micro-sized particles composition with the addition of CTAB in 1 wt%, 3 wt%, and 5 wt% composition.

2.2.Preparation of Carbon Water-based Microfluid

Carbon microparticles were done by the two-step method in laboratory grade obtained from Sigma-Aldrich. The preparation of carbon microparticles was carried out by milling of carbon powders for 15 hours at 500 rpm using planetary ball mill in Department of Metallurgy and Material Faculty of Engineering Universitas Indonesia, Depok. This process was originally aimed to decrease carbon powder to nano-sized. The next method was the dispersion of microparticles to distilled water with the addition of CTAB surfactant in 1 wt%, 3 wt%, and 5 wt% weight composition. Sonification was done using an ultrasonic cleaner for 280 seconds in room temperature to achieve greater dispersion and stability of carbon microparticles and surfactant in the fluid before followed by quenching [20].

2.3. Experimental Methods

After immersed in quench medium, samples of steel were washed by water then prepared by metallographic procedure then followed by etching in Nital containing 3ml nitric acid and 97 ml ethyl alcohol for five seconds after polishing. Observation of microstructures was carried out using an optical microscope at 100x magnification and hardness was obtained using a Vickers hardness test using 300gf diamond indenter load for 10 seconds. The result of the indentor size in samples was observed with a microscope then the calculation is done to the obtained final hardness value. Carbon powder characterization was done at Centre for Material Processing and Failure Analysis (CMPFA) Universitas Indonesia using Field Emission Scanning Electron Microscope (FE-SEM) and Energy Dispersive X-ray (EDX).

3. Results and Discussion

3.1. Carbon Microparticles Characterization

Carbon microparticles characterization is done using FE-SEM and EDX Test. Figure 1 shows the imaging result of FE-SEM examination of carbon microparticles in (A) 500x magnification and (B) 1000x magnification. The objective of this examination is to observe the particle size of the laboratory-grade carbon powder used in this study. The measurement of average size of carbon microparticles resulting in 15.08 μm.

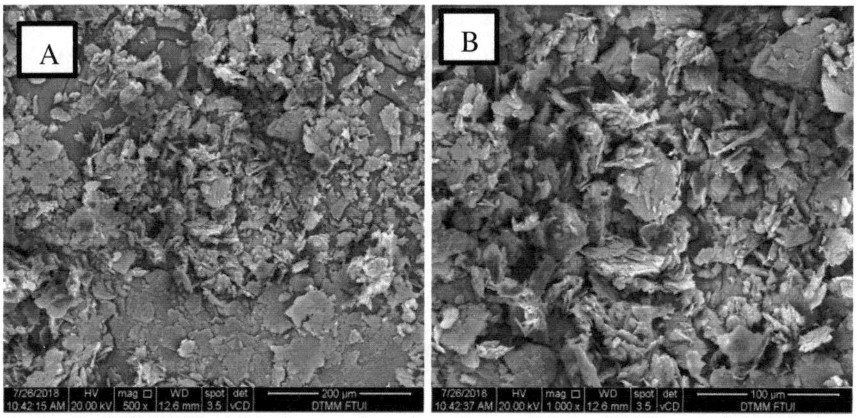

Figure 1. The result of FE-SEM for laboratory-grade carbon microparticles at (A) 500x and (B) 1000x magnification.

EDS test result is shown in Table 1. The objective of EDS is to obtain the composition of carbon microparticles and the purity of the powders. It is shown that the carbon composition is 94.16 wt.% and the rest of the powder is oxygen in 5.84 wt%. No impurities are detected from the result of EDS examination.

Table 1. The composition of laboratory-grade carbon microparticles from EDS. (%wt.)

Laboratory Grade Carbon Powder	C	O
	94.16	5.84

3.2. Microstructures Observation

Microstructures of S45C steel is shown in Figure 2. Figure 2A shows the microstructures of steel immersed in the pure distilled water. Figure 2B shows the microstructures in 0.1 wt.% carbon, then followed by the addition of CTAB 1% in 2C, CTAB 3% in 2D and CTAB 5% in

2E. Figure 2F shows the microstructure in 0.5 wt% carbon, with the addition of 1% CTAB in 2G, 3% CTAB in 2H and 5% CTAB in 2I.

Martensitic structures can be observed explicitly 2C-E and 2G-I marked by the darker area in brown color [21]. The greatest area of martensitic structure can be observed at 2C in 0.1wt% carbon and 1% CTAB. The comparison can be observed clearly to 2A and 2B which fewer brownish/darker area appear. The martensitic structure also has a fewer amount at 2B with 0.1 wt,% carbon than in 2F in 0.5wt% composition. It can be 0concluded that the addition of carbon can affect the greater amount of martensite achieved. Black spot captured in microstructure images can be explained with carbon sputtering out of the steel containing medium carbon composition by decarburization in high temperature due to the reaction of carbon reacts with oxygen or hydrogen in the furnace when heat treated thus the loss of carbon atoms in the surface happened to result on a black spot in microstructure images [22-23].

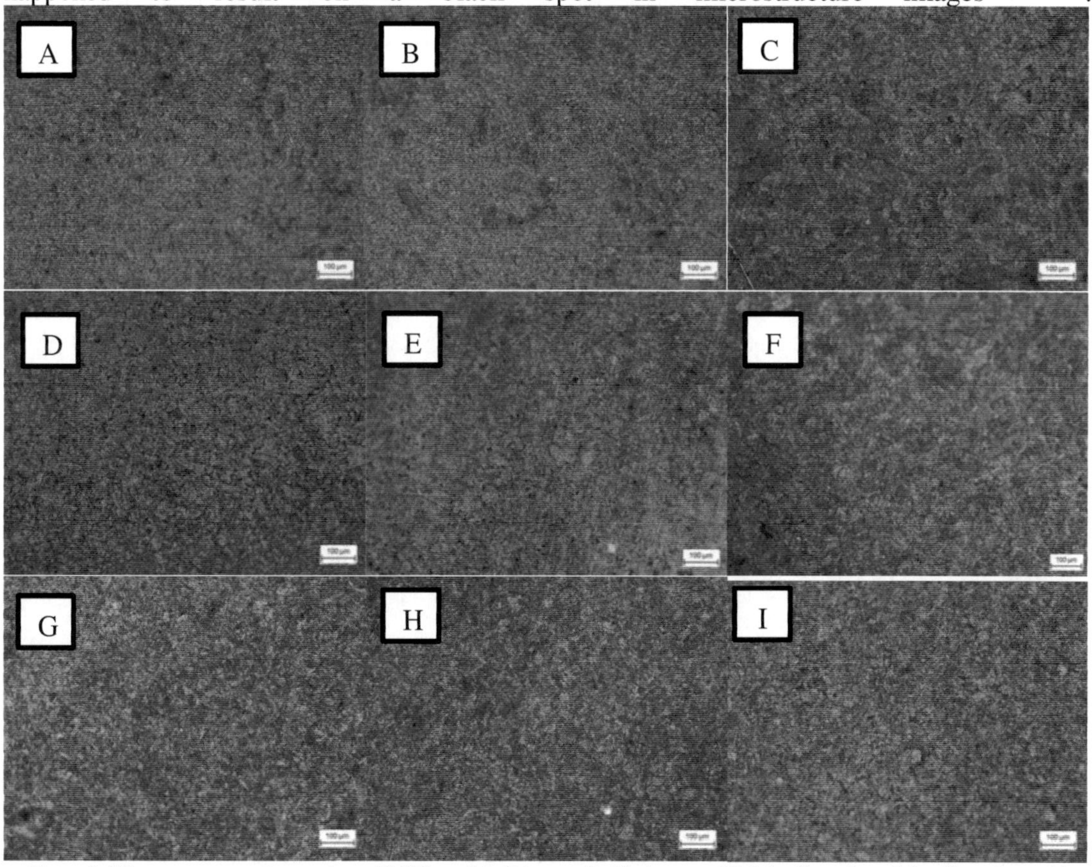

Figure 2. Microstructures observation of S45C steel in (A) pure distilled water, (B) 0.1wt% carbon, 0.1 wt% carbon with CTAB addition of (C) 1%, (D) 3% (E) 5%, (F) 0.5 wt% carbon 0.5 wt% carbon with CTAB addition of (G) 1%, (H) 3% (I) 5% in 100x magnification.

3.3.Hardness Values

The result of the hardness value is shown in Table 2. The hardness of the material in carbon 0.1wt% and 0.5wt% unexpectedly to be the same value which is 457.4 HV. The highest hardness value is 665.3 HV in 0.1wt% carbon composition with 1% CTAB addition, followed by 614.7 HV in carbon 0.5wt% with 3% addition. It can be stated that the optimum addition of the surfactant is 1% in 0.1wt% carbon microparticles and 3% in 0.5wt% carbon microparticles. The addition of surfactant in higher concentration seem to improve hardness value. This might due to the ability to reduce agglomeration. While the addition of surfactant more than optimum concentration can reduce hardness value.

IOP Conf. Series: Materials Science and Engineering **547** (2019) 012018 doi:10.1088/1757-899X/547/1/012018

Table 2. The hardness value of S45C steel in various components of microfluid.

CTAB Composition		Hardness (HV)
Carbon 0.1%	0%	457.4
	1%	665.3
	3%	569.7
	5%	578.9
Carbon 0.5%	0%	457.4
	1%	508.3
	3%	614.7
	5%	554.8

Figure 3 shows the graph of hardness value in various concentration of carbon and the addition of CTAB surfactant. The optimum CTAB concentration in 0.1wt% carbon is 1 wt% and 3 wt% in 0.5 wt % carbon microparticles. There is no close correlation between the carbon 0.1wt% and carbon 0.5 wt% that can be concluded. Both of the composition of microparticles yield a different result of hardness value depending on the addition of CTAB. The optimum concentration of CTAB and the highest hardness value obtained differ from one another. Yet it can be concluded that the addition of surfactant improved the hardness value compared to 0% use of surfactant.

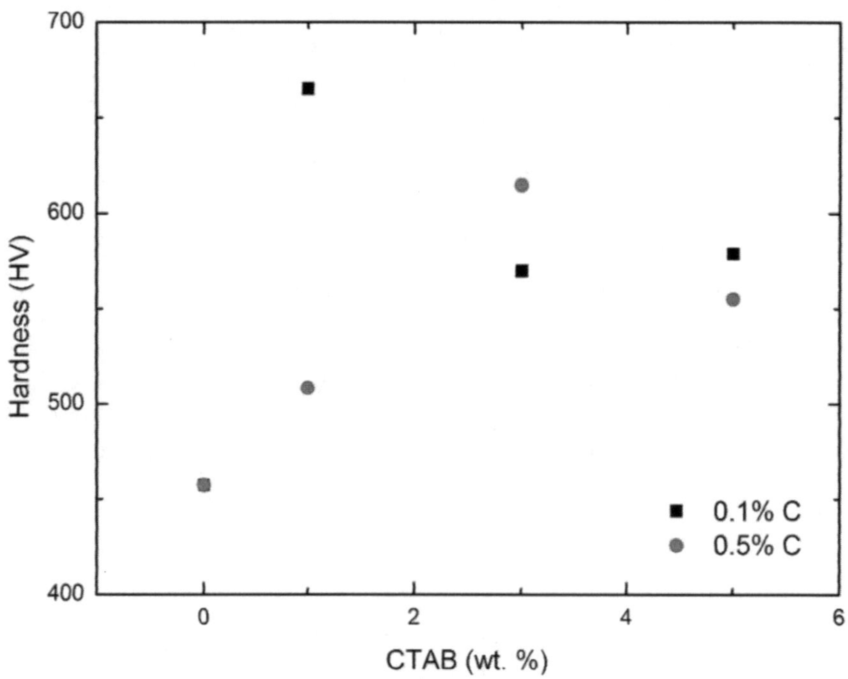

Figure 3. The visual graph of hardness value S45C steel immersed in various components of microfluid.

4. Conclusions

Cetyl trimethyl ammonium bromide (CTAB) was used as a cationic surfactant along with carbon microparticles in a water-based fluid as microfluid quenchants in order to investigate the effect of the surfactant on S45C microstructures and hardness. Carbon microparticles were used in 0.1wt% and 0.5wt% composition along with the addition of 1 wt%, 3 wt% and 5 wt% CTAB. Heat treatment by quenching produced martensitic structure marked by brown-colored area. Steel immersed in 0.1wt% carbon microparticles and 1% CTAB surfactant showed the greatest area of martensite. It also proved by the highest hardness value obtained which is 665.3 HV, followed by immersion in 0.5wt% carbon microparticles composition and 3% CTAB resulting in 614.7 HV. The addition of CTAB surfactant seem to improve the hardness of the material respectively. This improvment was probably due to the CTAB surfactant effect on the agglomeration reduction of carbon microparticles in a water-based microfluid.

5. Acknowledgments

The authors would like to thank Ms. Ghiska Ramahdita for the fruitful discussion, and gratefully acknowledged the financial support from Hibah PITTA 2018 funded by DRPM Universitas Indonesia No. 2555/UN2.R3.1/HKP.05.00/2018.

6. References

[1] Babu, K., Arularasan, R., & Srinath Ramkumar, S. (2017). Quenching performance of AISI 1010 in CNT nanofluids. *Materials Today: Proceedings*, *4*(10), 11044–11049. https://doi.org/10.1016/j.matpr.2017.08.065

[2] Gu, X., Zhang, F., Li, Y., Zhang, J., Chen, S., Qu, C., & Chen, G. (2018). Investigation of cationic surfactants as clean flow improvers for crude oil and a mechanism study. *Journal of Petroleum Science and Engineering*, *164*(September 2017), 87–90. https://doi.org/10.1016/j.petrol.2018.01.045

[3] Salamah, A., Phan, C. M., & Pham, H. G. (2015). Dynamic adsorption of cetyl trimethyl ammonium bromide at decane/water interface. *Colloids and Surfaces A: Physicochemical and Engineering Aspects*, *484*, 313–317. https://doi.org/10.1016/j.colsurfa.2015.08.010

[4] Das, P. K., Mallik, A. K., Ganguly, R., & Santra, A. K. (2018). Stability and thermophysical measurements of TiO2(anatase) nanofluids with different surfactants. *Journal of Molecular Liquids*, *254*, 98–107. https://doi.org/10.1016/j.molliq.2018.01.075

[5] Bílková, K., & Gulbrandsen, E. (2008). Kinetic and mechanistic study of CO2corrosion inhibition by cetyl trimethyl ammonium bromide. *Electrochimica Acta*, *53*(16), 5423–5433. https://doi.org/10.1016/j.electacta.2008.02.096

[6] Chen, L., & Xie, H. (2010). Properties of carbon nanotube nanofluids stabilized by cationic gemini surfactant. *Thermochimica Acta*, *506*(1–2), 62–66. https://doi.org/10.1016/j.tca.2010.04.016

[7] Soror, T. Y., & El-Ziady, M. A. (2003). Effect of cetyl trimethyl ammonium bromide on the corrosion of carbon steel in acids. *Materials Chemistry and Physics*, *77*(3), 697–703. https://doi.org/10.1016/S0254-0584(02)00129-3

[8] Kim, S., Tserengombo, B., Choi, S. H., Noh, J., Huh, S., Choi, B., … Jeong, H. (2018). Experimental investigation of dispersion characteristics and thermal conductivity of various surfactants on carbon based nanomaterial. *International Communications in Heat*

and Mass Transfer, *91*(December 2017), 95 102. https://doi.org/10.1016/j.icheatmasstransfer.2017.12.011

[9] Ueki, Y., Aoki, T., Ueda, K., & Shibahara, M. (2017). Thermophysical properties of carbon-based material nanofluid. *International Journal of Heat and Mass Transfer*, *113*, 1130–1134. https://doi.org/10.1016/j.ijheatmasstransfer.2017.06.008

[10] Choi, T. J., Jang, S. P., & Kedzierski, M. A. (2018). Effect of surfactants on the stability and solar thermal absorption characteristics of water-based nanofluids with multi-walled carbon nanotubes. *International Journal of Heat and Mass Transfer*, *122*, 483–490. https://doi.org/10.1016/j.ijheatmasstransfer.2018.01.141

[11] Xing, M., Yu, J., & Wang, R. (2016). Experimental investigation and modeling on the thermal conductivity of CNTs based nanofluids. *International Journal of Thermal Sciences*, *104*, 404–411. https://doi.org/10.1016/j.ijthermalsci.2016.01.024

[12] Das, P. K., Islam, N., Santra, A. K., & Ganguly, R. (2017). Experimental investigation of thermophysical properties of Al2O3–water nanofluid: Role of surfactants. *Journal of Molecular Liquids*, *237*, 304–312. https://doi.org/10.1016/j.molliq.2017.04.099

[13] Sánchez-Coronilla, A., Martín, E. I., Navas, J., Aguilar, T., Gómez-Villarejo, R., Alcántara, R., … Fernández-Lorenzo, C. (2018). Experimental and theoretical analysis of NiO nanofluids in presence of surfactants. *Journal of Molecular Liquids*, *252*, 211–217. https://doi.org/10.1016/j.molliq.2017.12.140

[14] Khaleduzzaman, S. S., Mahbubul, I. M., Shahrul, I. M., & Saidur, R. (2013). Effect of particle concentration, temperature, and surfactant on surface tension of nanofluids. *International Communications in Heat and Mass Transfer*, *49*, 110–114. https://doi.org/10.1016/j.icheatmasstransfer.2013.10.010

[15] Xuan, Y., Li, Q., & Tie, P. (2013). The effect of surfactants on heat transfer feature of nanofluids. *Experimental Thermal and Fluid Science*, *46*, 259–262. https://doi.org/10.1016/j.expthermflusci.2012.12.004

[16] Tiara, A. M., Chakraborty, S., Sarkar, I., Ashok, A., Pal, S. K., & Chakraborty, S. (2017). Heat transfer enhancement using surfactant based alumina nanofluid jet from a hot steel plate. *Experimental Thermal and Fluid Science*, *89*(March), 295–303. https://doi.org/10.1016/j.expthermflusci.2017.08.023

[17] Almahfood, M., & Bai, B. (2018). The synergistic effects of nanoparticle-surfactant nanofluids in EOR applications. *Journal of Petroleum Science and Engineering*, *171*(May), 196–210. https://doi.org/10.1016/j.petrol.2018.07.030

[18] Gnyusov, S. F., Rotshtein, V. P., Mayer, A. E., Astafurova, E. G., Rostov, V. V., Gunin, A. V., & Maier, G. G. (2017). Comparative study of shock-wave hardening and substructure evolution of 304L and Hadfield steels irradiated with a nanosecond relativistic high-current electron beam. *Journal of Alloys and Compounds*, *714*, 232–244. https://doi.org/10.1016/j.jallcom.2017.04.219

[19] Ulset, E. T., Kosinski, P., & Balakin, B. V. (2018). Solar steam in an aqueous carbon black nanofluid. *Applied Thermal Engineering*, *137*(December 2017), 62–65. https://doi.org/10.1016/j.applthermaleng.2018.03.038

[20] Estellé, P., Cabaleiro, D., Żyła, G., Lugo, L., & Murshed, S. M. S. (2018). Current trends in surface tension and wetting behavior of nanofluids. *Renewable and Sustainable Energy Reviews*, *94*(November 2017), 931–944. https://doi.org/10.1016/j.rser.2018.07.006

[21] Esmaeilzadeh, P., Bahramian, A., & Fakhroueian, Z. (2011). Adsorption of anionic, cationic and nonionic surfactants on carbonate rock in presence of ZrO2nanoparticles. *Physics Procedia*, *22*, 63–67. https://doi.org/10.1016/j.phpro.2011.11.009

[22] Soleimani, H., Baig, M. K., Yahya, N., Khodapanah, L., Sabet, M., Demiral, B. M. R., & Burda, M. (2018). Impact of carbon nanotubes based nanofluid on oil recovery efficiency using core flooding. *Results in Physics*, *9*, 39–48. https://doi.org/10.1016/j.rinp.2018.01.072

[23] Gangadevi, R., Vinayagam, B. K., & Senthilraja, S. (2018). Effects of sonication time and temperature on thermal conductivity of CuO/water and Al2O3/water nanofluids with and without surfactant. *Materials Today: Proceedings*, *5*(2), 9004–9011. https://doi.org/10.1016/j.matpr.2017.12.347

IOP Conf. Series: Materials Science and Engineering **547** (2019) 012019 doi:10.1088/1757-899X/547/1/012019

On Impact Problems Of Projectile Using Meshless Peridynamics Analysis

Mas Irfan P. Hidayat[1]*
[1]Department of Materials and Metallurgical Engineering-FTI, Institut Teknologi Sepuluh Nopember, Kampus ITS Keputih Sukolilo 60111, Surabaya

*E-mail: irfan@mat-eng.its.ac.id

Abstract. In this study, impact problems of projectile are investigated using meshless peridynamics analysis. Two impact cases of rigid projectiles having different stiffness values against aluminum target are examined numerically. The problems are simulated in LAMMPS software environments. For each case, numerical parameters affecting perforation patterns of target body are discussed in detail. This work focuses on the effects of numerical parameters of indenter stiffness (k) and critical bond stretch parameter alpha on the perforation patterns of the target for various impact velocities. Numerical results show that larger perforation is produced with the increase of k value of indenter. In addition, it is observed that the impacted plate appears to be more deflected, if greater value of α is employed.

Keywords: impact, projectile, peridynamics, LAMMPS

1. Introduction

Fracture of solids is main concern in engineering and industrial applications. Such material failure is important because integrity of a structure or component during its service is affected by the event of failure [1-3]. Amongst modes of failure is fracture due to impact, which is mainly characterized by collision particularly with an indenter of high speed. From numerical viewpoint, simulation of fracture is a challenging task. This is due to the fact that the initiation site/time of failure and failure direction in a material need to be determined accurately. On the other hand, how to realistically and efficiently simulate the material failure is difficult as well. In finite element (FE), process of re-meshing is frequently needed to adapt with new configuration of structure due to the failure. Obviously, the task would be cumbersome and time consuming [4]. Recently, a class of numerical techniques called as meshfree methods is intensively pursued for efficient modeling of fracture of materials [5-8].

In this study, impact problems of projectile are investigated using meshless peridynamics analysis. Two impact cases of rigid projectiles having different stiffness values against aluminum target are examined numerically. The problems are simulated in LAMMPS [9]. For each case, numerical parameters affecting perforation patterns of target body are discussed in detail. This work focuses on the effects of numerical parameters of indenter stiffness (k) and critical bond stretch parameter (α) on the perforation patterns of the target.

2. Materials and Methods

In classical continuum mechanics, differential notion is extensively used to represent equilibrium condition of media under the influence of forces/displacements exerted to it. Peridynamics is a reformulation of the classical continuum mechanics equations having a distinct advantage in treating discontinuities in solids. By employing non-local interaction concept, peridynamics resorts to formulation based upon integration rather than differentiation. As a result, it can naturally handle fracture events without necessity to use deformation gradients [10-13].

$$\rho(\mathbf{x})\ddot{\mathbf{u}}(\mathbf{x},t)=\int_{H_x}\mathbf{f}\left(\hat{\mathbf{u}}(\mathbf{x},t)-\mathbf{u}(\mathbf{x},t),\hat{\mathbf{x}}-\mathbf{x}\right)dV_x+\mathbf{b}(\mathbf{x},t)\quad \text{for}\quad \mathbf{x}\in\Omega\quad\text{and}\quad t\in[t_0,\infty] \tag{1}$$

where: Ω is the problem domain, ρ is material density, $\mathbf{f}$ is the pairwise force function in peridynamic bond connecting material points $\hat{\mathbf{x}}$ and $\mathbf{x}$, $\mathbf{u}$ and $\ddot{\mathbf{u}}$ are the displacement and acceleration vector, respectively, $\mathbf{b}$ is the body force, t and t_0 are time and initial time, respectively, H_x is horizon region, a local region in which the integral is defined. In addition, δ is defined as radius of the region, commonly taken as circle/disk in 2D or sphere in 3D.

Furthermore, a pairwise potential ϕ exists for a microelastic material such that:

$$\mathbf{f}(\xi,\eta)=\frac{\partial\phi(\xi,\eta)}{\partial\eta} \tag{2}$$

where: $\xi=\hat{\mathbf{x}}-\mathbf{x}$ is the relative position and $\eta=\mathbf{u}(\hat{\mathbf{x}},t)-\mathbf{u}(\mathbf{x},t)$ is the relative displacement.

The micropotential is defined as:

$$\phi(\xi,\eta)=\frac{1}{2}cs^2\|\xi\| \tag{3}$$

$$s=\frac{\|\eta-\xi\|-\|\xi\|}{\|\xi\|} \tag{4}$$

where: c constant and s bond stretch or strain.

Noting that with K the bulk modulus and δ the horizon, c for 3D models is given by:

$$c=\frac{18K}{\pi\delta^4} \tag{5}$$

Damage is defined as breakage of material bonds. For a prototype microelastic brittle (PMB) material, the pairwise function is given by:

$$\mathbf{f}(\xi,\eta)=g(\xi,\eta)\frac{\eta+\xi}{\|\eta+\xi\|} \tag{6a}$$

$$g(\xi,\eta)=\begin{cases}cs(t,\xi,\eta)\mu(\xi,\eta), & \|\xi\|\le\delta\\ 0 & \|\xi\|>\delta\end{cases} \tag{6b}$$

$$\mu(\xi,\eta)=\begin{cases}1 \text{ if } & s(t',\xi,\eta)<\min\left(s_0(t',\xi,\eta),s_0(t',\xi',\eta')\right)\text{ for all } 0\le t'\le t\\ 0 & \text{otherwise}\end{cases} \tag{7}$$

where: μ is the history-dependent scalar boolean function. $s_0(t,\xi,\eta)$ is a critical stretch, determined by parameters of critical bond stretch s_{00} and α, such that:

$$s_0(t,\xi,\eta)=s_{00}-\alpha s_{\min}(t,\xi,\eta),\quad s_{\min}(t)=\min_{\xi}s(t,\xi,\eta) \tag{8}$$

3. Results and Discussion

Aluminum cylindrical plate impacted by a rigid sphere indenter is considered in this study. The indenter is located at 0.0051 m from the plate and its velocity is varied from 200 to 400 m/s. Material properties for the aluminum plate are given in Table 1 below. For plotting the simulation results, OVITO [14] and MATLAB [15] soft-wares are employed.

Table 1. Properties of aluminum plate and parameters of simulation

Properties/parameters	**Values**
Density (ρ)	2700 kg/m3
Bulk modulus (K)	69E9 Pa
Poisson ratio (ν)	0.33
Fracture toughness (K_{Ic})	24 MPa√m
Horizon (δ)	0.0015
Critical bond stretch parameter 1 (s_{00})	0.00627
Critical bond stretch parameter 2 (α)	0.0 and 0.25
Indenter stiffness k	1E17 and 1E12
Time step	1E-7 sec

3.1. Influence of parameter k

Simulation results of perforation patterns on the plate are presented in Fig. 1. The stiffness value k is set to be 1E17, and two α values as given in Table 1 are employed.

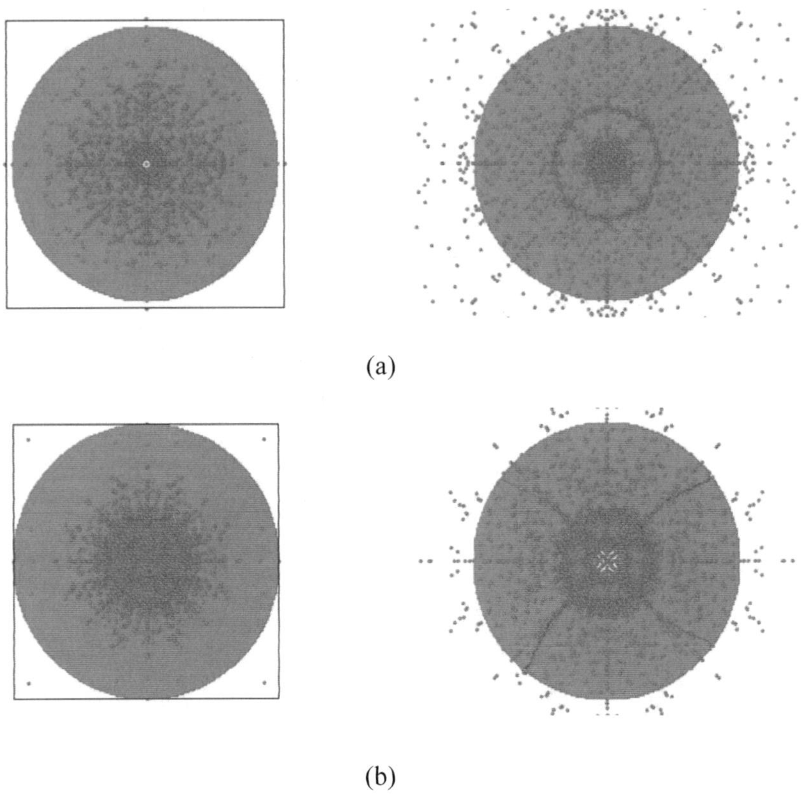

Fig. 1. Perforation patterns on the plate with: (a) $\alpha = 0$, and (b) $\alpha = 0.25$, taken at simulation steps of 1000 and 1800, respectively (velocity = 300 m/s and k = 1E17).

Fig. 2 shows the results of perforation patterns on the plate with the stiffness value k is set to be 1E20. Hence, the current indenter is harder than the previous one. From the results, it is seen that greater k value gives larger impact and perforation, for the same impact velocity.

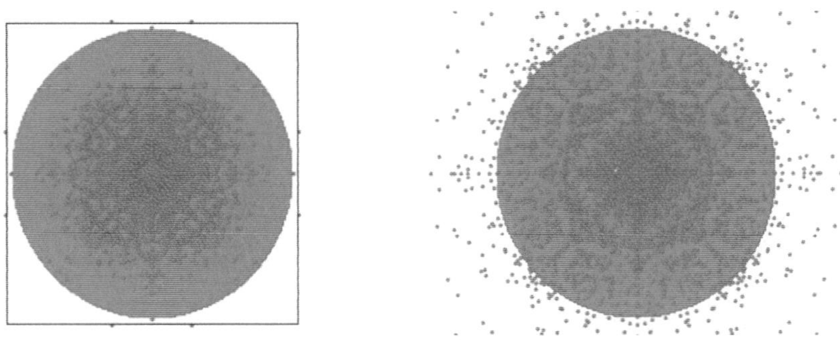

(a)

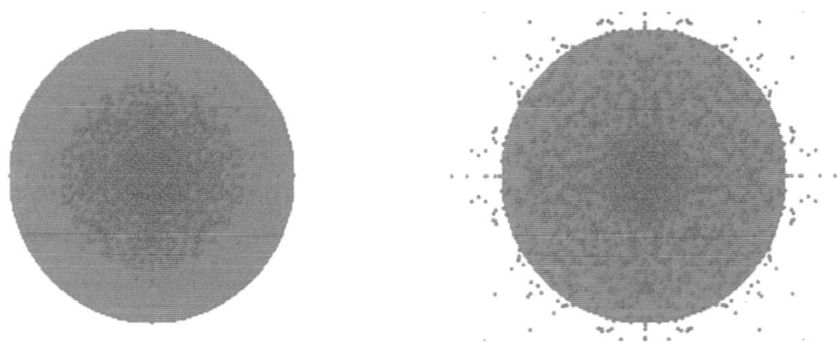

(b)

Fig. 2. Perforation patterns on the plate with: (a) $\alpha = 0$, and (b) $\alpha = 0.25$, taken at simulation steps of 1000 and 1800, respectively (velocity = 300 m/s and k = 1E20).

In addition, development of pressure during the simulation steps for various values of velocities and α is shown in Fig. 3. Note that the results are consistent with respect to the increase of k value.

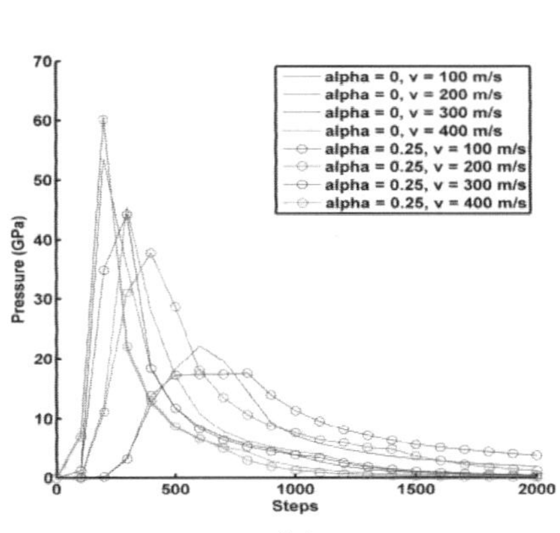

(a)

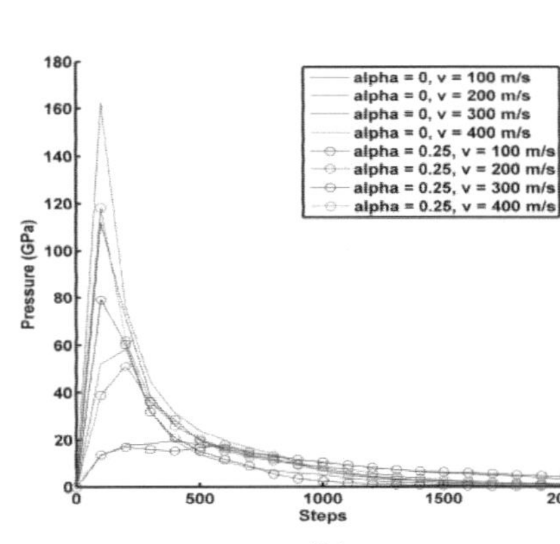

(b)

Fig. 3. Pressure during time steps for various impact velocities and α values with: (a) k = 1E17, and (b) k = 1E20.

3.2. Debris development

Development of debris from step to step is presented as well here, as shown in Figs. 4-7. The obtained results consistently show that greater debris development is obtained with the increase of indenter stiffness value. Moreover, more interesting discussion may be casted concerning with the influence of parameter α to perforation development, as in the following section.

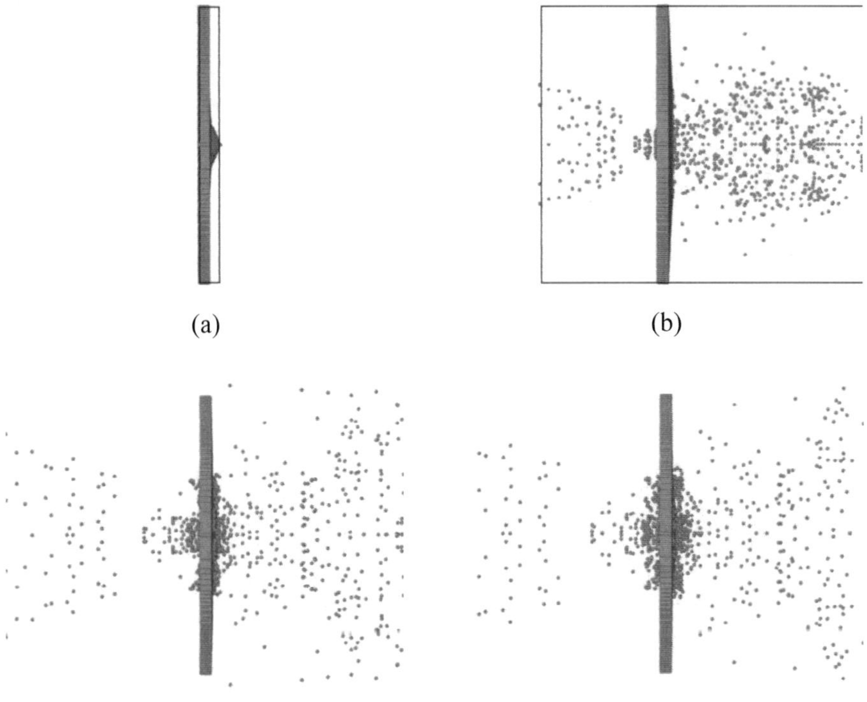

(a) (b)

(c) (d)

Fig. 4. Debris development at several simulation steps: (a) 200, (b) 800, (c) 1400, and (d) 1600 (the impact velocity = 300 m/s, $\alpha = 0$, k = 1E17).

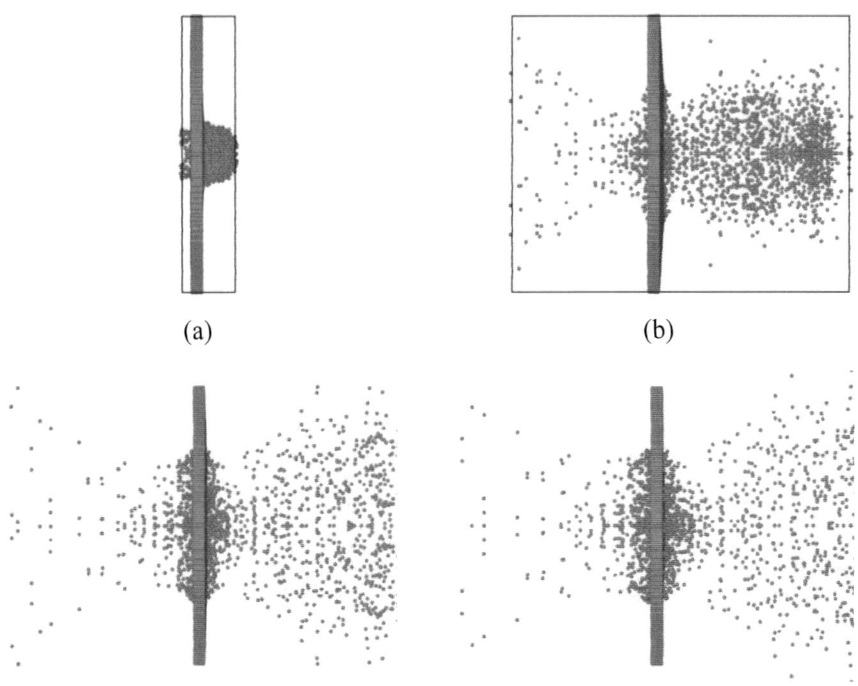

(a) (b)

(c) (d)

Fig. 5. Debris development at several simulation steps: (a) 200, (b) 800, (c) 1400, and (d) 1600 (velocity = 300 m/s, $\alpha = 0$, k = 1E20).

IOP Conf. Series: Materials Science and Engineering **547** (2019) 012019 doi:10.1088/1757-899X/547/1/012019

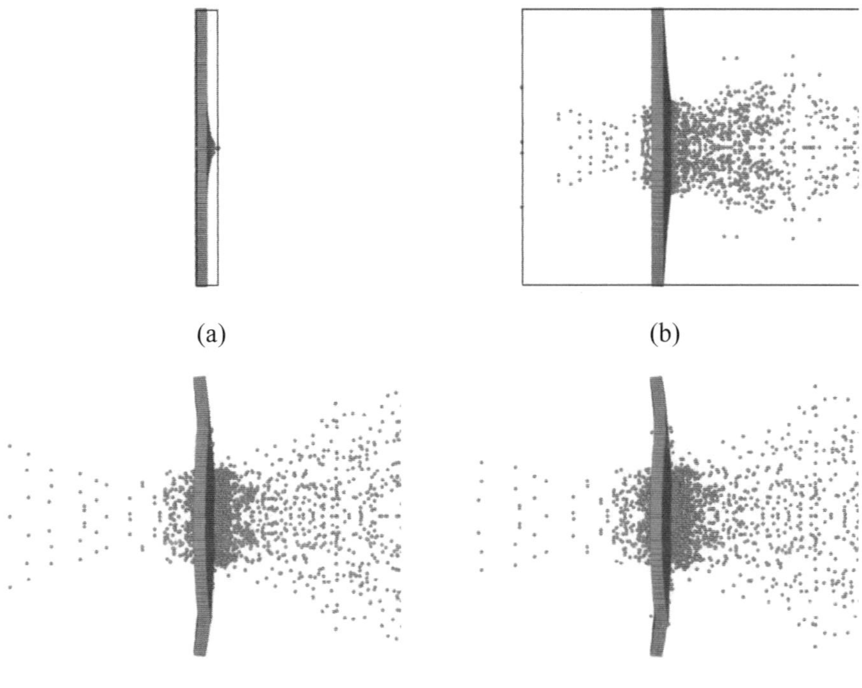

(a) (b) (c) (d)

Fig. 6. Debris development at several simulation steps: (a) 200, (b) 800, (c) 1400, and (d) 1600 (the impact velocity = 300 m/s, $\alpha = 0.25$, k = 1E17).

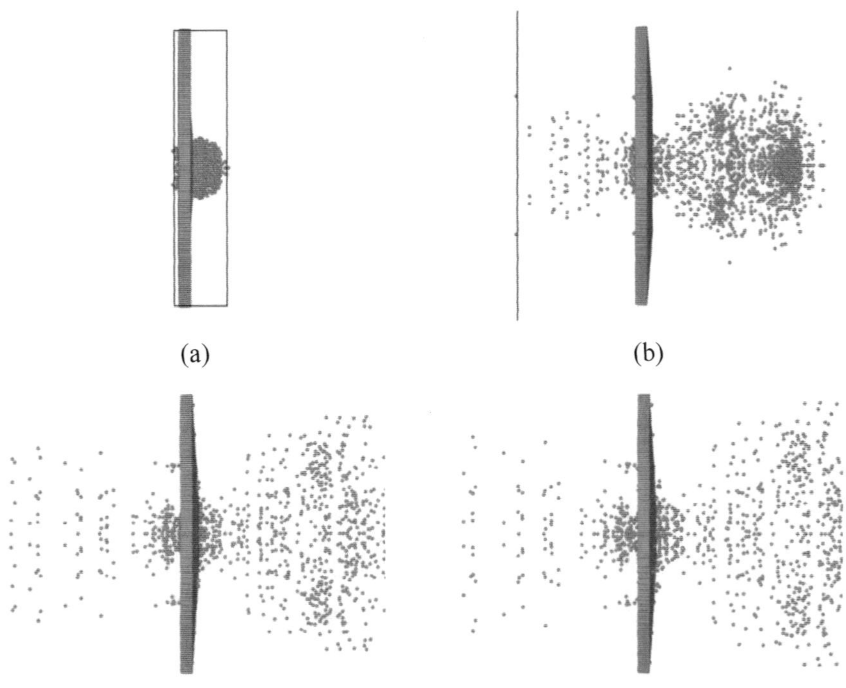

(a) (b) (c) (d)

Fig. 7. Debris development at several simulation steps: (a) 200, (b) 800, (c) 1400, and (d) 1600 (velocity = 300 m/s, $\alpha = 0.25$, k = 1E20).

3.3. Influence of parameter

Observing the obtained simulation results, it can be said that while larger perforation and range of debris is produced with the increase of k value of indenter, the variation of parameter α seems to give somewhat different effects on the debris intensity. For lower k value, debris is observed to be more localized near the plate for greater α, as shown by Figs. 4 and 6. On the other hand, for higher value of k, debris is more localized near the plate for lower value of α ($\alpha = 0$), as depicted in Figs. 5 and 7.

Apart from that, nonetheless, consistent results are observed by increasing α value that the impacted plate appears to be more deflected. This observation can be seen by checking the shape of impacted plate at the final time step, as shown by Figs. 6 and 7, and Fig. 8 below.

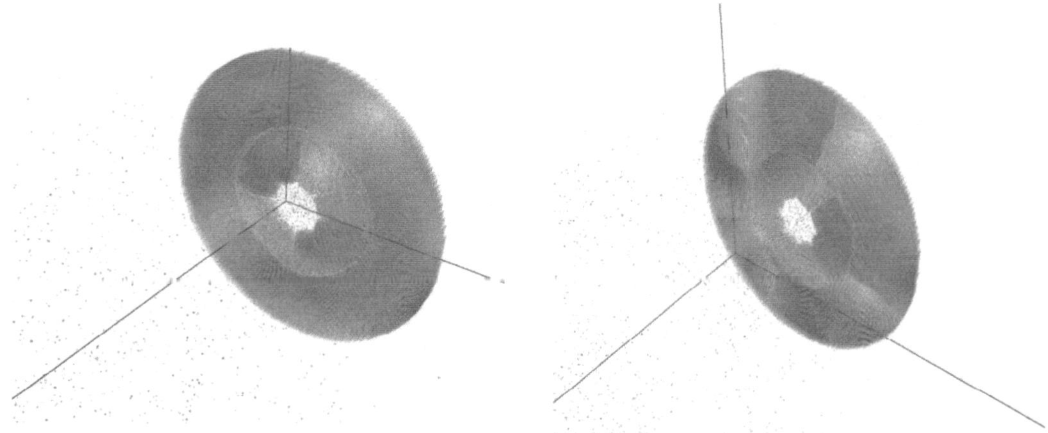

(a) (b)

Fig. 8. Fracture surface and shape of the impacted plate at time step of 2000 for: (a) $\alpha = 0$, and (b) $\alpha = 0.25$ (velocity = 300 m/s and k = 1E20).

4. Conclusions

In this study, analysis of peridynamics has been presented for impact problems of projectile on aluminum target. Effects of indenter stiffness (k) and critical bond stretch parameter (α) on the produced perforation patterns are discussed for various impact velocities. Numerical results show that larger perforation is produced with the increase of k value of indenter. In addition, it is observed that the impacted plate appears to be more deflected, if greater value of α is employed. Evaluation of impact problems with different projectile characteristics would be subjects of future study.

5. Acknowledgements

This work was supported by the Penelitian Dana Lokal ITS Surabaya (No. 1325/PKS/ITS/2018) and partially funded by the Research Grant of PTUPT (128/SP2H/PTNBH/DRPM/2018) Ministry of Research, Technology and Higher Education of the Republic of Indonesia (Kemenristek DIKTI).

References

[1] Reifsnider, K.L. ed., 1991. *Fatigue of Composite Materials*. Amsterdam: Elsevier.
[2] Harris, B. ed., 2003. *Fatigue in Composites*. Cambridge: Woodhead Publishing Ltd.
[3] Schijve, J. ed., 2009. *Fatigue of Structures and Materials*. Netherlands: Springer.

[4] Griebel, M. and Schweitzer, M.A., 2000, "A particle partition of unity method for the solution of elliptic, parabolic, and hyperbolic PDEs", *SIAM Journal of Sciencetific Computing*, 22, pp. 853–890.

[5] Lucy, L.B. (1977), "A numerical approach to the testing of the fission hypothesis", *Astron. J.*, Vol. 82, pp. 1013–24.

[6] Gingold, R.A. and Monaghan, J.J. (1977), "Smoothed particle hydrodynamics: theory and application to non-spherical stars", *Mon. Not. R. Astron. Soc.*, Vol. 181, pp. 375–89.

[7] Atluri, S.N. and Shen, S.P. (2002), *The Meshless Local Petrov-Galerkin (MLPG) Method*, Tech Science Press, USA.

[8] Liu, G.R. (2009), *Meshfree Methods: Moving Beyond the Finite Element Method*, 2nd ed., CRC Press, USA.

[9] Parks, M.L., Seleson, P., Plimpton, S.J., Lehoucq, R.B. and Silling, S.A. (2010), *Peridynamics with LAMMPS: A User Guide v0.2 Beta*, SANDIA REPORT, Sandia National Laboratories, USA.

[10] Silling, S.A. (2000), "Reformulation of elasticity theory for discontinuities and long-range forces", *J. Mech. Phys. Solids*, Vol. 48(1), pp. 175–209.

[11] Silling, S.A. and Askari. E. (2005), "A meshfree method based on the peridynamic model of solid mechanics", *Comput. Struct.* Vol. 83(17), pp. 1526–35.

[12] Bobaru, F., Yang, M., Alves, L.F., Silling, S.A., Askari, E. and Xu, J. (2009), "Convergence, adaptive refinement, and scaling in 1D peridynamics", *Int. J. Numer. Meth. Engng.* Vol. 77(6), pp. 852–77.

[13] Oterkus, S. and Madenci, E. (2012), "Peridynamic analysis of fiber-reinforced composite materials", *J. Mech. Mater. Struct.* Vol. 7(1), pp. 45–84.

[14] Stukowski, A. (2017), *OVITO User Manual*, Germany.

[15] Mathworks (2013), MATLAB User's Guide (R2013a), USA.

Effect of Pulse Parameters on The Particle Size of Copper Powder Electrodeposition

S Wahyudi[1,2,*], S Soepriyanto[3], M Z Mubarok[3], Sutarno[4]
[1] Doctoral program of Mining Engineering-Faculty of Mining and Petroleum, Institut Teknologi Bandung, Bandung, Indonesia
[2] Department of Material & Metallurgical Engineering-Faculty of Engineering, Institut Teknologi Sains Bandung, Bekasi, Indonesia
[3] Department of Metallurgical Engineering-Faculty of Mining and Petroleum, Institut Teknologi Bandung, Bandung, Indonesia
[4] Department of Metallurgical Engineering- Faculty of Engineering, Universitas Jenderal Achmad Yani, Bandung, Indonesia

*Corresponding author (E-mail): solehwahyudi@students.itb.ac.id

Abstract. In this paper, copper powder fabrication has been made via electrolysis method using $CuSO_4$ and H_2SO_4 in aqueous media. Electrodeposition process of copper powder on a metallic surface is done by applying to continue current and pulse current with rectangular pulse waveform. Characterization of copper powder was carried out by SEM to confirm particle size and shape and EDX to determine the copper powder composition. In electrodeposition by applying to continue current to produce the copper powder with particle shape was dendritic, and the elemental composition of the powder was Cu and O. Regarding average particle size, pulse electrodeposition process produces the copper powder with average particle size up to 0.85μm, 72.6% smaller than if apply to continue current.

Keywords: Copper powder, Electrodeposition, Pulse current, Particle size

1. Introduction

Copper is a metal that has excellent optical, electrical and thermal conductivity, magnetic and catalytic properties which in the form of nano-scale powder is widely applied to microdevices, electrical and thermal conductivity, paint, catalyst and lubrication [1,2,3]. Electrolysis is a method that has been commercially applied to produce copper powders. The advantages of copper powder produced by electrolysis method compared with copper oxide reduction method, precipitation, and atomization are high purity, low oxygen content and a wide range of apparent density [4].

The mechanism of the formation of copper powders by electrolysis method occurs on the surface of the cathode under conditions of low copper ion concentrations and high current density than the limiting diffusion current density. Under these conditions, the electrodeposition process of copper powder occurs simultaneously with the evolution of hydrogen gas [4]. The copper powder produced by direct current electrolysis method still produces the powder with dendrite structure and high of particle size. The dendrite structure resembles a tree, which has stems and branches [5]. The dendritic structure formed can result in a non-uniform shape of powder and has a high specific surface area. To control the formation of dendritic structures and reduce the particle size of the powder can be done by shortening the growth time of the nuclei that have been formed. The technique can be applied with a pulse electrolysis [6-10].

In this study, the type of pulse wave applied is a rectangular pulse wave shape, as shown in Figure 1. The regulated pulse parameters are peak cell voltage (E_p), frequency (f), and duty cycle (γ) in order to know the pulse on time or pulse time (T_{on}), the pulse off time or pause time (T_{off}), the period of one cycle of pulse (θ), and average cell voltage (E_m).

$$f = \frac{1}{T_{on} + T_{off}} \quad (1)$$

$$\gamma = \frac{T_{on}}{T_{on} + T_{off}} \quad (2)$$

$$\theta = T_{on} + T_{off} \quad (3)$$

$$E_m = E_p \cdot \frac{T_{on}}{T_{on} + T_{off}} \quad (4)$$

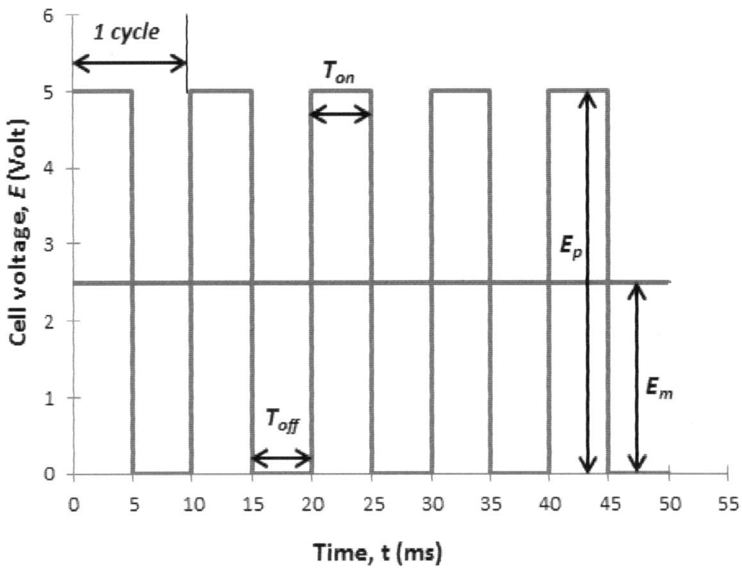

Figure 1. Pulse electrolysis with rectangular waveform at 100 Hz of frequency and 50% of duty cycle.

By adjusting the pulse parameters, conditions of the morphology and the size of the copper powder particles can be controlled. The reasons are since on the T_{off} condition where there is the relaxation on the interface occurred will allow the replenishment of concentration in the diffusion layer; and electroless adsorption, desorption, and recrystallization reactions on the surface. While on the T_{on} condition will occur nucleation [11].

In this study, the comparison of copper powder produced by direct current electrolysis and pulse electrolysis in order to examine as well as the effect of frequency and duty cycle of the pulse on morphology and the particle size of the resulting powder.

2. Materials and Methods

Electrolyte solution containing $CuSO_4.5H_2O$ with copper concentration of 0.01 M and an H_2SO_4 solution with a concentration of 1 M were placed into a single compartment of a two-electrode cell, as is shown in Figure 2. The total volume of the electrolytic bath solution was 100 ml. The anode electrode was a titanium grade 1 metal sheet (25 x 20 mm) and the cathode electrode was a copper 99,99% Cu plate. The deposition was carried out by employing (a) DC power supply and (b) DC-pulse power supply at room temperature (27 ± 1 °C) without stirring.

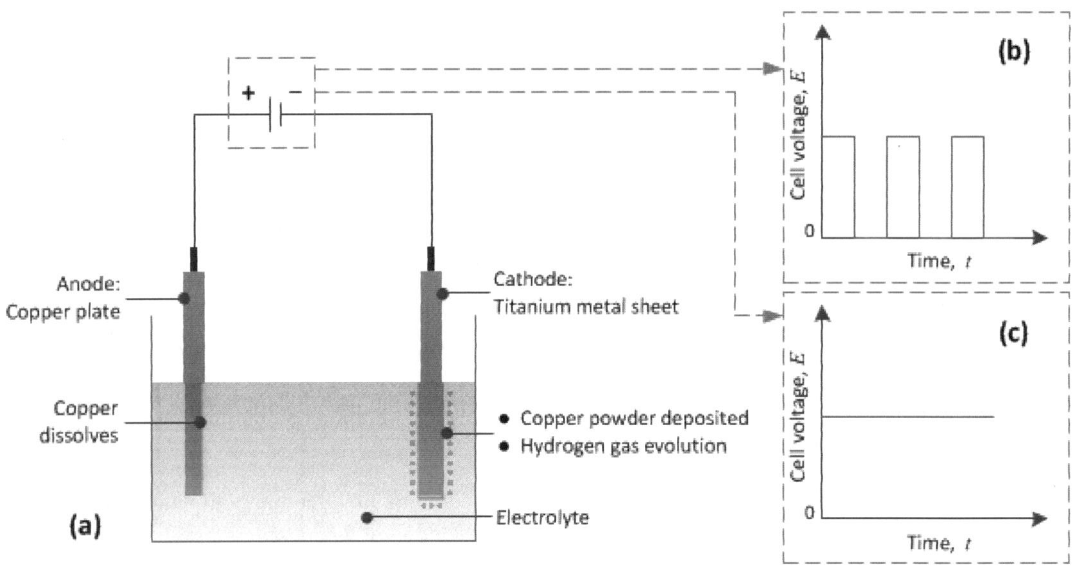

Figure 2. Schematic of (a) the experimental cell, (b) potentiostatic curve for pulse electrolysis, and (c) potentiostatic curve for DC electrolysis used in the experiment.

The cell voltage for DC electrolysis was 4 Volt. The peak cell voltage for pulse current electrolysis was 4 Volt. The pulsed waveform was rectangular with duty cycle variations of 30%, 50%, and 70% and frequency variations of 100, 1000, and 5000 Hz were used in the investigation. The duration of the deposition was set as 30 s to obtain an amount of powder that was sufficient for use in various analytical examinations. The duty cycle and frequency values as pulse electrolysis parameters with rectangular waveforms will produce different ton and toff values. Summaries of the pulse on time (T_{on}), the pulse off time (T_{off}), the period of one cycle of pulse (θ) in pulse electrolysis based on duty cycle and frequency values can be seen in Table 1.

The deposited powder on the cathode was washed slowly and several times using distilled water and then immersed in alcohol immediately. The morphology and composition of the copper powder were determined by using a scanning electron microscope (SEM) and energy-dispersive X-ray analyzer (JEOL Model JED 2300), respectively.

Table 1. Summaries of the pulse on time (T_{on}), the pulse off time (T_{off}), the period of one cycle of pulse (θ) in pulse electrolysis based on duty cycle and frequency values.

γ (%)	f (Hz)	θ (ms)	T_{on} (ms)	T_{off} (ms)
30	100	10	3	7
30	1000	1	0.3	0.7
30	5000	0.2	0.06	0.14
50	100	10	5	5
70	100	10	7	3

The particle size analysis of the SEM captured was performed by using ImageJ 1.52a, image-processing program [12]. Each image was converted to binary 8-bit type of image then using threshold adjustment, the contrast was modified under the B&W setting. Next, converted unit of length of image from pixel to micrometer with the 'scale set'. Check the dialog box of 'area' at set measurement to measure the area of selection in square pixels or in calibrated square units (μm^2) if *Analyze* ▶ *Set Scale…* was used to spatially calibrate the image. The final step was to utilize the 'analyze particles' function then set the show function to 'mask' to display the 8–bit binary image containing filled outlines of the measured particles (Fig. 3) and check the dialog box of 'display result' to produce the measurements of area (A in μm^2) for each particle.

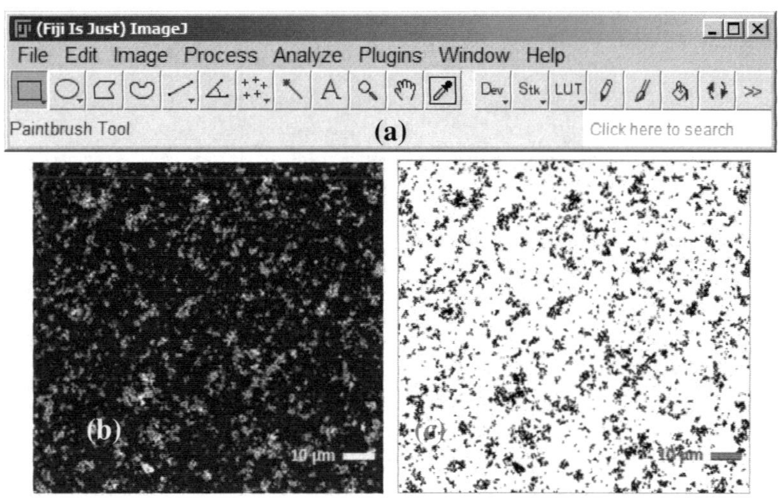

Figure 3. (a) The ImageJ toolbar; (b) Original SEM image and (c) the 8–bit binary image of copper powder.

IOP Conf. Series: Materials Science and Engineering **547** (2019) 012020 doi:10.1088/1757-899X/547/1/012020

The result then saves in files with csv format. With the assumption that the shape of particle is circular, the value of particle size (D in μm) of copper powder can be calculated with the equation below:

$$D = 2 \cdot \sqrt{A/\pi} \tag{5}$$

By using MS Excel, the csv format file of the result can be calculated the average particle size and size distribution of particle size of copper powder.

3. Results and Discussion

During the electrodeposition process of the copper powder, the applied cell voltage was kept constant, and hydrogen gas was formed on the surface of the cathode in addition to copper powder. The effect of duty cycle and frequency variation on pulse electrolysis on the morphology, particle size and size distribution of copper powder will be shown. Then, compared with the copper powder that produced by DC electrolysis.

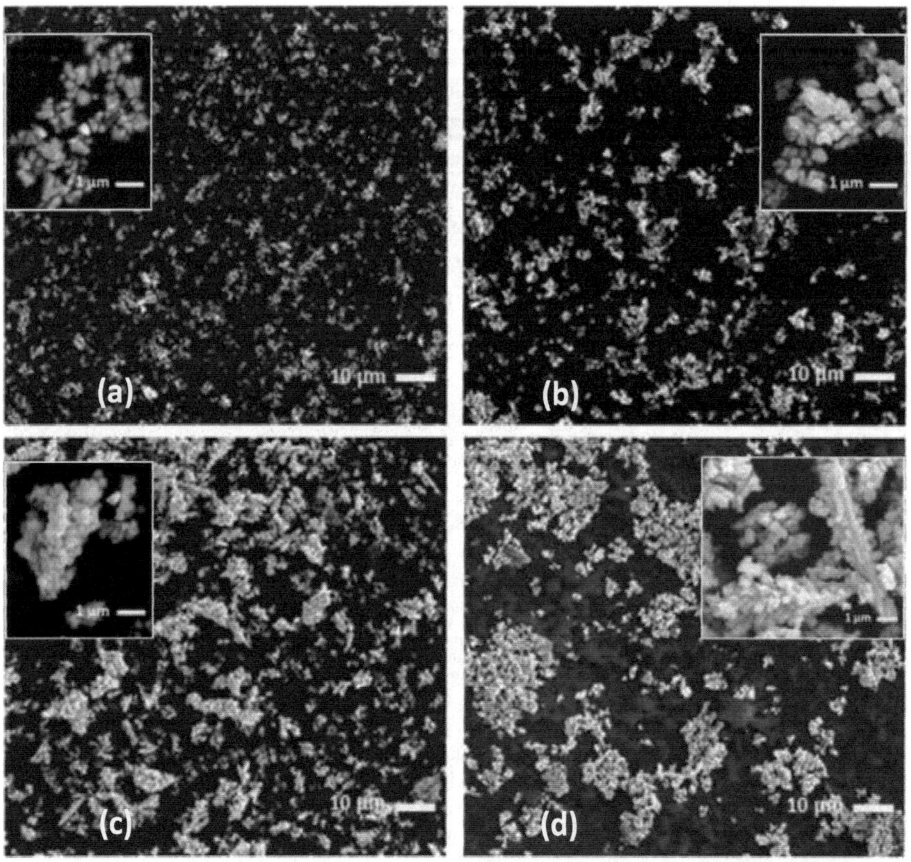

Figure 4. SEM image of copper powder with different duty cycle (a) 30%; (b) 50%; (c) 70%, in pulse electrolysis comparison with (d) SEM image of copper powder in DC electrolysis. The insert show the higher magnification SEM images of the copper powder structures formed.

3.1. Effects of duty cycle on particle size

Pulse electrodeposition of copper powder with duty cycle variation has been done. Figure 4 shows the results of SEM micrograph of copper powder in each duty cycle variation and Figure 5 shows the average particle size of copper powder in each duty cycle variation. As shown in Figure 4, the structure of copper powder particles decreases with decreasing duty cycle. Details of the shrinking of the copper powder structure are shown in the insert image. The decrease in the average particle size of the copper powder was confirmed by the 'analyze particle' results from ImageJ (Fig.5).

Based on Table 1, at fixed frequency, the pulse off time (T_{off}) will be longer if the duty cycle is smaller. T_{off} condition where there is the relaxation on the interface will occur the replenishment of concentration in the diffusion layer; and electroless adsorption, desorption, and recrystallization reactions on the surface [11].

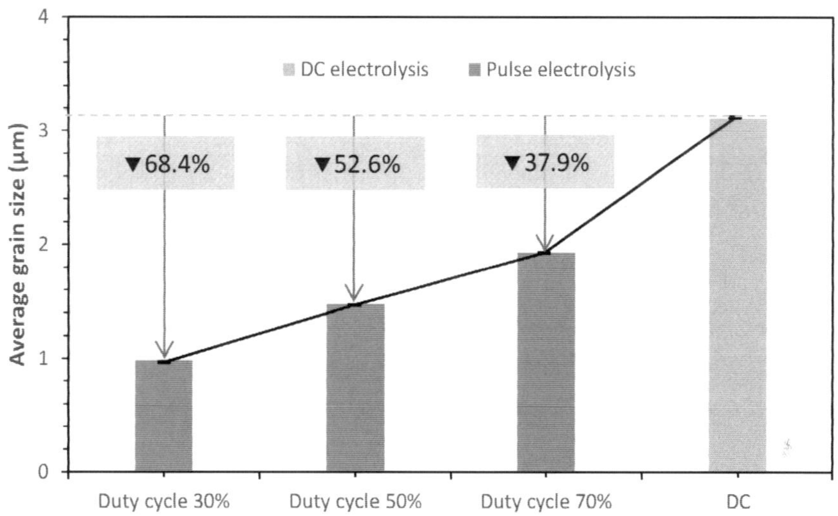

Figure 5. Effect of duty cycle on the average particle size of copper powder compared with the average particle size of copper powder produced from DC electrodeposition.

Compared with the results of pulse electrodeposition, it appears that the copper powder structure resulting from DC electrodeposition (Fig. 4.d) is greater. The copper powder produced by direct current electrolysis method still produces the powder with dendrite structure and high of particle size. The dendrite structure resembles a tree, which has stems and branches [5]. The dendritic structure formed can result in a non-uniform shape of the powder and has a high specific surface area. The average particle size of copper powder at 70% duty cycle was successfully decreased by 37.9% from that obtained by DC electrodeposition and decreased by 68.4% in a 30% duty cycle.

Based on Figure 6, the size distribution of copper powder in DC electrodeposition and pulse (at 70% duty cycle) is more fluctuating and has a longer tail than in pulse electrodeposition with duty cycle were 30% and 50%. By lowering the duty cycle it can decrease the particle size and the size distribution becomes more evenly distributed.

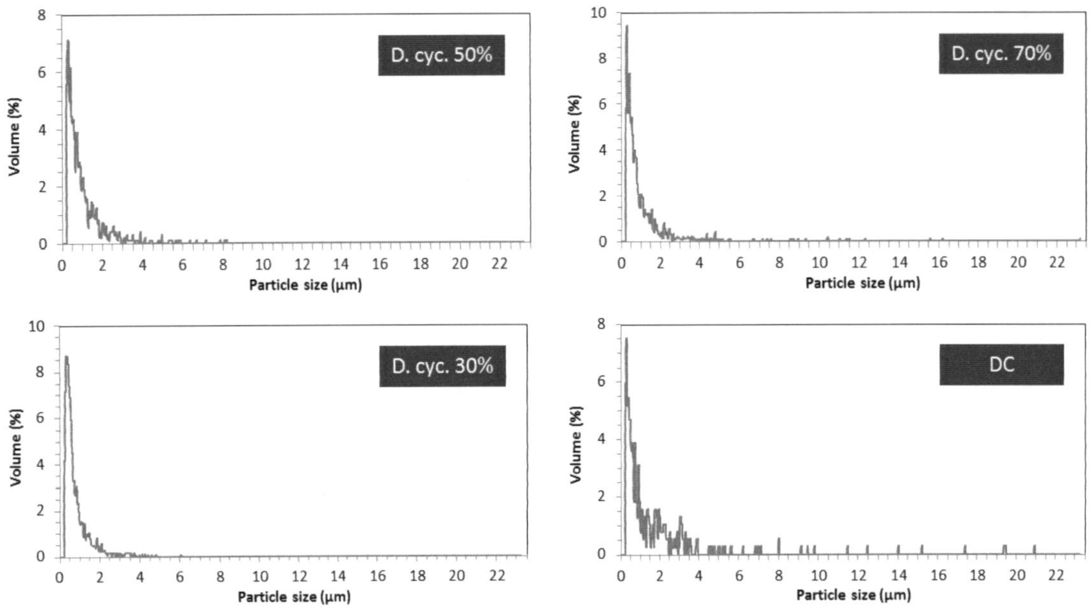

Figure 6. Effect of duty cycle on the particle size distribution of copper powder compared with the particle size distribution of copper powder produced from DC electrodeposition.

3.2. Effects of frequency on particle size

Figure 7 shows the SEM micrograph of the copper powder deposits produced from the pulse electrodeposition on the frequency variation. While in Figure 8 shows the average particle size of copper powder. The average particle size of copper powder will decrease as the pulse frequency increases [10].

In this experiment, the average particle size of copper powder decreased from 981 µm at a frequency of 100 Hz to 852 µm at a frequency of 1000 Hz. When pulse on time (T_{on}), the higher of frequency will be more often formed new nuclei on the surface of the cathode and the time required for growth becomes smaller. Under the same conditions, when pulse off time (T_{off}), is call relaxation time, will also be more frequent so adsorption of copper ion will be more frequent and particle growth more often inhibited. High frequencies can produce high instant overpotential during one pulse cycle so that the nucleation rate can be increased [10,11,13]. Therefore, the growth of the particles is inhibited and the morphology of the dendrite structure diminishes.

However, the average particle size of copper powder at a frequency of 5000 Hz indicates an increase. In the pulse electrodeposition known the transition time. The transition time (τ_t) is the time needed to switch from the end of the pulse time (T_{on}) to the pause time (T_{off}) where if the pulse time (T_{on}) is shorter than the transition time so the chances of producing an increasing powder subtle still exists because the growth of the deposit starts after the transition time. A dimensionless criterion, $k = T_{on}/\tau_t$ has been introduced for the time deposits of the pulse to the transition time in which no new dendrites appear [6].

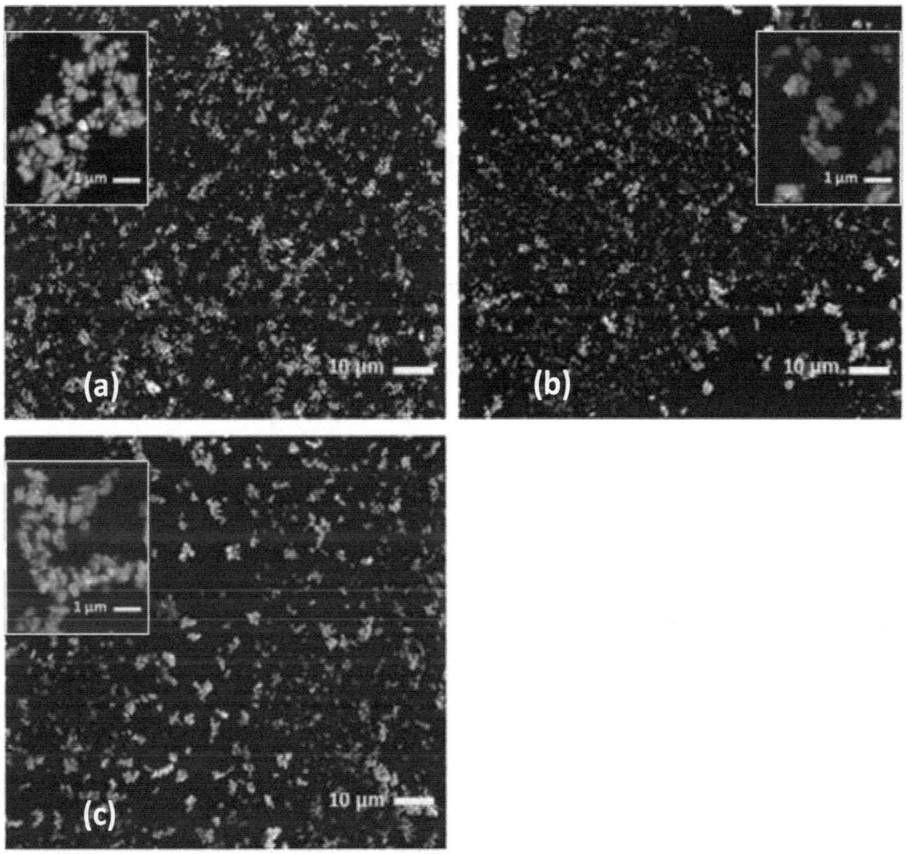

Figure 7. SEM image of copper powder with different frequency (a) 100 Hz; (b) 1000 Hz; (c) 5000 Hz. The insert show the higher magnification SEM images of the copper powder structures formed.

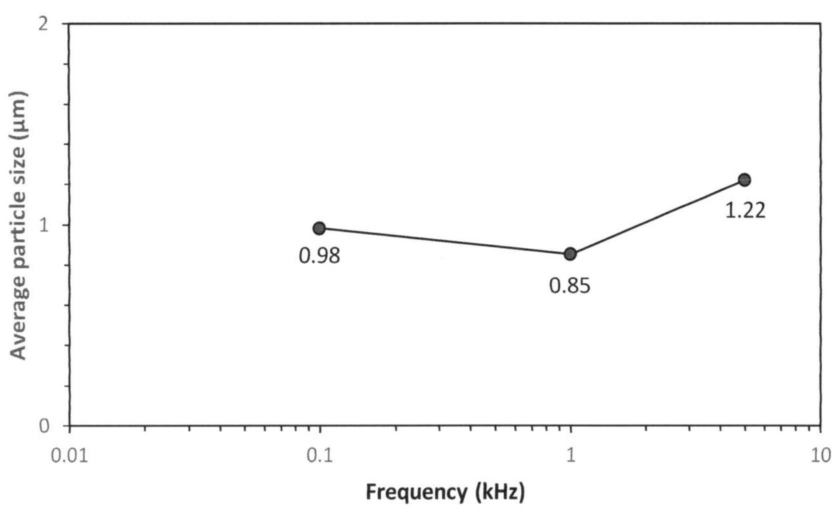

Figure 8. Effect of pulse frequency on the average particle size of copper powder.

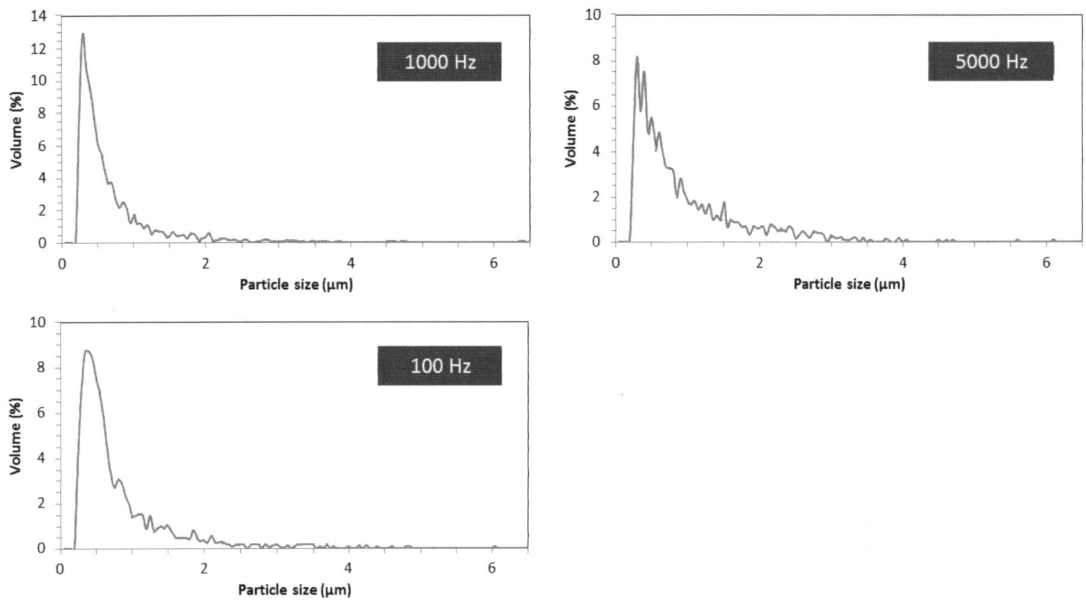

Figure 9. Effect of pulse frequency on the particle size distribution of copper powder.

The Sand's equation [6], the transition time equation in the case of direct current is:

$$I\sqrt{\tau}/S = zFC_0\sqrt{\pi D_0}/2 \; = \; \text{constant} \qquad (6)$$

where I is current, τ is transition time, S is the electrochemically active surface area, C_0 is the ion concentration in the solution bulk, D_0 is the diffusion coefficient of ions.

As the thickness of the deposit grows, electrochemically the active surface area (S) will increase; for I = constant, transition time (τ) will increase for each subsequent pulse [6]. Increasing the size of the copper powder particles occurring at a frequency of 5000 Hz is possible because the transition time is shorter than the pulse time (T_{off}).

Based on Figure 9, the size distribution of copper powder at a frequency of 5000 Hz is more fluctuating and has a longer tail, similar to DC electrodeposition. Compared with frequencies 100 Hz and 1000 Hz, by lowering the pulse frequency below 1000 Hz can decrease the particle size and the size distribution becomes more evenly distributed. Current distribution in pulse electrolysis tends to be smoother than in DC. In pulse electrolysis, the T_{off} is the time of interruption where the nucleus growth stops and the concentration of metal ions lead to equilibrium so that it reduces the thickness of the diffusion layer. In addition, there are also the T_{on} that take place very short. This condition limits the growth time of the nucleus. The mechanism of alternating T_{off} and T_{on} in pulses encourages nucleation and limits growth [14, 15].

3.3. Structure and morphology of copper powder

The SEM micrograph of morphological evolution of copper powder produced under the conditions of 0.01 M copper concentration, 1 M of H_2SO_4, 4 Volt of peak cell voltage at pulse

IOP Conf. Series: Materials Science and Engineering **547** (2019) 012020 doi:10.1088/1757-899X/547/1/012020

electrolysis, room temperature, and a 30-sec. in pulse electrodeposition then compared with DC electrodeposition were shown in Figure 10. The morphology of particles exhibited dendritic growth. This could be attributed to the hydrodynamic conditions surrounding the cathode [7]. The dendrite structure of copper powder in DC electrodeposition is the greatest compared to the dendrite structure of copper powder in the pulse.

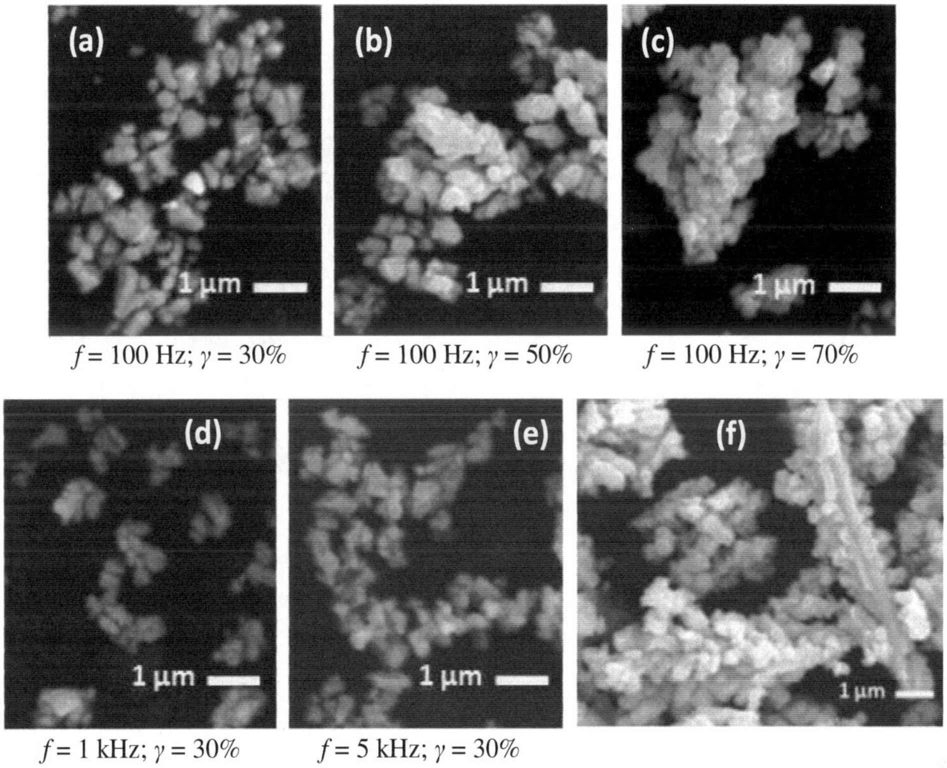

Figure 10. SEM of the morphological evolution of copper dendrites in (a-e) pulse electrolysis and (f) DC electrolysis.

To control the formation of dendritic structures and reduce the particle size of the powder can be done by shortening the growth time of the nuclei that have been formed. The technique can be applied with pulse electrodeposition [6-10]. This is in line with the results obtained in this study in which DC electrolysis yields average particle size of 3.11 µm (Fig. 5) whereas in the pulse can effectively decrease average particle size and decrease the size of the copper powder dendrite structure. The average particle size of copper powder from the pulse resulted between 0.85-1.93 µm (Fig. 5 & Fig. 7).

3.4. Chemical composition of copper powder

Figure 11 shows the EDX analysis of the copper powder produced under the same conditions as those described above. EDX analysis shows that the powder comprised copper and a small amount of oxygen. The oxygen may have been produced as a result of environmental moisture oxidation during drying and storage, as oxidation of particles is difficult to control in these stages [7]. The presence of carbon is attributed to the double-coated conductive carbon tape used to mount the sample.

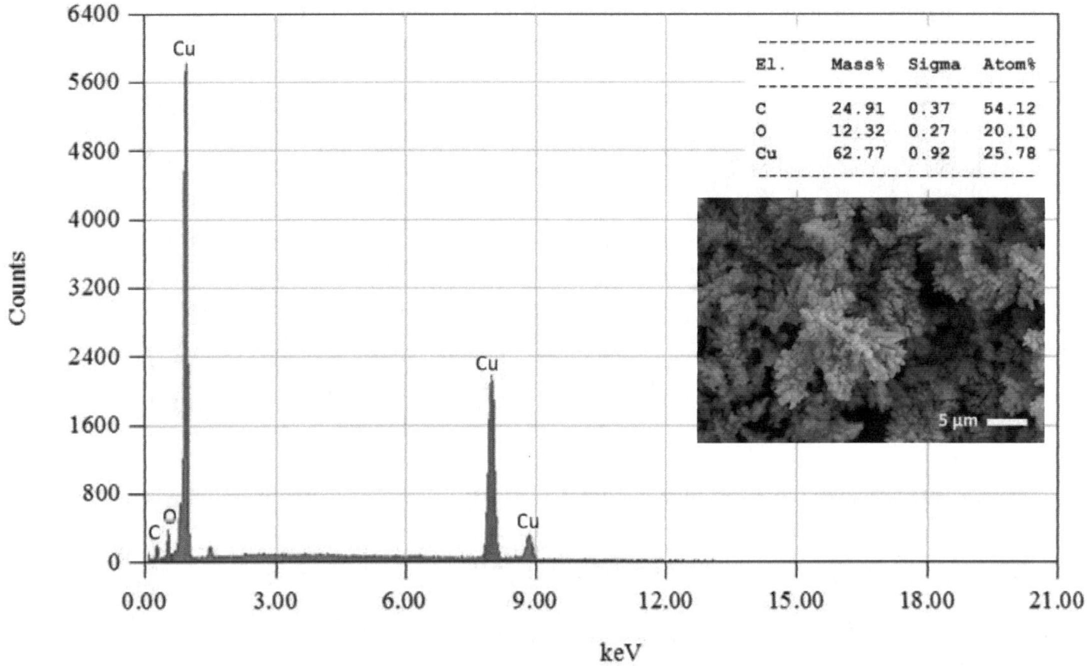

Figure 11. EDX analysis of copper powder.

4. Conclusions

Copper powder was prepared by DC electrodeposition in copper sulfate solution without any additives then compared by pulse electrodeposition with different duty cycle and frequency. Duty cycle and frequency on pulse electrodeposition affect the average particle size, size distribution and morphology of copper powder. The average particle size of the copper powder was decreased with decreasing duty cycle and increasing pulse frequency. However, the average particle size of copper powder will reach a minimum point on a certain larger pulse frequency.

The copper powder produced in both methods has a dendritic structure. However, dendritic structure and particle size of copper powder can be controlled by pulse electrodeposition. Regarding average particle size, pulse electrodeposition produces the copper powder with average particle size up to 0.85µm, 72.6% smaller than if apply to continue current.

5. Acknowledgements

Our participation in IC-DAEM 2018 is supported by the Ministry of Research, Technology and Higher Education of the Republic of Indonesia, and LPDP, for the doctoral scholarship program BUDI-DN 2016 (Grant number: 20161141011049).

IOP Conf. Series: Materials Science and Engineering **547** (2019) 012020 doi:10.1088/1757-899X/547/1/012020

6. References

[1] Bakthavatsalam R, Ghosh S, Biswas R K, Saxena A, Raja A, Thotiyl M Ottakam, Wadhai S, Banpurkarc A G and Kundu J 2016 *RSC Adv.* **6** 8416

[2] Wahyudi S, Soepriyanto S, Mubarok M Z, Sutarno 2018 *IOP Conf. Ser.: Mater. Sci. Eng.* **395** 012014

[3] Mubarok M Z, Aji L N and Wahyudi S 2017 *Prosiding SENAMM X 2017* pp 623 (e-ISBN: 978-602-51621-1-4)

[4] Orhan G and Gezgin G G 2012 *J. Serb. Chem. Soc.* **77** 651

[5] Pavlović M G, Pavlović L J, Maksimović V M, Nikolić N D and Popov K I 2010 Int. J. Electrochem. Sci. **5** 1862

[6] Ostanina T N, Rudoi V M, Nikitin V S, Darintseva A B and Ostanin N I 2017 *Russ. Chem. Bull.* **66** 1433

[7] Nekouie R K, Rashchi F and Joda N N 2013 *Powder Technology* **237** 554

[8] Nekouie R K, Rashchi F and Ravanbakhsh A 2013 *Powder Technology* **250** 91

[9] Nomura K, Shibata N and Maeda M 2002 *J. Electrochem. Soc.* **149** F76

[10] Yao Y, Zhao M, Zhao C and Wang X 2013 *J. Electrochem. Soc.* **160** D553

[11] Nusbaum T, Rosen B A, Gilcadi E and Eliaza N 2015 *J. Electrochem. Soc.* **162** D250

[12] Ferreira T and Rasband W 2012 *ImageJ User Guide* IJ 1.46r ImageJ/Fiji

[13] Nikolic N D and Popov K I 2012 *Electrochemical Production of Metal Powders, Modern Aspects of Electrochemistry* (New York: Springer Science+Business Media) chapter 3 pp 125-185

[14] Puippe J Cl and Ibl N 1980 *J. App.l Electrochem.* **10** 775

[15] Shanthi C, Barathan S, Jaiswal R and Arunachalam R M 2009 *IJEMS* **16** 128

IOP Conf. Series: Materials Science and Engineering **547** (2019) 012021 doi:10.1088/1757-899X/547/1/012021

The Effect of Variation of Nano SiC Reinforcement Particle Addition to Mechanical Properties of Mg/Nano SiC Composite by Stir Casting Method

A Z Syahrial[1*], and N D Hannan[1]
[1]Metallurgical and Materials Engineering Department, Universitas Indonesia, Depok, Indonesia

E-mail: anne@metal.ui.ac.id

Abstract. Composites are advanced material which is being developed that contains of two materials or more in order to improve the mechanical properties. Magnesium composite is a suitable material beside aluminium and its alloy to be applicated as vehicle body structure due to its lightweight and its low density. Thus, the vehicle body structure with lightweight and good mechanical properties can be achieved. In this research, magnesium composite is fabricated by mixing magnesium as composite matrix and nano SiC as reinforcement particle with the variation of the volume fraction 0.05 %v_f, 0.10 %v_f, 0.15 %v_f, 0.20 %v_f, and 0.25 %v_f with stir casting as fabrication methods. To characterize the composite product, several testing were done, which were chemical composition characterization, metallographic observation, SEM-EDS characterization, XRD characterization, hardness testing, wear testing, and impact testing. The result of this research shows that with nano SiC addition as reinforcement particle will increase mechanical properties of magnesium composite. The best mechanical properties of magnesium composite is shown by the addition of 0.15 %vf nano SiC. The result shows the brinnel hardness number of this composition is 39 HB and 0.0052 mm3/m wear rate. Mechanical properties of magnesium composite are increased due to several strengthening mechanism such as grain size reduction and Orowan strengthening mechanism.

Keywords: Magnesium composite, nano SiC, stir casting

1. Introduction

Fuel efficiency on the vehicle is one of the most developed subject in automotive technology. By improving fuel efficiency, the fuel consumption of a vehicle will be lower so as to reduce the costs incurred for fuel and can save the natural resources to fuel the vehicle. The mass of vehicles plays an important role towards the value of fuel efficiency. The higher the mass of the vehicle, the engine will require more fuel to drive the vehicle. Reduction of 10% vehicle load can reduce fuel usage by 6-8%. [1]. Car frame is one part of the vehicle whose mass can be reduced.

Magnesium is a metal that has the lowest density. Since magnesium density is only 1.74 g / cm^3 [2], magnesium can be used as a lighter car frame than a car-made aluminium and steel frame. So in terms of lightness, magnesium is a metal that is suitable to be used as raw material for car

frame with high fuel efficiency. Moreover the availability of magnesium in nature is very abundant because magnesium is the most abundant element of the sixth order on earth (2.7% of the earth's crust composition is magnesium) [3].
To create a car frame with better mechanical properties, magnesium composite can be used instead of magnesium because magnesium composite has the better hardness, stiffness, and wear rate [4]. In composites, there are at least two types of materials used. Both materials act as matrix and reinforcement. The matrix is magnesium. One of the reinforcement that can be used is the nano-sized silicon carbide ceramic particles (SiC) [5]. Ceramic reinforcement has isotropic properties, excellent mechanical properties, chemical inertness properties, stable at high temperatures, and its expansion coefficient can be controlled [6] so that ceramic particles are considered attractive for metal matrix composite fabrication.

2. Materials and Methods

2.1.Materials

In this research, pure magnesium was used as the matrix and nano sized SiC particles with 0.05, 0.10, 0.15, 0.20, and 0.25% volume fraction were used. The term to be used for composite sample according to the volume fraction is sample 1, 2, 3, 4, and 5. The pure magnesium that used in this research were containing 99.8 % magnesium and another element. The other element composition that contained in the pure magnesium can be seen at Table 1.

Table 1. Chemical Composition of pure magnesium ingot

Al (%)	**Be** (%)	**Cu** (%)	**Mn** (%)	**Zn** (%)	**Ag** (%)	**Ca** (%)	**Cd** (%)	**Sn** (%)	**Sr** (%)	**Mg** (%)
<0.005	0.0001	<0.001	0.0466	0.0377	<0.001	0.0023	<0.001	0.0421	0.0007	Bal.

2.2.Experimental Methods

Pure magnesium and nano SiC particle are cleaned in order to remove inclusion particle. After being cleaned, pure magnesium is weighted and inserted into tilting furnace at 800 oC temperature. In parallel, nano-SiC are preheated in order to remove H20 from the powder. Preheating process are being done at muffle furnace at 1000 oC in one hour temperature. Crucible and mold are also being preheated in muffle furnace at the same temperature in thirty and two minu tes in order to avoid thermal shock. After preheat process, insert nano SiC particle into molten magnesium and then stir it for 1 minutes to get homogenous mixture. Then, pour the molten composite into the crucible and then transfer the molten composite into the mold which has been preheated before. Open the mold after the composite are solidified. Solidified composite sample is being used for characterization.

2.3.Characterization

In order to characterize the composite product, several testing were done. There were physical characterization which contain hardness testing, impact testing, and wear testing. Characterization like chemical composition characterization, metallographic observation, SEM-EDS characterization, and XRD characterization were also being conducted.

IOP Conf. Series: Materials Science and Engineering **547** (2019) 012021 doi:10.1088/1757-899X/547/1/012021

3. Result and Discussion

3.1.Chemical Composition

OES (Optical Emission Spectrometry) was performed to analyze the elements contained in the material quantitatively. The purpose of the test is to identify and know the chemical composition contained in the magnesium as-cast sample and the magnesium nano SiC composite sample to know the effect of the casting process on the metal characteristics of the casting. SiC were not observed because SiC is a compound. OES characterization can only observe an element.
According to characterization result, there are some additional elements in the materials tested. This is because in the casting process are not completely clean of impurities so that elements other than magnesium are mixed into the matrix.

Table 2. Chemical compotition of pure magnesium as a matrix and Composite Materials

Elements	Content (Vf%)					
	As-cast	Sample 1	Sample 2	Sample 3	Sample 4	Sample 5
Mg	99.8	99.8	99.8	99.7	99.8	99.8
Al	< 0.005	0.0177	0.0162	0.0169	0.0126	0.0146
Be	0.0001	0.0002	0.0002	0.0002	0.0002	0.0002
Cu	< 0.001	0.0139	0.0148	0.0131	0.0077	0.0059
Mn	0.0466	0.0347	0.0341	0.0462	0.0452	0.0449
Zn	0.0377	0.076	0.048	0.0410	0.0411	0.0419
Ag	< 0.001	< 0.001	< 0.001	< 0.001	< 0.001	< 0.001
Ca	0.0023	0.0021	0.0023	0.0028	0.0025	0.0023
Cd	< 0.001	0.0014	0.0014	0.0015	0.0014	0.0016
Sn	0.0421	0.0640	0.0644	0.0639	0.0632	0.0635
Sr	0.0007	> 0.0008	> 0.0008	> 0.0008	> 0.0008	> 0.0008

3.2.SEM and EDS

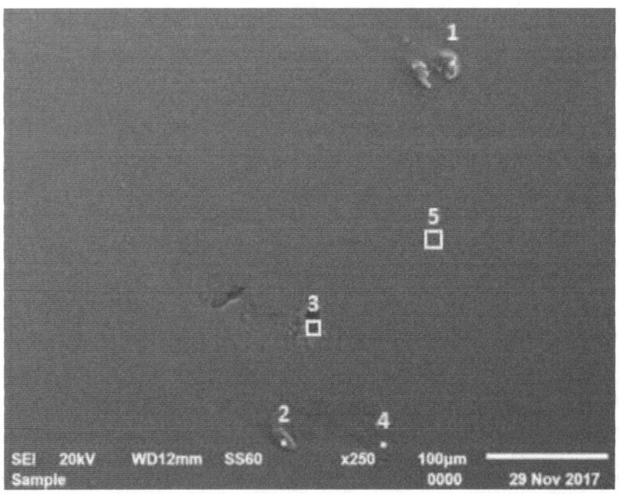

Figure 1. SEM observation result of composite material

SEM and EDS tests were conducted using magnesium composites sample 3 to represent all of the magnesium composite samples. At 250x magnification SEM image was being taken in five locations and can be seen in Figure 1.

Table 3. Chemical composition of phase based on SEM-EDS examination

Location	Content (wt%)					Possible Phase
	Mg	Al	Si	C	O	
1	18.53	0.28	39.46	32.08	9.55	MgO, SiC
2	57.80	1.81	0.73	4.37	35.10	MgO
3	23.26	11.33	0.11	38.26	26.54	MgO, Al_2O_3
4	16.60	0.55	26.40	36.76	19.27	MgO, SiC
5	97.21	0.00	0.00	1.74	0.83	Mg

Location 5 of the EDS test results is the magnesium matrix due to the high magnesium levels at that point and on the SEM result image, the magnesium matrix are represented with grey area. Magnesium content at location 1, 3, and 4 are low when compared with location 5. At location 1 and 4 there are high amounts of Si and C. High levels of Si and C it is possible that at that point there is a SiC thereby which mean there is a SiC content in this sample.
In addition there is an element of oxygen at most of the locations which indicates the creation of rust in the form of MgO. This is because magnesium is a highly reactive metal and very easily oxidized. There are some Al2O3 compound that observed in location 3 which indicates there are some contamination from casting process.

3.3.XRD

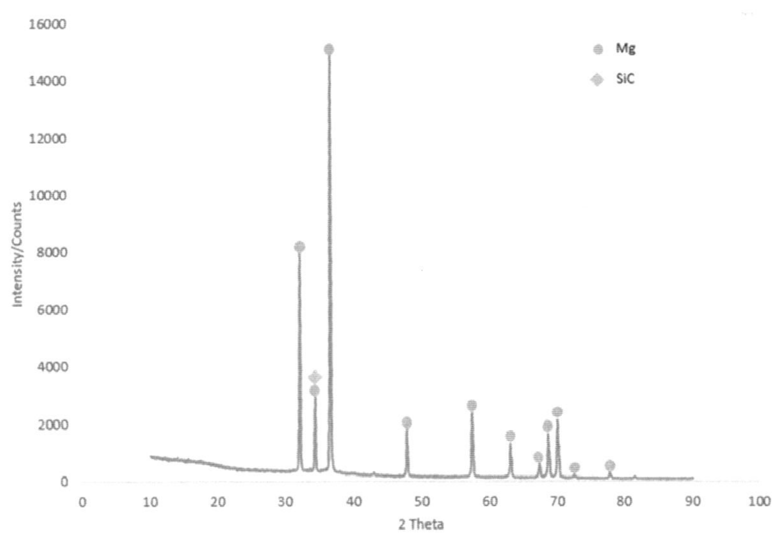

Figure 2. XRD observation result of composite material

XRD test was conducted magnesium composites sample 3 to represent all of the magnesium composite samples. From the data processing, there were 11 peaks at 02θ *i.e.* 32.0974, 34.3334, 36.5434, 47.7494, 57.3434, 62.9594, 67.2754, 68.6014, 69.9264, 72.4754, and 77.9354 can be seen at Figures 2. In the graph of XRD data reading, the magnesium phase is observed in all peaks. This indicates that magnesium as the matrix is the dominant phase in the magnesium composite. Nano SiC reinforcement particle is also seen at the second peak at XRD reading result. The presence of reinforcement particles in magnesium composites will make magnesium composites much better than as-cast magnesium.

3.4. Microsturcture

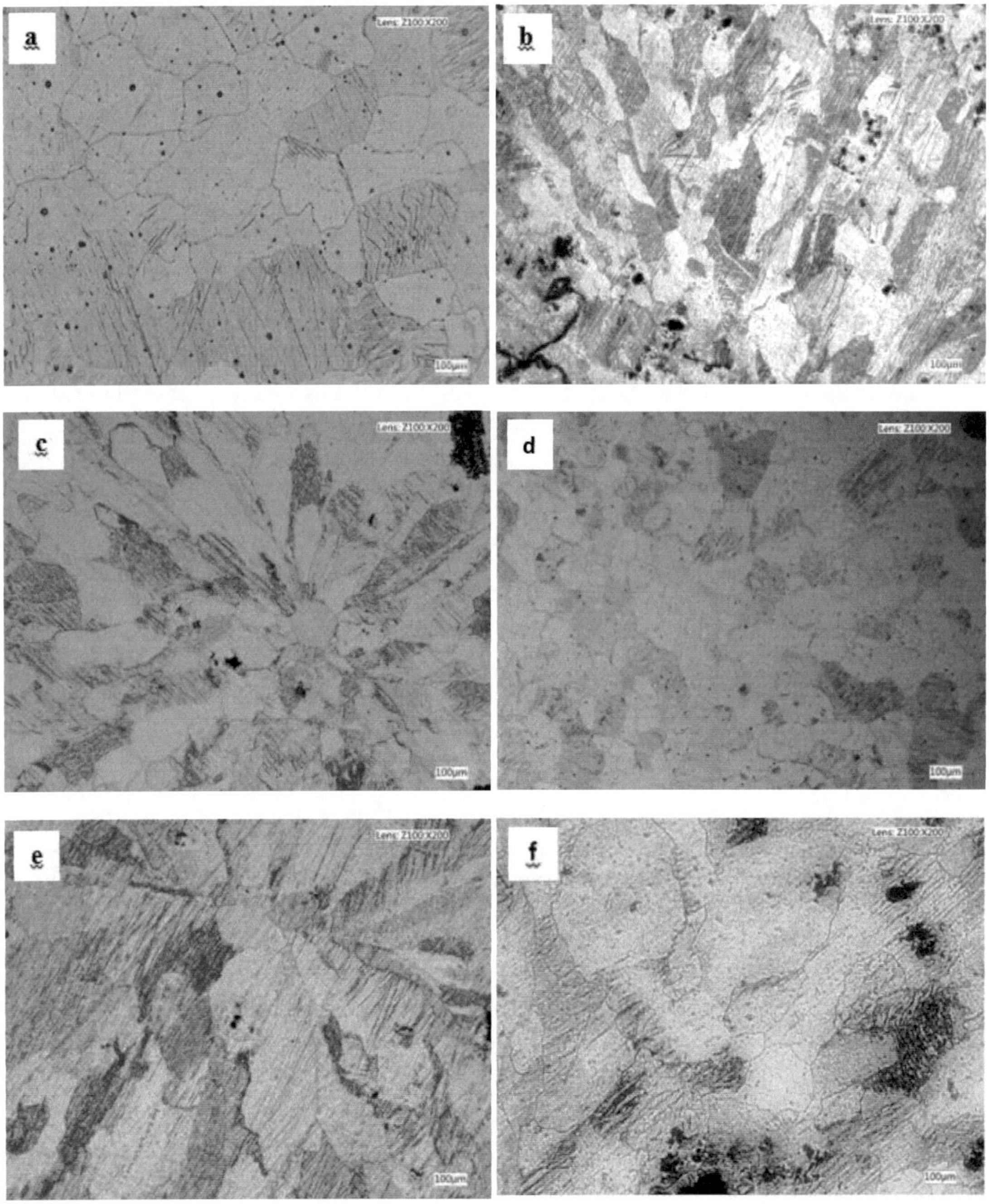

Figure 3. Microstructure images of (a) as-cast, (b) sample 1, (c) sample 2, (d) sample 3, (e) sample 4, and (f) sample 5 of magnesium composites

Microstructural observation of magnesium magnesium composites are being conducted by using digital optical microscopes with magnification 200 times The addition of nano SiC reinforcement particles will lead to a decrease in magnesium grain size as the volume fraction is added [7]. Because the nano SiC reinforcement particles act as a grain refiner. nano SiC reinforcement particles will usually occupy the grain boundary of magnesium and will cause pinning effect which will cause grain growth obstruction due to the presence of reinforcement particles so that the magnesium grains will be smaller [8].
The average grains size of sample 1, 2, 3, 4, and 5 are 201 μm, 198 μm, 174 μm 248 μm, and 264 μm. In this study there is a decrease in grain size on the sample of 1 to 3. However, in sample 4 and 5, the size of the grains become re-enlarged. This could be happened because the greater the addition of volume fraction of reinforcement particles, then the possibility of the reinforcement particle experiencing agglomeration will be greater. Agglomeration of reinforcement particles causes uneven distribution of reinforcing particles in the composite so that the pinning effect of the reinforcing particles against magnesium grains will not be optimal [8].

3.5. Mechanical Properties

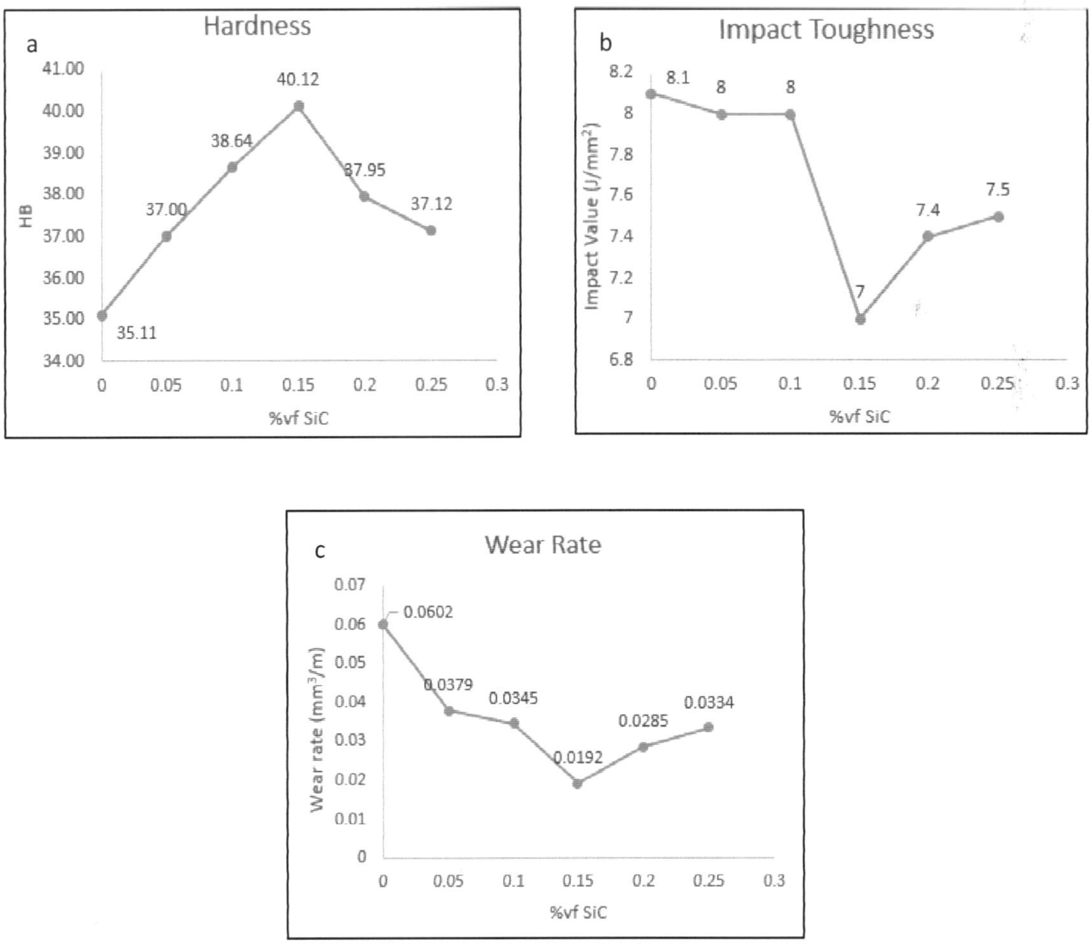

Figure 4. Effects of nano SiC particles addition on the (a) hardness, (b) impact toughness, (c) wear rates characteristics of composite materials

In this experiment, there is an increase in the hardness value of magnesium as-cast to magnesium composites. In magnesium composites, the more reinforcement particles increase the hardness value of the composite. This is because the higher levels of reinforcement particles, the grain size on the matrix will decrease because magnesium matrix grain growth were increased by reinforcement particles.
The smaller the grain size of the matrix then the compressive resistance will be increased. Additionally, the added nanoparticles increase the area of the amplifier so that the matrix bonds will become larger and give rise to Orowan defense mechanism where the deformation resistance will be better because the reinforcing particles act as a barrier to deformation [9]. The optimum hardness value is in sample 3 with 40.12 HB but the value will be decreased in sample 4 and sample 5 due to nano SiC agglomeration. Nano SiC in composites with larger volume fraction of reinforcement particle are easier to be agglomerated [10]. Agglomeration of the reinforcing particles causes a decrease in interface strength between the matrix and the reinforcement particles. Thus, the composite hardness will be decreased [11].
The addition of the nano SiC volume fraction will make the material harder. As the hardness increases, the material properties tend to change from ductile to brittle as a result of decreased ductility [12]. The more material gets brittle then it can decrease the energy that can be absorbed when fracture occurs. This results in a decrease in impact strength because the material will break as less energy is absorbed. The impact toughness value can be seen at figure 4 where sample 3 have the lowest impact toughness with 7 J/mm^2 impact resistance because sample 3 have the highest hardness if compared with another sample.
The reduction of the wear rate occurs with the addition of the nano SiC reinforcement particle to magnesium because reinforcement particles have a contribution in increasing the hardness of the material. The optimum wear rate is in sample 3 with 0.0192 mm^3/m wear rate because the harder the material will result in less abrasion to that material [13]. Increasing the volume fraction of the reinforcement particles can also increase the protection of the matrix surface from contact with other materials because the interface between the matrix and reinforcement particles are increased thus the strength of the interface will prevent the composite from peeling due to the friction of other materials [14].

4. Conclusion

Nano SiC reinforcement particles addition to magnesium matrix shows that magnesium composites have the better mechanical properties expecially in hardness value and wear rate. Nano SiC particles addition to magnesium composite will increase composite hardness with optimum value in sample 3 with 40.12 HB hardness value because the distribution of reinforcing particles are evenly distributed in the matrix which can increase deformation resistance and reduce the grain size. There is a decrease in hardness value in sample 4 and sample 5 due to the agglomeration of reinforcing particles. The hardness value affects the impact toughness and wear rate of magnesium composite. The sample with higher hardness value will have the lower wear rate and impact toughness.

5. Acknowledgements

Authors would like to thanks to Universitas Indonesia for financial support through HIBAH PITTA which help so much in this research.

6. References

[1] U.S. Department of Energy, Lightweighting Material: 2010 Annual Progress Report, DOE/EE-0577 pp. 1. 2011.

[2] Gupta, M., Sharon, N.M.L. Magnesium, Magnesium Alloys, dan Magnesium Composites. John Wiley & Sons, Inc. 2011.

[3] Clark, J. B., Zabdyr, L., Moser, Z, Phase Diagrams of Binary Magneisium Alloys, ASM International, Metals Park, Ohio, USA, 1988.

[4] Paramsothy, M., Hassan, S. F., Srikanth, N., Gupta, M. Simultaneous Enhancement of Tensile/Compressive Strength and Ductility of Magnesium Alloy AZ31 Using Carbon Nanotubes, J. Nanosci. Nanotechnol. 10 pp. 956-964. 2010.

[5] Dey, A., Pandey, K.M. Magnesium Metal Matrix Composite – A review. 2015. Rev. Adv. Mater. Sci. 42. Pp 58-67. 2015.

[6] Paramsothy, M, Chan, J., Kwok, R., Gupta, M. The Effective Reinforcement of Magnesium Alloy ZK60A Using Al2O3 Nanoparticles, J. Nanopart. Res. 13. pp. 4855-4866. 2011.

[7] Shen, M.J., Wang, X.J., Ying, T., Wu, K., Song, WJ. Characteristics and mechanical properties of magnesium matrix composites reinforced with micron/submicron/nano SiC particles. Journal of Alloys and Compounds volume 686 pp. 831-840. 2016.

[8] Matin, A., Saniee, F.F., Abedi, H.R. Microstructure and mechanical properties of Mg/SiC and AZ80/SiC nano-composite fabricated through stir casting method. Materials Science & Engineering A volume 625. Pp 81–88. 2015.

[9] Aravindan, S., Rao, P.V., Ponappa, K. Evaluation of Physical and Mechanical Properties of AZ91D/SiC Composites by Two Step Stir Casting Process. Journal of Magnesium and Alloys 3 (1), pp. 52-62. 2015.

[10] Muhammad, W.N.A.W., Sajuri, Z., Mutoh, Y., Miyashita, Y. Microstructure and Mechanical Properties of Magnesium Composites Prepared by Spark Plasma Sintering Technology. Journal of Alloys and Compounds. Volume 509 pp. 6021-6029. 2011.

[11] Sreekumar V, Babu NH, Eskin D, Fan Z. Structure–Property Analysis of In-situ Al–MgAl2O4 Metal Matrix Composites Synthesized Using Ultrasonic Cavitation. Mater Sci Eng: A. Volume 628 pp.30–40. 2015.

[12] Ye, H.Z., Liu, X.Y. Review of recent studies in magnesium matrix composites. Journal of Material Science volume 39 pp. 6153 – 6171. 2004.

[13] Archard J.F. Contact and rubbing of flat surfaces. Journal of Applied Physics, 24: pp. 981-988. 1953.

[14] Sharma, S.C., Anand, B., Krishna, M. Evaluation of Sliding Wear Behaviour of Feldspar Particle-Reinforced Magnesium Alloy Composites. Wear Journal. Volume 241 pp. 33-40. 2000.

IOP Conf. Series: Materials Science and Engineering **547** (2019) 012022 doi:10.1088/1757-899X/547/1/012022

Corrosion Behavior Of New Type Titanium Alloy As Candidate For Dental Wires In Artificial Saliva On Fluctuating Temperatures

Jon Affi[1], Fadhli Ihsan[1], Hidayatul Fajri[2], Gunawarman[1*]
[1]Mechanical Engineering Department, Faculty of Engineering, Andalas University, Padang, Indonesia
[2]Biomedical Science, Faculty of Medicine, Andalas University, Padang, Indonesia
*Corresponding author (E-mail): gunawarman@eng.unand.ac.id

Abstract. Titanium alloys are widely used for implant material due to high biocompatibility properties. It is well known that the corrosion resistance of titanium alloys is better than other metallic materials. However, if the alloys are applied for unknown condition such as wire in orthodontic application, the fluctuating temperature is important to be considered as parameter which influences on corrosion rate. In this study, the effect of fluctuating temperature on corrosion behavior of TNTZ was investigated. Two conditions of TNTZ were used in this study, i.e., Ti-29Nb-13Ta-4.6Zr *aging treatment* (AT) and TNTZ *solution treatment* (ST). For comparison, the commercial Ti6Al4V ELI was also used in this study. The corrosion rates of the samples were characterized in artificial saliva on fluctuating temperatures range between 10^0C and 50^0C by immersion testing. The results showed that the lowest corrosion rate for each variation of temperature was TNTZ ST. The corrosion process seems to decrease the hardness of TNTZ up to 26 VHN. The result also showed that the corrosion rate of TNTZ is much lower than that of Ti6Al4V in all cycles. Moreover, the hardness of Ti6Al4V decreases more significantly i.e., up to 77.2 VHN. The artificial saliva with fluctuating temperature induces corrosion process and decreases the hardness of the alloys.

Keywords: TNTZ, fluctuating temperatures, artificial saliva, corrosion rate, hardness.

I. Introduction

Titanium alloy has been used widely in the medical application such as orthodontic and orthopedic.[1]–[3] Compared with other traditionally metallic biomaterials, such as stainless steel and cobalt-chromium, the alloys have superior biocompatibility, lower modulus young, and enhanced corrosion resistance.[1] The attempting to upgrade the biocompatibility of titanium alloys in research is growing due to low biomedical performance of titanium still founds in some cases such as corrosion product attached at the tissue, implant breaks after long period's implantation, and etc. Several Ti alloys have been developed, such as Ti-6Al-4V and Ti-6Al-7Nb for their enhancement wear resistance and improvement mechanical strength[2], [3], Ti15Zr4Nb4Ta and Ti29Nb13Ta4.6Zr with better corrosion resistance and mechanical properties, [4], Ti25Ta25Nb and Ti25Ta25Nb3Sn related to their on cytotoxic property [5].

The corrosion rate of biomedical metal is more often quantified using simulated (artificial) physiological solutions which model the human body fluid. Generally, the simulated solutions consist of salts which are almost similar to actual physiological solution content. Therefore, the using of simulated solution for evaluation of corrosion performance is accepted. Corrosion resistance of the metals is generated by continues oxides which is covered surface against oxygen. However, the contact of actual physiological solution with metals causes wear at surface and lead localized breakdown. Utilizing of Ti alloys as biomaterial implants has shown tendencies to break after long term using in replacing the hard tissue. [6]. In the cavity of the mouth, the using of metal for dental material exhibiting higher degradation degree if the environment with fluctuating temperature, pH varieties, high humidity, stress, and microorganism [7]. The effect of metals corrosion behavior contact with fluids may be different between them. The temperature variation may also effect on the corrosion rate of metals. In order to verify the corrosion rate of the biomedical implant at fluctuating temperature, the metals were immersed in the artificial saliva for certain temperature in hot and cold alternately by different time scheme.

II. Material and Methods

Two types titanium alloy were used in this experiment namely TNTZ and commercial Ti6Al4V ELI. TNTZ was provided by condition as solution treatment (ST) and as aging treatment (AT). The commercial Ti6Al4V ELI was also prepared as a standard of comparison test pieces. All corrosion samples were cut into disc shape with the thickness around 1-3 mm and diameter around ±15 mm so those suitable to be flatten using rotating of sand paper on polishing machines up to #2000 grids. The samples were mirror polished by using suspended of alumina powder in water at cloth polishes. The artificial saliva was prepared according to "Afnor" solution with composition shown in Table 1. The temperature of solution was set to hold at 10°C and at 50°C alternately.

Table 1. Chemical composition of "Afnor" artificial saliva. [7]

Elemen	NaCl	KCl	KH_2PO_4	$NaHCO_3$	NaH_2PO_4	KSCN	UREA
g/l	0,7	1,2	0,2	1,5	0,26	0,33	0,13

All samples were immersed for 42 days (6 weeks) with three scheme hot-cold cycles as shown in Figure 1. In the scheme 1, the test pieces were held for 21 days at 50°C and 21 days at 10°C. In the scheme 2, the test pieces were held for 7 days at 50°C and for 7 days at 10°C and the cycle was repeated up to 42 days by 3 cycles. In the scheme 3, the test pieces were held for 3 days at 50°C and 3 days at 10°C and it was also repeated consecutively up to 42 days (7 cycles).

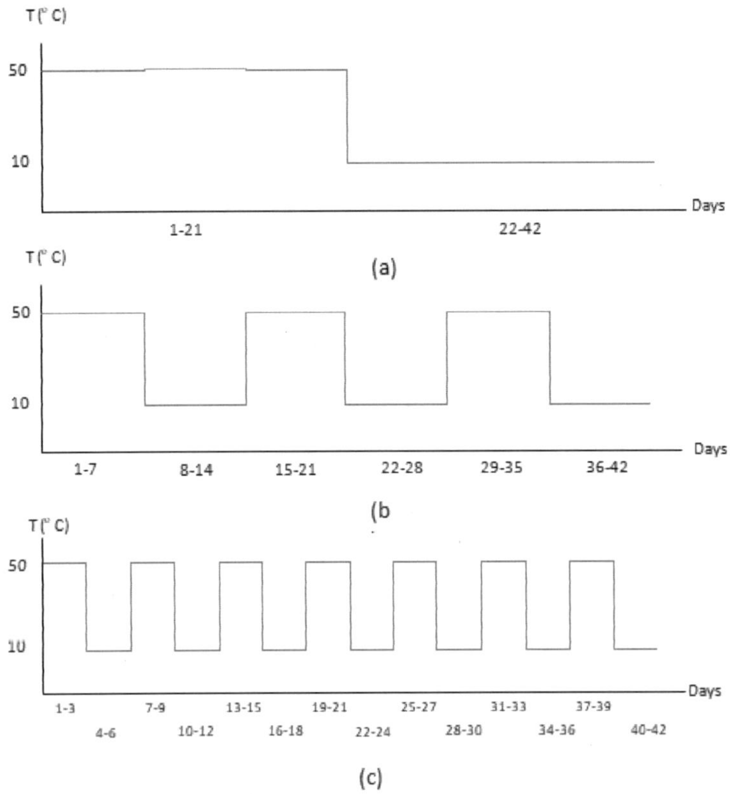

Figure 1. Scheme of testing variation; (a) 1 cycle, (b) 3 cycles, and (c) 7 cycles

Artificial saliva was put on the bottle with controlled environment. The low temperature of 10°C and the high temperature of 50°C were maintained by using the combination of a circulating heating simulator (Vivo Circulator iTherm RT2) and small ice blocks put on bottle surrounding. If the temperature drops below 10°C, the heating simulator will operate to increase temperature. Meanwhile, if temperature higher than 10°C, the heating simulator will operate by circulating ice blocks mixed in water to reduce temperature. Similarly, the high temperature of 50°C has resulted from the heater of the heating simulator.

The corrosion rate of samples was calculated by weight loss methods according to ASTM G1, Standard Practice for Preparing, Cleaning, and Evaluating Corrosion Test Specimens.

Characterizations of samples were performed by using *Scanning Electron Microscope* (SEM, S-3400N Type II). The initial condition of samples surfaces before corrosion was also evaluated using SEM for preliminary data. The hardness of samples before and after corrosion were evaluated with V*ickers hardness tester* (Shimadzu HMV-2).

III. RESULT AND DICUSSION

3.1. Effect of immersion cycles hot-cold on corrosion rate

Figure 2 and Figure 3 show the quantitative corrosion rate of titanium alloy in "*afnor*" artificial saliva. Figure 3 was constructed from data in Figure 2 using formulation for corrosion rate based on ASTM G1 i.e:

$$Cr = (K \times W)/(D \times A \times T)$$

Where :	Cr	= Corrosion rate (miles per year)
	Constant for unit conversion	= 3.45×10^6
	W	= weight losses due to corrosion (g)
	D	= Density of metal (g/cm^3)
	A	= Surface area (cm^2)
	T	= Time (hours)

There is no significant corrosion rate difference between those types of titanium alloys. Both as solution treatment and an aging condition of TNTZ elucidated similar corrosion rate after 42 days. The decline in the weight of the samples occurred due to formation metallic ion that suspended in the solution. As seen in Figure 2, weighing degradation is comparable with hot-cold immersion cycles i.e. more cycles resulted more weight reduction. It occurred for all types of titanium alloys. As seen in the Figure 2, the low corrosion rate occurred on samples with low cycles. Meanwhile, the samples with a high number of cycles have more weight reduction after completing 6 weeks immersed in the solution. Based on this result, the corrosion rate of dental wire would be affected by how often someone consumes meal at hot or cold temperature.

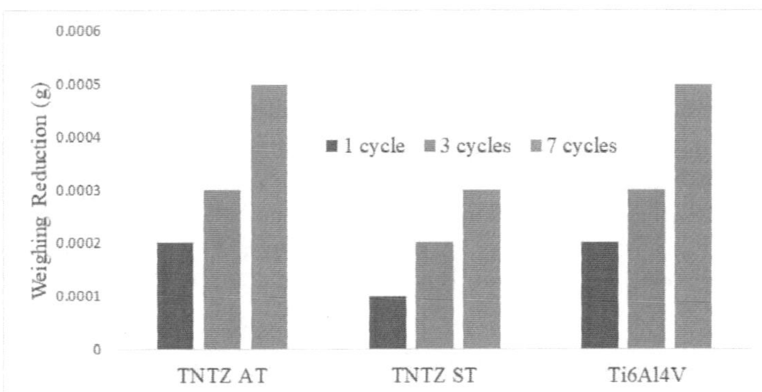

Figure 2 Weight Reduction of Titanium Alloys in "Afnor" Solution after 42 days

From Figure 2 is also revealed that weight reduction of Ti6Al4V Eli relatively similar to that of TNTZ AT for all cycles. However, TNTZ ST has the lowest corrosion rate than the TI6Al4V has. The results are agreed with previous study[8], [9]. It was found that the corrosion rate of TNTZ ST is lower than corrosion rate of the Ti6Al4V after being tested by potentiodynamic methods at the constant temperature. This is due to present of Niobium and Zirconium elements in TNTZ which are the highly passivating metals. [10] The elements improve the corrosion resistance by the formation of more oxides at the surface of titanium such as ZrO_2, TiO, Nb_2O_3, and Ta_2O_5. It strengthens the TiO_2 passive film layer from oxygen attack. Moreover, non-toxic and non-allergic elements of Ti, Ta, Nb, and Zr support good biocompatibility with the tissues.[11], [12]

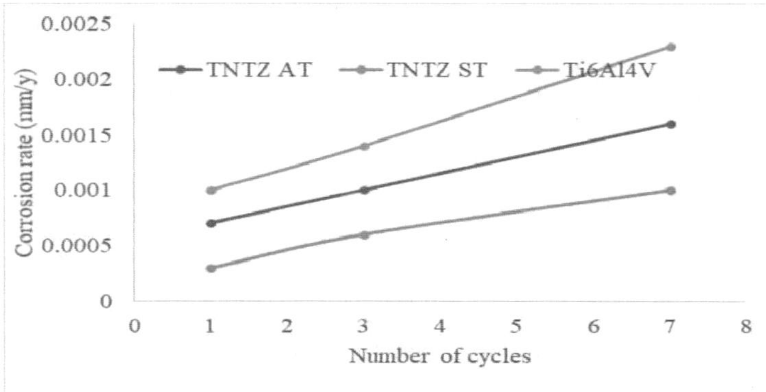

Figure 3 Corrosion Rate of Titanium Alloys in "Afnor" Solution after 42 days.

3.2 Observation of the surface morphology

Observation of surface corrosion were performed using SEM with 1000x magnification. As shown in Figure 4, the corrosion indication is marked as color differentiation, scrapping, or pitting in the surface of samples. The first row of the Figure 4 is condition of as polished samples for TNTZ ST, TNTZ AT and Ti6Al4V ELI respectively. As polished of TNTZ ST is homogenous distribution of single β phase. On the other hand, TNTZ AT and Ti6Al4V produce different surface morphology. Both of metal appears with other particle distributed on matrices β. As shown in Figure 4, the size of the second phase of Ti6Al4V seems to be larger than the second phase of TNTZ AT.

Along with increasing the cycle, crevice and pitting corrosion advanced in surface area. Samples of TNTZ ST with 1 hot-cold immersion cycle have induced pitting corrosion partially. An increasing in the number of cycles has promoted higher surface degradation where has shown as the pitting modes (Figure 4.b) changes into crevice of corrosion (Figure 4.c). Finally, the samples with 7 cycles yield more area of surface corrosion (Figure 4.d). Most of the surfaces have corroded. Similarly, the step of surface degradation also occurred on TNTZ AT samples but the corrosion appears worse than TNTZ ST samples as shown in Figure 4.g with larger crevice trace. Ti6Al4V samples revealed higher corrosion rate than TNTZ (Figure 4j- Figure 4l). There are some deep pitting and crevice at the surface of 3 and 7 cycles samples.

3.3 Hardness of samples

The hardness testing was performed in order to evaluate the effect of the corrosion process on the samples surface. The hardness of all samples showed in Figure 5. The initial hardness of samples before immersion in artificial saliva for TNTZ ST, TNTZ AT, and Ti6Al4V are 176.2 VHN, 332 VHN, and 346.8 VHN respectively. This result appropriates with the natural properties of the metal which content single or multiphase. TNTZ AT and Ti6Al4V consist of two-phase in the microstructure and thereby they have higher in hardness. The microstructure of multiphase has obtained increasing the strength through second phase strengthening mechanism

.

The Figure 5 shows that the sharp drop of hardness belonging to Ti6Al4V in which the total hardness reduction was is 77.2 HVN. Meanwhile, the lowest hardness reduction of TNTZ ST was 26 HVN. It proved that the solution treatment condition has better corrosion resistance than the aging condition or multiphase microstructure.

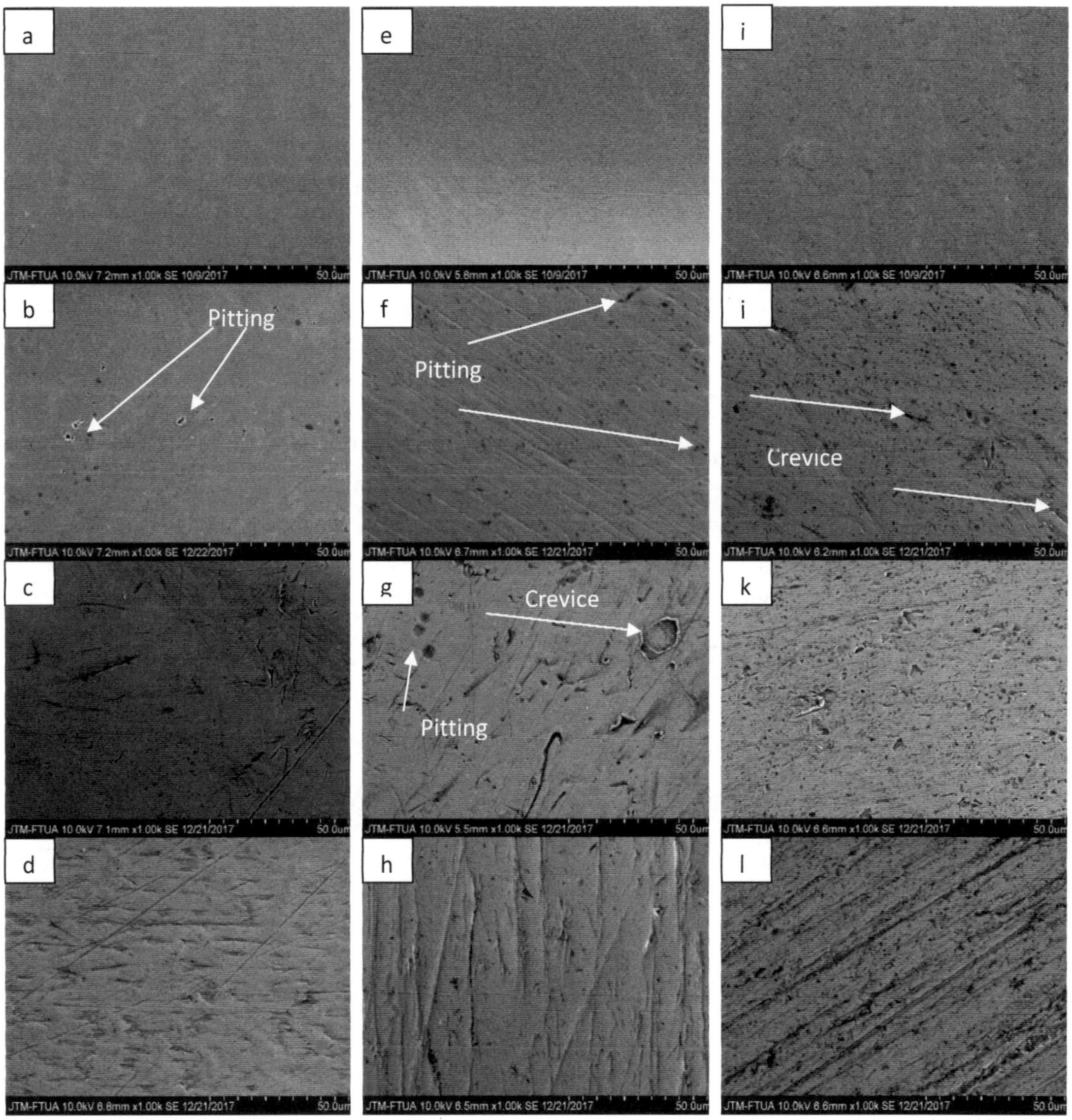

Figure 4 SEM results of observation samples, the note (a, b, c, d) are TNTZ ST samples. (e, f, g, h) are TNTZ AT samples and (i, j, k, l) are Ti6Al4V. The first row showed the surface feature before immersion. The second, third and fourth row is the surface feature of 1 cycle, 3 cycles, and 7 cycles immersion, respectively.

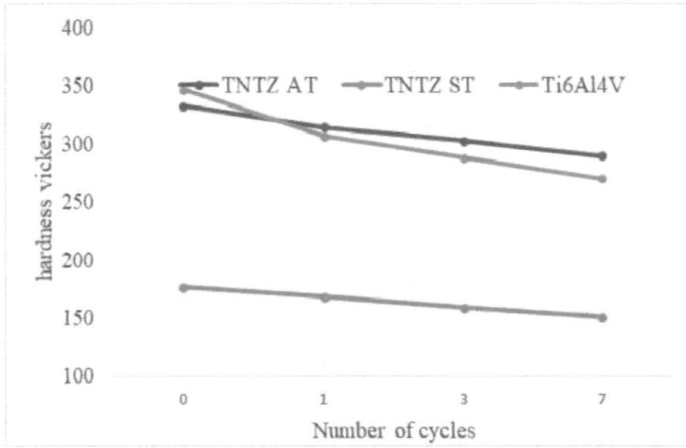

Figure 5 Hardness of samples before and after immersion

IV.CONCLUSIONS

Based on the result, we may conclude that:

1. The corrosion rate of implantable titanium alloys was influenced by hot-cold immersion cycles in artificial saliva. The corrosion rate was proportional with number of hot cold immersion cycles although they were immersed in similar periods (42 days).
2. The titanium alloys with solution treatment has better corrosion resistance than titanium multiphase for all hot-cold immersion cycles.
3. The lowest hardness reduction after corrosion has also occurred at solution treatment of titanium alloys.

ACKNOWLEDGEMENT

This work was support by "Skim Claster Riset Guru Besar" of Andalas University with contract number 19/UN.16.17/PP.RGB/LPPM/2018.

REFERENCES

[1] M. Long and R. HJ, "Titanium alloys in total joint replacement - A materials science perspective," *Biomater*, vol. 19, pp. 1621–1639, 1998.

[2] M. F. Semlitsch, H. Weber, R. M. Streicher, and R. Schön, "Joint replacement components made of hot-forged and surface-treated Ti-6Al-7Nb alloy," *Biomaterials*, vol. 13, no. 11, pp. 781–788, 1992.

[3] D. Iijima, T. Yoneyama, H. Doi, H. Hamanaka, and N. Kurosaki, "Wear properties of Ti and Ti-6Al-7Nb castings for dental prostheses," *Biomaterials*, vol. 24, no. 8, pp. 1519–1524, 2003.

[4] A. Fukuda *et al.*, "Bone bonding bioactivity of Ti metal and Ti-Zr-Nb-Ta alloys with Ca ions incorporated on their surfaces by simple chemical and heat treatments," *Acta Biomater.*, vol. 7, no. 3, pp. 1379–1386, 2011.

[5] M. R. Seixas, C. Bortolini, A. Pereira, R. Z. Nakazato, K. C. Popat, and A. P. R. Alves Claro, "Development of a new quaternary alloy Ti-25Ta-25Nb-3Sn for biomedical applications," *Mater. Res. Express*, vol. 5, no. 2, 2018.

[6] B. G. Pound, "Corrosion behavior of metallic materials in biomedical applications. I. Ti and its alloys," *Corros. Rev.*, vol. 32, no. 1–2, pp. 1–20, 2014.

[7] K. Małkiewicz, M. Sztogryn, M. Mikulewicz, A. Wielgus, J. Kamiński, and T. Wierzchoń, "Comparative assessment of the corrosion process of orthodontic archwires made of stainless steel, titanium–molybdenum and nickel–titanium alloys," *Arch. Civ. Mech. Eng.*, vol. 18, no. 3, pp. 941–947, 2018.

[8] D. D. Giat Mana, "Perilaku korosi TNTZ dan Ti-6Al-4V dalam Lingkungan Artficial Salifa," Andalas University, 2015.

[9] A. Satria Putra Amin, "Karakteristik korosi titanium beta (ti-Beta) dalam air ludah buatan pada tempertur konstan 36°C," Andalas University, 2017.

[10] Y. I. Yoshimitsu Okazaki, K. Kyo, and T. Tateishi, "Corrosion resistance and corrosion fatigue strength of new titanium alloys for medical implants without V and A1," *Mater. Sci. Eng. A*, vol. 213, pp. 138–147, 1996.

[11] M. Abdel-Hady Gepreel and M. Niinomi, "Biocompatibility of Ti-alloys for long-term implantation," *J. Mech. Behav. Biomed. Mater.*, vol. 20, pp. 407–415, 2013.

[12] M. Niinomi, "Recent metallic materials for biomedical applications," *Metall. Mater. Trans. A*, vol. 33, no. 3, pp. 477–486, 2002.

IOP Conf. Series: Materials Science and Engineering **547** (2019) 012023 doi:10.1088/1757-899X/547/1/012023

Performance of Vanadium Doped and Carbon Bamboo/ Carbon Black Coated Lithium Iron Phosphate for Battery Cathode

A Z Syahrial[1], W C Pawito[1], N Sofyan[1], W M F Rahmatulloh[1]

[1]Departement of Metallurgical and Materials Engineering, Universitas Indonesia, Depok, 16424, Indonesia

Email: anne@metal.ui.ac.id

Abstract. Synthesis of lithium iron phosphate ($LiFePO_4$) via wet chemical followed by a hydrothermal method has been carried out. The preparation of $LiFePO_4$ was begun with the precursor of LiOH, $NH_4H_2PO_4$, and $FeSO_4.7H_2O$ mixed stoichiometrically. After the synthesis, $LiFePO_4$ was doped using vanadium and then coated using two types of carbon sources, i.e. carbon black blended with activated carbon pyrolyzed from bamboo, through a solid-state reaction. The materials were mixed using a ball-mill and subsequently characterized using a thermal analyzer (STA) to determine the sintering temperature. The result shows that $LiFePO_4$ formation temperature is at 639^oC. The sintering process was performed for 4 hours and the characterization was done using X-ray diffraction (XRD) and electron microscope (SEM) equipped with energy dispersive x-ray spectroscopy (EDX). The electrochemical impedance spectroscopy (EIS) testing was performed to show the conductivity of the formed materials. XRD result showed that $LiFePO_4$/V/C phase has formed with an olivine structure, while the SEM result showed fair distribution and small particle size with some agglomerated microstructures. The EIS results showed that carbon coating on the active material increases the conductivity, whereas the addition of vanadium increases the conductivity of up to 5 wt.% vanadium but decreases at 7 wt.% vanadium.

Keywords: LiFePO4, hydrothermal, bamboo carbon, carbon black, battery cathode

1. Introduction

Currently, environmentally friendly energy sources such as water, wind, and solar radiation has been widely used. To store this kind of energy, a material with a large capacity such as lithium ion battery is required [1]. One kind of lithium ion battery active materials is lithium iron phosphate ($LiFePO_4$). This material has received much attention due to its simple olivine-based structure, in addition to its environmentally friendly, low material cost, good chemical stability, high theoretical capacity of 170 mAh/g and good potential (~ 3.45 V vs. Li/Li+) [2]. However, there are some drawbacks of this material, such as low intrinsic electronic conductivity (about 10^{-9}-10^{-11} S/cm) and low diffusion coefficients (about 10^{-11}-10^{-13} cm^2/s) depending on Li+ concentration and characterization method used [3]. Because of that, there has been considerable development in improving the electrochemical performance of $LiFePO_4$ cathodes, among them with conductive materials to improve its electrical conductivity [4], wherein the carbon layer is a promising and effective method [5]. The effect of carbon coating alone, however, decreased its tap density, therefore doping with the supervalent cation is commonly used to improve tap density [6].

In this study, a combination of carbon layer source (activated carbon from bamboo and carbon black) was performed. Activated carbon from bamboo as a reducer, mixed with raw materials, forming a mixture of precursors. Carbon black acts as a carbon coating source, added in the second step, wrapping the precursor particles. The addition of a supervalent cation element such as vanadium is also performed to form a number of active electrochemical compounds that can intercalate/deintercalate reversible lithium.

2. Experimental

2.1 $LiFePO_4$ cathode synthesis

Firstly, LiOH, $NH_4H_2PO_4$, and $FeSO_4.7H_2O$ powder were prepared stoichiometrically. The powders were dilute using distilled water up to 30 mL into a separate beaker glass. Further, the three solutions were mixed and stirred for homogenization in a magnetic stirrer for 30 minutes. The solution was then move to a Teflon autoclave and was heated at 180°C for 20 hours. The solution was move to a beaker glass to settle until the solids were separated from the solution. The solid was filtered and washed with deionized water to obtain a pH of 5. After all the process was completed, the precipitate was put in an oven at 80°C for 18 hours. Finally, the powder was ready for the next treatment.

2.2 $LiFePO_4$/V/C fabrication

The previously prepared $LiFePO_4$ powder was mixed with activated carbon pyrolyzed from bamboo, and carbon black, and H_4NO_3V with a composition of 92:2:2:5% wt. $LiFePO_4$ and H_4NO_3V powder were inserted into a ball miller. After the milling process, activated carbon powder was then fed into the $LiFePO_4$ and H_4NO_3V powders. The same procedure was carried out for other percentage variations of 0, 3, 5, and 7 wt.% of H_4NO_3V. After all the process of milling were set, thermal test on 4 variations of the sample was carried out. Thereafter, the materials were calcined at 750°C for 4 hours. The samples were then subjected to characterization using SEM, XRD, and EIS.

2.3 Battery fabrication and testing

All of these manufacturing processes were performed in a glove box on the split test cell. The electrode sheets were prepared and circular cut using a disc puncher. The material was arranged in sequence from below; bottom casing, cathode, separator, anode, and top casing. Further, the battery was filled with $LiPF_6$ electrolyte solution and tested using split test cell to obtain a potential difference in the component.

3. Results and discussion

3.1 Hydrothermal Synthesis Phase Formation $LiFePO_4$

Figure 1 shows STA test results from the $LiFePO_4$ cathode. It appears that the heat flow increases with the addition of temperature. This is because the sample continues to oxidize due to the absence of nitrogen or argon gas so that the test conditions are not inert. Then, on another graph or on the TGA there is a continuous weight increase at 10.71 at 242°C to 11.12 mg at 493°C. Then, rise slowly to 11.91 mg when the temperature is 639.18°C. Then it drops to 11.01 mg at 959.7°C.

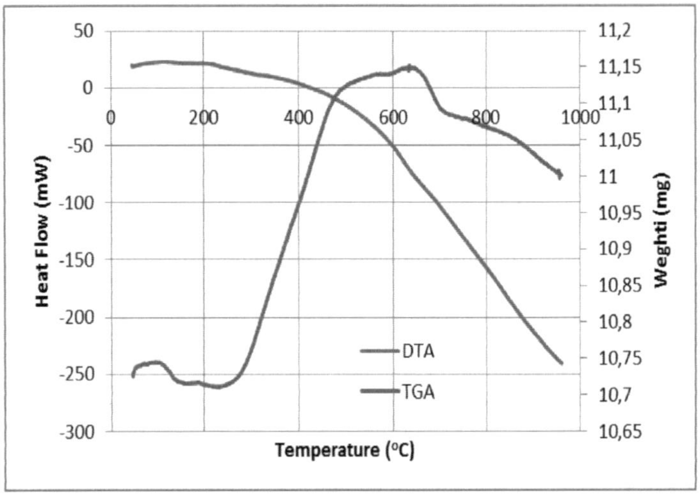

Figure 1. STA Graph with a heating rate of 10°C/min and a gas flow of 20 ml/min

3.2 Diffraction Patterns

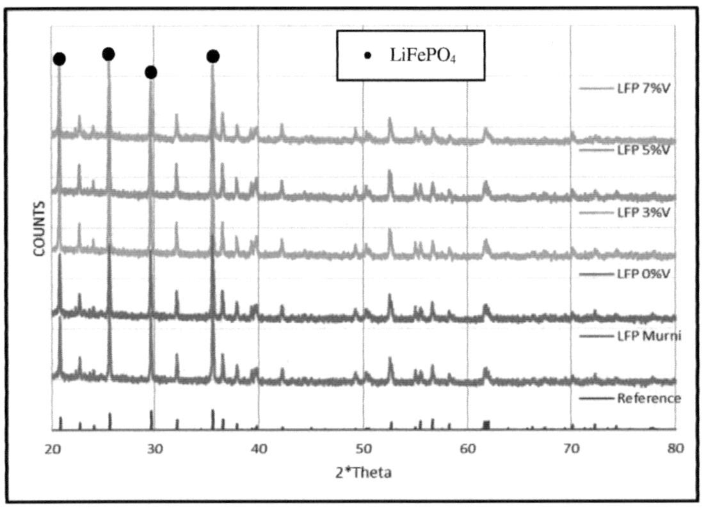

Figure 2. The comparison graph of the $LiFePO_4$ cathode diffraction pattern with various variables

It can be seen in Fig. 2 that the diffraction pattern of the $LiFePO_4$ cathode sample with the various variables of carbon coating difference and the addition of vanadium having a peak corresponding to the reference PDF2 diffraction pattern is the olivine structure with the orthorhombic Pmnb space group. However, a shift in the $LiFePO_4$/ V 7% sample was not significant due to the formation of a new phase $Li_3V_2(PO_4)_3$ in which the amount of vanadium addition would result in a peak shift. This vanadium content can only be seen with a TEM (Transmission Electron Microscopy) due to its very little percentage which only serves as a doping on $LiFePO_4$.

3.3 Morphological and Distribution of LiFePO4

Figure 3(a) is a result of a pure $LiFePO_4$ cathode having angular crystal structure and agglomerated by heterogeneous sizes. This is because the effect of sintering temperature is high enough to change the amorphous structure into crystalline so that the surface area is

IOP Conf. Series: Materials Science and Engineering **547** (2019) 012023 doi:10.1088/1757-899X/547/1/012023

smaller and the particles are stick together and form particles of large size. For Fig. 3(b) EDS results where the three spectra show the same result, composed $LiFePO_4$ compounds. But there is little difference in the amount of its element content. In the first spectrum there is a considerable amount of Fe content and on the third spectrum the content of the oxygen element sufficiently affects the color of the image.

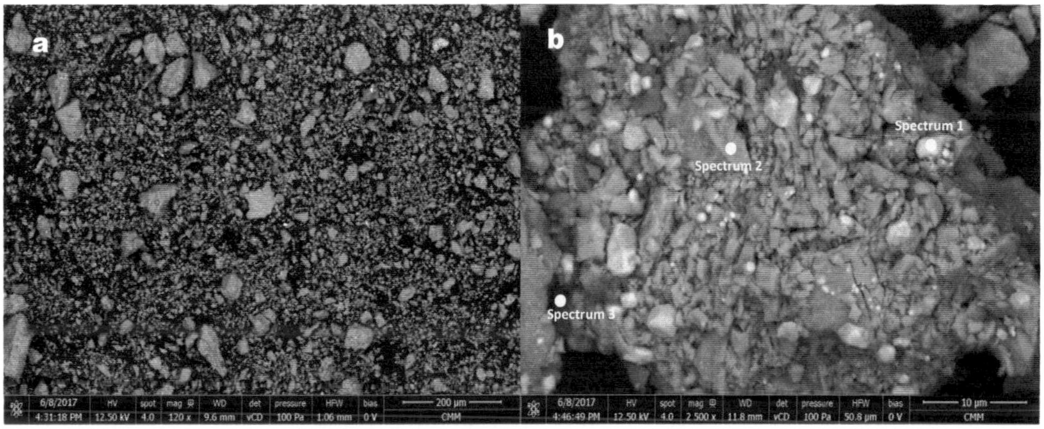

Figure 3. SEM results of the synthesized $LiFePO_4$ cathode and EDS point with magnification (a) 120X (b) 2500X

Figure 4. SEM $LiFePO_4$ cathode with carbon without vanadium doping and EDS point with magnification 2500X

Figure 4. SEM results of $LiFePO_4$ cathode material of carbon coating without vanadium doping. At 2500X magnification, there are some irregular shapes but a fairly homogeneous size, a small white sphere. Figure 6. EDS results where spectrum 1 and 2 show the presence of carbon that does not coat $LiFePO_4$. Spectrum 3 shows $LiFePO_4$ coated by carbon.

Figure 5. SEM $LiFePO_4$ cathode with carbon and doping 5% vanadium and EDS point with magnification 2500X

Figure 5. shows SEM results of $LiFePO_4$ cathode material with carbonaceous coating and 5% vanadium doping show reduced particle size and more evenly where vanadium reduces the carbon content to decrease the valence from V^{5+} to V^{3+} [6]. Figure 8. shows no significant difference where both spectra indicate $LiFePO_4$ which has been coated by carbon. However, in spectrum 2 it does not show the presence of vanadium in $LiFePO_4$ whereas in spectrum 1 there is little vanadium content.

3.4 EIS characterization of $LiFePO_4$ electrodes

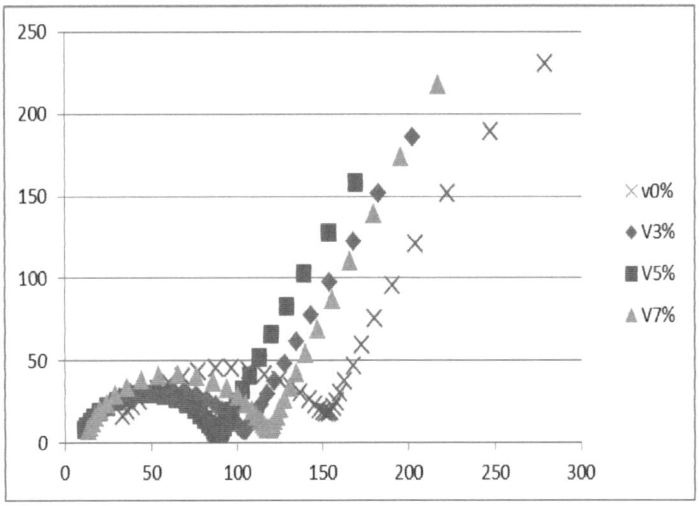

Figure 6. EIS test results on LFP samples with various variables

Figure 6 shows LiFePO4 sample results without vanadium doping (LFP-V0%) has a greater resistance value, which is 119 Ω. For samples of $LiFePO_4$ with vanadium doping (LFP-V3%, LFP-V5%, and LFP-V7%) had a smaller resistance value of 92 Ω, 77 Ω, and 105 Ω, respectively. This can be proven by the diameter of semicircle on the graph. In theory, the addition of vanadium will increase the conductivity of the active material, but the excessive addition of LFP-V7% may carry an impurity phase such as $Li_3V_2(PO_4)_3$ and block the diffusion channel Li^+, which will adversely affect the electrochemical performance [7]. Table 1. shows the comparison between carbon black or activated carbon from bamboo with using a mixture of both [8,9]. As a result, LFP-V0% with an activated carbon mixture of bamboo and carbon black has a larger conductivity value and has a larger Li rate as well, but at LFP-V5% the results look somewhat different.

Table 1. Comparison of total resistance and conductivity with different carbon coatings and vanadium doping

Sample	Rct (Ω)	t (cm)	A (cm^2)	σ (S/cm)
Carbon Black				
LiFePO4/C/0V	158	0.008	2.01	2.52 x 10-5
LiFePO4/C/5V	59	0.008	2.01	6.75 x 10-5
Activated Carbon Bamboo				
LiFePO4/C/0V	197	0.008	2.01	2.02 x 10-5
LiFePO4/C/5V	91	0.008	2.01	4.37 x 10-5
Mixed Carbon Black and Activated Carbon Bamboo				
LiFePO4/C/0V	119	0.008	2.01	3.34 x 10-5
LiFePO4/C/5V	77	0.008	2.01	5.17 x 10-5

4. Conclusion

The carbonization process of bamboo was successful and produced high crystallinity so that when mixed with carbon black yielded a low resistivity value of about 119 Ω of vanadium 0% addition. While the addition of vanadium 5% showed higher resistance value than using only carbon black and lower than using only bamboo activated carbon alone. The SEM test results show smaller and homogeneous particle size although some of the agglomerated particles are compared with pure $LiFePO_4$.

5. Acknowledgement

We would like to thank to Ministry of Research and Higher Education Republic of Indonesia for funding this research through Partial Hibah PITTA No. 822/UN2.R3.1/HKP. 05.00/2017, and Hibah Stranas Institusi No. 545/UN2.R3.1/HKP05.00/2018

6. Reference

[1] Yuan, L. X., Wang, Z. H., Zhang, W. X., Hu, X. L., Chen, J. T., Huang, Y. H., & Goodenough, J. B. (2011). Development and challanges of LiFePO4 cathode material for lihium-iron batteries. *Energy & Enviromental Science*. 4, 269-284.

[2] Bi, Z., Zhang, X., He, W., Min, D., & Zhang, W. (2013). Recent advances in LiFePO4 nanoparticles with different. *RSC Advances*. 3, 19744.

[3] Zhang, Y., Huo, Q., Du, P., Wang, L., Zhang, A., Song, Y., … Li, G. (2012). Advances in new cathode material LiFePO4 for lithium-ion batteries. *Synthetic Metals*. *162*(13), 1315–1326.

[4] Long, Y. F., Su, J., Cui, X. R., Lv, X. Y., and Wen, Y. X. (2015). Enhanced rate performance of LiFePO4/C by co-doping titanium and vanadium. *Solid State Sci*. vol. 48, pp. 104–111.

[5] Wang, Y., Mei, R., & Yang, X. (2014). Enhanced electrochemical properties of LiFePO4/C synthesized with two kinds of carbon sources, PEG-4000 (organic) and Super p (inorganic). *Ceramics International*. *40*(6), 8439–8444.

[6] Zhang, L.-L., Liang, G., Ignatov, A., Croft, M. C., Xiong, X.-Q., Hung, I.-M., … Peng, Y. (2011). Effect of Vanadium Incorporation on Electrochemical Performance of LiFePO 4 for Lithium-Ion Batteries. *Chemical Engineering*. *115*(27), 13520–13527.
[7] Ming, C., Leng-Leng, S., Hua-Bin, Y., Tie-Zhen, R., Gaohui, D., Zhong-Yong, Y. (2015). Vanadium-doping of LiFePO4/carbon composite cathode materials synthesized with organophosphorus source. *Electrochimica Acta*, vol. 167: p. 278-286
[8] Sofyan, N., Putro, D. Y., Zulfia, A. (2016). Performance of Vanadium-Doped LiFePO4/C Used as Cathode for Lithium Ion Battery, *International Journal of Technology* Vol 7 No 8 932-9406.
[9] Sofyan, N., Alfaruq, S., Zulfia, A., Subhan, A., Characteristics of vanadium doped and bamboo activated carbon coated $LiFePO_4$ and its performance for lithium ion battery cathode, J. Kimia Kemasan, Vol. 40 No. 1 April 2018, in press.

Development Of Short Carbon Fiber Reinforced Aluminium Matrix Composites By Low Pressure Infiltration Process

Xuan Meng[1], Yongbum Choi[2]*, Kazuhiro Matsugi[2], Wentao Hu[3] and Wenchang Liu[3]

[1] Department of Mechanical Science and Engineering, Graduate School of Engineering, Hiroshima University, 1-4-1 Kagamiyama, Higashi-Hiroshima, 739-8527, Japan
[2] Division of Materials and Production Engineering, Graduate School of Engineering, Hiroshima University, 1-4-1 Kagamiyama, Higashi-Hiroshima 739–8527, Japan
[3] State Key Laboratory of Metastable Materials Science and Technology, Yanshan University, Qinhuangdao, 066004, China.

*Corresponding author, E-mail: ybchoi@hiroshima-u.ac.jp

Abstract. Short carbon fiber reinforced aluminium based matrix composites were fabricated by a low-pressure infiltration process. This study is amid for development of short carbon fiber reinforced aluminium matrix composites. Short carbon fibers of 10 vol.% was used to preparing preforms for the low-pressure infiltration process. Afterwards, pure Al and Aluminium alloy under temperature of 1073K were infiltrated into the preforms under an applied pressure of 0.4MPa. Microstructure and Vickers hardness as well as microstructure–property relationships of these materials were studied. Microstructure observations indicated that: the SiO_2 binder was coating on the surface of short carbon fiber and distributed at the corner of carbon fibers in preforms. Short carbon fibers were homogeneously distributed in the matrix. Furthermore, properties of short carbon fiber reinforced pure Al and A336 alloy composites showed that: the relative density was 95.8% and 97.2%, respectively. Hardness of short carbon fiber reinforced pure Al composite increased by 76% comparing to the base pure Al, and short carbon fiber reinforced A336 alloy composite increased by 10% comparing to the A336 alloy.

Keywords: Aluminium matrix composite, Short carbon fiber, Low-pressure infiltration, Microstructure, Mechanical properties.

1. Introduction

Aluminium matrix composites are now attracting extensive attentions, due to their major advantages in both processing and final properties such as high strength and high thermal property[1, 2]. Carbon fiber (CF) has been widely used as very important reinforcement to fabricate high performance aluminium and aluminium alloys composite[3]. Consideration of carbon fiber as reinforcement for aluminium matrix composite are low density, high specific strength and stiffness, low coefficient of thermal expansion (CTE), and high electrical/thermal conductivity[4-6]. Therefore, the carbon fiber reinforced Al matrix composites have the great expectation to be applied as functional and structural materials in the future.

Pure Al possesses high thermal conductivity and electrical conductivity, whereas, the low mechanical property limits its application. The development of short carbon fiber reinforced pure Al composite aimed to improve the mechanical property for using as functional materials. A336 alloy was selected because of good fluidity and presence of Si which improves the alloy casting ability, the Mg which improves the wettability of reinforcement by the matrix and the Ni as well as Cu which improve tensile strength and hardness of the alloy. The development of short carbon fiber reinforced A336 alloy composite aimed to improve the thermal conductivity and mechanical property of composite by combining the high thermal conductivity of carbon fiber and the high mechanical property of A336 alloy, for which be used for structural materials. In the previous research, it concluded that carbon fibers exhibit high reactivity with Al and Al alloys, causing harmful interfacial reaction between CF and Al, forming undesirable interfacial reaction product as Al_4C_3 which limits the applications of CF/Al matrix composites. Carbon fibers and Al have poor wettability, so, it is essential to improve the wettability and to prevent formation of interfacial reactions[7, 8]. A new fabricating method named low pressure infiltration (LPI), which developed by Y. B. Choi[9] is gaining extensive attentions due to the advantages of low cost, well dispersion and volume control of reinforcement and energy saving. In the present study, SiO_2 binder was using for surface coating of carbon fibers for improving the wettability with Al and prevent the harmful interfacial reaction, short carbon fiber reinforced pure Al and A336 alloy composites were fabricated by LPI process. Microstructure and Vickers hardness as well as microstructure–property relationships of these materials were investigated. This study focuses on the development of multifunctional carbon fiber reinforced aluminium matrix composites by low pressure infiltration process.

2. Materials and Experiment Process

A1070 with aluminium content of 99.7% and high thermal conductivity (237W/(m • K)) and A336 alloy (Al-13%Si-1.5%Ni-1.3%Cu-1.3Mg) with high tensile strength (195MPa) are used as matrixes in this experiment. Carbon fiber (K13D2U, Mitsubishi Plastics, Inc. length of 2.5mm, aspect ratio of 230) with 10 vol. % and 8 mass % SiO_2 binder (Nissan Chemical Industries, Ltd.) are used for fabricating preform. Purpose of SiO_2 binder is to cross-link short carbon fibers and surface coating of short carbon fibers for wettability with Al. The size of preform is ϕ30 x h20mm. The preform is heat-treated for sintering at 1433K in Argon for SiO_2 binder. Low pressure infiltration process conditions are applied pressure of 0.4MPa, molten alloy of 1073 K and holding time of 3.6 ks. Microstructures of preforms and composites are revealed by Optical Microscope (OM, Metal Microscope OPTIPHOT-2, Nikon Corporation, Japan) and Scanning Electron Microscope (SEM, TOPCON SM-520, Japan). To evaluate the element distribution of the interlayer between carbon fiber and aluminium matrix and to prove the product composition of interlayer, Electron Probe Micro-Analyzer (EPMA, JXA-8900, Japan) is conducted. Furthermore, properties of composites are investigated by Vickers hardness test and density test, respectively. Vickers hardness test is performed by load of 5 kg_f for 10 s, the density of each composites is measured by Archimedes method.

3. Results and Discussion

3.1 Observation and characterization of carbon fiber reinforced Al composites

Fig. 1 shows SEM images of as-received carbon fiber and carbon fiber inside preform. In fig. 1 (a), the as-received carbon fibers presented surface without impurities. Fig. 1 (b) was observed that SiO_2 binder was distributed at the surface of carbon fiber and bridging between fibers. The thick SiO_2 layer wrapped carbon fiber would help the preform exhibit larger fiber adhesion. Therefore, the role of SiO_2 binder was bridging and coating fibers.
Fig. 2 shows OM and SEM images of short carbon fiber reinforced pure Al and A336 alloy composites. In fig. 2 (a) and (b), short carbon fibers were distributed evenly in both aluminium

matrix composites, indicating good dispersion of short carbon fibers in preform which could improving infiltration process. Defects were found between fibers in the matrix, defects content of short carbon fiber reinforced pure Al composite was higher than that of short carbon fiber

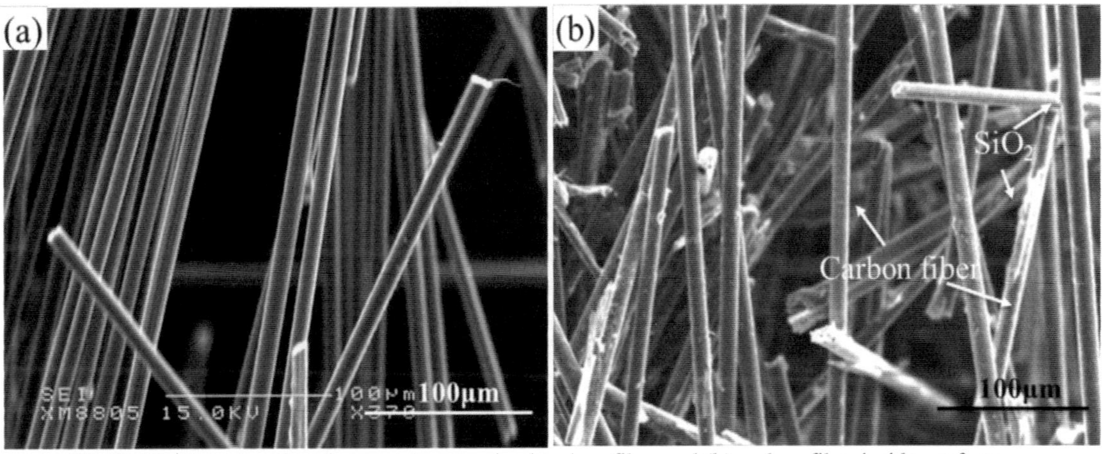

Fig. 1. SEM images of preforms: (a) as-received carbon fiber and (b) carbon fiber inside-preform.

reinforced A336 alloy composite. Defects content varied with matrix which related to the fluidity of matrix. The generation of defects was resulted by not only poor wettability of short carbon fiber by aluminium matrix but also the solidification of matrix before contact to short carbon fibers, cause the defects to remain in the composites. Defects directly influence on relative density of composites. To further analysis the microstructure of composites, SEM images of composites were shown in fig. 2 (a') and (b'), in both of images, the interlayers between fiber and matrix were smoothly, fibers were almost covered with oxide layers. While in fig. 2 (b'), intermetallic compounds of Al_3Ni and Si phase were generated in the matrix.

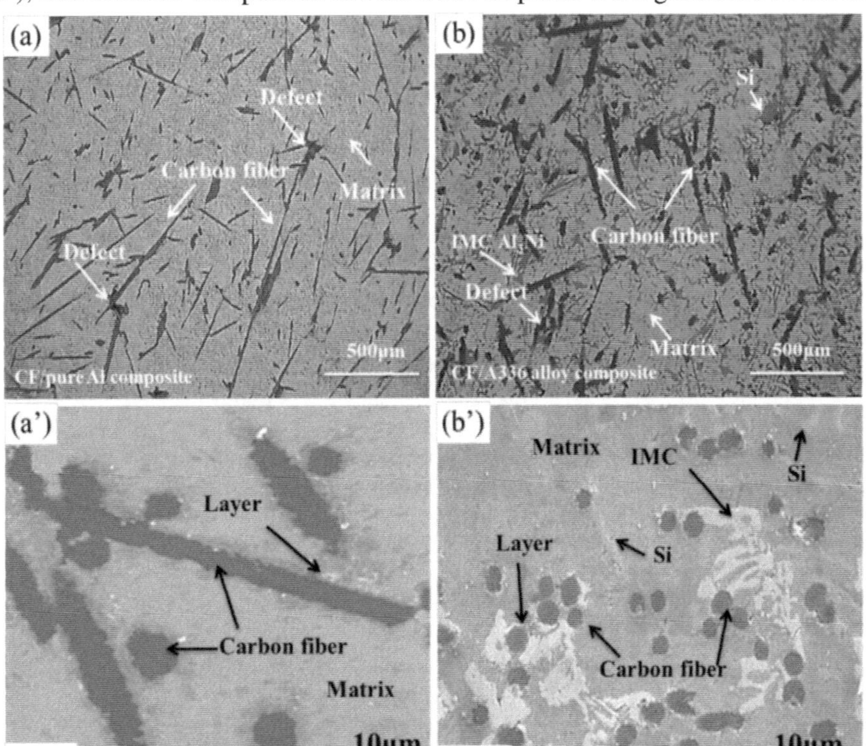

Fig. 2. OM and SEM images of the composites: (a), (a') short carbon fiber reinforced pure Al composite and (b), (b') short carbon fiber reinforced A336 alloy composite.

To characterize element distribution of the composite and the composition of the oxide layer, area mapping and point analysis by EPMA of short carbon fiber reinforced A336 alloy composite were shown in Fig. 3 Oxide element was mainly distributed in the interlayer between carbon fiber and matrix, Si phase was precipitated from the matrix during infiltration process. The point analysis revealed that the oxide layer mainly contained Al_2O_3 and SiO_2, the generation of Al_2O_3 could be explained by the reaction between Al matrix and SiO_2 binder as follow[8]:

$$4Al + 3SiO_2 \rightarrow 2Al_2O_3 + 3Si \tag{1}$$

Laurent V[10] concluded that the wettability of SiO_2-Al system transformed from nonwetting system of 150° to wetting system of 68° at melting temperature of 1173K. Therefore, the SiO_2 binder coating on the fibers is not only be a cause of preform strength improvement, but also improved the wettability between Al and fibers by reacting with matrix. Moreover, the improvement of reactivity between molten Al alloy and SiO_2 binder can lead to full densification of aluminium matrix composite, thus, the reaction is also looking forward to increasing the relative density of the composites.

3.2 Vickers hardness and Relative density of carbon fiber reinforced Al composites

Fig. 4 shows Vickers hardness and Relative density of composites. Hardness of pure Al and A336 alloy was 19.1 HV and 67.5HV, respectively. The hardness of short carbon fiber reinforced pure Al composite was 33.6 HV, 76% higher than the pure Al. The hardness of short carbon fiber reinforced A336 alloy composites was 74.3HV, showed an increasement of 10% to the A336 alloy. The increasement of hardness was caused by the addition of short carbon fiber reinforced phase which acted a pivotal part in dispersion strengthening.

The relative density of short carbon fiber reinforced pure aluminium and A336 alloy composite was 95.8% and 97.2%, respectively. Fig. 4 (a) shows more defects from entrapped gases in short carbon fiber reinforced pure aluminium composite which resulting in lower relative density

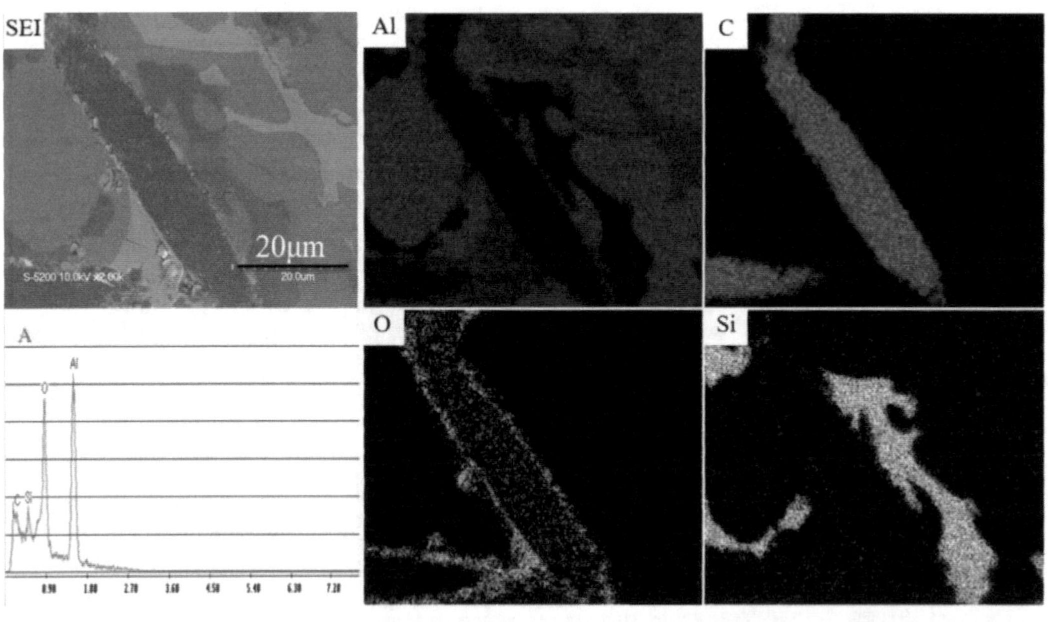

Fig. 3. Results of area mapping and point analysis by EPMA of short carbon fiber reinforced A336 alloy composite.

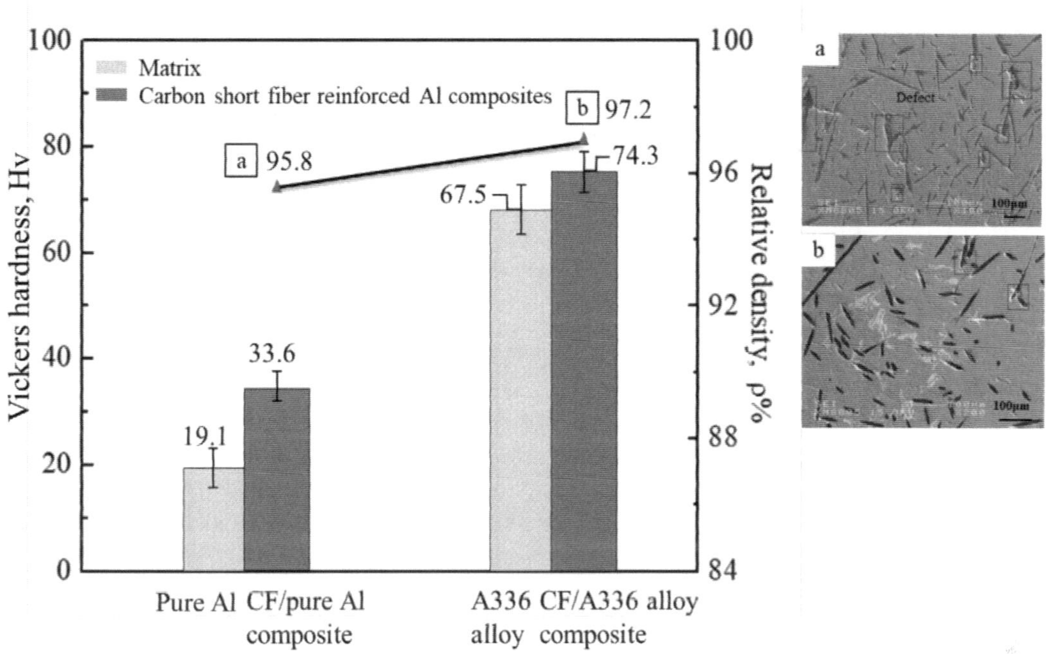

Fig. 4. Vickers hardness and Relative density of composites: (a) short carbon fiber reinforced pure Al composite and (b) short carbon fiber reinforced A336 alloy composite.

and poor wettability of the short carbon fiber by matrix. The carbon fiber preform could not be fully infiltrated because of higher viscosity of pure Al matrix resulted in higher porosity being remained in the composite. Fig. 4 (b) shows less defects in short carbon fiber reinforced A336 alloy composite, the higher relative density resulted from the high Si content of A336 alloy matrix, which caused an increased rate of infiltration by decreasing the viscosity of matrix and enhancing the wettability of short carbon fiber by the aluminium alloy.

4 Conclusions

In the present study, for developing multifunctional carbon fiber reinforced aluminium matrix composites, short carbon fiber reinforced pure Al and A336 alloy composites were fabricated by low-pressure infiltration method. The significant conclusions are summarized as follow: Pure Al and A336 Al alloy were used as matrix, on the one hand, to develop short carbon fiber reinforced pure Al and A336 alloy composites for using as functional materials and structural materials, on the other hand, to analysis the effect of element addition of A336 alloy compare with pure Al. Defects content in short carbon fiber reinforced pure Al composites was higher than that of short carbon fiber reinforced A336 alloy composites. High Si content of A336 alloy matrix caused an increased rate of infiltration by decreasing the viscosity of matrix resulted in less defects in the composite. Relative density of short carbon fiber reinforced pure Al and A336 alloy composite was 95.8% and 97.2%, respectively. Vickers hardness of short carbon fiber reinforced pure Al and A336 alloy composite was 76% and 10%, respectively, higher than pure Al and A336 alloy.

Acknowledgements
This work was supported by JSPS KAKENHI Grant Number 18K03839.

5 References

[1] S.T. Mavhungu, E.T. Akinlabi, M.A. Onitiri and F.M. Varachia, 2017, Aluminium Matrix Composites for Industrial Use:Advances and Trends, Procedia Manufacturing, 7, 178-182.

[2] B. Ralph, H.C. Yuen and W.B. Lee, 1997, The processing of metal matrix composites - an overview, Journal of Materials Processing Technology, 63, 339.

[3] J. D. Dong, C.Y. Jia and M.Q. Wang, 2017, Improved mechanical properties of carbon fiber-reinforced epoxy composites by growing carbon black on carbon fiber surface, Composites Science and Technology, 149, 75-80.

[4] E. Fitzer and L.M. Manoch, 1998, Carbon Reinforcements and Carbon/Carbon Composite, Springer, pp. 344.

[5] S. Chand, 2000, Review Carbon fibers for composites, Journal of Materials Science, 35, 1303–1313.

[6] S. J. Park and M.S. Cho, 2000, Effect of anti-oxidative filler on the interfacial mechanical properties of carbon–carbon composites measured at high temperature, Carbon, 38, 1053-1058.

[7] Y. P. Tang, L. Liu, W. W. Li, B. Shen and W. B. Hu, 2009, Interface characteristics and mechanical properties of short carbon fibers-Aluminium composites with different coatings, Applied Surface Science, 255, 4393-4400.

[8] M. H. Lee, Y. B. Choi, K. Sugio, K. Matsugi and G. Sasaki, 2014, Effect of Aluminium Carbide on unidirectional Cf-Al Composites fabricated by low pressure infiltration process, Composites Sciences and Technology, 97, 1-5.

[9] Y. B. Choi, K. Matsugi and G. Sasaki, 2013, Development of Intermetallic Compounds Reinforced Al Alloy Composites Using Reaction of Porous Nickel and Aluminium, Materials Transactions, 54, 595-598.

[10] V. Laurent, D. Chatain and N. Eustathopoulos, 1991, Wettability of SiO_2 and oxidized SiC by aluminium, Materials Science Engineering. A, 135, 89–94.

IOP Conf. Series: Materials Science and Engineering **547** (2019) 012025 doi:10.1088/1757-899X/547/1/012025

Relationship Between Heat Treatment And Mechanical Properties Of Al-Si-Mg Alloy

Yuki Namikawa[1], Kenjiro Sugio[1]*, Gen Sasaki[1], Junji Tabata[2] and Nobuyuki Fuyama[3]

[1]Graduate School of Engineering, Hiroshima University, Higashi-Hiroshima, Japan
[2]Hiroshima Aluminum Industry CO., LTD., Akitakata, Hiroshima, Japan
[3]Western Hiroshima Prefecture Industrial Reserch Institute, Kure, Hiroshima, Japan

*Corresponding author, E-mail: ksugio@hiroshima-u.ac.jp

Abstract. The relationship between heat treatment and mechanical properties of Al-Si-Mg alloy was investigated. Solution treatment at 808K for 6 hours, preliminary aging at room temperature for 60 hours and artificial aging at 473K for 1 hours were conducted. Not preliminary aged samples were prepared for comparison. The strength improved and the elongation decreased with increasing the amount of Mg. This trend was seen both preliminary aged and not preliminary aged samples. However, preliminary aged samples showed relatively high tensile strength and elongation comparing with not preliminary aged samples. EPMA analysis revealed that the precipitation of Mg_2Si in preliminary aged samples was larger than that of not preliminary aged samples, and which possibly contributed to the improvement of strength of the preliminary aged samples. Thermodynamic calculation showed that Mg_2Si phase was not undissolved at the solution temperature, 808K, and these undissolved Mg_2Si particles might influenced on mechanical properties negatively.

Keywords: Al–Si-Mg alloy, heat treatment, mechanical properties

1. Introduction

In recent years, global warming is a serious problem, and countermeasures against it are made in various fields. Particularly in the automobile industry it is strongly required to regulate automobile exhaust gas and improve fuel economy. Then, the aluminum material has a specific gravity of one third as compared with the steel material and has good workability. So, it is expected to be an effective material for weight reduction of the car body. At the actual production process, heat treatment is often applied after casting to improve the mechanical properties[1,2]. From the viewpoint of energy saving, if the influence of heat treatment on cast material is clarified, a more efficient production system should be prepared. Therefore, in this study, we aimed to investigate the influence of composition and heat treatment on mechanical properties by applying heat treatment to Al-Si-Mg based alloy which has both good ductility and strength[3].

2. Experimental procedure

2.1. Material preparation

In this study, Al-4wt.%Si-xwt.%Mg alloy with different heat treatment / composition was used as test materials. Procedures for preparing each sample are shown below. The composition is shown in weight percent. The master alloys, 99.99%Al, Al-10% Mg, Al-4%Si-0.5%Mg and Al-24.3%Si, were cut, weighted and blended in a predetermined composition. The samples were fabricated by gravity casting and listed in Table 1. The weighted master alloys were put into an alumina crucible and heated in an electric furnace to 1073 K and casted. Two types of heat treatment, non-pre aged and pre aged, were applied in this study as shown in Fig. 1. After casting, samples were maintained at 808 K for 6 hours as a solution treatment, followed by water cooling with water at room temperature. For pre aged ample, preliminary aging treatment at room temperature was conducted at 303 K for 60 h and then an artificial aging treatment was conducted at 473 K for 1 hour in an oil bath.

Table 1 Prepared samples in this study.

Heat treatment	Composition
non-pre aged	Al-4%Si-0.4%Mg
	Al-4%Si-0.6%Mg
	Al-4%Si-0.8%Mg
pre aged	Al-4%Si-0.4%Mg
	Al-4%Si-0.8%Mg
	Al-4%Si-1.2%Mg

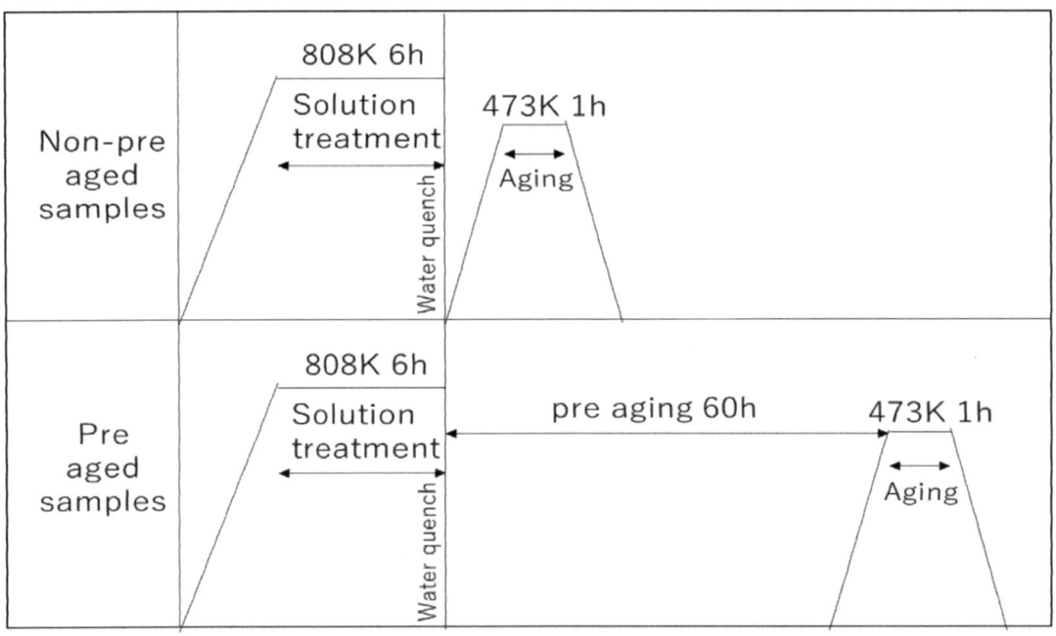

Fig.1 Heat treatment procedure of samples.

2.2. Evaluation method

A tensile test was conducted to measure tensile strength and elongation of samples. The tensile speed was 1 mm/min. The elongation was determined from the change in gauge distance. The samples were polished until the surface became a specular surface, and microstructural observation by SEM and elemental analysis by EPMA were carried out in order to investigate the relationship between the microstructure and tensile properties of the samples. JEOL JXA8900 was used with the acceleration voltage of 15kV.

3. Results and Discussion

3.1. Mechanical properties

Fig. 2 shows mechanical properties of samples. Although the ultimate tensile strength of the pre-aged Al-4%Si-0.8%Mg slightly decreased comparing with Al-4%Si-0.4%Mg, Al-4%Si-1.2%Mg shows a high strength exceeding 300MPa. However, the elongation of Al-4%Si-1.2%Mg greatly decreased. The strength of the non-pre aged samples increased with the amount of Mg and the elongation decreased. When comparison between non-pre aged and pre aged Al-4%Si-0.4%Mg is made, pre aged sample showed high ultimate tensile strength, as shown in Fig. 2(a), and this means that pre aged heat treatment affected positively. On the other hand, when comparison between non-pre aged and pre aged Al-4%Si-0.8%Mg is made, pre aged sample show low ultimate tensile strength, and this means that pre aged heat treatment affected negatively. Elongations of pre aged samples was higher than that of non-pre aged samples, and this means that pre aged heat treatment affected positively in elongation.

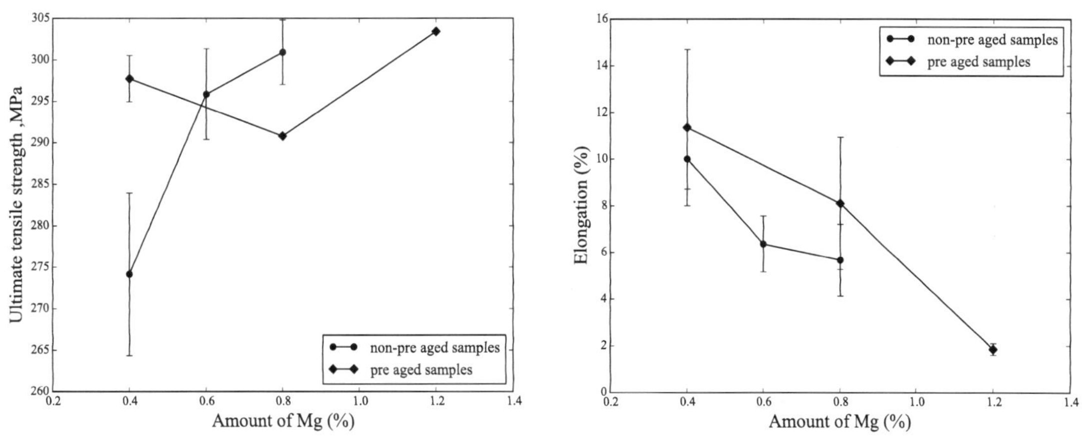

Fig.2 Mechanical properties of samples, (a)Ultimate tensile strength and (b)Elongation.

3.2. Structure observation

Magnesium silicide (Mg_2Si) is commonly formed in Al-Si-Mg based alloys[4]. Precipitation of Mg_2Si increased with increasing the amount of Mg in both pre aged and non-pre aged samples. However, precipitation of Mg_2Si in pre aged samples was larger than that of non-pre aged samples. In particular, Mg_2Si precipitates in pre aged Al-4%Si-0.8%Mg was formed along grain boundaries. Increase of precipitation of Mg_2Si possibly contributed to the improvement of the strength of pre aged samples.

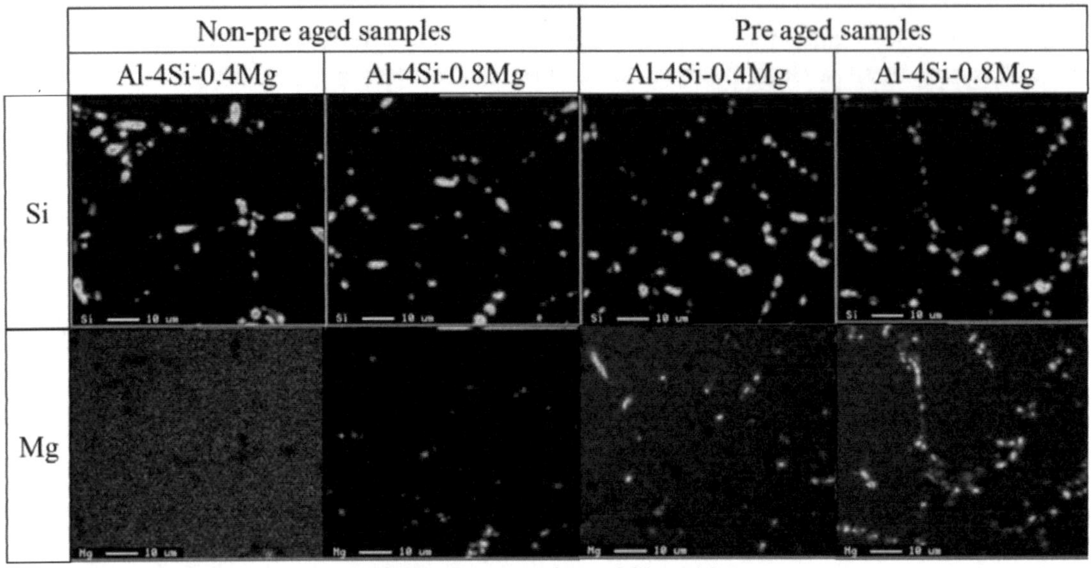

Fig.3 EPMA mapping of Si and Mg.

3.3. Phase calculation in each alloy

In this study, we used a thermodynamic calculation software called JMatPro to investigate the correlation between the proportion of each phase in the sample and the temperature. The result of the Al-4%Si-0.4%Mg alloy is shown in Fig. 4. It can be confirmed that the temperature at which all of the Mg and Si is solid solution is 773K or more. This means that a solution treatment should be conducted between 773K and solidus temperature, 833K. The difference between solid solution temperature and solidus temperature was defined as ΔT, and Fig. 5 shows the relationship between ΔT and amount of Mg. ΔT decreased with increasing amount of Mg. In case of Al-4%Si-0.6%Mg, the temperature at which all the Mg and Si is solid solution is 813K or more, and it is close to the solidus temperature. In case of Al-4%Si-0.8%Mg, the Mg_2Si phase does not solve until solidus temperature and it is considered that the unsolved Mg_2Si phase remained at 808K which was the solution temperature of this experiment. In case of Al-4%Si-1.2%Mg, considerable amount of unsolved Mg_2Si phase remained, and it had a negative influence on elongation.

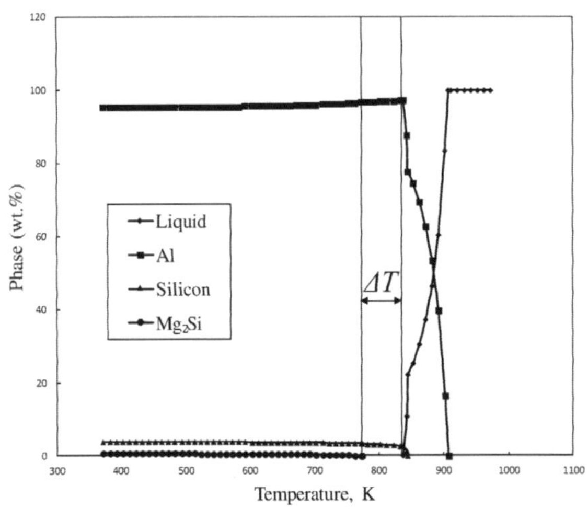

Fig.4 Percentage of each phase of Al-4%Si-0.4%Mg alloy calculated by JMatPro.

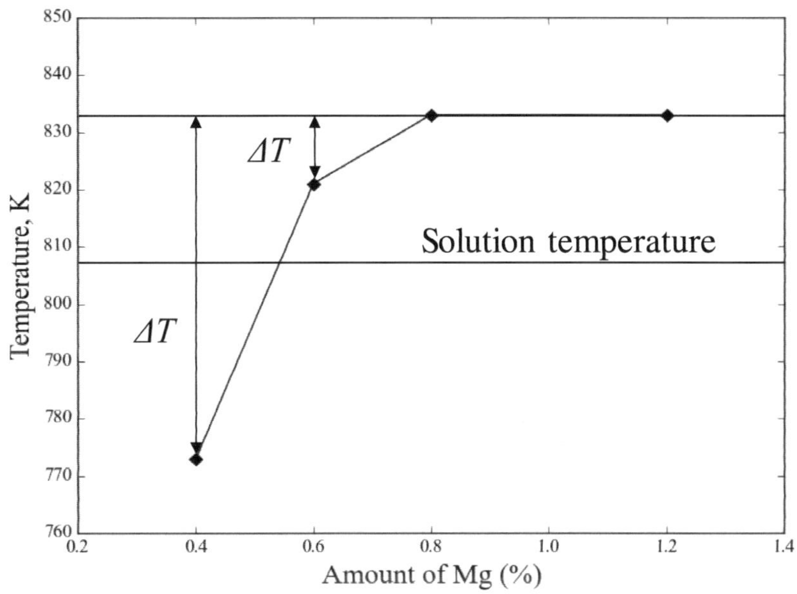

Fig.5 Relationship between ΔT and amount of Mg.

4. Conclusions

The influence of composition and heat treatment on mechanical properties of Al-4%Si-x%Mg alloy was investigated. The results are summarized as follows.

1. The strength improved and the elongation decreased with increasing the amount of Mg. This trend was seen in both non-pre aged and pre aged samples.

2. When comparing non-pre aged and pre aged samples, pre aged samples showed relatively high tensile strength and elongation.

3. EPMA analysis revealed that the precipitation of Mg_2Si in pre aged samples was larger than that of non-pre aged samples, and which possibly contributed to the improvement of strength of the pre aged samples.

4. The difference between solid solution temperature and solidus temperature decreased with increasing amount of Mg in Al-4%Si-x%Mg alloy. When amount of Mg was larger than 0.6%, the unsolved Mg_2Si phase possibly remained at 808K which is the solution temperature of this experiment.

5. References

[1] FURUI M, ISHIKAWA T, IKENO S, MIURA M, SAIKAWA S and SAKAI N 2011 *Journal of Japan Institute of Light Metals* Vol.61 No.1 pp 9–13

[2] TSUCHIDA S and YOSHIDA H 1989 *Journal of Japan Institute of Light Metals* Vol.39 No.8 pp 587–606

IOP Conf. Series: Materials Science and Engineering **547** (2019) 012025 doi:10.1088/1757-899X/547/1/012025

[3] KITAOKA S, UJIKURA C and KAMIO A 1988 *Journal of Japan Institute of Light Metals* Vol.38 No.7 pp 426–446

[4] YAMADA H 1957 *Journal of Japan Institute of Light Metals* Vol.7 No.4 pp 45–50

Effect of Gelatin On The Corrosion Resistance Of Cr-ZrO_2 Coatings In 3% NaCl Solution

Asep Ridwan Setiawan[1], Aditianto Ramelan[1]*, Riska Rachmantyo[1], Febi Luthfiani[2]
[1]Materials Engineering Department, Faculty of Mechanical and Aerospace Engineering, Institut Teknologi Bandung, Indonesia
[2]Graduated Students, Materials Engineering Department, Faculty of Mechanical and Aerospace Engineering, Institute Teknologi Bandung, Indonesia

*Corresponding author (E-mail): ramelan@material.itb.ac.id

Abstract. In this work, Cr-ZrO_2 composite coating were prepared by electrodeposition method on aluminium substrate from chromic acid based electrolytes containing gelatin. The effect of gelatin on the corrosion resistance of Cr-ZrO_2 coatings were studied in 3 wt.% sodium chloride solutions. Potentiodynamic polarization studies were performed using a Versastat 3 (Princeton Applied Research) electrochemical system with a three-electrode cell. Results shows that the corrosion resistance of Cr-ZrO_2 coating on Al substrate was improved by the addition of gelatin to the coating solution. The addition of gelatin shifted the potentiodynamic polarization curved in the noble direction. The corrosion potential changed positively from -778 mV to -259 mV. The corrosion rate of Cr-ZrO_2 coating decreased from 4.44±0.93 mpy to 0.13±0.01 mpy with the addition of 3g/L gelatin.

Keywords: electrodeposition, corrosion, gelatin, corrosion rate.

1. Introduction

Aluminum and its alloys are characterized by their high strength-to-weight ratio, good corrosion resistance, high thermal and electrical conductivity, good appearance and good fabrication characteristics. These properties make aluminum a very important material for applications in the aircraft and automotive industries [1]. However, the surface properties of aluminum need to be improved because it has a relatively low hardness and poor wear resistance [2-5]. Hard chromium coatings were often used to improve the surface performance of aluminum, thanks to its low frictional coefficients, high hardness, good corrosion resistance and good aesthetic appearance [6-11].

Recently, significant interest has been growing in the fabrication of Cr composite coatings, where ceramic particles are incorporated into a Cr layer. Jasim et al. [12] showed that the hardness of the Cr-ZrO_2 composite coating can increases with increasing zirconia concentration in the coating. The adddition of 4g/L zirconia into Cr layer improves the hardness up to 1017 HV0.2. Similar works by Hashimoto et al [7] showed that the addition of zirconia particles, ZrO_2, as much as 2g/L in Cr solution succeeded in increasing the hardness of the chromium layer on a steel substrate up to 1200 HV. However, the authors could not find any work related on the deposition of Cr directly on the aluminum substrate. Based on these fact, recently, Luthfiani [13] have been trying to deposit Cr-ZrO_2 particles on the aluminium substrate and evaluate its mechanical properties. Gelatin additive agent was used to improve the dispersion ability of ZrO_2 in the Cr plating solution. The addition of 1-wt% gelatin on the electrolyte suspension was found to improve the hardness of Cr-ZrO_2 layer up to 1220 VHN. In this work, the effect of gelatin on the corrosion resistance of Cr-ZrO_2 coatings will be studied in 3 wt.% sodium chloride solutions.

2. Materials and Methods

Aluminum series 1100 were selected as a substrate. The Al specimens were cut into coupon (20 x 10 x 1 mm) and mechanically polished with various size of abrasive paper up to 2000 grit. The area of specimens exposed to the plating solution was limited to the size of 10 x 10 mm. Afterwards the coupons were degreased with 50% HNO_3 for 1 minute at room temperature, followed by rinsing, and immersion in zincate solution (525 g/L NaOH and 98 g/L ZnO) for 2 minutes. Finally, the specimens were rinsed, and ready for electrodeposition process. Electrodeposition experiments were carried out at 45-50 °C in chromic acid electrolytes, under a constant current density of 0.8 A/cm^2 and deposition time of 120 minutes. The content of nano ZrO_2 in the solution was fixed at 3 g/L to improve the hardness of Cr layer. The gelatin content was varied from 0 to 3 g/L to improve the uniformity/homogeneity of the coating. The details of bath composition and plating parameters are shown in Table 1.

Table 1. The composition of the experimental electrolytes

Electrolyte composition		
	Chrome coatings	composite coatings
CrO_3	250 g/L	250 g/L
H_2SO_4	2.5 g/L	2.5 g/L
ZrO_2 particles		3 g/L
Gelatin		0, 1, 3 g/L

Microstructure observation and thickness of Cr-ZrO_2 coating was evaluated from its cross-section, using optical microscope (OM). Potentiodynamic corrosion test was carried out in a 3 wt% NaCl solution at room temperature with a Verstatat 3 electrochemical system using a three-electrode cell configuration. The schematic experimental setup can be observed in Figure 1. Cr-ZrO_2 composite coating was set as the working electrode (WE). Graphite rod was set as the counter electrode (CE), and saturated calomel electrode (SCE) was set as the reference electrode (RE). The measurement was conducted in the range of -0.3V (initial potential) to +0.2V (final potential). The corrosion potential (E_{corr}) and corrosion current density (I_{corr}) of the produced coatings were determined from the intersection of the cathodic and anodic curves by using tafel extrapolation method. After the test, the corrosion rate of specimen was calculated according to the following equation:

$$[CR]_{mpy}=[0.131] \times I_{corr} \times EW/\rho \qquad [1]$$

Where CR, Icorr, EW and ρ are the corrosion rate in mpy (mils/year), corrosion current density (μA/cm2), equivalent weight and density in g/cm3.

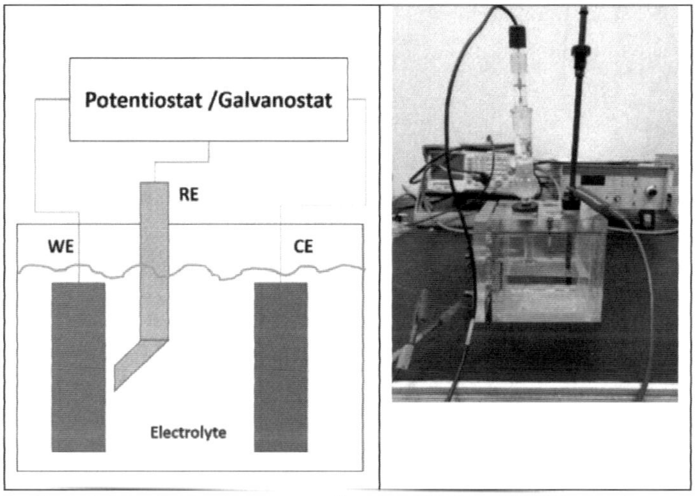

Figure 1. (a) schematics three electrode connection method and (b) actual experimental setup for corrosion test.

3. Results and Discussion

3.1 Coating thickness

The coating thickness of pure Cr, Cr-ZrO_2 with or without gelatin addition in the bath are shown in Figure 2.

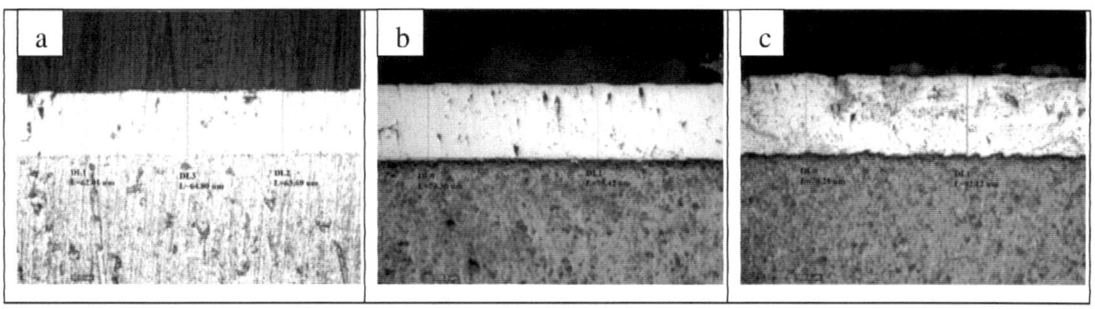

Figure 2. Cross section of (a) Cr coating, (b)Cr-ZrO2 coating and (c) Cr-ZrO2 coating with gelatin

In general, the cross-section micrograph shows a homogeneous and compact Cr and Cr-ZrO_2 coating without microcracks. The addition of nano ZrO_2 particles in the electrolytes slightly increase the thickness of Cr coating from 67 μm to 76 μm. The addition of gelatin in the electrolytes increases the thickness of Cr-nano ZrO_2 coating from 76 μm to 88 μm. Several researchers [14-15] have been reported that the addition of additive agent into electrolytes may improve the deposition rate and adherence of the coating into the substrate. Gelatin is a hydrolyzed form of collagen, which is sometimes added to electroplating baths to control the deposition rate, crystallization, leveling and brightness of the deposit [16]. During electrodeposition, gelatin were found to modify the characteristics and enhance the electrodeposition efficiency and/or corrosion resistance of Zn–Ni deposits [17-19]. In this work, it seems that gelatin affected the Cr-ZrO_2 deposition process through preferential adsorption on

IOP Conf. Series: Materials Science and Engineering **547** (2019) 012026 doi:10.1088/1757-899X/547/1/012026

the surface of the cathode. As a result, the deposition rate of Cr-ZrO is higher with the presence of gelatin in the electrolyte, resulted in thicker coating.

3.2 Corrosion resistance

Tafel polarization curves of Cr-ZrO_2 electrodeposited coating at different gelatin concentration in the plating bath are shown in Fig. 3. According to these figures, the corrosion potential and corrosion current density of as deposited Cr-ZrO_2 coating (3 wt%) is about -778 mV and 4.66 µA/cm2, respectively. It can be observed that by increasing the concentration of gelatin in the plating bath, the corrosion potential of Cr-ZrO_2 coating shifts to the noble direction, from -778 mv to about -259 mV. In this condition, the corrosion current density decreases from 4.66 µA/cm2 to 0.11 µA/cm^2.

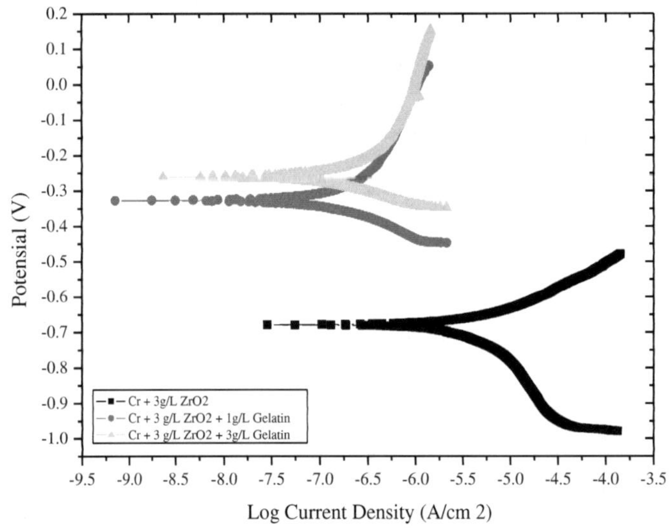

Figure 3. Potentiodynamic polarization curves of The Cr-ZrO_2 coating with or without gelatin

The change of corrosion potential, corrosion current density, and corrosion rates of Cr-ZrO_2 coating vs. gelatin concentration in the plating bath are also presented in Table 2. It can be concluded that the produced Cr-ZrO_2 coating at higher gelatin concentration in the plating bath exhibit better corrosion resistance on 3 % NaCl solution. The corrosion rate of Cr-ZrO2 coating decreases from 4.44±0.93 mpy to 0.13±0.01 mpy with the addition of 3g/L gelatin in the plating bath.

Table 2. Coating corrosion parameter obtained from potentiodynamic polarization curves

Coating system	**Corrosion rate (mpy)**	**E_{corr} (mV)**	**I_{corr} (µA)**
Cr + 3g/L ZrO_2	4,44 ± 0,93	-778	4,66
Cr + 3g/L ZrO_2 + 1g/L Gelatin	0,15 ± 0,02	-428	0,22
Cr + 3g/L ZrO_2 + 3g/L Gelatin	0,13 ± 0,01	-259	0,11

The higher corrosion resistance of Cr-ZrO_2 coating with the increase of gelatin concentration in the plating bath can be attributed to the:

1. Increase of ZrO content in the Cr matrix. The addition of gelatin to the plating bath may also promoted the formation of a relatively stable suspension of ZrO_2 particles. The adsorbed gelatin layer on the surface of ZrO_2 particles provides electrostatic stabilization for the suspension, preventing particle agglomeration [20]. As a result, the addition of gelatin improve the co-deposition of ZrO_2 particles in chrome matrix.
2. Formation of a uniform and highly protective passive film at the surface of Cr-ZrO_2 coating. Adsoption of gelatin into the Cr-ZrO_2 layer promotes the formation of passive films [21-22].
3. Lack of defect, such as grain boundaries or porosity in the coating. The presence of gelatin in the bath may covering the microcrack/porosities available at the Cr-ZrO_2 coating surface [13].

4. Conclusions

Based on our investigation about fabrication of Cr-ZrO_2 coating by means of electrodeposition, we summarize the primary results as follows. Cr-ZrO_2 coating were successfully formed during electrodeposition process. The addition of gelatin in the plating bath improves the deposition rates and corrosion resistance of Cr-ZrO_2 coatings. The corrosion rate of Cr-ZrO_2 coating decreases from 4.44±0.93 mpy to 0.13±0.01 mpy with the addition of 3g/L gelatin in the plating bath.

5. Acknowledgements

This research was conducted with the support from Institut Teknologi Bandung in the frame of the P3MI program.

6. References

[1] Dong, H 2010. *Surface engineering of light alloys Aluminium, magnesium and titanium alloys.* (Washington DC: Woodhead Publishing Limited)
[2] G.E. Totten 2003 *Handbook of Aluminium: Physical Metallurgy and Processes* (CRC Press)
[3] ASM Handbook Vol. 5 Surface Engineering. p797
[4] Deqing W, Ziyuan S and Tangshan K 2005 *Sur. Coat. Technol.* **191**(2) 324
[5] Wan S, Li D, Zhang G, Tieu AK and Zhang B 2017 *Tribol Int* **106** 10
[6] Zhenkun Yang. 2011 *Alternatives to hard chromium plating on piston rods* (Karlstads Universitet:Sweden)
[7] Hashimoto K, Sugio K, Sasaki G, Setiawan AR and Ramelan A 2016 *Mech. Eng. J* **3**(2) 15
[8] Barbato RS, Ponce FJ, Jara VM, Cuevas SJ and Egaña AR 2008 *J. Chilean Chem. Soc.* **53** 1429
[9] Svenson, E 2006 *Dura Chrome Hard Chromium Plating.* (USA: Plating Resources, Inc. Cocoa).
[10] Wu C, Hong Y, Chen W, Chen J, Yuan M and Liao XA 2016 **298** 83
[11] Biberger J, Füßer H-J, Klaus M, Genzel C. 2017 *Wear* **376** 1502
[12] Jasim, T. 2015 *J. Sci. Eng. Res.* **6**, 7
[13] Luthfiani, F 2018 *Study the Effect of Gelatin on Corrosion Rate on deposition of Cr-Nano ZrO2 Coatings*, (Thesis, Bandung Institute of Technology).
[14] Schrieber, R 2007 *Gelatine Handbook Theory and Industrial Practice.* (Germany: WILEY-VCH Verlag GmbH & Co.)

[15] Basics of Metal Matrix Composites 2006 *Metal Matrix Composites: Custom-Made Materials for Automotive and Aerospace Engineering*, ed K.U Kainer (Wiley-VCH Verlag GmbH & Co) pp 1–54

[16] American Society for Metals 1984 *Aluminum: Properties and Physical Metallurgy*, ed J.E. Hatch (ASM) pp. 30–35

[17] Torralba J., da Costa C and Velasco F 2003 *J. Mat. Proc. Techno.* **133**(1) 203

[18] Lurie S, Volkov-Bogorodskiy D, Solyaev Y, Rizahanov R and Agureev L 2016 *Comp. Mat. Sci.* **116** 62

[19] U. Pramod, K 2015 *Int. J. Thin. Fil. Sci. Tec.* **4**(2) 147

[20] Afifi SE, Ebaid AR, Hegazy MM and Barakat AK 1992 *JOM* **44**(1) 32

[21] Soares ME, Souza CAC and Kuri SE 2006 *Sur. Coat. Technol.* **201**(6) 2953

[22] Pedroza GAG, de Souza CAC, Carlos IA and de Andrade Lima LRP 2012 *Sur. Coat. Technol.* **206**(11) 2927

Fabrication of Cr based electrodeposited composite film using nano ZrO_2 particles on aluminum substrate

Kota SATO[1], Kenjiro SUGIO[1]*, Yongbum CHOI[1], Gen SASAKI[1], Asep Ridwan SETIAWAN[2] and Aditianto RAMELAN[2]
[1]Graduate School of Engineering, Hiroshima University 1-4-1 Kagamiyama, Higashi-Hiroshima, Hiroshima 739-8527, Japan
[2]Department of Materials Engineering, Faculty of Mechanical and Aerospace Engineering, Institute of Technology Bandung, West Java, Indonesia

*Corresponding author (E-mail): E-mail: ksugio@hiroshima-u.ac.jp

Abstract. In this study, the coating method suitable for producing high-quality film on aluminum substrate was investigated by using different surface treatment to each specimen before electrodeposition. Microstructure of coating cross section was observed with Electron Prove Micro Analyzer (EPMA). Vickers hardness of coating surface was measured. The measurement of polarization behavior of aluminum in CrO_3-H_2SO_4 electrolyte with ZrO_2 addition was carried out to investigate corrosion resistance. The addition of ZrO_2 particles increased Vickers hardness of Cr coating. The existence of Zr in the coating was confirmed by point analysis from coating cross section. Polarization point was shifted higher by dispersing ZrO_2 particles into the plating bath, and this means that corrosion resistance was improved by the addition of ZrO_2 particles.

Keywords: EPMA, Polarization behavior, Vickers hardness, Electrodeposited Cr, Zirconia particle

1. Introduction

In recent years, materials that have high hardness and excellent abrasion resistance are required in many industries. In advanced industries such as aerospace and automobile industry, aluminum is widely used alternative to steel and ferrous alloy. However, aluminum is inferior in mechanical properties such as hardness and abrasion resistance. One of the methods for improving these properties is electrodeposition. Hard Cr coating has excellent hardness and lubricating performance, corrosion resistance, and is considered to be the best option for achieving reduction of friction, minimization of wear and prevention of wear [1]. Aluminum is lightweight and hard Cr coating is durable. By combining these, application to new products becomes possible. Cr generally has a hardness of 65-70 HRc (840-1076HV [2]), it gives the coating surface durability to parts seizing, abrasion and abrasion. Hard Cr coating is produced by electrodeposition in a solution containing chromium oxide (VI) and a catalyst anion at an appropriate ratio. For applications to products it is used for items requiring high abrasion resistance and corrosion resistance such as tool cutting surface, piston rings, shock absorbers, cylinder liners, crankshafts and aircraft landing gear. It is known that lifetime of a piston ring which hard Cr coating has been applied is about 5 times longer than that of a material without coating [3]. When mechanical properties beyond this hard Cr coating are required, it is considered from the past study that composite coatings with ceramic particles will have better mechanical properties than coatings only with Cr. For example, it has been found that Cr-Al_2O_3 coating, which can be produced when rare metals are included in a hexavalent chromium bath, has 2 to 6 times better wear resistance compared with Cr coating [4]. Additionally, it is known that Cr-Al_2O_3/SiC coating has high corrosion resistance [5]. Furthermore, it has been found that composite coatings obtained by electrodeposition with ultrasonic agitation in a trivalent chromium bath containing multi-walled carbon nanotubes are superior in hardness, toughness and anti-friction properties compared with a coating only with Cr [6].

2. Materials and Methods

Fig.1 shows a schematic drawing of experimental equipment. Electrodeposition was carried out for 4 hours with a current density of 0.80 A/cm^2 and with stirring of electrolyte at about 100 rpm using YO60LX1200 manufactured by Acopian Corporation as a power source. UX369C manufactured by Heles was used as an ammeter and TY720 manufactured by Yokogawa was used as a voltmeter. A lead plate was prepared for an anode. For comparison, both an aluminum plate and a low-carbon steel plate were used as a cathode. A hole with a diameter of 2.0 mm was drilled in a cathode specimen beforehand so as to pass through a copper wire as a conductor to the sample. After passing the wire, the hole was closed with a screw so that the wire and the sample contacted properly to conduct electricity. The electrolytes consisted of CrO_3 (250 g/L), H_2SO_4 (2.5 g/L), and the added amount of ZrO_2 particles was varied from 0 g/L to 5.0 g/L. The particle size of ZrO_2 was 60-105 nm.

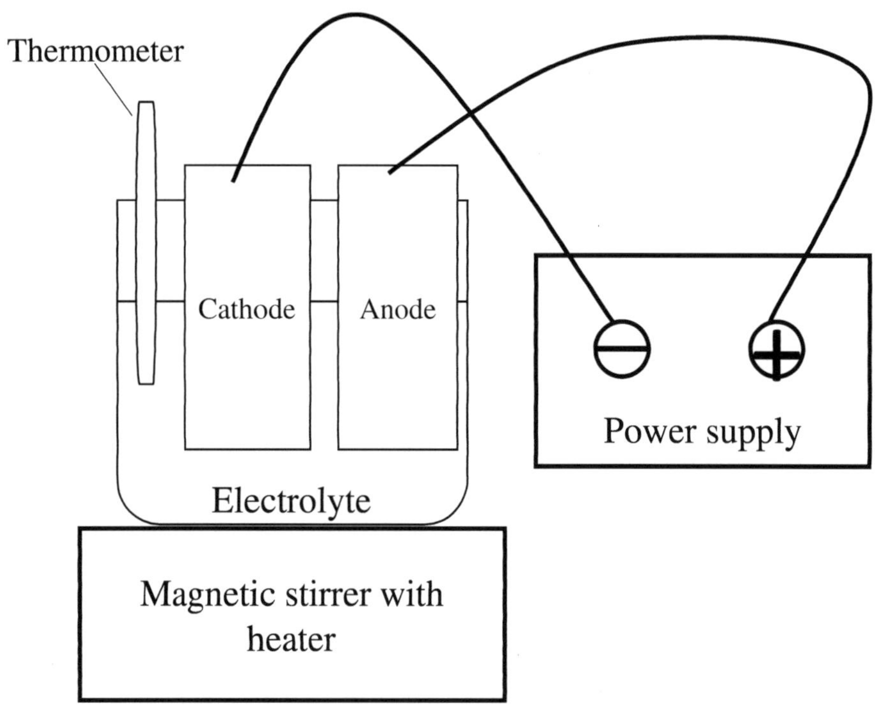

Figure 1 A schematic drawing of experimental equipment

Table 1 Each surface treatment up to electroplating

	A	B	C
1	Polishing	Polishing	Polishing
2	Etching	Acid dip	Acid dip
3	Rinsing	Rinsing	Rinsing
4	Acid dip	Zincate without stirring	Zincate with stirring
5	Rinsing	Rinsing	Rinsing
6	Electroplating	Electroplating	Electroplating

Three different procedures (A, B, C) of surface treatments were used for each cathode plates before electrodeposition. Table 1 shows each surface treatment before electroplating. In the procedure A, after polishing the specimen by waterproof abrasive paper from #600 to #1500, etching was carried out for 180 seconds in an aqueous sodium hydroxide solution (35 g/L, 50 °C) and washed with water. Acid activation was carried out for 180 seconds in an aqueous solution of sulfuric acid (15 wt%, 80 °C), and after that, the specimen was washed with water and electrodeposition was performed. In the procedure B, the specimen was polished in the same way as in the procedure A and immersed in an aqueous solution of nitric acid (50 vol%, room temperature) for 60 seconds, rinsed with water. After that, zincate treatment was conducted in a given solution (525 g/L NaOH, 98 g/L ZnO, room temperature) for 120 seconds without stirring. Moreover, the specimen was washed and electroplated. In the procedure C, all the treatment in the same way in the procedure B was carried out; however, zincate treatment was conducted with stirring of the solution. After the optimum surface treatment was evaluated, the temperature of the bath was varied (40±5, 50±5, 60±5°C) using the suitable surface treatment. Moreover, the amount of ZrO_2 particles was varied as well (0, 1.0, 3.0, 5.0 g/L) after the suitable bath temperature was determined.

Coating surface was observed by EPMA (JEOL JXA8900). Vickers hardness was examined by Micro Vickers Zwick Roell manufactured by Zwick. Ten times indentation was carried out on the surface of the specimens with load of 0.5 kgf and holding time of 10 seconds. In the measurement of polarization behavior, the specimen was mounted with thermosetting resin and polished by waterproof abrasive paper from #600 to #2000 and further polished using abrasive cloth. The cyclic voltammetry measurements were carried out by VersaSTAT-300 (Princeton™ Applied Research) and the potential scanning rate was 0.50 m/s. A saturated calomel electrode was used as the reference electrode. For the sample, an uncoated aluminum plate (2.1 cm^2) was used. In order to connect the sample to the measurement device, the sample was spot-welded with aluminum wire. CrO_3 (250 g/L), H_2SO_4 (2.5 g/L), ZrO_2 (0, 5.0 g/L) were prepared as the electrolytes. A stirrer was used when ZrO_2 particles were included in the solution.

3. Results and Discussion

Fig. 2 shows the surfaces of the coatings electrodeposited at the bath temperature of 50±5°C by using different procedures A, B, and C, respectively. In making comparison between the coating on the steel plate (Fig. 2 (a)) and the aluminum plate (Fig. 2 (b)) used the procedure A, peeling of the coating was observed in the latter specimen. Therefore, it is considered that the Al_2O_3 film on the aluminum surface adversely affected film formation [7]. The smoothest and uniform coating was obtained on the substrate used the procedure C (Fig. 2 (d)) in which zincate treatment with stirring was carried out. From this result, it is considered that the procedure C is suitable as a procedure for obtaining a smooth and uniform coating film. Subsequently, using the procedure C, coatings were produced by changing only the temperature of the plating bath.

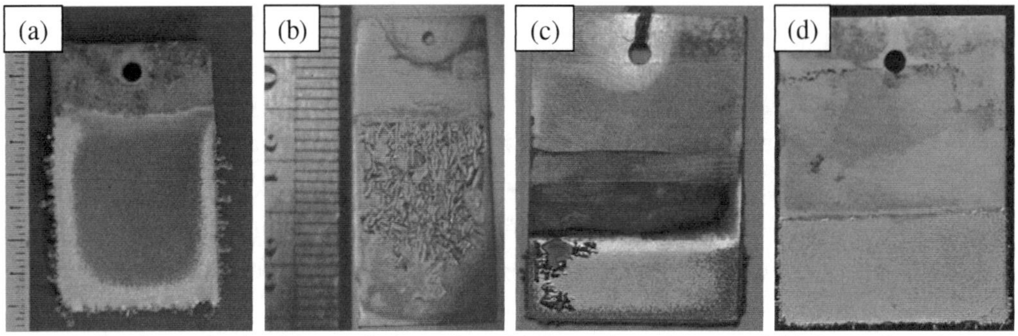

Figure 2 Coatings obtained by each procedure (a) Procedure A on steel (b) Procedure A on aluminum (c) Procedure B on aluminum (d) Procedure C on aluminum

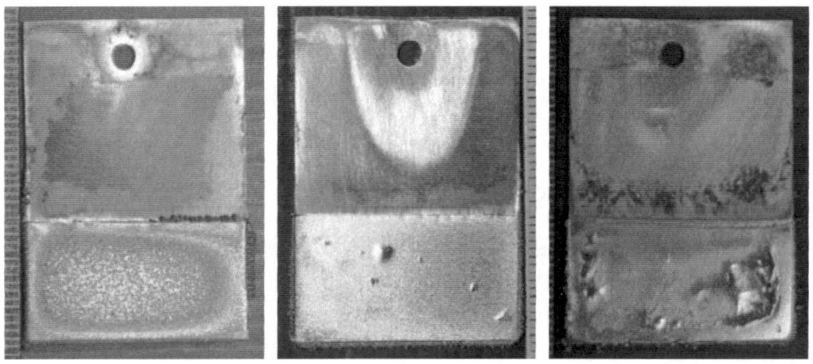

Figure 3 The specimen electroplated at (a) 40±5°C (b) 50±5°C (c) 60±5°C

Figure 3 shows the surface of the coating electrodeposited at each bath temperature (40±5, 50±5, 60±5°C). When electrodeposition was carried out, it was found that the evaporation amount of the bath increased as the bath temperature was raised. Coating obtained by electrodeposition at 40±5°C (Fig. 3 (a)) was not uniform. In the specimen plated at 60±5°C (Fig. 3 (c)), peeling of the film was observed. From this result, electrodeposition at 50±5°C is considered to be suitable for obtaining a smooth and uniform coating film.

Figure 4 shows SEM images of the cross section of the specimen electroplated at each bath temperature (40±5, 50±5, 60±5°C) and the thickness of the coating film is shown in Table 2. No difference was observed in the morphology of the coating cross section and we found that the film thickness was thinner as the temperature of the coating bath was higher. Since the thickness of general hard Cr coating is in the range of 2.54-254 μm [8], the thickness of the coating from 89 μm to 107 μm obtained in the present work seems to be appropriate.

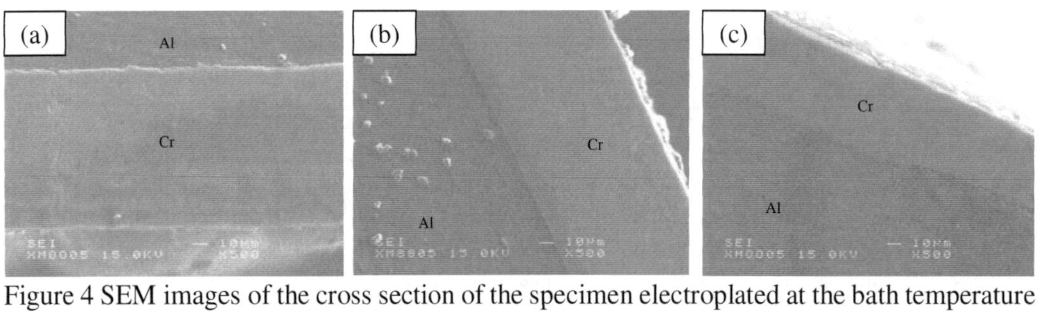

Figure 4 SEM images of the cross section of the specimen electroplated at the bath temperature of (a) 40±5°C (b) 50±5°C (c) 60±5°C

IOP Conf. Series: Materials Science and Engineering **547** (2019) 012027 doi:10.1088/1757-899X/547/1/012027

Table 2 Thickness of the coating at each bath temperature

Temperature of the plating bath [°C]	Thickness of the coating [μm]
40±5	107
50±5	96
60±5	89

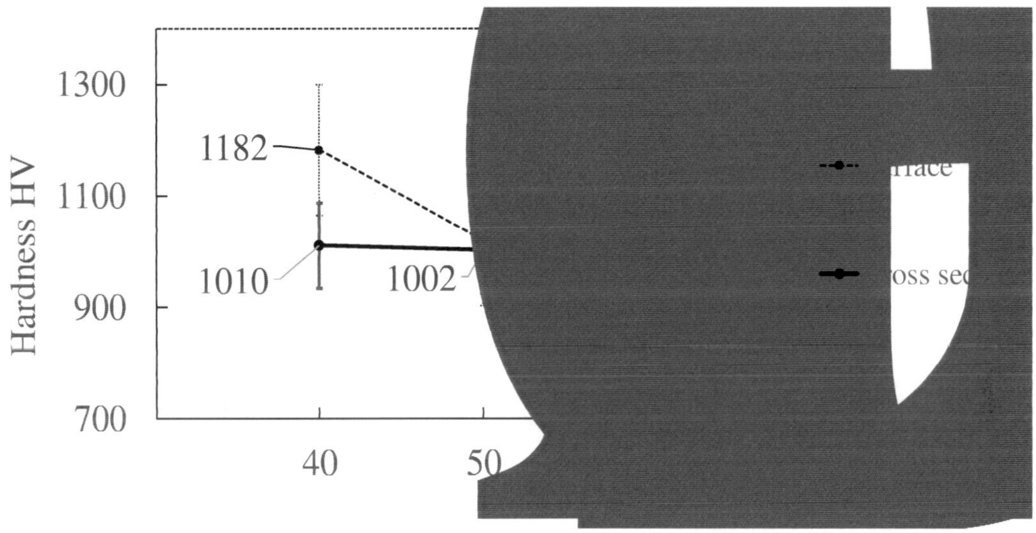

Figure 5 Vickers Hardness on the specimen which plated in the bath with different temperature

Figure 5 shows the result of hardness measurement on each specimen electrodeposited at different coating bath temperature. When the hardness was measured at the coating surface, it was obvious that the lower the hardness in the higher the temperature of the plating bath. Meanwhile, when it was measured at the cross section, hardness decreased slightly. From the graph shown in Fig. 5, it is considered that suitable temperature of the bath is 40±5°C in order to obtain a film with the highest hardness; however, from the results in Fig. 3, it is assumed that electrodeposition with bath temperature of 50±5°C is optimal for producing a smooth and uniform coating with high hardness.

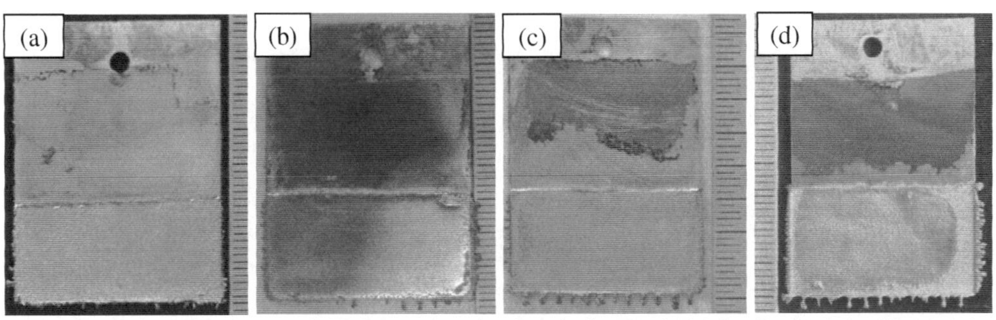

Figure 6 Coated surface on each ZrO_2 content in the plating bath (a) 0 g/L (b) 1.0 g/L (c) 3.0 g/L (d) 5.0 g/L

Figure 6 shows the surface of the coating electrodeposited in the bath with different ZrO_2 content (0, 1.0, 3.0, 5.0 g/L). Obviously, peeling of the coating film was not observed in any of

IOP Conf. Series: Materials Science and Engineering **547** (2019) 012027 doi:10.1088/1757-899X/547/1/012027

the samples. There is a possibility that a smooth and uniform coating can be obtained when ZrO_2 particles were dispersed into the plating bath.

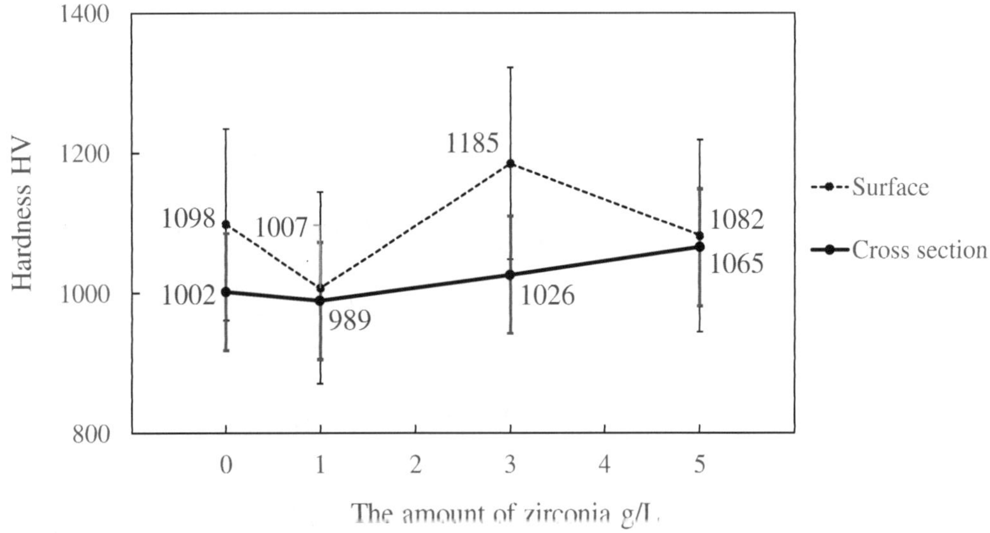

Figure 7 Vickers Hardness on each content of ZrO_2 in the plating bath

Figure 7 shows the result of hardness measurement on each specimen electrodeposited in the bath with different ZrO_2 content. From the result of measurement at the cross section, as the amount of ZrO_2 particles in the bath increased, the hardness of the coating increased. When the hardness was measured at the surface of the coating, no clear relationship was found between each value. It is considered that the hardness of the substrate affected the measurement at the coating surface.

Figure 8 shows EPMA mapping image at the cross section of Cr coating on aluminum plated in the bath which contains 5.0 g/L of ZrO_2 and the content of Zr in the coating obtained by point analysis is shown in Table 3. Although the existence of Zr was not confirmed by the mapping image, from the result of point analysis, Zr was detected as 1.235 wt% and 0.6924 atom% at a certain point in the coating film near the boundary with the aluminum substrate (circled in the composition image, CP, in Fig. 8). It is assumed that ZrO_2 particle dispersed in the coating film affected the increase of the hardness.

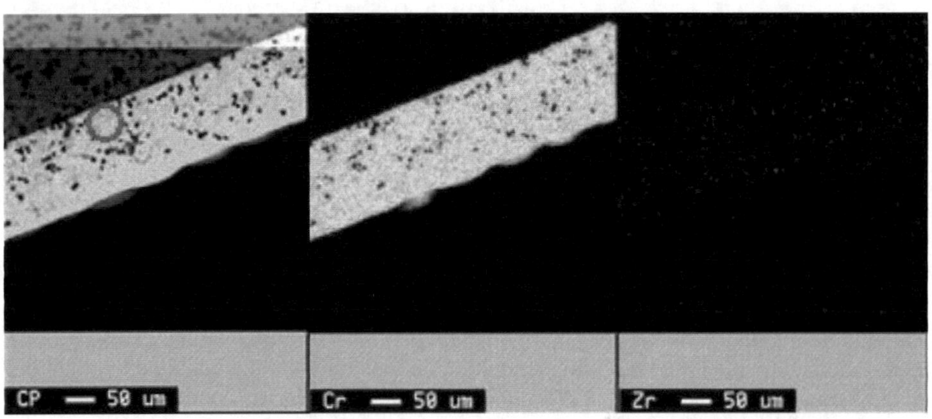

Figure 8 EPMA mapping image at the cross section of Cr coating on aluminum plated in the bath which contains 5.0 g/L of ZrO_2

Table 3 The content of Zr in the coating obtained by point analysis

Element	Wt %	Atom %
Zr	1.235	0.6924

Figure 9 shows the polarization curves of each ZrO_2 amount (0, 5.0 g/L). The corrosion current density (i_{corr}) increased in the solution containing ZrO_2. The corrosion potential (E_{corr}) was larger in the solution containing ZrO_2. In previous study, corrosion tends to occur as the corrosion potential is closer to the cathode side [9]. The material corrodes easily when polarization potential is minus. Hence, it is assumed that the addition of ZrO_2 into the plating bath improved the corrosion resistance.

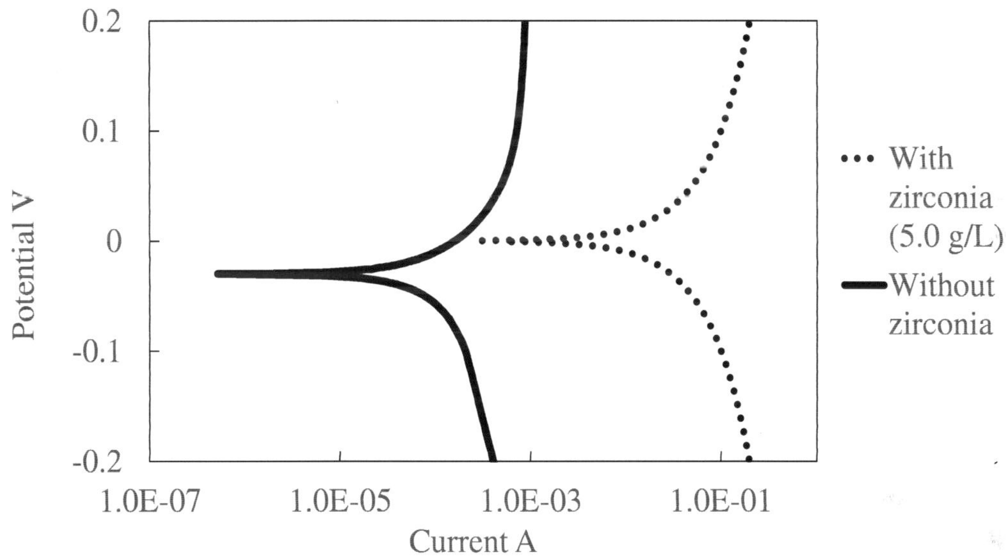

Figure 9 Polarization curves at each ZrO_2 density

4. Conclusions

In this study, the coating method suitable for producing high-quality film on aluminum substrate was investigated by using different surface treatment to each specimen before electrodeposition. Procedure C (polishing, acid dipping, and zincate treatment accompanied by stirring is performed on the sample before electrodeposition) was suitable to obtain a smooth and uniform film on an aluminum substrate by electrodeposition. Although the temperature of the plating bath for producing a film with the highest hardness by electrodeposition was 40±5°C, the optimum temperature for producing a smooth and uniform coating film was 50±5°C. A smooth and uniform film was obtained by electrodeposition with addition of ZrO_2 particles into the plating bath, and hardness of the coating increased with increasing the amount of ZrO_2 particles. The polarization potential increased when ZrO_2 particles were included in the plating bath. It means that corrosion resistance of the coating with ZrO_2 particles was improved.

5. References

[1] Chem Processing, Inc. "Hard Chrome Plating" available at: http://www.chemprocessing.com/page.asp?pageid=64& (accessed 2 December 2016)
[2] NDT Resource Center "Hardness Conversion Chart" available at:

IOP Conf. Series: Materials Science and Engineering **547** (2019) 012027 doi:10.1088/1757-899X/547/1/012027

https://www.nde-ed.org/GeneralResources/HardnessConv/HardnessConvPrint.pdf (accessed 5 December 2016)

[3] Catherine M C, James A S, and Fred A S Jr. *ASM Metals Handbook* **05** 1994 Surface Engineering pp 684–6

[4] Sun K, Hu X, Zhang J and Wang J Electrodeposited Cr-Al_2O_3 composite coating for wear resistance 1996 *wear* **196** , pp 295–7

[5] Jifeng G and Jinping S Preparation and characterization of the electrodeposited Cr-Al_2O_3/SiC composite coating 2011 *Applied Surface Science* **257** pp 9643–8

[6] Bin L, Zhixiang Z and Yimin L Mechanical properties of hard Cr-MWNT composite coatings 2009 *Surface & Coatings Technology* **203**, pp 3610–3

[7] U.S.Chrome "Chromium Plating on Aluminum" available at: http://www.uschrome.com/chromium-on-aluminum (accessed 5 December 2016)

[8] POETON 2017 "Surface treatments to reduce wear, friction and corrosion" available at: http://www.poeton.co.uk/chromium.htm (accessed 17 January 2017)

[9] Arenas-Martínez L.F and García-Cerecero G. Hydrogen Sulfide Corrosion of Weld Regions in API X52 Steel 2012 *Ingeniería Investigación y Tecnología* **volumen XIII (número 4)** pp 473–8

The Influence of Chitosan Concentration on Polyelectrolytes Complexes (PECs) of Chitosan – Poly-2-Acrylamido-2-Methylprophane Sulfonic Acid (PAMPS) as Potential Drug Carrier in Pulmonary Delivery Application

Arie Wibowo[1,2,*], Mumtaziah[1], Systi Adi Rachmawati[1], Fitriyatul Qulub[1], Lia Amelia Trisna Wulan Asri[1], Akfiny Hasdi Aimon[3], Rochim Suratman[1]

[1]Department of Materials Engineering, Faculty of Mechanical and Aerospace Engineering, Institut Teknologi Bandung

[2] Research Center for Nanoscience and Nanotechnology, Institut Teknologi Bandung

[3]Department of Physics, Faculty of Mathematics and Natural Sciences, Institut Teknologi Bandung

*Corresponding author: ariewibowo@material.itb.ac.id

Abstract. Pulmonary drug delivery system is one of potential way to effectively treat lung-based disease, such as tuberculosis, because this route offers a possibility to reduce systemic toxicity and achieving higher drug concentration at the main site of infection. One of potential drug carrier for this purpose is chitosan based polyelectrolytes complexes (PECs) because micron-sized particles with good biocompatibility, non-toxic and antibacterial properties, that was inherited from chitosan, could be achieved. In this research, PECs were prepared by simple mixing method between chitosan solution and poly-2-acrylamido-2-methylprophane sulfonic acid (PAMPS) solution. The influence of chitosan concentration on obtained PECs were studied by varying chitosan concentration from 0.1 to 2.0% w/v. Based on dynamic light scattering (DLS) and zetasizer results, increasing chitosan concentration lead to increase the particle size from 1.1 μm to 10.7 μm and their net surface charge of particles were changed from -4,2 mV to 32.8 mV. Considering that particle size of pulmonary drug carrier should be around 1-3 μm and preferable possesses positive net surface charge, PECs prepared from chitosan solution 0.5% w/v might be potentially useful as drug carrier for pulmonary drug delivery system.

Keywords: chitosan, PAMPS, polyelectrolytes complexes, pulmonary drug delivery system

1. Introduction

Tubercolosis (TB) is an infectious disease that was caused by mycobacterium tuberculosis bacteria and classified as one of the top 10 causes of death worldwide. Currently, the most common method to treat TB are using anti-TB drugs via oral and injectable intramuscular treatment [1]. However, both of them still have a problem with drug's efficacy since the drug still subjected to first-pass metabolism [1,2]. Therefore, an alternative strategy to deliver anti-TB drugs with better drug efficacy is needed.

Nowadays, pulmonary drug delivery systems (PDDS) have been widely investigated as attractive strategy for the treatment of pulmonary-related diseases because it could directly deliver anti-TB drugs to the lungs. Thus, it is expected that side effects of systemic administration could be reduced and therapeutic efficacy could be enhanced [1-6]. For effective delivery of drugs using this strategy, drug carrier with aerodynamic diameter between 1 and 3 μm are required for successful deposition in deep-lung [3]. One of attractive materials as drug carrier is polyelectrolytes complexes (PECs) because of its simplicity, low toxicity, high drug loading efficiency, and desired structural stability [7-9]. Previously, we have been successfully prepared PECs from chitosan as positively charged polymers and Poly-2-Acrylamido-2-Methylprophane Sulfonic Acid (PAMPS) as negatively charged polymers by simply mixing of their water-based solution [9]. Chitosan are appealing natural polymers for biomedical applications due to their non-toxic, good biocompatibility, biodegradability, and antimicrobial activity [10-12]. Besides their application as biomaterials, chitosan can also be used as chelating agent for heavy metal recovery [11,12], solid state batteries [12,13], etc. In this experiment, chitosan-PAMPS based PECs were prepared by simple mixing and various chitosan concentration (0.1; 0.5; 1.5 and 2.0% (w/v)) were used to know the influence of chitosan concentration to particles size of obtained PECs.

2. Materials and Methods

Chitosan (with degree of deacetylation (DD) 77% [9]) was purchased from CV. Biochitosan Indonesia. Before used, molecular weight of chitosan and PAMPS were roughly determined by dilution viscometry method using Mark-Houwink-Sakurada equation (Eq. 1) [14,15].

$$[\eta] = K\, M_v^{\alpha} \qquad (1)$$

where K and α are Mark-Houwink constants that depend on the polymers – solvent interaction. This characterization step is essential for assigning the possible range of PECs charge ratio.

Then, PECs were prepared by simple mixing of water-based solution chitosan and PAMPS following procedures that has been describe previously [9]. To detect the present of colloid and qualitative predict particle size of colloid in solution of obtained samples, Tyndall effect were observed by exposing laser pointer light to colloidal solution samples. Then, 20 mL of colloidal solution of obtained samples were characterized by Dynamic Light Scattering (DLS) to measure their particle size and Zetasizer to measure their surface charge using Delsa™ Nano Series. The particle size measurement was carried out with accumulation time 30 s at scattering angle 165°. The zeta potential measurement was conducted at 24.9 °C with scattering angle 34.1° and average electric field 7.45 V/cm.

3. Results and Discussion

3.1. Molecular weight measurement analysis

The molecular weight (M_v) of chitosan and PAMPS were calculated using Eq. (2.1) by considering K and α values of chitosan and PAMPS in their solvents. The value of K and α for chitosan in acetic acid 1% are 4.74x10^5 dL/g and 0.72 respectively [14]. While the value of K and α for PAMPS in sodium chloride 0.1 N are 1.67x10^5 dL/g and 0.88 respectively [15]. The calculated M_v of chitosan and PAMPS can be seen in Table 1.

Table 1. Molecular weight from dilution viscometry method

Polymers	M_v [g/mol]	R^2
Chitosan	231,736	0.88
PAMPS	568,238	1.00

The composition of these two polymers could be predicted in order to fabricate the PECs with positive net charge. Regarding to Table 1, the molecular weight of PAMPS is roughly 2.5 times higher than molecular weight of chitosan. Since DD of chitosan that was used in this experiment is about 77%, it could roughly predicted that neutral PECs from PAMPS and chitosan could be formed if the amount of chitosan is about 3.2 times than PAMPS. Considering this prediction, chitosan concentration were varied from 0.1% w/v (ratio of chitosan/PAMPS = 1; PAMPS-rich composition) to 2.0% w/v (ratio of chitosan/PAMPS = 20; chitosan-rich composition) to obtain some insight about characteristic of obtained PECs as a function of chitosan concentration.

3.2. Visual observation of colloidal solution samples

Visual observation of Tyndall effect of samples solution after exposing with laser pointer light can be seen in Fig.1. Generally, bigger the particle size induced light is being more dispersed. According to visual observation, it was predicted that as chitosan concentration increased, the bigger the particle size.

Figure 1. Visual observation of Tyndall effect of samples solution with various chitosan concentration: a) 0.1% w/v, b) 0.5% w/v, c) 1.5% w/v, d) 2.0% w/v, after exposing with laser pointer light

3.3. Particle size and zeta potential analysis

Particles size and zeta potential results of obtained samples can be observed in Fig. 2a. and 2b respectively.

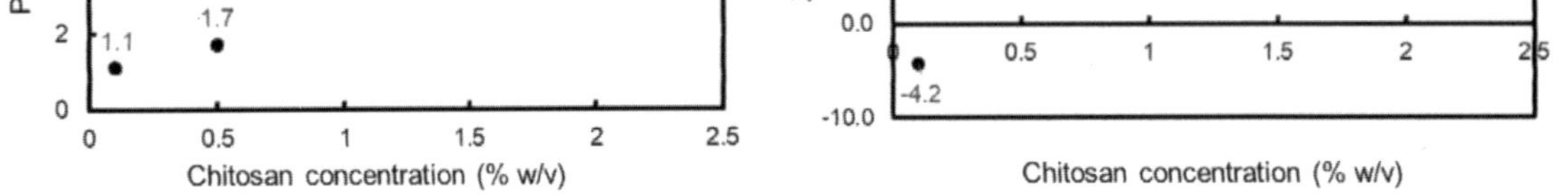

Figure 2. a) Particles size and b) zeta potential results of colloid in samples solution as a function of chitosan concentration

IOP Conf. Series: Materials Science and Engineering **547** (2019) 012028 doi:10.1088/1757-899X/547/1/012028

As can be seen in Fig 2.a, only samples with chitosan concentration 0.1 and 0.5% w/v (with particle size 1.1 and 1.7 μm respectively) are potential as drug carrier for PDDS application because their particles size are within the range of good carrier for PDDS (1-3 μm) [3]. While samples with chitosan concentration 1.5 and 2.0% w/v might not suitable as drug carrier for PDDS application because their particles size are higher than 3 μm. Zeta potential data (Fig. 2.b) showed that net surface charge of samples with chitosan concentration 0.1 and 0.5% w/v are -4.2 and 16.0 mV respectively. Considering that positively charge biomaterials possesses better antimicrobial effect than negatively charge biomaterials [10], it is expected that sample with chitosan concentration 0.5% w/v are better candidate than 0.1% w/v.

Characteristic of PECs from their oppositely charged polymers are following the nature of their forming polymers. PAMPS are naturally flexible and able to be folded [16], while chitosan are naturally rigid and cannot be folded easily [17]. As can be seen in Fig. 2a and 2b, PECs with negative charge in their surface (chitosan concentration 0.1% w/v) are smaller than PECs with positive charge (chitosan concentration 0.5-2.0% w/v). At chitosan concentration 0.1% w/v, the dominant polymers in their composition are PAMPS (which is indicated by their negative charge in their surface). As the consequences, most of polymers in this composition are flexible and could be folded to form small particles of PECs. At chitosan concentration 0.5-2.0% w/v, chitosan are more abundant than PAMPS in the PECs (which is indicated by their positive charge in their surface). Thus, particles size of PECs increased as chitosan concentration increased because most of polymers in this composition are rigid and difficult to be folded. The possible PECs formation and their characteristic regarding to their charge ratio is illustrated in Fig. 3 below.

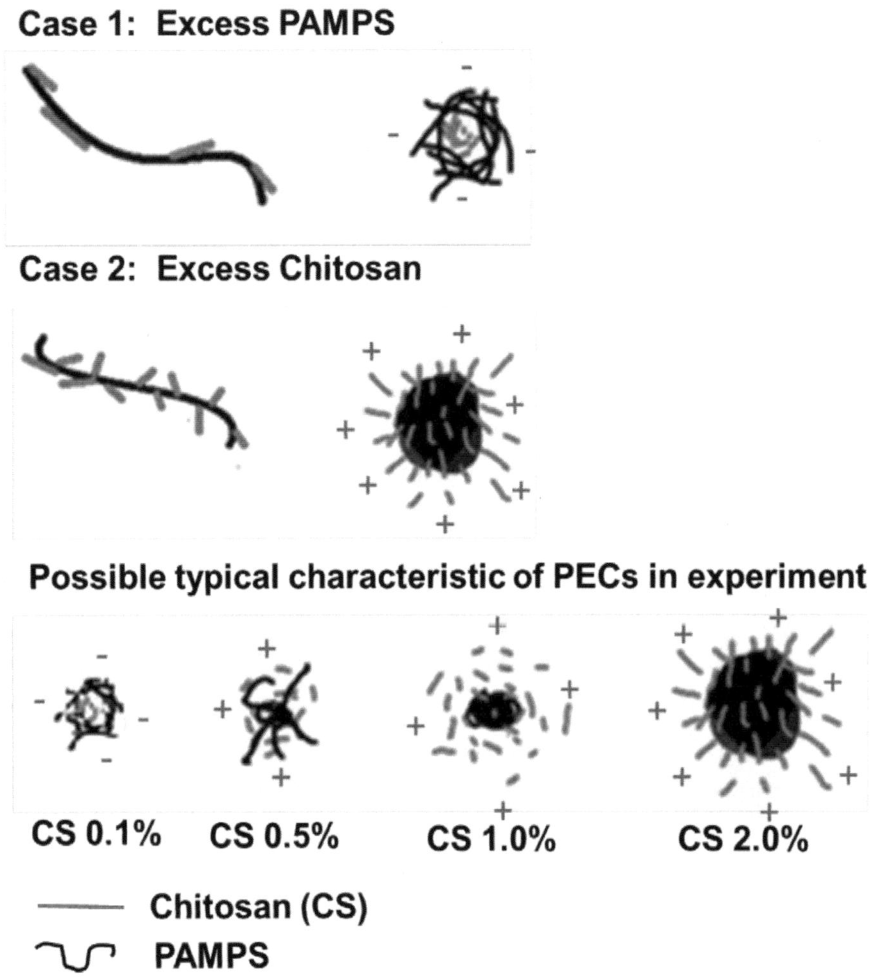

Figure 3. PECs formation model based on PAMPS and chitosan

4. Conclusions

Formation and characteristic of chitosan-PAMPS based PECs has been studied by varying chitosan concentration from 0.1 – 2.0% w/v. PECs that prepared from 0.5% w/v of chitosan concentration might be the most promising for pulmonary drug delivery system because their particle size (1.1 um) are within the range of good carrier for PDDS (1-3 μm). Also, obtained PECs possesses positive charge in their surface (16.0 mV) might possesses better antimicrobial effect against bacteria that are useful for tuberculosis treatment.

5. Acknowledgements

This research was funded by Overseas Research Collaboration Fund from Indonesian Ministry of Research, Technology and Higher Education 2018

6. References

1. Pham, D. D.; *International Journal of Pharmaceutics,* **2015**, *478,* 517
2. Wan, F.; Moller, E. H.; Yang, M.; Jorgensen, L.; *Drug Discovery Today: Technologies,* **2012**, *9 (2)*, e141
3. Yang, W.; Peter, J. I.; William III, R. O.; *International Journal of Pharmaceutics,* **2008**, *356,* 239
4. Sung, J. C.; Pulliam, B. L.; Edwards, D. A.; *Trends in Biotechnology,* **2007**, *25(12),* 563
5. Bailey, M. M.; Berkland, C. J.; *Medicinal research Reviews,* **2009**, *29*, 196
6. Mansour, H. M.; Rhee, Y.-S.; Wu, X.; *International Journal of Nanomedicine,* **2009,** *4*, 299
7. Luo, Y.; *International Journal of Biological Macromolecules* **2014**, *64*, 353– 367
8. Zhang, L., Wang, J., Ni, C., Zhang, Y., Shi, G., *Materials Science and Engineering C,* **2016**, *58*, 724 – 729
9. Wibowo, A.; Indrawan, R. F.; Triadhi, U.; Aimon, A. H.; Iskandar, F.; Ardy, H.; *Materials Research Express*, **2018**, *5*, 024005 (DOI: 10.1088/2053-1591/aaac86)
10. Rabea, E. I.; Badawy, M. E.-T.; Stevens, C. V.; Smagghe, G.; Steurbaut, W.; *Biomacromolecules*, **2003**, *4(6)*, 1457 – 1465
11. Zargar, V.; Asghari, M.; Dashti, A.; *ChemBioEng Reviews*, **2015**, *2(3)*, 204-226
12. Kumar, M. N. V. R.; *Reactive and Functional Polymers*, **2000**, *46(1)*, 1-27
13. Asri, L. A. T. W.; Nuruddin, A.; Hidayatullah, S.; Simatupang, M.; Wibowo, A.; Septawendar, R.; Purwasasmita, B. S.; *Journal of Australian Ceramics Society*. **2017**, (DOI 10.1007/s41779-017-0103-1)
14. Kasaai, M. R.; Arul, J.; Charlet, G.; *Journal of Polymer Science Part B: Polymer Physics*, **2000**, *38*, 4325-4332
15. Tan, J. S.; Fisher, L. W.; Markus, P.; ACS National Meeting in Philadelphia, division of organic coating and plastic Preprints, **1992**, *35(1)*, 348
16. Setoyama, H., Murakami, Y., Inoue, K., Iwata, H., Kitamura, H., Shimada, T., Kaji, H., Ikada, Y., Imamura, M., *Transplantation Proceedings*, **1999**, *31*, 2818-2822.
17. Terbojevich, M.; Cosani, A, Conio, G.; Marsano, E.; Bianchi, E.; *Carbohydrate Research*, **1991**, *209*, 251-260

Effect of Aging Time on Characteristics of ADC12/ SiC Composite with Addition of Sr Modifier and Al-5TiB Grain Refiner

Anne Zulfia Syahrial[1]*Muhammad Salman[1] Budi Wahyu Utomo[2]
[1]Department of Metallurgy and Materials, Faculty of Engineering, University of Indonesia, Kampus Baru Universitas Indonesia, Depok 16424, Indonesia
[2]PT Astra Honda Motor, Jl. Raya Kalimantan Blok AA Kawasan Industri MM2100 Cikarang Barat, Indonesia

*Corresponding author (E-mail): anne@metal.ui.ac.id

Abstract. An aluminium composite is developed as a light material with high strength for transportation application. Aluminium ADC 12 matrix was reinforced with SiC and added with Al-5TiB grain refiner and Sr modifier which is fabricated by stir-casting. The composites then were T6 treated for the various aging time from 2, 6, 10, 15 and 24 hours. The composites were characterized including tensile strength, impact resistance, hardness testing, wear resistance, density and porosity as well. The microstructural analysis to obtain the microstructure and phases changed after T6 using SEM-EDS and confirmed by XRD. Results show that T6 treatment has affected the morphology of primary and eutectic Mg_2Si, as well as precipitation of metastable β'-Mg_2Si from the aging process. T6 heat treatment has improved mechanical properties for all composites compared to as-cast composites. The highest ultimate tensile strength value and hardness are shown to be 144.5 MPa and 64.467 HRB after T6 for 6 hours at 170°C.

Keywords: *Aluminum composite, ADC 12, micro SiC, modifying agent Sr, grain refiner TiB, artificial aging*

1. Introduction

Aluminium composite is a potentially excellent material for many purposes, especially transportation and aerospace applications, due to its lightweightness and high strength [1]. Aluminum ADC12 alloy is an interesting choice as the Al composite matrix, due to its corrosion resistance, excellent thermal conductivity, as well as its mechanical strength relative to other aluminium alloys. Reinforcement used for the composite is silicon carbide particles, due to its ability to increase tensile strength, hardness, elasticity modulus, wear resistance, as well as thermal shock resistance, of the ADC12 alloy. The composite is also added grain refiner Al-5Ti-B and modifier Sr to form a good microstructure to improve mechanical properties further [2]. The composite is also given T6 heat treatment in order to further enhance the mechanical properties of the material [3]. This work is carried out in order to study the effect of various aging durations on the physical and mechanical properties of the composite material.

Metal-matrix composites are generally fabricated through means of stir casting, powder metallurgy, squeeze casting, preform infiltration, and ultrasonic vibration [4]. In this study, aluminium/SiC composites are fabricated using the stir casting method, due to simplicity as well as inexpensiveness to apply the technique. From this study, the resulting material is expected to have excellent mechanical properties, high fatigue resistance, excellent creep resistance, and practicality for field applications.

2. Materials and Methods

Aluminium ADC12 alloy is used as the matrix of the composite; SiC particles are used as reinforcement. Sample fabrication is carried out by stir casting method. SiC particles are first heated at 900°C for one hour in order to deagglomerate the SiC particles. Aluminium ADC12 is melted and degassed with argon to reduce porosity caused by hydrogen present in the melt, then added magnesium to increase the wettability of SiC particles to aluminium melt. Preheated SiC particles are added slowly while stirred with a pneumatic stirrer for 1-2 minutes. Before casting, the metal mold is preheated to avoid thermal shock, temperature drop, as well as to eliminate water content. The composite melt is poured into the mold immediately after stirring, then once the composite is set the product is taken out of the mold.

T6 heat treatment is carried out after composite fabrication. The samples are solution heat-treated at a temperature of 490°C for 60 minutes, then quenched using water at room temperature. The samples are the artificially aged at a temperature of 170°C for 2 hours, 6 hours, 10 hours, 15 hours, and 24 hours respectively.

Samples are characterized to obtain the characteristics of the ADC12/SiC composite as well as to study the effects of different aging durations. Characterizations carried out include chemical composition characterizataion by optical emission spectroscopy (OES), microstructural observation using optical microscope (OM), X-ray diffraction characterization (XRD), scanning electron microscopy enhanced with energy dispersive spectroscopy (SEM-EDS), density testing, tensile testing, hardness testing, and impact resistance testing.

3. Results and Discussion

Optical emission spectroscopy results on ADC12/SiC composite samples are shown in Table 1. It is shown from the results that the main composition of the composite is Al-Mg-Si. Excess Mg comes from the addition of 5 %wt Mg as a wetting agent for SiC particles towards aluminium matrix. The presence of Mg and Si indicates that phase formation that is predicted to occur on the ADC12/SiC composite is the formation of Al, eutectic Si, and Mg_2Si phase [5]. The results also show the incompatibility between actual Sr content and the supposed 0.03 %wt addition where Sr barely reaches 0.0064 %wt, due to an occurrence of fading of Sr particles. Meanwhile, TiB content data shows insignificant addition from pure ADC12, with 0.02 addition of %wt Ti and 0.01 %wt addition of B compared to a total of 0.15 %wt addition, due to the occurrence of fading on TiB so that Ti content on end product is lower than supposedly added. The presence of significant amounts of Fe on the material allows the possibility of formation of intermetallic phase β-AlFeSi which could decrease the mechanical properties of the material [6].

Table 1. Results of OES on ADC12 and ADC12/SiC composite.

Element	**Al**	**Si**	**Fe**	**Cu**	**Mn**	**Mg**	**Cr**	**Ni**	**Zn**	**Ti**	**Sr**	**B**
ADC12	84.8	10.5	0.864	2.33	0.223	0.221	0.0389	0.0759	0.637	0.0462	<0.0001	<0.0005
ADC12/SiC	73.91	9.837333	1.454333	3.442	0.096367	5.161333	0.155533	0.151633	1.035667	0.0822	0.0064333	0.017215

Microstructure examination is carried out using an optical microscope at magnifications 100, 200, 500 and 800x using HF as an etchant for 8-10 seconds. Figure 1 shows the microstructure of ADC12 aluminium alloy and ADC12/SiC composites given T6 heat treatment. The columnar shape of the grains means that grain size can be compared by measuring the width of the columnar grains, shown in Figure 1, and compared in the graph shown in Figure 2. From there primary Mg_2Si, pointed out in Figure 1, is spread throughout the sample surface. The preliminary chemical composition characterization shows that the composition of this sample consists of 73.9 %wt Al, 9.84 %wt Si, 5.16 %wt Mg, and 1.45 %wt Fe. Thus to predict the phases formed the Al-Mg-Si and Al-Si-Fe ternary diagrams are examined.

Determining phases formed in the composite is theoretically done by examining the ternary phase diagrams of Al-Mg-Si and Al-Si-Fe systems and observing the solidification path based on the composition from the spectroscopy results. Furthermore, examination of the quaternary phase diagrams of the Al-Si-Fe-Mg and Al-Si-Cu-Mg systems is carried out. From there the following reactions are obtained [7].

Table 2 Reactions that occur based on ternary Al-Si-Mg and Al-Fe-Si systems [7].

Reaction	Temperature (°C)
$L \rightarrow Mg_2Si + L_1$	647
$L_1 + Mg_2Si_p \rightarrow (Mg_2Si + Al)_E + Mg_2Si_p$	555
L → α-Al AlFe5Si → Al + β-Al5FeSi + Si	567

From the Al-Fe-Si-Mg quaternary phase diagram, the reaction L → (Al) + (Si) + Mg_2Si + $Al_9FeMg_3Si_5$ is predicted, meanwhile from the Al-Si-Cu-Mg quaternary phase diagram, the reaction L → Q-$Cu_2Mg_8Si_6Al_5$ + (Si) + θ-Al_2Cu + (Al) is predicted [8].

Heat treatment on Al-Mg-Si alloys could transform the morphology of primary Mg_2Si phases becoming smoother and more globular or more octahedral. During solution treatment, Mg and Si from sharp corners of the flake-like primary Mg_2Si phases dissolve more quickly so that the sharp edges erode. Furthermore, dendrite fracture of the flake-like Mg_2Si helps smoothen the Mg_2Si primary phases and make it more globular-shaped [9]. Primary Mg_2Si phases shaped near-octahedral and globular will increase mechanical properties of Al-Mg-Si alloys. Furthermore, heat treatment also causes Mg and Si from eutectic Mg_2Si phases to dissolve into the Al matrix and diffuse into locations with lower atomic concentration so that fragmentation of the long-shaped eutectic Mg_2Si phases occurs, to become more dissolved, diffused evenly, distributed evenly and also more round-shaped [10].

IOP Conf. Series: Materials Science and Engineering **547** (2019) 012029 doi:10.1088/1757-899X/547/1/012029

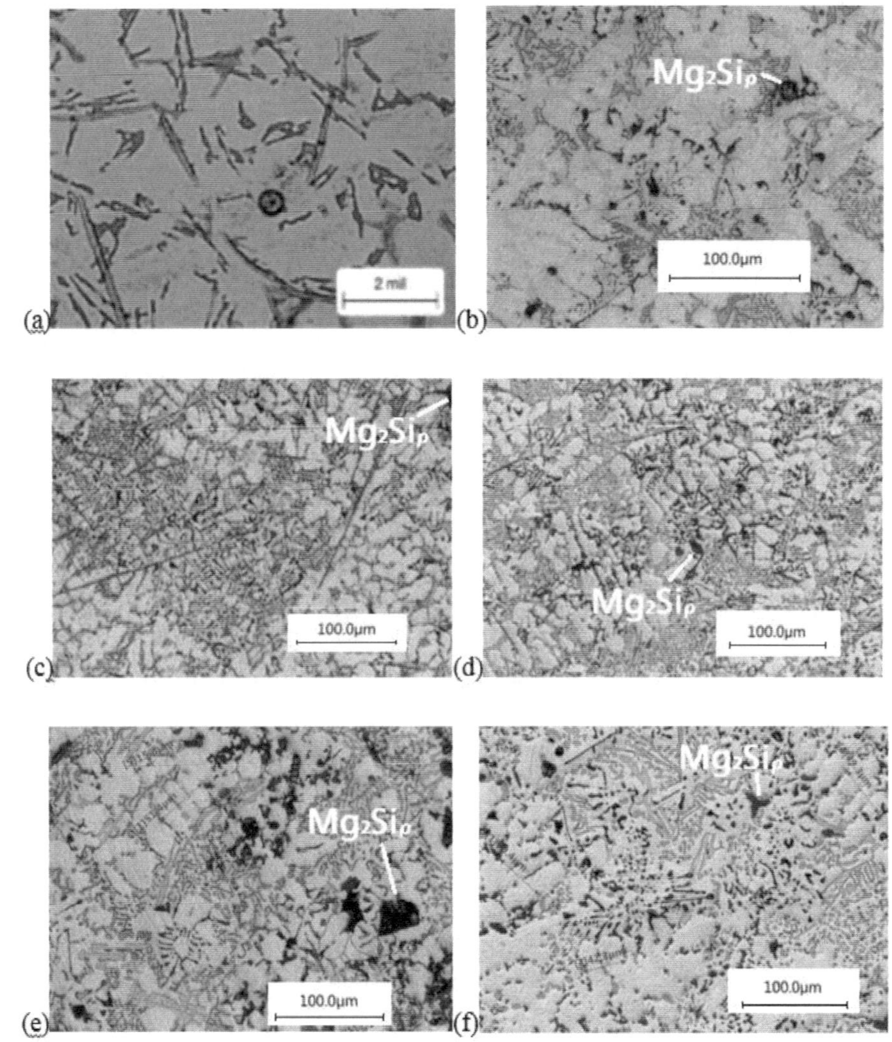

Figure 1 Micrograph of ADC12/SiC composite given T6 heat treatment of (a) ADC12 (b) 2 hours (c) 6 hours (d) 10 hours (e) 15 hours and (f) 24 hours using OM magnification 500x, HF etchant for 8-10 seconds.

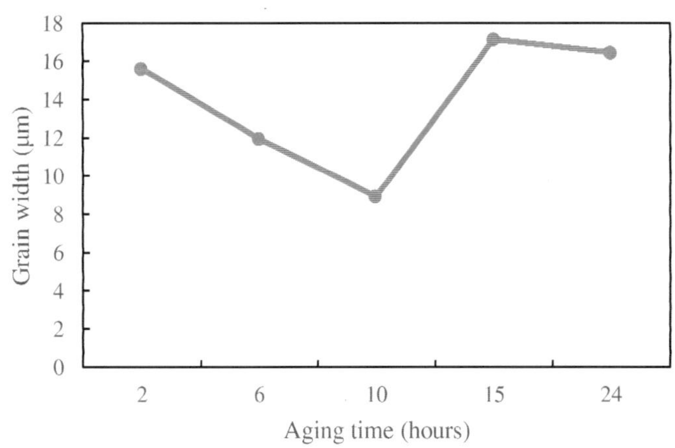

Figure 2 Columnar grain width of ADC12/SiC composite samples

XRD examination results on the optimum ADC12/SiC composite sample as shown in Figure 3 shows the formation of magnesium oxide (MgO) on the composite. MgO is formed from the reaction caused by high Mg content on the material which will react with SiC particles oxidized forming a SiO_2 layer, and the following reaction occurs: $SiO_2 + 2Mg \rightarrow 2MgO + Si$. Furthermore, the presence of spinel ($MgAl_2O4$) is indicated, which shows the formation of an interface between the matrix and the reinforcement of the composite. Another indication is the formation of Mg_2Si, a compound capable of increasing mechanical properties of the composite due to its high hardness, low thermal expansion, and high elasticity modulus [11].

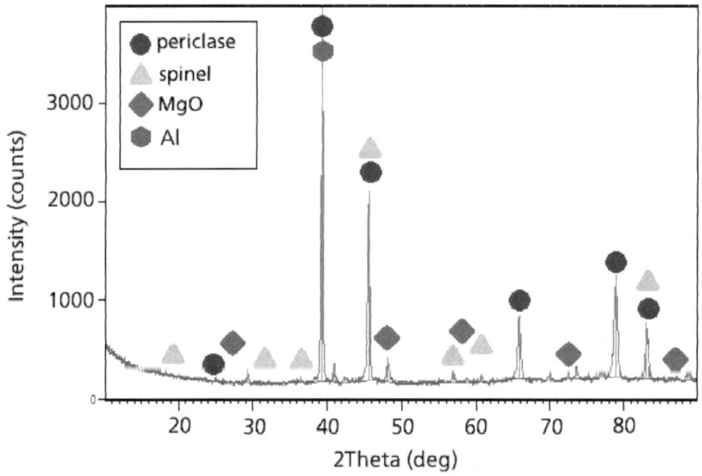

Figure 3 XRD result of the ADC12/SiC composite with the addition of Sr and Al5TiB and T6 heat treatment for 6 hours.

SEM/EDS characterization indicates the formation of primary and secondary Mg_2Si phases, as well as MgO phases. On point 056 in Figure 4(a) it is shown to contain Fe on the grey needle-shaped phase, and also contains Cu. This corresponds with the theory that Fe often coexists with Cu elements, as in the solidification reaction in the Al-Si-Cu-Fe quaternary system on the temperature of 525°C [7]. At this point, Al_2Cu and Al_5FeSi are formed, as well as Q-phase or $Cu_2Mg_8Si_6Al_5$. This phase is formed when Cu content exceeds 0.3%, and dissolved Mg_2Si will be replaced by $Cu_2MgSi_6Al_5$. This phase is formed on aluminium with dominant Si-Mg or Si-Cu alloying elements. With heat treatment on Al-Si-Mg-Cu systems, β' (metastable Mg_2Si) and Q-phase precipitates will be formed with a rod-like morphology.

a) 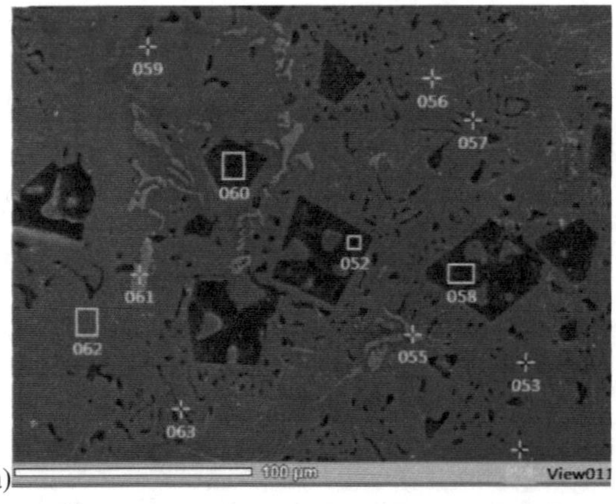

b) 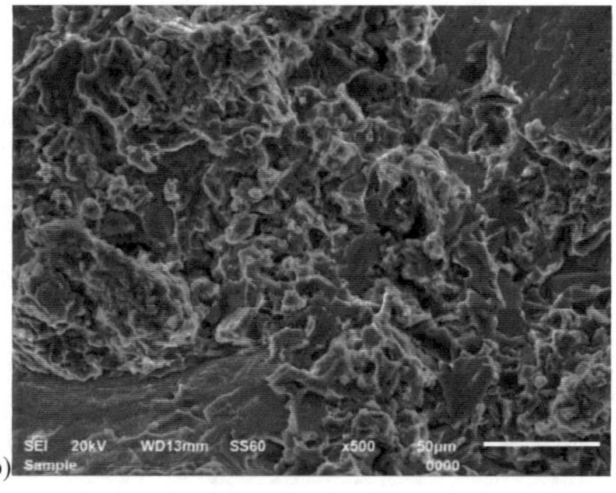

Figure 4 (a) SEM/EDS micrograph on the composite sample (b) SEM fractography of the composite sample

SEM microscopy is also carried out to observe the fracture zone of the impact resistance test samples. Figure 4(b) shows a mix of irregular large dimple fractures and cleavage fractures, which could be said as a quasi-brittle fracture [12]. Quasi-brittle fractures typically occur on composites consisting of ductile and brittle materials, so that failure will result in two fracture types: dimple (ductile fracture) and cleavage (brittle fracture).

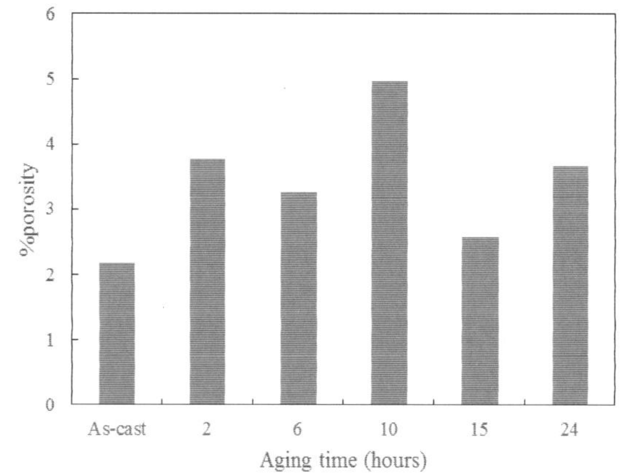

Figure 5 Porosity content of T6-treated ADC12/SiC composite samples

The porosity of composite samples is determined by calculation of actual density of composite samples compared to the theoretical density of the composite based on the composition of the composite. Figure 5 shows the result of calculation of porosity percentage of the composite samples. Porosity on aluminium composites could be caused by, to name a few, hydrogen evolution, gas entrapment, as well as shrinkage during solidification. Hydrogen entrapped in aluminium melt comes from atmospheric water vapour which binds with aluminium melt which then dissociates into hydrogen gas and dross [13]. The formed hydrogen gas becomes entrapped as the aluminium solidifies, and becomes porosity on aluminium matrix.

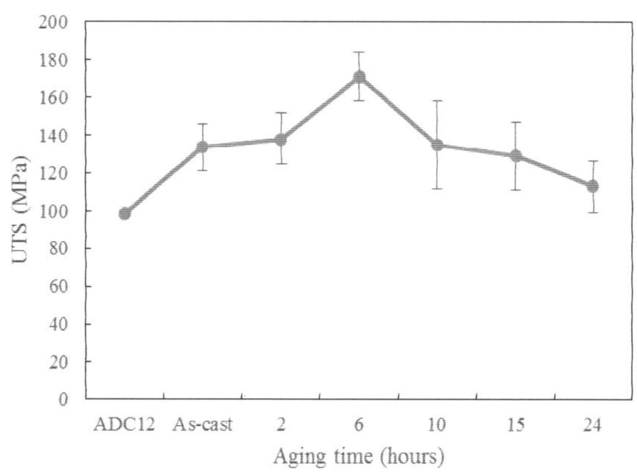

Figure 6 UTS of ADC12/SiC composites at different aging durations.

The ultimate tensile strength of ADC12/SiC composite as shown in Figure 6 is shown to increase compared to unreinforced ADC12 alloy, showing the effect of SiC reinforcement on the ADC12 matrix, due to the superior elasticity modulus of SiC. It is also seen that T6 heat treatment for 6 hours increases UTS more significantly than 2 hours of treatment. This shows that T6 heat treatment rises UTS of the composite, due to precipitation of semi-coherent β' (metastable Mg_2Si) precipitates and metastable Q-phase ($Cu_2Mg_8Si_6Al_5$) precipitates, as well as morphological modification of primary and binary Mg_2Si phases [14]. Excessive aging time will cause overaging which will reduce the tensile strength of the material. This is shown on the chart trend where from aging at 10 hours onwards show a decrease in UTS. Excessive aging causes the formation of precipitates incoherent with the matrix, where $\beta' \rightarrow \beta'' \rightarrow$ equilibrium β-Mg_2Si, which causes the precipitates to lose its strengthening ability [15].

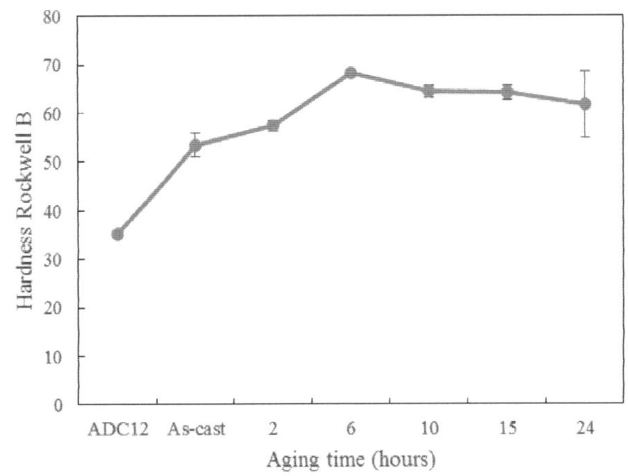

Figure 7 Rockwell B Hardness of ADC12/SiC composites at different aging durations

Rockwell B hardness of the ADC12/SiC composite samples is shown as in Figure 7 to increase compared to unreinforced ADC12 alloy, due to the high hardness and elasticity modulus of the SiC particles as reinforcement. The hardness value of the samples are shown to increase with a heat treatment up to the 6 hours of aging, then decreases after aging for 10 hours onwards. Aging treatment causes the formation of semi-coherent precipitates that will harden the composite material. Based on the maximum hardness as shown on the graph, optimum aging time for the composite in this study is 6 hours. Aging exceeding the optimum aging time will cause overaging, where the formed semi-coherent precipitates will become incoherent with the matrix lattice which causes the material hardness to decrease.

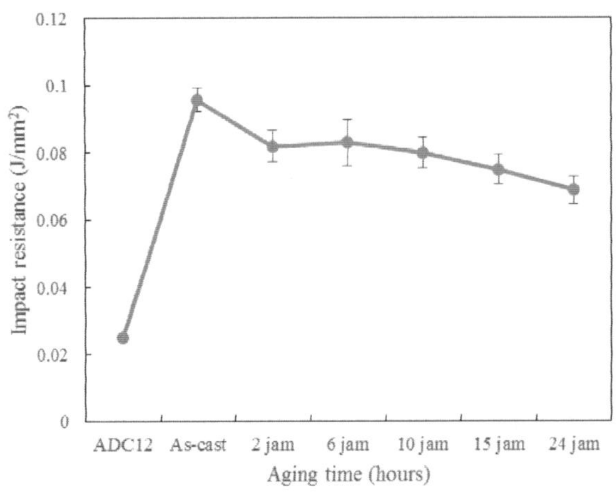

Figure 8 Impact resistance of ADC12/SiC composite samples at various aging durations.

Impact resistance of the ADC12/SiC composite material is shown as in Figure 8 to decrease after given heat treatment. Heat treatment for 6 hours is shown to increase the impact resistance of the sample compared to samples treated for 2 hours, afterwards impact strength is shown to decrease starting from 10 hours of treatment onwards. Impact strength is shown to increase due to the hindrance of dislocation motions by the formation of semi-coherent precipitates so that the material becomes tougher. The longer the heat treatment, the larger the precipitates will form so that it could hinder the motion of dislocations, by means of either precipitate tearing or Orowan looping [16]. Excessively long aging time will cause overaging, where the formed semi-coherent precipitates become incoherent with the matrix so that it can no longer hinder the motion of dislocation within the composite, as seen in the impact resistance chart.

4. Conclusions

The study on the effect of T6 aging time on the characteristics of ADC12/SiC composite modified with Sr and grain-refined with Al5TiB has shown morphological changes on primary and binary Mg_2Si phases as well as eutectic Si phases. SEM-EDS characterizations show the formation of primary and binary Mg_2Si phases, as well as the possibility Q ($Cu_2Mg_8Si_6Al_5$) phase formation. XRD examination shows the configuration of MgO, spinel ($MgAl_2O_4$) formed on matrix-reinforcement interfaces, and Mg_2Si in primary and binary forms, as well as metastable β'-Mg_2Si precipitates. T6 heat treatment is shown to have improved mechanical properties, including tensile strength, hardness, and impact resistance, due to the formation of β'-Mg_2Si and Q-$Cu_2Mg_8Si_6Al_5$ precipitates as well as morphology modification of primary and eutectic Mg_2Si phases, up to an optimum time of 6 hours. Aging exceeding 6 hours are shown to decrease mechanical properties due to the overaging phenomenon, where the formed precipitates have grown too large and incoherent with the matrix to hinder the motion of dislocations and thus lose its strengthening properties.

5. Acknowledgements

The Authors would like to express their gratitude for financial support from the Ministry of Research and Technology and Higher Education, Republic of Indonesia, through PTUPT Grant with contract number: 487/UN2.R3.1/HKP05.00/2018

6. References

[1] R. T. Mousavian, D. Brabazon, and A. F. Boostani, "Fabrication of aluminium composites reinforced with nano- to micrometre-sized SiC particles," *Materials and Design*, vol. 89, pp. 58–70, 2016.

[2] M. A. Simanjuntak and S. Abda, "Karakterisasi Komposit Matriks Logam AlSiC pada Produk Kanvas Rem Kereta Api," *e-Dinamis*, vol. 6, no. 2, 2013.

[3] D. Maisonnette, M. Suery, D. Nelias, P. Chaudet, and T. Epicier, "Effects of heat treatments on the microstructure and mechanical properties of a 6061 aluminium alloy," *Materials Science & Engineering A*, vol. 528, no. 6, pp. 2718–2724, 2011.

[4] M. J. Shen, X. J. Wang, T. Ying, K. Wu, and W. J. Song, "Characteristics and mechanical properties of magnesium matrix composites reinforced with micron / submicron / nano-SiC particles," vol. 686, 2016.

[5] N. A. Belov, D. G. Eskin, and A. A. Aksenov, "Alloys of the Al-Cu-Si-(Mg-Fe) System," *Multicomponent Phase Diagrams*, 2005.

[6] S. Ji, W. Yang, F. Gao, D. Watson, and Z. Fan, "Effect of iron on the microstructure and mechanical property of Al-Mg-Si-Mn and Al-Mg-Si diecast alloys," *Materials Science & Engineering A*, vol. 564, pp. 130–139, 2013.

[7] L. F. Mondolfo, *Aluminium Alloys: Structure and Properties*. Butter Worths, 1976.

[8] K. Chang, S. Liu, D. Zhao, Y. Du, L. Zhou, and L. Chen, "Thermochimica Acta Thermodynamic description of the Al – Cu – Mg – Mn – Si quinary system and its application to solidification simulation," *Thermochimica Acta*, vol. 512, no. 1–2, pp. 258–267, 2011.
[9] A. Malekan, M. Emamy, J. Rassizadehghani, and M. Malekan, "Effect of Isothermal Holding on Semisolid Microstructure of Al–Mg_2Si Composites," *ISRN Metallurgy*, 2012.
[10] Z. Li, C. Li, Y. Liu, L. Yu, Q. Guo, and H. Li, "Effect of heat treatment on microstructure and mechanical property of Al–10%Mg_2Si alloys," *Journal of Alloys and Compounds*, vol. 663, pp. 16–19, 2016.
[11] J. Jahromi, A. Dehghan, and S. Malekjani, "Effect of Optimum Amount of Sr and Sb Modifiers on Tensile, Impact, and Fatigue Properties of A356 Aluminum Alloy," *Journal of Science and Technology*, vol. 28, 2004.
[12] Tongmin Wang, Yuanping Zheng, Zongning Chen, Yufei Zhao, and Huijun Kang, "Effect of Sr on The Microstructure and Mechanical Properties of In Situ TiB_2 Reinforced A356 Composite," *Materials and Design*, vol. 64, pp. 185–193, 2014.
[13] J. E. Gruzleski and B. M. Closset, *The Treatment of Liquid Aluminum-Silicon Alloys*. Illinois: American Foundrymen's Society, Inc., 1990.
[14] A. Zulfia and M. K. Ajiriyanto, "Effect of Artificial Aging (T6) on Microstructure of Al-AC8H / Al_2O_3 MMC Produced by Stir Casting Route," vol. 330, pp. 1552–1555, 2011.
[15] W. D. Callister, *Materials Science and Engineering: an Introduction*, 8th ed. New York: John Wiley & Sons, 2007.
[16] R. R. Ambriz and D. Jaramillo, "Mechanical Behavior of Precipitation Hardened Aluminum Alloys Welds, Light Metal Alloys Applications.," 2014. [Online]. Available: https://www.intechopen.com/books/light-metal-alloys-applications/mechanical-behavior-of-precipitation-hardened-aluminum-alloys-welds. [Accessed: 03-Jul-2018].

Effects Of Ionic Immersion Time On Optical Properties Of Cu_2ZnSnS_4 (CZTS) Deposited by Successive Ionic Layer Adsorption And Reaction (SILAR)

Asywendi Rukini[1]* Alvian Reza[2] Badrul Munir[3]
[1]Universitas Teknologi Sumbawa
[2]IBM
[3]Universitas Indonesia

*Corresponding author (E-mail): asywendi@gmail.com

Abstract. Cu_2ZnSnS_4 (CZTS) thin films fabricated by successive ionic layer adsorption and reaction (SILAR) method is a promising combination to actualize a low cost thin films Solar Cell (TFSC). Ionic exchange and rinsing time are critical point for ionic layer formation in SILAR method. The purpose of this study is to investigate the influence of cationic and anionic immersing time on the band gap energy of CZTS. Immersing time was varied into 10, 20, 30 and 40 seconds for both deposited and annealed samples, which annealing was conducted on 300°C for two hours. Deposited results showed a linear increase on band gap energy as cationic immersing time extended. Whereas, nonlinear increment with optimum band gap 1.35eV at 30 seconds immersion happened for anionic time variation samples. Annealed CZTS with cationic time variation also showed a linear increasing band gap. Meanwhile, annealed with anionic time variation showed a backward tendency of the deposited samples which the band gap decrease nonlinearly as anionic immersing time lengthened.

Keywords: Semiconductor, band gap energy, CZTS, SILAR

1. Introduction

Theoretically, solar energy provide 7000 times amounts of whole world electrical energy needs, yet in real utilization solar cell only supplies 0,1% of world's demands [1]. Earlier development solar cell requires high production cost. To produce 1 watt electricity during 12 hours sun rays expose with intensity about 18% still costs more compare to gas or coal to convert its calorie into 1 watt of electrical energy. Although the newest production cost has been reduced through various modifications of process [2][3], the relatively low efficiency of energy conversion compare to calorie conversion of coal into electrical energy is still a major problem. In addition, high cost of raw materials and expensive technology contribute to the high initial cost of solar panel. Therefore, researchers focus on developing high efficiency solar cell with low-cost method and affordable raw material.

Cu_2ZnSnS_4 or CZTS is a potentially low cost quaternary thin films which each element of CZTS is earth abundant. Cu, Zn, Sn, and S are relatively non-toxic and cheaper than the rare earth such Ga and In. Katagiri reported that the band gap energy of CZTS is 1,4-1,5 eV with absorption coefficient greater than 10 cm which suits to be applied as semiconductor on solar cell [4][5]. CZTS thin films have been developed via various method either physical or chemical route, yet wet chemical is the most suitable on low cost thin films solar cell concept. Successive ionic layer adsorption and reaction or known as SILAR is a modified version of chemical bath deposition (CBD) method which is considered as one of promising method to produce thin films in the industrial-scale [6] due to its simple procedures and inexpensive equipment. With numbers advantages SILAR does not need high quality substrate and a set of expensive vacuum system [6][7]. The combination of simple method and cheap based semiconductor assumed to be relatively reduce the production costs.

The key of SILAR process is ionic exchange time, where the length of the ionic immersing and rinsing time process is a critical point in the formation of ionic layers. Unnecessary precipitation growth need to be avoided in SILAR in order to keep bulk less condition for thin film [6]. The present study intends to explore the effect rinsing time and ionic soaking time on CZTS semiconductor optical properties. By controlling immersion variable, the study expected to gain information on optimal treatment to achieve band gap energy of CZTS near to 1,4-1,5 eV.

2. Materials and Methods

The composition of the precursors refers to the experiment conducted by Patill and Mali et al. [7][8]. Concentration of the cationic precursors was 0,2 M of $CuSO_4$, 0.01 M of $ZnSO_4$, 0.02 M of SnSO4 and for anionic precursor was 0.16 M of Na_2S. Those precursors deposited on SS304 substrate with immersion area are 4cm x 4cm. Variable immersion time for each sample can be seen in Table 1 and repeated in 40 cycles as shown in Fig 1. Deposited samples with cationic soaking variable addressed as K1, K12, K3 and K4 while the annealed one marked with K1a, K2a, K3a, and K4a which same rules on anionic variable. Annealed was conducted in atmosphere an oven at a temperature of 300°C for 180 minutes or 2 hours.

Chemical compound on the surface of the samples was examined by X-Ray Diffractometer Shimadzu 700 (Cu Kα, $\lambda = 1.5418$ Å) scanned form 20° to 90° (2θ). Study of optical absorbtion of CZTS was carried out by Shimadzu UV-2410 instrument within wavelength range of visible light 350 – 800 nm. While surface morphology was characterized using field emission type scanning electron microscope (FE-SEM) S-4800, Hitachi Co., Japan.

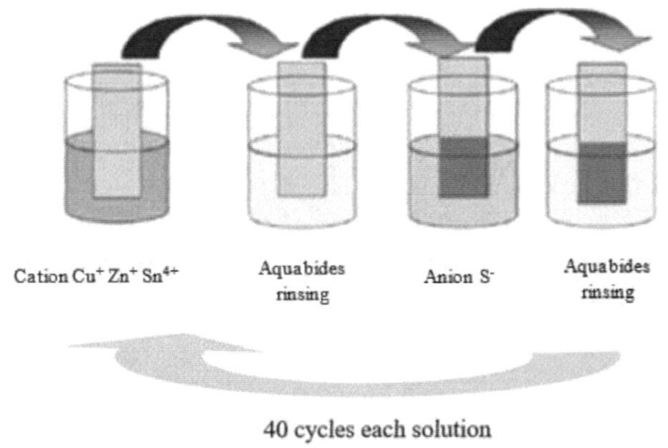

Fig. 1. Schematic of repeated immersion

3. Results and Discussion

Visual examination

Visual examination on deposited K samples went darker from brown to black as the cationic immersion time increase (Fig. 2). While surface of anionic sample are going darker from dark brown to black until 30 seconds immersion, yet after 40 seconds soaked the surface color turn lighter to brown again. Anionic samples show the same tendency with previous studies lead by Shinde [9][10]. This tendency can be attributed to slight decrease in film thickness where it is due to the formation of outer porous layer or the film may have developed stress which tends to cause delamination that resulting in peeling off the film after the film reaches at maximum thickness [8]

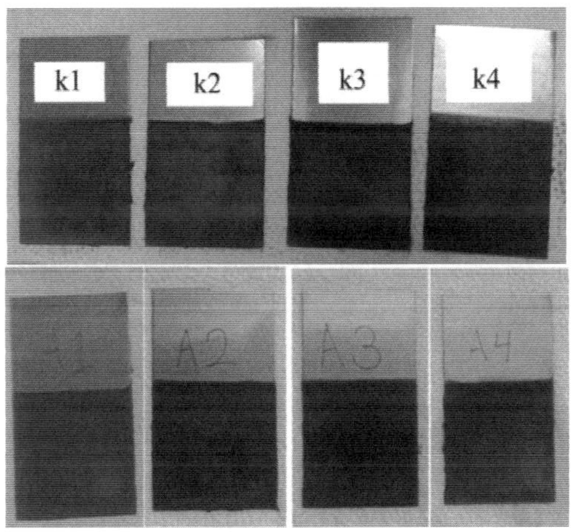

Fig. 2. As deposited sample's surface cationic samples (top) and anionic samples (bottom)

Optical absorption studies

Variation of absorbance values (α_o) versus wavelength (λ) of deposited and annealed samples can be observed in Fig 3. The optical band gap energy is estimated by using equation 1. Where α_0 is absorption coefficient and E_g is band gap energy. Since CZTS is a semiconductor with a direct transition, therefore the values n = ½. With both sides with the rank constituent quadratic equation then obtained equation 2. By extrapolating the curve to zero ($\alpha = 0$) energy gap values obtained from each sample.

$$\alpha = \frac{\alpha_0 (hv - E_g)^n}{hv} \quad (1)$$

$$(\alpha hv)^2 = \alpha_0^2 (hv - Eg) \quad (2)$$

Fig. 3a and 3b show band gap energy of cationic samples. Deposited K samples show an increasing band gap energy as the cationic immersion time increase from 1.37eV to 1.41eV after 40s. K samples with anneal treatment show the same tendency, the band gap energy increases as the cationic immersion time increase. These band energy level of annealed samples surpasses the range of CZTS band energy due to the formation of ZnS structure as it shown in XRD study. The band gap energy is expected to be decrease after anneal treatment, therefore the band gap increment was the contribution of ZnS which has energy band gap above 3eV.
As for anionic samples in Fig 3c and 3d, the band gap energy of deposited A samples showed nonlinear increasing band gap as the anionic immersion time increase. After soaked for 30s the energy band gap A samples reached optimum value at 1.35eV and decrease to 1.25eV after 40s.

IOP Conf. Series: Materials Science and Engineering **547** (2019) 012030 doi:10.1088/1757-899X/547/1/012030

Deposited A samples showed such low energy band gap likelihood due to formation of other substance which have lower band gap energy such as Cu_2SnS_3 (0,9 eV) or Cu_xS (1,2 eV). In the other hand, anionic samples with anneal treatment showed nonlinear decreasing band gap as the immersion time increase which A3a samples hit the lowest band gap energy at 1.3eV. Otherwise A2a and A4a showed such nearly ideal band gap energy, 1.4eV and 1.55eV respectively.

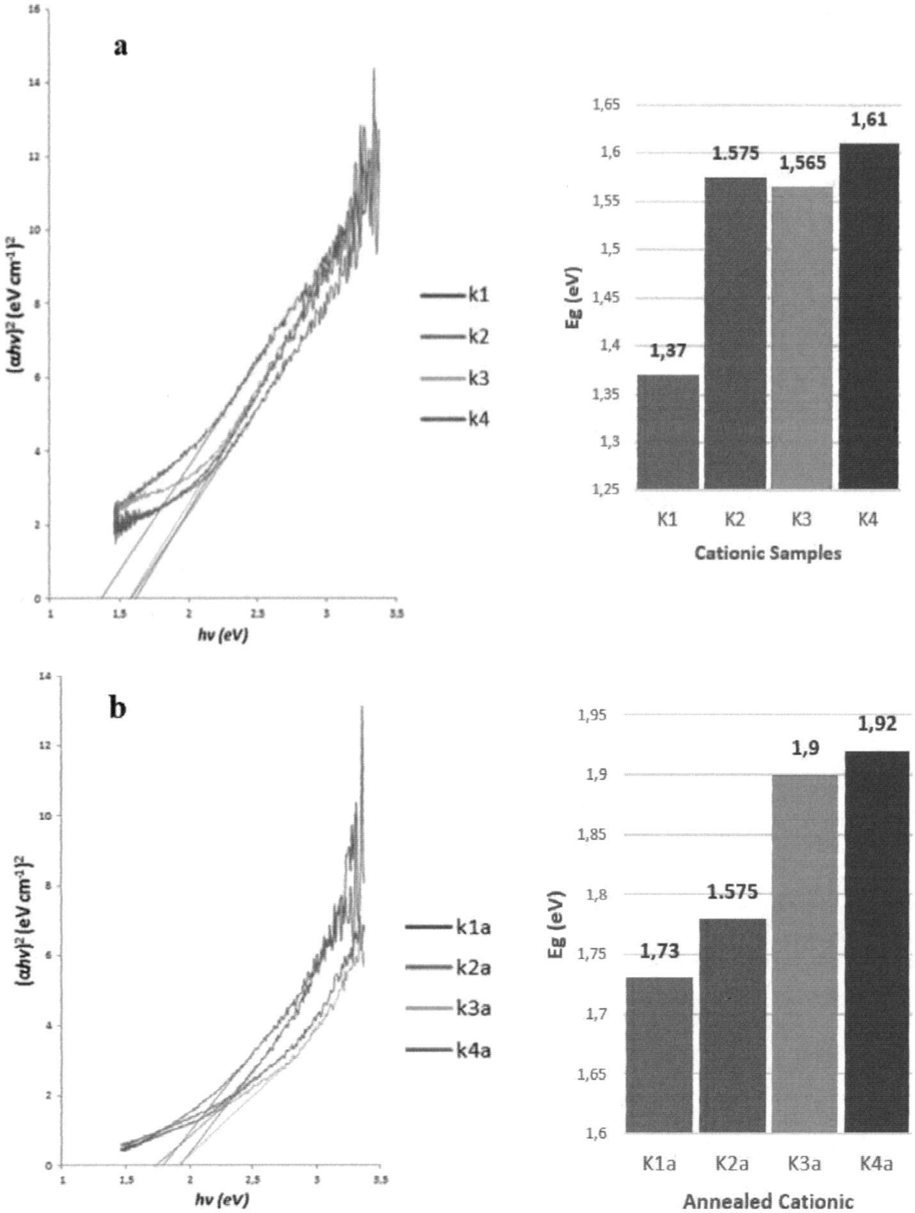

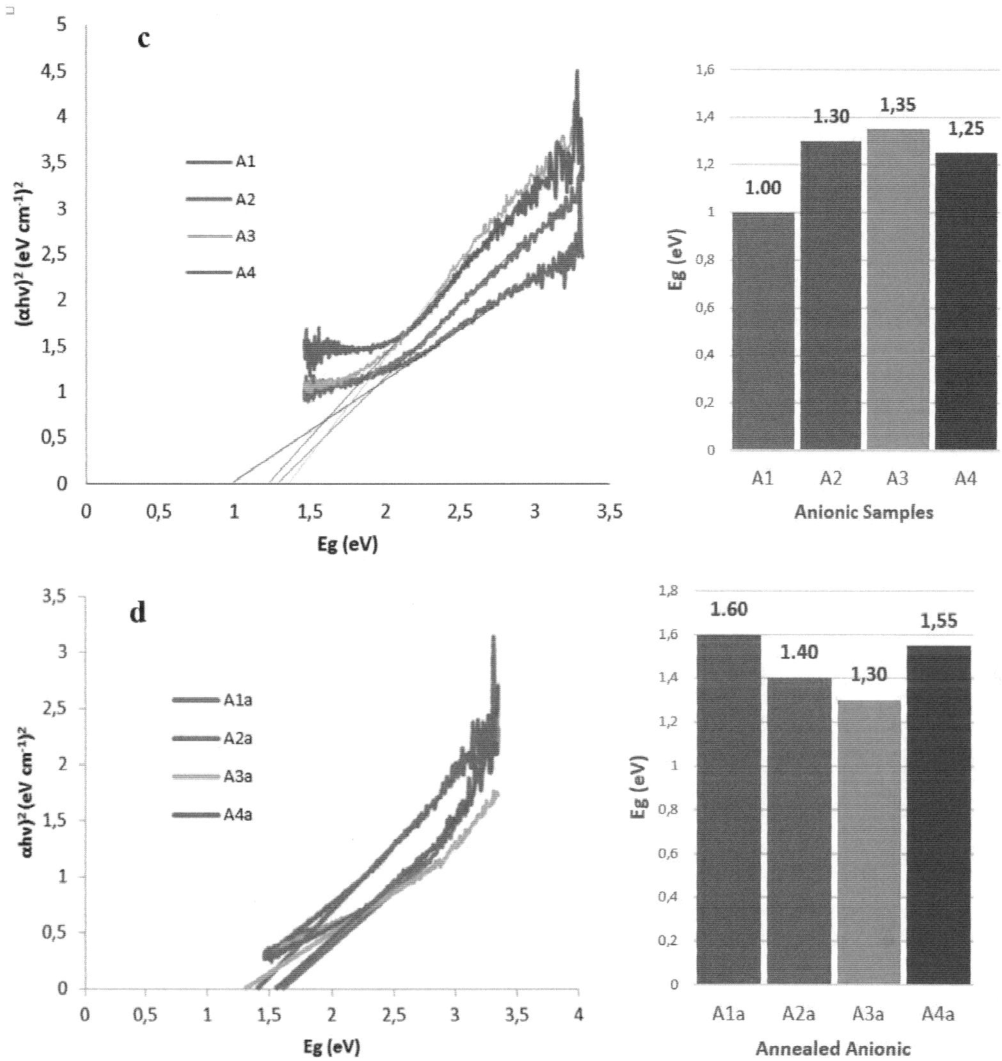

Fig. 3 Band gap energy of a) cationic as deposited sample b) cationic annealed sample c) anionic as deposited sample d) anionic annealed sample with each immersion time 10 seconds, 20 seconds, 30 seconds, and 40 seconds.

Structural and Morphological Studies

XRD analysis and SEM observation ware carried on the sample with the lowest, highest and optimum band gap energy. Which for deposited cationic samples K1 and K4 and for anionic samples A2 and A3 were selected. As for annealed treated samples K1a, K4a, A2 and A3a were also observed. XRD result for K1, K1a, K4 and K4a are shown in Fig. 4 which the deposited sample K1 show no indication of CZTS peak due to amorphous phase and K4 showed that Zns phase were precipitated on the sample after soaked for 40s. Samples K1a and K4a apparently show (1 1 2) plane in 25 degree 2θ after anneal treatment. Elemental study for A2, A2a, A3 and A3a can be observed in Fig. 3b where the peak of deposited CZTS films are still quite amorphou, thus the diffraction peak of (1 1 2) plane is insignificant and another compound that were expected to be observed such as Cu_2SnS_3 or Cu_xS could not be seen. While Fig. 4 shows a peak of (1 1 2) plane. The other significant peaks that are observed are SS304 peaks. Fig. 5 shows scanning electron micrographs of as deposited CZTS at 3000 times magnification and Fig. 4b shows annealed CZTS surface with same magnification. It is observed from both surface that CZTS films are not densely packed on substrate surface. Round grains are observed from as deposited CZTS and flakes grains from annealed CZTS. Anneal treatment changes the grain shape of CZTS from round shape into flakes due to the evaporation of hydrate and water content loss.

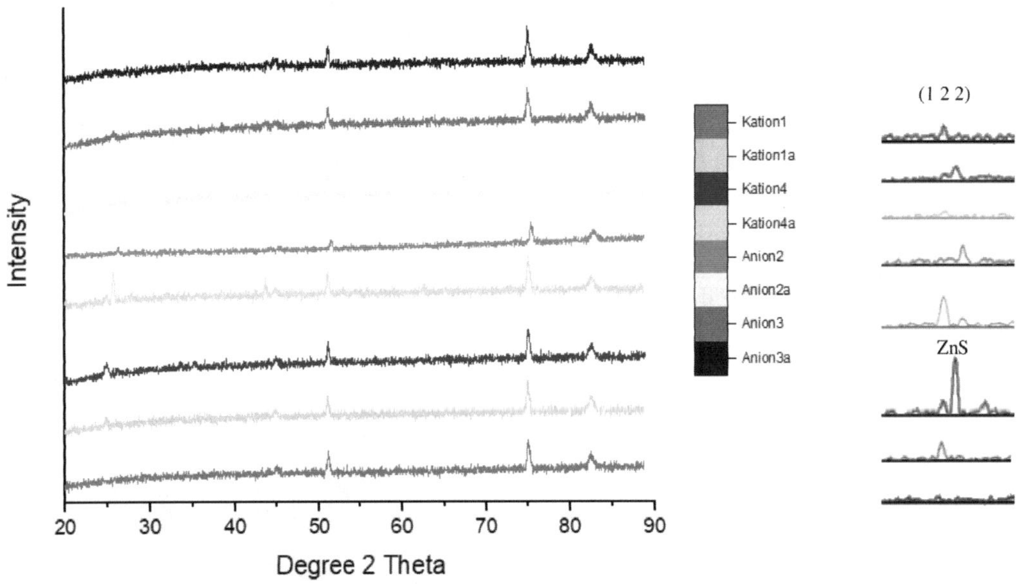

Fig. 4. XRD pattern of as deposited and annealed CZTS on SS304 substrate

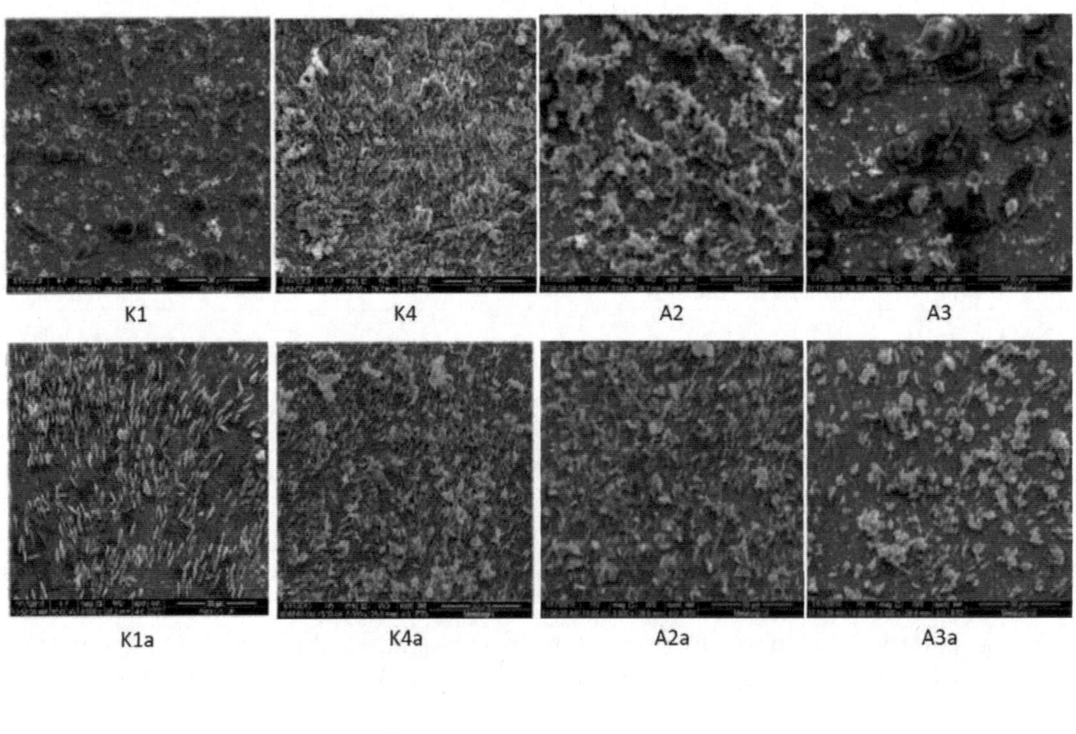

Fig. 5. SEM images of a) as deposited (left to right K1, K4, A2, A3) b) annealed CZTS (left to right K1a, K4a, A2a, A3a at 3000x magnification

4. Conclusions

Increasing of cationic immersion time for deposited and annealed K sample increase the band gap energy of CZTS, yet this linear increasing tendency is not essentially noteworthy since by keep increasing the variable the band gap will be wider. In the other hand as deposited sample show non-linear increment of band gap energy with optimum band energy achieve at 1.35 eV for deposited anionic sample. While annealed anionic sample also show non-linear decreasing of band gap energy with the opposite trend of deposited samples which A3a sample being the lowest (1.3 eV) and A2a reach ideal value (1.4eV). In other word, the optimum condition anionic samples hold key role in SILAR process.

5. References

[1]: 2008 Solar Technologies Market Report, US Department of Energy, NREL Report no. TP-6A2-46025, DOE/GO-102010-2867, 2010, pp. 1–131

[2] International energy Agency and Nuclear Energy Agency (2015) Projected Cost of Generating Electricity, France. ISBN 978-92-64-24443-6.

[3] International Renewable Energy Agency (2018) Renewable Power Generation costs in 2017, Abu Dhabi. ISBN 978-92-9260-040-2.

[4] Katagiri, H., Sasaguchi, N., Hando,S., Hoshino, S., Ohashi, J., Yokota, T. (1997). *Preparation and Evaluation* Cu_2ZnSnS_4 lapisan tipis*s by Sulfurization of E-B Evaporated Precursors,* Solar Energy Materials and Solar Cells 407-414. Elsevier.

[5]: Ito, K dan Nakazawa, T. (1989). *Photovoltaic Science and Engineering* , Proc. 4th Int. Conf, p.341

[6]: Pathan, H.M. dan Lokhande,C.D. (2003) *Deposition of Metal Chalcogenide* Lapisan tipis*s by Succesive Ionic Layer Adsorption and Reaction (SILAR) Method*, Bull. Mater. Sci, Vol 27, pp. 85-111. Indian Academy of Science

[7]: Mali, S.S., Patil B.M., Betty C.A., Boshale, P.N., Oh Young W., Jadkar S.R., Devan R.S., Ma Yuan-Ron, Patil, P.S. (2012). *Novel Synthesis of Kesterite* Cu_2ZnSnS_4 *Nanoflakes by Succesive Ionic Layer Adsorption and Reaction Technique : Characterization and Application,* Electrochimica Acta 216-221. Elsevier.

[8]: Mali, S.S., Patil B.M., Betty C.A., Boshale, P.N., Oh Young W., Jadkar S.R., Devan R.S., Ma Yuan-Ron, Patil, P.S. (2012).*Synthesis and Characterization of* Cu_2ZnSnS_4 *Thin Films by SILAR Method,* Journal of Physics and Chemistry of Solids 735-740. SciVerse ScienceDirect

[9]: Shinde, N.M., Dubal, D.P., Dhawale, D.S., Lokhande, C.D., Kim, J.H., Moon,J.H. (2012). *Room Temperature Novel Chemical Synthesis of Cu_2ZnSnS_4 (CZTS) Absorbing Layer for Photovoltaic Application*, Material Research Buletin 302-307. SciVerse ScienceDirect

[10]: Shinde, N.M., Lokhande, C.D., Kim, J.H., Moon,J.H. (2012). *Low Cost and Large Area Novel Chemical Synthesis of Cu_2ZnSnS_4 (CZTS) Films*, Journal of Photochemistry and Photobiology A: Chemistry 14-20. SciVerse ScienceDirect

IOP Conf. Series: Materials Science and Engineering **547** (2019) 012031 doi:10.1088/1757-899X/547/1/012031

Nanotube Formation And Surface Evolution On Hydrothermally Treated Ti In Alkali Solution

Asywendi Rukini[1]* Kim Sukyoung[2]
[1]Departemen Teknik Metalurgi, Universitas Teknologi Sumbawa
[2]Department of Materials Science and Engineering, Yeungnam University

*Corresponding author (E-mail): asywendi@gmail.com

Abstract. The present study intends to check which nanotube formation mechanism match the surface evolution on bulk Ti surface after soaked in hydrothermal treatment. Polished cp-Ti disks were hydrothermally immersed in various concentration NaOH for 1h, 2h, and 4h at 220°C. The SEM observations and XRD studies showed a sequence transformation from polished Ti surface into titanium oxide with plate shape and leave-like structure in the beginning. Bigger plates or leaves oxide then evolved into sodium titanate nanorods that finally became nanotubes as the immersing time lengthened. Apparently self-assembly and partial dissolution were the most applied mechanism to explained the sodium titanate formation. Plate and leaves-like oxide mainly formed by dissolution of Ti metal followed by precipitation of TiO_6 in self-assemble manner as the resultant of reaction between $Ti(OH)_3^+$ and OH^-. While, nanorod and nanotube structure were produced by partial dissolution of leaves structure followed precipitation of sodium titanate by reaction of Na^+ with $HTiO_3^-$ generated from OH- reaction with Ti metal and TiO_6.

Keywords: sodium titanate nanotube, hydrothermal, formation mechanism, dissolution-precipitation, self-assembly

1. Introduction

Titanate nanotube is widely known for its electrical properties which extensively applied in various application such as semiconductor[1][2] and photocatalysis[3] due to its high amount of surface area and superior physico-chemical properties. When titanate is synthesized via single step hydrothermal route in NaOH without calcining, ion exchange or HCl washing, sodium titanate can be generated [4][5]. In medicine applications, sodium titanate layer has been reported as bioactive layer due to its ion exchangeable surface making bone-like apatite deposited easily during immersion in SBF (simulated body fluid) solution [6][7]. Other studies also found that nanowire and nanofiber sodium titanate could form a self-assembly scaffold that is expected to allow living tissues to have ingrowth access [8]. These previous studies mentioned suggested nanostructure titanate as a potential coating layer in biomaterial especially hard-tissue application. Sodium titanate nanotube also performs anti-bacterial properties by performing novel photocatalysis reaction that can reduce the number of *Staphylococcus aureus* [7] (with or without UV radiation) and *Escherichia coli* [9] under UV radiation.

Despite its broad potential application, discussion over the formation mechanism of titanate nanotube is still being explored ever since the first time titanate nanotube was synthesized via hydrothermal treatment by Kasuga in 1998[10]. There are couple acknowledge formation mechanisms proposed in explaining nanotube formation without template-assisted. First postulate is rolling mechanism which explains that nanotubes are formed by rolled nano-sheets in the continuation of sheets exfoliation from the precursor [11]-[15]. The second statement is self-assembly which involving dissolution and precipitation process. Dissolution of outer parts of precursor occur as reaction with active component of basic solution, until precipitation begins, the structure will assemble accordingly to form nanotube [16][17]. Rolling mechanism can technically explain the process nanotube development synthesized from powder precursors. Yet, for non-powder precursor, it tends to be difficult to elaborate delaminating and scrolling process in structure development. Both rolling and self-assembly mechanisms involve ion exchange step in nanotube formation, which studies found that nanotube will not be generated without acid washing step [18]. Apparently, nanotube can still be generated in single step hydrothermal treatment only the nanotube has sodium in titanate compound [4][5]. Regarding to this, another formation mechanism were proposed by Wenjun D. et al. hypothetically explain sodium titanate nanotube formation on the surface of titanium mesh, which Wenjun D. presented upward-downward mechanism that involving co-mechanism between corrosion and self-assembly. Nanotube structure grows upward at the tip nanowire by self-assembly whilst nanowire vertically keeps growing downward on corrosion region [7].

Each mechanism brought up such a supportive evidence; therefore further experiments are needed to verify the empirical value of proposed mechanism. The present study successfully synthesizing sodium titanate nanotube on bulk titanium disk surface via single step hydrothermal treatment. Sodium nanotubes were obtained through single alkali treatment without ion exchange or acid washing precursor. By observing morphology evolution, this study intends to predict the chronological reactions and to configure which formation mechanism fits.

2. Materials and Methods

Sample preparation

C.p. Ti disks grade 2 with diameter 15 mm and thickness 2 mm were polished on SiC paper #400, #800, #1200, and #2000 in sequence. Polished c.p Ti disks were ultrasonicated in acetone, ethanol, and distilled water for 20 minutes each then hydrothermally immersed in three different concentrations of NaOH; 0.5M, 1M and 2M at 220°C, for 1h,2h and 4h. Hydrothermal treatment was carried out in stainless steel-Teflon line vessel where four disks were immersed in 10mL of NaOH solution. After treatment the disks were sonicated in distilled water for 15 minutes and dried in room temperature.

Characterizations

Chemical compound on the surface of the samples was examined by X-Ray Diffractometer for thin films, MPD-Pro. PANalytical Co. Ltd., Germany (Cu Kα, λ = 1.5418 Å) scanned form 20° to 70° (2θ). While surface morphology and elemental composition were characterized using field emission type scanning electron microscope (FE-SEM) S-4800, Hitachi Co., Japan equipped with energy dispersive X-ray spectroscopy (EDS) and field emission type transmission electron microscope (FE-TEM) Tecnai G2 F20 S-TWIN. IR study was also carried out to observe the hydroxyl (OH) group on nanotubes structure using Thermo Scientific Smart iTR Nicolet iS10 from wavelength 570- 4000 cm^{-1} in reflectance mode.

3. Results and Discussion

Morphology and composition studies

Various morphologies on Ti surface after treated in NaOH at 220°C can be observed in Table 1. After 1h immersed in 0.5 M, a uniform plates structure formed on Ti surface which the structure grew bigger and turned into a leaves-like shape on several places when immersion time increase into 2h. After 4h the leaves-like structure dominates with some rod and short wire structures appeared in between. Samples treated in 1M concentration for the morphology appeared relatively similar with samples immersed in 0.5M, only leaves-like structure formed early after 2h. Leaves structure completely changed into rod or short wire shape after 4h. While samples immersed in 2M NaOH show leaves-like structure appeared just after 1h and transformed into rod structure after 2h. After 4h immersed in 2M concentration, the rod structure grow longer into nanowire structure with the diameter of wire around 25 nm – 50 nm. Further observation on nanowire by TEM imaging in Fig. 1 shows that nanowires are happened to be multilayer nanotubes with lateral growth direction and wall thickness about 5-10 nm. XRD analysis in Fig. 2 showed only octahedral oxide peaks when plate-like or leaves-like structure appeared. As rod structure appeared in between leaves structure, sodium titanate peaks were detected with stronger intensity as immersion time gets longer. Although the morphology completely turns into randomly arranged nanotubes on samples treated in 2M concentration for 4h, octahedral oxide peaks still showed along with sodium titanate strong peaks.

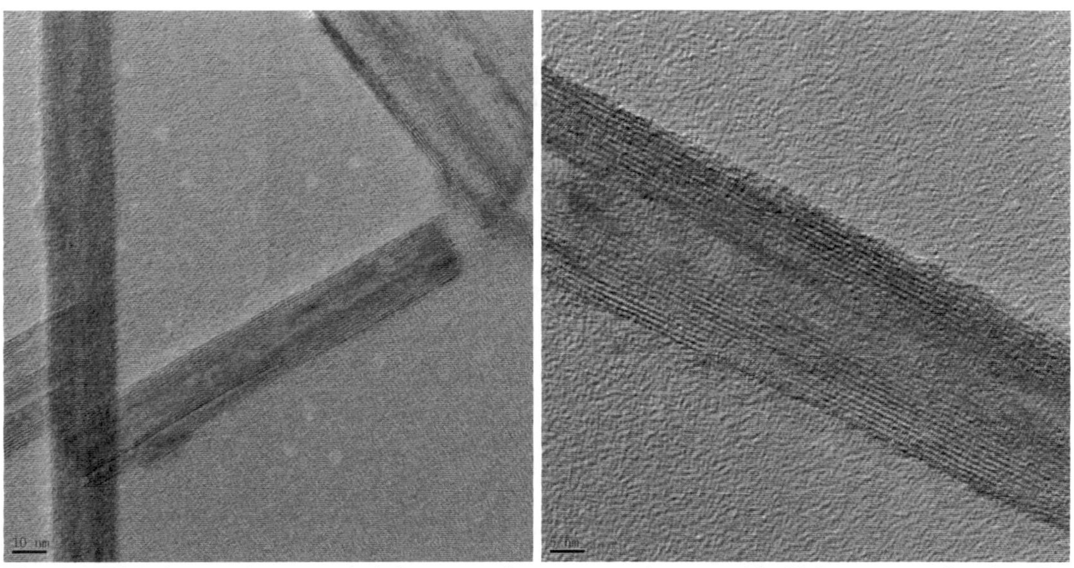

Fig. 1. TEM observation revealed the nanowires on sample treated in 2M concentration for 4h are nanotubes

IOP Conf. Series: Materials Science and Engineering **547** (2019) 012031 doi:10.1088/1757-899X/547/1/012031

Time	Concentration		
	0.5M	**1M**	**2M**
1h	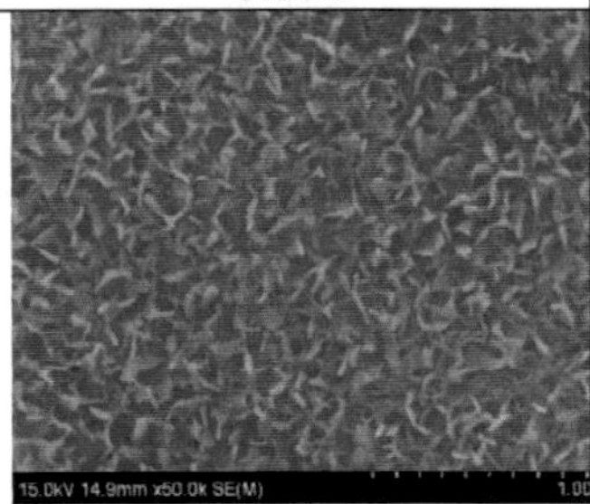	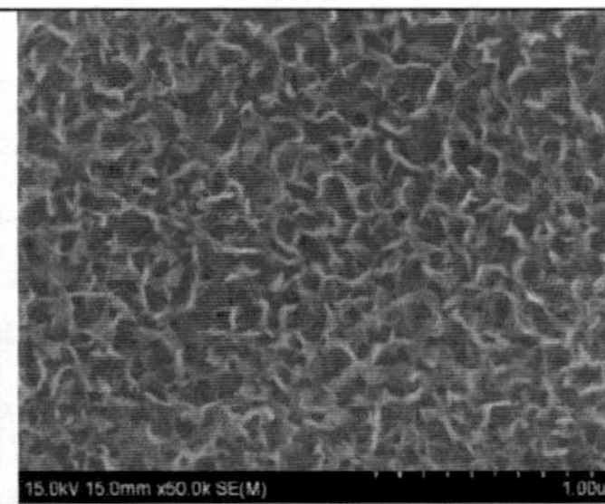	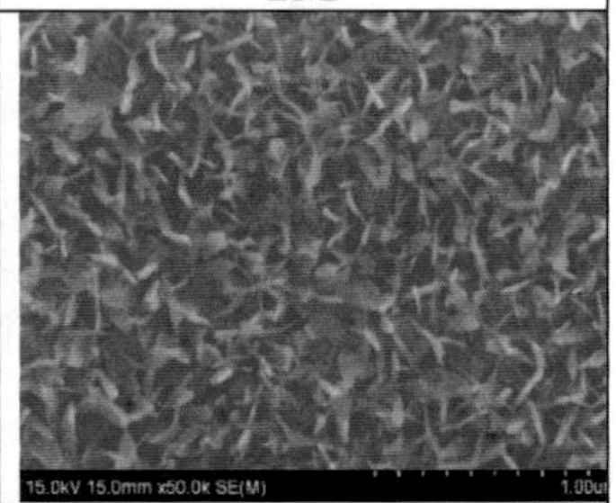
2h	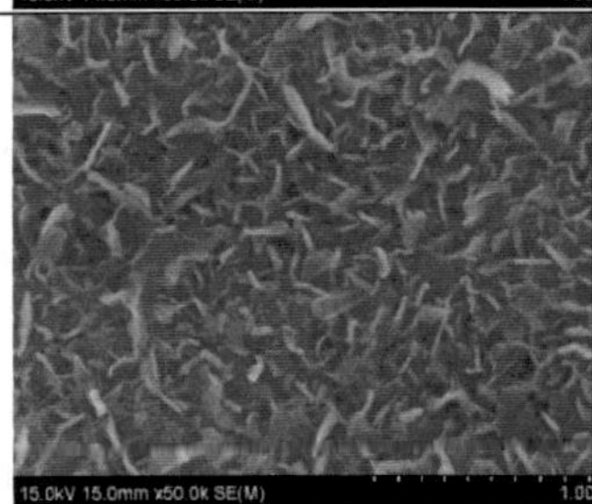	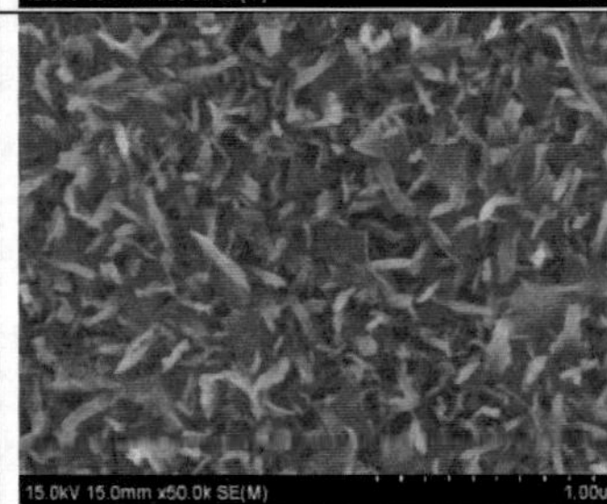	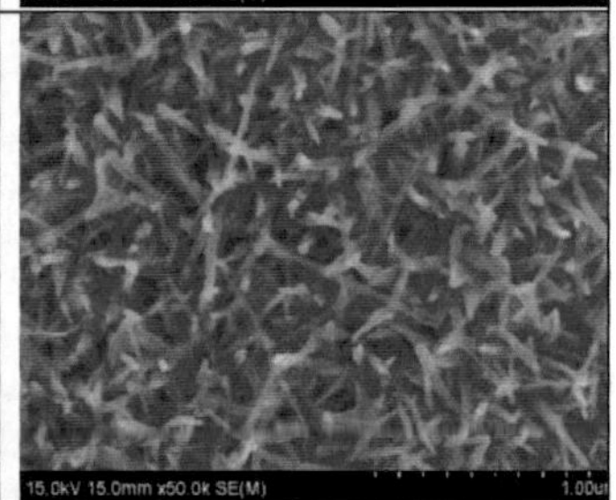
4h	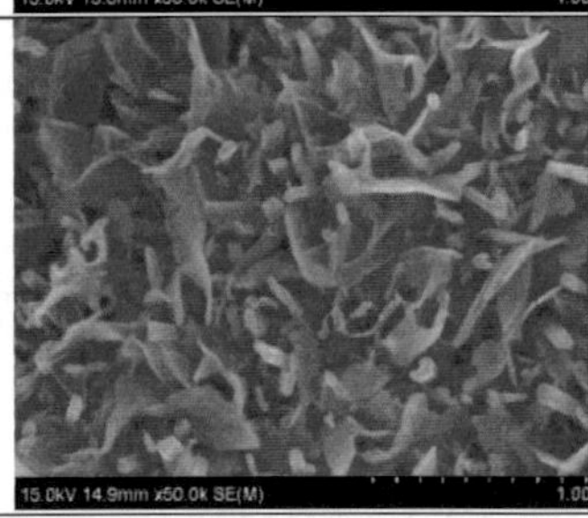		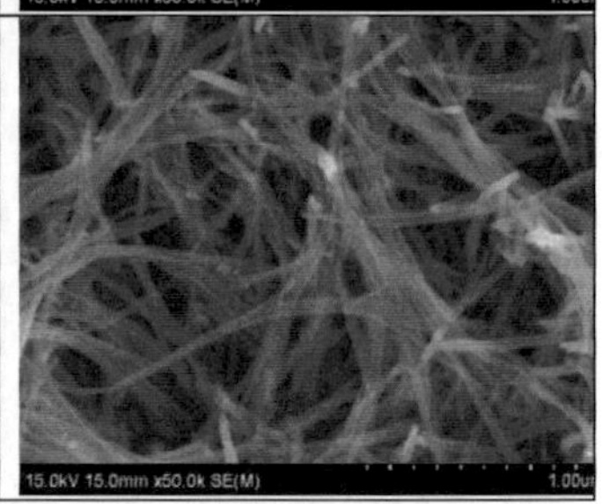

Table 1. SEM observation of NaOH treated samples at 220°C with various concentration and immersion time

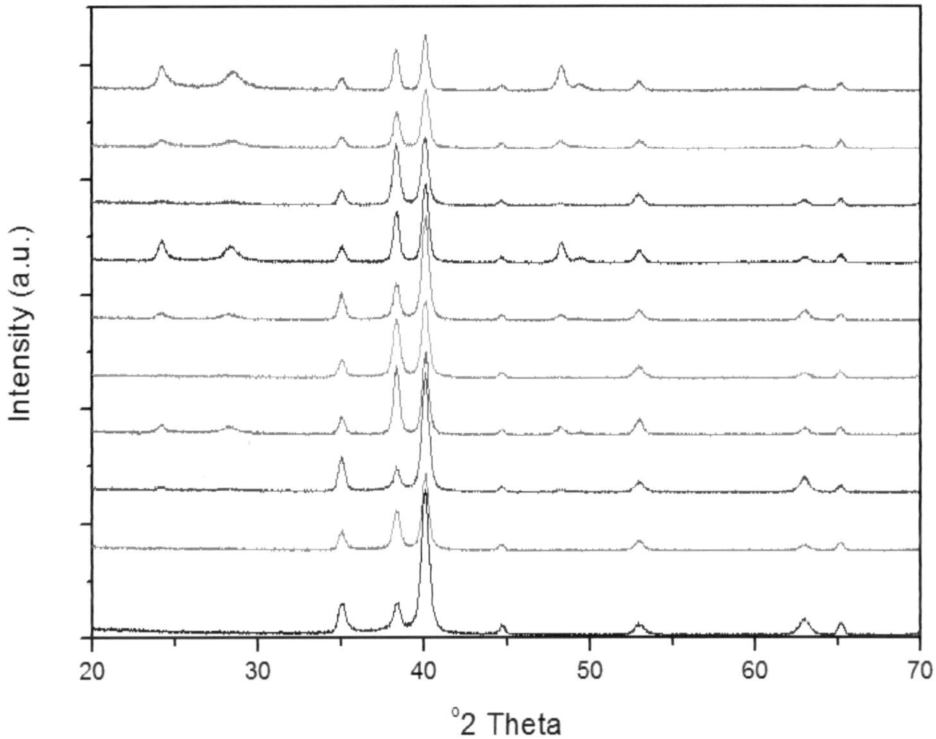

Fig 2. XRD pattern of NaOH treated samples with various concentration and immersion time treated at 220°C

Hyrothermal reaction and structure formation

SEM observation of morphology evolution on the present work cannot strongly support the rolling mechanism. In rolling mechanism the exfoliation and roll process were logically accepted when the precursors utilized is powder, yet to have the plate or leaves-like structure exfoliate or peel off from the disk to rolled up and then re-attach to disk is unmatched explanation for the morphologies observed in Table 1. While, upward and downward mechanism seems more acceptable in explaining how the rod shape growths into long wire and nanotubes formed by self-assembly. However, upward and downward mechanism does not completely satisfy explanation how the plate and leaves-like structure appeared and disappeared during the observation.

Under hydrothermal condition there are some different expectations of reaction type reported [15] such as; Decomposition, dissolution, corrosion, etching; Crystal growth of inorganic compounds; Synthesis of new phases; Leaching of ores in metal extraction; Preparation of microcrystallites with well-defined sized and morphology for specific application. By assuming that either the plate structure or nanotubes formation follows self-assemble mechanism means there are several dissolution and precipitation processes involved as illustrated in Fig. 3. The present study proposed partially dissolve mechanism on sodium titanate nanotube formation. The mechanism consists of two stages of dissolution and precipitation processes.

IOP Conf. Series: Materials Science and Engineering **547** (2019) 012031 doi:10.1088/1757-899X/547/1/012031

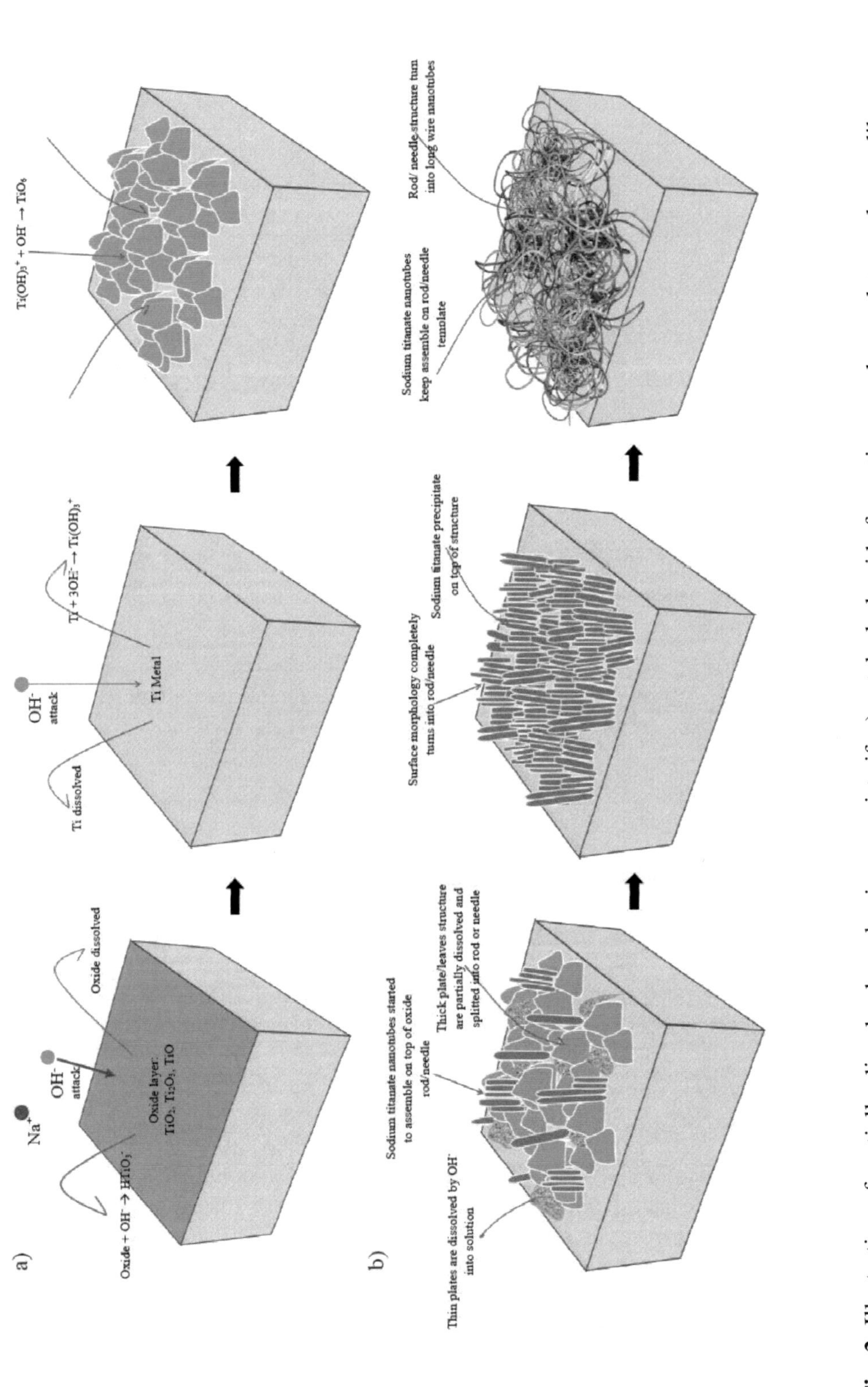

Fig. 3. Illustration of partially-dissolved mechanism consists if; a) octahedral oxide formation, produces plates to leaves-like structure; b) sodium titanate formation, involves partial dissolve and splitting process

IOP Conf. Series: Materials Science and Engineering **547** (2019) 012031 doi:10.1088/1757-899X/547/1/012031

The first dissolution-precipitation produces fine plate octahedral oxide that continuously grows into leaves-like structure. The plate structure formation most probably started with the dissolution of the thin passive oxide layers (TiO, TiO_2 and Ti_2O_3) on Ti by the corrosive attack of hydroxyl group (OH^-). Hyrogen metatitanate ($HTiO_3^-$) from Eq. 1-3 reacted with sodium ion could formed sodium titanate ($Na_xTi_yO_z$) deposit (Eq. 8), yet sodium titanate nanotubes were still not precipitated in this stage due to the small amount of passive oxide layer. Studies on titanium oxides [16][17] confirmed that after one day of exposure to air the oxide thickness reach up to 3.17 nm, this amount was probably not enough to initiate precipitation in the first stage of dissolution-precipitation process.

$$TiO + 2OH^- \rightarrow HTiO_3^- + \tfrac{1}{2} H_2\,(g) \quad (1)$$

$$TiO_2 + OH^- \rightarrow HTiO_3^- \quad (2)$$

$$Ti_2O_3 + 3OH^- \rightarrow 2HTiO_3^- .H_2O \quad (3)$$

After the passive layers were done, the hydroxyl group will attack Ti metal and formed Ti hydroxide ion (Eq. 4). The precipitation started to occur once hydroxide ion ($Ti(OH)_3^+$) that react with hydroxyl group produced titanium octahedral oxide (TiO_6) (Eq. 5) and quickly reached supersaturated point.

$$Ti + 3OH^- \rightarrow Ti(OH)_3^+ + 4e^- \quad (4)$$

$$Ti(OH)_3^+ + 6OH^- \rightarrow TiO_6 + 3H_2O \quad (5)$$

In second stage both of octahedral oxides with the plate and rod structure exist on titanium surface. In this point these structures might compete for their existence to grow. Since thermodynamically plate and leaves-like structure are less stable than needle and rod structure, they tend to dissolved or transformed. Fig. 4 showed some rods appeared in between of various thicknesses plate structure. The plates with enough thickness will probably splitting into rods while the thin one will partially dissolved by hydroxide into hydrogen metatitanate. After the partial-dissolve process happened, octahedral plate and leaves structure disappeared left the needle or rod structure completely dominated Ti surface. Rod or needle structure itself seems to be the template for nanotubes to assembly in this case.

$$TiO_6 + OH^- \rightarrow HTiO_3^- + 2O_2 \quad (6)$$

Sodium titanate precipitation process relates to the continuation of Eq.4. Before sodium titanate started to assembly on the top of rod or needle structure, titanium dioxide hydrate with hydrogen gas formed first. The pressure generated by H_2 gas could be one of the hindrance factors why the oxide hydrate was not precipitate. Instead of precipitating, titanium dioxide reacted with hydroxide producing hydrogen metatitanate. Further reaction between hydrogen metatitanate and sodium ion (Na^+) generated sodium titanate. Hydrogen metatitanate generated from both partial-dissolves process and dissolution of oxide hydrate produced the self assemble sodium titanate nanotube on the top of rod/needle structure. Partial-dissolves process also explains the existence of the octahedral oxide peaks when plate strucutre disappeared and titanium surface fully covered with of nanotubes. The octahedral oxide plates were only dissolved partially with some part of them transform into template for nanotubes growth.

$$Ti(OH)_3^+ + e^- \rightarrow TiO_2.H_2O + \frac{1}{2} H_2 (g) \quad (7)$$

$$TiO_2.H_2O (aq) .+ OH^- \rightarrow HTiO_3^- + H_2O \quad (8)$$

$$HTiO_3^- + Na^+ \rightarrow Na_xTi_yO_z(OH)_w \quad (9)$$

Eq. 9 implied that sodium titanate nanotubes were associated with hydroxyl group in final product. IR spectra on Fig. 5 showed OH broad peak from wave number 3100 to 3600cm^{-1} with the maximum percent reflectance at 3400 cm^{-1}. Another OH group vibration also appeared at 1623 cm^{-1} on the surface of titanate nanotubes, this peak is identified as molecular water bending [18].

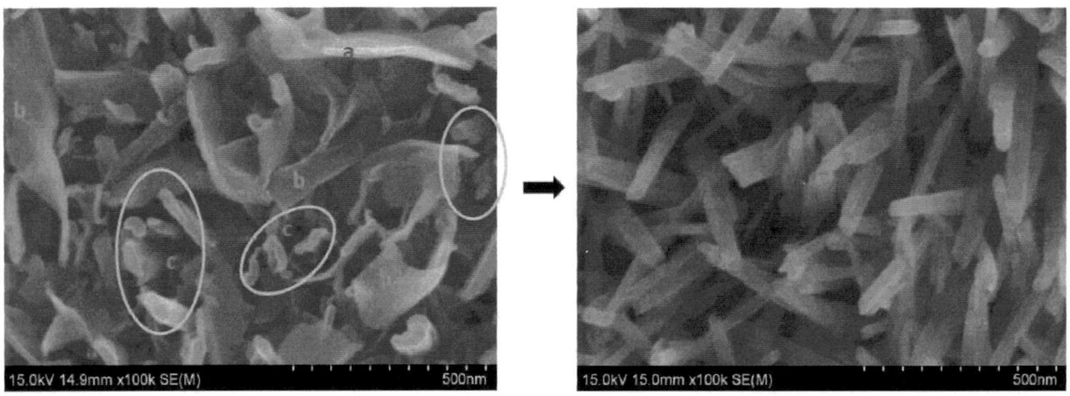

Fig 4. Transformation from plate/leaves-like structure to rod/needle structure involves partial dissolves process a) some thick plate or leaf structures split into rod b) the thin plates or leaves that is potentially disappeared c) splinted rod

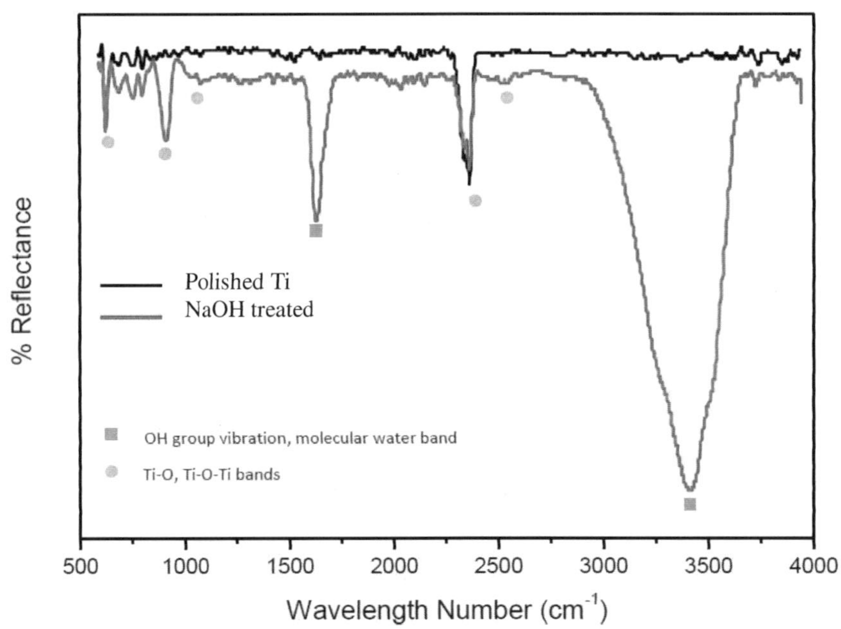

IOP Conf. Series: Materials Science and Engineering **547** (2019) 012031 doi:10.1088/1757-899X/547/1/012031

Fig. 5. IR reflectant spectra of polished Ti and sodium titanate nanotubes in wave number range from 570 to 400

4. Conclusions

Self-assembly with two stages dissolution and precipitation process is the most suitable in explains the sodium titanate nanotube formation. The process is started with octahedral oxide plate structure formation, followed by the octahedral oxide plate partially-dissolve and annihilation then ended up with sodium titanate nanotube self-assembly with the oxide rod/needle as template. However, predicted reactions were purely based on morphology observation and elemental studies therefore further works are needed to confirm.

5. References

[1] P. Roy, S. Berger and P. Schmuki : Angewandte Chemie International Edition (2011) 50 pg. 2904
[2] Y. Ohsaki, N. Masaki, T. Kitamura, Y. Wada, T. Okamoto, T. Sekino, K. Nihara, S. Yangida : Phys. Chem (2005) 7 pg. 4157
[3] A. Fujishima, K. Honda : Nature (1972) 238 pg. 37
[4] Q. Chen, G.H Du, S. Zhang, L.M. Peng : Acta Crystallograph (2002) 3 pg.50
[5] C.W. Peng, MRichard-Plouet, T.Y. Ke, C. Y. Lee, H.T. Chin, C. Marhic, E. Puzenat, F. Lemoigno and L. Brohan : Chem. Mater (2008) 20 pg. 7228
[6] H. M. Kim, F. Miyaji, T. Kokubo, Nakamura T. : Journal of Biomedical Materials Research (1996) vol. 32, no. 3, pg. 409–417

[7] W. Dong, T. Zhang, J. Epstein, L. Cooney, H. Wang, Y. Li, Y-B. Jiang, A. Cogbill, V. varadan, Z.R. Tian: Chem. Mater. 19 (2007) pp. 4454-4459

[8] H. Zhao, W. Dong, Y. zheng, A. Liu, J. Yao, W. Tang, B. Chen, G. Wang, Z. Shi : Biomaterials 32 (2011) pg. 5837-5846
[9] Y. Mitsunori, I. Yuko, N. Iwao, M. Tomohiro, T. Toshio, W. Takamori, H. Takao: Hindawi Publishing Corporation Journal of Nanomaterials Volume 2013, Article ID 476585, 9 pages
[10] Kasuga T., M. Hiramatsu, A. Hoson, T. Sekino, K. Nihara: Langmuir 14 (1998) pp. 3160-3163

[11] T. Kasuga, M. Hiramatsu, A. Hoson, T. Sekino, K. Niihara: Advanced Materials 11 (1999) pp. 1307-1311

[12] Nakahira A., T. Kubo, C. Numako: Inorganic Chem. 49 (2010) pp. 5845-5852

[13] Haiqiang Lu, L. Li, J, Zhao, Z. Zhu: Chem. Phys. Letters 508 (2011) pp. 258-264

[14] Tsai C.C. and Teng H.,: Langmuir 24 (2008) pp. 3434-3438

[15] A. Nakahira, T. Kubo, C. Numako: Inorg. Cem (2010) 49 pg. 5845

[16] W. Chen, X. GuoS. Zhang, Z. Jin; Nanoparticle Res (2007) 7 pg. 1173

[17] J. Xie, X. Wang, Y. Zhou; J. Mater Sci Tech (2012) 28 pg. 488

[18] Tsai C.C. and Teng H.: Chem. Mater (2004) 16 pg. 4352

[19] K. Byrappa and Mashahiro Yoshimura. *Handbook of hydrothermal technology: A technology for crystal growth and material processing*. Noyes Publication/ William Andrew Publishing LLC (2001) United States.

[20] Sittig,C., M. Textor, ND. Spencer, M. Wieland, PH. Valloton: Journal of Materials Science - Materials in Medicine 10 (1999) pp. 35-46

[21] Nekati Harmankaya,*Titanium Oxide and Bone Anchorage Role of the Complement System, and Delivery of Osteoporosis Drugs from Mesoporous TiO_2*. University of Gothenburg, Gothenburg 2003.

[22] Maira A.J. J.M. Coronado, V. Augugliaro, K.L. Yeung, J.C. Conesa, J. Soria: Journal of Catalysis 202 (2001) 413-420.

The Study of Thermal Decomposition of Rice Husk in Silica Production: The Effect of Hydrochloric Acid Leaching

D Dhaneswara[1], J Fajar Fatriansyah[1], A Kusuma Wardana[1], A Nurul Haqoh[1], S Aida Khairunnisa[1]

[1]Metallurgy and Materials Engineering Departement, Faculty of Engineering, Universitas Indonesia

Abstract. The effect of hydrochloric acid leaching on thermal decomposition process and morphology of the rice husk in silica production have been investigated. Amorphous silica can be obtained by thermal decomposition of rice husk in acid solution. The aim of this study is to investigate the stages of thermal decomposition of rice husk and its morphology. In this work, thermal decomposition of rice husk has been investigated in neutral and acid solution (hydrocloric acid 1%v/v, 3%v/v). The thermal decomposition study was performed at 225°c, 450°c, 700°c. The thermal behavior were investigated using DSC while the microstructure and properties of rice husk were characterized by SEM, FTIR, XRD.The results showed that the pre-treatment of the rice husk with acid leaching accelerated thermal decomposition of organic components at slight lower temperature as showed by differential scanning calorimetry (DSC) and scanning electron microscope (SEM) analysis. This knowledge may provide better insight in order to obtain high grade amorphous silica.

Keywords: rice husk, thermal decomposition, acid leaching.

1. Introduction

Rice Husk is a byproduct of the milling process of paddy into rice. They are become waste in many rice producing countries, including in Indonesia. Based on data released by the Central Bureau of Statistics (BPS), rice production in 2016 amounted to 79.14 million tons [1]. About 20%-22% of total rice production is the rice husk[2]. Rice husk have no commercial interest because it has poor nutrient and used as animal food. In addition, rice husk is often disposed of using open burning, leading to environmental pollution. This will be an environment threat if it is notwellrecycled. However, rice husk consists of lignin (22 wt. %)[3], celluloses (55-60wt. %) and inorganic components (20-25wt.%)[4, 5]. Rice husk used in various application such as fuel for boiler, electricity generation and feedstock for bio-fuel production. In particular, the high amorphous silica in the range of 20-25% wt. % has attracted the concentration of researcher for its potential use as a raw material for the production of silicon-based materials.Sintering the rice husk produces the rice husk ash which is mainly contains silica. This silica has been shown to be a good material for the synthesis of very pure silicon, silicon carbide and other high-grade silica. Silica is a basic raw material widely used in electronics, semiconductors, ceramics and several industries[5]. A research demonstrated that silica from rice husk can be used to synthesis SBA-15 and SBA-3 mesoporous materials [6] which have high surface area and applicable in various applications such as catalyst, membrane and adsorption Cu, Cd, Zn, Pb in the case of SBA-15 [7-10].

IOP Conf. Series: Materials Science and Engineering **547** (2019) 012032 doi:10.1088/1757-899X/547/1/012032

Several methods have been used to obtain silica from rice husk. Zhang, et al [11] obtained superfine silica with a diameter 20-200 nm by incineration, while Wu (1996) used hydrothermal decomposition. Controlled combustion could be a convenient method to remove impurities for the release of silica from rice husk.The stages of thermal decomposition of rice husk could help on the process of manufacturing silica from rice husk. However, the mechanism of thermal decomposition hasreceived much less attention. In this paper, thermal decomposition of rice huskis investigated. In addition, the rice husk is treated with acid leaching to remove impurities. The variation of reflux and concentration of hydrochloric acid are used to investigate the stages of thermal decomposition of rice husk.

2. Materials and Methods

2.1 Raw materials

The rice husk was obtained from Paddy Mill Industry,Leuwiliang, Bogor and was dried for 8 hours under sun. NaOH, HCl, distlilled water was obtained from C.V. Mitra Bersama and used as it is without any further purification.

2.2 Acid solution preparation

1%v/v HCl is obtained by diluting 13,51 ml of 37%v/v HCl in 486.49 ml of water. 3%v/v HCl is obtained by diluting 40.54 ml of 37%v/v HCl in 459.46 ml of water

2.3 Acid leaching

The rice husk is leached with hydrochloric acid using reflux technique. 25 gram of rice husk were added into three-neck rounded flask. Then, hydrochloric acid solution was poured. After that,they were refluxedfor 2 hours at temperature of 100 °C.

2.4 Rice husk ash preparation

RHA was obtained after controlled combustion of rice husk that has been leached at 250°C, 400°C, and 700°C for 6 hours. Before that, the rice husk is dried using oven at 110°C for 2 hours.

2.5 The characterization

Rice husk was leached with hydrochloric acidand its ash thermal decomposition behavior wasanalysed using Differential Scanning Calorimetry (DSC) and Thermo Gravimetric Analyzer (TGA).The chemical composition of rice husk ash was characterized using Fourier Transform Infra-Red (FTIR), Scanning Electron Microscope (SEM) and Energy-dispersive X-ray spectroscopy(EDS).

3. Results and Discussion

The thermograms of rice husk are shown in Fig. 1.It is shown that there are three peaks which can be recognized as stages in the decomposition.There is a small endothermic curve that followed by exothermic shift (first peak). The second peak is exothermic and followed by endothermic shift at 380°C. The third peak is on 690-700°C which was followed by exothermic shift.

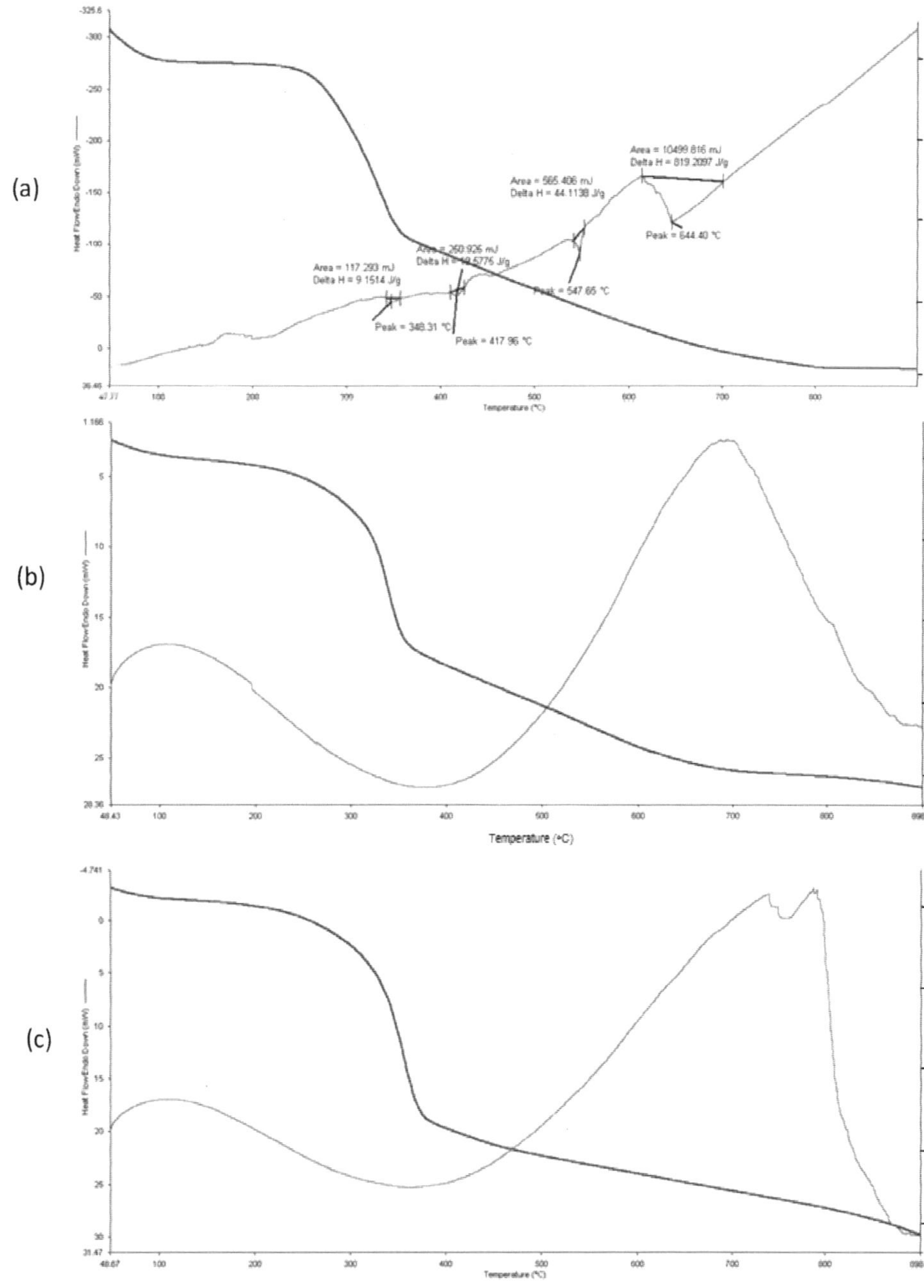

Fig. 1 DSC and TGA of rice husk: (a) non treatment, (b) leached by HCl 1%(v/v) and (c) leached by HCl 3%(v/v)

The initial 5% weight loss below 110°C shown in fig 1, and followed by rapid weight loss ~56% started at 300°C. And at 700°C, the weight loss reaches ~83%. The first peak which followed by 5% weight loss is observed as loss of absorbed water.It is reported that the DSC of cellulose shows exothermic peaks [12-14]. The second peak which is exothermic peak is followed by rapid weight loss indicates the decomposition of cellulose. The decomposition of cellulose is reported [4] to overlap with the decomposition of lignin.The weight of rice husk continued to loss at the third peak that has observed as an ongoing decomposition of lignin and another organic constituent.

Fig. 2 shows weight losses of rice husk versus temperature. The samples were leached

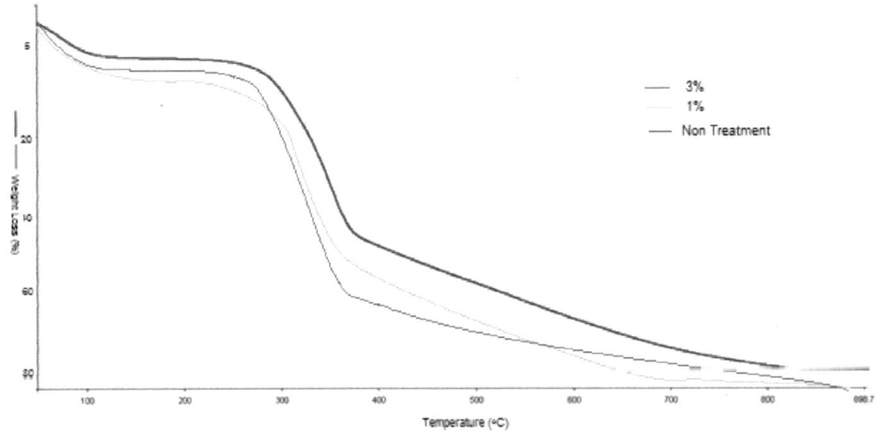

Fig. 2 TG curve of rice husk ash

in hydrochloric acid 1%v/v, 3%v/v and sample non treatment. It is clear that rice husk that wereleached by hydrochloric acid has higher weight loss than rice husk non treatment. This indicates that the organic constituent like cellulose and lignin are decomposed more effectively due to the reduced impurities in the rice husk.The inorganic impurities may raise the decomposition temperature. Duringacid leaching, inorganic impurities were lost. That statement can be inferred from the fig 3(a) and (b) which show that as we increase HCl concentration, some dirt and surface features which represent inorganic impurities in the surface were disappeared.With the loss of those impurities, the organic constituent could decompose at lower temperature.

Table 1 shows the silica content of rice husk ashes, after combustion in a furnace for 6 hours at 700°C. Rice husk that was leached by 3 %(v/v) HCl has higher silica content than 1%(v/v) due to the loss of organic and inorganic impurities.

Table 1
Silica content of rice husk ash

Sample	Temperature	SiO2 (wt%)
RHA leached by 1%(v/v) HCl	250	47.25
	450	80.52
	700	85.22
RHA leached by 3%(v/v) HCl	250	59.58
	450	81.69
	700	95.92

Fig. 3 Electron micrograph of rice husk leached by HCl (a) 1% before combustion, (b) 3% before combustion, (c) 1% at 250°C, (d) 3% at 250°C, (e) 1% at 450°C, (f) 3% at 450°C, (g) 1% at 700°C, and (h) 3% at 700°C.

Fig. 3 present electron micrograph of rice husk obtained from the combustion. Fig 3a and b shows the epidermis from the rice husk. Bumpy outer epidermis is clearly visible. In Fig 3c and d visible epidermis is still attached to silica.Similar microstructural features have been reported [15] for silica samples released from rice husk. It is indicate that the organic constituent has not been decomposed completely. Futhermore, Fig 3e and f shows the haphazard structure of silica deposit. It proved that the cellulose ,lignin and another organic substances has been decomposed. The same structural features can be seen in Fig 3g and h. It shows the silica has a relatively high purity.

(b)

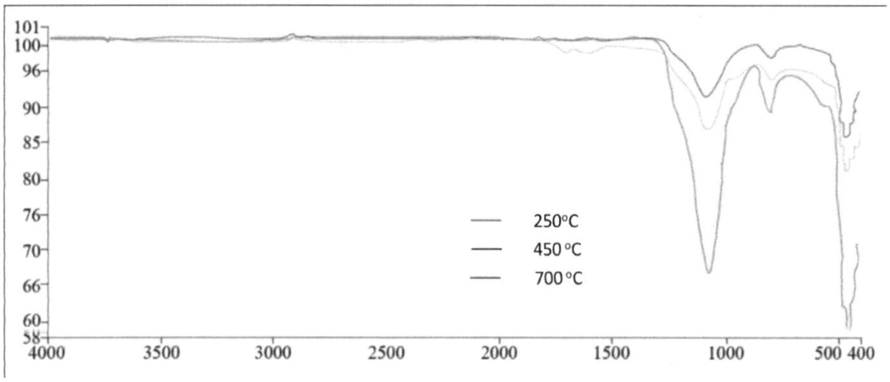

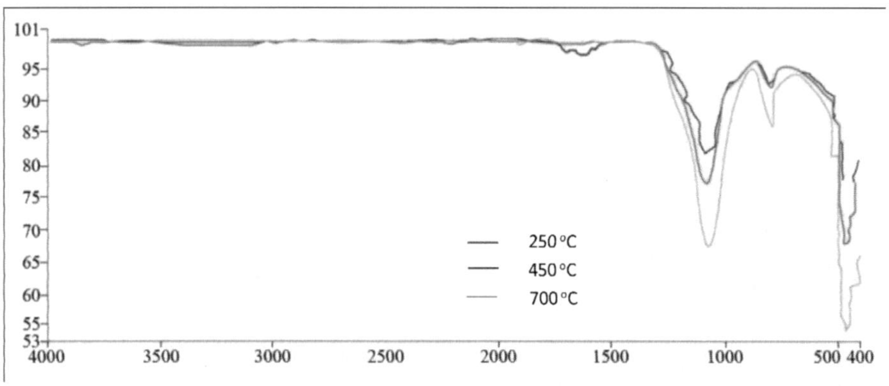

Fig. 4 Weight loss - Temperature curve of rice husk (a) leached by HCL 1%(v/v) and (b) leached by 3%(v/v)

FTIR characterization was carried out to examine the functional group characteristicscontained in the silica produced from RHA. The spectra were obtained at wavenumber 4000 – 400 $cm^{-1}$and the results are shown in Fig. 4. The observed band at 470 cm^{-1} is associated with a network of O-Si-O bond-bending [16]. Increases in intensity of that band indicate an increases in silica purification. The peak appearing at ~1720 cm^{-1} corresponds to the C=O stretching of p-coumaric acids of lignin and the acetyl groups of hemicelluloses [11]. And the band at ~1620 cm^{-1} could be represent to either C=C vibration or to O-H stretching of adsorbed water [17]. The decreases in intensity of that band shows the progres of decomposition and their disappearance is an indication that the decomposition of organic substances (cellulose and lignin) is completed.

4. Conclusion

This study has investigated the effect of hydrochloric acid leaching on thermal decomposition process. Silica with purity ~96% can be obtained from combustion of rice husk that was leached by hydrochloric acid. The rice husk that is leached by hydrochloric acid are decompose more effectively due to the loss ofthe inorganic content.Lignin and cellulose of rice husk are leached by hydrochloric acid are completely decomposed at temperatures ≥ 450°C. The process will be more effective when using hydrochloric acid with a higher concentration.

5. Acnowledgement

This research is funded by Directorate of Research and Community Services (DRPM), Universitas Indonesia, through Hibah PTUPT 2019.

References

[1]Indonesia, Central Bureau of Statistic,http://st2013.bps.go.id/dev2/index.php (2018)

[2] H. A. Saudi, S. M. Salem, S. S. Mohammad, A. G. Mostafa and M. Y. Hassaan, Am. J. Phys. and App.**3(3)**, 97 (2015)

[3] James, J., & Rao, M. S. (1986). Silica from rice husk through thermal decomposition. *Thermochimica Acta*, *97*(C), 329–336.

[4] Hui, C., Lei, Z., Xitang, W., &Shujing, L. (2015). Preparation of NanoporousTiO 2 / SiO 2 Composite with Rice Husk as Template and Its Photocatalytic Property. *Rare Metal Materials and Engineering*, *44*(7), 1607–1611. https://doi.org/10.1016/S1875-5372(15)30101-6

[5] Liou, T. (2004). Preparation and characterization of nano-structured silica from rice husk, *364*, 313–323. https://doi.org/10.1016/j.msea.2003.08.045

[6] Jullaphan, O., T. Witoon, and M. Chareonpanich. "Synthesis of mixed-phase uniformly infiltrated SBA-3-like in SBA-15 bimodal mesoporous silica from rice husk ash." *Materials Letters* 63.15 (2009): 1303-1306.

[7] Dhaneswara, D., Delayori, F., Utami, S., Kuskendrianto, F. R., Abdurrahman, M. H., & Fatriansyah, J. F. (2018). The Adsorption of Copper by SBA-15-CPTMS Mesoporous for Water Remediation. In E3S Web of Conferences (Vol. 65, p. 05007). EDP Sciences.

[8] Dhaneswara, D., Fatriansyah, J. F., Mahagnyana, A. B., Delayori, F., Putranto, D. A., & Anwar, S. U. A. A. (2018, January). The Role of Modification SBA-15 Mesoporous Silica with CPTMS in Cd Adsorptions. In *IOP Conference Series: Materials Science and Engineering* (Vol. 299, No. 1, p. 012063). IOP Publishing.

[9] DhaneswaraD., PutrantoD.A., BachtiarM., FatriansyahJ. F.,(2018). Selective Adsorption of CPTMS-SBA-15 towards Zinc and Cadmium for Liquid Waste Remediation. In E3S Web of Conferences (Vol. 65, p. 05008). EDP Sciences.

[10] FatriansyahJ. F.,DhaneswaraD., SitumorangF. W., Mahagnyana A. B., DelayoriF., S. U. A. Anwar, KusumawardhaniD. P.,The Comparative Study of Pure Mesoporous Silica SBA-15 and CPTMS-SBA-15 Adsorption of Pb Heavy Metal, In E3S Web of Conferences (Vol. 65, p. 05016). EDP Sciences.

[11] Zhang, Z., Zhu, M., & Zhang, D. (2018). A Thermogravimetric study of the characteristics of pyrolysis of cellulose isolated from selected biomass. *Applied Energy*, *220*(October 2017), 87–93.

[12] F. Shafizadeh and A.G.W. Bradbury, J. Appt. Polym. Sci., 23 (1979) 1431.

[13] W.K. Tang, in R.C. Mackenzie (Ed.), Differential Thermal Analysis, Vol. 2, Academic Press, London, 1972, p. 523.

[14] M.V. Ramiah, J. Appl. Polym. Sci., 14 (1970) 1323.

[15] P. Bartha and E.A. Huppertz, Proc. Int. Conf. Rice By-product utilisation, Vol. 1, Rice husk utilisation, Valencia, Spain, 1974, pp. 89-98.

[16] Shahnani, M., Mohebbi, M., Mehdi, A., Ghassempour, A., & Aboul-Enein, H. Y. (2018). Silica microspheres from rice husk: A good opportunity for chromatography stationary phase. *Industrial Crops and Products*, *121*(April), 236–240.

[17] Zhao, J., Xiuwen, W., Hu, J., Liu, Q., Shen, D., & Xiao, R. (2014). Thermal degradation of softwood lignin and hardwood lignin by TG-FTIR and Py-GC/MS. *Polymer Degradation and Stability*, *108*, 133–138.

Air Jet Erosion Study on Stir Casted Metal Matrix Composite Aluminum-SiC

M. Longalayuk[1], C. Ruskandi[2], B. H. Setiamarga[3]*
[1,3]Materials Engineering Study Program, Faculty of Mechanical and Aerospace Engineering, Institute of Technology Bandung.
[2] Foundry Technology Department, Bandung Manufacturing Polytechnics.

*Corresponding author (E-mail): mardilongalayuk@gmail.com; c2p_r@yahoo.com; budi.setiamarga@itb.ac.id

Abstract. For equipment that operates in high-speed abrasive media, erosion will be the major factor for its failure. Metal Matrix Composite (MMC) Aluminum-Silicon Carbide has been developed for improving the lifetime for such application.
In this research, MMC Al-SiC has been made using Al-7075 scrap with 5%, 10%, and 15% SiC powder addition. This MMC was manufactured using stir casting technique with 1% Magnesium addition. Optical metallography has been done to know the SiC distribution homogeneity. Erosion resistance of the MMC was studied using Air Jet Erosion Test with Alumina powder as abrasives.
The microstructural study showed that MMC Al-SiC has been successfully manufactured although the SiC particle distribution has not been fully homogeneous yet. Air Jet Erosion Testing revealed that the addition of 5% SiC powder was optimal in increasing its erosion resistance. The SiC addition from 10 to 15% did not improve its erosion resistance significantly.

Keywords: Metal Matrix Composite; Al-7075; SiC; Stir Casting; Air Jet Erosion Test

1. Introduction

Technology development in materials has made a way to help reducing working cost and increasing efficiency in industries. Metal Matrix Composite (MMC) Al-SiC is a composite material with aluminum matrix reinforced with silicon carbide (SiC) particles [1]. This composite has been used in energy and aviation industries because it is relatively easy to produce, light weight and has good mechanical properties.

Good MMC Al-SiC has been manufactured with stir casting process using Aluminum Al7075, SiC powder grit 220 and the addition of 1 wt.% of magnesium powder [2, 3]. Even though stir casting is the simple way to produce composite, this process has some drawbacks that reduce the quality of product, i.e. the presence of porosity and particle agglomeration. There are some factors that influence the MMC product quality, such as interface reaction, wettability, and porosity. At the interface region, reaction between SiC powder and molten aluminum must be controlled to avoid unnecessary reaction that can produce more harmful compound like Al_4C_3 [4]. Wettability is the tendency of a certain fluid to spread on a solid surface. To increase the wettability of the fluid, there are three main things that must be accomplished which are increasing surface energy of solid particle, decreasing surface tension of fluid, and decreasing interfacial energy of solid-fluid [5]. Porosity is a void or an opening space between grains or trapped gas in grains in microstructure. The presence of porosity in manufacturing products cannot be avoided but can be reduced. Porosity was formed by the presence of impurities of hydrocarbon or entrapped gas because of vigorous stirring [6].

In some applications, parts made from MMC Al-SiC may operate in environment that contains high-pressure fluid. This high-pressure fluid may bring particles that can cause erosion on the composite and may reduce its life-time. It is our aim in this research to investigate the effect of SiC content to the erosion resistance of MMC Al-SiC.

2. Materials and Methods

Before the stir casting process, 50 µm SiC powder was heated at 1050^{0}C for 8 hours to develop thin SiO_2 oxide layer at the surface of SiC particles. This SiO_2 oxide layer will prevent the formation of Al_4C_3 whenever molten Al reacts with the surface of the SiC particles [7]. This brittle Al_4C_3 will reduce the interface bond strength of the MMC. The stir casting process was done by mixing aluminum Al7075 scrap and oxidized SiC powder with the addition of 1 wt.% magnesium powder. SiC and 1 wt.% magnesium was preheated before introduced to the molten aluminum at 700°C, then stirred for one minute outside the furnace. The melted composite was then reheated before poured into a 33 mm diameter cylindrical steel container with 120 mm height. This sample was then cut and characterized for optical metallography to know the homogeneity of the MMC Al-SiC sample. The observation was done for the un-etched samples at nine different locations as described in Figure 1.

Resistance to erosion was studied with air jet erosion tester that was in the Metallurgy and Materials Laboratory, Institute of Technology, Bandung. The 2.5 cm × 2.5 cm specimen were placed in a jet erosion testing machine with 45° contact angle, then the erodent particle of 50 µm alumina powder was impacted to the specimen with fluid velocity of 70 m/s. The erosion test was conducted for 20 minutes with 2 grams/minute of erodent discharge. The difference in weight before and after the erosion test will then be determined for different percentage of SiC. The weight loss was then divided by specimen density to obtain volume loss in cubic millimeter. Erosion resistance will be presented in the erosion value that is the volume loss divided by total mass of alumina erodent that used to erode specimen. The erosion jet testing is schematically presented in Figure 2.

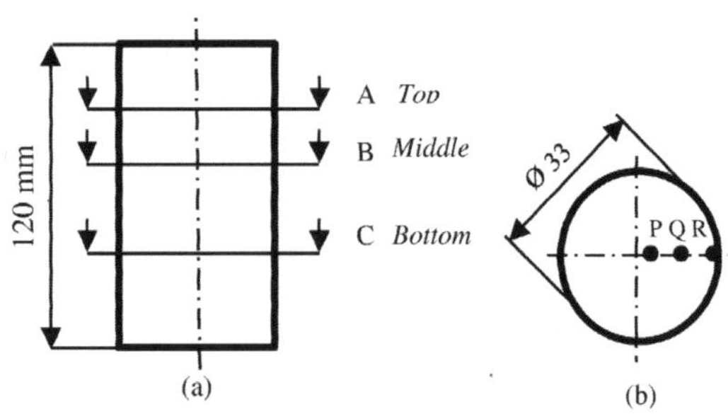

Figure 1 Location of observation on cylindrical sample (a) side view and (b) cross section view

\

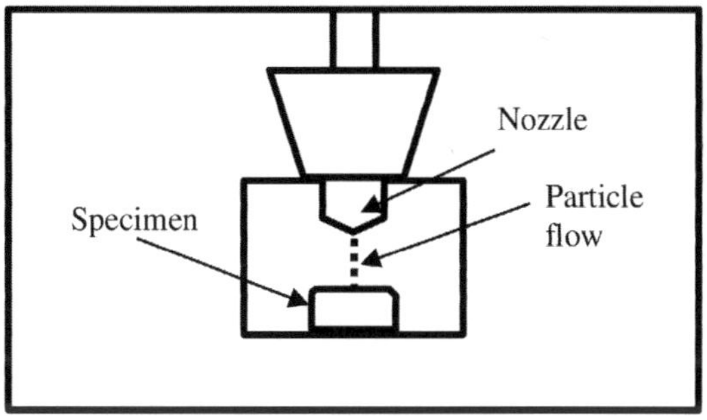

Figure 2 Air jet erosion test scheme

3. Results and Discussion

The result of optical metallography of the un-etched sample of MMC Al-15 wt.% SiC in Figure 3 shows the homogeneity of SiC distribution. It presents the microstructure of the nine observation locations at the cross sections of the cylindrical sample as schematically shown in Figure 1. It can be demonstrated that the SiC particles have been mixed nicely in the cylindrical sample from the upper part to the lower part of the cylinder although the distribution of SiC particles for each location were not homogeneously distributed. In the mixing process, SiC powder will tend to form clusters following the flow of the melted aluminum. Because the stirring process was done outside of the furnace, there was not a lot of time to do the mixing before the temperature reached the liquidus temperature of the Al-7075 and solidified. This will result in the SiC particles distribution that was still locally clustered, although the SiC particles still could be distributed all over the cylinder. The SiC particles apparently were quite intact at the aluminum matrix due to the addition of 1 wt.% Mg. Without the addition of magnesium, it has been proven that the SiC powder would tend to float at the surface of the molten aluminum due to the high surface energy of the molten aluminum. The addition of 1 wt.% magnesium has made the SiC powder not to float when the stirring process was done. This shows that the magnesium addition has reduce the surface tension of aluminum, reducing the interfacial energy between molten aluminum and SiC particles, and improving the surface energy of the solid, in accordance with the investigation for MMC Al-SiC by Singh et.al. [5] This happened due to the formation of $MgAl_2O_4$ at the surface of the oxidized SiC particle which improved the wettability of the SiC particles towards aluminum.

The amount of SiC particles in Figure 3 was counted to know the homogeneity of the SiC distribution from the upper part to the lower part of the cylinder. The result is presented in Figure 4. In general, the outer cylinder has lower density of SiC particle. SiC particles tend to be at the center of the cylinder. After stirring, the molten MMC was casted to the steel cylinder die. It will solidify starting from the outer part that was closer to the steel wall. Due to the surface energy of the molten metal, the SiC particles will tend to flow up to the surface. This can only happen at the aluminum that has not been solidified yet. Therefore, the SiC particle then will be pushed upwards and try to move to the area that was still melted at the center of the cylinder. This explains why the SiC particles to be more at the center and less at the outer part of the cylinder. This also explains why the upper part of the cylinder has more SiC particles than the lower part.

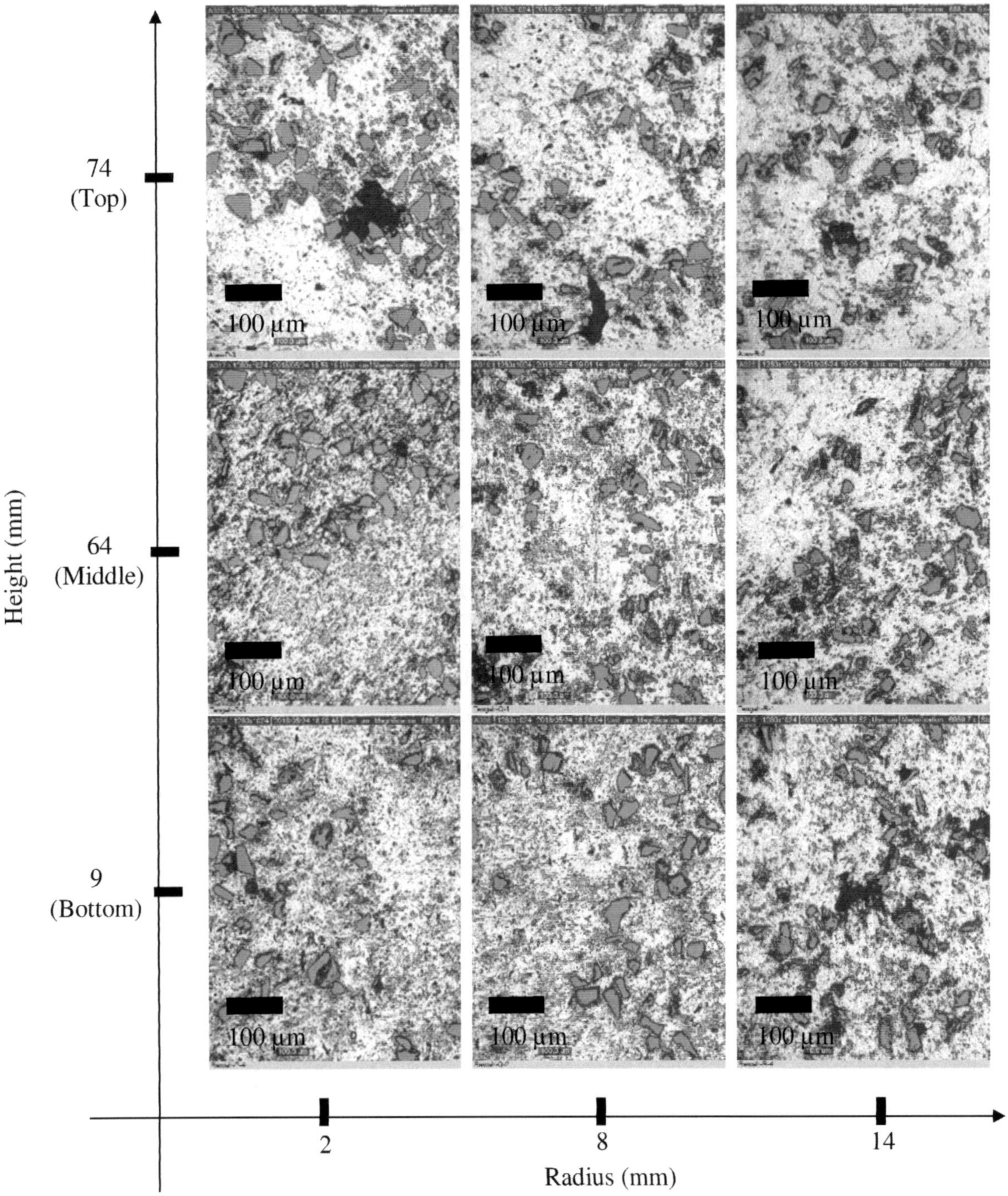

Figure 3 The distribution of SiC particle in MMC Al-15 wt.% SiC

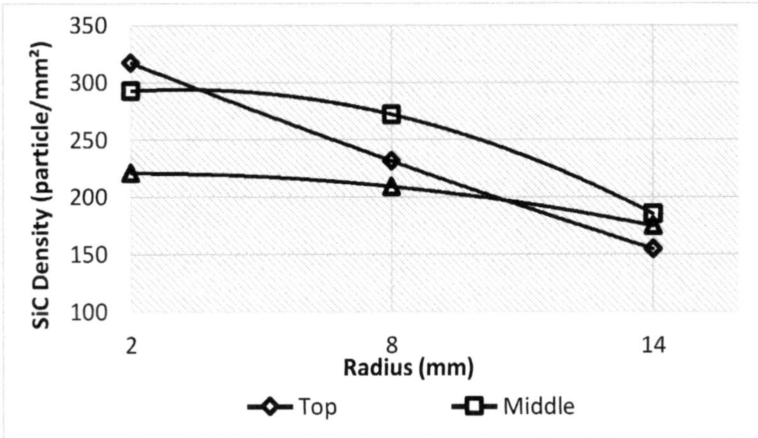

Figure 4 Particle density of SiC in Aluminum matrix

Figure 5 presents the result of the jet erosion testing for four different addition of SiC powder to aluminum, i.e. 0 wt.%, 5 wt.%, 10 wt.%, and 15 wt.% SiC. The result shows that the addition of SiC will decrease the erosion value which means improving the erosion resistance of MMC Al-SiC. The SiC particles apparently act as a shield to the jet flow of fluid that contains the alumina erodent in jet erosion testing. This will protect the aluminum matrix to the erosion wear due to the jet flow. However, the addition of more than 5 wt.% SiC will not make a statistically significant difference in erosion value. This is due the presence of the porosities that tend to be more at higher percentage of SiC addition. These porosities will make the SiC to be easier to be pulled out when it was impacted by the jet flow. The porosities will be getting bigger and the weight loss will be more. This results in the reduction of the erosion value such that there is no statistically significant difference in erosion value after 5 wt.% SiC.

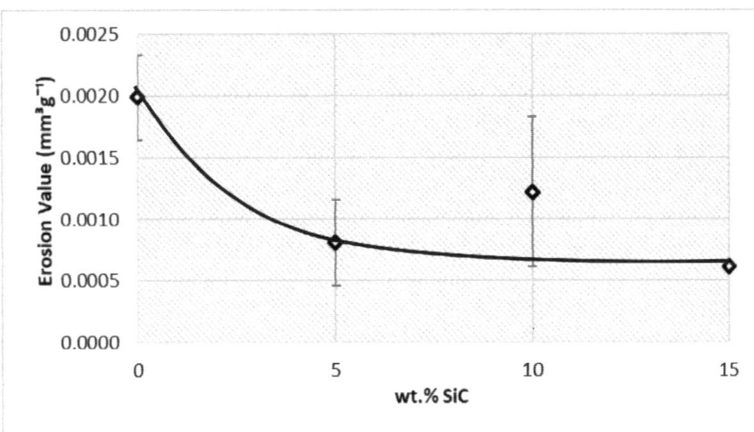

Figure 5 Erosion value variation from air jet erosion test

4. Conclusions

Metal Matrix Composite Al-SiC can be produced using the stir casting procedure with 1 wt.% magnesium addition. Even though SiC particle was not distributed homogenously in Al matrix, the addition of SiC powder in aluminum matrix makes MMC Al-SiC to have better erosion resistance. The test showed that the optimum erosion resistance can be accomplished with 5 wt.% SiC. In the future study to ensure the homogeneity of SiC particle distribution, it is suggested to increase the stiring time of the molten MMC using an externally heated container.

5. References

[1] T. W. Clyne and P. J. Withers, An Introduction to Metal Matrix Composite, E. A. Davis and I. M. Ward, Eds., Oakleigh, Melbourne: Press Syndicate o the University of Cambridge, 1993.

[2] E. B. Kohara, "Study of the Effect of Aluminum Types and SiC Powder Oxidation in the Making of Al-SiC MMC," Intitute of Technology Bandung, Bandung, 2016.

[3] P. K. Sudarsono, "Study of the Effect of SiC Concentration and Magnesium Addition on Hardness of Aluminium Matrix Composite," Institute of Technology Bandung, Bandung, 2017.

[4] Amitesh and V. C. Kale, "Aluminium Based Metal Matrix Composite for Aerospace Application: A Literature Riview," *IOSR Journal of Mechanical and Civil Engineering,* pp. 31-33, November-Desember 2015.

[5] V. K. Singh, S. Chauhan, P. Gope and A. Cahudhary, "Enhancement of Wettability of Aluminum Based Silicon Carbide Reinforced Particulate Metal Matrix Composite," *High Temperature Materials Processes,* p. 2, 2004.

[6] X. G. Chen and E. S., "Formation of Gas Porosity in Aluminium Alloys," *American Foundry Society Transactions,* Vols. 92-94, pp. 673-682, 1993.

[7] J. Hashim, M. Looney and M. S. J. Hashmi, "The Enhancement of Wettability of SiC Particles in Cast Aluminium Composites," *Journal of Material Processing Technology,* no. 119, pp. 329-335, 2001.

The Effect of Adding Synthesis Silica from Rice Husk as Cement Substitution for High Strength Concrete

J Fajar Fatriansyah[1], D Dhaneswara[1], S Aida Khairunnisa[1], D Kusumawardhani[1]

[1]Metallurgy and Materials Engineering Department, Faculty of Engineering, Universitas Indonesia

Abstract. The synthesis of silica from agricultural waste was conducted. The thermal decomposition was conducted at 700°C which yields white-brown silica powder. Acid mechano-chemical method was done to reduce the size particle of silica and homogenize the distribution particle. The variables used were the rotational speed of planetary ball-mill at 22 rpm and the addition of acid (HCl 1M). Under controlled burning, the highly reactive ash (silica products) that is produced can be used as a supplementary cementing material or in the production of high strength concrete (HSC). The samples were tested during its 28th day to know its compressive strength.

Keywords: Silica, Rice husk, High strength concrete

1. Introduction

Rice is one of the most consumed staple food in the world [1]. Rice Husk (RH) is the outer layer of the paddy crop. It is enormously available throughout the world wide as an agro waste [2]. Rice husk weight that produced is 20%-22% from the total weight of rice produced.[3] In 2015, Indonesia produced about 75 million tons of paddy,[4] while in worldwide 740 million tons of paddy produced[5]. The enormous amount of rice husk produced will increase each year since the increase of global demand for the paddy for years ahead [6]. Rice husk is surface is hard because high content of silicon, therefore it is hard to decomposed by bacteria. This waste can possibly threatening for environment if it is not treated well [7].

Rice husk is widely used in the industries as a fuel. Due to burning of rice husk, its ash form has produced known as Rice Husk Ash (RHA) [3]. Recycling different types of ashes for cement production has gained increasing attentions worldwide in a bid to close the waste loop. It minimizes waste landfilling and meanwhile produces useful secondary materials with reduced costs [8]. RHA mainly contains silica as a major component, due to which RH as well as RHA has used as a prime reinforcing agent in order to improve the mechanical properties of alloys [3]. Rice husk gives a good quality of pozzolanic ash containing silica, which is considered a good substitute for cement [9-10].

RHA addition in concrete could increase compressive strength and durability, decrease porosity on concrete matrix and reduce the carbon dioxide emission. Partial replacement of Portland cement with rice husk ash improve the workability and mechanical resistance also give advantage in economic view with lower construction cost needed [11-15].

Rice husk contains amorphous silica about 20-25 wt. %. Silica crystallinity in RHA can transform depend on the combustion process, it will transform into crystalline state if heated in high temperatures [15]. Controlled combustion of rice husk can produce amorphous silica with large surface area. Amorphous silica is obtained from rice husk combustion at 700°C, and it will transform into crystalline state with increase of temperature more than 850°C [17]. Pozzolanic nature of RHA depends on RHA particle size, surface area of the ash, and crystallization state of the silica [18]. Small particle size, large surface area and amorphous phase are ideal condition to get the best pozzolanic nature [19-21]. Planetary ball mill can be used to reduce the particle size of RHA. The mechanical force when steel ball colliding with RHA will crush the particle to the smaller size.

Costa et.al [22] reported that chemical treatment for RHA using hydrochloric acid could produce higher content of silica and lower content of the impurities. The silica that produced with this treatment dominantly on amorphous state. Chemical treatment also could affect the particle size of RHA because acid can react with the RHA and diminish the intercalation between the particles.

In this study, we tried to evaluate the effects of RHA addition in cement according to some experimental background of what influence theses admixtures may have on concrete composition. We present different kinds of RHA addition, followed by some tests to examine the influence of additions on basic properties of concrete. The variable used are rotational speed of planetary ball-mill (22 rpm) and 1M hydrochloric acid.

2. Experimental

2.1 Materials

Acid Solution Preparation

IOP Conf. Series: Materials Science and Engineering **547** (2019) 012034 doi:10.1088/1757-899X/547/1/012034

The Hydrochloric acid was obtained from CV Mitra Bersama and used without further purification. 1 M hydrochloric acid was obtained by diluting 20.72 ml of 37% HCl in 229.28 ml of water.

Rice Husk Ash Preparation

Raw rice husk ash was obtained from Paddy Mill Industry, Leuwiliang, Bogor. This wet rice husk was dried for 8 hours under the sunlight then burned in controlled temperature of 700°C for 5 hours. The rice husk ash then milled to reduce its particle size. The process of rice husk ash preparation can be seen on **Figure 1.**

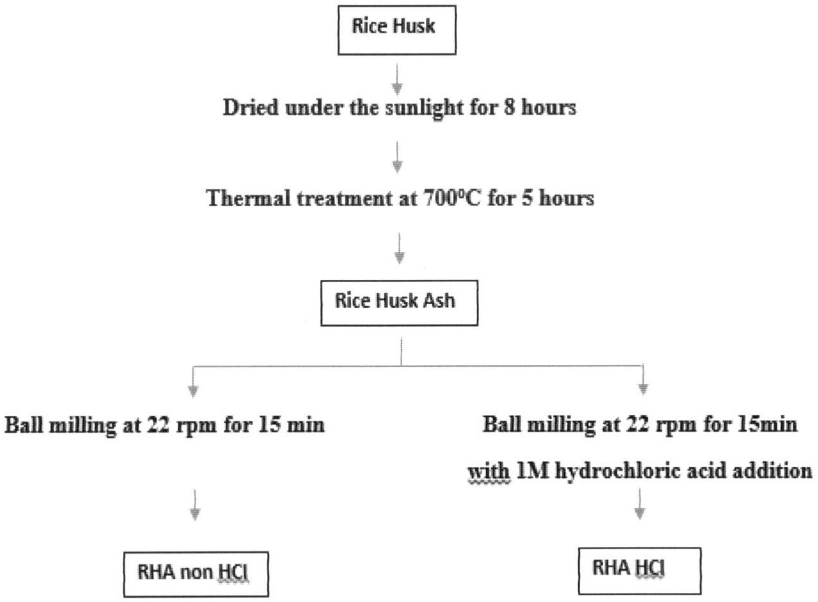

Figure. 1 The Process of Rice Husk Ash Preparation

The addition of RHA non-HCl and RHA HCl will improved the compressive strength of concrete depend on its chemical composition. The chemical composition of each RHA added are given in **Table 1.**

	Pure Silica	RHA non HCl	RHA HCl
Chemical analysis (wt. %)			
MgO	1.51	1.55	1.43
SiO2	82.5	88.13	86.71

P2O5	1.89	1.69	1.58
Cl	0.62	0.37	0.56
K2O	4.75	4.14	3.92
CaO	1.74	1.7	1.49
Fe2O3	3.95	2.32	4.21
NiO	0.58	-	-
ZnO	2.35	-	-
Ag2O	0.11	0.12	0.11

Cement

The cement we used for this study is ordinary Portland cement type I from PT Indocement. This cement specifications conform to ASTM C 150 (Standard Specification for Portland Cement)

Aggregates

There were 2 types of aggregates used for this study. The fine aggregate is sand from Bangka province. The sand particle size range is 0 to 5 mm sieving grade. The coarse aggregate is gravel from Rumpin with particle size range from 10 to 20 mm.

2.2 Concrete Cube

Concrete

The concrete consist of aggregates (fine aggregate or sand and coarse aggregates or gravel), cement and RHA as supplementary cementious materials. Mix all of the materials evenly and slowly add water into the dry materials so it can bind all of the four materials. The composition of each materials in the making of concrete can be seen in **Table 2.**

Table 2. Composition of Concrete Making Materials

Concrete	RHA (g)	Cement (g)	Sand (g)	Gravel (g)	Water (g)
C - Normal	0	483	973	1200	267
C - RHA non HCl	18.5	464.5	973	1200	267
C - RHA HCl	18.5	464.5	973	1200	267

Mold

Cube molds that were used is 100mm x 100mm x 100mm made from wooden board. Before the concrete is scooped into the molds, the molds must be lightly coated with a mold release agent to ensure that the concrete does not stick to the mold and makes it easier to remove the concrete cube. When using a 100 mm mold, the concrete sample is scooped into the mold in 2 equal layers (50 mm) and compacted between each layer manually using a compacting rod. Each layer compacted is tampered using certified compacting rod, and the side of the molds is tapped usig a hammer. Tampering and tapping used to removes trapped air in the concrete and allows compaction of the sample.

Storage

The cube concrete then being labelled to identify each of the cubes and to record how they have made, before being stored in dry environment at a temperature range of 20 ± 5 degrees Celsius for 24 hours.

2.3 Concrete Cube Test

After 28 days, where the concrete shows almost at 99% of its strength, the cubes being tested. The cubes are tested on the face which perpendicular to the molding face. The compressive machine exerts a constant progressing force on the cubes till they fail. The maximum compressive strength of the concrete will be shown once the concrete reach its failure. The failure of the concrete was indicated by the change of its shape which has been altered due to compression.

3. Result and Discussion

Concrete is used mostly for structural purposes such as foundations, columns, beams and floors and therefore must be capable in taking the loads that will be applied. One of the methods of checking its fit for purpose is to carry out a concrete cube test which measures the compressive strength of the concrete cubes.

Compressive strength of concrete depends on many factor such as water-cement ratio, cement strength, quality of concrete material, etc. Test for compressive strength is carried out either on cube or cylinder. In this experiment, we're using cube as a sample shape for concrete. Type of specimens that used for cube test was 100mm x 100mm x 100mm.

3.1 Particle Size of Rice Husk Ash

In this experiment, we get to know the effect of adding acid (1M HCl) while grinding to the particle size of RHA which presented in **Table 3.** It was found that the rice husk ash milled together with 1M hydrochloric acid had a smaller particle size compared to rice husk ash that was milled without 1M hydrochloric acid. RHA with high fineness can improve the compressive strength [23]. In addition, its corresponding to the compressive strength that was shown at **Table 5**, where C-RHA HCl has higher compressive strength than C-RHA non HCl. Compare to pure silica, both experiment show smaller particle size.

Table 3. Data Obtained from the Particle Size Analysis

	Particle size (µm)
Pure Silica	51.02
RHA non HCl	30.16
RHA HCl	23.24

3.2 Surface Area of Rice Husk Ash

In this experiment, we got the comparison of surface area of pure silica, RHA non HCl and RHA HCl by using BET analysis which presented in **Figure 2**. The graphic present show the increase of surface area in RHA non HCl and RHA HCl compare to the surface of the pure silica. And its corresponding to the result of the particle we got, where the smaller particle size have larger surface area, which will increase the compactness of the concrete to increase its compressive strength.

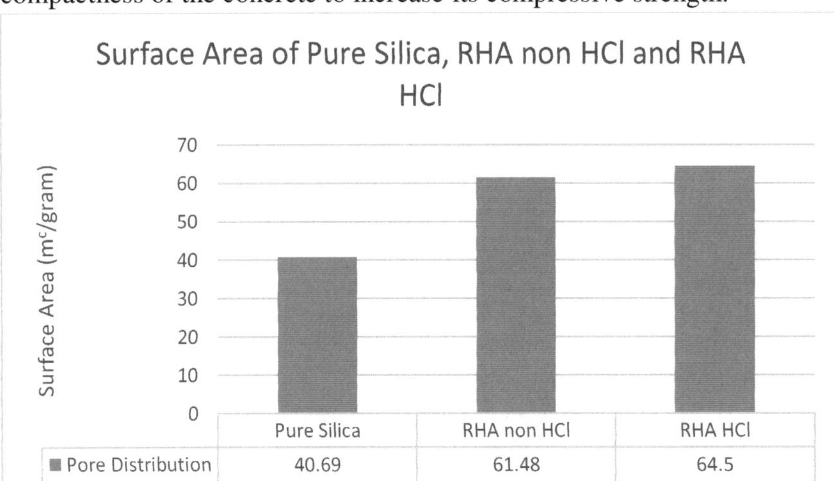

Figure 2. Surface Area of Pure Silica, RHA non HCl, and RHA HCl

3.3 Compressive Strength

Compressive strength of all concrete cubes with different replacement of RHA at 28 days are presented in **Table 4**. It was found that the concrete with RHA replacement had higher compressive strength compared to the normal concrete (without any replacement of RHA). Beside from the filling effect because of its particle size, pozzolanic reactivity of RHA make outstanding contribution to the development of compressive strength [24]. Reactivity of RHA usually associated to its high content of silica and its very large surface area governed by the porous structure of the particle [18]. The highest amount of amorphous silica is achieved when RHA is burned between 500 and 700°C [25]

Table 4. Data Obtained from the Compressive Strength Test

Concrete	Compressive Strength (MPa)	Compressive Strength Average (MPa)
C - Normal	27.23	26.38
	24.14	
	27.77	
C - RHA non HCl	27.05	28.25
	29.53	
	28.17	
C - RHA HCl	30.4	28.53
	26.41	
	28.78	

4 Conclusions

The replacement of 10% cement with RHA, increased the compressive strength at 28 days by 7-8% compare to the normal concrete. The particle size of RHA used as replacement was reduce by 23% using acid mechano-chemical with addition of 1M hydrochloric acid when ball milling compare to ball milling without the addition of 1M hydrochloric acid. The compressive strength of concrete increase as the particle size of RHA reduced, and the surface area increase to strengthen the bond within the particle. Thus, the replacement of cement with RHA using acid mechano-chemical has higher compressive strength than the replacement of cement with RHA without the addition of 1M hydrochloric. Further study on the effect of rice husk ash on concrete can be done by more trials and samples to increase the consistency of the data obtained in the experiment. Different addition of acid/alkali when milling the rice husk ash other than what has been used in this paper, can also be explored.

References

[1] FAO, 2015. Rice Market Monitor XVI, pp. 1e14.

[2] Tiwari, S., & Pradhan, M. K. (2017). Effect of rice husk ash on properties of aluminium alloys: A review. *Materials Today: Proceedings*, *4*(2), 486–495. https://doi.org/10.1016/j.matpr.2017.01.049

[3] H. A. Saudi. S. M. Salem, S. S. Mohammad, A. G. Mostafa and M. Y. Hassaan, Am.J.Phys. and App.3(3), 97 (2015)

[4] Indonesia Central Bureau of Statistic, https://www.bps.go.id/dynamictable/2015/09/09/865/produksi-padi-meurut-provinsi-ton-1993-2015.html (accessed 2018)

[5] Food and Agriculture Organization of the United Nation. *Rice Market Monitor*. December 2015. Vol XVIII.1

[6] P.A. Seck, A. Diagne, S. Mohanty, M.C.S. Wopereis, Crops that feed the world 7: rice, Int. Soc. Plant Pathol. 4 (2012) 7–24.

[7] Zhang, H., Ding, X., Chen, X., Ma, Y., Wang, Z., Zhao, X., 2015. A new method of utilizing rice husk: consecutively preparing d-xylose, organosolv lignin, ethanol and amorphous superfine silica. J. Hazard. Mater. 291, 65e73. https://doi.org/10.1016/j.jhazmat.2015.03.003.

[8] Yin, K., Ahamed, A., & Lisak, G. (2018). Environmental perspectives of recycling various combustion ashes in cement production – A review. *Waste Management*, *78*, 401–416. https://doi.org/10.1016/j.wasman.2018.06.012

[9] Zhong Yuxia, Wang Xianli, Hu Xiangxiu, Chen Xiaoyu, Zhnag Rui, "Study on the Effects of Rice Husk Quantity and Particle Size on The Compressive Strength of Concrete Based on Experiment," 7th International Conference on Intelligent Computation Technology and Automation (ICICTA), 2014.

[10] Swaminathen, A.N., "Indian Rice Husk Ash – Improving the strength and durability of concrete: A review," International Conference on Current Trends in Engineering and Technology (ICCTET), 2013.

[11] N. Yalçin, V. Sevinç, Studies on silica obtained from rice husk, Ceram. Int. 27 (2001) 219–224, https://doi.org/10.1016/S0272-8842(00)00068-7.

[12] B. Chatveera, P. Lertwattanaruk, Durability of conventional concretes containing black rice husk ash, J. Environ. Manag. 92 (2011) 59–66.

[13] D. Chopra, R. Siddique, Kunal, Strength, permeability and microstructure of self compacting concrete containing rice husk ash, Biosyst. Eng. 130 (2015) 72–80.

[14] R. Kishore, V. Bhikshma, P.J. Prakash, Study on strength characteristics of high strength rice husk ash concrete, Procedia Eng. 14 (2011) 2666–2672.

[15] S.H. Sathawane, V.S. Vairagade, K.S. Kene, Combine effect of rice husk ash and fly ash on concrete by 30% cement replacement, Procedia Eng. 51 (2013) 35–44, https://doi.org/10.1016/J.PROENG.2013.01.009

[16] Othman Ali, I., Hassan, A.M., Shaaban, S.M., Soliman, K.S., 2011. Synthesis and characterization of ZSM-5 zeolite from rice husk ash and their adsorption of $Pb2^+$ onto unmodified and surfactant-modified zeolite. Sep. Purif. Technol. 83, 38e44. https://doi.org/10.1016/j.seppur.2011.08.034.

[17] A. Mehta, and Dr. R. P Ugwekar, Int. J. Eng. Res. And App. 5, 43 (2015)s

[18] Rukzon, S.; Chindaprasirt, P.; Mahachai, R. Effect of grinding on chemical and physical properties of rice husk ash. Int. J. Miner. Metall. Mater. 2009, 16, 242–247.

[19] N. Yalçin, V. Sevinç, Studies on silica obtained from rice husk, Ceram. Int. 27 (2001) 219–224, https://doi.org/10.1016/S0272-8842(00)00068-7.

[20] M.N. Al-khalaf, H.A. Yousift, Use of rice husk ash in concrete, Int. J. Cem. Compos. Light Concr. 6 (1984) 241–248, https://doi.org/10.1016/0262.

[21] Agarwal, S. K. (2006). Pozzolanic activity of various siliceous materials. Cement and Concrete Research, 36(9), 1735–1739. https://doi.org/10.1016/j.cemconres.2004.06.025

[22] Santana Costa, J. A., & Paranhos, C. M. (2018). Systematic evaluation of amorphous silica production from rice husk ashes. *Journal of Cleaner Production*, *192*, 688–697. https://doi.org/10.1016/j.jclepro.2018.05.028

[23] Xu, W., Lo, T. Y., Wang, W., Ouyang, D., Wang, P., & Xing, F. (2016). Pozzolanic reactivity of silica fume and ground rice husk ash as reactive silica in a cementitious system: A comparative study. *Materials*, *9*(3), 1–14. https://doi.org/10.3390/ma9030146

[24] V.T.A. Van, C. Rößler, D.D. Bui, H.M. Ludwig, Rice husk ash as both pozzolanic admixture and internal curing agent in ultra-high performance concrete, Cem. Concr. Compos. 53 (2014) 270–278.

[25] Liou, T.H. Preparation and characterization of nano-structured silica from rice husk. Mater. Sci. Eng. 2004, 313, 313–323.

Failure Analysis of Superheater Boiler Tube SA 213 T12

Husaini Ardy[1]* **Deri Andika Bangun**[1]

[1]Materials Engineering Department, Bandung Institute of Technology, Indonesia

*Corresponding author (E-mail): husaini@material.itb.ac.id

Abstract. Superheater tubes SA213 T12 (OD = 57 mm, WT = 6 mm) has been in service for 11 years in a supercritical coal fired boiler (315 MW). Tubes operating temperature are 430°C and pressure is 11 MPa. Failure occurred on tube no. 20, row 2. Three tubes specimens from the same row (failed tube, tube no. 3, and no. 4), were drawn for laboratory examinations. Vickers microhardnesses of failed and no. 3 tubes are the same (140 VHN), while tube no. 4 hardness is 170 VHN. These hardness values are much lower compared to the new tube (192 VHN). The lower hardness value of failed tube is due to complete decomposition of pearlite colonies in inner and outer side of failed tube to ferrite and spheroidized cementite. Failure of tube is characterized by thin lip rupture which indicated that the tube circumferential deformation occur gradually following spheroidization process. Failure by intergranular crack occurred along grain bondaries when the stress on the tube wall (hoop stress) exceeding material strength at service temperature. Tube no. 3 and 4 are partially spheroidized, mostly on the innerside. The severity of microstructure damage is related to the flow direction of flue gas, from tube no. 1 to no.2, no.3 and so on. Tube no. 1 was survived because it was protected by a front cover that will limit heat transfer from the flue gas. The previous failure analysis of similar boiler tube revealed that the hardness of the bulging tube is about 140 VHN which is the same as the hardness of the failed tube (no. 2) and no.3. Tube no. 3 is in the early stage of damage and shall be replaced because the hardness is much lower than 170 VHN. The effect of flue gas flow direction not only affects the hardness and microstructure but also deposit thickness. Deposit thickness in tube no. 2, 3, and 4 are 149, 123, and 100 μm respectively. Thicker deposit in tube no. 2 is also contributed to premature failure of this tube compared to the others. The local temperature of tube wall will be higher because thick deposit will reduce heat transfer between flue gas and steam inside the tube.

Keywords: SA213 T12, superheater tube, creep damage, spheroidization, deposit thickness

1. Introduction

Superheater tube is one of important components in the power boiler system. Steam for power generation was produced by heating the tube containing water or steam by flue gas from the burning coal. Due to their nature of service, superheater tube will be exposed to high temperature and coal debries during service. Failure of superheater tube might occur when it has been exposed continuously to high temperature for long period of time. Several failure mechanisms might occur to the tube, such as creep, spheroidization, graphitization, overheating, and wear [1]. Creep is time-dependent degradation of material under stress and high temperature. Creep damage can be identified through several characteristics such as long exposure time, thick crack lip, microvoids, and intergranular crack. Spheroidization of cementite is another type of high temperature failure that arises from decomposition of pearlite into ferrite and spherical cementite that will reduce material strength substantially. Spheroidization occurs above 600°C and can be identified by disappearance of perlite colonies in microstructure. Similar to spheroidization, graphitization is degradation due to decomposition of pearlite into ferrite and graphite at a temperature lower than 600°C followed by intergranular crack along graphite sites on the grain boundaries [2]. This paper will present the root cause analysis of repetitive failure of superheater tube in a 315 MW coal fired boiler. First failure was found in 2014 [3], and recent failure in 2018 (after 11 years of service). Superheater tube operating temperature is 430 °C and pressure of about 11 MPa.

2. Materials and Methods

Superheater tube material is ASME SA213 T12, outer diameter and thicknesses are 57 and 6 mm respectively. Three tube samples were drawn from tube no 20; row 2, 3, and 4 (Figure 1). Tube row 2 is the failed tube and tubes row 3 and 4 are un-failed tubes near row 2. Superheater tube material identification was carried out by chemical analysis, tensile test, microhardness test, and microstructure analysis. Chemical compositions of each tube were examined using optical emission spectroscope ARL 3460. Tensile tests were conducted using 20 tonnes Tarno Grocky universal tensile testing machine. Hardness tests were conducted using Zwick Digital Vickers Microhardness Tester at 600 gram load. Microstructures were observed using Nikon Eclipse MA2000.

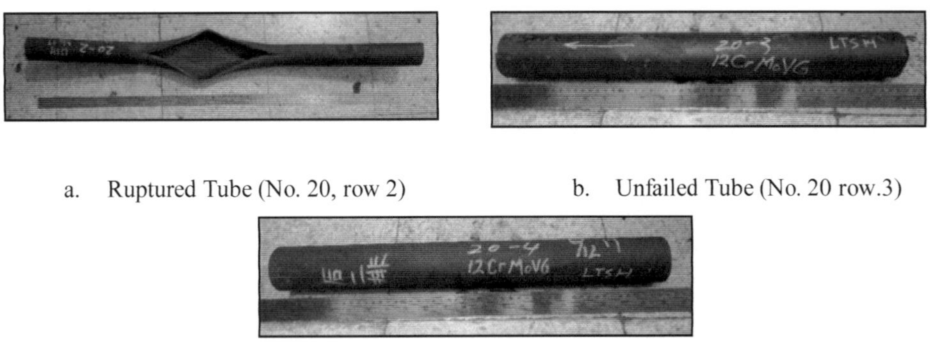

a. Ruptured Tube (No. 20, row 2) b. Unfailed Tube (No. 20 row.3)

c. Unfailed Tube (No. 20 row 4)

Figure 1. Pipe samples for laboratory examinations

3. Results and Discussion

3.1. Tube Material Properties

The average results from three examinations of chemical compositions on each tube sample and their comparison to ASME and China (GB) Standards are shown in **Table 1**. Chemical compositions of ruptured and unfailed tubes fulfil 12Cr1MoV specification.

Table 1. Chemical Composition (weight %) of Tubes

Element	Testing Results			Specifications [4, 5]	
	Ruptured Tube No. 20 row 2	Unfailed Tube No. 20 row 3	Unfailed Tube No. 20 row 4	ASME SA213 T12 Specs.	GB 8162 12Cr1MoV Specs.
C	0.12	0.14	0.14	0.05 – 0.15	0.08 – 0.15
Si	0.19	0.20	0.19	≤0.50	0.17 - 0.37
P	0.0035	0.0050	0.0049	≤0.025	≤0.035
S	0.013	0.021	0.019	≤0.025	≤0.035
Mn	0.56	0.56	0.54	0.30 - 0.61	0.40 - 0.70
Ni	0.061	0.044	0.043	---	≤0.30
Cr	0.93	0.95	0.94	0.80 - 1.25	0.90 - 1.20
Mo	0.28	0.30	0.29	0.44 - 0.65	0.25 - 0.35
V	0.19	0.18	0.18	----	0.15 - 0.30
Cu	0.16	0.13	0.12	----	---
Fe	Balance	Balance	Balance	Balance	Balance

Tensile test was carried out according to ASTM A 370 standard. Result of tests and its conformation to ASME and GB standard is shown in **Table 2**.

Tensile strength of failed tube (No. 20 row 2) is much lower compare to that of unfailed tube but slightly above minimum requirement. The lower value of tensile strength is due to material degradation after long time exposure; but the yield strength is still above specification. Tensile and yield strengths of two good tubes (No. 20 row 3 and 20 row 4) are still fulfils ASME SA213 T12 and GB specification [3,4].

Table 2. Tensile Properties of Tube

Properties	Testing Results			Specifications [4, 5]	
	Failed Tube No. 20 row 2	Unfailed Tube No. 20 row 3	Unfailed Tube No. 20 row 4	SA213 T12	GB 8162 12Cr1MoV
Tensile Strength, MPa	474	542	567	≥415	470 - 640
Yield Strength, MPa	349	418	412	≥205	≥255
Elongation, %	39	34	30	≥30	≥21

Table 3. Hardness Test Result

Sample	Average Hardness (VHN)	SA213 T-12 Specification [4]
Hardness data in December 2017		≤ 170
Tube No. 20 row 2	140	
Tube No. 20 row 3	141	
Tube No. 20 row 4	170	
Hardness data (mid-thickness) in June 2014		
Good Tube No. 10.4	192	
Bulging Tube No. 12.4	139	

The hardness of tubes in row 2 (failed) and 3 (un-failed) are below specification. The lower values of hardness indicated that the material properties have been degraded. The hardness of tube row 4 (un-failed) is in the lower specification limit.

The current result will be compared to the previous data taken in June 2014 of the new and bulging tubes. The hardness of new tube is 192 VHN and the bulging one is 139 VHN. It is clear that the hardness of 140 VHN of un-failed tube (row 3) is closed to the hardness of the bulging tube, therefore this tube shall be replaced to prevent unexpected failure during service. The hardness value of 140 VHN can be used as reference for tube replacement.

Microstructure examinations will be conducted to verify the degree of degradation in material by observing the status of pearlite in microstructure. Microstructures of tubes were taken from the longitudinal section of the failed tube (row 2), and two good tubes (row 3 and 4). Specimen of the failed tube was taken from location far from the bursting part (near one of the edge in **Figure 1a**).
Figure 2 shows the microstructure of tubes at the mid-thickness. The microstructure of good tubes (**Figure 1b** and **c**) consists of ferrite and pearlite. Pearlite structures are partially degraded but still present. Pearlite phases in failed tube have been decomposed completely (**Figure 1a**) to ferrite and spheroidized cementite. **Figure 3a** shows the optical micrograph of failed tube at high magnification. Some microvoids on grain boundaries are evident (arrow signs). Scanning electron microscope (SEM) examination was also performed on the same sample to obtain a clearer picture of microvoids on the grain boundaries.

Figure 2-4 shows the comparison of microstructure of the tubes inner surface. It can be seen in **Figure 4** that the failed tube has the thickest deposit among them. The damage in pearlite structure will occur in shorter time with increasing deposit thickness because of higher local temperature.

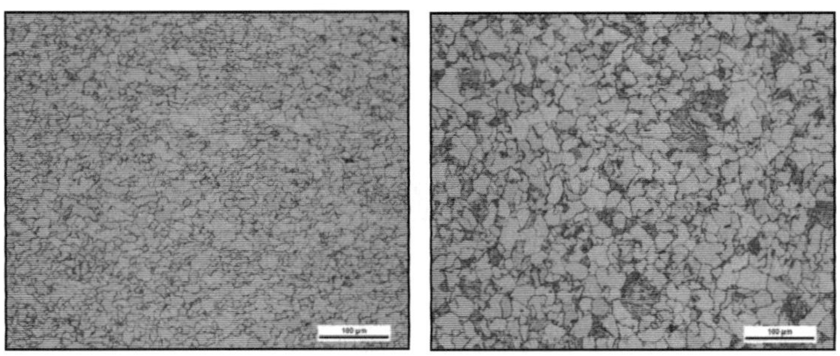

a. Tube 20 row 2 b. Tube 20 row 3

c. Tube 20 row 4

Figure 2. Optical micrographs of tubes at mid thickness

a. Optical micrograph

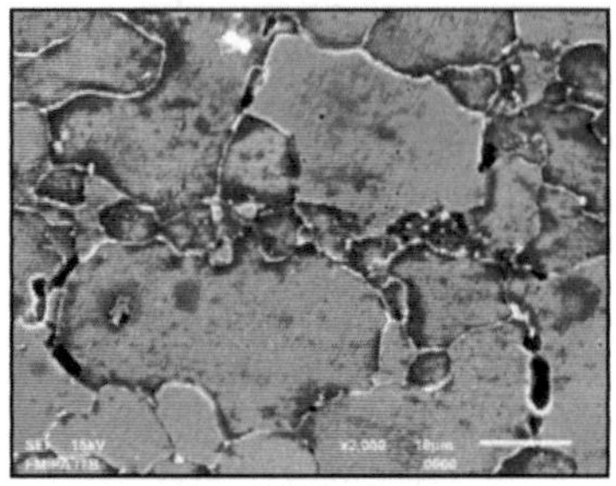

b. SEM micrograph

Figure 3. Microstructure of microvoids at grain boundaries (arrow signs)

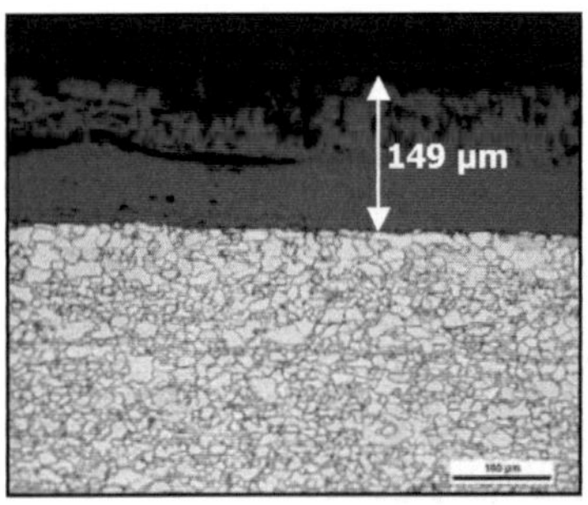

a.Tube No. 20 row 2

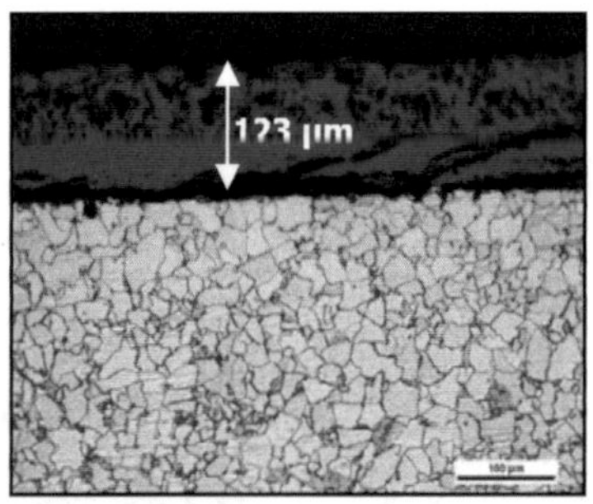

b. Tube No. 20 row 3

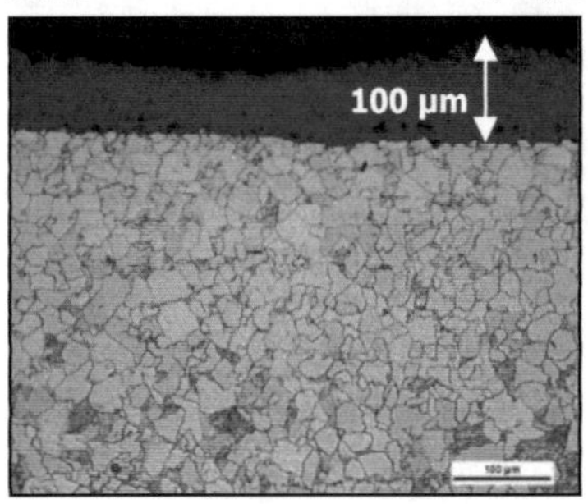

c. Tube No. 20 row 4

Figure 4. Microstructure of tubes at inner surface

3.2. Failure Mechanism

Tube shows a thin lip rupture types of failure. Thin lip rupture is a characteristic of stress rupture failure that was initiated by bulging. Bulging itself is a time dependent process where spheroidization process decreased material strength gradually. Rupture occurred when the stress on the tube wall (Hoop stress) is higher than material strength at temperature. Microstructure of failed tube has been examined to obtain data about microstructure degradation and crack propagation mode. Two locations were observed, near the fractured lip and its opposite side (180° from the fractured lip).

Figure 5 shows the microstructure of failed tube at the fractured lip. **Figure 5a** shows the remaining deposit on the inner surface, most of the deposit has been wiped out by the steam flow during rupture. Crack propagation along grain boundaries (intergranular crack) are shown in **Figure 5b** to **d**. Intergranular cracks occurred along the grain boundries. It can also be seen that pearlite structure vanished completely due to spheroidization process (**Figure 5d**).

Figure 6 shows the microstructure at the opposite of the crack (180° from the crack lip). There is no intergranular cracks found on the mid-thickness section, however the pearlite structure also disappear completely by spheroidization process (**Figure 6d**).

Hardness of pipes at the fractured lip and its opposite side have been examined using Vickers microhardness to confirm the degradation of microstructure because the damage or transformation of pearlite into ferite and graphite will decrease the hardness significantly. The average hardness of fractured lip is 130 VHN and the opposite side is 152 VHN. The lower hardness near the fractured lips confirmed the effect of microstructure degradation to decrease the hardness and strength.

a. Inner surface b. Inner surface, fractured lip

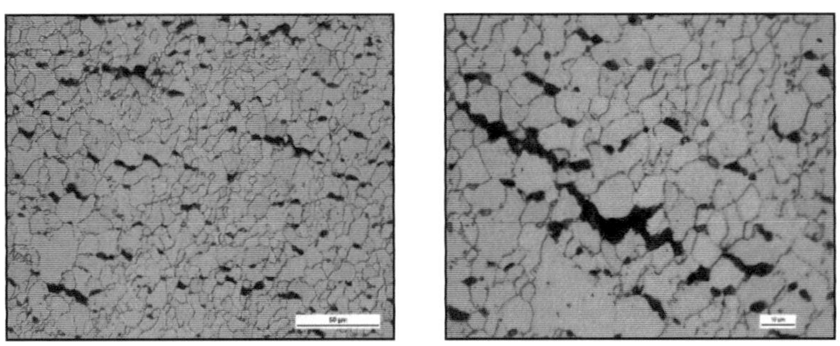

c. Mid-thickness d. Mid-thickness at higher magnification

Figure 5. Microstructure at the cracked tube

a. Inner side b. Outer side

c. Mid-thickness, low magnification

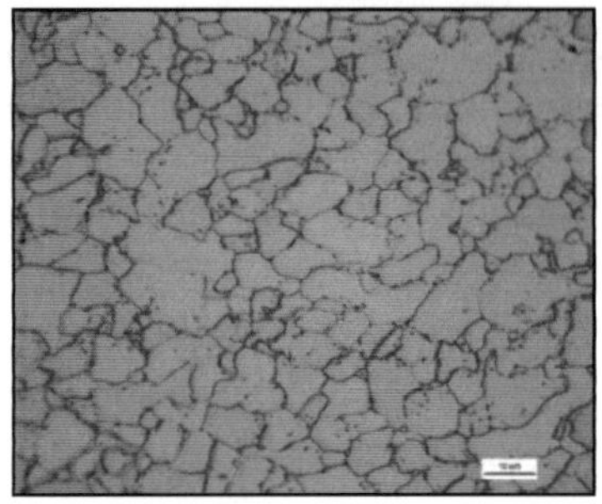

d. Mid-thickness, high magnification

Figure 6. Microstructure of tube at opposite side of crack.

4. Conclusions

Failure of low temperature superheater tube is caused by spheroidization after about 11 years of service. Spheroidization damage in tube No. 20 row 2 has been accelerated by the present of thick deposit on the inner side. Observation of deposit thickness revealed that tube No. 20 row 2 has the thickest deposit compared to that of tube No. 20 row 3 and 4. Thicker deposit will promote higher local temperature because deposit will decrease the heat transfer between flue gas and steam.

The tube rupture is related to the microstructure damage by mean of pearlite tranformation to ferrite and carbides and followed by the decrease in hardness. Pearlite colonies in tube No. 20 row 2 has completely damaged and the hardness decrease to 140 VHN which is the same as the hardness of the bulging tube (previous failure in 2014). Microstructure of pipe No. 20 row 3 also affected and the hardness decreases to 141 which indicate that this material is entering the early stage of creep and shall be replaced immediately to prevent unexpected failure. Microstructure of pipe No. 20 row 4 was not affected significantly and the hardness of 170 is far above the bulging hardness. The current findings of tube hardness (140 VHN) which is supported by previous case (failure in 2014) can be used as reference hardness for tube replacement without conducting time consuming destructive test.

5. Acknowledgements

The authors are grateful to PT. Sumber Segara Primadaya for the research funding, tube samples, and service data.

6. References

[1] 2002 *ASM Handbook: Failure Analysis and Prevention Vol. 11* ed. Becker W T, Shipley R J (Ohio: ASM International) p 1540

[2] French D N 1993 *Metallurgical Failures in Fossil Fired Boilers* 2nd ed (New York: John Wiley) p 115

[3] Ardy H 2014 *Failure Analysis of Low Temperature Superheater Tube* internal report (Bandung: LAPI-ITB)

[4] ASTM 2017 *Standard Specification for Seamless Ferritic and Austenitic Alloy-Steel Boiler, Superheater, and Heat-Exchanger Tubes* (Harbour Drive: ASTM International)

IOP Conf. Series: Materials Science and Engineering **547** (2019) 012035 doi:10.1088/1757-899X/547/1/012035

[5] Wegst C W 1995 *A Key to Steel* (Marbach: Verlag Stahlschussel Wegst GMBH) p 95

Physical Aging of $Zr_{65}Cu_{17.5}Ni_{10}Al_{7.5}$ Glassy Alloy

Daisman P.B. Aji*, Triyono, Yusep Mujalis
Department of Mechanical Engineering, Universitas Trisakti
Jl. Kyai Tapa 1, Grogol, Jakarta, Indonesia 11440

*Corresponding author (E-mail): daisman.aji@trisakti.ac.id

Abstract. Physical aging of $Zr_{65}Cu_{17.5}Ni_{10}Al_{7.5}$ glassy alloy has been studied after holding its samples with variation of time and temperature of annealing. The results showed that the enthalpy and entropy decreased with time according to a non-exponential kinetics. The Differential Scanning Calorimetry (DSC) heating scan after annealing was analyzed using Tool-Narayanaswamy-Moynihan (TNM) model. The analysis showed that a set of parameters that fitted the heating scan data for one annealing time did not fit the heating scan data for another annealing time. This is explained due to approximations used in the model and the effect of Johari-Goldstein (JG) relaxation to the physical aging.

Keywords: *Physical Aging*; *Glassy Alloys*; *Glass Transition*; *Calorimetry*; *Johari-Goldstein Relaxation*

1. Introduction

Structural relaxation on aging of glasses has been studied extensively mainly because it changes the properties of a glass during its use. When a glass is heated toward its glass-softening temperature or when a glass is cooled, all its physical properties change not only with temperature but also with time. Moreover, when a glass is kept at a certain temperature, its properties change with time. Therefore, the properties of a glass depend on its thermal history. By studying the structural relaxation of a glass, one can predict the changes in properties of a glass with time. Change in properties during structural relaxation of glasses has been generally studied [1-4]. Based on the study of structural relaxation in multicomponent metallic glasses, there is a distinction between the topological short range order (TSRO) and the chemical short range order (CSRO) [5]. The change of TSRO and CSRO of metallic glasses is due to annealing or ageing. X-ray and electron diffraction measurements have been used to gain insight of this concept on the atomic diffusion processes occurring either during the isothermal annealing or during heating toward T_g. TSRO is a slow process which begins after the completion of the fast CSRO process. From the study of structural relaxation in several Pd- and (Fe, Ni)- based metal-metaloid glasses and Zr-(Cu,Fe,Ni) binary and ternary metallic glasses, at low temperatures structural relaxation occurs by local and medium range rearrangement of the atoms with weak bonding. In contrast, at high temperatures it occurs by the long range cooperative regrouping of metal and metalloid atoms [6].

Johari and coworkers [7] have studied several Ni-based metallic glasses and organic and inorganic glasses. They found that there is no difference between structural relaxation in metallic glasses and other types of glasses when the distribution of self-diffusion times is considered. They showed that structural relaxation of metallic glasses can be simulated by the same stretched-exponential relaxation function of diffusion times and non-linearity of relaxation behavior as used for polymeric and inorganic glasses. This similarity led to the conclusion that structural relaxation of metallic glasses instead involves changes in both TSRO and CSRO. They concluded that the origin of the relaxation is seen in the time- and temperature-dependent regain of the equilibrium state. Those local groups of atoms, whose diffusion is fast enough to allow loss of enthalpy and entropy on their approach to an equilibrium configurational state of low energy on annealing, absorb heat to reach their new equilibrium configuration state of higher energy at a higher temperature on heating at a certain rate. Hence each mode of atomic diffusion in the structure has its own 'mini glass-transition temperature', and reversible relaxation is a reflection of a broad distribution of times arising from temporal and spatial variations in the atoms' environment.

The study described here is intended to further examine the various conclusions reached in earlier studies with both technological and academic viewpoint.

2. Materials and Methods

The bulk metallic glass (BMG) of composition $Zr_{65}Cu_{17.5}Ni_{10}Al_{7.5}$ were prepared by (i) arc melting of an accurately weighed mixture of pure elements under a titanium-gettered Argon atmosphere and (ii) transferring the melt to a water-cooled copper crucible. To homogenize the melt, the glassy solid was melted two to four times. Plate-like specimens of thickness 1 - 2 mm, width 5 - 8 mm, and length of 50 -100 mm and rod specimens of diameter 3 - 4 mm and length of 50 - 100 mm were produced by suction casting in a copper mold. Their glassy nature were characterized by diffraction methods and calorimetry.

For the DSC measurement, the metallic glass samples were cut from the specimens using diamond cutting saw with oil cooling to keep the specimens at ambient temperature. Before cutting the specimens, the surfaces of the specimens were ground using a sand paper so that the dirt and oxide formed on the surfaces were removed. After the cutting, the samples were cleaned by using ethanol. Aluminum pans were also cleaned by using ethanol. The samples contained in the Al-pans were accurately weighed in 10 to 15 mg amounts. The shape of the samples were made such that the surface of the samples has a good thermal contact with the pans and to obtain further better contacts, the samples were sealed in the pans by crimping them. During the DSC measurement, 99.9% pure argon gas was used as purge gas. The normalized DSC output divided by their mass did not differ by more than 0.5%. Thus the effect of the samples' mass on the measured values were negligible. Repeat measurements showed that the metallic glass sample or its components did not chemically react or form an alloy with the aluminum pan over the temperature range of the study.

As in a typical study of structural relaxation, the as-cast sample of $Zr_{65}Cu_{17.5}Ni_{10}Al_{7.5}$ glass was first heated at a rate of 20 K/min in the DSC from 50 °C to 500 °C. This led to its crystallization. The scans obtained are shown in Figure 1. Figure 1 shows two exotherms at high temperature corresponding to the crystallization of the sample. In Figure 1, a broad exotherm was observed due to the structural relaxation on first heating of the as-cast sample over the 50 °C range. From this scan, we obtain the onset crystallization temperature T_x, as 424 °C for heating at a rate 20 K/min, and the temperature range of the supercooled ultraviscous liquid of about 380 to 420

°C. From these scans, we obtained the T_g of 325 °C for the heating rate of 20 K/min and ΔC_p as 19 J/mol K. For the isothermal annealing experiments, the experiments were done for a fixed annealing temperature and different annealing times. The annealing temperatures are 280, 285, 290, 295, and 300 °C. The annealing times for each annealing temperature are 5, 10, 20, 40, 80, 160, and 320 minutes.

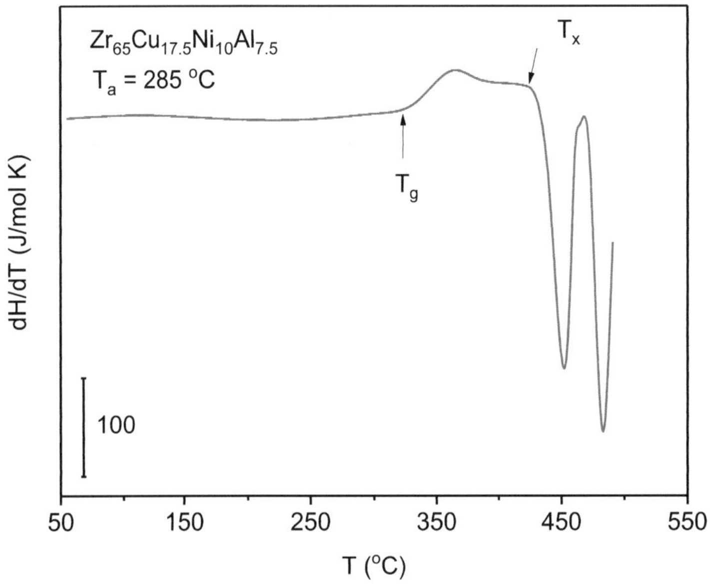

Figure 1. DSC scan for $Zr_{65}Cu_{17.5}Ni_{10}Al_{7.5}$ glass at 20 K/min rate.

In the isothermal annealing experiment, a new as-cast sample was heated from 250 °C (initial temperature, T_0) to 400 °C (upper temperature limit, T_h) at 20 K/min and cooled from 400 °C to 250 °C at 20 K/min. After this, the sample was heated and cooled again over the same temperature range and the same heating and cooling rate as for the first scan. The second heating scan was obtained for the unannealed sample. After the second cooling, the sample was heated at heating rate of 100 K/min to the annealing temperature, 280 °C, held for 5 minutes at 280 °C, and then cooled at 100 K/min to 250 °C. The sample then heated at a heating rate of 20 K/min to 400 °C and cooled at 20 K/min to 250 °C. The heating scan obtained was the heating scan after annealing at 280 °C for 5 minutes. The DSC scanning was repeated for the annealing times of 10, 20, 40, 80, 160, and 320 minutes. Experiments were done also for annealing temperatures of 285, 290, 295, and 300 K with other new as-cast samples for the same annealing times as given above.

3. Results and Discussion

An isothermally annealed or structurally relaxed glass generally shows an endothermic feature or an 'overshoot' in the DSC scan. The reversibility of this effect on thermal cycling between T_0 and T_h indicates that the underlying atomic diffusion processes are reversible in nature. The DSC scans, of the annealed and unannealed samples of $Zr_{65}Cu_{17.5}Ni_{10}Al_{7.5}$ bulk metallic glass obtained during heating at 20 K/min after annealing at 285 °C for annealing times of 5, 10, 20, 40, 80, 160, and 320 minutes are shown in Figure 2. The plot for annealed sample shows an excess endothermic heat flow or an overshoot. The dH/dt measured in the scan was multiplied by molecular weight and divided by the mass of the sample and by the heating rate, q_h, and thus converted to (dH/dT) in units of J/mol K. The difference between the annealed DSC scan and unannealed DSC scan, i.e., $[(dH/dT)_{an} - (dH/dT)_{un}]$ is plotted against T in Figure 3 for annealing temperatures of 285 °C. The plots of (dH/dT) and $[(dH/dT)_{an} - (dH/dT)_{un}]$ for other annealing temperatures are not shown in this report.

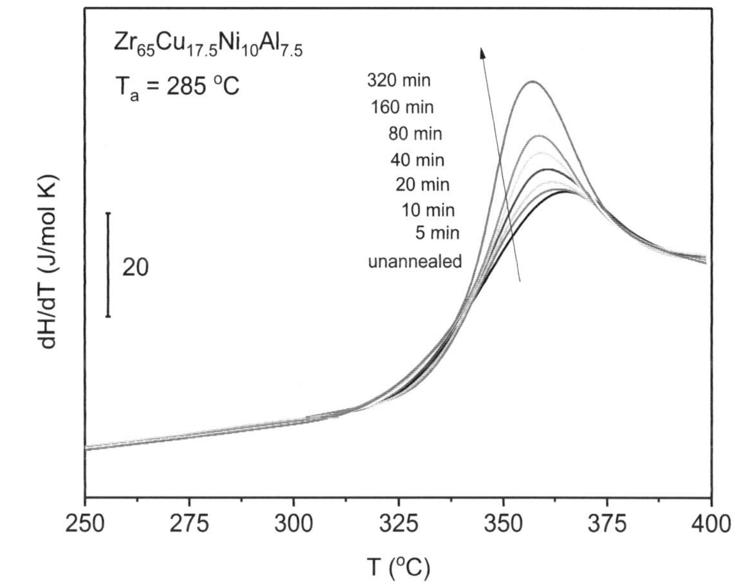

Figure 2. DSC scan for $Zr_{65}Cu_{17.5}Ni_{10}Al_{7.5}$ glass at 20 K/min rate after annealing at 285 °C for annealing times of 5, 10, 20, 40, 80, 160, and 320 min.

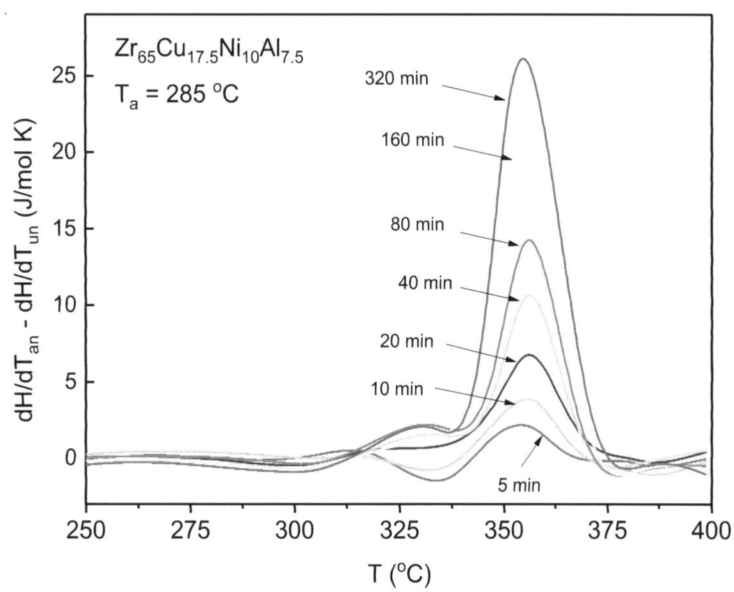

Figure 3. The plots of $(dH/dT)_{an} - (dH/dT)_{un}$ for $Zr_{65}Cu_{17.5}Ni_{10}Al_{7.5}$ glass at 20 K/min rate after annealing at 285 °C for annealing times of 5, 10, 20, 40, 80, 160, and 320 min.

In these isothermal annealing experiments, a great amount of time and effort was made to repeat the measurement for different samples so that no crystallization occurred. If the sample showed crystallization, *dH/dt* in its ultraviscous liquid state was lesser than before, and all data were rejected. In most of the experiments after annealing 10 or 20 minutes at fixed annealing temperature the samples partially crystallized.

A rapid and large magnitude of structural relaxation in metallic glasses, as found for $Zr_{65}Cu_{17.5}Ni_{10}Al_{7.5}$, is observed when an as-cast sample or a metallic glass kept for a long period at ambient temperature is heated more slowly than it was cooled from the liquid state. Thermodynamically, its state during heating approaches a structure of lower energy and entropy. Usually, in the production of a metallic glass, a high cooling rate is required to convert its liquid state to glassy state. The high cooling rate during the glass formation in a mold creates also stored-strains in the specimen across its thickness, because of the different rate of cooling in its bulk. When a sample of this specimen is heated, enthalpy, entropy as well as some of the elastic energy stored in the sample are released. These effects appear as an exothermic feature in the DSC scan, as is seen in the first heating scan in Figure 1. After the first heating of the as-cast sample from its glass to its supercooled liquid, any further relaxation that occurs on thermal cycling is referred to as reversible relaxation. In this reversible relaxation, the physical properties are recovered after subjecting the glass to a variety of thermal histories.

During the course of structural relaxation, the enthalpy and entropy of a glass decrease. The decrease in enthalpy and entropy is a combined effect of two occurrences, first, loss of the configurational enthalpy and entropy as a result of localized atomic diffusion in the disordered structure, and second, changes in the phonon frequencies and the corresponding density of states [7]. Contributions from these two effects can be seen in the exothermic and endothermic features of the DSC scan shown in Figure 2. In the figure, the features show generally how the recovery of the configurational enthalpy and entropy by local atomic diffusion and by changes in phonon behavior produce an endothermic feature when a sample, after having been annealed is heated to its ultraviscous liquid state.

Structural relaxation can be appropriately described in the terms of changes in the thermodynamic state functions without using a model. For that purpose, the decrease in enthalpy on annealing, $d\Delta H_a$, may be written as the sum of several contributions [7],

$$d\Delta H_a = \left(\frac{\partial \Delta H}{\partial T_a}\right)_{t_a, q_h, q_c} dT + \left(\frac{\partial \Delta H}{\partial t_a}\right)_{T_a, q_h, q_c} dt_a + \left(\frac{\partial \Delta H}{\partial q_h}\right)_{T_a, t_a, q_c} dq_h + \left(\frac{\partial \Delta H}{\partial q_c}\right)_{T_a, t_a, q_h} dq_c \quad (1)$$

where $d\Delta H_a$ is the difference between the enthalpy of the material before and after the annealing experiment. The prefix Δ is used to maintain that quantities determined by experiments are [*H* - *H*(0 K)] and [*S* - *S*(0 K), where *H*(0 K) and *S*(0 K) are the enthalpy and entropy of a material at 0 K. It should be noted also that *H*(0 K) decreases on annealing as the structure densifies, as does *S*(0 K) because the randomness of the zero-point energy changes.

The first term on the right hand side of equation (1) represents the enthalpy loss with respect to temperature for constant annealing time, t_a, heating rate, q_h, and cooling rate, q_c, and the second term represents this loss with respect to t_a, for constant T_a, q_h, and q_c. The third term represents the enthalpy loss over a certain temperature range with respect to the heating rate, for constant t_a, T_a, and q_c and the fourth term represents the enthalpy loss over a certain temperature range

with respect to the cooling rate, for constant t_a, T_a, and q_h. Because q_h and q_c for the experiments are kept constant, the magnitudes of the third and fourth terms are zero.
The corresponding decrease in entropy is given by [7],

$$d\Delta S_a = \left(\frac{\partial \Delta S}{\partial T_a}\right)_{t_a,q_h,q_c} dT + \left(\frac{\partial \Delta S}{\partial t_a}\right)_{T_a,q_h,q_c} dt_a + \left(\frac{\partial \Delta S}{\partial q_h}\right)_{T_a,t_a,q_c} dq_h + \left(\frac{\partial \Delta S}{\partial q_c}\right)_{T_a,t_a,q_h} dq_c \qquad (2)$$

where the terms have the same meaning as in equation (1). These two equations describe how *H* and *S* of a glass will change during annealing or during aging if the temperature fluctuates at a certain rate.

Effects of the Annealing Temperature
To investigate the effects of the annealing temperature for a constant annealing time, we first determine the first term on the right hand side of equations (1), and (2), i.e., the enthalpy and entropy loss when a vitrified sample is annealed at a selected T_as for a fixed period. The magnitude of $d\Delta H_a$ of equation (1) was determined from the area of the difference between the annealed DSC scan and unannealed DSC scan, i.e., $[(dH/dT)_{an} - (dH/dT)_{un}]$ plotted on a linear temperature scale using equation (3) shown in Figure 4 and $d\Delta S_a$ was determined from the area of the difference curve plotted on a logarithmic temperature scale using equation (4).

$$d\Delta H_a = \int_{T_1}^{T_2}\left[\left(\frac{dH}{dT}\right)_{\text{annealed}} - \left(\frac{dH}{dT}\right)_{unannealed}\right] dT \qquad (3)$$

$$d\Delta S_a = \int_{T_1}^{T_2}\left[\left(\frac{dH}{dT}\right)_{\text{annealed}} - \left(\frac{dH}{dT}\right)_{unannealed}\right] d\ln T \qquad (4)$$

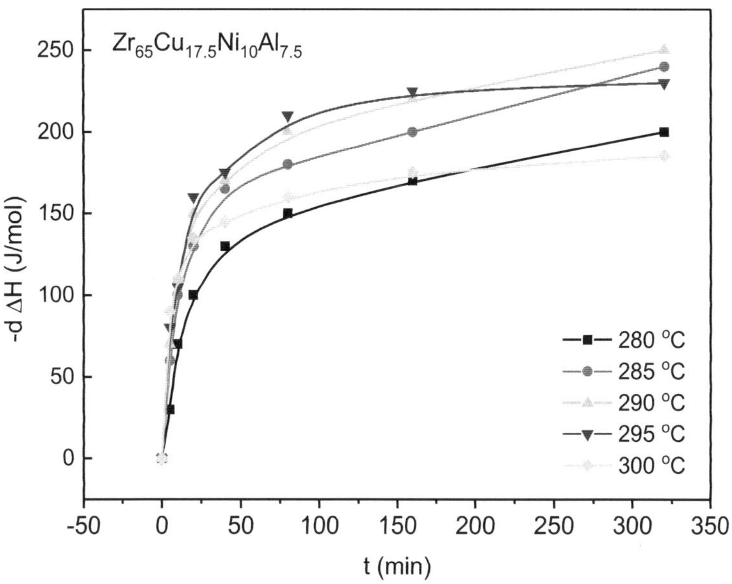

Figure 4. The enthalpy loss for $Zr_{65}Cu_{17.5}Ni_{10}Al_{7.5}$ glass after annealing for annealing times of 5, 10, 20, 40, 80, 160, and 320 min at 280, 285, 290, 295, and 300 °C.

The area under the peaks in the difference curves in Figure 3 was integrated against T with a baseline at $dH/dT = 0$ and this gave the value of the enthalpy recovered on heating the annealed sample, which should be the same as the enthalpy lost on structural relaxation during isothermal annealing at a chosen T_a and t_a, plus any other enthalpy lost on heating when the sample does not crystallize or chemically react during cooling, annealing, or heating. To determine the entropy decrease on structural relaxation, the difference $[(dH/dT)_{an} - (dH/dT)_{un}]$ can be constructed against ln(T) instead of against T. The peak area in the difference curve of such a plot yielded the entropy recovered on heating, which should be the same as the entropy decrease on structural relaxation at a chosen T_a and t_a. The enthalpy loss on annealing of $Zr_{65}Cu_{17.5}Ni_{10}Al_{7.5}$ at a fixed annealing temperature and different annealing times is plotted against time in Figure 4. Since H and S of an equilibrium liquid at a temperature substantially above its T_g is independent of the path by which it is reached from the glassy state, we can determine the enthalpy and entropy at all temperatures between T_a and $T_{(equilibrium)\ liquid}$ from the relation,

$$\Delta H(T) = \Delta H(T_a) - \int_{T_{ann}}^{T} \left(\frac{dH}{dT}\right) dT \tag{5}$$

$$\Delta S(T) = \Delta S(T_a) - \int_{T_{ann}}^{T} \left(\frac{dH}{dT}\right) d\ln T \tag{6}$$

where (dH/dT) is determined from DSC. Thus the $\Delta H(T)$ and $\Delta S(T)$ plots against T for all t_a and T_a condition would merge in the equilibrium liquid state. The values of $\Delta H(T_a)$ and $\Delta S(T_a)$ are taken from Figure 2 and the starting of the integration is from $T_{(equilibrium)\ liquid}$ of 250 °C.

Theoretical Consideration

The observations of a similarity of structural relaxation in glassy polymers and in metallic glasses [7] indicate that either the conceptual and mathematical treatments in terms of a two-level system used for metallic glasses may be useful for the glassy state in general, or that, the concept and mathematical models which are used currently for interpreting the structural relaxation in other glasses be extended to metallic glasses. It is expected that the atomic level processes in all glasses be similar, and that, during the annealing of a glass, localized diffusion leading to a spontaneous decrease in H and S of a glass is a reflection of the spontaneous change in either the overall glass structure to a lower configurational energy state, or of only those group of atoms in it that have been trapped in a deep energy minimum. In one general model, which is independent of the molecular or atomic details, developed by Perez and coworkers [8], the high energy sites in configurational space are termed defects. These defects are the local groups of atoms where potential energy correspond to a point on either the attractive side of the potential energy minimum or the repulsive side of it. There are also alternative representations referred to as "rugged energy landscape" models containing numerous minima of varying depth. The potential energy landscapes are now regarded also as time-dependent.

Variations in the atomic environment in a glass structure, in either representation, leads to a distribution of diffusion coefficients, which may be expressed as a distribution of mainly the activation energy, E, in the Arrhenius equation, $D = D_0 \exp(-E/kT)$, where D_0 is the pre-exponential term and k the Boltzmann constant. On cooling at a certain rate, those atoms which diffuse too slowly to achieve their lowest configurational enthalpy and entropy in the glass structure, become kinetically frozen. As cooling causes a progressively rapid decrease of the self-diffusion coefficient, the majority of the atoms in a glass structure becomes kinetically frozen-in just below T_g. The fraction of the atom population that ceases to contribute to a thermodynamic function during the cooling of a glass leads to further development of a kinetic instability with respect to those parts of the structure that fail to attain a thermodynamic equilibrium with decrease in kT. This appears as a gradual decrease in C_p, H and S of a glass over a temperature range extending to 0 K. On isothermal annealing, only those local configurations in which atomic diffusion is sufficiently rapid, approach an equilibrium structure within a given annealing time. On heating an annealed glass, self diffusion in the local regions becomes faster according to the same Arrhenius equation as on cooling. And C_p, H and S become equal to the sum of contributions from all states that achieve configurational equilibrium on heating plus the vibrational contributions. For each structural state that had attained a configurational equilibrium and reached its T_f, defined as the temperature at which a glass is kinetically stable with respect to its equilibrium liquid, on heating at a certain rate, the slopes of the H and S curves rapidly increase once the temperature exceeds T_f, goes through a point of inflexion and the curves meet the curves for the configurationally equilibrated state from below the equilibrium state curve.

In the usual DSC scan, when an annealed glass is heated towards its equlibrium liquid state, this recovery appears as an overshoot at $T > T_g$, before the equlibrium liquid state is reached. The area of the overshoot is a measure of the H and S lost during the annealing. Even when the majority of the atoms in a glass structure at $T < T_g$ are kinetically frozen-in, the rapid approach towards an equilibrium of a small, but increasing number of locally diffusing atoms produces a small overshoot at a temperature determined by its diffusion rate and the rate of heating. In this view, the endothermic peak observed on heating an annealed glass is a reflection of the sum of a multitude of C_p overshoots, each corresponding to a "mini glass-liquid transition" of the localized groups of atoms or molecular segments.

In previous studies on physical aging of several metallic glasses where the effect of the distribution of relaxation times was investigated in the glassy state [7,9-14], the glasses were cooled from $T < T_g$, annealed and then heated back to the same $T < T_g$. Johari and coworkers [7,9] observed small endothermic peak in the DSC scan. However, such endothermic was not observed in the DSC scan [10-14] which is similar to the result in the present work. The different findings for this thermally cycling in a T range below T_g can be explained by the competition between two processes: α-relaxation and JG relaxation. The endothermic peak observed in Refs. 7 and 9 was attributed to the unfreezing of the faster relaxation modes in the distribution of the α-relaxation process [15,16]. If the rate of the JG relaxation occurs is comparable to these faster modes, its process strength would decrease on annealing and on heating, it would be recovered [15,16]. The recovery would be observed as an overshoot at $T > T_g$ and attributed to the α-relaxation. On contrary, when the rate of the JG relaxation is close to that of the α-relaxation, recovery of the JG relaxation configurations would be accompanied by the recovery of the α-relaxation contributions. Therefore the JG relaxation in a DSC scan can be observed only if its relaxation rate is much faster than that of the α-relaxation and its magnitude relatively large.

Modeling

To examine whether or not the reversible relaxation in metallic glasses may be attributed to a distribution of relaxation time in the same manner as for other glasses, we qualitatively simulate its feature by using the TNM equations for the structural relaxation of organic and inorganic polymers as done by Hodge [17]. As mentioned earlier, the main observations of this study are: (i) the broad temperature range of structural relaxation, and (ii) the appearance of an endothermic feature on annealing. The kinetics of atomic and segmental motion involves a broad distribution of relaxation times which can be represented in two ways: (i) a set of a separate relaxation mechanisms for the various relaxation time components of $g(\tau)$ which represents the distribution of relaxation times, and (ii) an empirical nonexponential relaxation function.

The relaxation function for the irreversible change in the structure is written in terms of a distribution of relaxation times. This distribution of relaxation time is written in the form of a relaxation function,

$$\phi(t) = \exp\left[-\left(\frac{t}{\tau}\right)^{\beta}\right] \tag{7}$$

where $\phi(t) = \dfrac{H(t) - H(\infty)}{H(0) - H(\infty)}$, ($H(t)$ is the value of enthalpy at time t, $H(\infty)$ is the value of the enthalpy at equilibrium (infinite time), $H(0)$ is the value of the enthalpy at initial time (the value at $t = 0$)), τ is characteristic relaxation time, β is a parameter that expresses the breadth of the distribution of relaxation time ($0 < \beta \le 1$). A single relaxation time is denoted by $\beta = 1$. Equation (2.13) is equivalent to a set of separate relaxation mechanisms for the various relaxation time components of $g(\tau)$ which represents the distribution of times:

$$\phi(t) = \int_0^{\infty} g(\tau)\exp\left(-\frac{t}{\tau}\right)d\tau \tag{8}$$

for $\int_0^{\infty} g(\tau)d\tau = 1$, and,

$$\tau(T, T_f) = A\exp\left[\left(\frac{x\Delta h^*}{RT}\right) + \left(\frac{(1-x)\Delta h^*)}{RT_f}\right)\right] \tag{9}$$

where $\phi(t)$ is the relaxation function ,Δh^* is the activation energy, T_f is fictive temperature, A is a parameter equal to τ when both T and T_f are formally infinity, and $x < 1$ is an empirical parameter referred to as the nonlinearity parameter. For $x = 1$, equation (9) becomes the Arrhenius equation.

The DSC scans of our experimental data were qualitatively simulated by using the parameters: ln A = -76, x = 0.29, β = 0.69 and Δh^* = 408,000 J/mol. The simulated curves are shown in Figure 5 for the annealed sample at 285 °C for annealing times 5, 10, 20, 40, 80, 160, and 320 minutes. In the simulation, a glass was cooled at 100 K/min from 407 °C to 250 °C and heated to 285 °C at 100 K/min then held at 285 °C for a certain annealing time. The heating curve of annealed glass thus could be obtained. These parameters should describe the curves for all the samples, whether annealed or unannealed.

It is evident in Figure 5 that a good fit is obtained for the sample after annealing for 5 minutes, and somewhat satisfactory fit is obtained for the sample annealed for 10 minutes. But the fit is quite unsatisfactory for the samples that were annealed other annealing times. The discrepancy between the simulated results and the experimental data for the metallic glasses seems to support the premise that only for a few cases structural relaxation of a metallic glass may be described in terms of a distribution of relaxation times. The above-given comparison of simulation of the DSC scans for metallic glasses, even though they fail to predict the structural relaxation or annealing over a long time period, shows that the structural relaxation observed in metallic glasses, whether irreversible or reversible, has a broad distribution of atomic diffusion time. Formalisms used for glasses in general can be used to describe the structural relaxation in metallic glasses.

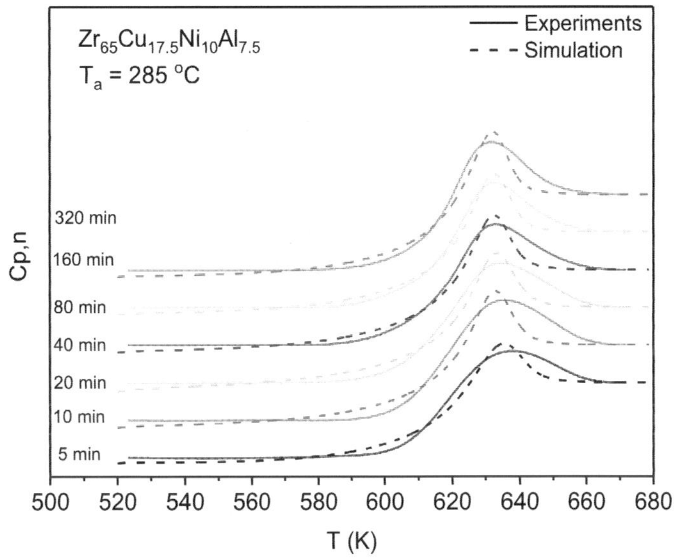

Figure 5. TNM simulations of the DSC scans for $Zr_{65}Cu_{17.5}Ni_{10}Al_{7.5}$ glass after annealing at 285 °C for annealing times of 5,10, 20, 40, 80, 160, and 320 minutes.

The simulations also have been performed previously for several Zr-based metallic glasses [10-14]. The simulation results were similar to our finding where results indicated that the glasses had a broad glass transition region. According to Hammond et. al. [11], Wunderlich's model may be used to explain that structural relaxation in these glasses involved the cooperative motion of a single bead. Similar to previous studies of other Zr-based metallic glasses [10-14], relaxation behavior was well described by the TNM model. Qiao et. al. [14] further elaborated that the values of β and relaxation times were sensitive to the composition of the metallic glasses. The enthalpy relaxation that we observed in the present work were sensitive to both annealing time and temperature, which is in a good agreement with the results relative to other metallic glasses [11,14].

4. Conclusions

The structural relaxation studies of metallic glasses presented in this section leads to the following conclusions:

1. There are two types of relaxation observed in a metallic glass. The first relaxation is irreversible relaxation and the second is reversible relaxation. The irreversible relaxation is observed as an exothermic feature during the first heating of the as-cast sample. The exotherm is due to the very large enthalpy frozen-in during the high cooling rate in the preparation of the metallic glass and its release during the slower heating rate of the first heating in the DSC.
2. Isothermally annealing of a sample at a certain temperature and a certain time produces also endothermic effect during the sample's heating, which are the reversible relaxations. The

peaks area is larger than that for the unannealed sample. For a fixed annealing temperature but different annealing times, the peaks area continuously increases as annealing time increases. These features correspond to the enthalpy and entropy loss during annealing.

3. Simulation of the heat capacity curves obtained from the DSC scans in terms of a nonexponential and nonlinear relaxation concept for metallic glasses fit satisfactorily for some isothermal annealing experiment results but do not do so for other experimental results.

5. Acknowledgements

The authors are grateful for the grant in this research from the Faculty of Industrial Technology, Universitas Trisakti.

6. References

[1] Inoue, A., Kato, A., Zhang, T., Kim, S.G., and Masumoto, T., *Mater. Trans. JIM* **32** (1991) 609.

[2] J.W.P. Schmelzer, V.M. Fokin, A.S. Abyzov, E.D. Zanotto, I. Gutzow, *Int. J. Appl. Glass Sci.* **1**, 16 (2010).

[3] E. Donth, *The Glass Transition: Relaxation Dynamics in Liquids and Disordered Materials* (Springer, Berlin, 2001).

[4] Aji, D.P.B. and Johari, G.P., *Enthalpy and entropy change on structural relaxation of* $Mg_{65}Cu_{25}Tb_{10}$ *glass, Thermochimica Acta* 503-504 (2010) 121-131.

[5] Balanzat, E., Stanley, J.T., Marry, C., and Hillairet, J., *Acta Metall.* **33** (1985) 785.

[6] Inoue, A., Masumoto, T., and Chen, H.S., *J. Mater. Sci.* **20** (1985) 4057.

[7] Johari, G.P. and Shim, J.G., *J. Non-Cryst. Solids* **261** (2000) 52.

[8] Johari, G.P. and Perez, J., *Mol. Phys.* **83** (1994) 235.

[9] Shim, J.G. and Johari, G.P., *Philos. Mag. B* **79** (1999) 565.

[10] Zappel, B. and Sommer, F., *Mater. Sc. and Eng.* **A179/A180** (1994) 283.

[11] Hammond, V.H., Houtz, M.D., and O'Reilly, J.M., *J. Non-Cryst. Solids* **325** (2003) 179.

[12] Wang, X.D., Jiang, J.Z., and Yi, S., *J. Non-Cryst. Solids* **353** (2007) 4157.

[13] Evenson, Z. and Busch, R., *J. Alloys and Comp.* **509S** (2011) S38.

[14] Qiao, J.C. and Pelletier, J.M., *Intermetallics* **19** (2011) 9.

[15] Johari, G.P. and Goldstein, M., *J. Chem. Phys.* **53** (1970) 2372.

[16] Casalini, R. and Roland, C.M, *Phys. Rev. Lett.* **102** (2009) 035701.

[17] Hodge, I.M., *J. Non-Cryst. Solids* **169** (1994) 211.

Effect of Sodium Nitrate-Sodium Fluoride Ratio as Degasser in Al-7Si-2Cu Casting Product

Donanta Dhaneswara, Jaka Fajar Fatriansyah, Naila Hanandhira[1]*, Agatha Rona Basa, M. Reza Firmansyah.
[1]Departement of Metallurgical and Materials Engineering, Faculty of Engineering, Universitas Indonesia, Kampus UI, Depok, Indonesia 16424

*naila.hanandhira@ui.ac.id

Abstract. Effect of Sodium Nitrate-Sodium Fluoride Ratio as degasser in Al-7Si-2Cu Casting product has been investigated. Degassing is one of the methods used in the casting process to remove gases, such as hydrogen gases, in the molten metal. The most commonly used degassing method is by injecting an inert gas such as argon. In this experiment, a conventional degassing method with degasser-based sodium nitrate-sodium fluoride was used with changes in sodium nitrate to sodium fluoride ratio variables are 3:5, 4:4, and 5:3. The type of material used is Al-7Si-2Cu material with additional scrap. The material melted first at 800°C, then the degasser was added into the furnace and held for 3 minutes. High temperature was used to melt the material due to the solubility of hydrogen gases in liquid metal at high temperature is high. The molten metal then poured into the dies at approximately 690°C. The casting process results are then prepared for mechanical testing, such as tensile test, impact test, and hardness test, and microstructure testing. The results show that at the ratio of 1:1.5 (normal ratio), the porosity was lower (0.8 at average) and the mechanical strength was higher (1.2 at average).

Keywords: Casting. Degassing. Porosity. Sodium.

1. Introduction

Aluminum is one of the most used material because of it's light weight and high ductility properties made it compatible to use in casting product. Unfortunately, pure aluminum has a low strength and some low mechanical properties, therefore it needs to be alloyed with other elements such as silicon and copper. The addition of alloying elements helps to increase its mechanical properties and adjust to the requirement needed on the final product. For example, addition of Si will improve the cast ability of the materials, lessen the possibility of hot tearing, lowering the thermal expansion coefficient, and improving it's wear resistance[1]. Meanwhile, addition of Cu increases the strength, corrosion resistance, creep resistance, weld ability, machinability, and also make it possible to undergo a heat treatment process[2]. These alloys are commonly use in high speed rotating parts, pistons, structural aerospace components, fuel pumps, and other machine parts[1,3].

In order to produce a sound casting product using aluminum alloy, there are factors that needed to be considered such as melting temperature, pouring temperature, and turbulence during pouring. If one of those factors is neglected, it might cause some defects such as porosity in the final product. Porosity itself usually formed during solidification process in which the hydrogen gas was soluble and trapped within the molten metal. There are multiple studies which shows, that the presence of gas porosity in a casting product could significantly lower its mechanical properties such as impact toughness, elongation, and tensile strength[4,5].

To lower the amount of hydrogen gas soluble in the molten metal, there are several methods the industry can use. One of which commonly use method are fluxing and degassing method. Some research has been done using the fluxing method, especially using cover flux with NaCl and KCl as its main composition[6]. As to degassing method itself, there is a cheaper yet still compatible method which is using tablet degasser, mostly containing C_2Cl_6. In this study, the tablet degasser used is mixed degasser based sodium nitrate ($NaNO_3$) and sodium fluoride (NaF) The expected result is by using this degasser, it will decrease the amount of porosity by binding the H_2 gas out. In addition, the ideal composition of the degasser will also be studied by changing the ratio amount of $NaNO_3$ and NaF used.

2. Materials and Methods

2.1 Making of $NaNO_3$ and NaF Based Degasser

The mixed degasser used is containing 18%wt $NaNO_3$ in powder forms, 40%wt NaF, and the other 42%wt consists of NaCl, NH_4Cl(KCl), Na_2SO_4, and dye. All of them were then mixed with some water using a mixer machine. The stirring process then conducted for approximately 30 minutes to help homogenized the mixture. After that, the final mixture then processed and formed into the desirable shape, in this case it's a cylindrical tablet shape, using a printing machine. Lastly, the tablet was put into the oven at 100°C for 1 hour to evaporate all of the moistures inside.

Some changes were made in order to conduct this study, those are the ratio amount of $NaNO_3$ and NaF used. The ratio of sodium nitrate to sodium fluoride used are 3:5, 4:4, and 5:3.

2.2 Characterization of Al-7Si-2Cu

By using Optical Emission Spectroscopy (OES) material testing, the chemical composition of Al-7Si-2Cu were obtained and was shown in Table 1. It could be seen that the main component is Al with majority alloying element of Si and Cu.

Table 1. Chemical Composition of Al-7Si-2Cu

Al	Si	Fe	Cu	Mn	Mg	Zn	Cr
89.1	7.06	0.350	2.30	0.178	0.246	0.495	0.0220

The casting procedure started with melting the materials with additional scrap using an electric furnace at 800°C until its melted completely. While the materials were melting, the mold was coated with zircon and thinner mixture then preheated at 400°C for 30 minutes. The degasser itself also was preheated using a muffle furnace at 100°C for 1 hour to make sure there was no moisture left. After all the aluminum alloy melt completely, the preheated degasser then dipped into the furnace using a coated plunger. It needed approximately 3 minutes to do the degassing process until the degasser was completely dissolved. Next, the degassed molten metal was poured into the metal mold at 690°C and into the porosity testing cup. The schematic casting design can be seen in Fig. 1. The poured metal then underwent a solidification process, while the one in the porosity testing cup was put into a vacuum condition at a constant pressure of 100mmHg for 315 seconds.

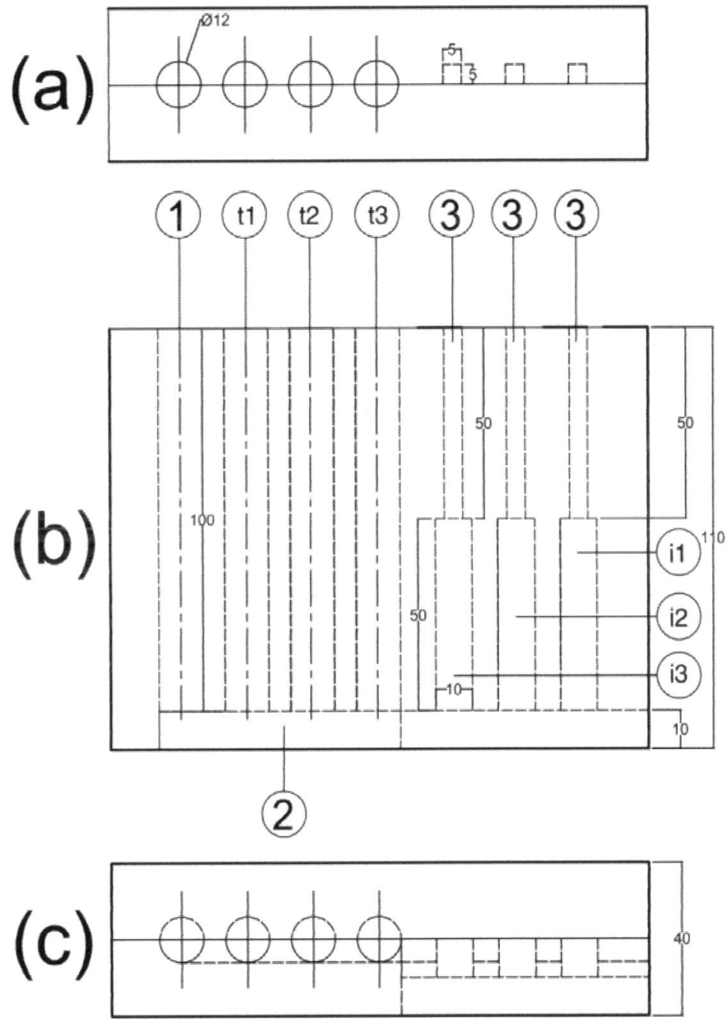

Fig. 1. Schematic of casting design. (a) Top View ; (b) Front View ; (c) Bottom View ; (1) Down Sprue ; (2) Runner ; (3) Gas Tunnel ; (t1, t2, t3, t4) Tensile Specimens ; (i1, i2, i3) Impact Specimens.

The mechanical testing specimens were made using a metal mold design shown in Figure 1. The tensile specimens were obtained from the t1, t2, and t3, while the impact specimens were obtained form i1, i2, and i3. On the mold itself, the down sprue was placed on the far left side with the diameter of 12mm which can be seen as number 1, is where the molten metal was poured. The runner which is shown as number 2, help the molten metal flow and fill in all the specimen's casting. Meanwhile the gas tunnels were placed on the right top side in order to remove gasses from within the metal mold, can be seen as number 3. After the casting process was finished, the specimen then prepared and observed to help to determine the mechanical properties of the casting products.

Tensile test, impact test, hardness test, and microstructure test were then conducted to help characterize the mechanical properties of Al-7Si-2Cu as cast. In order to do the tensile test, GOTECH AI-7000 LA 10 machine was used with the JIS Z2241 standard. The impact test was conducted using frank impact testing machine, with the charpy method. For the hardness test, LCB3100 LECO brinnel machine with 10mm diameter of steel balls indenter was used with the force parameter of 1000 kgf.

For the porosity testing sample, after they were solidified in a vacuum condition, they were quenched in water to fasten the solidification of the remaining liquid metal in the center. The sample then cut in half, prepared, and observed to investigate the microstructure and the amount of porosity left in the solid casting products.

3. Results and Discussion

3.1 Mechanical Properties of Casting Al-7Si-2Cu with Degasser

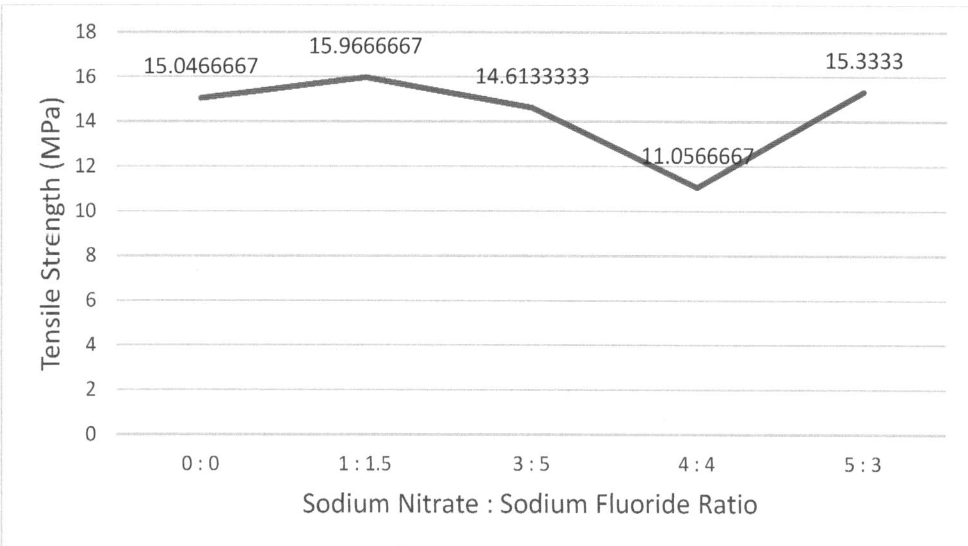

Fig. 2. Effect of Ratio $NaNO_3$ and NaF degasser on Tensile Strength of Al-7Si-2Cu

Fig. 2. above shows the tensile test result according to the amount of degasser use in the casting process. The ratio of $NaNO_3$ and NaF also has an impact towards the tensile strength of Al-7Si-2Cu as cast. From the figure above, it could be seen that by using a 1:1.5 ratio of $NaNO_3$ to NaF, the highest tensile strength was obtained. Meanwhile, by using a 4:4 ratio, it resulted in the lowest tensile strength, even lower than the sample without using any degasser. Therefore, for the tensile test result, the degasser of 1:1.5 ratio works the best on removing the hydrogen gasses and increasing the tensile strength.

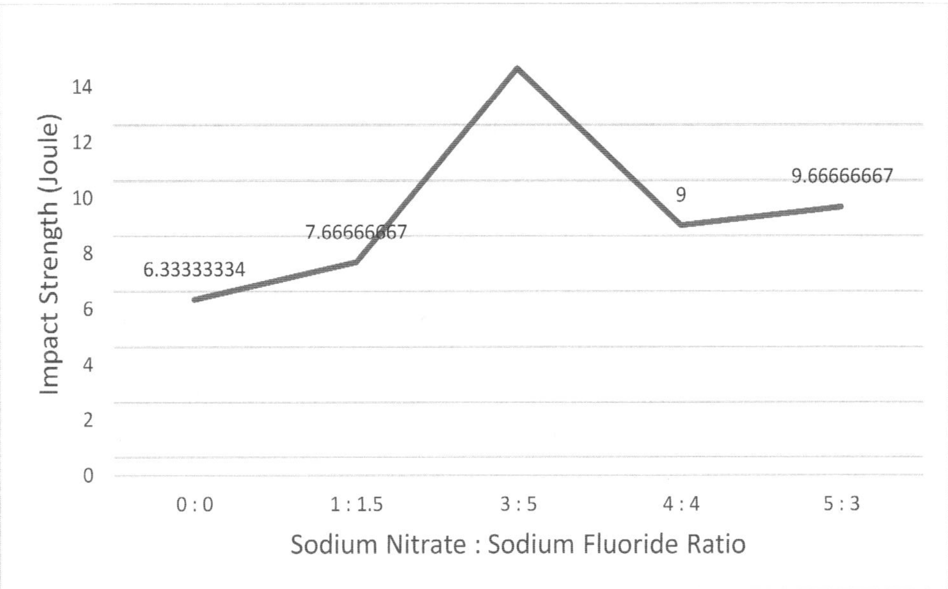

Fig. 3. Effect of Ratio $NaNO_3$ and NaF degasser on Impact Strength of Al-7Si-2Cu

Figure 3 represents the effect of $NaNO_3$ to NaF ratio on the impact strength of Al-7Si-2Cu as cast. The result is quite different from the tensile test result, because in here it was found that by using a 3:5 ratio of $NaNO_3$ to NaF, it has the highest impact strength. The lowest impact strength was obtained from the sample without using any degasser. It is true, since by not using any degasser at all, the amount of hydrogen gas soluble in the molten metal must be in a high number, therefore it's decreasing the ability of the sample to withstand any sudden load.

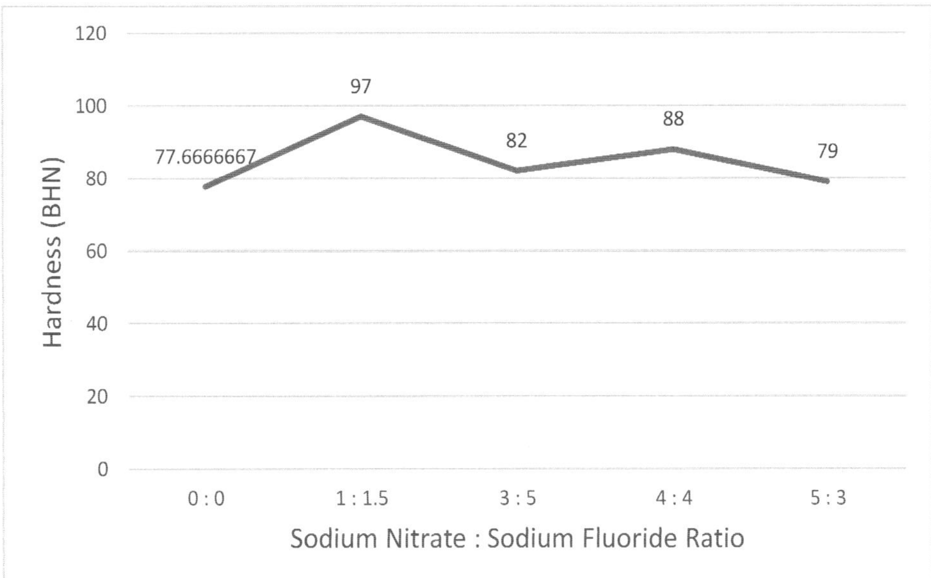

Fig. 4. Effect of Ratio $NaNO_3$ and NaF degasser on Hardness of Al-7Si-2Cu

Figure 4 above shows the hardness test results of the Al-7Si-2Cu as cast without degasser and by using degasser. The highest hardness was obtained from a sample using a 1:1.5 $NaNO_3$ to NaF ratio with 97 BHN, it is the same as the tensile test result. The lowest hardness end result was obtained from the sample without using any degasser. In this case, without any help of degasser, the more the hydrogen gas are soluble within the molten metal, therefore the porosity is higher. A high amount of porosity could decrease the sample's mechanical properties.

3.2 Microstructural Observation and Phase Identification of Al-7Si-2Cu

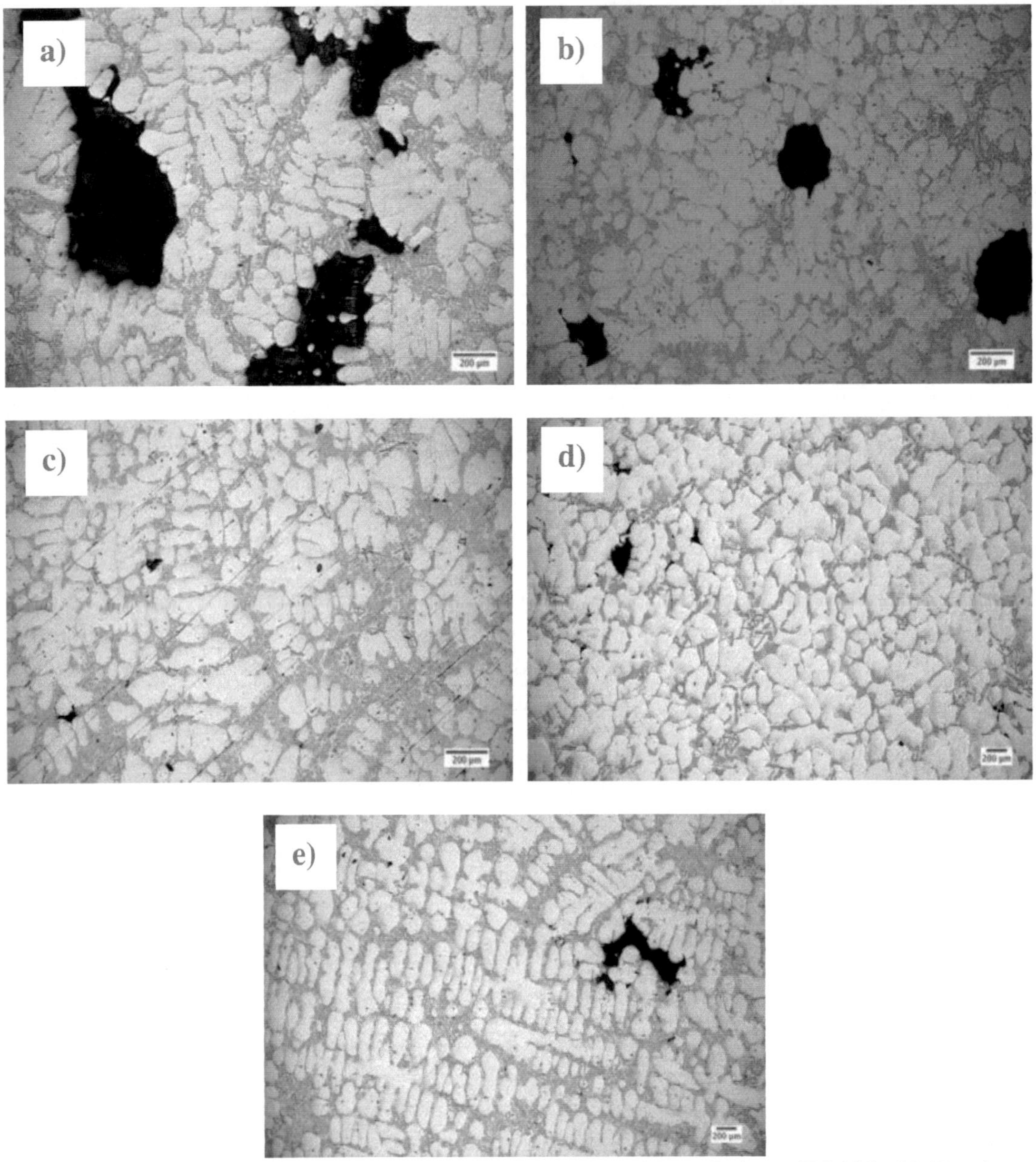

Fig 5. Optical micrograph of as-cast alloy Al-7Si-2Cu with variation of $NaNO_3$:NaF ratio: (a) 0:0 ; (b) 1:1.5 ; (c) 3:5 ; (d) 4:4 ; and (e) 5:3.

It is already known that the mechanical properties of aluminum casting alloys depend on microstructural characteristics such as the grain size, dendritic cell size, and the size and morphology of the β-Al_5FeSi phase, as well as on the pores and inclusions[4]. Figure 5 represents the optical micrograph results for all the ratio. It could be seen from above there are some obvious black area which is the gas porosity trapped within the microstructure. Predominant phase of α-Aluminum in a dendritic form is shown by the white area. In this particular as-cast, the orientation and morphology of α-Aluminum is very disorderly. Accurate identification of the relatively coarse iron rich intermetallic phases commonly found in aluminum–silicon casting alloys is important, because some of these phases are associated with reduced mechanical properties[7]. In a very low concentration, Cu is soluble in α(Al) and is a major constituent in the intermetallic phase $CuAl_2$ (θ-phase). After the precipitation of $CuAl_2$ at α-grain boundaries in Al-Si-xCu alloy, the microstructure becomes brittle and fails intergranularly[8] and this was proven by the fractograph of the impact test results.

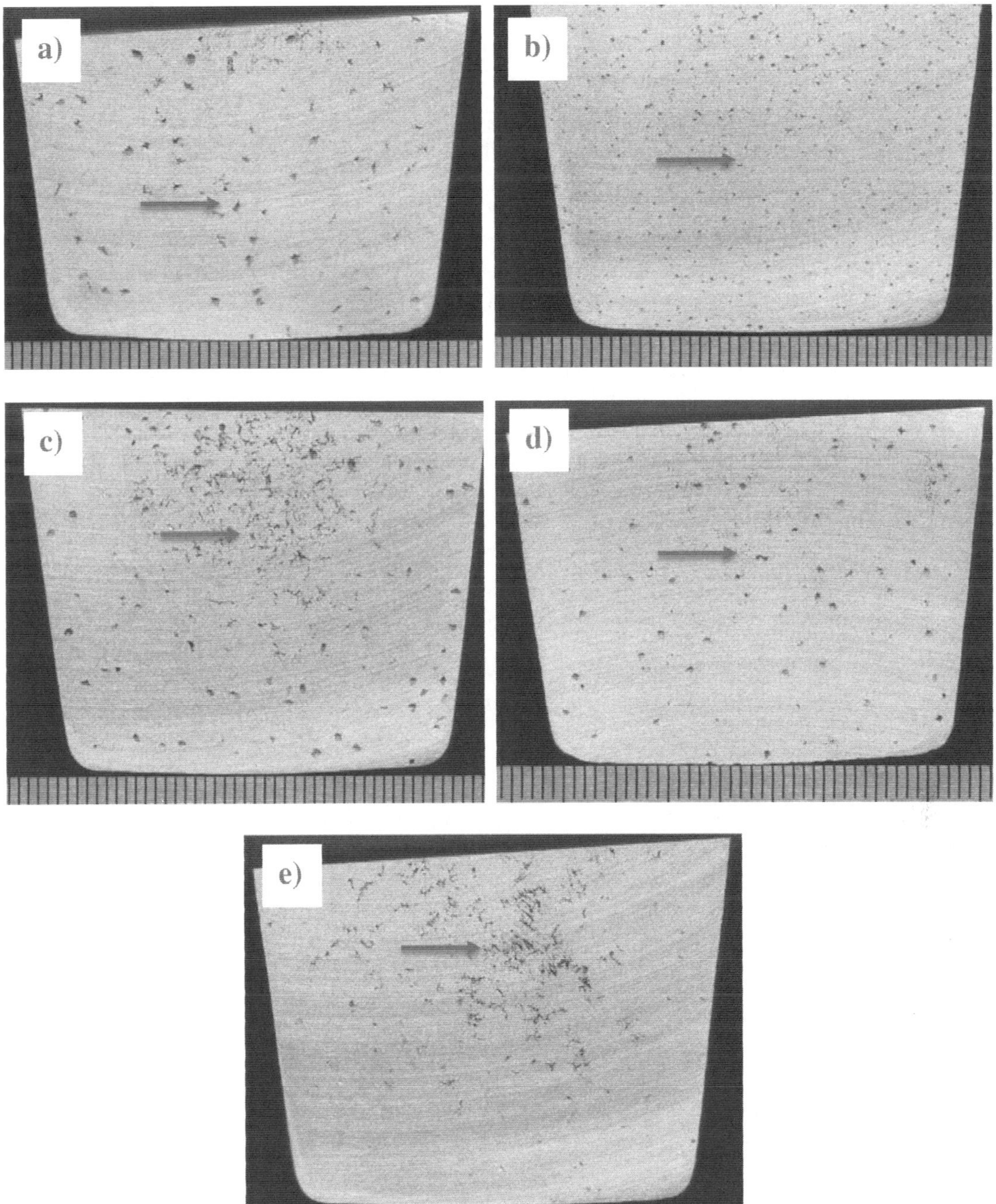

Fig 6. Macrostructure of as-cast alloy Al-7Si-2Cu with variation of $NaNO_3$:NaF ratio: (a) 0:0 ; (b) 1:1.5 ; (c) 3:5 ; (d) 4:4 ; and (e) 5:3.

Figure 6 shows the macrostructure appearance of Al-7Si-2Cu with all ratio. It could be seen that every cast result has small porosities from the gas entrapment, but the 1:1.5 ratio shows finer and more dispersed porosities. This result corresponds with the mechanical properties result that 1:1.5 ratio also has the highest value on tensile strength and hardness properties. By resulting in finer and more dispersed porosities, this means that the mixed degasser works best in 1:1.5 ratio, that this is the effective ratio for the degasser to binds the H_2 gas out.

3.3 Effect of Sodium Nitrate and Sodium Fluoride Addition

The simplest method to remove dissolved hydrogen is to hold the metal for some time, allowing for some degassing. For a faster degassing process, it can be achieved by gas purging, the application of a vacuum, tableted flux degas- sing, or mechanical stirring[9].The mixed degasser used in this study was mainly based on $NaNO_3$ and NaF. By equation (i) and (ii), it is believed that each of them has an effect regarding to the amount of soluble hydrogen gas.

$$2\,NaNO_3 + H_2 \rightarrow 2\,Na + 2\,HNO_3 \quad \text{(i)}$$
$$2\,NaF + H_2 \rightarrow 2\,Na + 2\,HF \quad \text{(ii)}$$

Within the degasser used itself, all of the compound will later on produce nitrogen gas when it got submerged into the molten aluminum. Those nitrogen gasses then bound with the H_2 to form bubbles which resurfaced on the molten metal. These compound also has the ability to remove impurities within the molten metal and carry it to the surface which will make a disposal dross.

Sodium fluoride is actually a common ingredient to use, which formed a ternary eutectic with KCl and NaCl with a melting point of 607°C. A low melting point was needed in order to improve the fluidity[8]. Addition of NaF is studied to be increasing the hydrogen dissolution into the slag formed on the surface[10]. $NaNO_3$ and NaF are both salt fluxes that has a function to help wet the oxide inclusion. This resulted in enabling the removal of hydrogen that are trapped within the molten aluminum. By removing the H_2 gas, it is also reducing the amount of porosity that might get trapped during solidification. Thus, it could be resulting in a more sound casting products.

4. Conclusions

The use of a 1:1.5 ratio of $NaNO_3$:NaF degasser improved mechanical properties of Al-7Si-2Cu the best, due to the least amount of gas porosity based on the microstructure test results. Varying the amount of $NaNO_3$ and NaF has their own impact on the mechanical properties such as a higher impact strength but with a lower hardness, therefore it did not provide the best mechanical properties.

5. Acknowledgements

The authors gratefully acknowledge financial support provided by Hibah PITTA with contract No: 2421/UN2.R3.1/HKP.05.00/2018.

6. References

[1] Zamani, M. (2015). *Al-Si Cast Alloys-Microstructure and Mechanical Properties at Ambient and Elevated Temperature*
[2] Talamantes-Silva, M. A., Rodríguez, A., Talamantes-Silva, J., Valtierra, S., & Colás, R. (2008). Characterization of an Al-Cu cast alloy. *Materials Characterization*, *59*(10), 1434–1439.
[3] Robert S.G. (2000). High Performance Alloys – Trans Tech Publications LTD,

USA.

[4] Mugica, G. W., Tovio, D. O., Cuyas, J. C., & González, A. C. (2004). Effect of Porosity on the Tensile Properties of Low Ductility Aluminum Alloys. *Materials Research*, *7*(2), 221–229.

[5] Moscovitch, N., Gertsberg, G., Nagar, N., Lautzker, M., Fantetti, N., & Bronfin, B. (2006). The effect of porosity on the microstructure and mechanical properties of die cast Mg alloys, *2006*(724), 1–6.

[6] Shih, T.-S., & Wen, K.-Y. (2005). Effects of Degassing and Fluxing on the Quality of Al-7%Si and A356.2 Alloys. *Materials Transactions*, *46*(2), 263–271. http://doi.org/10.2320/matertrans.46.263

[7] Kral, M. V., McIntyre, H. R., & Smillie, M. J. (2004). Identification of intermetallic phases in a eutectic Al-Si casting alloy using electron backscatter diffraction pattern analysis. *Scripta Materialia*, *51*(3), 215–219.

[8] Zeren, M., Karakulak, E., & GÜmÜ, S. (2011). Influence of Cu addition on microstructure and hardness of near-eutectic Al-Si-xCu-alloys. *Transactions of Nonferrous Metals Society of China (English Edition)*, *21*(8), 1698–1702.

[9] Utigard, T. A., Friesen, K., Roy, R. R., Lim, J., Silny, A., & Dupuis, C. (1998). The properties and uses of fluxes in molten aluminum processing. *Jom*, *50*(11), 38–43.

[10] Chung, S. H., Park, J., Park, S., Chang, W., & Sohn, I. (2013). Effect of NaF and CaO / SiO 2 on the Hydrogen Dissolution in the CaO – SiO 2 – FeO t Based Flux System. *ISIJ International*, *53*(1), 27–33.

Study of Si Surface Adsorption Towards Hydrogen Molecule

D Dhaneswara[1] J Fajar Fatriansyah[1] M Bachtiar Yusuf[1]* M Hanif Abdurrahman[1] F Riski Kuskendrianto[1]
[1]Metallurgy and Materials Engineering Department, Faculty of Engineering, Universitas Indonesia

*Corresponding author (E-mail): muhammad.bachtiar@ui.ac.id

Abstract. Recent approaches to address more efficient method on storing hydrogen by using adsorbent materials has been done. The hydrogen adsorption on silicon has been studied through molecular dynamics simulations and experiment by researchers. We conducted molecular dynamics simulation using a Lennard-Jones potential to demonstrate the hydrogen adsorption capability of silicon surface (001) and (111) with various temperature applied. The amount of hydrogen adsorbed by silicon surfaces are higher as entropy of the system decreases. Without considering entropy, Si (111) has higher adsorption capability due to its lower energy surface than Si (001).

Keywords: hydrogen storage; silicon; molecular dynamics simulation; Lennard-Jones potential; adsorption

1. Introduction

The availability of energy is decreasing as time goes by, transcendence of technology and human growth, it causes a dependency in energy sources. The energy that is a main source for supplying daily needs came from the fossil fuel. One of the energy sources that mostly used for life support is the energy from fossil fuel. Fuel energy has a big role to run various of daily needs such as industries and transportation. By that time, an alternative fuel is needed for the future energy demand when those fossil fuels become unavailable, more importantly for the transport sector. Among hydrogen appears to be the best choice due to the highest energy density per unit mass (120 MJ/kg), non-toxic molecule, colorless no environmental implications, so it is suitable to use as a fuel for motor vehicles. Even so, the using of hydrogen will be facing a problem in distribution and storing process due to its volatility [1].

To face the problem in the using of hydrogen, adsorption is the best method. Using that method, hydrogen will be saved in a porous material. Basically, gas adsorption happens because van der Waals's force between gas molecules and adsorbent molecules that causes gas molecules trapped in a pore structure. One of the porous materials that can be used is using adsorption method in silicon. Silicon has a large surface and a big volume of pores, so it is suitable for hydrogen's storage media [1-2]. The advantage of using the technique of hydrogen storage using adsorption method is that it is more secure and cheaper in the application [2].

Adsorption of gases on the silicon surface has been studied. Adsorption properties of different types of gases in silicon studied by measuring pressure to identify decomposition products of the adsorbed gas. For the adsorption of hydrogen on silicon is quite a lot of study about it such as the terms of mechanisms, factors that affect, and the nature of adsorption. The H_2 desorption mechanism and H_2 interactions on the Si surface have been studied in. Factors that affect the adsorption of hydrogen on silicon have been widely explored. Temperature and pressure factors have been studied. The hydrogen atoms will adsorb prudentially on adatoms at low temperatures whereas the rest atoms are the first to become hydrogen terminated at elevated temperatures [3-4]. In another research, the amount of adsorption increases with the decreasing of the temperature and increasing of the pressure [3]. The effects of kinetic energy, dynamic lattice distortions, static lattice distortions, molecular vibrations, surface coverage and the angle of incidence of H_2 have been studied in [5].

The location and geometry of Si also affect this adsorption process. The properties of the respective lattice distortion is clear for Si (111)7×7. In this system to make lower the adsorption barrier, the backbonds of the silicon adatoms must be widely stretched in the transition state [6]. In another research the adsorption geometry changed from phase 2 x 1 monohydride to phase 1 x 1 dihydride with increased exposure to hydrogen [7]. From the difference in hydrogen chemisorption energy between the adsorption area on the terraces and at the steps of a vicinal Si (001) surface the equilibrium hydrogen position of the various surface sites can be derived [7-8]. Another factor like saturation of the atomic bonds between silicon atoms on this surface by hydrogen have been studied in [6]. In other paper a minor influence in adsorption is tunneling, molecular vibrations, and the structural details of surface. In enhancing the adsorption quantity, the structure of the silicon is modified as in nano-size. The combined ab-initio quantum mechanical calculations and Canonical Monte Carlo (CMC) simulations were developed to investigate the ability of hydrogen adsorption on silicon nanotubes. Studies on the configuration or structure of nanotubes of silicon and the structural and electronic properties of these nanostructures were also developed [8].

The hydrogen adsorption to the silicon surface that involved potential interaction between H_2-H_2 and Si-H_2 was considered as simple physical adsorption. Relatively low attractive force between adsorbate with adsorbent surfaces induced physical adsorption, so that adsorbate moved to another part of adsorbent surface [9]. In this study, we conducted molecular dynamics simulation towards the hydrogen interaction with Si (111) and Si (001) by Lennard-Jones function. The (001) and (111) Si surfaces are preferred due to its relatively feasible preparation with good quality characterization [10]. A research done by a simulation of molecular dynamics using a Lennard-Jones potential to demonstrate the hydrogen molecule adsorption capability of silicon surface with various temperature applied. The scope of the discussion in the present review is limited to recent advances in hydrogen storage using silicon adsorbents that have received significant attention from the hydrogen storage.

2. Materials and Methods

In this paper we study the gas adsorption of Si atoms using LAMMPS (Large-scale Atomic/Molecular Massively Parallel Simulator) software. The arrangement of Si atoms was constructed using VMD (Visual Molecular Dynamics) software. The interaction of hydrogen molecule and silicon atom was estimated by simply employing Lennard-Jones function with potential parameter known. The intermolecular potential function of hydrogen molecule was also estimated with known parameter, while intramolecular interaction such as vibration and rotation activity and quantum effect of hydrogen were neglected [11].

Atom	σ (Å)	ε (kJ/mol)
H	0	0
Si	3.385	2.4522

Table 1. Parameter used to conduct Lennard-Jones interaction [12].

$$U = 4\varepsilon\left\{\left(\frac{\sigma}{r}\right)^{12} - \left(\frac{\sigma}{r}\right)^{6}\right\} \tag{1}$$

Here, σ is length parameter of Lennard-Jones potential and r is distance between two molecules. Cross parameter of different pair, termed as i and j particles formulated using Lorentz-Berthelot mixing function. The simulation ran with probabilistic concept, a system composed of interacting atoms which are given initial coordinates and orientations as a set. The successive random displacement of the atoms generated the initial set.

$$\varepsilon_{ij} = \sqrt{\varepsilon_{ii}\varepsilon_{jj}} \tag{2}$$

$$\sigma_{ij} = \frac{(\sigma_{ii} + \sigma_{jj}}{2} \tag{3}$$

Temperature was set to 248K, 273 K, and 298 K for silicon surface adsorption towards hydrogen molecule which is higher than temperature of nitrogen in liquid state. Temperature condition was set to be higher than 77 K was based on its feasibility as hydrogen storage utilization [13]. Nevertheless, the critical temperature where hydrogen can be liquefied is 33.19 K and considered as safe to be stored and transported. The gravimetric storage capacity ρ_w of Si calculated as the amount of hydrogen adsorbed where N are number of atoms or molecule and m are atomic relative masses.

$$\rho_w = \frac{N_H.m_H}{N_{Si}.m_{Si}+N_H.m_H} \tag{4}$$

Formulation to determine the amount of hydrogen input on certain pressure condition defined in van der Waals equation state which are modification of ideal gas law:

$$(P + a\left(\frac{n}{V_m}\right)^2)(V_m - n.b) = nRT \tag{5}$$

where a (2.44×10^{-7} m^6.atm/mol^2) and b (2.661×10^{-5} m^3/mol) represent van der Waals coefficient for hydrogen, V_m is volume of a hydrogen molecule (m^3/molecule), R is universal gas constant (8.314 J/mol.K) [14].

3. Results and Discussion

The initial condition were 60 hydrogen atoms (30 molecules) which were placed outside of Si surface. This initial pressure condition represents roughly 8 atm. The simulations run in molecular time scale of (10000, 25000, 50000 and 100000). In this molecular simulation, one-time step represents roughly 1 femtosecond. The results were given in table (1), (2) and (3) for the temperature of 248 K, 273 K and 298 K respectively. The results were also plotted in figure 1 (a) and (b) for surface of Si (001) and Si (111).

Si Surface	**Number of Atoms Adsorbed per Run Timestep**			
	10000	**25000**	**50000**	**100000**
Si (111)	6	9	17	32
wt%	*0.06652*	*0.09978*	*0.18847*	*0.35477*
Si (001)	6	10	14	29
wt%	*0.07236*	*0.1206*	*0.16884*	*0.34973*

Table 2. Number of hydrogen atoms adsorbed at 248 K.

Si Surface	**Number of Atoms Adsorbed per Timestep**			
	10000	**25000**	**50000**	**100000**
Si (111)	6	10	17	30
wt%	*0.06652*	*0.11086*	*0.18847*	*0.33259*
Si (001)	4	10	14	26
wt%	*0.04824*	*0.1206*	*0.16884*	*0.31356*

Table 3. Number of hydrogen atoms adsorbed at 273 K.

Si Surface	**Number of Atoms Adsorbed per Timestep**			
	10000	**25000**	**50000**	**100000**
Si (111)	6	9	15	29
wt%	*0.06652*	*0.09978*	*0.1663*	*0.32151*
Si (001)	3	10	14	24
wt%	*0.03618*	*0.1206*	*0.16884*	*0.28944*

Table 4. Number of hydrogen atoms adsorbed at 298 K.

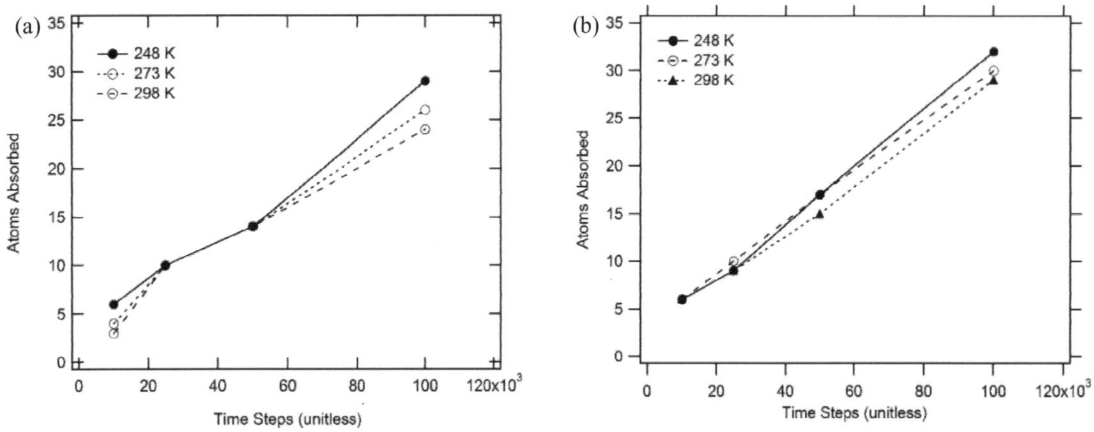

Figure 1 (a) The hydrogen molecule adsorption of Si surface (001) (b) Si surface (111).

From the results, the hydrogen adsorption is temperature dependent. As the temperature increases, the hydrogen adsorption decreases according to statistical thermodynamics of adsorption that prefers adsorption at lower temperature. Physical adsorption is done in exothermic condition, so it worked better at lower temperature as heat is liberated and entropy of the system decreases due to its decreased microstate number and freedom of hydrogen molecules movement [15]. This shows that our simulation is in accordance with the real phenomena. This curve shows saturation of adsorption which is usually demonstrated in BET (Brunauer, Emmett and Teller) adsorption measurement. Our results also showed that the Si surface (111) has higher adsorption capacity in comparison with Si surface (001). This result is related to the stability of (111) surface in comparison with (001) surface. Surface energy as thermodynamic property of Si indicate which surface indices has relatively better stability between each Si atoms [16]. Without considering entropy, Lu et al., has shown that the Si (111) surface has lower energy in comparison with the surface of (001) which lead to more hydrogens can be adsorbed physically without disturbed by Si-Si interaction [17].

4. Conclusions.

In this simulation, we had successfully calculated hydrogen adsorption of Si surface of (001) and (111) using LAMMPS and VMD as builder. We demonstrated that Si surface of (111) has higher adsorption capability in comparison with Si surface of (001) due to the lower surface energy and condition which has lower temperature, silicon tend to adsorb more hydrogen as entropy of the system decreases. In general, our simulations capture the real phenomenon of hydrogen adsorption; saturation and the temperature dependence of hydrogen adsorption.

5. Acknowledgements

We gratefully acknowledge this part of work as this was supported by DRPM (Universitas Indonesia Research Directorate) under Hibah PITTA Contract No. 2420/UN2.R3.1/HKP.05.00/2018 grant. There is no conflict of interest between the authors and the institution regarding the publication of this article.

6. References

[1] Centre, M. S. (2006). Dissociative adsorption of molecular hydrogen on silicon surfaces, *61*, 465–526. https://doi.org/10.1016/j.surfrep.2006.08.002

[2] Law, B. Y. J. T., Francois, E. E., Laboratories, B. T., & Hill, M. (1956). ADSORPTION OF GASES ON A SILICON SURFACE, *28*(6).

[3] Wu, C. J., & Carter, E. A. (1991). Adsorption of hydrogen atoms on the Si (lOO) -2 x implications for the H2 desorption mechanism surface :, *185*(I).

[4] Razado-colambo, I., Zhang, H. M., Uhrberg, R. I. G., Hansson, G. V, Razado-colambo, I., Zhang, H. M., … Hansson, G. V. (2005). STM study of site selective hydrogen adsorption on Si (111) 7 × 7, (111). https://doi.org/10.1103/PhysRevB.71.235411

[5] Chitsazan, A., Monajjemi, M., Aghaei, H., & Sayadian, M. (2017). Neutral Gases Adsorption with Hydrogen on Silicon Nanotubes: A Fuel Cell Invistegation.

[6] Dürr, M., & Höfer, U. (n.d.). Reaction dynamics of molecular hydrogen on silicon surfaces - importance of lattice degrees of freedom, (1), 1–24. https://doi.org/10.1103/PhysRevB.54.5978

[7] Sakurai, L. H. and W. X. and K. M. and T. H. and T. (1992). FI-STM study of hydrogen adsorption on Si(100) surface. *Acta Physica Sinica (Overseas Edition)*, *1*(3), 191. Retrieved from http://stacks.iop.org/1004-423X/1/i=3/a=005

[8] Pehlke, E., & Kratzer, P. (1999). Density-functional study of hydrogen chemisorption on vicinal Si (001) surfaces. *Physical Review B*, *59*(4), 2790–2800.

[9] Med, A. I. (1960). Studies." (28922) V. T. MADDAIAH, A. A. KURNICKS, (1), 6 8.

[10] Chen, H. (2013). *Atomic and molecular adsorptions of hydrogen and oxygen on silicon nanotubes: An ab initio study* (Order No. 3568834). Available from ProQuest Dissertations & Theses Global; SciTech Premium Collection. (1418802092).

[11] Balilehvand, S., Hashemianzadeh, S. M., Razavi, S., & Karimi, H. (2012). Investigation of hydrogen and methane adsorption/separation on silicon nanotubes: A hierarchical multiscale method from quantum mechanics to molecular simulation. *Adsorption*, *18*(1), 13–22. https://doi.org/10.1007/s10450-011-9375-x

[12] Mashayak, S. Y., & Aluru, N. R. (2012). Coarse-Grained Potential Model for Structural Prediction of Confined Water. *International Journal Chemistry Theory Computation,* 8, 1828-1840. https://dx.doi.org/10.1021/ct200842c

[13] Mahdizadeh, S. J., & Goharshadi, E. K. (2014). Hydrogen storage on silicon, carbon, and silicon carbide nanotubes: A combined quantum mechanics and grand canonical Monte Carlo simulation study. *International Journal of Hydrogen Energy*, *39*(4), 1719–1731. https://doi.org/10.1016/j.ijhydene.2013.11.037

[14] Bratu, P., Brenig, W., Gro\S, A., Hartmann, M., Höfer, U., Kratzer, P., & Russ, R. (1996). Reaction dynamics of molecular hydrogen on silicon surfaces. *Phys. Rev. B*, *54*(8), 5978–5991. https://doi.org/10.1103/PhysRevB.54.5978

[15] Sing, K. (2001). The use of nitrogen adsorption for the characterization porous materials. *Colloids and Surfaces A: Physicochem. Eng. Aspects*, 3(9), 187-188.

[16] Liu, F., Hohage, M., Lagally, M.G. (1999). Encyclopedia of Applied Physics. Wiley-VCH Verlag GmbH London, 321.

[17] Lu, G. H., Huang, M., Cuma, M., & Liu, F. (2005). Relative stability of Si surfaces: A first-principles study. Surface science, 588(1-3), 61-70.

IOP Conf. Series: Materials Science and Engineering **547** (2019) 012039 doi:10.1088/1757-899X/547/1/012039

Development of Steel Matrix Composites Used for Metal Die

Gen Sasaki[1]*
[1]Graduate School of Engineering, Hiroshima University, Higashi-Hiroshima, Japan

*Corresponding author (E-mail): gen@hiroshima-u.ac.jp

Abstract. In this study, we show the development of the steel matrix composites used for metal die. Matrix used is 40CrMoV5 (ISO standard) steel, which is used as press and die-casting dies. The dispersants used are vapor grown carbon nanofiber (VGCF), which is one of carbon nanofiber (CNF) and titanium die-bride (TiB_2) particle powder. After mixing 40CrMoV5 powder and the dispersants by ball milling, the composites was prepared by spark plasma sintering (SPS) process. For VGCF, 1.9 and 3.8 vol. % was added to 40CrMoV5 powder in order to obtain the composites. The thermal conductivity of the VGCF/40CrMoV5 alloy composites improved about 20% compared with the 40CrMoV5 monolithic alloy. The tensile strength of the composites with 1.9 vol. % VGCF is about 1980MPa, which is higher than the monolithic alloy with 1250 MPa. For TiB_2 particle, 4.0, 8.0 20.0 vol. % powders was added to 40CrMoV5 powder. By adding 8.0 vol. % TiB_2, the thermal conductivity improved about 40% compared with the monolithic alloy, but tensile strength decreased to 612 MPa in 8.0 vol. % VGCF composites, which are caused by the aggregation of TiB_2 powders. It seems the strength of the composites will improve by controlling the dispersibility of TiB_2 powders.

Keywords: Steel composites, Sintering, Thermal conductivity, Mechanical property, Microstructure

1. Introduction

Die-cast process is used in the automobile industry mainly because of high dimensional accuracy, smooth casting surface, high productivity and easy catting process. Above all, the aluminum alloy die-casting accounts for 95% of total amount of die-cast production. On the other hand, Al products used in vehicle has been substituted from wrought product to casting product because of high cost performance in recent years. Especially, die-cast process has been grate-watched, because the die-cast process is able to make high performance products. Die-cast process is as follows; after coating the die surface by lubricant, die was clamped. Then the molten metals was poured in cavity with high pressure. After cooling and opening the mold, we can obtained the cast products. Repeat of rapid cooling and heating of metal die generate the thermal stress. The extrusion pressure and mold clamping lead the mechanical stress to metal die. The degradation of life time of die-cast mold and surface quality of die-cast products is caused by the defect such as the heat check, the crack and the melting generated during operation. In order to improve the productivity, cost performance and stability of the product quality for die- cast process, the improvement of the life time is important. In order to improve

the life time of die, the improvement of the thermal conductivity and the mechanical properties is required because of the degradation of thermal stress and the toughening.
The objective in this study is to relax the thermal stress generated in metal die and to improve the life time of metal die by using the metal matrix composites (MMC). Especially, we try to improve the thermal conductivity without the degradation of mechanical properties for conventional steel die. We used the dispersant with high thermal conductivity and small size under sub-micron for composites. Small sized dispersants seems to affect to the improvement of the strength and toughness of metal die because of the refinement of the microstructure and the improvement of grain boundary strength.

2. Experimental Procedure

Two type of carbon nanofiber (CNF) and ceramic particle dispersed steel composites was prepared in this study. Starting material used as matrix is 40CrMoV5 steel alloy powder with average diameter 70μm (Mitsubishi steel MFG Co., Ltd.). Table 1 shows the chemical composition of 40CrMoV5 used in this study. Dispersants used are vapor grown carbon nanofiber (VGCF) and TiB_2 particles. VGCF is one of carbon nanofiber (CNF), which was developed by Showa Denko Co., Ltd. in Japan and forms 150 nm in diameter and 10-20 μm in length. VGCF has high thermal conductivity as 1200W/mK along fiber direction. TiB_2 particle forms equiaxial with average diameter 2.62 μm, which is developed by Japan New Metals Co., Ltd. Then, these powders was mixed by V shape type ball mixer, which is conventional mixing method. Each powers was inserted in stainless vessel with ethanol and mixed under 50 rpm in rotational speed of mixer. Volume fractions of VGCF were 1.9 and 3.8 vol. %. Volume fractions of TiB_2 were 4.0, 8.0 and 20.0 vol. %. After mixing, the composites were prepared by spark plasma sintering (SPS) process under the condition of 1173K and 1273K in sintering temperature, 50 MPa in applied pressure, 0.6ks in keeping time of pressure and $<10^{-2}$Pa in atmosphere. Relative density and microstructure of the composites was evaluated by Archimedes method, optical microscopy, SEM and EPMA. Thermal conductivity and tensile strength were measured by steady state method and tension tester, respectively.

Table 1 Chemical composition of 40CrMoV5 steel powders used in this study.

	C	Si	Mn	P	S	Cu	Ni	Cr	Mo	V	Fe
40CrMoV5[wt.%]	0.51	1.09	0.06	0.008	0.007	0.03	0.03	4.63	1.03	1.06	Bal.

3. Results and Discussion

Fig. 1 shows the microstructure of VGCF dispersed 40CrMoV5 steel composites and TiB_2 particle dispersed 40CrMoV5 steel composites. Fig. 1 (a) is 1.9 vol. % VGCF/ 40CrMoV5 steel composites. Dark dots and lines were observed at the boundary of sintered steel particle, which is aggregation of VGCFs. Fig. 1 (b) and (c) is 4.0 vol.% and 8 vol. % TiB_2/ 40CrMoV5 steel composites, respectively. Dark area in these figs. is traces of falling particles by polishing. It seems the dispersed nanofibers or particles segregated at the boundary of sintered steel particle. Sintered steel particle forms equiaxed, which forms maintain the as-received particle.
As VGCF and TiB_2 may react with Fe thermodynamically, we have to check the reactivity of the composites during sintering. Fig. 2 shows the mapping of C, Fe and Cr element around the dark dots in fig. 1 (a) observed by EPMA, which is 1.9 vol. % VGCF/ 40CrMoV5 steel composite. The composites was fabricated at 1173K in sintering temperature. The density of the composites was over 98%. Fe and Cr elements was not detected in the area shown strong carbon element. It means VGCF did not react with Fe and Cr to form these carbides. This composites prepared by sintering at 1173K. But, most VGCF react with Fe in the composites

prepared by sintering at 1273K. It seems 1173 K is suitable temperature to obtain dense composites without reaction between VGCF and Fe.

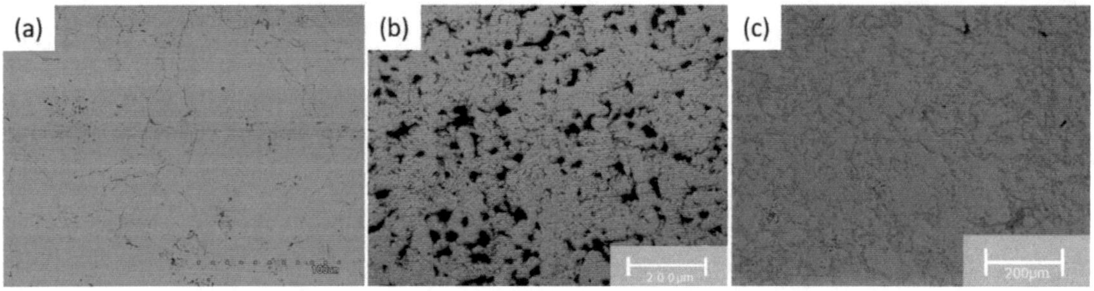

Fig. 1 Microstructure of (a) 1.9 vol. % VGCF/ 40CrMoV5 steel composites, (b) 4.0 vol. % TiB_2 / 40CrMoV5 steel composites and (c) 8.0 vol. % TiB_2 / 40CrMoV5 steel composites.

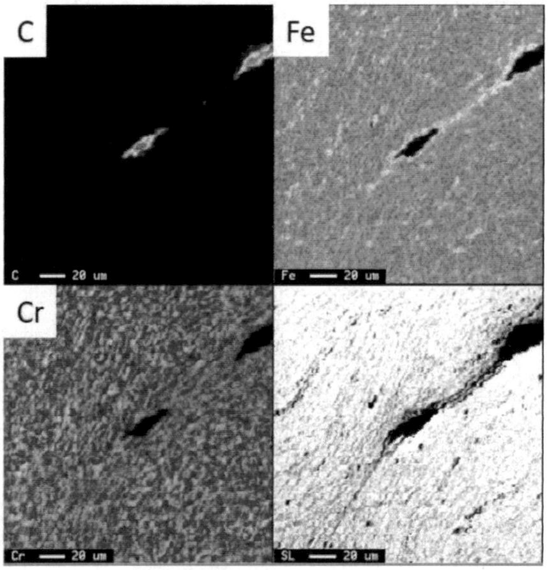

Fig. 2 EPMA mapping of 1.9 vol. % VGCF/ 40CrMoV5 steel composite.

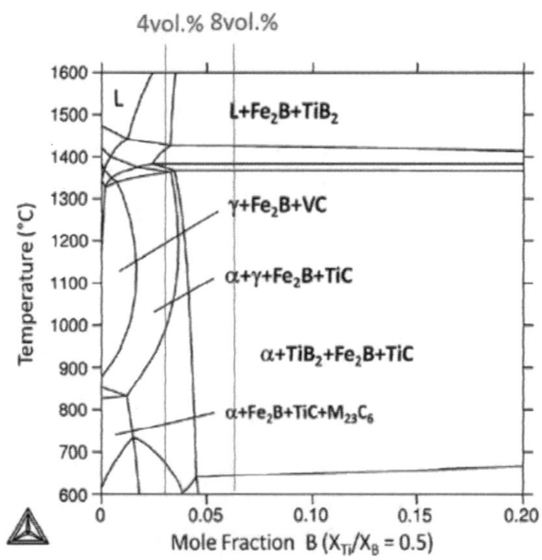

Fig. 3 Equilibrium phase diagram between 40CrMoV5 alloy and TiB_2 [2].

Fedrizzi et al. [1] shows TiB_2 react with iron element in 40CrMoV5 alloy as following reaction.

$$TiB_2 + \alpha\text{-Fe}(\uparrow C) \rightarrow TiC + Fe_2B + \alpha\text{-Fe}(\downarrow C) \quad (1)$$

Furthermore, Fedrizzi et al. [1] shows the equilibrium phase diagram between 40CrMoV5 alloy and TiB_2 by using Thermo-Calc® software and TCFe3 database [2]. Fig. 3 shows this diagram. Molar fraction of boron in horizontal axis in Fig. 3 for 4 vol. %, 8 vol. % and 20 vol. % TiB_2/ 40CrMoV5 alloy composites are 0.035, 0.069 and 0.15. This diagram shows the stable phase at 1273K in 4 vol. % TiB_2/ 40CrMoV5 alloy composites are $\alpha+\gamma+TiB_2+Fe_2B$, and TiB_2 phase is unstable phase. But in 8 vol. % and 20 vol. % TiB_2/ 40CrMoV5 alloy composites, stable phases are $\alpha+TiB_2+Fe_2B+$ TiC, and TiB_2 phase remains as stable phase. As the thermal conductivity of TiB_2, TiC and Fe_2B are 66.4 W/mK, 16.7W/mK and 20.7W/mK, it seems the generation of TiC and Fe_2B lead to degrade the thermal conductivity of the composites.

Fig.4 shows XRD analysis of 20 vol. % TiB_2 /40CrMoV5 composites. Two peaks are observed at around 34 degree and 44 degree. The former is TiB_2 peak, and the latter is sum of Fe and TiB_2 peaks. This results shows many TiB_2 remains in composites. It seems TiB_2 phase remains in 8 vol. % TiB_2 /40CrMoV5 composite. But obvious TiB_2 was not observed in XRD result of 8 vol. % TiB_2 /40CrMoV5 composite.

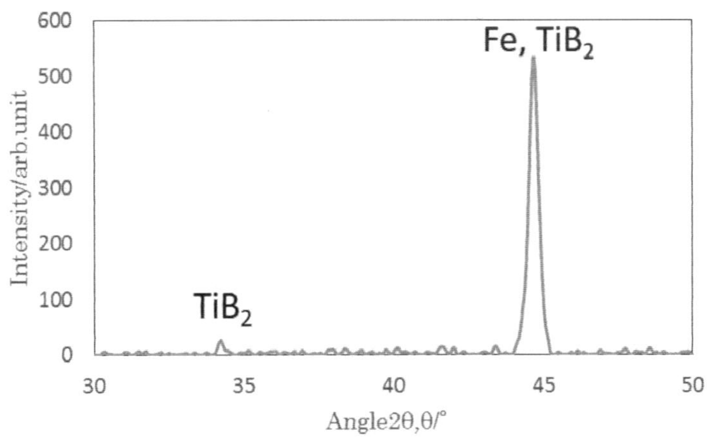

Fig. 4 XRD analysis of 20 vol. % TiB_2 /40CrMoV5 composites.

Fig. 5 shows the thermal conductivity of monolithic 40CrMoV5 alloy, VGCF dispersed 40CrMoV5 alloy composites and TiB_2 disperse 40CrMoV5 alloy composites. Thermal conductivity of monolithic 40CrMoV5 alloy is about 20.0 W/ mK. By adding 1.9 vol. % VGCF, the thermal conductivity increased to 25.0 W/ mK. On the other hand, 4 vol. % TiB_2 /40CrMoV5 alloy composites was 18.0 W/mK, which is caused by the generation of TiC and Fe_2B and dissipation of TiB_2 by reaction during sintering. But by adding TiB_2 content to 8 vol. %, the thermal conductivity increased to 34 W/ mK, which is caused by the remains of TiB_2 in composites. Consequently, the addition of VGCF and TiB_2 is effective for the improvement of the thermal conductivity of monolithic 40CrMoV5 alloy.

The keeping of strength for composites is important for practical use of metal die. Fig. 6 shows the tensile strength of the monolithic alloy and the composites. By adding 1.9 vol. % VGCF, the tensile strength of the composites improved dramatically from 1250MPa to 1980MPa. It seems the strength of 40CrMoV5 particle boundary in composites was improved by the bridging of 40CrMoV5 particles by VGCF. On the other hand, by adding TiB_2, the strength of 40CrMoV5 alloy degraded. For 4 vol. % and 8 vol. % TiB_2 addition, the strength of the composites was decreased as 1213 MPa and 612 MPa, respectively. The degradation of the strength seems to be caused by the stress concentration at aggregation area of TiB_2 particles.

On the other hand, the elongation of TiB / 40CrMoV5 is improved by increasing TiB contents, which is caused by the grain boundary strengthening by boron element, which is supplied from the reaction of Fe and TiB_2. In order improve the strength of the composites, the microstructure control is very important.

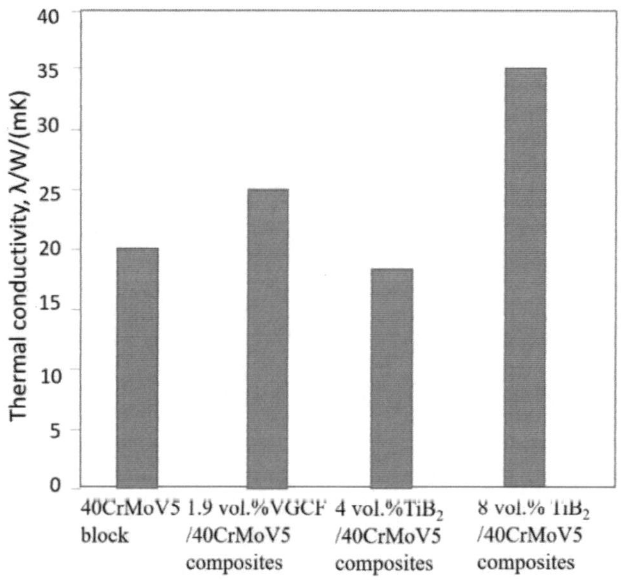

Fig. 5 Thermal conductivity of monolithic 40CrMoV5 alloy, VGCF dispersed 40CrMoV5 alloy composites and TiB_2 disperse 40CrMoV5 alloy composites

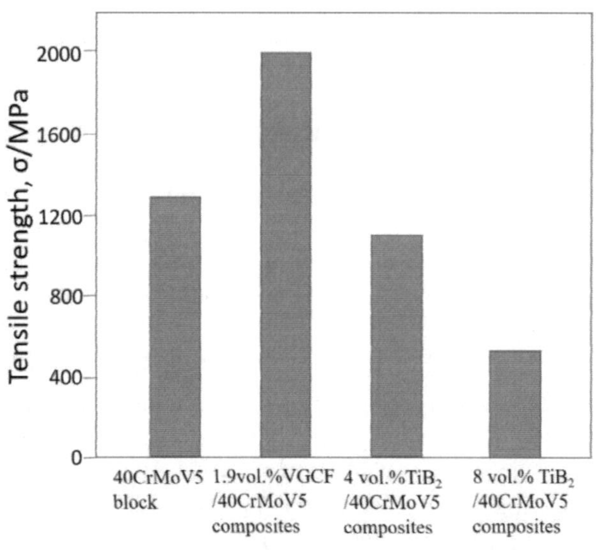

Fig. 6 Tensile strength of monolithic 40CrMoV5 alloy, VGCF dispersed 40CrMoV5 alloy composites and TiB_2 disperse 40CrMoV5 alloy composites.

4. Conclusions

In order to obtain high performance metal die with high thermal conductivity and high strength, 40CrMoV5 alloy based composites was prepared by SPS process. Two type's materials was investigated as dispersant of the composites, which is VGCF and TiB_2 particles. For VGCF, the suitable sintering temperature was 1173K because of few reaction and dense composites. The thermal conductivity and the tensile strength of 1.9 vol. % VGCF/ 40CrMoV5 alloy composites was better than the monolithic 40CrMoV5 alloy block. On the other hand, Addition of 4vol. % TiB_2 degraded the thermal conductivity and tensile strength because of the reaction of Fe and TiB_2, but by increasing TiB_2 contents more, thermal conductivity of the composites increased because of the suppress of the reaction.

5. Acknowledgements

This study was partly supported by JSPS KAKENHI Grant Number 18K03840.

6. References

[1] FEDRIZZIA A, PELLIZZARIA M, ZADRAB M and MARIC E 2013 *Materials Characterization* Vol. 86 pp 69-79
[2] SUNDMAN B, JANSSON B and ANDRESSON J-O.1985 *The Thermo-Calc databank system. Calphad* 9:153–90.

Synthesis and Characterization of Ca-doped $Li_4Ti_5O_{12}$ Using $CaCO_3$ from Chicken Eggshell as a Dopant for Lithium-Ion Battery Anode Material

B Priyono[1]*, D K Ibrahimi[1], A Z Syahrial[1], H Jodi[2], A Subhan[3], M R Nugraha[1]
[1]Department of Metallurgy and Materials, Faculty of Engineering, Universitas Indonesia, Depok, 16424, Indonesia
[2]Center for Science and Technology of Advanced Materials, BATAN, PUSPITEK, Tangerang, Banten, 15310, Indonesia
[3]Center for Research of Physics, LIPI, PUSPITEK, Tangerang, Banten, 15310, Indonesia

*bambang.priyono@ui.ac.id

Abstract. In order to improve the performance of $Li_4Ti_5O_{12}$ (LTO) anode, this research was focused on Ca^{2+} ion doping as a substitute to Li^+ ion to form $Li_{4-x}Ca_xTi_5O_{12}$ with values of x=0, 0.05, 0.075, and 0.125 using solid-state reaction. The Ca^{2+} ion source was $CaCO_3$ which synthesized from the chicken eggshell. The LTO was prepared by a solid-state method using TiO_2 Degussa. The pristine LTO and Ca-doped LTO sample powder was characterized by XRD, SEM, and were also tested its electrochemical performance by EIS, CV and CD. The $CaCO_3$ dopant characterization results showed $CaCO_3$ in calcite polymorph as the main phase, with agglomerated fine particulate morphology and high purity. Characterization of LTO sample powder with XRD revealed that dopant Ca successfully enter the structure of LTO spinel, with maximum addition level x=0.05, which excessive addition led to $CaTiO_3$ impurity forming. SEM result showed all Ca-doped LTO have almost similar morphology, which was agglomerated particulate. Ca-doped LTO samples have smaller particle size compared to pristine LTO. Electronic conductivity improvement was spotted at all of Ca-doped LTO sample, with $Li_{3.75}Ca_{0.125}Ti_5O_{12}$ (LCaTO-3) showed the lowest charge transfer resistance of 29.5 Ω. $Li_{3.75}Ca_{0.125}Ti_5O_{12}$ (LCaTO-3) also had the highest initial discharge capacity of 168.2 mAh/g. Nevertheless, in high rate application, the best performance was showed by $Li_{3.85}Ca_{0.075}Ti_5O_{12}$ (LCaTO-2) with the discharge capacity of 30.2 mAh/g at 12 C, which capacity retention percentage of 21.43% compared to its discharge capacity at 0.2 C.

Keywords: $Li_4Ti_5O_{12}$ anode, solid-state, Ca-doped, dopant $CaCO_3$, chicken eggshell waste

1. Introduction

Lithium-ion (Li-ion) battery is one of the most commonly used in today's technology. Increased demand for lithium batteries has increased over time, coupled with a decrease in the cost of manufacturing. The conventional LiB currently are generally made with graphite as anode material. Graphite has advantages include relatively low fabrication costs and large capacity (372 mAh/g) [1]. However, the safety level is still low in the use of graphite, and it is still the primary concern. This issue is due to the voltage operations close to 0 V (vs. Li/Li^+) at a time when the process of Li^+ insertion, so the metal lithium deposits formed on the surface of the carbon electrode is dendritic, commonly referred to as solid-electrolyte interface (SEI). SEI formed at potential under 0.8 V (vs. Li/Li^+) this will reduce the amount of lithium in the electrolyte, so that cause a decrease in the specific capacity [2].

Along with the rapid advancement of technology, other materials were developed to accommodate the shortcomings of graphite anodes, one of which is the titanate of lithium

($Li_4Ti_5O_{12}$). The anode of lithium titanate (LTO) spinel has the specific smaller capacity (175 mAh/g) compared to graphite, but have higher operating potential (1.55 V vs. Li/Li^+) so it is safer to operate [3]. In addition to the above characteristics, LTO has other advantages, such as the nature of zero strain material LTO in the lithiation-delithiation process which have an impact on the results of the stability of cycles and good reversibility [4]. However, anode LTO has some disadvantages in addition to specific capacity is low, i.e. the coefficient of electronic conductivity is low (10^{-13} S/cm) and the coefficient of diffusion of ions Li^+ which is also low (10^{-8} cm ^{2}/s) that affect the rate of performance [5].

Based on the above conditions, improved performance anode LTO can be done in several ways, ranging from lower particle size, coat the surface of the particles with conductive materials such as carbon and silver, as well as doping by using ion metal and non-metal [3]. Increased performance and capacity can also be made by combining LTO with anode material combines with a higher value of the specific capacity.

The increase in conductivity is also one of the ways to improve the performance of LTO. Conductive phase such as carbon or FRS is usually added in the matrix so that the conductivity of LTO increases. Also, the treatment of doping can also be done, such as using metal ions (Ag^+, Mg^{2+}, Zn^{2+}, Sr^{2+}, Sn^{2+}, etc.) and other non-metallic ions (F^-, Br^-, etc.). Doping on LTO is proven to increase the electronic conductivity effectively [6]. One of the interesting research is doping LTO using the atom of calcium. Generally, the source of dopant atoms in calcium derived from CaO or $CaCO_3$.

On the other hand, the $CaCO_3$ turns out to be is a constituent component of natural materials such as shells and egg shells [7]. The said two ingredients can be processed and used as $CaCO_3$ from a natural source to serve as the dopant of the LTO.

2. Materials and Methods

Calcium carbonate was used in this research is prepared with the same method as the previous research [8]. Preparation process begins with obtained ingredients of chicken eggshells are then cleaned of debris physically using the water until clean. Once cleaned, chicken egg shells are then crushed and pounded until fine powder-shaped. Then, the powder obtained is dried at a temperature 100°C for 12 hours. The results obtained are calcium carbonate derived from nature.

The main ingredients needed are a titanium oxide TiO_2 as a source of titanium, lithium carbonate (Li_2CO_3) as a source of lithium, and $CaCO_3$ prepared from chicken eggshells as a dopant source. The calculation was made so that the final result was obtained by the stoichiometry of $Li_{4-x}Ca_{2x}Ti_5O_{12}$ with variable values x of 0, 0.05, 0075, and 0125. The TiO_2 used is TiO_2 nano-sized Degussa P25 commercial size ~ 20-30 nm, while Li_2CO_3 was Merck Sigma Aldrich. The composition of reactants used customized with stoichiometric forming $Li_4Ti_5O_{12}$, except of Li_2CO_3 where added in-excess of 4% to accommodate the loss of the Li^+ during the high sintering temperature. The addition of dopant is then carried-out by mixing $CaCO_3$ from eggshells along with Li_2CO_3 and TiO_2.

Table 1 shows the composition used in this research, in which the amount of dopant $CaCO_3$ adapted to the variable substitution for lithium ion doping $Li_{4-x}Ca_{2x}Ti_5O_{12}$.

Table 1. Formulation and sample code

Formulation	Moles of Ca doping (x)	Moles of Li	Material	Sample Code
	0	4	$Li_4Ti_5O_{12}$	LTO
$Li_{4-x}Ca_xTi_5O_{12}$	0.05	3.90	$Li_{3.90}Ca_{0.05}Ti_5O_{12}$	LCaTO-1
	0.075	3.85	$Li_{3.85}Ca_{0.075}Ti_5O_{12}$	LCaTO-2
	0.125	3.75	$Li_{3.75}Ca_{0.125}Ti_5O_{12}$	LCaTO-3

Once weighed, all reactants are introduced into the chamber of the High Energy Vibrating Ball Mill (MSKK-SFM-3 High-Speed Vibrating Ball Mill-1200 rpm) for mixing so that the mixture is homogeneous. Milling was done during the 45-minute pauses every 10-15 minutes stopped to avoid overheating in the chamber because of the process of high-energy, rotation and release particles that stick to the wall of the chamber. Powder milling result is then ejected from the chamber to be calcined at a temperature of 800°C for 6 hours.

The process of manufacturing the batteries begins with the preparation of materials active. The active material is crushed with a mortar until smooth then sifted with a 140 mesh sieve. Active material already sifted is then weighed to measure that is specified by the stoichiometric. Next up is the making of the slurry, i.e. by dissolving active anode materials of $Li_4Ti_5O_{12}$, PVDF, and acetylene black in comparison with 8:1:1 in a solution of NMP. Further mixing with magnetic stirrer sized for 2 hours until homogenized slurry is formed.

The slurry then coated on a sheet of a current collector of Cu foil. This coating process used a special tool that is doctor-blade. After slurry exposed on a certain thickness, with results of the coating is heated in the appliance at a temperature of 80°C. The results of coating is then transferred to the oven at a temperature of 120°C and kept for 24 hours.

Further cutting of sheet forming a circle with dimensions that match the size of the coin battery half-cell. Other cell components also prepared as a counter electrode (lithium foil), $LiPF_6$ as an electrolyte, and 2300 Celgard as a separator. Everything is placed in the glove box with argon gas atmosphere and moisture content of less than 10 ppm. Coin cell battery components assembled in the glove box. A pressure of 10 MPa was applied for compacting of the battery arrangement. Next, the battery left for 24 hours to be ready to do the testing and performance.

3. Results and Discussion

XRD Analysis

Figure 1 shows the XRD diffractogram of the samples. The XRD characterization results were processed using the program *Plus HighScore X'Pert* refers to ICDD PDF2 database. The diffraction peak of recording range trail from 10° up to 80° (2θ). Diffraction XRD peak data charts of each sample can be seen in Figure 1. The phase Component tries to found through matching with the reference code is $Li_4Ti_5O_{12}$ (00-049-0207), TiO_2 rutile (01-072-1148), and $CaTiO_3$ (01-072-1192). It also seen how the influence of addition of doping ions of Ca^{2+} on the characteristics of the diffraction patterns of a sample.

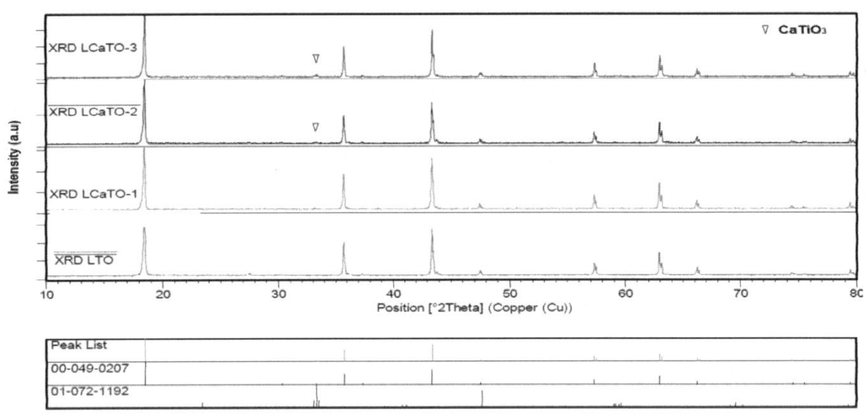

Figure 1. Diffraction patterns LTO, LCaTO-1, LCaTO-2 and LCaTO-3.

All of the four samples are characterized, the majority of the main peak is the diffraction patterns from Li_4TiO_{12} spinel. Diffraction peak (2θ) appear at the angle 18.45°, 37.29°, 43.47°, 47.43°, 57.45°, 63.08°, 66.14°, 74.42° and 79.41° which is the main constituent of diffraction peak patterns of phase LTO spinel. Through matching with reference code JCPDS no: 00-049-0207 on a sample LTO without doping, the sample LCaTO-1, LCaTO-2 and LCaTO-3 gained a score of 89, 91, 83, 78, respectively. Thus, it indicates that compound of lithium titanate (LTO) was successfully synthesized using methods of the synthesis of solid-state doping, and the addition of Ca^{2+} into LTO does not change the structure of the host of the LTO.

The peak of $CaTiO_3$ shows at the 33.21° from the sample of LCaTO-2 and LCaTO-3. The peak matches with JCPDS no: 01-072-1192 with common name calcium titanate. The formation of calcium titanate is due to the calcium ion doping has exceeded the maximum doping amount. Further, from XRD test results above, it is found that the amount of TiO_2 rutile which is very few in the four samples, the samples are either doped or without doping. Matching for the phase2 rutile TiO_2 use the JCPDS reference code: 01-072-1148 on LTO without doping, LCaTO-1, LCaTO-2, and LCaTO-3 row obtained a score of 4, 0, 3, and 4. The existence of this phase of TiO_2 rutile is not significant when compared to the phase of the LTO. So, it shows that the amount of TiO_2 powder rutile in very little. Some factors that become the cause is the use of calcined TiO_2 nano-sized and the temperature high enough. The use of TiO_2 nano-sized lead to spread of the material that is more equitable and homogeneous. Also, the high surface area of the use of TiO_2 nano-sized has the advantages of increasing the probability of contact between the titanium source with a lithium source, and with the spread evenly so the amount of TiO_2 reacting become more massive.

The size of the crystallites averages obtained through approach using Scherer's equation. The average value of the smallest size of crystallites is variable with a value of LTO 38.12 nm, followed by LCaTO-2, LCaTO-1 and LCaTO-3 consecutive 47.32, 52.78, and 59.66 nm. Thus, this LTO preparation using TiO_2 Degussa obtained a nanocrystalline particle.

SEM Analysis

Figure 2 shows the morphology of the LTO sample. It can be seen that the particles experience the agglomeration with various size distribution. Agglomeration occurs starting from the process of mixing using HEBM that is too long. Particle LTO mutually connects between each other occur because of high sintering temperature. Agglomerate in the picture also big. It is possible to occur because of the snapping and hardening powder on the wall of the chamber HEBM during the process of mixing. Characteristics of morphology sample are also porous. The calculation of the average particle size of the sample LTO using software image-J shows the value of 0.809 μm with a standard deviation of 0.155 μm.

Figure 2. The SEM result of LTO.

Figure 4 shows the morphology of the samples (a) LCaTO-1 (b) LCaTO-2, and (c) LCaTO-3. Similar to the morphology at the sample of LTO, LCaTO-1 also shows the agglomeration particles and porous microstructures as seen in Figure 3 previously. However, the agglomerate formed not as big as that of the sample LTO without doping. Calculation of particle size using software image-J produces average particle size worth 0.548 µm with a standard deviation of 0.125 µm. The results of SEM samples LCaTO-2 shows the characteristics of agglomerate particles and porous as seen in the results of SEM other variables. Calculation of average particle size produces an average of 0.347 µm with standard deviation 0.052 µm, where this value is the average value of the smallest compared to other variables. On a scale of 5 µm particle size that looks smooth and pores that much formed. The results of SEM samples LCaTO-3 still shows the morphology of agglomerate particles and porous. If viewed on a scale of 20 µm, the agglomerate formed to have the shape and size of a fairly homogeneous. Meanwhile, the size of the particles a bit more rounded than the other samples. The average particle size is obtained through measurements using image-J software is 0.622 µm with standard deviation 0.124 µm. From this finding, the Ca-doping could reduce the agglomeration size in general.

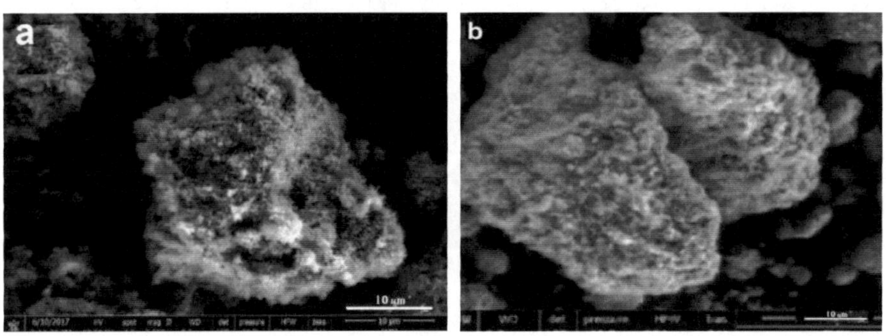

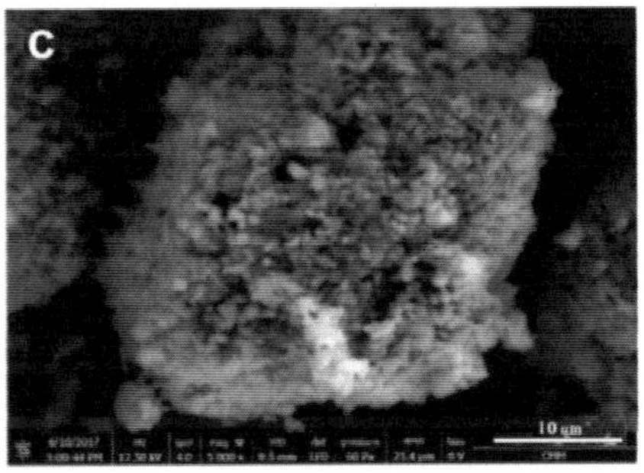

Figure 3. The results of SEM of samples (a) LCaTO-1 (b) LCaTO-2 (c) LCaTO-3.

EIS Analysis

The test results of the EIS can be seen in Figure 4. The value of R_{ct} smallest displayed by LCaTO-3 with a value of 29.5 Ω, followed by LCaTO-1 with a value of 68.8 Ω, and LCaTO-2 with 84.4 Ω. Sample LTO without doping results R_{ct} highest, i.e.: 86.7 Ω.

With reverse relations as already described above, then the information can be drawn that the LTO without doping have the lowest conductivity among other samples. In general, it can be inferred that the ion doping treatment of Ca^{2+} in the LTO can lower the resistivity, which means it also increases the value of conductivity.

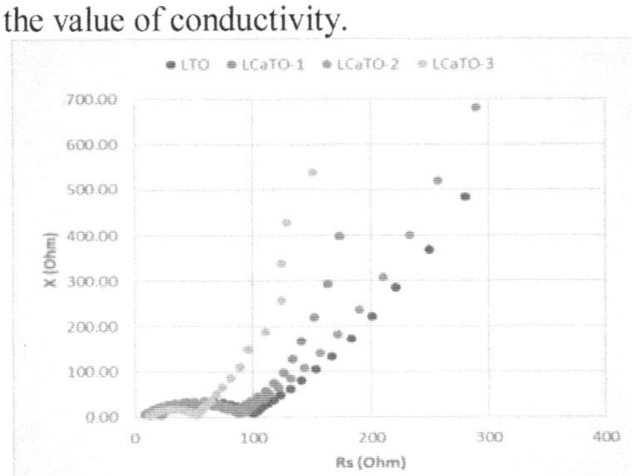

Figure 4. The combined Graph EIS for each sample.

Table 2. The value of the charge transfer resistivity of each sample

Sample	R_e (Ω)	R_{ct} (Ω)
LTO	14.3	86.7
LCaTO-1	20.1	68.8
LCaTO-2	8.6	84.4
LCaTO-3	12.9	29.5

EIS test results can be seen in Figure 5. In general, it could be concluded that the treatment of Ca^{2+} ion doping on LTO can lower the resistivity, which is about increasing the value of conductivity. Referring to the XRD diffraction data of test results, the shift of the diffraction peaks on doped samples showed Ca^{2+} influx into the LTO structure replaces multiple Li^+ ion site. Ca^{2+} influx into the host structure LTO to replace Li^+ ions will contribute more electrons. Compensation charge of substitutions cause a transition from Ti^{4+} to Ti^{3+}, where the presence of additional electron donation will increase the electrical conductivity due to the increased

IOP Conf. Series: Materials Science and Engineering **547** (2019) 012040 doi:10.1088/1757-899X/547/1/012040

concentration of electrons. Ti^{3+} ions play a role more effectively in the increase and decrease in polarization electrode conductivity[11][12].

The presence of impurities in the form of CaTiO3 also possible to affect the electronic conductivity doped samples. Their impurity dopant showed that the number of atoms that made it into the lattice structure of the LTO had passed the optimum limit. In several studies such as doping atom doping Gd and Ce on the structure of LTO, the impurity phase will create barriers charge transfer becomes larger so that the lower the value of conductivity[13] [14], But in this study, a decrease in conductivity does not occur. One indication of the cause of the high levels of conductivity is CaTiO3 characteristics that have perovskite crystal structure and an n-type semiconductor [15]. In a study on the characterization of CaTiO3 perovskite ceramics found variations in the electronic conductivity of 1.4 x 10-9 to 4.2 x 10-11 S / cm, where the value is still higher than the value of the electronic conductivity LTO (10-13 S / cm) [16].

Cyclic Voltammetry Analysis

The Figure 5 shows CV graph in the peak results of cathodic and anodic generated by each sample is not much different from each other. Cathodic peak and anodic from the sample LTO are at 1.785 and 1.405 volts, respectively. The working voltage is obtained from average voltage of oxidation and reduction on LTO sample at 1.595 V.

The LCaTO-1 sample Cathodic voltage value and anodic of 1.759 V and 1.434 V, with a working voltage of 1.5965 V. On sample LCaTO-2, cathodic and anodic voltage value of 1.424 V and 1.776 V consecutively, with working voltage 1.6 V. while at the doping with the highest concentration of Ca i.e., sample LCaTO-3, anodic and cathodic voltage value of 1.739 V and 1.453 V, with 1.596 V working voltage.

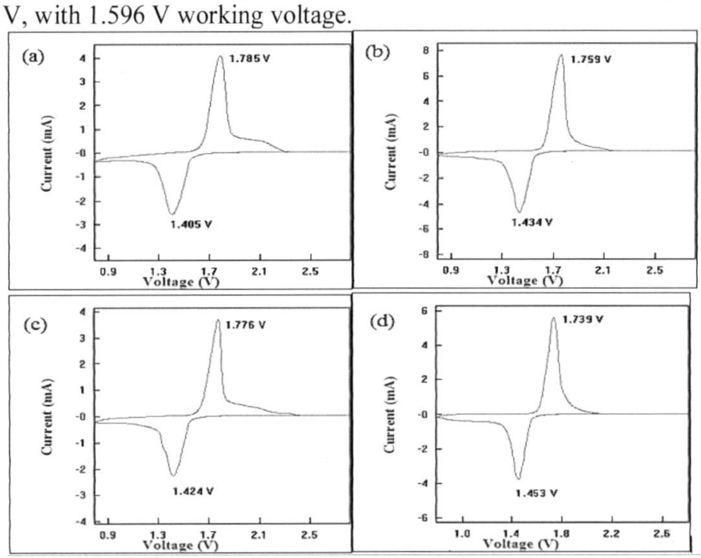

Figure 5. Graph of cyclic voltammetry sample LTO, LCaTO-1, LCaTO-2 and LCaTO-3.

All samples show the voltage value is not much different from each other, either cathodic voltage, anodic or working voltage cells. Working voltage on all samples show the value that almost approaches the working voltage theoretical LTO, which is 1.55 V. This shows that the phase is the phase of the primary constituent of the LTO from the powder of the active material.

View from the perspective of polarization, there was a decrease in the level of polarization on the atomic addition of Ca doping in the process. Difference of cathodic voltage and anodic is one of the parameters of the polarization level electrode electrochemical cell. The lowest level of polarization is owned by LCaTO-3 sample, with a value of voltage difference 0.286 V. then followed by LCaTO-1, LCaTO-2, and LTO with consecutive values 0.325 V, 0.352 V, and 0.38 V. Decrease in the level of polarization is closely related with the electronic conductivity factor, where the polarization of the lowest represented by the sample with the highest electronic conductivity, i.e. LCaTO-3.The trend of his descent is also the same with the existing trend of increased electronic conductivity in the analysis of the test results of the EIS in advance. It is aligned with the findings stating that the Ca atom doping into the structure polarization levels decline supports LTO electrode so that the kinetics reaction better [17].

Charge and discharge Analysis

Figure 6 shows the curve of the charge/discharge in C-rate. Almost all samples show the plateau in the form of a straight line in ~ 1.6 volts, which is by the characteristics of charge and discharge materials spinel LTO [2]. On any increase in C-rate decline in capacity. This phenomenon could be explained that the greater the flow rate causes a reaction that occurs is not comprehensive to all parts of the active material. On high C-rates, a lithiation reaction happens more on the surface. From this downturn, it will be seen how the retention capacity of the sample, i.e. the percentage of the sample capacity at a certain rate if compared to the capacity at the smallest C-rate (0.2 C).

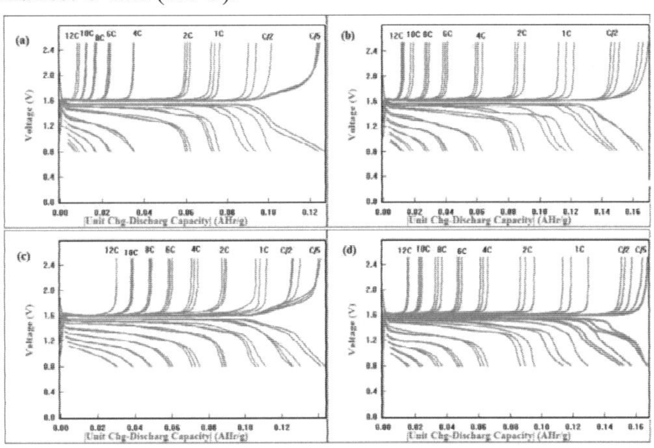

Figure 6. Charge/discharge curve of the sample (a) LTO (b) LCaTO-1 (c) LCaTO-2 (d) LCaTO-3.

Figure 7 shows the initial capacity rate obtained at the C-rate 0.2. Generally, Ca-doping is proven to increase the specific capacity value of LTO. The highest capacity at 168.2 maH/g obtained at doping of 0.125 mol Ca – 3.75 mol Li (LCaTO-3). This result corresponds to the charge-transfer resistivity (R_{ct}) value obtained from the EIS testing result. The Ca-doping will decrease the charge-transfer resistivity (R_{ct}) value and thus, increasing the specific capacity of the sample [18].

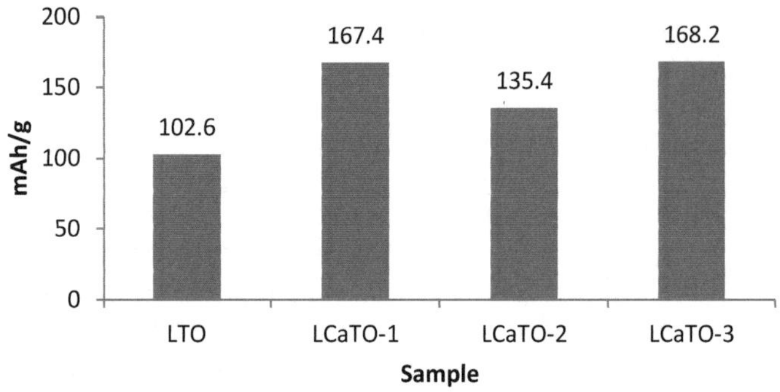

Figure 7. Initial Discharge capacity rate

Figure 8 shows the graph of C-rates of performance in each sample. It can be seen a reduced rate of specific capacity with the increasing rate is used. Sample LCaTO-1 and LCaTO-3 have the highest specific capacity value is indeterminate at rate of 0.2 C, i.e. 166.6 mAh/g Sample with specific capacity under them is LCaTO-2 value of 140.9 mAh/g, and the last is a sample LTO with capacities the lowest specific i.e. 123.7 mAh/g. With the increase in the rate of current use, the specific capacity gradually decreases. On the higher rate, the performance of the samples is best shown by the LCaTO-2 with capacity discharge 30.2 mAh/g at 12 C, and 39.2 mAh/g at 10 C. Sample LCaTO-3 has a capacity at the C-rate is high under LCaTO-2, i.e. worth 16.25 mAh/g at 12 C and 24.6 mAh/g at 10 C. Under LCaTO3, there is LCaTO-1 sample with a discharge capacity of 11.75 mAh/g at 12 C and 12.5 mAh/g at 10 C.

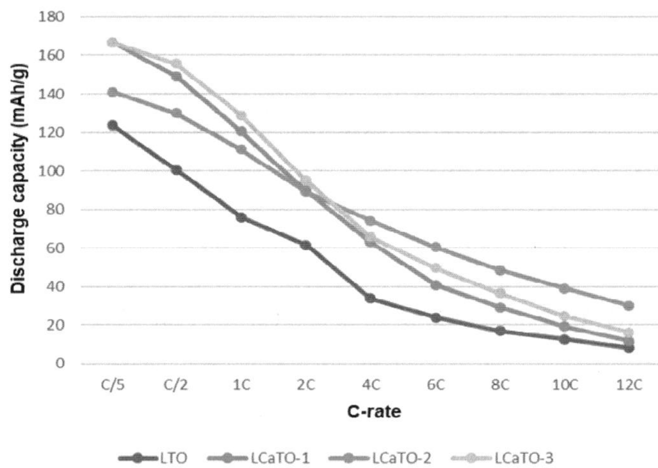

Figure 8. Discharge capacity Charts vs. *C-rates.*

Table 3. Percentage of capacity retention on different C-rate vs. capacity at the rate of 0.2 C

C-Rate	Percentage of capacity retention			
	LTO	LCaTO-1	LCaTO-2	LCaTO-3

C/2	81.33	89.44	92.19	93.34
1C	61.28	72.33	78.85	77.37
2C	49.80	53.72	63.31	57.02
4C	27.49	37.88	52.56	39.50
6C	19.32	24.61	42.87	29.77
8C	13.66	17.53	34.40	21.97
10C	10.19	11.52	27.75	14.77
12C	6.62	7.05	21.43	9.75

Discharge capacity retention percentage is highest at the rate implied by the high flow samples LCaTO-2. Initially at a rate of flow is low (0.5 C and 1 C), LCaTO-3 still show retention capacity with the highest percentage (93.34% vs. capacity at 0.2 C). However, starting in the current rate of 2 C to 12 C, LCaTO-2 shows the performance of the most high capacity retention among all samples. In the 12 C, sample LCaTO-2 shows the discharge capacity of 30.2 mAh/g, 21.43% of capacity at 0.2 C. This value is far greater than the sample LTO without doping which saw capacity 8.19 mAh/g, only 6.62% of the capacity of 0.2 C. Thi finding shows that the rate capability performance also increased with Ca doping [18].

In addition to retention capacity at the C-rate is high, coulombic efficiency data is also included in the test data processing charge-discharge. Figure 9 shows the graph of the coulombic efficiency of each sample in C-rate. The value of this coulombic obtained from the efficiency percentage of capacity discharge divided by the capacity charge on each C-rate. All samples showed high efficiency values almost reaching ~ 99%, with the lowest efficiency 98.80%. There are anomalies where the value of the efficiency that exceeds 100% on one sample. This happens because the application of cut-off voltage at 0.80 V, so it is possible there are still some lithium ion that is left on the previous cycle and migrating the next cycle so that the resulting value of the efficiency exceeding 100% because the discharge capacity. In general, high coulombic efficiency on all samples showed that doping does not influence negatively to coulombic efficiency LTO obtained due to the characteristic zero strain when the lithiation/delithiation process.

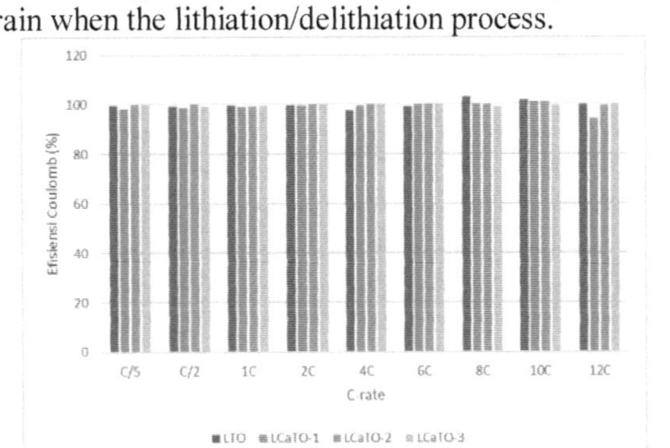

Figure 9. The graph of the coulombic efficiency vs. C-rates.

4. Conclusions

1. Synthesis of active anode material $Li_4Ti_5O_{12}$ doped using dopant Ca^{2+} from $CaCO_3$ of chicken eggshell successfully carried out using the method of solid-state.

2. Characterization via XRD testing indicates doping ion Ca^{2+} does not change the host structure of LTO spinel. Optimum doping levels ion Ca^{2+} can enter the structure of LTO shown by $Li_{3.90}Ca_{0.05}Ti_5O_{12}$, with the addition of doping impurities produce exceeds the value in the form of $CaTiO_3$. There is a difference in the size of the crystallites on average on a significant sample of LTO and $Li_{3.85}Ca_{0.075}Ti_5O_{12}$ probably caused by less evenly temperature distribution on a furnace that used.
3. SEM Results showed material synthesis through the method of solid-state using High Energy Ball Mill (HEBM) produces the characteristic morphology of the agglomerate particles and porous on all samples. The addition of doping on LTO exerts influence in the form of a decrease in the size of the particles, with $Li_{3.85}Ca_{0.075}Ti_5O_{12}$ as the smallest particles with the size of the sample.
4. Calcium Doping LTO using dopant $CaCO_3$ chicken eggshells enhance electronic conductivity in general. The highest conductivity indicated by $Li_{3.75}Ca_{0.125}Ti_5O_{12}$ charge transfer barriers with the lowest i.e. 39.5 Ω.
5. The CV test results showed that doping calcium ions on the LTO using dopant $CaCO_3$ chicken egg-shells exert influence on the decline in the level of polarization of electrodes that deliver impact reaction kinetics.
6. Calcium doping improves performance on high C-rates. The best performance is indicated by $Li_{3.85}Ca_{0.075}Ti_5O_{12}$ with discharge capacity totalling 30.2 mAh/g at 12 C along with the increase in electronic conductivity and decrease of particle size so that shortening the diffusion distance.

5. Acknowledgements

The authors would like to express gratitude to "The Direktorat Riset dan Pengabdian Masyarakat Universitas Indonesia (DRPM-UI)" for the financial support for this research under the grant of Riset Hibah no: PITTA/561/FT/2018.

6. References

[1] N. Nitta and G. Yushin, "High-Capacity Anode Materials for Lithium-Ion Batteries : Choice of Elements and Structures for Active Particles," 2013.

[2] B. Xiangchen Sun, Pavle V. Radovanovic, Cui, "Advances in spinel Li4Ti5O12 anode materials for lithium-ion batteries," *New Journal of Chemistry*, 2014.

[3] R. Cai, X. Yu, X. Liu, and Z. Shao, "Li4Ti5O12/Sn composite anodes for lithium-ion batteries: Synthesis and electrochemical performance," *Journal of Power Sources*, vol. 195, no. 24, pp. 8244–8250, Dec. 2010.

[4] A. Laumann, "Novel routes to Li4Ti5O12 spinel: Characterization and phase relations," *Electrochemistry*, 2010.

[5] T. Shu-Yi, F. Kuan Zong, and N. Chung-Ta, "Conductivity Enhancement and Thin-Film Processing of Li 4 Ti 5 O 12 (LTO) Spinel for Li Battery Applications Shu-Yi Tsai," vol. 68, no. 2, pp. 37–43, 2015.

[6] Q. Zhang and X. Li, "Recent developments in the doped-Li4Ti5O12 anode materials of lithium-ion batteries for improving the rate capability," *International Journal of Electrochemical Science*, vol. 8, no. 5, pp. 6449–6456, 2013.

[7] J. Neunzehn, T. Szuwart, and H. P. Wiesmann, "Eggshells as natural calcium carbonate source in combination with hyaluronan as beneficial additives for bone graft materials,

an in vitro study," *Head and Face Medicine*, vol. 11, no. 1, pp. 1–10, 2015.

[8] D. Setiawan, A. Subhan, and S. A. Saptari, "Ca-doped LTO using waste eggshells as Ca source to improve the discharge capacity of anode material for lithium-ion battery," *AIP Conference Proceedings*, vol. 030073, p. 030073, 2017.

[9] H. Yan, Z. Zhu, D. Zhang, W. Li, and Qilu, "A new hydrothermal synthesis of spherical Li4Ti5O12 anode material for lithium-ion secondary batteries," *Journal of Power Sources*, vol. 219, pp. 45–51, 2012.

[10] Y. Wang, C. G. Niu, L. Wang, Y. Wang, X. G. Zhang, and G. M. Zeng, "Synthesis of fern-like Ag/AgCl/CaTiO3plasmonic photocatalysts and their enhanced visible-light photocatalytic properties," *RSC Advances*, vol. 6, no. 53, pp. 47873–47882, 2016.

[11] Q. Zhang, S. Zhang, F. Ning, X. Lu, Y. Liu, and L. Nie, "Calcium Doping of Lithium Titanium Oxide Nanospheres : A Combined First-Principles and Experimental Study," vol. 510515, pp. 1–5, 2016.

[12] A. Nugroho, K. Y. Chung, and J. Kim, "A Facile Supercritical Alcohol Route for Synthesizing Carbon Coated Hierarchically A Facile Supercritical Alcohol Route for Synthesizing Carbon Coated," *The Journal of Physical Chemistry*, no. July, pp. 18–193, 2015.

[13] Q. Zhang, M. G. Verde, J. K. Seo, X. Li, and Y. S. Meng, "Structural and electrochemical properties of Gd-doped Li4Ti5O12 as anode material with improved rate capability for lithium-ion batteries," *Journal of Power Sources*, vol. 280, pp. 355–362, 2015.

[14] T. P. Zhou, X. Y. Feng, X. Guo, W. W. Wu, S. Cheng, and H. F. Xiang, "Solid-state synthesis and electrochemical performance of Ce-doped Li4Ti5O12 anode materials for lithium-ion batteries," *Electrochimica Acta*, vol. 174, pp. 369–375, 2015.

[15] G. Pfaff, "Synthesis of calcium titanate powders by the sol-gel process," *Chemistry of Materials*, vol. 6, no. 1, pp. 58–62, 1994.

[16] G. Gralik, A. E. Thomsen, C. A. Moraes, F. Raupp-pereira, and D. Hotza, "Processing and characterization of CaTiO 3 perovskite ceramics," pp. 53–57, 2014.

[17] Q. Zhang, C. Zhang, B. Li, S. Kang, X. Li, and Y. Wang, "Preparation and electrochemical properties of Ca-doped Li4Ti5O12 as anode materials in lithium-ion battery," *Electrochimica Acta*, vol. 98, pp. 146–152, 2013.

[18] Q. Zhang, S. Zhang, F. Ning, X. Lu, Y. Liu, L. Nie, C. Ouyang, and L. Zhang, "Calcium Doping of Lithium Titanium Oxide Nanospheres: A Combined First-Principles and Experimental Study," *Energy Technology*, vol. 5, no. 4, pp. 539–543, 2017.

The Effect of Service Time on Carbide Volume Fraction and Hardness of Catalyst Tube Materials

Maulidin Bastian[1] **Eko Sugiharto**[1] **Husaini Ardy**[2*]
[1]Department of Inspection, PT Pupuk Sriwidjaja Palembang, Indonesia
[2]Department of Material Engineering, Bandung Institute of Technology, Indonesia

*Corresponding author (E-mail): husaini.ardy@material.itb.ac.id

Abstract. The effect of service time on secondary carbides formation in reformer furnace tubes has been studied to obtain a relationship between carbide volume fraction, service time, and hardness. Tube material is Kubota KHR35CT which is part of ASTM A297 HP class. The study was performed by examinations of microstructural evolution and hardness. The microstructures were examined using optical and scanning electron microscopes of the new tube, and the other tubes that had been in service for 17,500; 96,300 and 113,800 hours at 850°C. Carbide volume fraction was measured using image analyzer. Carbide volume fraction of the new tube is 7.72%, and increased to 8.86% after 17,500 hours, 9.54% after 96,300 hours, and 10.95% after 113,800 hours. The increase in carbide volume fraction to 8.86 % did not affect the hardness substantially because the hardness only increases from 191 to 198 BHN. However, increasing exposure times to 96,300 and 113,800 hours decreased the hardness to 170 and 173 BHN respectively. These decreases in hardness may be corresponding to the coarsening and agglomeration of carbide particles due to longer exposure time at 850°C.

Keywords: catalyst tube, creep, degradation, microstructure, carbide precipitate.

1. Introduction

Cast alloys of ASTM A297 HP are widely used for radiant tube in reformer furnace to produce hydrogen gas as a result of endothermic reaction between light hydrocarbon and water vapor. The main degradation mode of reformer tubes is creep because they are being operated at high temperatures (up to 1000°C) and internal pressures. In some cases other degradation modes such as overheating and thermal shock [1,2] may also contribute to shorten the service life. Therefore, the tube material shall have high creep resistance and long service life.

The primary reformer in ammonia plant of Pusri Unit 3 has a capacity of 1100 MTPD and consists of 378 catalyst tubes of Kubota KHR35CT that arranged in nine harps of 42 tubes each. The tubes are vertically installed and hanged on the spring at the top and to expand downwards. The light hydrocarbon feeding gas and steam enters from the top end at a temperature of 400°C and pressure of 455 psig, and flowing down through the catalyst in individual tubes before coming out of the tubes at 850°C and pressure of 443 psig. The endothermic reforming reaction takes place in the catalyst tubes. The heat is supplied from 200 arch burners and 10 tunnel burners in the furnace. The heat is transferred to the catalyst tubes through radiation to maintain metal temperature at 850°C.

The microstructure of ASTM A297 HP alloy consists of austenitic dendrite surrounded by eutectic carbides. The primary carbides were precipitated at the grain boundaries during solidification and fine secondary carbides were precipitated in the matrix at service temperature [3,4]. High temperature strength of the tube was obtained from these precipitates [5,6]. Embrittlement and reduction of strength might occur when these precipitates coarsened or agglomerated. Further microstructure degradation may involve the formation of creep cavities, cavities coalescence, micro-cracking and final propagation of cracks [1,3].

The present work will discuss about the microstructure evolution and its effect on the hardness at various service time of reformer furnaces.

2. Materials and Methods

Nine tubes samples, consists of a new tube and eight used tubes from Primary Reformer Pusri Unit 3 Row-J harp with different service time are shown in Table. 1. Figure 1 shows the location of samples taken from each tube for laboratory analysis

Table1. Tube samples from primary reformer Pusri-3

Identity	Service time (hours)
New (N)	0
T8	17500
T11	17500
T2	96300
T12	113800
T18	113800
T22	113800
T37	113800
T42	113800

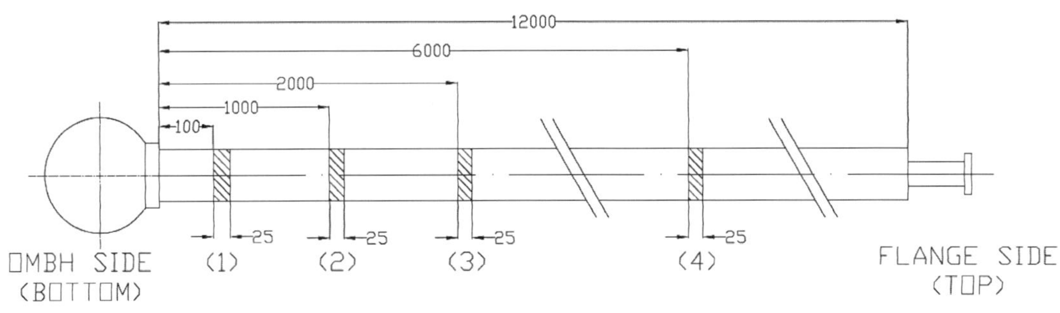

Figure 1. Tube dimensions and location of sampling

Tubes materials have been exposed to high temperature at various time, there is a possibility that some of the tubes has been degraded by creep. Safety limit for creep expansion on this catalyst tube is 6% [7]. To verify the status of each tube; circumferential dimensions were measured to obtain data about deviation from the new tube dimensions. Figure 2 shows the measurement data. It can be seen that most of the tube was expand less than 1%, however small portion of the tubes were expanded higher but still below 2%.

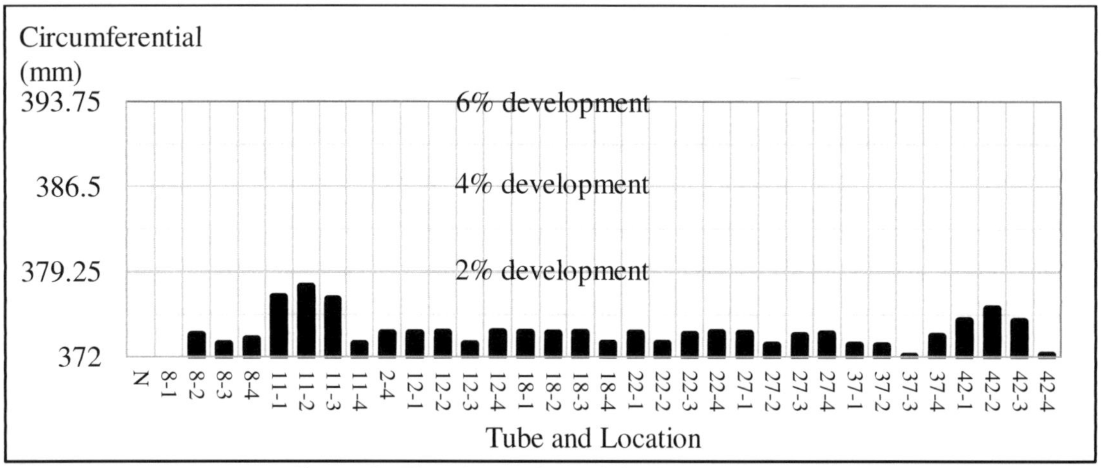

Figure 2. The circumferential tubes expansion

Chemical compositions of each tube were examined using optical emission spectroscope ARL 3460. Table 2 shows chemical composition of each tube and comparison to ASTM A297 HP and Kubota KHR35CT specifications. All of chemical composition fulfills ASTM A297 HP specifications except nickel content is slightly higher. Some of the tubes have carbon content slightly lower than minimum requirement for Kubota KHR35CT.

Table. 2. Chemical Composition of Tubes (weight %)

Service Time (hours)	Identity	%Cr	%Ni	%Nb	%Ti	%Mn	%Si	%P	%Mo	%C
	ASTM A297 HP	**24-28**	**33-37**	**-**	**-**	**2.0 Max**	**2,5 Max**	**<0,04**	**0,5**	**0,35-0.75**
	KHR35CT	**24-27**	**33-37**	**0,4-1.0**	**Add**	**2.0 Max**	**1,5 Max**	**<0,03**	**0,5**	**0,45-0.55**
Measured										
0	New	24.87	36.98	0.45	0.07	0.84	0.82	0.02	0.01	0.40
17500	T8	25.16	37.52	0.44	0.09	0.84	0.95	0.02	0.01	0.39
17500	T11	24.95	37.31	0.44	0.04	0.96	0.72	0.01	0.01	0.47
96300	T2	25.57	35.31	0.43	0.03	0.73	0.74	0.02	0.01	0.42
113800	T12	24.42	36.97	0.52	0.04	0.88	0.72	0.02	0.01	0.42
113800	T18	25.16	38.10	0.50	0.05	0.89	0.87	0.01	0.01	0.41
113800	T22	24.39	36.40	0.45	0.05	0.97	0.83	0.02	0.01	0.40
113800	T37	24.67	38.07	0.46	0.07	0.80	0.80	0.02	0.01	0.40
113800	T42	25.36	38.63	0.43	0.07	0.87	0.80	0.01	0.01	0.47

*%Fe: Rem

Microstructures of 2 × 1.4 × 2 cm specimens were performed on cross section perpendicular to the tube axis using light microscope Nikon Eclipse MA2000 and scanning electron microscope JEOL JSM 6510A. Volume fraction of carbides from un-etched specimens was quantified at outer, middle and inner wall section using Image-J software. Brinell hardness measurements were conducted by Mori Testing Machine using 1500 kN load following JIS Z 2243:2008 standard on specimens from the center thicknesses of tube wall.

3. Results and Discussion

3.1. Microstructure of as cast material

Kubota KHR35CT tubes were manufactured by centrifugal casting and the microstructures consist of dendrites and directional grains. Referring to Fe-Cr-Ni phase diagram [8], the primary phase (matrix) of tube sample is austenite. At temperature over 700°C, carbide-former element will strongly tie up carbon and there will be no free carbon detached from structure [3]. Chromium content is the highest among other carbide forming elements, therefore chromium carbides will be present extensively in the dendrites structure along grain boundaries. Niobium and titanium elements will form secondary carbides in the matrix [1,5,6,9,10]. Figure 3 shows the microstructure of as cast tube. It can be seen that the matrix material was free from secondary carbides precipitates.

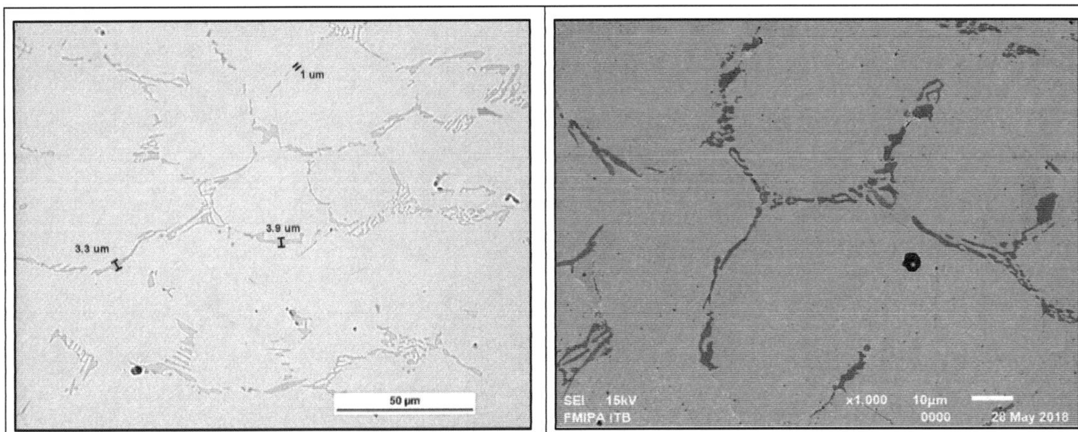

a. Light microscope image b. Scanning electron image

Figure 3. The microstructure of as cast tube.

3.2. Microstructure after long term services

Figure 4 shows optical and SEM micrographs of sample after 17,500 hours of services. Dendrites boundaries are thin and discontinuously ordered with fine precipitates in the matrix. On the SEM images (Figure 4), the grey area is chromium carbides, the black and white area are TiC and NbC respectively [1,9].

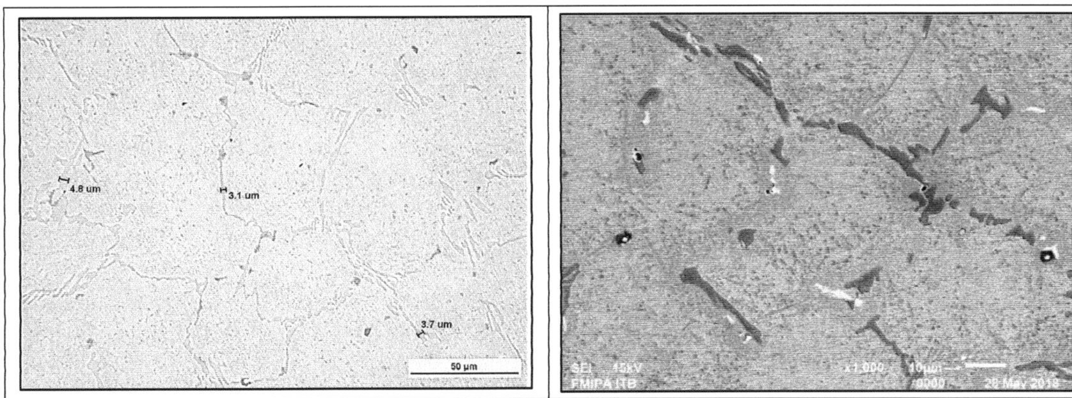

a. Light microscope image b. Scanning electron image

Figure 4. The microstructure of tube after 17,500 hours of service.

IOP Conf. Series: Materials Science and Engineering **547** (2019) 012041 doi:10.1088/1757-899X/547/1/012041

After 96,300 hours of service (Figure 5), precipitates in matrix are coarser than that of 17,500 hours. Volume fraction of fine precipitate Cr_7C_3 in matrix was decreased at this stage and thickening of dendrite boundaries by primary carbides was also evident. The thickening of primary carbides is due to the formation of $Cr_{23}C_6$ carbides along grain boundaries above 700°C [3,4]. The carbides formation reactions is,

$$23Cr_7C_3 \rightarrow 7Cr_{23}C_6 + 27C$$

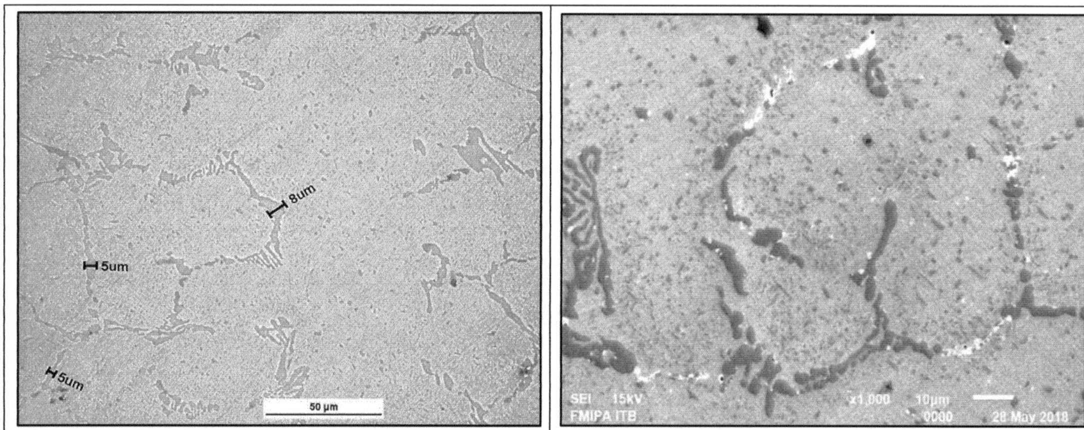

a. Light microscope image b. Scanning electron image

Figure 5. Microstructure of tube after 96,300 hours in service.

The microstructure after 113,800 hours of service is shown in Figure 6. It can be seen that the dendrite boundaries were thicker than that of 96,300 hours, and continuous. Thickening of the dendrites boundaries is due to coarsening and agglomeration of primary carbides. Precipitates in the matrix become less distributed, and also coarsening. At this stage, more white area of G-phase (Ni-Nb-Si) [6] is present on some part of the boundaries. The microstructures at several times of services in this paper are consistent with the observation by Almeida et. al [6] and Liu et. al [1].

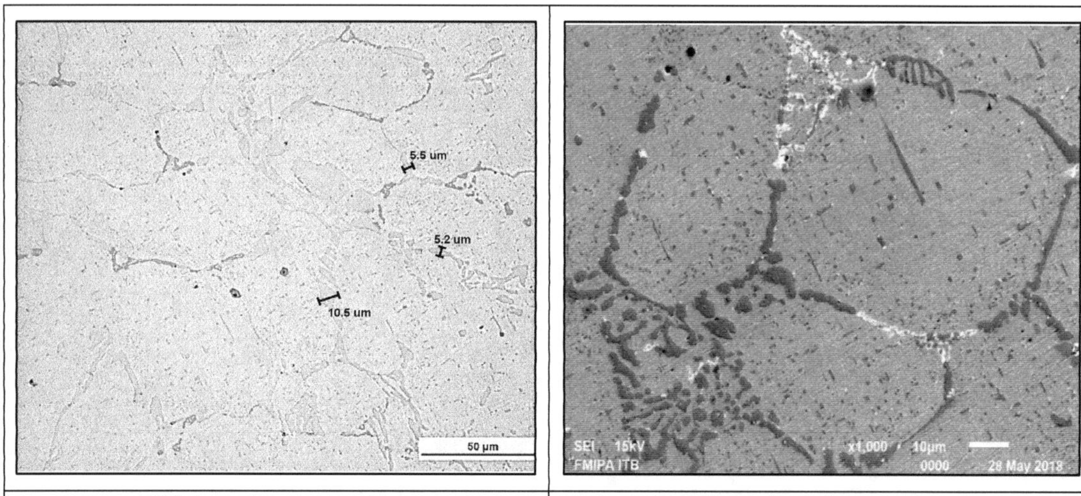

a. Light microscope image b. Scanning electron image

Figure 6. Microstructure of tube after 113,800 hours of service.

The voids are observed after long term services (Figure 7). The formation of voids indicated that creep deformation by grain boundary sliding mechanism were active. The presence of voids can be seen clearly at the dendrites boundaries of primary carbides. The same results are also observed by Wahab et.al [11].

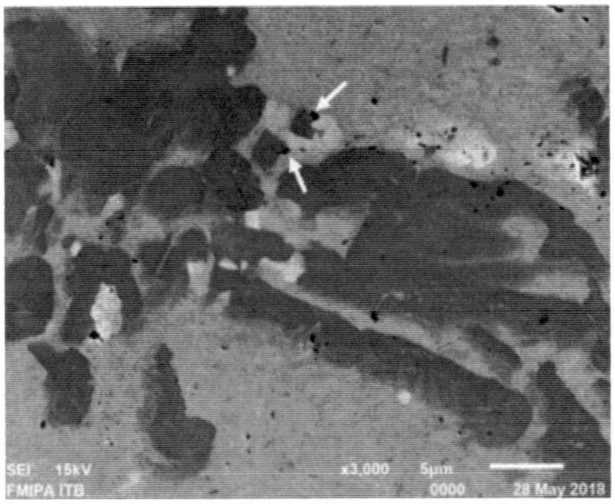

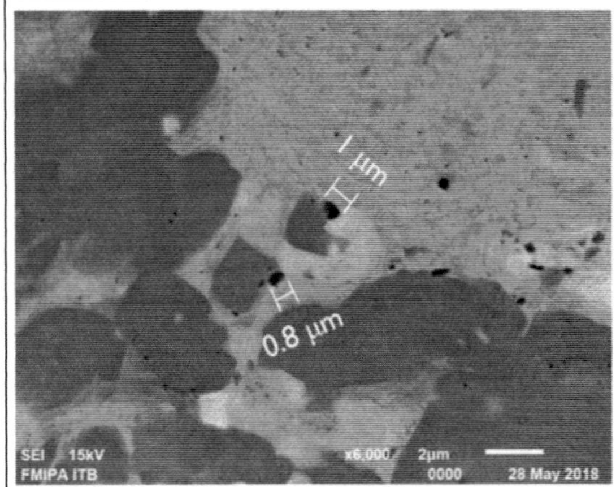

a. 3000x magnification	b. 6000x magnification

Figure 7. Voids by SEM imaging at 113800 hours of service time

3.3. *Effect of service time on carbides*

Precipitation of Cr_7C_3 in Fe-Cr-Ni alloy steel occurred primarily at dislocation in matrix at 500˚C [4] and slowly dissociated at 700°C. Above this temperature, formation of more stable $Cr_{23}C_6$, TiC and NbC in several locations such as dendrite boundaries, area near boundaries, and matrix occurred. Fine and unstable Cr_7C_3 precipitates have higher chromium concentration than coarse and stable $Cr_{23}C_6$ precipitates, so that diffusion of chromium atoms occurred from fine Cr_7C_3 to coarser $Cr_{23}C_6$ precipitates, until fine precipitate dissolve completely and coarse precipitate agglomerate. From diffusion criteria, time of service is proportional to cubic of precipitate radii (r^3) while the precipitate growth is inversely proportional to r^2 [4]. Fine precipitates not only have higher growth rate but also form at shorter time. When service time is very long enough, fine precipitate dissolve slowly and coarse carbides grow up, agglomerate and have lower growth of precipitate than fine precipitate. This can be seen in Figure8.

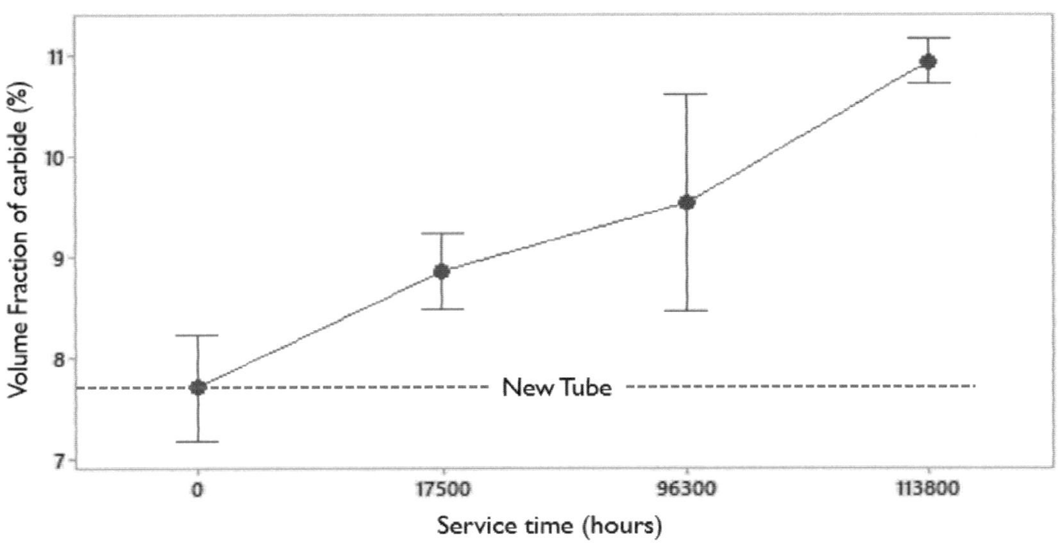

Figure 8. The plot of volume fraction of carbides against service time of tubes.

Volume fraction of carbides in the new tube is 7.72% and increased to 8.86% after tubes had been in service for 17,500 hours. Precipitates were fine and density was high especially in the matrix, as seen in Figure 4. After long time exposure, the carbide content was increased to 9.54% at 96.300 hours of service and to 10.95% at 113,800 hours of service respectively. Precipitates were coarse and density was low because nucleation rate was very low compare to growth rate (Figure 5-6).

3.4. Mechanical properties of tubes at various service time

Creep resistance of high temperature alloys rely upon development of carbides in the structure. Kubota KHR35CT has primary chromium carbides and also small additions of strong secondary carbide-forming elements, particularly titanium and niobium. Small amount of titanium and niobium will form very fine secondary precipitates in the matrix to hinder the dislocations movement.

Plastic deformation of material occurred because of dislocations movement. These dislocations on precipitation strengthened alloys could cut and-or bow at precipitates [4]. If precipitates are fine, wide interparticle spacing, and soft enough, dislocation will cut the precipitates. For coarse and hard precipitates, dislocation will be bowing just before passing precipitates to produce a dislocation loop and leaved a back stress (Orowan Stress). This stress is inversely proportional to inter-precipitate spacing. For large precipitate, this stress is low because interspacing is wider than fine precipitate.

During dislocation cutting, the stress will arise with increasing precipitate size and proportional to square root of precipitate size. When dislocation is bowing, the back stress will decreased with increasing precipitates size. These two mechanisms will present during precipitates-dislocation interaction as shown by the pink line in Figure 9. The maximum strengthening effect is achieved at transition from dislocation cutting to bowing mechanism at optimum size of precipitate (r^*).

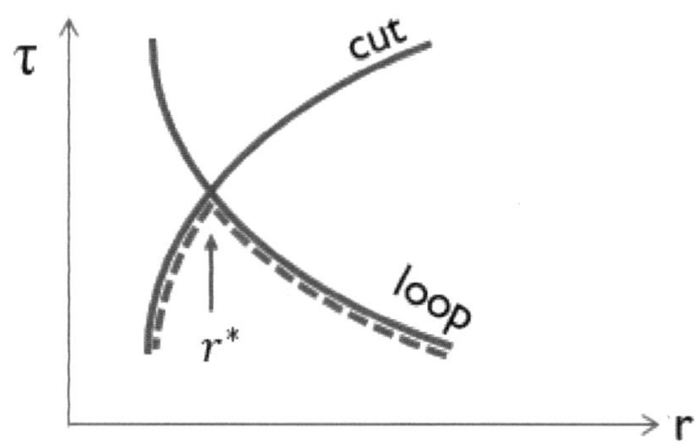

Figure 9. Transition of strengthening on precipitated alloys.

The effect of service time on material hardness is shown in Figure 10. At shorter service time (0 to 17,500 hours), the hardness only arises from 191 to 198 BHN. The precipitates resistance to dislocation motion is still lower due to low volume fraction of fine precipitates and relatively wide interparticle spacing. The carbides were coarsened and agglomerates (Figure 5-6) when the tube had been service for longer time (after 96,300 and 113,800 hours), where dislocation bowing is dominant and dislocation resistance decreased significantly as shown by decreasing in hardness from 198 to 170 BHN. It can also be seen that the hardness is almost constant at 170 BHN after 96,300 hours of service.

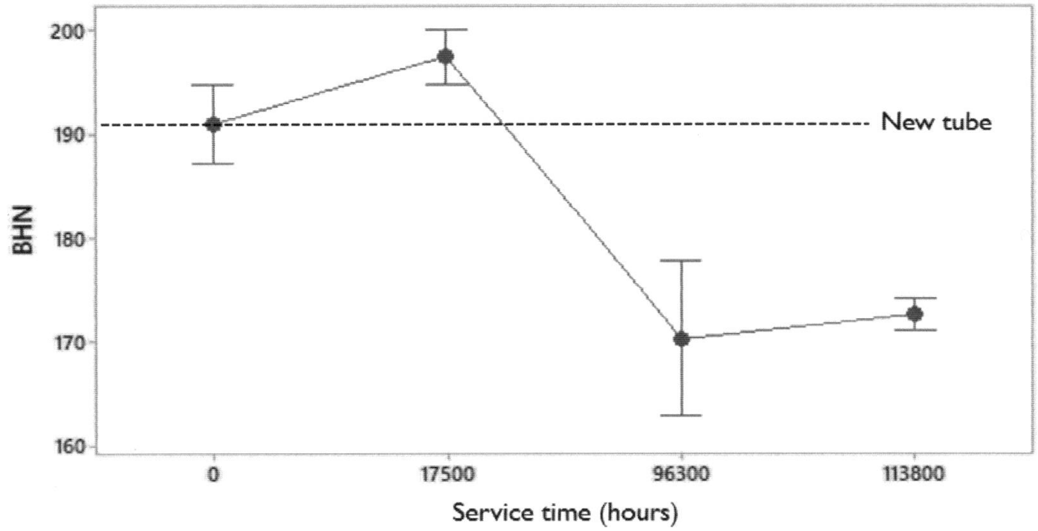

Figure 10. The effect of service time at 850°C on the hardness

4. Conclusions

a. Carbide volume fraction in new tube is 7.75% and increases linearly to 10.95% after 113,800 hours of service at 850°C.
b. The hardness of the new tube is 198 BHN and decreased to 170 BHN after 96,300 hours of service at 850°C.
c. Beyond 96,300 hours of service time, the hardness is almost constant at 170 BHN.

5. Acknowledgements

The author (MB) is grateful to PT Pupuk Sriwidjaja Palembang for financial support during graduate study at Materials Engineering Department, Bandung Institute of Technology.

6. References

1. Liu CJ, Chen Y. Variations of the microstructure and mechanical properties of HP40Nb hydrogen reformer tube with time at elevated temperature. Mater Des [Internet]. 2011;32(4):2507–12. Available from: http://dx.doi.org/10.1016/j.matdes.2010.08.031
2. Huber J, Jakobi D. Centricast Materials for High-Temperature Service. Nitrogen+Syngas. 2010;307:117–21.
3. Porter DA, Easterling KE, Sherif MY. Phase Transformations in Metals and Alloys. 3rd ed. Porter D, Easterling K, editors. London, New York: CRC Press; 2014. 522 p.
4. Dieter GE. Mechanical metallurgy [Internet]. SI Metric. Vol. 273, Journal of the Franklin Institute. McGraw Hill Book Company (UK) Ltd; 1962. 338 p. Available from: http://linkinghub.elsevier.com/retrieve/pii/S0016003262911456
5. Jasinsky W, Zawada P. The precipitation processes of Manaurite XM superalloy after 2000 hours of aging. Foundry Eng. 2008;8(1/2008):155–8.
6. De Almeida LH, Ribeiro AF, Le May I. Microstructural characterization of modified 25Cr-35Ni centrifugally cast steel furnace tubes. Mater Charact. 2002;49(3):219–29.
7. Sugiharto E. Remaining Life Estimation on Material ASTM A297 HP Mod (Nb/MA)

by Ultrasonic Attenuation Method and Creep Strain Measurement. Institut Teknologi Bandung; 2011.

8. ASM International Handbook Committee. The Materials Information Company. Technology [Internet]. 2001;2:3470. Available from: http://books.google.com.hk/books?id=eC-Zt1J4oCgC
9. Nunes FC, de Almeida LH, Dille J, Delplancke JL, Le May I. Microstructural changes caused by yttrium addition to NbTi-modified centrifugally cast HP-type stainless steels. Mater Charact. 2007;58(2):132–42.
10. Sourmail T. Precipitation in creep resistant austenitic stainless steels. Mater Sci Technol [Internet]. 2001;17(1):1–14. Available from: http://www.tandfonline.com/doi/full/10.1179/026708301101508972
11. Wahab AA, Hutchinson CR, Kral M V. A three-dimensional characterization of creep void formation in hydrogen reformer tubes. Scr Mater. 2006;55(1 SPEC. ISS.):69–73.

The Effect of Minimum Distance Between Bamboo Strengtheners on Flexural Strength of Bamboo Reinforced Cement

Aditianto Ramelan[1],* Riska Rachmantyo[1], Ahmad Yasier Hanif[1], and Firmansyah Sasmita[1]

[1]Department of Materials Engineering, Faculty of Mechanical and Aerospace Engineering, Bandung Institute of Technology, Ganesha 10, Bandung 40132, Indonesia

ramelan@material.itb.ac.id

Abstract. Various species of Bamboo is widely available in Indonesia. Usually it is used as pillars, roof frame, and for walls in a woven form and it can be used for steel substitute in reinforced concrete. In making bamboo as a strengtheners or reinforcement a proper volume fraction should be chosen. Besides that, the minimum distance between bamboo bars or inter-strengtheners distance must be determined so that the reinforcement effect on cement matrix takes place. Ten variations of mixtures were used in this research. The number and distance between reinforce bars and its arrangement perpendicular or parallel with loading direction. From the observation it is found that both of the vertically and horizontally arranged bamboo bars at minimum inter-strengtheners distance of 10 mm gave maximum reinforcement and the flexural strength are 16.1 ± 0.3 MPa and 15.8 ± 0.6 MPa. Whereas cement only sample gave 7.2 ± 0.6 MPa hence 2 folds increase. For distance less than 10 mm, the flexural strength is lower, even volume fraction is relatively high and it supposed to give higher strength, in fact this is due to composite reinforcement effect did not occur and this is belief due to a hindered hydration process.

Keywords. Cement, bamboo, composite, minimum inter-strengthener distance, flexural strength.

1. Introduction

There are about 154 types of bamboo and can be found abundant in Indonesia [1]. One of the usages of bamboo is to be used for masonry or house building and has it been used since long time ago. Usually used as column, roof frame or as a woven for partition. Bamboo has a high tensile strength that almost near to steel [2]. Besides it is cheaper than steel it also grows very fast i.e. between 3-5 years before it can be used for masonry. The disadvantage of bamboo is easily attacked by moths, less resistance to weather and easily get burnt off [3]. Bamboo can also be used to replace steel as reinforcement in a concrete. The amount of bamboo used should be optimized. Besides the volume fraction the minimum distance between strengtheners (inter-strengtheners distance) is very important so that the properties of a composite take effect. In this research besides observing the minimum distance between the strengtheners also direction of its arrangement was also observed to the effect to flexural strength of bamboo composite.

2. Method of Experiment

This research used Gombong bamboo (*Giganthocloa Pseudoarundinacea*) that is widely available in in West Java. The rectangular bamboo reinforced cement was made with an arrangement as shown in figure 1 and 2. Prior to cement casting the bamboo should be treated first with a polychloroprene solution to make sure the interfacial strength between bamboo and cement is high [4]. This research is an analogy to ACI 318-08 chapter 7 that is specifically for steel bar in a concrete. The three point bending test or flexural test was undertaken on the rectangular specimen. The bamboo is arranged vertical and horizontal in the cement rectangular sample perpendicular or parallel to the load. The rectangular specimen could consist of 2, 3 and 4 bamboos strengtheners as shown in figure 1. Two sizes of bamboo bar or strengtheners were used which are 5x5 mm and 7x7 mm. The sizes of the rectangular were kept the same for the specimens, hence the more the bamboo added distance between them became narrower, this is to observe the effect of strengtheners distances on the composite behaviour. The length off specimen is 290 mm with a rectangular shape 50 x 50 mm. To maintain a fixed distance between bamboos a stiff cable ties was used, hence on cement casting the arrangement will keep in its predesigned configuration.

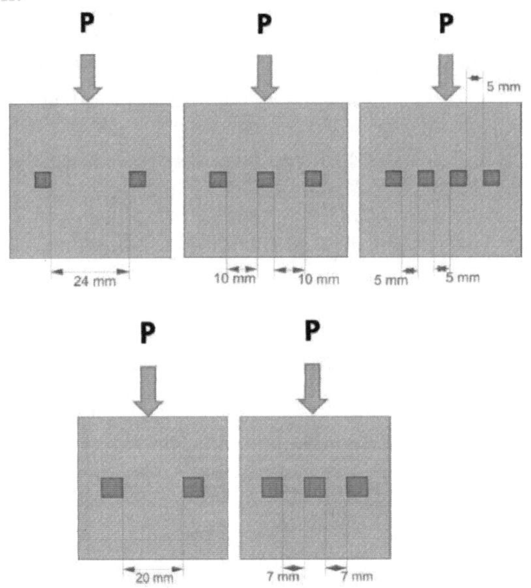

Figure 1. Cross sectional BRC with horizontally arranged bamboo strengtheners shown as squares. Distance between bamboos is shown on the above schematic.

The three points bending test on rectangular specimens were carried out in the Metallurgical and Materials Engineering Laboratory of Faculty of Mechanical and Aerospace Engineering, Bandung Institute of Technology. The tests were carried out by using a hydraulic compression machine Tarnogrocki with a loading speed of 7 mm/min and the distance between supports is 17 mm and specimen were necessary measured for further calculation.

3. Results and Discussion

The specimens were tested on a three point bending test or flexural test. The results are as shown in table 1, table 2.

Table. 1. Flexural strength of cement only specimen.

Speciment Number	Flexural Strength (MPa)	Flexural Strength Average (MPa)	Stdev
1	8.0		
2	7.1		
3	6.7	7.2	0.6
4	7.6		
5	6.5		

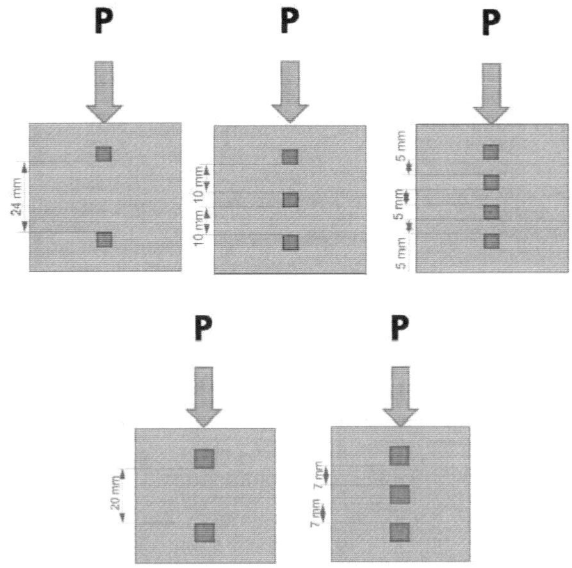

Figure 2. Cross sectional BRC with vertically arranged bamboo strengtheners shown as squares. Also written in small letters The istance between bamboos.

In table 1 it is shown that the flexural strength of a cement only rectangular sample is 7.2 ± 0.6 MPa. The hydration process took place to form the CSH and the specimens were cured for 28 days. In table 2 is shown the flexural strength of vertically arranged BRC under three point bending or flexural test.

Table. 2. Flexural strength of a vertically arranged BRC

Type	Specimen	Flexural Strength (MPa)	Average	stdev
2 Bamboo, distance 24 mm ($BRCv_{24}$)	1	11.4		
	2	11.5	11.9	0.7
	3	12.7		
3 Bamboo, distance 10 mm ($BRCv_{10}$)	1	15.8		
	2	16.2	16.1	0.3
	3	16.3		
4 Bamboo, distance 5 mm ($BRCv_{5}$)	1	11.4		
	2	12.6	12.3	0.8
	3	12.9		
2 Bamboo, distance 20 mm ($BRCv_{20}$)	1	11.8		
	2	12.2	12.5	0.9
	3	13.5		

IOP Conf. Series: Materials Science and Engineering **547** (2019) 012042 doi:10.1088/1757-899X/547/1/012042

3 Bamboo, distance 7 mm (BRC_{v7})	1	12.4	13.5	1.1
	2	13.7		
	3	14.5		

In figure 3 it is shown the comparison between semen rectangular and vertically arranged BRC.

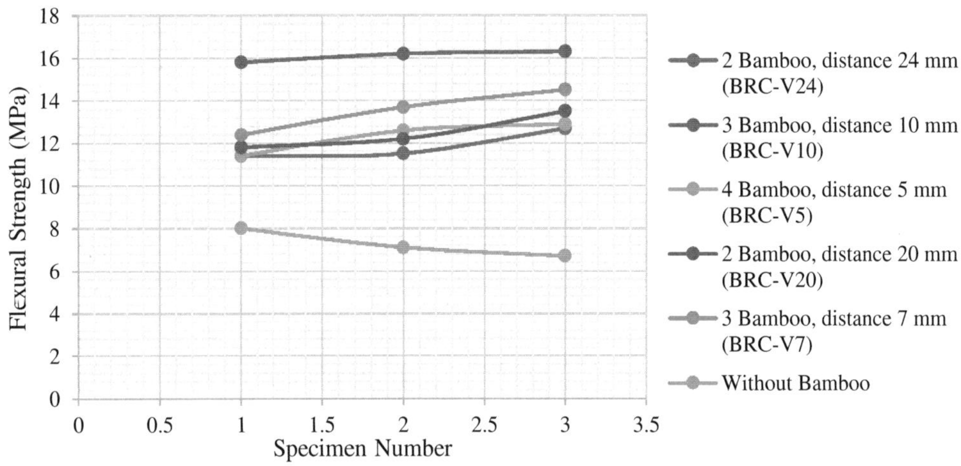

Figure 3. Flexural Strength curve on vertically arranged BRC.

In figure 3 the addition of bamboo as a strengtheners to a cement only rectangular sample will increase its flexural strength from 7.2 ± 0.3 MPa to maximum 16.1 ± 0.3 MPa. This shows a composite strengthening effect in two folds. Cement or concrete has a resistance to compressive stress rather than a tensile stress. On bending the surface of the specimen undergoes compression, shear in the middle and tension at the bottom. The stress experienced at the top will be at maximum and 0 at the neutral axis. If the load is high at the top hence the bottom of the specimen will experience a high tensile stress, once crack occur the specimen will continue to break. In a BRC the existence of bamboo as a strengtheners will compensate the tensile stress on the bottom part of specimen and at once it can also prevent the crack to propagate further. Hence the flexural strength increased as shown in figure 3. In figure 3 the vertically arranged BRC with 10 mm distance between strengtheners (BRC_{v10}) has the highest strength of 16.1 ± 0.3 MPa. In BRC with 24 mm and BRC with 20 mm inter-strengtheners distances namely BRC_{v20} and BRC_{v24} both have 2 bamboos strengtheners. The strengthener size are respectively 5x5x290 and 7x7x290 mm. According the rule of mixture the volume fraction of bamboo the BRC_{v20} has a higher flexural strength than BRC_{v24} 12.5 and 11.9 MPa respectively. BRC_{v10} and BRC_{v7} have the same amount of bamboo strengtheners which is 3 but differ in size which is 5x5 mm and 7x7 mm and distance between bamboos which is 10 and 7 mm. Supposed that the flexural strength of BRC_{v7} is higher than BRC_{v10} according to volume fraction. In fact it is observed as the other way around the flexural strength of BRC_{v10} is higher than BRC_{v7} which are 16.1 and 13.5 MPa, hence the inter-strengtheners distance has a dominant effect on the composite properties. This also occurred to BRC_{v5} that has higher volume fraction than the BRC_{v10} and supposed to have a higher flexural strength too. In fact the observation shows that the flexural strength of BRC_{v5} is 12.3 MPa which is lower than the BRC_{v10} which is 16.1 ± 0.3

MPa. From the observation above we can conclude that it is not only strengtheners volume fraction has a dominant effect on flexural strength of a composite but also the inter-strengtheners distance will determined the mechanical properties of the cement composite. At an inter-strengtheners distance of 5 and 7 mm the composite strengthening do not occur. In the case of concrete with steel reinforcement the individual distance between strengtheners should not be less than the diameter of the strengtheners if is uniform and the biggest diameter it is non uniform or 5 mm bigger than the coarse aggregate [5]. In concrete making according to ACI 318-08 chapter 7, the clean distance between parallel strengtheners on the same layer should not be smaller than the dimension of the strengtheners or 25 mm [6]. In case of the distance between strengtheners less than the minimum distance this might hinder the cement paste to enter the gap between strengtheners completely. This will cause the formation of pores or gap with no cement in between hence will decrease the flexural strength of the specimen. Besides tested with a vertical arrangement the BRC is also tested with a horizontal bamboo arrangement as shown in figure 1 and the results is tabulated in table 3.

Table. 3. Flexural strength of a horizontally arranged BRC (BRCh).

Type	Specimen	Flexural Strength (MPa)	Average	stdev
2 Bamboo, distance 24 mm ($BRCh_{24}$)	1	13.0	14.2	1.0
	2	14.7		
	3	14.9		
3 Bamboo, distance 10 mm ($BRCh_{10}$)	1	15.1	15.8	0.6
	2	16.2		
	3	16.2		
4 Bamboo, distance 5 mm ($BRCh_5$)	1	12.7	13.3	0.6
	2	13.4		
	3	13.9		
2 Bamboo, distance 20 mm ($BRCh_{20}$)	1	12.8	13.3	0.4
	2	13.5		
	3	13.7		
3 Bamboo, distance 7 mm ($BRCh_7$)	1	12.6	13.7	0.1
	2	14.1		
	3	14.5		

In table 3 it is shown that the horizontal arrangement BRC with inter-strengtheners distance 10 mm or $BRCh_{10}$ has the highest flexural strength of 15,8 ± 0.6 MPa. This result is similar with the the vertically arranged BRC with 10 mm inter-strengthener distance or $BRCv_{10}$.

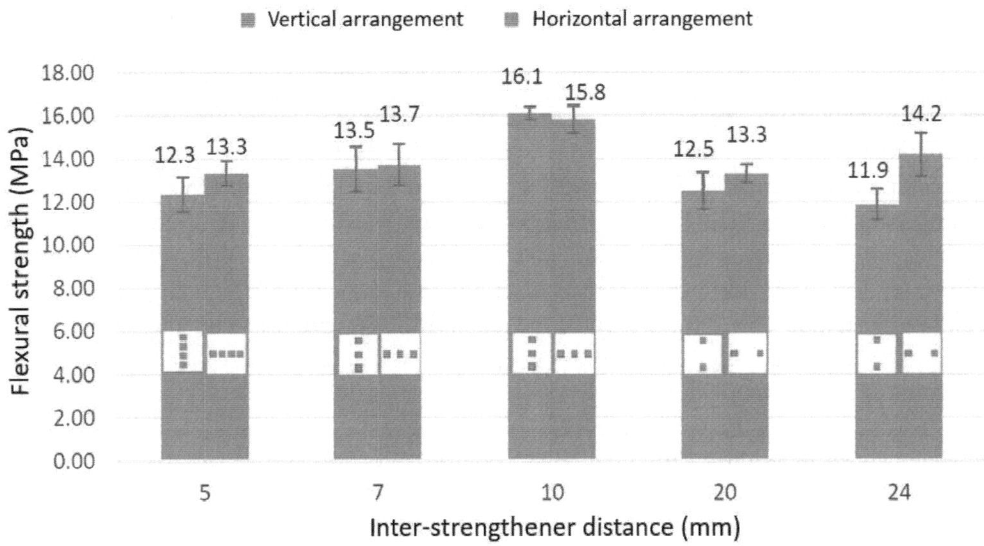

Figure 4. Flexural strength of bamboo reinforce cement with vertically arranged strengthener or BRCv and and horizontally arranged BRCh.

In figure 4 it is shown that the flexural strength of all BRCh is higher than the BRCv except for 10 mm distance between strengtheners. The flexural strength of BRCh has a higher value relatively to BRCv, this might be due to the bigger load bearing area. The area under the load for bamboo with vertical arrangement strengtheners is smaller in compare with the area under load on horizontal arrangement, which were 5 x 290 mm and 5 x 290 x by number of bamboo bars or strengtheners. Hence the stress experienced by the bamboo will become smaller too. In figure 1 it is shown the thickness of cement above the bamboo on horizontal arrangement will be thicker in comparison with the vertical one. The thickness of the cement above bamboo arrangement will determine the flexural strength and also the crack propagation behaviour once a crack initiate on the BRC's surface.

4. Conclusions

The flexural strength of bamboo reinforced cement (BRC) with horizontal and vertical arrangement of bamboo bars or strengtheners is not only determined by the volume fraction of the components but also depending on the minimum inter-strengtheners distance. As shown above as the inter-strengtheners distance of 10 mm the effect of components on composite strengthening still occurred and giving a value of 16.1 ± 0.3 MPa for $BRCv_{10}$ and 15,8 ± 0.6 MPa for $BRCh_{10}$ less than 10 mm the flexural strength will decrease. The minimum value of 10 mm could be proposed to ACI for standard in making bamboo reinforced cement as stated in ACI 318-08 chapter 7 for steel bar reinforcement. The hydration state of the cement matrix between strengtheners strongly affects the flexural strength hence minimum inter-strengtheners distance became the main issue.

5. References

[1] E.A, Widjaja. 1995. *Gigantochloa nigrociliata, gigantochloa pseudoarundinacea, and gigantochloa robusta.* Backhuys Publishers Leiden 114-120.

[2] K. Ghavami. 2005. *Bamboo as reinforcement in structural concrete elements.* Journal of Cement and Concrete Composites Vol. 27 637-649.
[3] W. Liese. 1985. *Anatomy and Properties of Bamboo.* Institute of Wood Biology and Wood Preservation of the Federal Research Centre for Forestry and Forest Product. Germany.
[4] A. Ramelan, R. Rachmantyo, MK Bijaksana, F. Sasmita. 2015. *Karakteristik antarmuka komposit semen berpenguat bambu Gombong (BRC).* Nasional Seminar Proceeding in Material and Metallurgy VIII 160-164.
[5] B.C. Punmia. 2007. *Limit state design of reinforced concrete.* New Delhi : Laxmi Publications.
[6] ACI 318-08. 2008. American Concrete Institute.

The Tensile Strength of Petung Bamboo Fiber Reinforced Epoxy Composites: The Effects of Alkali Treatment, Composites Manufacturing, and Water Absorption

Gunawan Refiadi[1]* Yusi Syamsiar[2] Hermawan Judawisastra[3]
[1]*Mechanical Engineering Vocational Education Study Program, College of Teacher Training and Education (STKIP) Sebelas April, Jl. Anggrek Situ 10 Sumedang 45323, Indonesia*
[2]*Textile Chemistry Department, Polytechnic of Textile Technology, Jl. Jakarta 31 Bandung 40272, Indonesia*
[3]*Material Engineering Department, Faculty of Mechanical and Aerospace Engineering, Bandung Institute of Technology (ITB), Jl.Ganesha 10 Bandung 40132, Indonesia*

*gunawan_refiadi@stkip11april.ac.id

Abstract. Natural fiber application due to its competitiveness attracted many research in green composites. However, the tensile properties of natural fiber composites might be influenced by several factors such as treatment of the fibers, manufacturing processes, and water absorption levels. In this research, we use petung bamboo fibers as natural fibers candidate for a good reinforcement in green composites. The study focused on the tensile property of bamboo fiber reinforced epoxy composites due to alkali treatment, manufacturing process, and water absorption. The composites were made with three variations: the fibers were varied without and with alkaline treatment (5% NaOH); manual lay-up method and hot press were applied as manufacturing variation; moisture content in the composites was varied in dry condition and after water absorption treatment. Tensile testing and Scanning Electron Microscope (SEM), were performed to improve tensile strength of composite and fiber-matrix interface quality. The alkali treatment has less effect to the tensile strength than of the volume fraction gain and the void content. Composite manufacturing by hot press has a significant effect to the improvement of tensile strength up to 37% compare to the hand lay-up method. Water absorption up to 8.8% had decreased the tensile strength of composites up to 29%. The alkali treatment on the bamboo fibers had effectively reduced water absorption into the composite as well as the tensile strength reduction.

Keywords: petung bamboo, alkali treatment, composites manufacturing, water absorption

1. Introduction

Polymeric composites such as CFRP (Carbon Fiber-Reinforced Polymer) and GFRP (Glass Fiber-Reinforced Polymer) had abundantly produced due to their advantages as well as high mechanical specific performances. Unfortunately, the drawback of those materials is lack of sustainable environment due to pollutant expel both to the air and the land. However, this limitation in turn opens up other material research opportunities such as green composites. The composites use lignocellulosic structure natural resources as reinforcement such as ramie, coconut, hemp, jute, and bamboo. In tropical country such Indonesia, bamboo has a potential advantage as an abundance resource of natural fibers. In addition, bamboo have low-cost, fast growth, renewable, CO_2 free, flexibility, as well high specific strength and stiffness compared to glass fiber [1][2]. Several considerable research dealing with bamboo were reviewed [3] and in Indonesia petung bamboo (Dendrocalamus asper) in the form of stems, slats, and fibers was studied also [4]–[6]. However, petung bamboo fibers as well as the other lignocellulose materials still have several weaknesses and problems. High variability in properties and high moisture absorption [7] affected the interfacial bonding quality between fibers and matrix. Meanwhile, composite materials in its application could be affected by hygroscopic conditions. To face the problem, several chemical methods such as alkalization, graph copolymerization, and coupling agents have been proposed to improve both interfacial quality and hydrophobing of the fibers [8], [9]. Studies of water absorption effects on bamboo fiber-reinforced polyester through fibers surface modifications through alkali treatment [6] have been reported. In terms of composite manufacture methods, hand layup and hot press were commonly used to fabricate natural fiber composites considering time and cost [11]. However, the effects of combination between manufacturing methods, alkali treatment and water absorption to tensile properties of petung bamboo fiber reinforced composite has not been investigated yet. This study is aimed to investigate tensile strength of petung bamboo fiber-reinforced epoxy composites resulting from alkali treatment, composites manufacturing, and water absorption.

2. Materials and Methods

2.1 Materials

Supplied from Hutan Penelitian Bambu in Arcamanik, Bandung (West Java, Indonesia), petung bamboo stems then extracted into 100~300 mm fiber bundles by chemo-mechanical methods. The resulting fibers then variably prepared both by non-alkali (NA) and alkali method using 5% v/v NaOH solution (A). Resulted before [12], both treated fibers (NA and A) has density of 1.05 and 1.11 g/mm^3, respectively. The 5% alkali was found as the optimum solution resulted maximum bamboo fiber strength [12] and least water uptake in bamboo composites [10]. The matrix used in this study was Lycal GLR 1011 Part A epoxy resin with curing agent GLR 1011 Part B with density of 1.13 g/mm^3.

2.2 Composites Manufacturing and Testing

Prior to composites manufacturing by hand lay-up (**LU**) and hot press (**HP**) methods, bamboo fibers were oven dried 24 hours – 110°C to achieve dry condition. After petung bamboo fiber aligned unidirectionally on the molding surface, we poured epoxy mixing. For HP method, pressure, temperature, and holding time were set on 10 MPa 60°C, and 1 hour, respectively. With the LU process, impregnation were apllied manually using specific roll at 25°C. Demolding then performed after the composite 24 h cured.

Composites characterization were evaluated physically by volume fraction and density measurements. Fiber volume fraction (V_f) and void volume fraction (V_v) calculated using eq. (1) and eq. (2) as per ASTM D-3171 based on bamboo petung fibers and epoxy densities. We measured composite density by pycnometry method.

$$V_f(\%) = \left[\frac{(w_b/\rho_b)}{(w_b/\rho_b) + (w_m/\rho_m)}\right] x\ 100\% \qquad \text{.......... (1)}$$

$$V_v(\%) = 1 - \rho_c[(w_b/\rho_b) + (w_m/\rho_m)] x\ 100\% \qquad \text{.......... (2)}$$

IOP Conf. Series: Materials Science and Engineering **547** (2019) 012043 doi:10.1088/1757-899X/547/1/012043

where W^b, W^m, are weight fractions of bamboo fibers, epoxy matrix and ρ^b, ρ^m and ρ^c are densities of bamboo fibers, epoxy matrix, composites, respectively.

Water absorption tests were performed to the tensile specimen as per ASTM D570-98 but modified as per previous research [6]. in this study, we use water absorption by two intervals of immersion. Firstly, weighing the samples in boiling water 60 minutes for every 10 minute. Secondly, with 30 minute intervals from the minutes of 60 up to 240. The water uptake in the composites was weighted by electronic balance with accuracy 10^{-4}. Moisture content in the composite samples then calculating by weight difference using Eq.3.

$$W_t(\%) = \frac{W_1 - W_0}{W_0} x\ 100\% \qquad \text{.......... (3)}$$

where W_t total water-absorbed in composites, W_0 and W_1 are composites weighs on H0 and H1 respectively. Tensile testing of composites were conducted on both H0 and H1 samples per ASTM D3039. We use Tensilon machine equipped by a 5 kN load cell to get the tensile property. Besides, SEM (*Scanning Electron Microscope*) analysis using JEOL GSM-636OLA type were also conducted after tensile test to evaluate fracture surface of the composites.

3. Results and discussion

3.1 Physical Properties of Petung Bamboo-Reinforced Epoxy

Table 1 and Figure 1(a) and (b) show the fiber volume fraction and the void volume fraction of petung bamboo-reinforced epoxy under varying processing conditions. The fiber volume fraction was 22% up to 28% and the void volume fraction ranges from 0% to 6%. The higher pressure with uniform distribution in hot press processing method resulted in higher fiber content as shown in Figure 1(a). Alternatively, the composite fabricated by manual lay-up had a higher void content as shown in Figure 1(b).

Table 1. Fiber and void volume fractions of petung bamboo fiber-reinforced epoxy

Manufacture Processes	V_f (%)		V_v (%)	
	NA	A	NA	A
LU	24	22	1	6
HP	28	28	1	0

V_f – fiber volume fraction, V_v – void volume fraction; NA – Non-Alkalized; A – 5% Alkalized

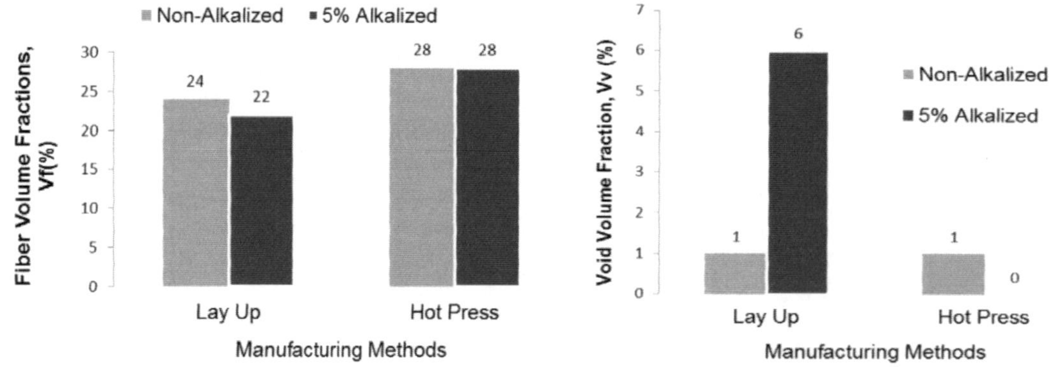

(a) (b)

Figure 1. Fiber volume fraction, (a) and void volume fraction, (b) of composites as per manufacturing methods and alkali treatments. [LU: Hand Layup, HP: Hot Press]

The high void content in composite primarily due to several factors such as air bubbles entrapped during matrix mixing and composite fabrication. The air from the process of manual lay-up, and the volatile component from the heat process due to chemical reactions [13]–[15]. The void content resulted in both of hot press and lay up methods are the same which is only 1.0%. After the alkali treatment the void content become 6% and 0% in both of the hand lay-up and hot press methods, respectively. Tightly controlled parameters in the hot press method through closed-loop system resulting in the lower void content whereas the open-loop controlled system in the hand lay-up method showed an anomaly result by increasing void content.

3.2. The Effects of Alkali Treatment on Tensile Strength and Water Uptake

Table 2 shows the influence of alkali treatments (0% and 5% NaOH) on tensile strength and water absorption of petung bamboo fiber reinforced epoxy composites were varied from 79 MPa up to 126 MPa while the water absorption gain was ranged from 3 up to 8.8%.

Table 2. The tensile properties and water absorption of petung bamboo fiber reinforced epoxy composites before and after alkali treatment.

Manufacture Process	Tensile Strength (MPa)		Water absorption (%)	
	NA	A	NA	A
LU	83	79	5,15	3,79
HP	126	126	8,82	6,57

NA – Non-Alkalized; A – 5% Alkalized

Figure 2 shows that the alkali treatment effect in the hot press (HP) method did not increase the composite's strength as stated in the literature [16]. Moreover, using lay up (LU) method, the effect of alkali treatment showed slight reduction of tensile strength due to high void content, 6% (see Table 1). The void causes stress concentrations and tend to weaken the fiber-matrix bonding [17]. The tensile strength of hot-pressed composites seem to be more dominated by high fiber volume fraction, 28%V_f, and lower void content than that of the alkali effect.

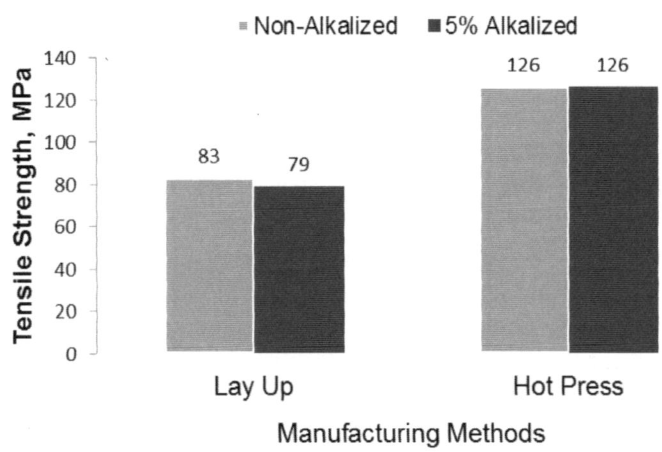

Figure 2. Tensile properties of petung bamboo fiber reinforced epoxy before and after alkali treatment as per hand layup (LU) and hot press (HP) methods. [NA: Non-Alkalized, A: 5% Alkalized]

Figure 3 shows water absorption reduction after fiber alkali treatment on lay-up (LU) and hot press (HP) composites. The higher fiber volume fraction in the hot press composites lead to higher water uptake than that of the lay-up composites. This uptake was due to the inherent hygroscopic properties of bamboo fibers as lignocellulosic materials[16] which contain polar hydroxyl groups produced by both cellulose and hemicellulose content.

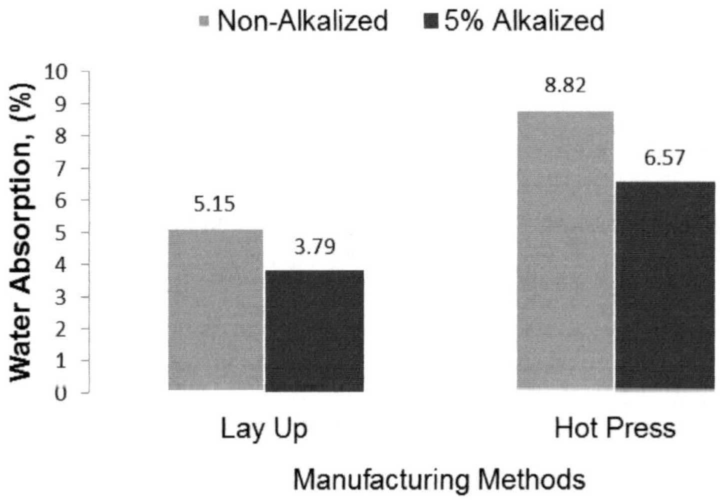

Figure 3. Water absorption characteristics of composites based on alkali treatment. [NA: Non-Alkalized, A: 5% Alkalized]

Both composites have 26% reduction of water uptake after the alkali treatment. This was due to the removal of hemicellulose and lignin content during the treatment [16]. In the previous report [10] hemicellulose was responsible for water absorption of the fibers. Alkali treatment makes the fibers less hydrophilic as the number of hydrophilic hydroxyl groups reduces by react which 5% NaOH [17], [18]. Therefore, reduction of water absorption enhanced the water resistance of composites. Besides, alkali treatment increases fiber fibrillation and therefore decreases fiber diameter, which in turn improve the effect of surface area for contact with the matrix [16]. This might be contributing to decrease water absorption due to bamboo fibers more encapsulated by the epoxy matrix.

3.3. Effect of Manufacture Process on Tensile Strength

Figure 4 shows the comparison of tensile strength of composites resulting from manual lay-up (LU) and hot press (HP) processing. The highest tensile strength of composites was obtained by using hot press method. It can be seen that hot press processing has improved the tensile strength of composites up to 37%. The combination of higher fiber volume fraction and lower void content improved tensile strength, see Table 1. The higher fiber volume fraction in the hot press (HP) was achieved by homogenous temperature and steadier pressure distributions. In this method temperature and pressure are tightly controlled in closed-loop system. Conversely, processing parameters in hand lay-up (LU) controlled manually through the operator skills level which is the open-loop system. Dealing with the lower void content in the hot pressed composite, this could be achieved due to uniformly high pressure – forced out the entrapped void – supported by the molding mechanism.

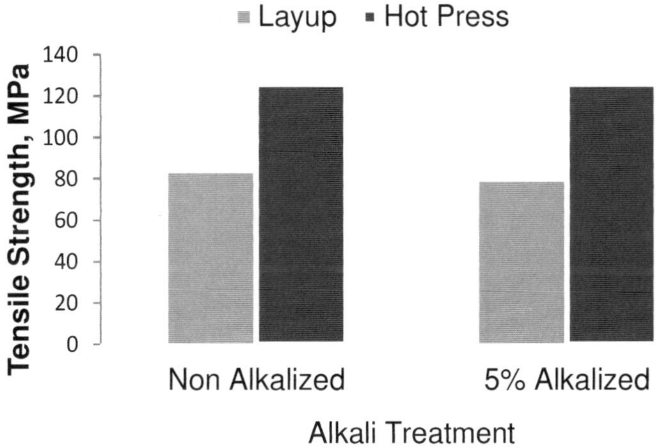

Figure 4. The comparison of the tensile strength by using manual lay-up (LU) and hot press (HP) processing

Figure 5 shows the composites fracture surfaces resulting from both the layup and hot press processes. Lower extent of fiber pullout in HP composites (Fig. 5b) compared to LU composites confirms the improvement of fiber-matrix interface, resulting in higher tensile strength. This improvement might be resulted from the combined temperature and pressure energy supply in HP method.

(a) (b)

Figure 5. Comparison of the fracture surface of composites manufactured by (a) Hand Lay Up and (b) Hot Press

3.4. Effect of Water Absorption on Tensile Strength

Table 3 and Figure 6 depicted the effect of water absorption on tensile strength of petung bamboo fiber-reinforced epoxy composites based on alkali treatment and manufacturing processes. The absorption of water from 3.79% up to 8.8% leads to the degradation of fiber-matrix interface [18]. The result was a reduction of composites tensile strength from 17% up to 29% (see Table 3). Besides, water composites immersion swelled the hydrophilic bamboo fibers resulting in micro cracking of the brittle matrix[19]. As water, penetrating into the interface through the micro cracks fiber-matrix debonding occurs and leads to composite failure [20]. High strength reduction on both non-alkalized composites (LU-NA and HP-NA) as high as 25% and 24% were due to high water absorption 5.15% and 8.8%, respectively. This water absorption made hydrogen bonds to the fibers and decreased fiber-matrix interconnections. Large strength reduction on the alkalized lay-up (LU-A) composites was dominated by high void content (see Fig. 1) than the water uptake in the composites. From Fig. 6 the best composite with the lowest strength reduction was resulting from alkalized hot press composites (HP-A). The composite shows only 17% tensile strength reduction, showing the effectiveness of alkalization to reduce water absorption and minimize the strength reduction.

Table 3. Tensile properties decrement of Petung bamboo fiber reinforced epoxy composites due to water absorption

Composite - Fiber treatments	**Tensile decrements (%)**	**Water uptake (%)**
LU-NA	25	5.15
LU-A	29	3.79
HP-NA	24	8.8
HP-A	17	6.57

NA – Non-Alkalized; A – 5% Alkalized

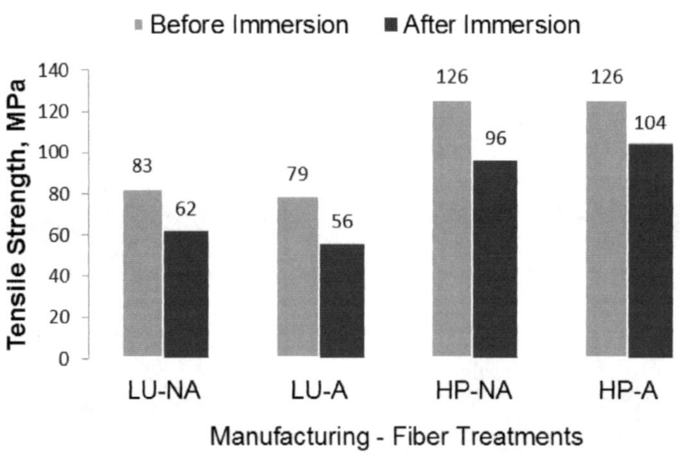

Figure 6. The effect of water absorption on tensile strength petung bamboo fiber reinforced epoxy composites.

Supporting tensile test results, SEM qualitative evaluation were added in Fig. 7. Fracture surfaces were based on fiber treatments, manufacturing processes, and water absorption conditions. Fracture surfaces of composites after water immersion (Fig. 7b, d, f, h) show larger extent of fibers pullout that related to poor fiber-matrix interface. These results corroborate with the tensile strength reduction of composites due to water absorption.

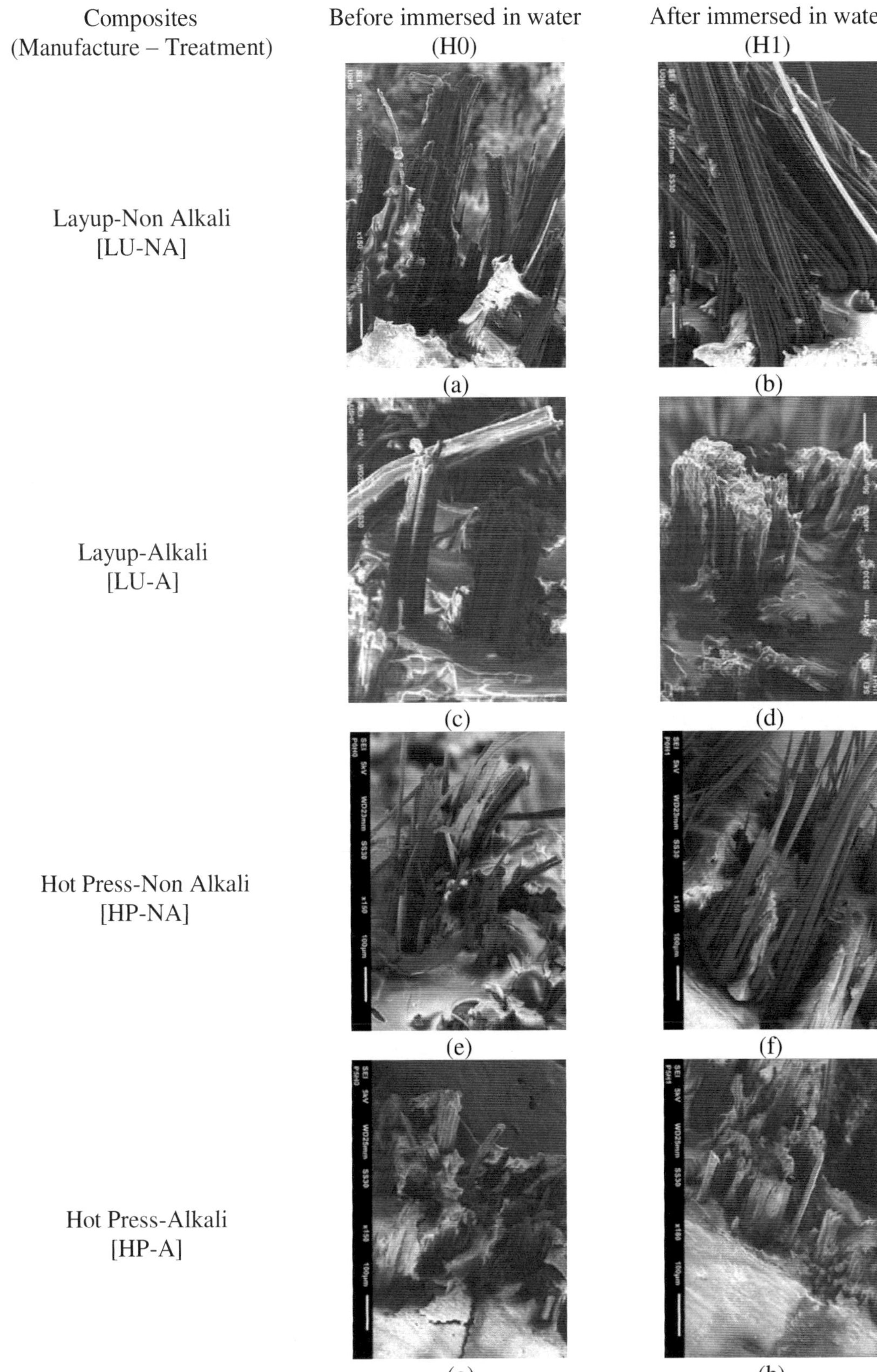
Composites
(Manufacture – Treatment)
Before immersed in water
(H0)
After immersed in water
(H1)
Layup-Non Alkali
[LU-NA]
(a)
(b)
Layup-Alkali
[LU-A]
(c)
(d)
Hot Press-Non Alkali
[HP-NA]
(e)
(f)
Hot Press-Alkali
[HP-A]
(g)
(h)

Figure 7. Comparison of composites fracture surfaces due to alkali treatment, manufacturing technique, and water absorption

4. Conclusion

The effects of alkali treatment, composite manufacturing, and water absorption on the tensile strength of petung bamboo fiber-reinforced epoxy composites have been studied. In this research, the alkali treatment has less effect to the tensile strength than of the volume fraction gain and the void content. Hot press composites manufacturing method has significant effect to the improvement of tensile strength up to 37% than that of lay-up method. Water absorption up to 8.8% decreased the tensile strength of composites up to 29%. The alkali treatment had effectively reduced water absorption and consequently minimized the tensile strength decrement.

Acknowledgements

The authors would like to express their gratitude and appreciation to Directorate General Research Strengthening and Development, Ministry of Research, Technology and Higher Education of the Republic of Indonesia (DRPM-Kemristekdikti) that have been founded the Inter-Universities Collaboration of Research (PEKERTI)

References

[1] T. HIROGAKI, E. AOYAMA, M. HUYNH, Y. NAKAMURA, K. OGAWA, and H. NOBE, "Hot press fabrication of hemisphere shell product made of bamboo fibers extracted with a machining center," J. Adv. Mech. Des. Syst. Manuf. Hot, vol. 9, no. 3, pp. 1–15, 2015.
[2] P. Chaowana, "Bamboo: An alternative raw material for wood and wood-based composites," J. Mater. Sciience, vol. 2, no. 2, pp. 90–102, 2013.
[3] H. P. S. Abdul Khalil, I. U. H. Bhat, M. Jawaid, A. Zaidon, D. Hermawan, and Y. S. Hadi, "Bamboo fibre reinforced biocomposites: A review," Materials and Design, vol. 42. pp. 353–368, 2012.
[4] S. Delgado, "The potential of bamboo in the design of polymer composites," Mater. Res., vol. 15, no. 4, pp. 639–644, 2012.
[5] H. Supomo et al., ""Analysis of the Adhesiveness and Glue Type Selection in Manufacturing of Bamboo Laminate Composite for Fishing Boat Building Material," Appl. Mech. Mater., vol. 874, pp. 155–164, 2018.
[6] H. Judawisastra, R. D. R. Sitohang, and M. S. Rosadi, "Water absorption and tensile strength degradation of Petung bamboo (Dendrocalamus asper) fiber — reinforced polymeric composites," Mater. Res. Express, vol. 4, no. 9, 2017.
[7] K. L. Pickering, M. G. A. Efendy, and T. M. Le, "A review of recent developments in natural fibre composites and their mechanical performance," Compos. Part A Appl. Sci. Manuf., vol. 83, pp. 98–112, 2016.
[8] M. Das and D. Chakraborty, "Evaluation of improvement of physical and mechanical properties of bamboo fibers due to alkali treatment," J. Appl. Polym. Sci., vol. 107, no. 1, pp. 522–527, Jan. 2008.
[9] A. W. van Vuure, L. Osorio, E. Trujillo, C. A. Fuentes, and I. Verpoest, "Long Bamboo Fibre Composites," in 18TH INTERNATIONAL CONFERENCE ON COMPOSITE MATERIALS, 2013, vol. 3, no. 3, p. 25.

[10] P. K. Kushwaha and R. Kumar, "Studies on the water absorption of bamboo-epoxy composites: The effect of silane treatment," Polym. - Plast. Technol. Eng., 2010.

[11] A. F. . et. al. Jusoh, "Natural Fiber Reinforced Composites: A Review on Potential for Corrugated Core of Sandwich Structures," in MATEC Web of Conference, 2016.

[12] G. Refiadi, N. Bayu, H. Judawisastra, and Mardiyati, "Jurnal selulosa," J. Selulosa, vol. 8, no. 1, pp. 9–20, 2017.

[13] G. Y, L. M, Z. Z, and S. Z. J, "Void formation model and measuring method of void formation condition during hot pressing process," Polym. Compos, vol. 31, no. 9, pp. 1562–71, 2010.

[14] e. a. K. H., "Effect of variation pressure, temperature, and vacuum-application time on porosity and mechanical properties of carbon fiber/epoxy," J. Compos Mater Compos., vol. 46, no. 16, pp. 1985–2004, 2011.

[15] Y. Li, Q. Li, and Z. Ma., "The void formation and their effects on the mechanical properties of flax fiber reinforced epoxy composites," Compos. Part A, 2015.

[16] A. U. M. Shah, M. T. H. Sultan, M. Jawaid, F. Cardona, and A. R. A. Talib, "A review on the tensile properties of bamboo fiber reinforced polymer composites," BioResources, vol. 11, no. 4, pp. 10654–10676, 2016.

[17] D. Kim, D. J. Hennigan, and K. D. Beavers, "Effect of fabrication processes on mechanical properties of glass fiber reinforced polymer composites for 49 meter (160 foot) recreational yachts," Int. J. Nav. Archit. Ocean Eng., vol. 2, no. 1, pp. 45–56, 2010.

[18] Z. Hanmin, L. J. J. Weibang, and N. Zhang, "Relation of modification and tensile properties of sisal fibers,"," Acta Sci Nat Uni Sunyatseni, vol. 35, pp. 53–57, 1996.

[19] A. Bismarck, I. Aranberri-Askargorta, J. Springer, T. Lampke, B. Wielage, A. Stamboulis, I. Shenderovich, and H. Limbach, "Surface characterization of flax, hemp and cellulose fibers; Surface properties and the water uptake behavior.," Polym Compos, vol. 23, pp. 872–894, 2002.

[20] G. Marom, "The role of water transport in composite materials," in Polymer Permeability, 1985, pp. 341–356.

Effect of Temperature on Corrosion behavior of Copper Silver Phosphorus Brazing Alloys in Chloride Containing Solution

Ekavianty Prajatelistia [1*], Andrieanto Nurrochman [2], Kyehyun Cho [3]

[1] Materials Science and Engineering Department, Institut Teknologi Bandung, Indonesia
[2] Mechanical Engineering Department, Sekolah Tinggi Teknologi Mandala, Indonesia
[3] School of Materials Science and Engineering, Yeungnam University, South Korea
*ekavianty@material.itb.ac.id, andrienurrochman@gmail.com, kcho@ynu.ac.kr

Abstract. The groups of phosphorus-containing brazing alloys are the most favored filler metals used for joining copper and copper-base alloys. These brazing alloys are expected to have the optimum corrosion resistance during the application. However, failures arise at the braze joint generally caused and initiated by the corrosion of the brazing alloys when used at higher temperature than the room temperature. Since most studies of the phosphorus containing brazing alloys were performed at room temperature, this work has been extended to investigate the effect of various temperature on the corrosion behavior. Electrochemical measurements were carried out to observed the temperature effect with the temperature variation of 25, 40, 55 and 70 °C in 3.5% NaCl electrolyte. As the results, the copper-silver-phosphorus brazing alloy was undergone a selective dissolution by the attack of Cl- to the copper element. The temperature dependence of the alloy showed in agreement with the relationship of 2.3RT/F that have similar manner on pure copper tafel region. The schematic diagrams of the corrosion process were then proposed. This work could be helpful to understand in the field practically because the application temperature is usually higher than that of room temperature in general.

Keywords: Brazing alloy, corrosion, high temperature corrosion

1. Introduction

The families of phosphorus-containing brazing alloys are the most favored when copper and copper-base alloys are brazed [1,2]. During the brazing process, phosphorus reacts when the alloys melts, producing copper metaphosphate slag. This feature is known as a self-fluxing function, therefore, usual fluxes may be ommited when phosphorus-containing brazing alloys are used for joining copper and copper base alloys. Moreover, the higher phosphorus content, generally with the maximum of 8.25 wt%, the better the alloy flowing into the joint gap.

Silver addition to a phosphorus-containing brazing alloy is well known not only improving its flow characteristic, but also increasing its ductility. Better properties in ductility possibly gained at the equal flow characteristic of the silver-free brazing alloy [3,4].

The phosphorus-containing brazing alloys with silver addition corrosion behaviour was studied by Valero-Gomez *et al.*[5] in lithium bromide solution at 25°C. The passive layer of the alloys was formed. The passive region showed a large range of inhibition even though the cracked layer were observed and occured from its started to form. Furthermore, copper-silver-phosphorus alloys were proved to have more noble potential in lithium bromide solution than the silver-free alloys. It also has been established that phosphorus-containing brazing alloys with silver addition formed a protective passive film on the surface of the alloys when exposed to water. Idrissi *et al.*[6] found that the film mainly made of cupruos oxide and calcium carbonate when studied the corrosion behaviour of the alloy in local water at room temperature.

Besides its excellences as a corrosion resistant, some failures of the brazed joint were often arisen. Sigler *et al.* [7] reported a failure in welding transformers that brazed with copper-silver-phosphorus brazing alloy. The water leaks of failed transformers were caused by the mechanism of sulfide-induce corrosion. Selective dissolution of copper was observed and postulated occurred by the attack of sulfide species initiated at the braze alloy. The cavity became larger due to the catalyzed attack of copper either in the braze or in the base material by the phosphoric acid that produced from the changes of water chemistry under the corrosion product of copper sulfide.

Furthermore, a leak in the clip-to-strand braze connection that initiated in the braze alloy has been reported by Joseph A. Worden *et al.* [8]. In the report, the corrosion mechanisms were proposed into two stages. At the first stage, the brazing alloy was attacked by the water, particularly at the phosphorus rich phase of the alloy. The existing voids at the brazing alloys give the opportunity for water to stagnate and initiated the corrosion. Afterwards, this void grows as the corrosion progresses. At the second stage, the corrosion of the void penetrates until reaches the copper as the base metal. Hence, the phosphoric acid that produced by the corrosion of the brazing alloy attacked the copper trough the void.

The copper-silver-phosphorus brazing alloys have much attention as consequences of some good properties when used for joining. However, the failures arise frequent at the braze joint. The failures are generally caused and initiated by the corrosion of the brazing alloys. Copper as the base material has been getting a large attention for their corrosion behaviour. Yet, less comprehensive information available for the corrosion behaviour of the brazing alloys, particularly in the chloride containing electrolyte.

The present study was intended to observed the behaviour of the copper-silver-phosphorus brazing alloy exposed in the 3.5% NaCl solution. In addition, since most studies of the phosphorus containing brazing alloys have been made at room temperature, this work was performed to study the temperature effect of copper-silver-phosphorus brazing alloy corrosion with temperature at 25, 40, 55 and 70°C. This work could be helpful to understand in the field practically because the application temperature is usually higher than that of room temperature in general.

2. Experimental Procedures

The materials used was commercial copper-silver-phosphorus brazing alloy with composition of 85% Cu, 15% Ag, 5% P, the materials were made with the exposed area approximately 1cm^2. The pure Copper of 99% purity was also utilized to compare the corrosion behavior. All the materials were supported by Doosan Heavy Industries & Construction Ltd. Prior each experiment the materials were mechanically polished successively with SiC abrasion paper from 600 to 2000 grits, and then mirror finished using Buehler polishing cloth wetted with diamond suspension sequentially from 9 to 1 μm. After polishing the materials were cleaned by ultrasonically degrease the coupon for 5 minutes in methanol or ethanol, and then rinse thoroughly in distilled water before dried.

The electrochemical data aquisition equipment was Princeton Applied Research potentiostat model 273A controlled with PowerSuite software, and the reference electrode used was silver/silver chloride (Ag/AgCl saturated KCl).

The experiments were conducted using three-electrode cell with a platinum mesh as the counter electrode adjacent to the working electrode (brazing alloy and copper).The condition was in open air atmosphere under 3.5% NaCl electrolyte. The electrolyte solution was prepared by using distilled water and analytical grade NaC1. The potentiodynamic experiments of the brazing alloy were performed at scanned potential from -0.1 mV_{SSCE} vs OCP to 0.4 V_{SSCE} with the scan rate of 0.1 mV/s at 25, 40, 55 and 70°C.

Cyclic voltammetry experiments were carried out with copper and brazing alloy at 25°C and the cyclic voltammograms were recorded starting from -0.8 mV_{SSCE} and reversed when reaching 1 V_{SSCE} back to the starting point at the scan rates of 10, 20, 30, and 40 mV/s.

Potentiostatic polarization experiments were performed on active-passive transition region potential of the brazing alloy that obtained from the potentiodynamic polarization curve. The corrosion behaviour of the alloy was evaluated by statically polarized the materials at each temperature of 25, 40, 55 and 70°C for 3600 s.

The corrosion products at each temperature resulted from the potentiostatic polarization experiments were identified by XRD examination. The morphologies of the corroded samples were analyzed by optical microscope and scanning electron microscopy (SEM, Hitachi S-4100).

3. Results

3.1. Potentiodynamic polarization

The potentiodynamic polarization of the copper-silver-phosphorus brazing alloys in 3.5% NaCl solution resulted in E vs log i curves and representated in figure 1.

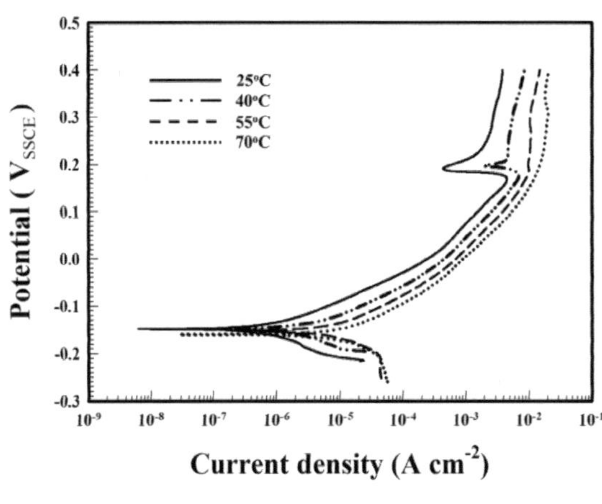

Figure 1. Potentiodynamic polarization curve of copper-silver-phosphorus brazing alloys in 3.5% NaCl solution at various temperature.

Tafel slope method were used to obtain the corrosion potential and corrosion current of the alloy. The corrosion potential (E_{corr}) shifted towards more negative value with respect to the increased of temperature and the corrosion current was also increased regarding to the increased of temperature. Moreover, the anodic tafel slopes were followed the changes of the temperature, at higher temperature the slope was steeper than that of lower temperature. The corrosion properties of the alloy that obtained by tafel slope method were summarized in table 1.

Table 1. Copper-silver-phosphorus eletrochemical parameters obtained from tafel slope method

T (°C)	E_{corr} (V_{SSCE})	I_{corr} (A)	η_a (mV/dec)	η_c (mV/dec)
25	-0.149	9.036×10^{-7}	58	76
40	-0.154	2.8×10^{-6}	61	66
55	-0.156	7.43×10^{-6}	70	57
70	-0.164	2.01×10^{-5}	73	68

The peak current obviously appeared in of the potentiodynamic polarization curve at 25°C. This peak has appeared up to 40°C with increased in current value as the temperature increased. While, the potentials were reatively unchanged. Further polarization above the peak current potential produced region with decreasing in current until touched the minimum current value. Subsequently, the current rapidly increasing until followed by a current plateau at the limiting current region.

3.2. Cyclic voltammetry

Cyclic voltammetry experiments were conducted at the scan rate varied from 10 mVs^{-1}, continued to 20 mVs^{-1}, 30 mVs^{-1} and up to 40 mVs^{-1}. The experiments resulted voltammograms of copper and copper-silver-phosphorus brazing alloys that exposed to the solution contained Cl- in an open air atmosphere. At the anodic region of copper voltammogram two peaks were showed in each scan rate, noted as A1, and A2 in figure 2. The A2 peak appeared like a shoulder at the small scan rate, then gradually more apparent with increasing scan rate. When the scan rate reached 40 mVs^{-1}, both of the anodic peaks almost have the same current intensity. Instead of the A2 peak, the A1 peak was obvious in any scan rate. Moreover, both of the anodic peak shifted towards positive values with increasing scan rate.

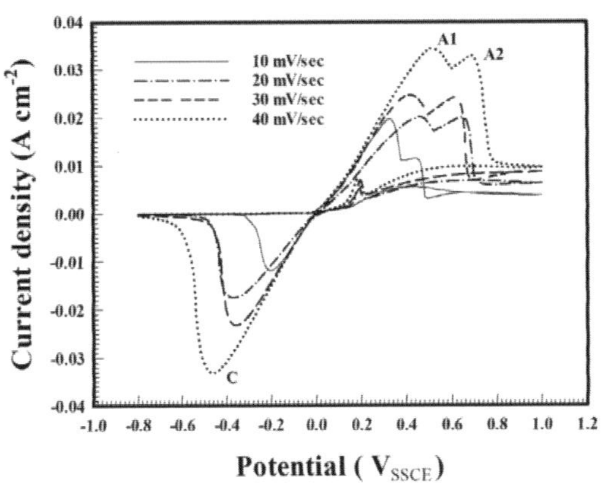

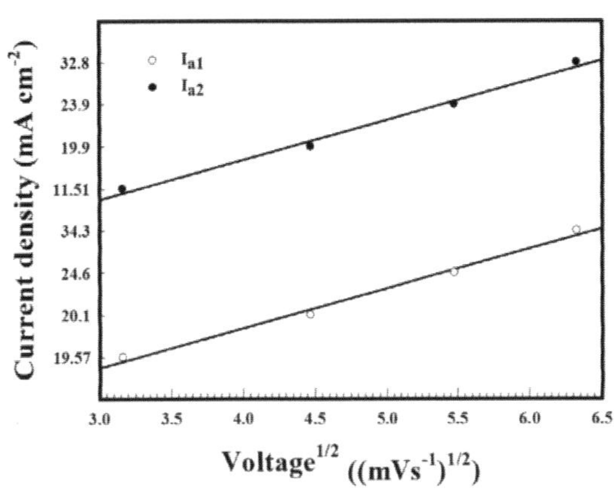

Figure 2. a) Cyclic voltammograms of Copper in 3.5% NaCl electrolyte at 25°C with the scan rates of 10, 20, 30, 40 mV/s, b) Copper anodic peak current versus square root of scan rate.

The cathodic peak of copper voltammogram , noted as C in the figure 2a, was obviously appeared at any scan rate. The peak increasing in current value with increasing scan rate. Linear dependence of I_a on the $\nu^{1/2}$ obtained by cyclic voltammetry were showed in figure 2b. All copper's cyclic voltammetry parameters are presented in table 2.

Table 2. Cyclic voltammetry anodic-cathodic peak of Copper in 3.5% NaCl solution at 25°C

Scan Rate	Ea1 Peak (V_{SSCE})	Ea2 Peak (V_{SSCE})	Ec Peak (V_{SSCE})	Ia1 Peak (A)	Ia2 Peak (A)	Ic Peak (A)
10 mV/sec	0.335826	0.45251	-0.208297	0.01957	0.01151	-0.01191
20 mV/sec	0.466369	0.652792	-0.371896	0.0201	0.0199	-0.01745
30 mV/sec	0.429262	0.634639	-0.358017	0.0246	0.0239	-0.0232
40 mV/sec	0.518960	0.688710	-0.460460	0.0343	0.0328	-0.0331

The voltammograms of copper-silver-phosphorus brazing alloy were shown with the potential value of cathodic peak in the range of copper's voltammograms potential and depicted in figure 3a. The anodic peaks were seen consisted of two peaks agreed with the typical manner with copper's voltammograms, unless the peak at higher potential poorly showed and more like shoulder than a peak.

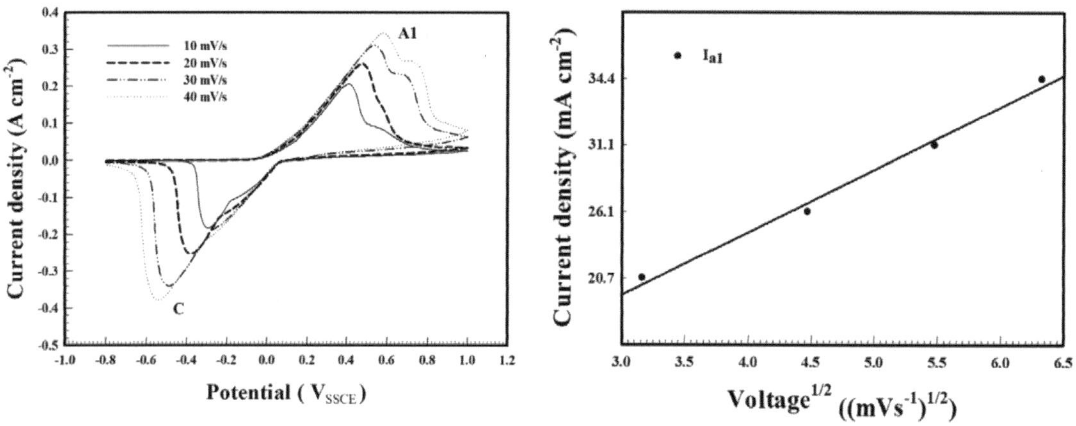

a) b)

Figure 3. a) Cyclic voltammograms of Copper-silver-phosphorus brazing alloy in 3.5% NaCl electrolyte at 25°C with the scan rates of 10, 20, 30, 40 mV/s, b) Copper-silver-phosphorus brazing alloy anodic peak current versus square root of scan rate.

The dependence of E_a and I_a on the $\nu^{1/2}$ of the alloy were evaluated, plots in figure 3b showed a high degree of linearity. All the cyclic voltammetry parameters of copper-silver-phosphorus brazing alloy are presented in table 3.

Table 3. Cyclic voltammetry anodic-cathodic peak of Copper-silver-phosphorus brazing alloy in 3.5% NaCl solution at 25°C

Scan Rate	Ea Peak (V_{SSCE})	Ec Peak (V_{SSCE})	Ia Peak (A)	Ic Peak (A)
10 mV/s	0.4130	-0.2907	0.0207	-0.0182
20 mV/s	0.4765	-0.3792	0.0261	-0.0253
30 mV/s	0.5367	-0.4887	0.0311	-0.0340
40 mV/s	0.5792	-0.5417	0.0344	-0.0378

3.3. Potentiostatic polarization

The active-passive transition of the copper-silver-phosphorus brazing alloy were tried to evaluate by potentiostatic polarization at each temperature of 25, 40, 55, and 70°C. The potential of the active-passive trantition region that chosen was 250 mV_{SSCE}. The curves in figure 4 represented that the current resulted become greater with increasing temperature. The current curve at 70°C showed some abruption as the consequences of less stability of the corrosion product on the surface, meanwhile the curve at 25°C showed the highest degree of its stability.

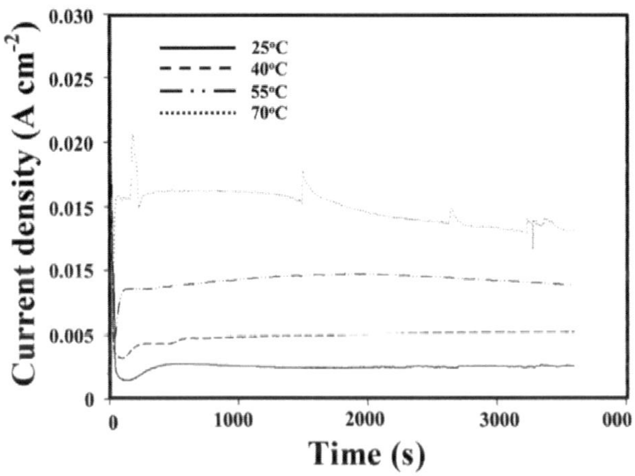

Figure 4. Potentiostatic polarization of Copper-silver-phosphorus brazing alloy with the potential of 250 mV in 3.5% NaCl at 25, 40, 55, and 70°C.

3.4. Characterization

X-ray diffraction was used to characterize the corrosion product that produced on the surface of the copper-silver-phosphorus brazing alloy as the result of potentiostatic polarization. In addition, the untested brazing alloy was observed to obtain the origin of the peak. The X-ray diffraction pattern of the corrosion products are summaryzed in figure 5. The Cu, Ag, and Cu_3P were identified as the main elements of the brazing alloy. When the alloy exposed to 3.5% NaCl at 25°C the existence of CuCl were confirmed. Indeed, the CuCl were found in each of the temperature. The higher the temperature the corrosion worsened and started to dissolving more element, hence the AgCl were detected at 55 and 70°C. In addition, the CuCl were futher appeared in each varied time when statically polarized.

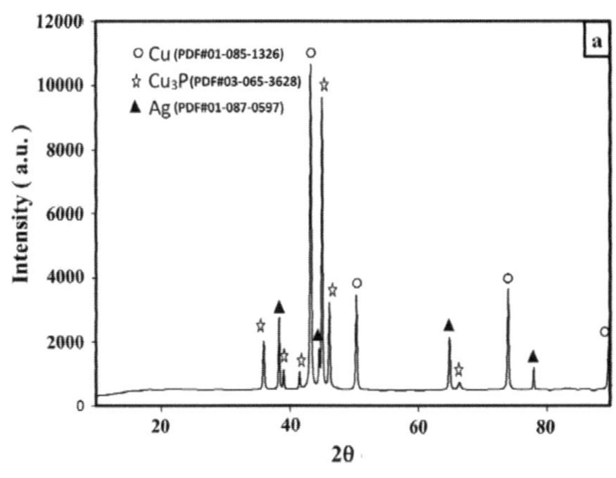

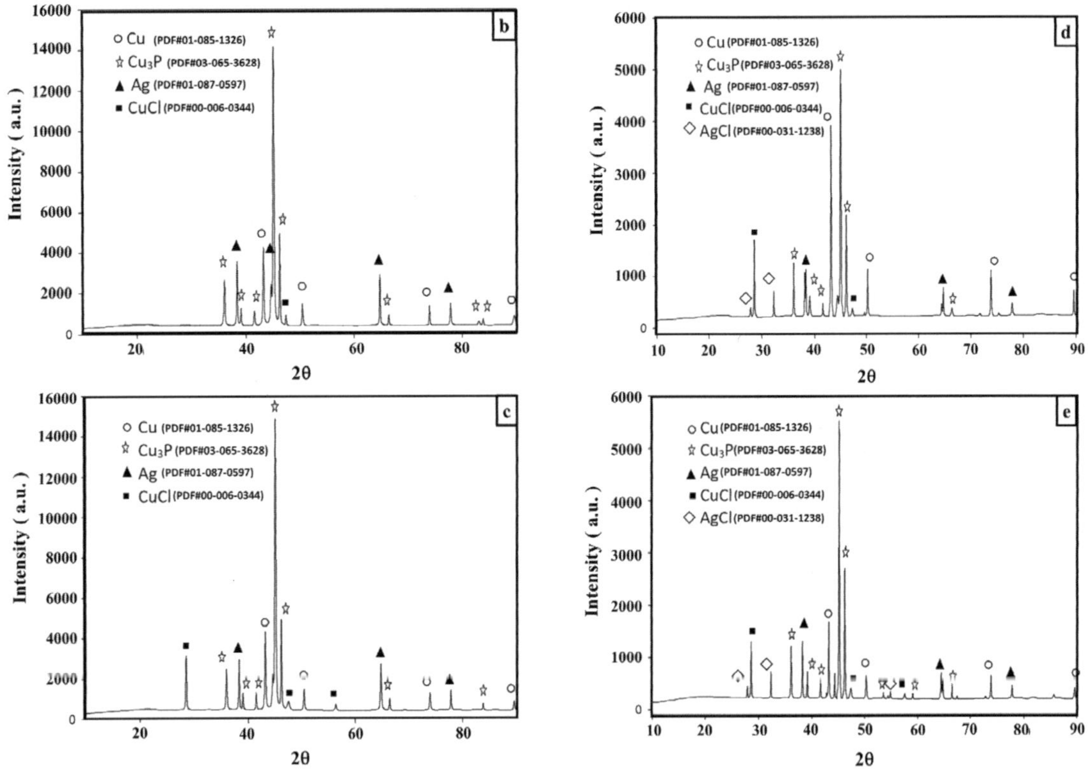

Figure 5. XRD examination of Copper-silver-phosphorus brazing alloys, a) untested, b) at 25°C, c) at 40°C, d) at 55°C, and e) at 70°C

SEM and EDS examination were done to analyze the corroded area in the microscopic scale. figure 6 showed SEM micrographs of the four brazing alloys based on the time of measurements. The quality of the corroded area observed were deteriorating with the increasing of temperature.

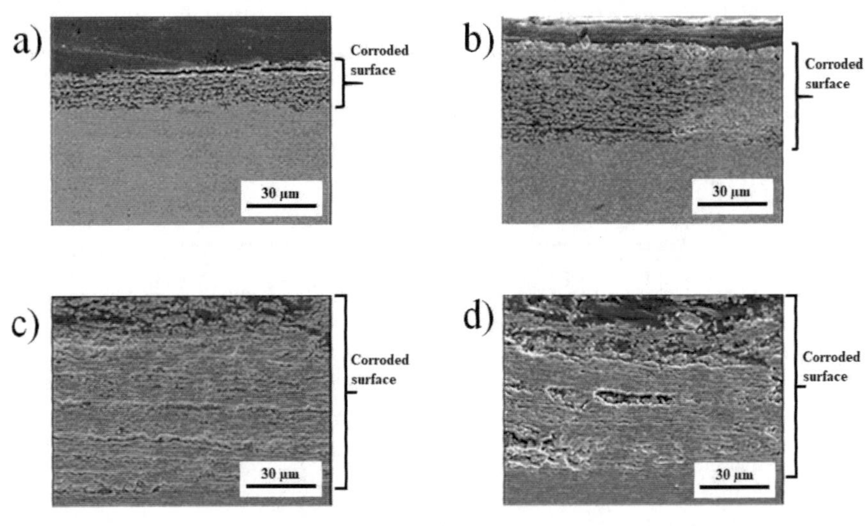

Figure 6. SEM images of the alloys's cross section with 1000x mag. after potentiostatic polarization at 250 mV(SSCE) in 3.5% NaCl solution at various temperature: a) 25°C; b) 40°C; c)55°C; d)70°C

4. Discussion

4.1. Electrochemical corrosion behavior

The anodic region of the alloy potentiodynamic curves showed the typical anodic behaviour of pure copper in chloride containing electrolyte [9-16]. The alloy anodic curves can be divided into three sections as shown in figure 1: the first section is the apparent tafel region. The second section is where the peak and minimum current appeared. The last section is where the limiting current region existed.

At the apparent tafel region in 25°C potentiodynamic curve, the anodic slope was 58 mV/decade. P. A. Lush and M. J. Carr [18] proposed that at low current density with the tafel slope close to 60 mV/decade for the pure copper in chloride containing electrolyte the dissolution was mass-transfer controlled. Hence, the dissolution of the alloy might have the similar manner. However, at high current density the tafel slope of the alloy change to close to 120 mV/decade. It was believed that the steep slope at high anodic current density may influenced by the existence of film on the surface caused the limited diffusion of the reacting species. Previous studies have reported such behavior on covered copper with porous film on the surface [18,19].

The anodic dissolution behaviour of pure copper in chloride containing electrolyte has been studied comprehensively by some authors [9-11, 14-16]. The dissolution of copper by the attack of Cl- was believed initiated with the adsorption of Cl- ions on copper surface and formed submonolayer of Cl adatoms that shielded the surface [15]. The dissolution was then continued with the formation of CuCl and/or $CuCl_2^-$ corrosion product. H. Lal and H. R. Thirsk [17] have suggested that the $CuCl_2^-$ were the limited diffusion species to the solution and CuCl were precipitated on the surface. The voltammograms of the alloy (figure 3) showed the first peak at the lower potential that corresponds to the precipitation of CuCl. The second peaks of the alloys voltammograms, unlike the copper (figure 2) as described by D. Trornans and Ru-hong Sun [11], the peaks were more like a shoulder.This shows that the rate of the dissolution of the alloy through $CuCl_2^-$ is more undesireable than in the pure copper. The relationship between I_a on the $v^{1/2}$ suggested that the anodic reaction was mainly diffusion controlled with the limmiting diffusion species is $CuCl_2^-$. The alloy's nature of microstructure was believed to be the main cause of the limited mass-transport processes. Since the dissolution occured by selective attack of copper element in the alloy, the other elements inhibit the corrosion product dissoluted to the solution.

In compliance with the cyclic voltammetry behavior, at the active-passive region of the 25°C potentiodynamic curve, the peak current at the minimum current density was contributed by the formation of CuCl that covered the surface. As the potential shifted to the more positive value the current density observed increased once again, this anodic behaviour was associated with a strong role of the $CuCl_2^-$ dissolution from the surface to the bulk solution.

4.2. The effect of temperature

The apparent tafel slopes were in agreement with the relationship of 2.3RT/F as firstly introduced by Lal and Thirsk [17] for pure copper in chloride containing electrolyte. Lee and Nobe [13] offered the similar relationship which indicate that in the apparent tafel region electrodissolution of Cu controlled by mixed mass-transfer and charge-transfer. The alloy tafel slopes behaviour dependencies to the temperature with such relationship of 2.3RT/F showed by the tendency of the E_{corr} and the anodic slope that follow the changes of the temperature. Shifted to more negative potential, and at higher temperature the slope was steeper than that of lower temperature.

The appearances of active-passive region were also gradually vanished as the temperature increased. The formation of CuCl is faster than the formation of $CuCl_2^-$ at lower temperature of 25 C and 40 C, thus the surface have a high possibility to be covered by precipitated CuCl that change the slope and obviously decreasing the current density. On the other hand, at higher temperature of 55 °C and 70 °C, the temperature not only improving the rate of the $CuCl_2^-$ formation but also faster the mass-tranport of $CuCl_2^-$ limited diffusion species to the solution. These circumtances make the CuCl cannot inhibit the corrosion process.

D. Starosvetsky *et al.* [15] reported on their work that the anodic behaviour of the active-passive region was a function of the passivity breakdown process that associated with incubation periode. In the similar approach, the alloys were statically polarized in the anodic potential of 250 mV_{SSCE} at each temperature. Static polarization curve showed that the incubation periode decreasing with temperature increased. In agreement with the result of D. Starosvetsky *et al.* the activation was stimulated by either with the applied potential or the temperature. The incubation periode seems to decrease with increasing temperature or increase in applied potential, and almost directly activated for dissolution at higher potential or temperature. Similar manner of the incubation periode was shown by the iron and steels, and agreed that this feature is the initial stage of pit nucleation [15]. The surface characterization of the alloys after potentiostatic experiments showed the tendency of pitting to occur become higher with increasing temperature. Some abruptions were also found in the potentiostatic curve at 70°C.

This incubation periode time was observed at different time statically polarized the alloy at the same potential of 250 $mV_{ssce.}$ The existence of the CuCl salt layer was identified at each varied times and temperature by the XRD SEM and optical microscope examination showed that the layer was in the form of selective dissolution region and this region increased in the depth of the attacked region with time and temperature. This feature suggested that the alloy was in the active state at the potential of 250 mV_{ssce} as either at the incubation periode time. From the electrochemical measurements, the schematic drawing proposed as in figure 7.

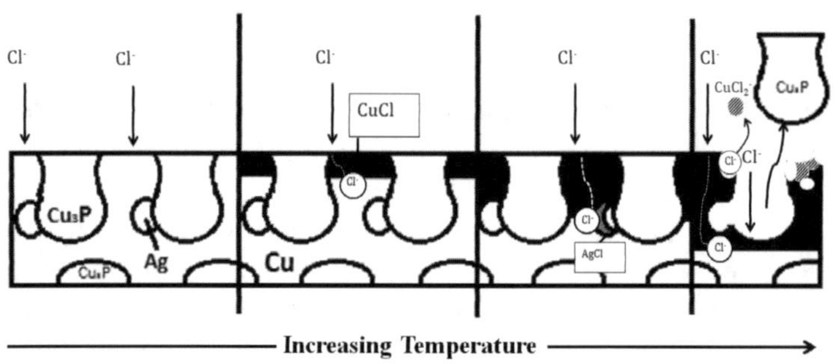

Figure 7. Schematic diagrams of Copper-Silver-Phosphorous Brazing Alloy for electrochemical corrosion with the effect of temperature.

5. Conclusion

The copper-silver-phosphorus brazing alloy was undergone a selective dissolution by the attack of Cl- to the copper element. The anodic behaviour of the alloy showed the typical manner of the pure copper anodic behaviour. The temperature dependence of the alloy showed in agreement with the relationship of 2.3RT/F that have similar manner on pure copper tafel region. The increasing of the slopes was more due to disturbed in balance between the charge-transfer and mass-transfer control. The temperature effect have similar role with the applied potential, which is the higher both applied, the easier the copper and brazing alloys activated. The morphologies of the corroded area of the alloy informed that the typical characteristic of pitting nucleation by the presence of incubation periode were in a good agreement.

REFERENCES

[1] ASM Handbook, "Welding brazing and soldering, ASM International," 1993, Vol. 6.
[2] Schwartz, M. M., "Brazing, 2ndEdition," ASM International, pp. 8–16, 2003.
[3] Roberts, Philip M., "Industrial Brazing Practice," CRC Press LLC, United States of America, 2004.
[4] Chu Hyon Cho, "Copper-phosphorus-strontium brazing alloy," US Patent Publication, Pub. No. : US 2012/0288401 A1, United States of Americaa, 2012.
[5] A. Valero-Gomez, A. Igual-Munoz, J. Garcia-Anton, "Corrosion and Galvanic Behavior of Copper and Copper-Brazed Joints in Heavy Brine LiBr Solutions," Corrosion (2006), in press.
[6] Idrissi, H., Millet, J. P., and Audisio, S., "Natural formation of a protective layer on copper and copper brazing in domestic water: structural characterization and electrochemical behavior." Journal of Materials Science 29: 5327–5332, 1994.
[7] D. R. Sigler, J. G. Schroth, Y. Wang, D. Radovic, "Sulfide-Induced Corrosion of Copper-Silver-Phosphorus Brazed Joints in Welding Transformers," Welding journal, 340-348, vol.86, 2007.
[8] Joseph A. Worden, Jorge M. Mundulas, "Understanding, Diagnosing, and Repairing Leaks in Water-Cooled Generator Stator Windings," GE Power Systems, Schenectady, NewYork, 2008.
[9] C. Deslouis, B. Tribollet, "Electrochemical behaviour of copper in neutral aerated chloride solution. I. Steady-state investigation", Journal of Applied Electrochemistry, Vol.18, pp. 374-383, 1988.
[10] Z. H. Gu, T. Z. Fahidy, "An experimental study of anodic dissolution of copper in aqueous sodium chloride via laser-induced image processing", Journal of Applied Electrochemistry , Vol.19, pp. 354-360, 1989.
[11] D. Trornans, Ru-hong Sun, "Anodic Polarization Behavior of Copper in Aqueous Chloride/Benzotriazole Solutions", J. Electrochem. Soc., Vol. 138, No. 11, 1991.
[12] G. Kear a, B.D. Barker b, F.C. Walsh c, "Electrochemical corrosion of unalloyed copper in Chloride media—a critical review," Corrosion Science, vol. 46, pp. 109–135, 2004.
[13] H. P. Lee, Ken Nobe, "Kinetics and Mechanisms of Cu Electrodissolution in Chloride Media," J. E, lectrochem. Soc., Vol. 133, No. 10, 2035-2043, October 1986.
[14] J. Crousier, L. Pardessus, J.P. Crousier, "Voltammetry Study of Copper in Chloride Solution, "Electrochemica Acta, Vol. 33. No. 8, pp. 1039-1042, 1988.
[15] D. Starosvetsky, O. Khaselev, M. Auinat, Y. Ein-Eli, "Initiation of copper dissolution in sodium chloride electrolytes", Electrochimica Acta, Vol. 51, pp. 5660–5668, 2006
[16] Boyu Yuan, Chao Wang, Liang Li, Shenhao Chen, "Real time observation of the anodic dissolution of copper in NaCl solution with the digital holography", Electrochemistry Communications, Vol. 11, pp. 1373–1376, 2009.
[17] H. Lal and H. R. Thirsk," The Anodic Behawiour of Copper in Neutral and Alkaline Chloride Solutions", J.Chem. Soc., pp. 2638, 1953.
[18] P. A. Lush and M. J. Carr, "Copper Dissolution in Sea Water Under Mixed Activation and Diffusion Control", Corrosion Science, Vol. 19, pp. 1079-1088, 1979.
[19] R.J.K. Wood, S.P. Hutton, D.J. Schiffrin, Corrosion Science, Vol. 30, pp. 1177–1201, 1990.

Young's Modulus Determination of Polyester and Epoxy by Means of Ultrasonic Pulse Echo Testing

Firmansyah Sasmita[1]* Thea Athalia Candra[1] Hermawan Judawisastra[1] Toni Agung Priambodo[2]
[1]Materials Engineering Department, Faculty of Mechanical and Aerospace Engineering, Institut Teknologi Bandung
[2]Center for Material Technical Products, the Ministry of Industry – Republic of Indonesia

*Corresponding author: firmasyah.sasmita@material.itb.ac.id

Abstract. The Young's modulus or elasticity modulus is material ability to resist elastic deformation when loaded. Thermosetting polymers have a wide range of modulus elasticity values. To determine the elasticity modulus value, materials need to be destructive tested, such as tensile test or bending test. But in industrial field, they avoided destructive test. Theoretically, the modulus of elasticity can also be determined by ultrasonic testing pulse echo method. This method has already applied successfully for metals material. The goal of this research was to evaluate probe used in longitudinal velocity measurement of polyester and epoxy and to determine Young's Modulus of polyester and epoxy by means ultrasonic pulse echo testing. This method was performed using three longitudinal probes that have different diameters, frequencies, and additional delay line. The specimens were varied with different thicknesses. The ultrasonic pulse echo method uses probe 3 with a diameter of 12.7mm, a frequency of 1.5MHz, and additional delay line is most suitable to measure the longitudinal velocity of polyester and epoxy with thickness of 2.82-36.72 mm. Probe 3 has a smaller diameter and additional delay lines thus eliminating the effect of dead zones. The elastic modulus error of thermoset polymer material through ultrasonic testing of pulse echo method compared to the mechanical modulus of elasticity is still very large, i.e. 102% for epoxy, and 159% for polyester. A large error is caused by attenuation and the use of the elasticity modulus equation from ultrasonic testing that still requires modification.

Keywords: Elastic Modulus, Ultrasonic Testing, Pulse Echo, Polyester, Epoxy, Longitudinal Velocity

1. Introduction

Thermosetting polymers, e.g. unsaturated polyesters and epoxies, are widely used in industries. Unsaturated polyesters are extensively used to produce structural parts with a glass-fiber reinforcement, while epoxies are employed for aeronautical and electronic applications [1]. Thermosetting polymers have a wide range of Young's modulus values which affected by its structure and dependent on molecular weight, molecular orientation, and density [2]. Young's modulus, also known as elastic modulus, is an important mechanical property of materials. This property describes the stiffness of materials or material's resistance to elastic deformation [3]. The elasticity of thermoset polymer elasticity can decrease due to degradation. This degradation generally caused by components exposed to ultraviolet light, high temperatures, and chemical liquids so that the polymer chain chain begins to break [4]

To determine the elastic modulus, materials need to be tested destructively, such as tensile test or bending test. Industrial field tries to avoid destructive tests as it will damage the test object and cannot be conducted in-situ. Ultrasonic testing is common non-destructive test technique that frequently used to detect defects in materials [5]. The modulus of elasticity theoretically can be determined by ultrasonic testing. Ultrasonic wave propagation in material has strong proportional relation to its elastic properties [6].

Two common methods of ultrasonic testing are pulse echo and through transmission [7]. Ultrasonic through transmission testing for measurement of polymers' elastic modulus and dynamic elastic modulus have been studied, however this method is not practical as it requires two sides of the test object, and immersed in liquid [8]. Ultrasonic pulse echo testing is more practical to be used in industries, especially in the real time monitoring of in service structures as it only requires one side of the test object. Determination of Young's modulus with ultrasonic testing has already applied successfully for metals with error less than 5% [9].

Error in determination of Young's modulus of polymers by means of ultrasonic testing can be caused by some factors such as attenuation and properties of materials i.e. density and viscoelasticity [9]. Attenuation is a factor that describe the decrease in ultrasound intensity with distance, normally expressed in decibels per unit length [10]. When sound travels through a medium, its intensity diminishes with distance, or in the other hand, material thickness. In idealised materials, sound pressure as signal amplitude is only reduced by the spreading of the wave. Ultrasonic attenuation is the decay rate of the wave as it propagates through material [11].Higher attenuation experienced by material can caused higher error in Young's modulus determination by ultrasonic testing also higher viscoelasticity of material. While higher density of material reduced error in Young's modulus determination with ultrasonic testing [12].

Dead zone and near field common phenomenon occurred in ultrasonic testing. Both will made difficulties in ultrasonic testing reading. Dead zone is the distance in material from the surface of the test specimen to the depth at which a reflector can first be resolved under specified conditions [5]. Length of dead zone depends on probe and ultrasonic test equipment. Meanwhile near field is the distance immediately in front of the transducer in which the ultrasonic beam exhibits complex and changing wave fronts [10].

To improve the inspection with ultrasonic testing changing frequency and diameter of the probe can be done. Also a zero impedance probe (ZIP), probe with delay line is commonly used. The primary function of a delay line is to introduce a time delay between the generation of the sound wave and the arrival of any reflected waves. This allows the probe to complete

IOP Conf. Series: Materials Science and Engineering **547** (2019) 012045 doi:10.1088/1757-899X/547/1/012045

its "sending" function before it starts its "listening" function so that near surface resolution is improved [13]. These will eliminate or reduce dead zone or near field factor.

Therefore, this research aims to evaluate the use of different probes to determine longitudinal velocity of polyester and epoxy and to determine Young's Modulus of polyester and epoxy by means ultrasonic pulse echo testing.

2. Materials and Methods

The polyester specimens were made from unsaturated polyester resin Yucalac ® C108-B with 2% weight curing agent. Eposchon A as primer epoxy resin and Eposchon B as hardener used to make the epoxy specimens with 1:1 weight ratio. Both thermosets were cured in room temperature. The thickness of polyester ultrasonic specimen used for the experiment are 2.8 mm, 11.6 mm, and 36.7 mm. While the thickness of epoxy specimens are 3.9 mm, 13.2 mm, and 33.7 mm. Material density measurement was conducted with KERN ABJ-NM/ABS-N device at 25°C using water media according to ASTM D 792 [14].

Ultrasonic test was conducted with GE USM 35X device with Pulse-Echo method to obtain the longitudinal wave velocity of materials. Sonotech® Ultragel II® was used as the couplant. Both showed in Figure 2.1.

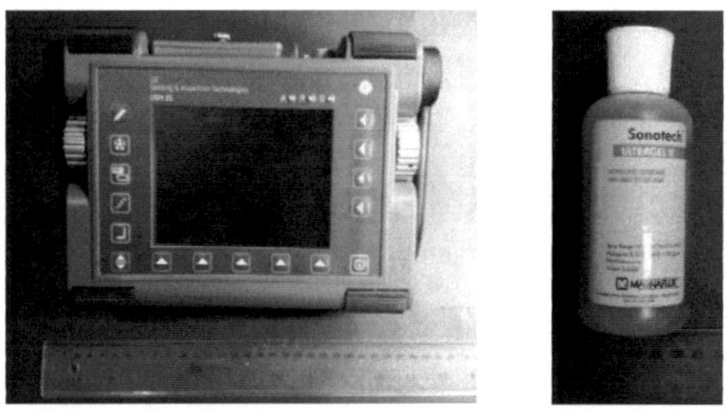

(a) (b)

Figure 2.1 (a) Ultrasonik device GE USM 35X and (b) couplant

There were 3 probes used for the ultrasonic test: probe 1 using 1 MHz frequency with 25.4 mm diameter; probe 2 using 1.5 MHz 12.7 mm; and probe 3 using 1.5 MHZ 12.7 mm with 13 mm thickness delay line. Table 2.1 exhibit the probes.

Table 2.1 Probes

Probe	Side view	Below View
Probe 1 (Ø 24 mm 1 MHz)	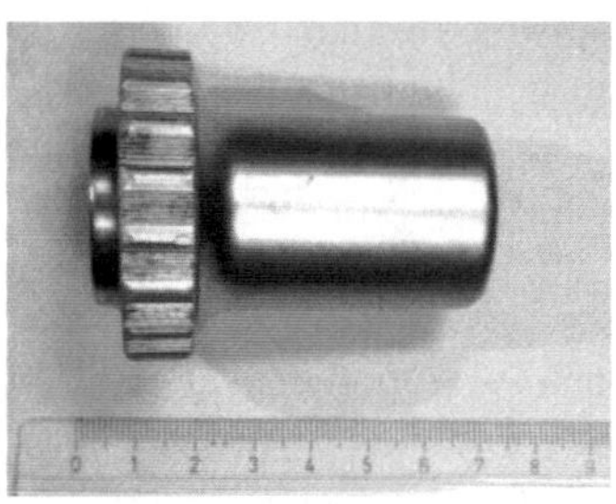	

Probe 2 (Ø 12,7 mm 1,5 MHz)	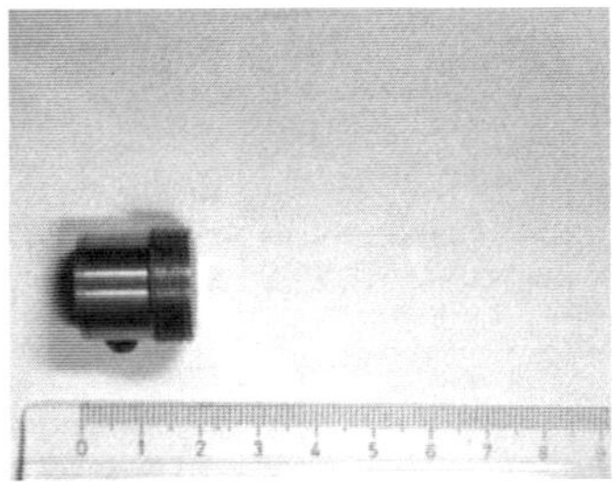	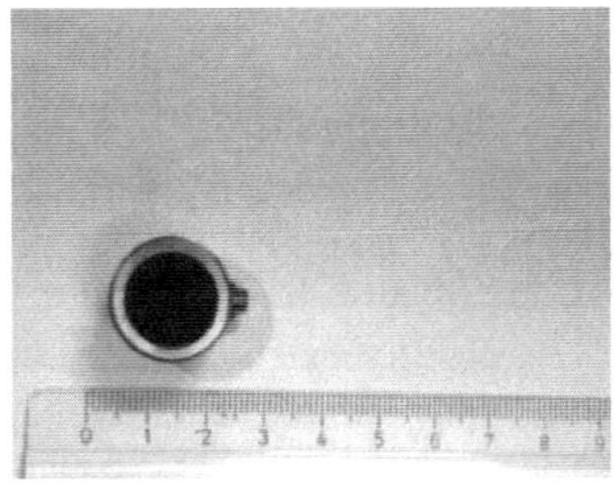
Probe 3 (Ø 12,7 mm 1,5 MHz, with delay line)	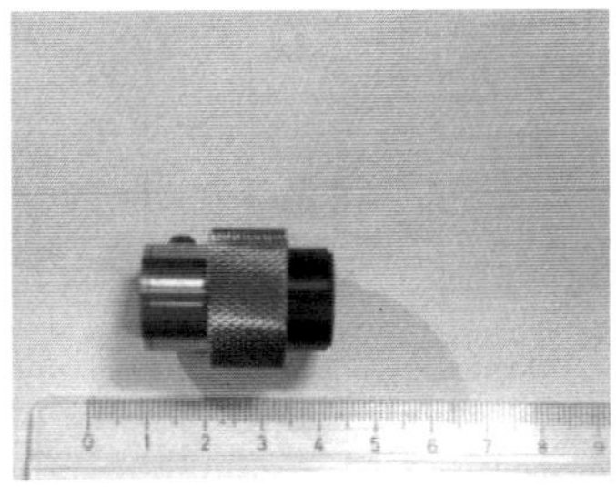	

Three times of longitudinal wave measurement in every section were conducted on polyester and epoxy by means of all three probes. The velocity is calculated based on ASTM E 494 [15]. The longitudinal wave velocity obtained from the measurement was used to obtain Young's modulus using **Equation 1**.

$$E = \frac{V_{long}^2 \times \rho \times (1+\nu)(1-2\nu)}{(1-\nu)} \qquad (1)$$

where V_l is longitudinal sound velocity, ρ is material density, E is modulus of elasticity, and μ is material Poisson's ratio.

The apparent attenuation (α) is determined by means of **Equation 2**, based on ASTM E 664 [16]. 36.72 mm polyester and 33.71 mm epoxy were used to obtain apparent attenuation for each material.

$$\mathrm{A} = \mathrm{A_0} \times \mathrm{e}^{-\alpha z} \qquad (2)$$

where A the amplitude that is reduced after the wave has passed the distance z, A_0 is initial amplitudo, α is attenuation coeficient, and z is distance travelled.

Tensile test was conducted to obtain the elastic modulus of polymeric specimens based on ASTM D 638 [17]. The specimens were type I for dimension with speed of testing was 5 mm/min. The Young's modulus obtained from tensile test shall be the reference and will be compared with the Young's modulus resulted from the ultrasonic test. Poisson's ratio of polyester and epoxy were from literature [18].

IOP Conf. Series: Materials Science and Engineering **547** (2019) 012045 doi:10.1088/1757-899X/547/1/012045

3. Results and Discussion

3.1 Physical Properties

Table 3.1 presents the average density of polyester used was 1.21 g/cm^3 with poisson's ratio value was 0.36. Meanwhile epoxy has 1.10 g/cm^3 density and 0.37 poisson's ratio rate. The densities achieved in this experiment within range of literature's density [19]. Poisson's ratio literature was used in this research.

Table 3.1 Polymers Density and Poisson's Ratio

Polymers	Average Density (g/cm^3)	Poisson's Ratio
Polyester	1.21	0.36 [18]
Epoxy	1.10	0.37 [18]

3.2 Probes Evaluation Results

Table 3.2 and **Figure 3.1** depicts the longitudinal velocity of polyester and epoxy specimens obtained from ultrasonic testing with different thicknesses and probes compared with literature [18, 20]. The measurement of longitudinal velocity using the third probe that used delay line, obtain better results for thin specimens (2.8 mm thickness polyester and 3.9 mm epoxy) than the other probes. The result of longitudinal velocity measurement using of delay line tend to be have stable values for each polymers, though there were thickness differences, unlike the first and second probe that did not use delay line material.

Table 3.2 Longitudinal velocity measurement result for polyester and epoxy with specimens thickness variation using different probe

Polymers	Specimen Thickness (mm)	Ultrasonic longitudinal velocity				
		Experiment			References	
		Probe 1 (1 MHz Ø 24 mm)	Probe 2 (1,5 MHz Ø 12,7 mm)	Probe 3 (1,5 MHz Ø 12,7 mm with delay line)	Lower limit	Upper limit
Polyester	2,84	1831±23	1904±11	2505±59	2290	2430
	11,64	2416±78	2367±27	2540±15		
	36,72	2518±26	2501±21	2551±14		
Epoxy	3,92	2146±22	2122±4	2306±16	2480	2680
	13,24	2228±61	2214±7	2330±12		
	33,71	2344±1	2297±24	2336±61		

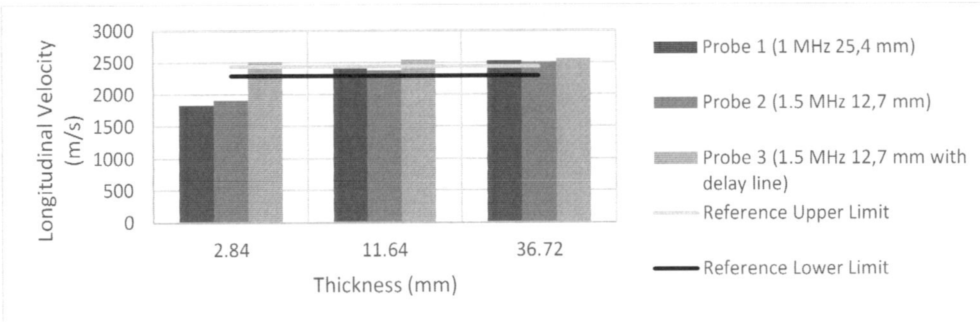

(a)

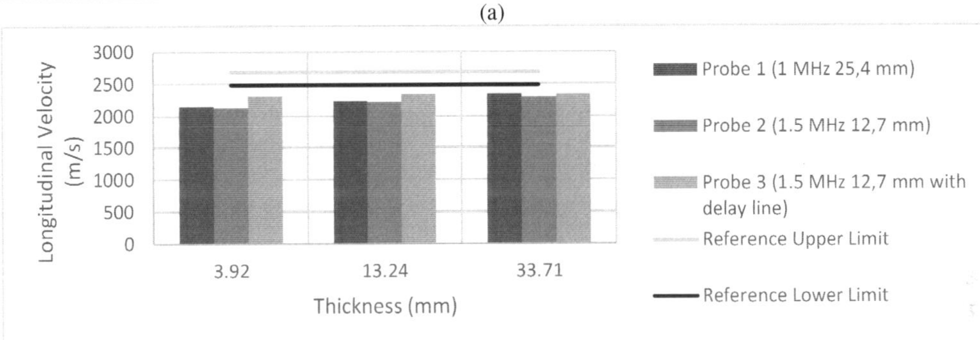

(b)

Figure 3 1 Longitudinal Velocity of (a) Polyester and (b) Epoxy from Reference and Ultrasonic Testing for Each Probe and Thickness

Longitudinal velocity measurements using probe 1 and probe 2 without delay line for specimens with thicknesses of 2.82-3.92 mm showed unfavorable results because they produced ultrasonic longitudinal velocity values that were much lower than in other thicknesses or measurements using probe 3 with delay line at the same thickness. Even so, measurements on specimens with thicknesses above 10 mm measuring using probe 1 and probe 2 without delay line have produced velocity values that approach the literature for polyester and epoxy.

The results of the ultrasonic longitudinal velocity measurements of each thermoset polymer using probe 3 with a delay line resulted in a relatively similar value for specimens with a thickness of 2.82 mm to 36.72 mm. Therefore, the use of probe 3 with delay line is considered the best compared to probe 1 or probe 2 which does not use delay line.

Table 3.3 exhibit the near field length on each probe and polymers. Probe 1 generates 60 – 66 mm near field length for polyester and 65 – 70 mm for epoxy. Probe 2 had lower near field length, 23 – 25 mm for polyester and 24 – 26 mm for epoxy. Probe 3 had lowest near field because of the substraction of deline line thickness, 10 – 12 mm for polyester and 11 – 13 mm for epoxy. Compared to **Figure 3.1**, near field seems not affected longitudinal velocity measurement by ultrasonic testing in this experiment because there are not significantly differences between measurement results below and upper near field.

Table 3.3 Near field length

Polymers	Near field length (mm) Probe 1 (1 MHz Ø 25,4mm)	Probe 2 (1,5 MHz Ø 12,7mm)	Probe 3 (1,5 MHz Ø 12,7mm dengan *delay line*)
Polyester	60 – 66*	23 – 25*	10 – 12**
Epoxy	65 – 70*	24 – 26*	11 – 13**

$*N = \frac{D^2 f}{4V}$, $**N = \frac{D^2 f}{4V} - 13mm\ (delay\ line)$

Besides near field, another phenomenon in ultrasonic testing is dead zone. The dead zone phenomenon will occur errors in ultrasonic reading for thin material so that it produces an incorrect longitudinal speed [10]. **Figure 3.2** shows dead zone displayed at ocilator screen.

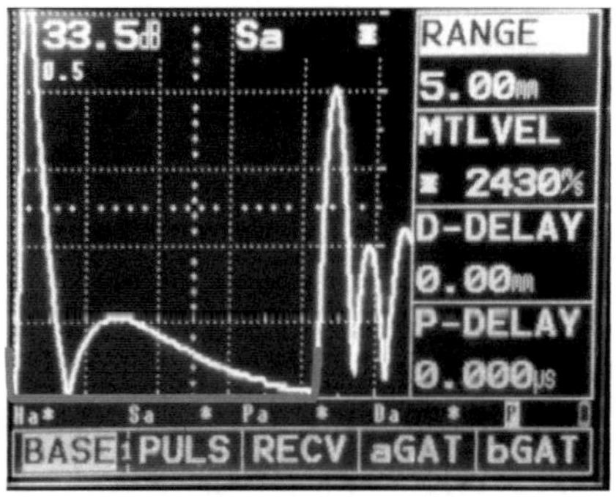

Figure 3.2 Dead zone displayed at ocilator screen

It is seen that the display area of the dead zone for this ultrasonic device is about 4 mm length. So that for specimens with a thickness less than the length of the dead zone, amplitude readings are not clearly visible when using probe 1 and probe 2 and causes errors in ultrasonic longitudinal velocity measurements. This is because on thin specimens, the entire thickness is the area of the probe to ringing down.

The effect of dead zone phenomenon will be lost when using probe 3 which has a delay line. This can be seen in the longitudinal velocity measurement with probe 3, in **Figure 3.1**, resulting in a more stable value (not too much different) even though there is a difference in specimen thickness.

3.3 Longitudinal Velocity Measurement Results

Table 3.4 shows the error of longitudinal velocity measurement of polyester and epoxy compared to literature, 4.18% for polyester and 6.29 for epoxy. **Table 3.4** also figures that Epoxy's apparent attenuation was 1.84 dB/mm, larger than polyester's 1.20 dB/mm.

Table 3.4 Longitudinal velocity, error, and attenuation on polymers and epoxy

Polymers	V_{Long} UT with ZIP (m/s)	Error of V_{Long} Measurement with ZIP (%)	Apparent Attenuation (α) (dB/mm)
Polyester	2532	4.18	1.20
Epoxy	2324	6.29	1.84

The larger error in longitudinal velocity measurement using ultrasonic testing commonly caused by larger apparent attenuation of materials [5]. Scattering, absorption and geometric are three parameters that effect the attenuation. Small discontinuities like boundaries between each crystal are the source of scattering [7]. In this case, there be influenced from density of materials, but the relationship of density and attenuation constant are inversely proportional as represent in **Table 3.1** and **Table 3.4**.

3.4 Young's Modulus Determination from Ultrasonic Testing

Table 3.5 figures that error of Young's modulus from ultrasonic testing compared to mechanical testing is 159% for polyester and 104% for epoxy. This large errors of calculation occurred even though errors of longitudinal velocity measurement with ultrasonic pulse echo testing with zero impedance probe were less than 7%, according to **Table 3.5**.

Table 3.5 Elastic Modulus from Ultrasonic and Mechanical Testing

Polymers	Young's Modulus UT (E_{UT}) (GPa)	Young's Modulus Tensile Testing (E_M) (GPa)	Error Elastic Modulus (%)
Polyester	4.62	1.78	159
Epoxy	3.36	1.65	104

Therefore **Table 3.6** shows that the result of polymer and epoxy Young's modulus calculation using longitudinal velocity from reference. This verify that **Equation 1** above is not applicable for polyester and epoxy used in this research and need to be modified.

Table 3.6 Elastic Modulus using Longitudinal Velocity References

Polymers	V_{Long} Reference (m/s)	Young's Modulus Calculation (GPa)	Young's Modulus Tensile Testing (E_M) (GPa)	Error Elastic Modulus (%)
Polyester	2290 [20]	3.78	1.78	112
	2430 [18]	4.25		138
Epoxy	2480 [20]	3.83	1.65	132
	2680 [18]	4.47		171

Table 4.6 proves that although using ultrasonic longitudinal velocities from the literature, the elastic modulus error is still very large, 112% -171%. This may be due to the formula that is used cannot be applied to thermoset polymers. Allegations related to this problem are the movement of thermoset polymer chains when exposed to dynamic loads on ultrasonic testing in contrast to chain movements when exposed to static loads on tensile testing [6]. This can occur because the variation of thermoset polymer movement is more diverse than metal. Variations in the movement of thermoset polymers are illustrated in Figure 4.11.

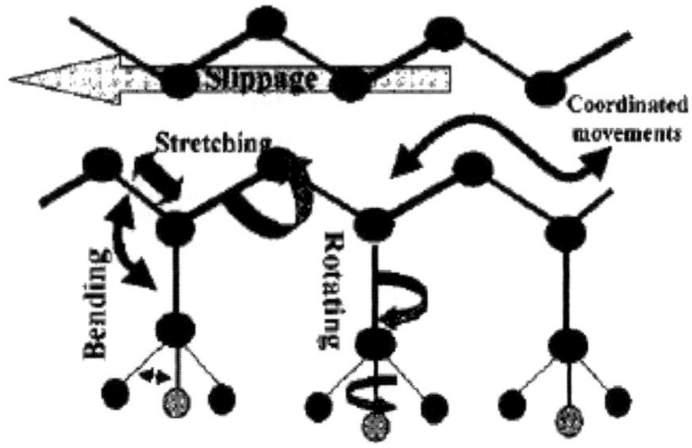

Figure 3.3 Crankshaft model to explain polymer chain movement [21]

Variations in the movement of the thermoset polymer chain make movement different for each dimension. Whereas in metals whose bonds consist of atoms, their movements tend to be the same in each direction in 3 dimensions, as illustrated in Figure 4.12 [3].

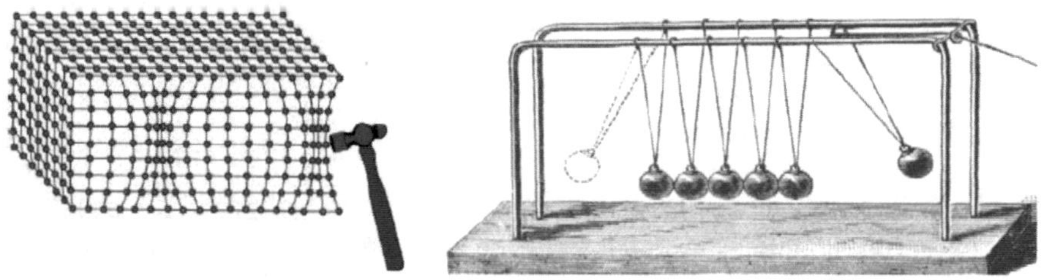

Figure 3.4 Atoms of metal movement illustration [3, 6]

So that standard equations can be used to determine the elastic modulus of thermoset polymers, then:

1. It is necessary to add the transversal direction velocity factor to the standard equation [6].
2. The standard equation requires other correction factors such as the nature of the thermoset polymer itself, for example density and attenuation [6].

4. Conclusions

The ultrasonic pulse echo method uses probe 3 with a diameter of 12.7mm, a frequency of 1.5MHz, and additional delay line is most suitable to measure the longitudinal velocity of polyester and epoxy with thickness of 2.82-36.72 mm. Probe 3 has a smaller diameter and additional delay lines thus eliminating the effect of dead zones. The elastic modulus error of thermoset polymer material through ultrasonic testing of pulse echo method compared to the mechanical modulus of elasticity is still very large, ie 102% for epoxy, and 159% for polyester. A large error is caused by attenuation and the use of the elasticity modulus equation (**Equation 1**) from ultrasonic testing that still requires modification.

5. Acknowledgements

The authors would like to express gratitude to ITB for funding this research under P3MI-KK FY 2018 project. The results discussed in the present paper where obtained within this project.

6. References

[1] Jean-Pierre Pascault, Henry Sautereau, Jacques Verdu, Roberto J. J. Williams, Thermosetting Polymers, New York, USA: Marcel Dekker, Inc., 2002.

[2] M. Chanda and S. K. Roy, Plastics Technology Handbook, 4th ed., Florida, USA: Taylor & Francis Group, LLC, 2006.

[3] W. D. C. Jr., Materials Science and Engineering: An Introduction, 8th ed., New Jersey, USA: John Wiley & Sons, Inc., 2010.

[4] K. L. Ho and A. L. Pometto, "Effects of Temperature and Relative Humidity on Polylactic Acid Plastic Degradation," *Journal of Polymers and the Environment,* vol. 07, 1999.

[5] D. Ensminger and L. J. Bond, Ultrasonics: Fundamentals, Technologies, and Applications, 3rd ed., Florida: CRC Press, 2012.

[6] J. Krautkramer and H. Krautkramer, Ultrasonic Testing of Materials, Berlin: Springer-Verlag, 1990.

[7] S. Gholizadeh, "A review of non destructive testing methods of composite materials," in *XV Portuguese Conference on Fracture*, Portugal, 2016.

[8] F. Lionetto and A. Maffezzoli, "Polymer Characterization by Ultrasonic Wave Propagation," *Advanced in Polymer Technology,* vol. 27, pp. 63-73, 2008.

[9] E. E. Franco, J. M. Meza, F. Buiochi, "Measurement of Elastic Properties of Materials by the Ultrasonic Through-Transmission Technique.," *Dyna rev.fac.nac.minas, Medellín,* vol. 78, p. 58–64, Aug. 2011.

[10] C. J. Hellier, Handbook of Nondestructive Evaluation, New York: The McGraw-Hill Companies, Inc., 2003.

[11] "NDT Resource Center," [Online]. Available: https://www.nde-ed.org/EducationResources/CommunityCollege/Ultrasonics/Physics/attenuation.htm. [Accessed 8th August 2018].

[12] J. Jarzynski, "Mechanisms of Sound Attenuation in Materials," in *In Sound and Vibration Damping with Polymers*, Washington, DC, American Chemical Society, 1990.

[13] "NDT Resource Center," [Online]. Available: https://www.nde-ed.org/EducationResources/CommunityCollege/Ultrasonics/EquipmentTrans/transducertypes.htm. [Accessed 8th August 2018].

[14] ASTM, ASTM D792 Standard Test Methods for Density and Specific Gravity (Relative Density) of Plastics by Displacement, Pennsylvania: ASTM International, 2013.

[15] ASTM , ASTM E 494 Standard Practice for Measuring Ultrasonic Velocity in Materials, Pennsylvania: ASTM International, 1995.

[16] ASTM, ASTM E 664 Standard Practice for the Measurement of the Apparent Attenuation of Longitudinal Ultrasonic Waves by Immersion Method, Pennsylvania: ASTM International, 1993.

IOP Conf. Series: Materials Science and Engineering **547** (2019) 012045 doi:10.1088/1757-899X/547/1/012045

[17] ASTM, ASTM 638 Standard Test Method for Tensile Properties of Plastics, Pennsylvania: ASTM Internasional, 2002.

[18] D. W. v. Krevelen, Properties of Polymers, 4th Edition, Oxford: Elsevier B.V., 2009.

[19] J. E. Mark, Polymer Data Handbook, Second Edition, New York: Oxford University Press, 2009.

[20] J. N. Kim, R. Tutwiler, D. R. Kwak, I. Park and C. Miyasaka, "Multilayer Transfer Matrix Characterization of Complex Materials with Scanning Acoustic Microscopy," in *Proceeding of SPIE Vol. 8694*, San Diego, 2013.

[21] J. McHugh, "Ultrasound Technique for the Dynamic Mechanical Analysis (DMA) of Polymers," Research gate, Berlin, 2008.

Preparation of Nanocellulose-Alginate Nanocomposites for Chlorhexidine Digluconate Drug Carrier

Angela Evelyna,[1*] Tiara Khumaira Astifanni,[1] Imelda Ruth,[1] Lia Asri,[2,*] Bambang Sunendar Purwasasmita[3,4]

[1]Faculty of Dentistry, Maranatha Christian University, Indonesia
[2]Materials Science and Engineering, Bandung Institute of Technology, Indonesia
[3]Advanced Materials Processing Group, Engineering Physics Study Program, Bandung Institute of Technology, Indonesia
[4]Research Center for Nanosciences and Nanotechnology, Bandung Institute of Technology, Indonesia.

*Corresponding authors: angela.evelyna@gmail.com; lia.asri@material.itb.ac.id

Abstract. Root canal treatment failures generally caused by microorganisms that have ability to invade dentinal tubules. Chlorhexidine is widely known to effectively eradicate broad spectrum bacteria that infected dentinal tubules. The delivery of chlorhexidine to the infected site of root canal still becomes an issue. In order to increase the stability and to regulate the release of chlorhexidine, encapsulation of chlorhexidine digluconate was conducted by employing nanocellulose and alginate nanocomposites. Nanocellulose was prepared from palm kernel cake using acid hydrolysis method. Transmission electron microscopy images showed the formation of whisker and fiber nanocellulose with the average diameter of 20 nm. Resulting nanocellulose was combined with alginate in the presence of Ca^{2+} crosslinker. Scanning electron microscopy images displayed sphere and oval morphologies of microcapsules with diameter of 500 nm. The microcapsules were further loaded with chlorhexidine digluconate 2% (w/v). In vitro drug release properties were evaluated in PBS at pH 7.4 (normal tooth environment) and at pH 5.5 (infected tooth environment). The UV-VIS results showed that chlorhexidine digluconate has higher release rate at pH 5.5 than at pH 7.4, suggesting that the microcapsule is a good candidate for the delivery of chlorhexidine digluconate to the infected root canal tooth environment.

Keywords: Root canal treatment, nanocellulose, chlorhexidine digluconate, microcapsules, alginate, drug delivery.

1. Introduction

The main purpose of root canal treatment in endodontic field is to disinfect the entire root canal system based on eradication of broad-spectrum of microorganisms [1]. Eradication of microorganisms from infected root canals is a complex task that involves the employment of various instrumentation methods, irrigation procedures and intra-canal medicaments. Preservation of microorganisms in dentinal tubules after root canal chemo-mechanical treatment has been reported [2]. The complex anatomy of the root canal and dentinal buffering ability lead to a complicated and challenging delivery of antimicrobial agents to the

infected site. Delivery systems of antimicrobial active agents are needed to guarantee efficient bacteria eradication.
Chlorhexidine (CHX) is a strong antiseptic biguanide that widely used in dentistry. CHX is a cationic biguanide comprising a positively charged hydrophobic and lipophilic molecule that show a broad-spectrum bactericidal and bacteriostatic properties [3]. The translation of controlled CHX drug delivery system in intracanal medicament is still limited. There is a need in lengthening the time release of CHX beyond the current delivery systems to improve bacterial reduction in infected root canal systems and to create a suitable environment for periapical healing. Encapsulation of CHX could offer a better medicament release control while further prolonging the delivery period of the bactericidal action in dentinal tissues.
Nanocellulose is a highly crystalline cellulose nanoparticle that can be employed as drug carrier to perform control drugs release [4, 5]. Formation of nanocellulose nanocomposites with other polymers such as alginate is important to increase the drug loading capacity. Drug carriers derived from nanocelulose-alginate nanocomposites showed pH-dependent and sustained drug release properties [6]. Therefore, the aim of this work was to obtain CHX controlled drug delivery system by encapsulating CHX in nanocellulose-alginate nanocomposite microcapsules. The release behavior of CHX was further evaluated.

2. Materials and Methods

Materials. Palm kernel cake (PKC) was obtained from the local market. Alginic acid sodium salt from brown algae was obtained from Fluka. Sulfuric acid was purchased from Smart Lab. Nitric acid, sodium sulfate, sodium hypochlorite technical grade 12% and sodium nitrite were obtained from Brataco. Chlorhexidine digluconate was obtained from PPH Cerkamed. Phosphate buffer saline pH 7.4 was obtained from Biogear. All chemicals were used without further purification.
Extraction of cellulose from PKC. PKC (75 g) was added into solution containing HNO_3 3.5% and$NaNO_2$, heated at 90°C for 2h. Resulting suspension was filtered and washed several times with water. The solid residue was treated with 750 ml of solution containing NaOH 2% and Na2SO3 2% at 50°C for 1h, followed by filtration and washing with water until pH suspension reached 7. The solid residue was subsequently bleached with 250 ml of NaOCl 1.75% at 100° C for 30 min. Resulting cellulose was washed with water for several times.
Isolation of nanocellulose. Cellulose (135 g, wet residue) was suspended in 500 ml of H_2SO_4 45% at 45°C for 45 min. Suspension was cooled down to RT by adding 500 ml of water and was stored at RT overnight. Resulting colloidal solution was washed with demineralized water and sonicated at pH 7. Colloidal solution was centrifuged at 10000 rpm for min 10 min.
Preparation of nanocellulose-alginate nanocompostes microcapsules. Nanocellulose-alginate nanocomposites was prepared according to Ning et al. [6] and and our previous work [7] with modified procedure. A hundred ml nanocellulose 2wt% was mixed with sodium alginate 1wt% and stirred for 15 min to result in homogeneous colloidal solution. Resulting nanocellulose-alginate solution was sprayed into 800 ml $CaCl_2$ 2wt%, followed by centrifugation at 10000 rpm. The residue was freezed dried overnight.
CHX drug loading. Nanocelullose-alginate nanocomposite microcapsules (0.75 g) were immersed in 200 ml of CHX 2wt% and stirred for 48h. Microcapsules loaded with CHX were separated by filtration.
***In vitro* CHX release test.** A twenty five mg of nanocellulose-alginate nanocomposite microcapsules was stored in cellulose dialysis membrane tubing (Sigma-Aldrich, MWCO 14000) and incubated in 50 mL PBS pH 7.4 at 37 °C. At a predetermined time interval, 5 mL of incubated solution was taken out to determine the amount of CHX, at the same time 5 mL of fresh PBS solution was added into the solution containing loaded drug to keep the total

volume of solution constant. The amount of rutin released during incubation with specified time was measured using UV VIS spectrophotometer. In vitro release test was also conducted in HCl solution at pH 5.5 using the same procedure.

Characterizations. FTIR measurements were recorded with KBr pellets on Prestige 21 Shimadzu at wavelength 4500 to 400 cm^{-1}. Spectra were measured at a resolution of 4 cm^{-1} with number of scan 40. Transmissions Electron Microscopy (TEM) images of nanocellulose were taken using HITACHI HT7700 with acceleration voltage of 80-100 kV. Sample was prepared by deposition of a small droplet of colloidal solution of nanocellulose in isopropyl alcohol on carbon coated copper micro grid (EM-japan) and allowed to air dry. Scanning Electron Microscopy (SEM) images were taken on Hitachi SU3500. The samples were dispersed in isopropyl ethanol, sonicated for 10 min, deposited on cover glass and dried in an oven. Afterwards the samples were gold-coated in a sputter coater MC-1000, Hitachi. X-Ray Diffraction (XRD) analysis, measured on Rigaku using Cu anode with wavelength of 1.5406 Å. The data were collected over the 2θ range of 10°-60°. UV-VIS measurement was performed on Shimadzu UVmini-1240 at wavelength of 255.5 nm.

Statistical Analysis. Comparison between *in vitro* CHX drug release at pH 7.4 and 5.5 was evaluated using Independent T-test SPSS program. The differences were considered as statistically significant at $p<0.05$.

3. Results and Discussion

3.1 Preparation of nanocellulose

Nanocellulose was prepared from PKC using acid hydrolysis method. Pretreatment of PKC was first conducted to open the bulk structure of PKC using nitric acid, resulting in a brown solid product. The hemicellulose content was further removed using sodium hydroxide solution, followed by dissolution of lignin substance in NaOCl solution. Resulting cellulose was indicated by the formation of white suspension that purified further with demineralized water. A high concentration of sulfuric acid (45 wt%) was used to hydrolyze resulting cellulose, leading to the cleavage of cellulose microfibrils. The acid esterifies some of surface hydroxyl groups of cellulose to produce nanocellulose comprising negatively charged sulfate groups that help to form a stable dispersed colloidal solution [8] (**Figure 1a**). XRD diffractogram of nanocellulose reveals the crystalline nanocellulose with characteristic peaks of native cellulose as shown by diffractions at 15°, 16.5°, 20.5°, 22.5°, 34°, and 35° (JCPDS#030289). The value of crystallinity index of nanocellulose accounted for 60%. It was calculated from the ratio of the height of the peak at 22.5° and the height of the minimum (Intensity of amorphous area, which is at about 18°) [9, 10].

TEM images of nanocellulose in Figure 1c and 1d display various structure of nanocellulose, prepared using the same acid hydrolysis procedure but different purification processes were applied. Nanocelullose in Figure 1c was obtained after immediate washing of nanocellulose to reach neutral pH of filtrate. This treatment resulted in whiskers nanocellulose with diameter around 20 nm and length of 200-300 nm (ratio diameter to length= 1:10 ~ 1:15). Additionally, nanofibers structure was also observed with diameter of 20-50 nm. Nanocellulose in Figure 1d was obtained after washing nanocellulose that was stored at pH 2-3 at 4°C around two weeks. Afterwards, nanocellulose was washed to reach pH 7. During storage, the hydrolysis continued to occur, causing further cleavage of glycosidic linkage [11] and finally resulted in spherical nanocellulose with diameter around 20 nm. Beside the sphere structure, agglomerated nanofibers were also found. It is suggested that various morphologies of nanocellulose was affected by different particle size of PKC starting material.

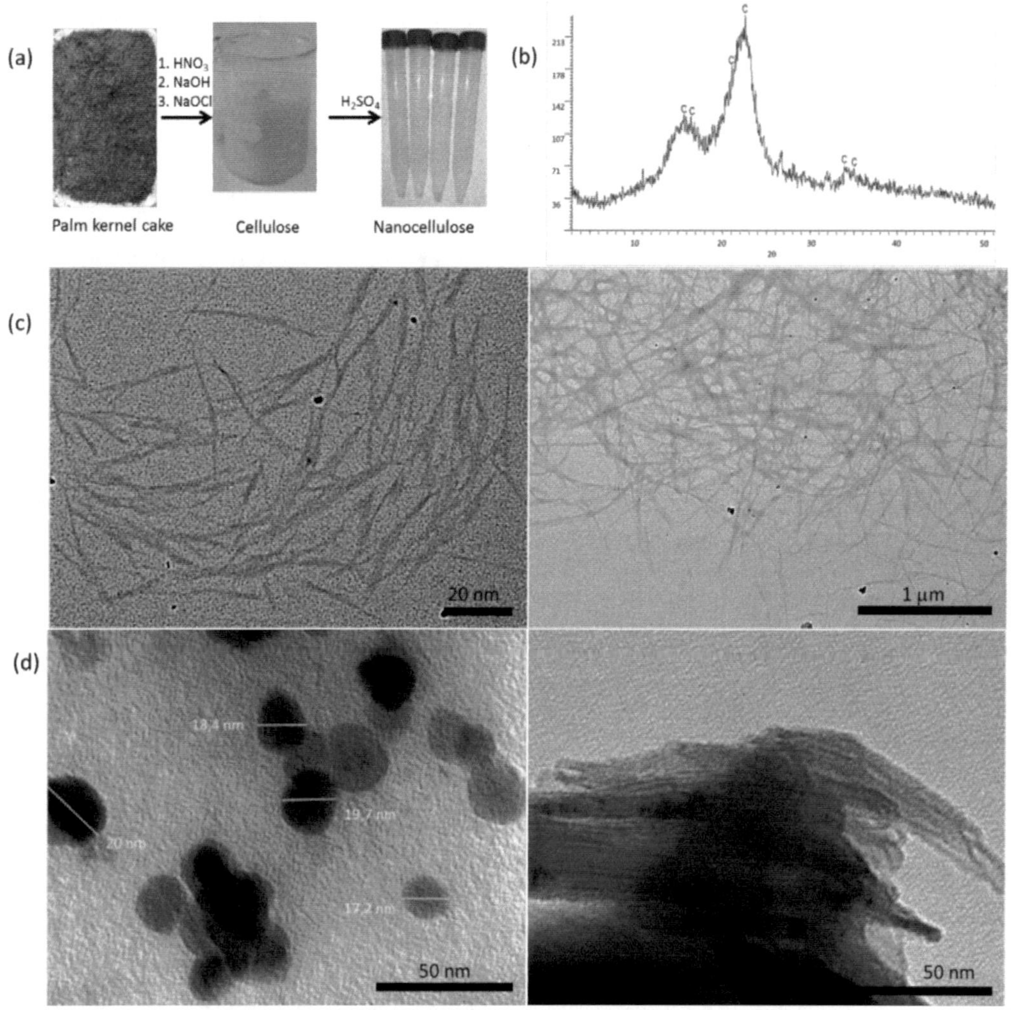

Figure 1. (a) preparation of nanocellulose from palm kernel cake (b) XRD diftractogram of nanocellulose (c) and (d) TEM images of nanocellulose with various morphologies.

3.2 Preparation of alginate-nanocellulose nanocomposite microcapsules

Microcapsules were prepared by combining nanocellulose and alginate. Nanocellulose was employed in order to stabilize the structure of alginate and to control the release properties of the microcapsule. Nanocellulose and alginate colloidal solution was sprayed into $CaCl_2$ solution to form cross-linked structure of alginate via ionic interaction between carboxylate group (COO^-) and Ca^{2+} ions, while nanocellulose was entrapped between alginate networks [7]. SEM images of nanocellulose-alginate nanocomposites depicted in Figure 2 exhibit sphere and oval shape with an average diameter of 500 nm and a length of 1.2 μm. Some inhomogeneous, irregular and agglomerated particles are still observed. This might be caused by combined spraying technique (rate of spraying, interval time etc.) and the concentration of sodium alginate which was probably too high, making alginate polymer chains easily to interact and to cross-link each other. These parameters (spraying rate, time interval and alginate concentration) during crosslink process need to be investigated further.

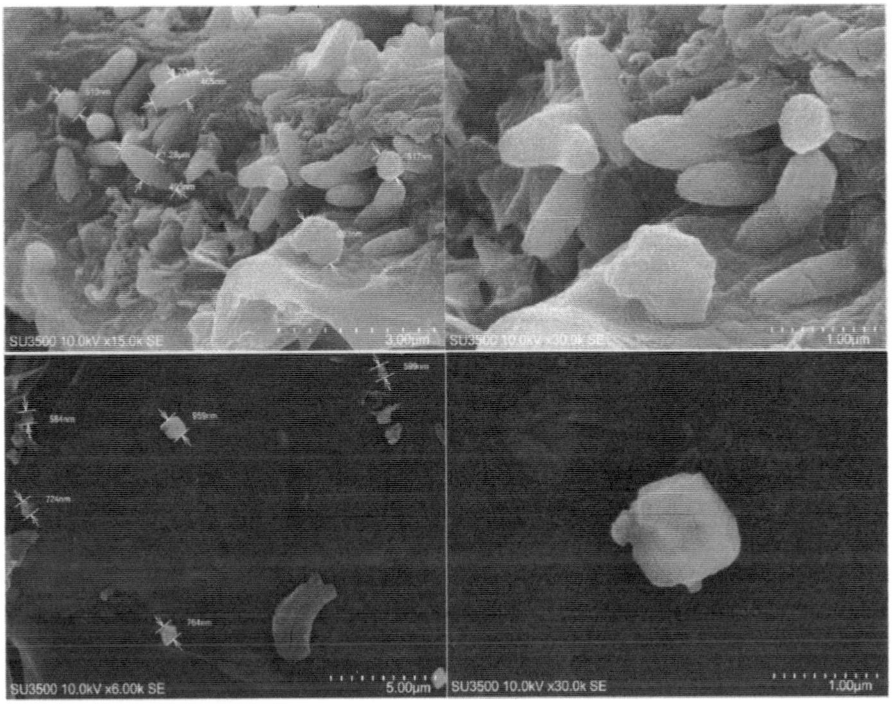

Figure 2. SEM images of nanocellulose-alginate nanocomposites microcapsules.

Nanocellulose-alginate nanocomposite microcapsule was used to entrap CHX antibacterial agent for root canal medicament application. CHX has been proved highly effective as an intracanal treatment by inhibiting the growth of *Enterococcus faecalis [12, 13]*. To enhance bacterial elimination from infected root canal systems and to make an ideal environment, it is highly desirable to prolong the release of CHX *in situ*. The root canal drugs should be able to penetrate the dentine tubules canal that has diameter varies between 1 and 4 micrometers [14, 15]. It was hypothesized that drug carrier prepared with a suitable size and a controlled release behavior could deliver the medicament deep into dentin tubules, allowing for sustained inhibition of bacteria in the root canal system. Notably, SEM results show that the size of nanocellulose-alginate microcapsules (500 nm-1.2 µm) are considered small enough for penetration inside the dentin tubules.

3.3 *In vitro* drug release properties

Entrapment of CHX was conducted by immersing nanocellulose-alginate microcapsules in CHX 2%. FTIR spectrum of nanocellulose-alginate nanocomposite loaded with CHX in Figure 3a indicates characteristic absorptions of these three compounds. The presence of CHX is shown by absorption at 1529 cm^{-1} and 1411 cm^{-1} (C=C aromatic functional groups), 3213 cm^{-1} (N-H group) and 2931 cm^{-1} (C-H stretching vibration). Characteristic absorptions of nanocellulose are shown by peaks at 1000-1249 cm^{-1}, attributed to C-O stretching vibration (overlapped with alginate functional groups), peak at 3213-3334 cm^{-1} from O-H hydrogen bonded stretching vibration (overlapped with peak of CHX and alginate). Alginate can be identified from the presence of absorptions at 1633 cm^{-1} and 1411 cm^{-1}, originating from asymmetric and symmetric stretching vibrations of carboxylate salt ions. The N-H functional groups from alginate appear as overlapped peak at 3213 cm^{-1}.

IOP Conf. Series: Materials Science and Engineering **547** (2019) 012046 doi:10.1088/1757-899X/547/1/012046

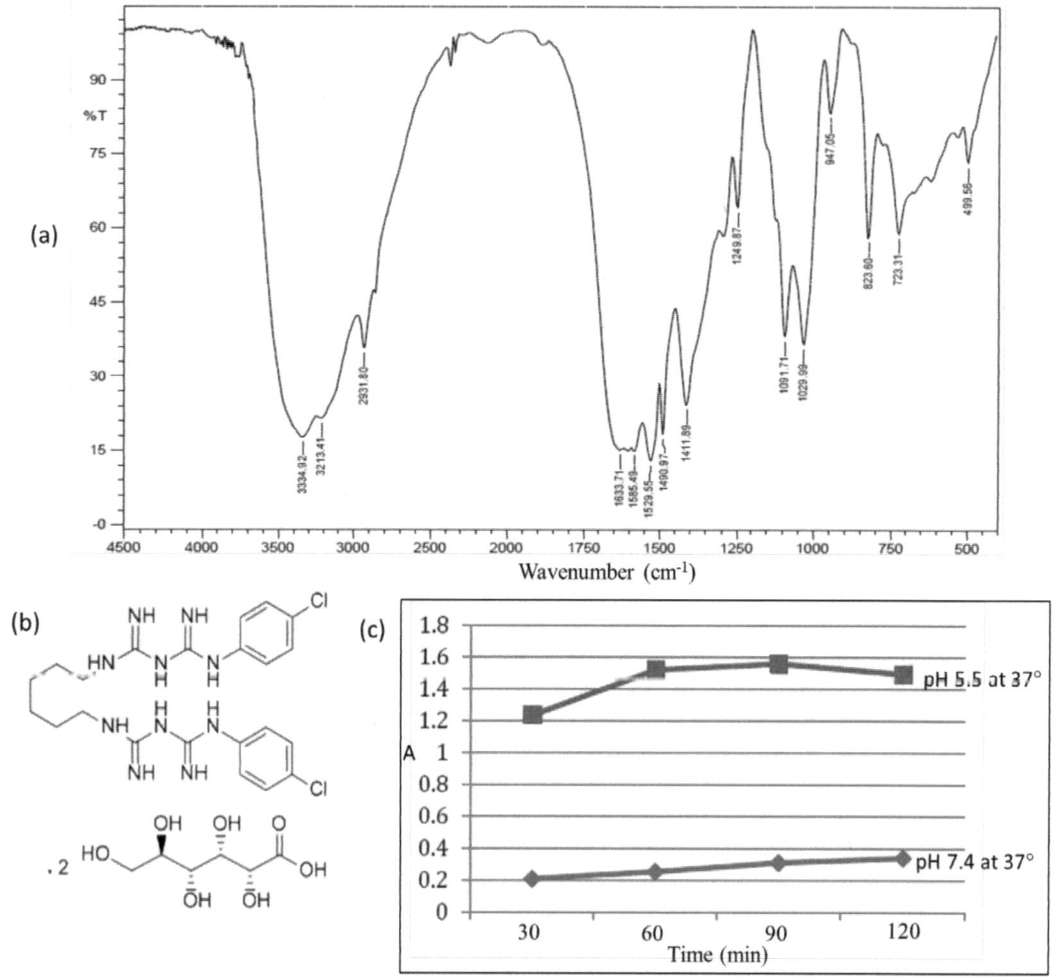

Figure 3. (a) FTIR spectrum of nanocellulose-alginate containing CHX (b) chemical structure of CHX (b) *in vitro* drug release of CHX at pH 7.4 and 5.5 at 37°C, conducted in triplicate.

In vitro drug release was evaluated at two conditions, at pH 7.4 in PBS solution to represent normal tooth environment and at pH 5.5 in HCl solution to mimic the infected tooth condition (Figure 3c). Saphiro Wilk homogeneity test and normality test results showed that p value (sig.) in all sample groups is more than 0.05. It can be concluded that the data is homogeneous and normal. Statistical analysis using Independent T-test showed that all sample groups had p value less than 0.05. Statistically H0 is rejected, confirming a significant difference between the release rate of CHX in normal condition (pH 7) and infected condition (pH 5.5). The release rate of CHX is much higher in an infected condition than a normal condition.

Lin et al. [6] reported that nanocellulose incorporated with alginate polysaccharides nanocrystals had maintained the integrity the nanocomposites, attributed to the presence of rigid nanocrystals that improved the stability and prolong the degradation. They showed that the microcapsules performed better stability at pH 1 (HCl solution) than at pH 6.8 and 7.4. In our case, the release profile was higher at pH 5.5 than 7.4, which is contrary to the results of Lin et al. This can be explained by the properties of CHX loaded into nanocomposites. It is envisaged that CHX comprising amine groups and hydroxyl groups are able to form stronger hydrogen bonding with alginate and nanocellulose compared with theophylline drug model employed by Lin et al. [6] This hydrogen bonding will be easily disrupted in the presence of acid (at low pH), resulting in higher release CHX than at pH normal 7.4. This results show that nanocellulose-alginate nanocomposites is a promising candidate for delivering CHX for root canal medicament.

4. Conclusions

In this work, nanocellulose-alginate nanocomposites were prepared as drug carrier for CHX entrapment. It is shown that size of resulting nanocomposites is suitable for tubuli dentin penetration, although some irregular and agglomerated particles were still observed. *In vitro* drug release evaluation of nanocellulose-alginate loaded with CHX demonstrated controlled release properties, showing higher release properties of CHX at pH 5.5 (infected condition) than pH 7.4 (normal condition). This result suggests that nanocellulose-alginate nanocomposites drug carrier is a potential candidate for delivery CHX to tubule dentin. *In vitro* antibacterial properties of the nanocellulose-alginate loaded with CHX will be evaluated in the future work.

5. Acknowledgements

We thank Viny Tanuwijaya for her assistance with SEM measurements. L.A.T.W.A acknowledges financial support from the Postdoctoral National-ITB Program, World Class University Grant, Indonesian Ministry of Education and Technology 2017.

6. References

[1] Siqueira Jr JFandRôças IN 2009 *Oral Surgery, Oral Medicine, Oral Pathology, Oral Radiology, and Endodontology* **107** 870-8.
[2] Nair P 2006 *International endodontic journal* **39** 249-81.
[3] Jenkins S, Addy MandWade W 1988 *Journal of Clinical Periodontology* **15** 415-24.
[4] Müller A, Ni Z, Hessler N, Wesarg F, Müller FA, Kralisch DandFischer D 2013 *Journal of Pharmaceutical Sciences* **102** 579-92.
[5] Lin NandDufresne A 2014 *European Polymer Journal* **59** 302-25.
[6] Lin N, Huang J, Chang PR, Feng LandYu J 2011 *Colloids and Surfaces B: Biointerfaces* **85** 270-9.
[7] Lia Amelia Tresna Wulan A, Amelia R, Muhammad Zulfan F, Muhamad IandBambang S 2018 *Materials Research Express*.
[8] Wang HandRoman M 2011 *Biomacromolecules* **12** 1585-93.
[9] Segal L, Creely JJ, Martin AEandConrad CM 1959 *Textile Research Journal* **29** 786-94.
[10] Park S, Baker JO, Himmel ME, Parilla PAandJohnson DK 2010 *Biotechnology for Biofuels* **3** 10.
[11] Rinaldi RandSchüth F 2009 *ChemSusChem: Chemistry & Sustainability Energy & Materials* **2** 1096-107.
[12] Varoni E, Tarce M, Lodi GandCarrassi A 2012 *Minerva stomatologica* **61** 399-419.
[13] Ercan E, Dalli MandDülgergil ÇT 2006 *Oral Surgery, Oral Medicine, Oral Pathology, Oral Radiology, and Endodontology* **102** e27-e31.
[14] Goldberg M, Kulkarni AB, Young MandBoskey A 2011 *Frontiers in bioscience (Elite edition)* **3** 711.
[15] Schilke R, Lisson JA, Bauß OandGeurtsen W 2000 *Archives of Oral Biology* **45** 355-61.

IOP Conf. Series: Materials Science and Engineering **547** (2019) 012047 doi:10.1088/1757-899X/547/1/012047

Elastic Modulus Determination of Thermoplastic Polymers with Pulse-Echo Method Ultrasonic Testing

Hermawan Judawisastra[1]*, Claudia[1], Firmansyah Sasmita[1], Toni Agung P[2]
[1]Materials Engineering, Faculty of Mechanical and Aerospace Engineering, *Institut Teknologi Bandung*, Bandung 40132, Indonesia
[2]Center for Material and Technical Product, the Ministry of Industry – Republic of Indonesia, Bandung 40132, Indonesia

* hermawan.judawisastra@material.itb.ac.id

Abstract. The determination of the elastic modulus of the structural thermoplastic polymers is especially required during in-service monitoring of the components. This elasticity modulus determination can simplify the inspection process and predict the lifetime of the polymer being used. The elastic modulus of the polymer can be determined by pulse-echo method testing ultrasonic which is easier to use than transmission method. This research aims to determine elastic modulus of thermoplastic polymers i.e. polymethylmethaacrylate (acrylic), polyamide (nylon) and low density polyethylene (LDPE) calculated from measurement of velocity and attenuation of ultrasonic pulse-echo. The research begins by looking at the effect of the use of three different probes to ultrasonic velocity readings. Then elastic modulus of material polymers from ultrasonic velocity was calculated using standard equation. The elastic modulus was compared with the elastic modulus from mechanical testing. A linear model elastic modulus from ultrasonic reading which consists of velocity and apparent attenuation was developed. The results were elastic modulus value obtained from ultrasonic velocity and standard equation had a profound error 98% to 158%, especially for polymer with an eminently low density and high ratio of viscous property to elastic property. The attenuation from ultrasonic reading was influenced by the viscoelastic properties and the elastic modulus from ultrasonic velocity reading was influenced by the density. A linear model for thermoplastic elastic modulus determination based on ultrasonic pulse echo testing has been developed and could reduced the error to 3.45%.

Keywords: ultrasonic, pulse-echo, elastic modulus, polymer, attenuation, viscoelastic

1. Introduction

Thermoplastic polymer is commonly used as a structural material, but it can be degraded by exposure to temperature and relative humidity [1]. As a result the elastic modulus of polymers decreased than lead to a structural failure. By monitor its modulus and predict the lifetime of polymer component, failure can be prevented. Polymer also has a wide range of elastic modulus because of its microstructure variation, such as molecular weight distribution, branching, fillers and blends [2]. Therefore it is important to characterize even monitor the elastic modulus of polymer.

Tensile testing or destructive testing is the most familiar method to determine elastic modulus. There are several disadvantages of destructive testing for example it needs arduous sample preparation and is impossible to use in the real time monitoring of in service structures. Meanwhile the field of non-destructive testing involves the identification without cutting apart or harming the object being tested. The most popular technique used is ultrasonic testing. It is able to use in the field and fast scan for repeatedly test. There are two approaches of ultrasonic: pulse echo and through transmission [3]. Elastic modulus and dynamic elastic modulus measurement with ultrasonic through-transmission method for polymers have been studied [4] and also wave velocity measurement in polymer with ultrasonic pulse-echo method has also been performed [5]. Pulse-echo is easier to use compare to transmission method and require only one side of the material. Both of these use high frequency sound waves on the order of 1-50 MHz [6].

The most commonly used indicators of material properties are velocity and attenuation or energy loss. The velocity depends on elastic modulus, poisson's ratio and density. The ultrasonic results are also influenced by the nearfield, diameter, and frequency of probe [7] Recent research study is about energy loss discover a few additional characteristics of a material. Scattering, absorption and geometric are three parameters that effect the attenuation. Small discontinuities like boundaries between each crystal are the source of scattering [3]. In polymeric materials, especially thermoplastics, there may be influenced from density [8] and viscoelastic properties [9]. The lost energy is related to viscous properties used for chain disentangle and secondary bonds dissolve which cannot return to their original positions. Besides that, the amplitude readings from ultrasonic have not interrupted by nearfield phenomenon [7].

Therefore, this research aims to determine elastic modulus of thermoplastic polymers i.e. polymethylmethacrylate (PMMA or acrylic), polyamide (PA or nylon) and low density polyethylene (LDPE) calculated from measurement of velocity and attenuation of ultrasonic pulse-echo with regard to the effect of nearfield probe, density and viscoelastic polymer, and develop a linear model elastic modulus from ultrasonic reading which consists of velocity and apparent attenuation.

2. Materials and Methods

Thermoplastic polymers were as-received PMMA, PA, and LDPE polymers that were given cutting and grinding treatment to create a smooth surface. Material density was conducted with KERN ABJ-NM/ABS-N device that used archimedes principle at 25°C and ethanol 99.8% was used as the reference density.

Dynamic mechanical analysis test was conducted to obtain the viscoelastic properties of specimens that were storage modulus (E'), loss modulus (E'') and tan delta. The test itself was based on ASTM D 5023 [10] with 3 point bending type, 10 Newton load and 100Hz frequency at room temperature. Dimension of specimen was 50 x 20 x 2mm. Storage modulus (E') is a measure of elastic response of polymers. It measures the stored energy. Loss modulus (E") is a measure of viscous response of polymers. It measures the energy dissipated as heat. Tan delta is the ratio of loss to the storage and called as damping. It related to how good a material will be at absorbing energy [11]. The value obtained from this test would be compared with the apparent attenuation and error elastic modulus from the ultrasonic test.

Tensile testing was conducted to obtain the elastic modulus of specimens. The test itself was based on ASTM D 638 [12] with speed of testing 5 mm/min and type I dimension. The value obtained from this test would be compared with the one from the ultrasonic test.

Ultrasonic test was conducted with GE USM 35X device with pulse-echo method to obtain the materials' longitudinal wave velocity. The probe used for the test were 1.5 MHz 12.7 mm diameter with delay line (probe 1), 1.5 MHz 12.7 mm diameter without delay line (probe 2), and 1 MHz 24 mm diameter (probe 3).

The nearfield zone of probe calculated with equation 1: where D is diameter probe (mm), f is frequency probe (Hz) and V is longitudinal wave velocity (mm/s). The nearfield of probe 1 and probe 2 is ± 20mm and nearfield of probe 3 is ±60 mm.

$$N = \frac{D^2 f}{4V} \tag{1}$$

Three times of longitudinal wave measurement in every section were conducted in every polymers and every probe. Measuring the velocity is based on ASTM E 494 [13] and apparent attenuation based on ASTM E 664 [14]. The sample code followed by materials- probe-sample thickness above (AN)/ below nearfield (BN). Table 1 showed the code of sample. The thickness samples of PMMA are 142, 40, 25, and 7 mm; PA are 98, 49, 25, and 16 mm; LDPE are 63, 30, 15, and 7 mm.

Table 1 Sample code (XX-XX-XX) of: materials – probe – sample thickness above/ below nearfield

	Thermoplastic Polymers		
	PMMA	**PA**	**LDPE**
Probe 1	PMMA-P1 -AN	PA-P1 -AN	LDPE-P1 -AN
	PMMA-P1 -BN	PA-P1 -BN	LDPE-P1 -BN
Probe 2	PMMA-P2 -AN	PA-P2 -AN	LDPE-P2 -AN
	PMMA-P2 -BN	PA-P2 -BN	LDPE-P2 -BN
Probe 3	PMMA-P3 -AN	PA-P3 -AN	LDPE-P3 -AN
	PMMA-P3 -BN	PA-P3 -BN	LDPE-P3 -BN

All of the longitudinal wave velocity obtained from the measurement would proceed to calculation process with equation 2: where E is elastic modulus, V_{long} is longitudinal wave velocity obtained from ultrasonic test, ρ is material's density, and v is material's Poisson's ratio. For this experiment, Poisson's ratio that were used for calculations were literature's Poisson's ratio.

$$E = \frac{V_{long}^2 \rho (1 + v)(1 - 2v)}{1 - v} \tag{2}$$

The apparent attenuation is determined by equation 3: where A_m and A_n are amplitudes of the mth and nth back reflection ($n > m$), and T specimen thickness.

$$\text{Apparent attenuation} = \frac{20 \log_{10} \frac{A_m}{A_n}}{2\,(n - m)T} \tag{3}$$

Linear regression was used to develop a model to find the best relationship between elastic modulus from mechanical testing as the dependent variable E_m and elastic modulus from ultrasonic testing E_{UT} and attenuation a as the independent variables. The correlation is calculated by Pearson method and parameter of linear equation is found by least square [15].

3. Results and Discussion

3.1 Physical, Mechanical and Viscoelastic Properties of Polymer

Table 2 showed the average of density, elastic modulus from tensile testing, viscoelastic properties which are storage modulus, loss modulus and tan delta, and Poisson's Ratio of three different types of thermoplastic polymers..

Table 2 Density, Elastic Modulus from Mechanical Testing, Viscoelastic and Poisson's Ratio of polymer

Polymers	Density (gr/cm^3)	Elastic Modulus from Tensile Test (GPa)	Viscoelastic Properties			Poisson's Ratio
			Storage Modulus (GPa)	Loss Modulus (GPa)	Tan Delta	
PMMA	1.19	1.99 ± 0.09	3.64	0.226	0.062	0.4 [16]
PA	0.95	1.06 ± 0.06	2.69	0.135	0.050	0.41 [17]
LDPE	0.89	0.29 ± 0.00	0.35	0.041	0.115	0.46 [17]

From the result, material polymers had a variation of density, elastic modulus, and viscoelastic properties and poisson's ratio. So that, these polymers could be used to find its influence in elastic modulus determination by ultrasonic testing.

The density of PMMA, PA and LDPE were 1.19 gr/cm^3, 0.95 gr/cm^3, 0.89 gr/cm^3 respectively. The elastic modulus of PMMA, PA and LDPE were 1.99 GPa, 1.06 GPa, 0.29 GPa respectively. Density had a linear relation with an elastic modulus of materials. The denser materials would have a higher elastic modulus.

The storage modulus of PMMA, PA and LDPE were 3.64 GPa, 2.69 GPa, 0.35 GPa respectively. There was a linear relation between elastic modulus from tensile test and storage modulus from dynamic mechanical analysis. The higher elastic modulus from tensile test the higher storage modulus from dynamic test. But the storage modulus is not equal to elastic modulus because material response differently in quasi static tensile testing and dynamic testing bending [18].

The loss modulus of PMMA, PA and LDPE were 0.226 GPa, 0.135 GPa, 0.041 GPa respectively. The energy absorb use for chain disentangle and secondary bond dissolve. Tan delta or the ratio of viscous and elastic of PMMA, PA and LDPE were 0.062, 0.050 and 0.115. LDPE has the highest ratio of viscous and elastic, because the amorf and branched structure of LDPE make high loss of energy [19].

3.2 Evaluation of Ultrasonic Velocity Reading

Table 3 showed the velocity of several types of thermoplastic polymers from ultrasonic reading probe 1, probe 2, and probe 3 compared to velocity from reference [20]. Probe 1 diameter 12,7 mm with delay line showed more consistent velocity readings because velocity reading is not influenced by nearfield. Because the length of nearfield from ±20 mm decrease with the length of delay line 13 mm so the length of nearfield becomes only ±7 mm. Moreover, probe 1 is highly damped and low frequency delay line transducers. The low frequency characteristics make this probe good for penetrating thick or thin and highly attenuative materials [21].

For probe 1 the difference in speed between the material's thickness above the nearfield and below nearfield is 0.04% for PMMA, 0.3% for PA, and 5.4% for LDPE. Meanwhile probe 2 and probe 3 showed big differences velocity reading for the sample thickness which is below nearfield. The velocity difference for probe 2 is 10% for PMMA, 6% for PA, and 25% for LDPE and the velocity difference for probe 3 is 4% for PMMA, 5% for PA. A big difference ultrasonic velocity in probe 2 and 3 compared to the velocity references was due to velocity reading with thickness below nearfield affected by nearfield. Nearfield is a region where wave interference phenomenon occurs, which will disrupt wave propagation and lead to misreading [22].

Table 3 Velocity of thermoplastic polymers from reference and Probe 1, Probe 2, Probe 3

Polymers	V_{long} reference (m/s)	Vlong **ultrasonic (m/s)**					
		Zip Probe 1 (Ø 12,7 mm with delay)		**Probe 2 (Ø 12,7 mm without delay)**		**Probe 3 (Ø 24 mm)**	
		N2 - AN	**N2 - BN**	**N2 - AN**	**N2 - BN**	**N3 - AN**	**N3 - BN**
PMMA	2690 [20]	2693	2684	2725	2443	2708	2604
PA	2400 [20]	2380	2379	2392	2243	2357	2230
LDPE	2237 [19]	2011	2126	2424	2984	-*	1856

*No sample thickness above 60mm

Therefore probe 1 diameter 12,7 mm 1,5 MHz with delay line are good for velocity reading for several thermoplastic polymer and several thickness. This probe will be used for calculation of apparent attenuation.

3.3 Comparison between Elastic Modulus from Mechanical Testing and Ultrasonic Testing

The results of elastic modulus from mechanical testing and elastic modulus from ultrasonic testing are shown in Table 4. Error of elastic modulus from ultrasonic E_{UT} is higher than elastic modulus from mechanical testing E_M. Error elastic modulus of PMMA is 98.5%, PA 119.5%, and LDPE 157.9%. Other research also showed that there are a profound error between elastic modulus from ultrasonic testing and elastic modulus from mechanical testing. Error can be caused by low density, attenuation and viscous elastic properties [8] [9].

Table 4 Elastic Modulus from ultrasonic testing and mechanical testing

Polymers	V_{long} Ultrasonic (V_{UT})(m/s)	Elastic Modulus UT (E_{UT}) (Gpa)	Elastic Modulus mechanical testing (E_M) (Mpa)	Error Elastic Modulus (%)
PMMA	2671	5.03	1.99	98.5
PA	2385	2.72	1.06	119.5
LDPE	1969	1.62	0.29	157.9

Figure 2 show the correlation between error elastic modulus calculation and density. The lighter polymer show the higher error elastic modulus, because the lower density polymer tends to have a bigger free volume and bigger viscous properties. Higher free volume makes the chain easier move between each other and did not return to their previous position. The displaced molecules have no tendency to slip back to their original positions, therefore these deformations are not recoverable absorb energy [23]. Meanwhile higher density of materials caused the waves propagate faster. This happened because there is less chain movements and less absorb energy [8]. Therefore the error elastic modulus in thermoplastic polymer is related to density.

The result from apparent attenuation ultrasonic testing and viscoelastic properties are shown in Figure 3. There is a strong correlation between attenuation from ultrasonic testing and viscoelastic properties. Attenuation is the combined effect of a number of parameters: absorption and scattering. Absorption is related to the viscoelastic properties, the more viscous polymers showed higher loss energy [24]. It is shown by tan delta, the ratio of loss modulus and storage modulus of viscoelastic polymer. Higher tan delta means loss energy or viscous property is higher compare to storage modulus or elastic property. The energy absorb use for chain disentangle and secondary bond dissolve. Meanwhile scattering happens because of small reflection of material micro interfaces, such as crystalline and amorphous zone in polymer or bulky pendent group [7].

The highest error and attenuation is from LDPE followed by PA. The energy of ultrasonic wave excites the long range movement of LDPE and PA chains, so there is big absorb energy from LDPE and PA. The error would come from highly energy absorb by the microstructure of LDPE which is consists of branched and amorphous chain [24]. PA is less amorphous and branched chain than LDPE so it has less attenuation. Meanwhile PMMA has 92 % optical transmission [25] and it is denser so it is less scatter the energy than the other polymers.

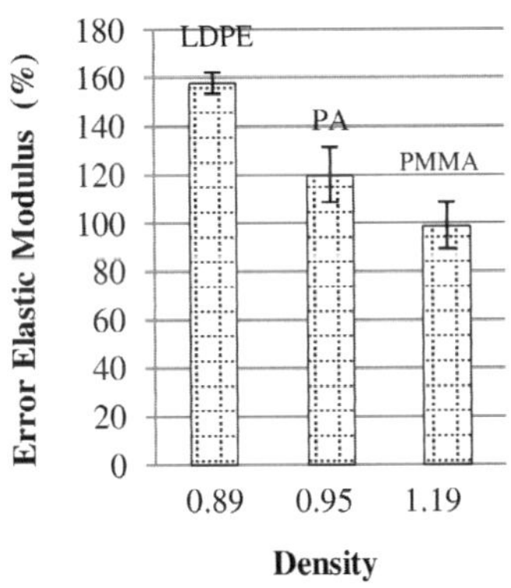

Figure 1 Error of elastic modulus and density

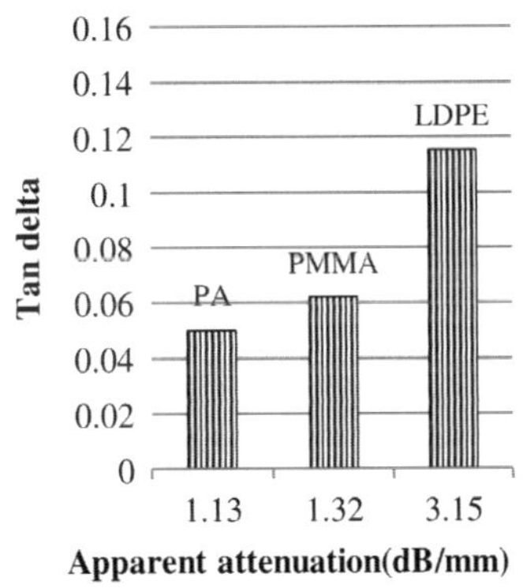

Figure 2 Viscoelastic properties and apparent attenuation

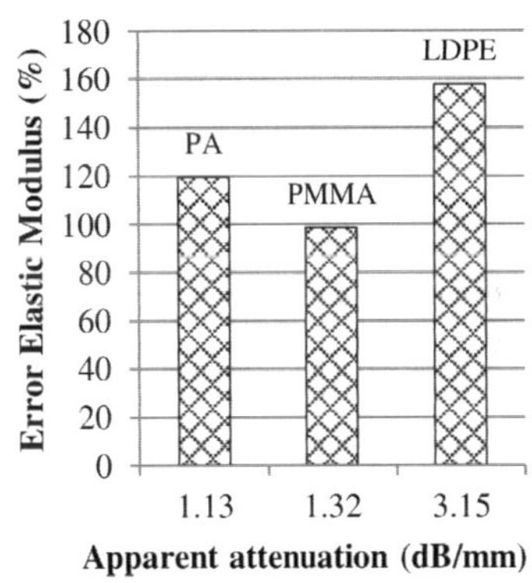

Figure 3 Error of elastic modulus and apparent attenuation

There is also a correlation nonlinear between apparent attenuation and error elastic modulus which is shown in Figure 4. The error value is greater if the apparent attenuation is greater. But there is a different trend in PA because attenuation is a combine effect of absorption and scattering [7].

Therefore the error of elastic modulus is still influenced by the viscoelastic and materials density. The viscoelastic properties have a strong correlation with the apparent attenuation.

3.4 Model of Thermoplastic Elastic Modulus Determination based on Ultrasonic Pulse Echo Testing

The correlation of modulus elastic from mechanical testing to modulus elastic from ultrasonic and attenuation is shown at Table 5. There were a strong linear correlation between elastic modulus from mechanical testing with elastic modulus and attenuation from ultrasonic testing. The model is shown by equation 4. Elastic modulus from ultrasonic testing affects elastic modulus from mechanical testing of 0.528. Attenuation affects elastic modulus from mechanical testing of 0.0128. While -0.161 is from constant parameters such as similar temperature [26], probe, and ultrasonic method [9].

Table 5 Correlation of modulus elastic from mechanical testing to modulus elastic from ultrasonic and attenuation

	Eut	**a**
Em	0.993	-0.771

$$\hat{E}\text{model}(GPa) = 0.0128 \times a\left(\frac{dB}{mm}\right) + 0.528 \times E_{UT}(GPa) - 0.161 \quad (4)$$

Figure 5 shows comparison of elastic modulus from ultrasonic velocity calculated using standard equation (E_{UT}), elastic modulus from tensile testing (E_M), and elastic modulus from E_{UT} and attenuation from ultrasonic testing calculated with model (E_{model}). The use of developed model successfully reduced the error of elastic modulus from 98.5%- 157.9% to 2.16% -5.52% for 3 types of thermoplastic polymers.

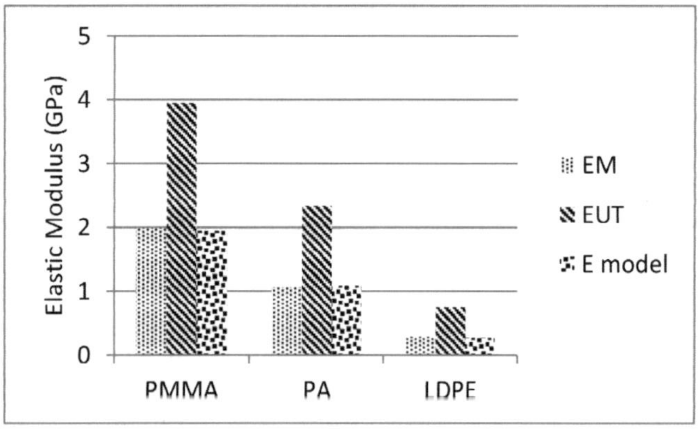

Figure 4 Comparison of elastic modulus of thermoplastic polymer from mechanical testing, ultrasonic, and model ultrasonic

3.4 Comparison between Elastic Modulus from Dynamic Mechanical Testing and Ultrasonic Testing

Table 6 Comparison between Elastic Modulus from Dynamic Mechanical Testing and Ultrasonic Testing

Polymers	Storage Modulus from DMA 3 point bending (GPa)	Elastic Modulus from Ultrasonik (E_{UT}) (Gpa)	Elastic Modulus from Tensile Testing (E_M) (Mpa)
PMMA	3.64	5.03	1.99
PA	2.69	2.72	1.06
LDPE	0.35	1.62	0.29

The difference between elastic modulus from ultrasonic E_{UT} and storage modulus from dynamic mechanical testing is smaller than elastic modulus from ultrasonic E_{UT} and elastic modulus from tensile testing. Elastic modulus from ultrasonic is near to the storage modulus from dynamic mechanical testing, because both ultrasonic testing and dynamic testing are describe dynamic mechanical behavior[4]. Ultrasonic wave is the particle making up the substance, and the restoring forces are the elastic bonds holding the substance together. While dynamic testing is sinusoidal load and response of material, so elastic and loss modulus can be determined. There was a slight difference between elastic modulus from ultrasonic and storage modulus from dynamic mechanical testing because the propagation of ultrasonic wave in polymers not only depend on elastic modulus, but also density and viscoelastic behaviors [4], which had been proved by experiment results in the previous section.

While the elastic modulus from 3 point bending was greater than elastic modulus from tensile testing because 3 point bending testing often results in a greater modulus of elasticity due to the smaller number of defects in the smaller test volume and there is a shear stress involvement [27].

4. Conclusions

The probe 1.5MHz 12.7 mm diameter with delay line or zero impedance probe (ZIP) results in more consistent reading for various thickness and enables the measurement of wave velocity and attenuation. The elastic modulus testing from ultrasonic on three types of thermoplastic polymers i.e. PMMA, PA and LDPE were still contained error from 98% up to 156%. Elastic modulus value obtained from ultrasonic testing from ultrasonic test had a profound error, especially for polymer with an eminently low density and high ratio of viscous property to elastic property. The attenuation from ultrasonic reading was influenced by the viscoelastic properties and the elastic modulus from ultrasonic velocity reading influenced by the density. A linear model for thermoplastic elastic modulus determination based on ultrasonic pulse echo testing has been developed and successfully reduced the error to 2.16 - 5.52% for three types of polymers.

5. References

[1] K. L. Ho and A. L. Pometto, "Effects of Temperature and Relative Humidity on Polylactic Acid Plastic Degradation," *Journal of Polymers and the Environment,* vol. 07, 1999.

[2] W. Cox, Understanding Rheology od Thermoplastic Polymers, TA Instruments, 1958.

[3] S. Gholizadeh, "A review of non destructive testing methods of composite materials," in *XV Portuguese Conference on Fracture*, Portugal, 2016.

[4] F. Lionetto and A. Maffezzoli, "Polymer Characterization by Ultrasonic Wave Propagation," *Advanced in Polymer Technology,* vol. 27, pp. 63-73, 2008.

[5] H. A. Afifi, "Ultrasonic Pulse Echo Studies of The Physical Properties of PMMA, PS, and PVC," *Polymer Plastics Technology and Engineering,* vol. 42, pp. 193-205, 2003.

[6] ASM Handbook volume 17, Nondestructive Evaluation and Quality Control, Metal Park, Ohio: ASM International, 2006.

[7] C. Hellier, Handbook of Nondestructive Evaluation, USA: McGraw Hill, 2003.

[8] F. G. R. d. Oliveira, "Relationship between density and ultrasonic velocity," *Bioresource Technology,* vol. 97, p. 2443–2446, 2006.

[9] T. L. Szabo, A Model for Longitudinal and Shear Wave Propagation in Viscoelastic Media, Acoustical Society of America, 2000.

[10] ASTM, ASTM D 5023 Standard Test Method for Plastics: Dynamic Mechanical Properties: In Flexure (Three Point Bending), ASTM International, 2001.

[11] Y. K. P. Menard, Dynamic Mechanical Analysis: A Practical Introduction, Second Edition.

[12] ASTM, ASTM 638 Standard Test Method for Tensile Properties of Plastics, United States: ASTM Internasional, 2002.

[13] ASTM , ASTM E 494 Standard Practice for Measuring Ultrasonic Velocity in Materials, Pennsylvania: ASTM International, 1995.

[14] ASTM, ASTM E 664 Standard Practice for the Measurement of the Apparent Attenuation of Longitudinal Ultrasonic Waves by Immersion Method, United States: ASTM International, 1993.

[15] R. E. Walpole and R. H. Myers, Ilmu Peluang dan statisitka untuk insinyur dan Ilmuwan edisi ke-4, Bandung: Penerbit ITB, 1995.

[16] K. Nitta and M. Yamana, "Poisson's Ratio and Mechanical Nonlinearity Under Tensile Deformation in Crystalline Polymers," [Online]. Available: http://cdn.intechopen.com. [Accessed 12 June 2018].

[17] Goodfellow, "Polyamide-Nylon-6," [Online]. Available: http://www.goodfellow.com. [Accessed 12 June 2018].

[18] H. F. Brinson and L. C. Brinson, Polymer Engineering Science and Viscoelasticity, New York: Springer, 2008.

[19] H. I. W. P. D. Davidse, "Sound Velocity and Young's Modulus in Polyethylene.," *Journal of Polymer Science,* vol. 59, pp. 389-400, 1962.

[20] J. E. Mark, Physical Properties of Polymer Handbook 2nd, New York, USA: Springer, 2007.

[21] G. i. technologies, "Ultrasonic Probes," [Online]. Available: http://www.ndtproducts.ca/products/files/UT%20Probe%20Cat.pdf. [Accessed 21 June 2018].

[22] D. K. P. a. S. Pandey, Ultrasonic A technique of Material Characterization in Acoustic Waves, Croatia: Sciyo, 2010.

[23] M. Chanda and S. K. Roy, Plastics Technology Handbook, New York: CRC, 2006.

[24] S. Brent, Plastics Materials and Processing, Ohio: Pearson Prentice Hall, 2006.

[25] H. a. D. E. Lin, "Optical and Mechanical Properties of Optically Transparent PMMA Composites," *Polymer Engineering and Science,* vol. 32, 1992.

[26] F. Massines, "Ultrasonic Characterization of Polymer Viscoelasticity Application to Polysterene," *Makromol Chem.,* pp. 121-137, 1989.

[27] F. Mujika, "On the effect of shear and local deformation in three point bending test," *polymer testing,* vol. 26, p. 869–877, 2007.

IOP Conf. Series: Materials Science and Engineering **547** (2019) 012048 doi:10.1088/1757-899X/547/1/012048

The Effect of Dibutyl phthalate plasticizer to Impregnation and Mechanical Properties of Biocomposite Sansevieria trifasciata/High Impact Polypropylene

Apang DjafarShieddieque*[1, 2 a)], Mardiyati[1, b)], Rochim Suratman[1, c)] and Bambang Widyanto[1, d)]

[1]Faculty of Mechanical and Aerospace Engineering, Bandung Institute of Technology, Ganesha 10, Bandung, 40132, Indonesia
[2]Faculty of Mechanical Engineering, Sekolah Tinggi Teknologi Wastukancana, Jln Cikopak No. 53, Purwakarta, 41151, Indonesia

[*] Coresponding Author: apang_djafar@yahoo.com

Abstract. The aim of this research was to investigate the effect of dibutyl phthalate plasticizer to impregnation and mechanical properties of biocomposite *sansevieria trifasciata*/high impact polypropylene and to analyze the morphology of interfacial bond between matrix and fiber. In this research, biocomposite *sansevieria trifasciata*/high impact polypropylene was manufactured by using a hot press method with temperature 180°C and pressure 140 bar for 10 minutes. Fibers were extracted from *sansevieria trifasciata* by using mechanical retting. In this experiment, the fibers weree treated by NaOH 3% (wt%) solution at temperature 100°C for an hour. The volume fractions of fiber used in this experiment were 0%, 15%, 25%, and 35% with unidirectional orientation. To improve impregnation of high impact polypropylene matrix, 2% dibutyl phthalate was added as plasticizer. Actual density and mechanical properties of biocompositesare conducted based on ASTM standard testing methods D792 and D3039.The results showed that tensile strength, modulus of elasticity and elongations of bio composite *sansevieria trifasciata*/high impact polypropylene with plasticizer were higher than without added plasticizer. The fracture analysis was investigated by scanning electron microscope (SEM). The highest tensile strength, modulus of elasticity and elongation of biocomposites were 65 Mpa, 2.31 GPa and 0.42% respectively. Their highest properties were obtained from bio composites with 35 % volume fraction of unidirectional alkaline treated fiberand 2% dibutyl phthalate. Morphological feature based on scanning electron microscope (SEM) observation showed that biocomposites with added plasticizer had better interface bond between matrix and fibers.

Keywords: Alkaline treatment, *Dibutyl phthalate*, High impact polypropylene, Plasticizer, *Sansevieriatrifasciata*

1. Introduction

Light weight vehicle has become very important nowadays due to depletion of the fossil fuel availability and increasing of CO_2 emissions [1–2]. To overcome those problems, the usage of lightweight materials in automotive to reduce vehicle weight which eventually reduce the fuel consumption and improve the performance as well [3]. Plastic has been developed as one of promising alternatives to replace the original materials [1]. However, plastic such as high impact polypropylene that has been developed has weakness due to its mechanical properties that still not suit up to the expected standard as material of vehicle components. Fibers have been used as reinforcement in plastic to improve the mechanical properties, named as composite. Natural fiber has been a better alternative than glass fiber as it offers a lower carbon footprint in manufacturing process and less energy required for production [2 – 4].Inthis research, *Sansevieria trifaciata* is used as to reinforce the high impact polypropylene matrix. *Sansevieria trifaciata* is common plant and can be found throughout Indonesia. To improve the mechanical properties of the composite, plasticizer has been added into the composite matrix (i.e polymer). Plasticizer is added toaid the polymer processing. The aim of this study was toinvestigate the effect of plasticizer addition on impregnation and mechanical properties of *Sansevieria trifasciata*/high impact polypropylene (STF/HIPP) biocomposites.

2. Materials and Methods

Materials

Sansevieria trifasciata leaves and high impact polypropylene were used in this research. The leaves were obtained from local residents in Citarum, Bandung, and high impact polypropylene was provided by PT Chandra Asri Petrochemical. *Sansevieria trifasciata* fibers were prepared from *Sansevieria trifasciata* leaves by mechanical retting and washed with water to remove leaves' component, then dried at room temperature. Plasticizer used in this study was dibutyl phthalate (DBP) from Sigma Aldrich. The chemical structure of DBP is shown in Fig.1

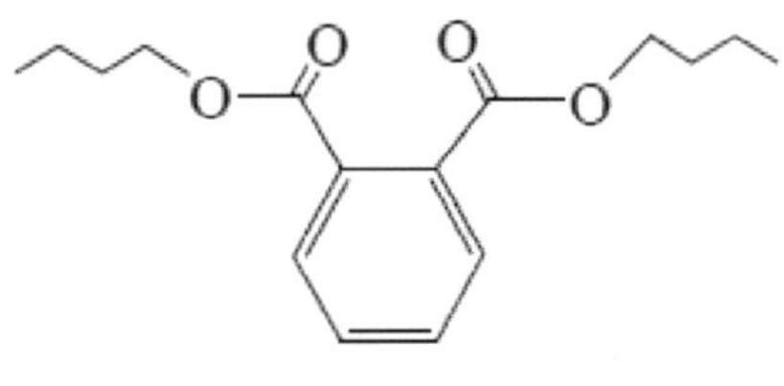
DBP ($C_{16}H_{22}O_4$)

Fig 1.The chemical structure of dibutyl phthalate (DBP) [5].

Alkaline Treatment

Sansevieria trifasciata fibers were treated by refluxing 10 grams of fibers in NaOH solution with ratio 1:100. The concentration of NaOH solution was 3% wt. Fibers were treated at 100°C for an hours. Then, the treated *Sansevieria trifasciata* fibers were washed with demineralized water and dried at room temperature about 1 day. The fibers orientation was set up into unidirectional orientation.

Manufacturing

In this research, *Sansevieria trifasciata*/high impact polypropylene (STF-HIPP) without and with 2%wt. of DBP were prepared. *Sansevieria trifasciata*/high impact polypropylene biocomposites were manufactured by using a hot compression molding at 180°C with the pressure at 140 bar for 10 minutes. Fibers orientations were unidirectional with volume fraction of fiber 0%,15%, 25% and 35%.

Testing Method

The mechanical properties of *Sansevieria trifasciata*/high impact polypropylene biocomposites (STF/HIPP) were examined by using universal testing machine method (Tensilon). The density of composite was determined according to ASTM D792, constituent contents were determined according to ASTM D3171 and the void content of composites were determined according to ASTM D2734-94. The morphologies of the fracture surface were observed using Scanning Electron Microscope (SEM).

3. Result and Discussion

Mechanical properties of Composites

The density of unidirectional *Sansevieria trifasciata*/high impact polypropylene (STF/HIPP) with 2 wt. % DBP and without DBP are presented in Table 1. The tensile strength, the stiffness and elongation of the biocomposites are shown in Fig, 2. The results show that the addition of fraction of fiber will increase the density of biocomposites and void as well. The amount of voids can lead the stress consentration that can cause to failure, but because the void difference obtained for each kind of composite is considered small, the effect can be neglected [6-8].

Table 1 Density and void percentages of STF/HIPP biocomposites

No	Sample	Fiber Volume Fraction (%)	Density (kg/m3)	Void (%)
1	STF/HIPP without plasticizer	15	985	3.14
2		25	989	4.40
3		35	992	5.80
4	STF/HIPP + 2 wt.% DBP	15	988	3.12
5		25	990	4.00
6		35	995	5.5

The mechanical properties of biocomposites are shown in Fig. 2-4. The tensile and stiffness properties of STF/HIPP biocompositesincrease with increasing the volume fraction of fiber. The addition of volume fraction of *Sansevieria trifasciata* fibers up to 35 % into high impact polypropylene matrix increases the tensile strength from 10.8 MPa to 63.9 MPa. Biocomposites STF/HIPP with 2wt. % DBP have slightly higher tensile and stiffness strength as shown in Fig. 2 and Fig. 3.

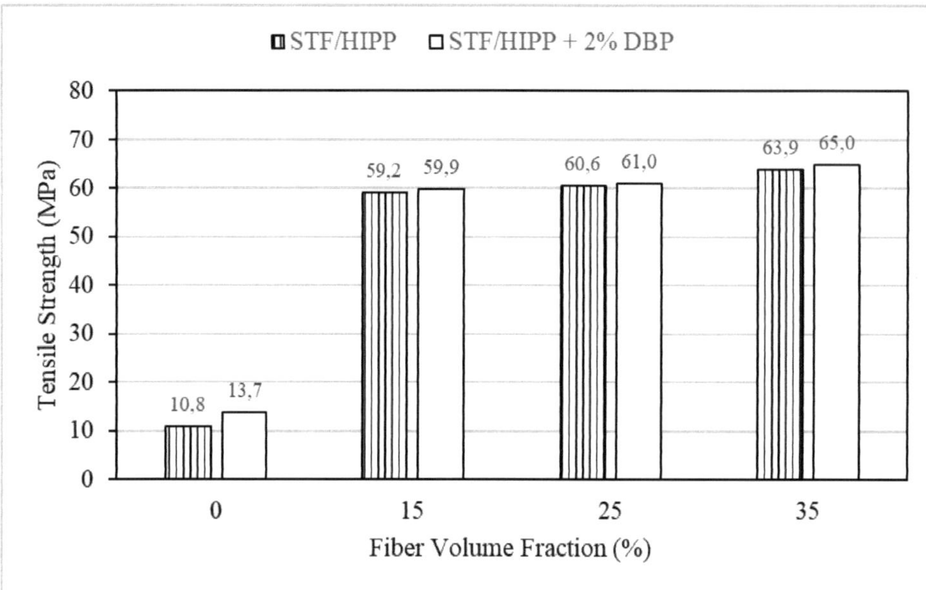

Fig. 2. Tensile Strength of Composites

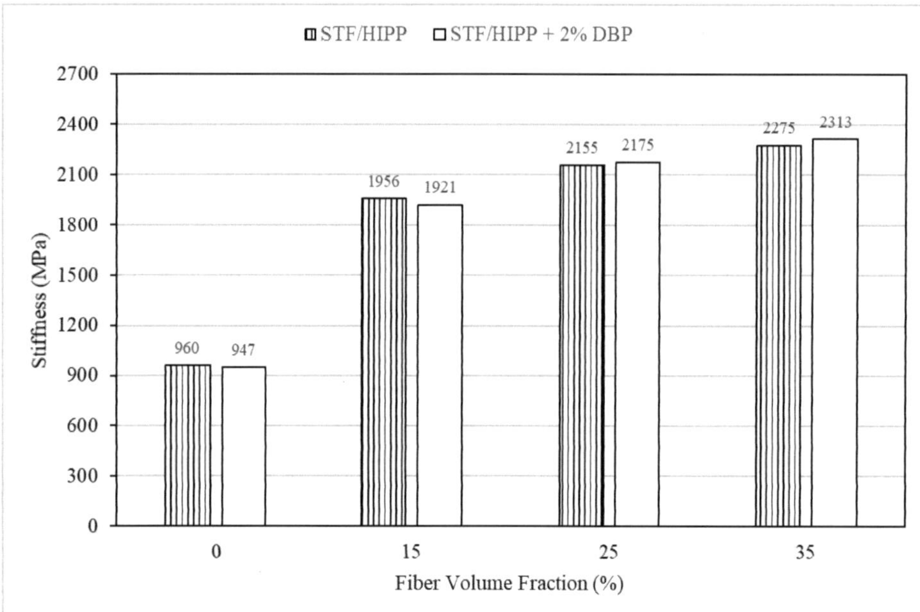

Fig. 3.Stiffness of Composites.

Figure 4 shows the elongation properties of STF/HIPP biocomposites. Biocomposites STF/HIPP with DBP addition has higher elongation compare with STF/HIPP biocomposites without DBP addition. In polymer processing, plasticizer is added to improve the film flexibility, reduce glass transition temperature, improve the impact strength, facilitate on molding, etc. [9]. Reducing the glass transition temperature will lead to improvement of ductility of polymer, therefore enhances the toughness and impact resistance of the polymer. In this study it is observed that addition small amount of DBP as plasticizer improve the elongation value for unidirectional STF/HIPP biocomposites with 15 % and 25 % vol. fraction of fiber. The elongation of unidirectional STF/HIPP biocomposites with 35 % vol. fraction of fiber shows slightly improvement that might be caused by the reinforcement effect of the fiber more dominant than the effect of plasticizer.

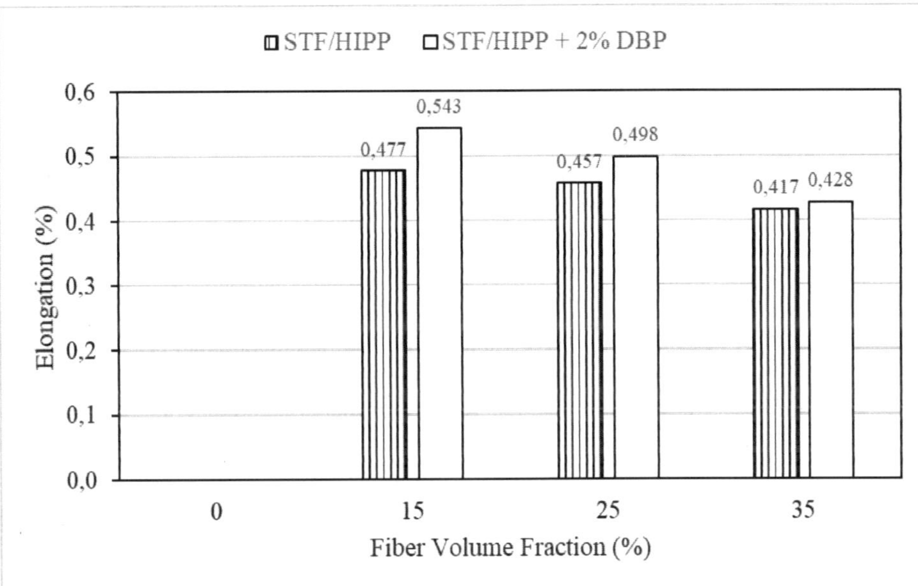

Fig. 4 Elongation of Composites.

Morphology of Fracture Surfaces

The morphology of the fracture surface of the unidirectional STF/HIPP biocomposites are shown in Fig. 5 and Fig. 6. There were gaps and cavities in all images of fracture surface that associated with the pull out of the fiber from the matrix. Pull out fibers are observed on the fracture surface of the unidirectional STF/HIPP biocomposite without DBP plasticizer and with addition of DBP plasticizer as shown in Fig. 5a and Fig 6a. However, the well bond *Sansevieria trifasciata* fibers are more likely found on the fracture surface of unidirectional STF/HIPP biocomposite with the 2 wt. % of DBP (Fig. 6a). The addition of DBP improves the impregnation of the matrix into fiber, results in increasing the mechanical properties as mentioned above. The less impregnation of matrix can be seen on existing of the gap or cavity around the fiber as a result of pull out in Fig. 5b, meanwhile the gap or cavity is not found around the pullout fiber of the unidirectional STF/HIPP biocomposites with DBP addition as seen in Fig. 6b. It is clear the addition of DBP can improve the impregnation matrix into fiber.

IOP Conf. Series: Materials Science and Engineering **547** (2019) 012048 doi:10.1088/1757-899X/547/1/012048

(a) (b)

Fig. 5 SEM images of fracture surface of unidirectional STF/HIPP

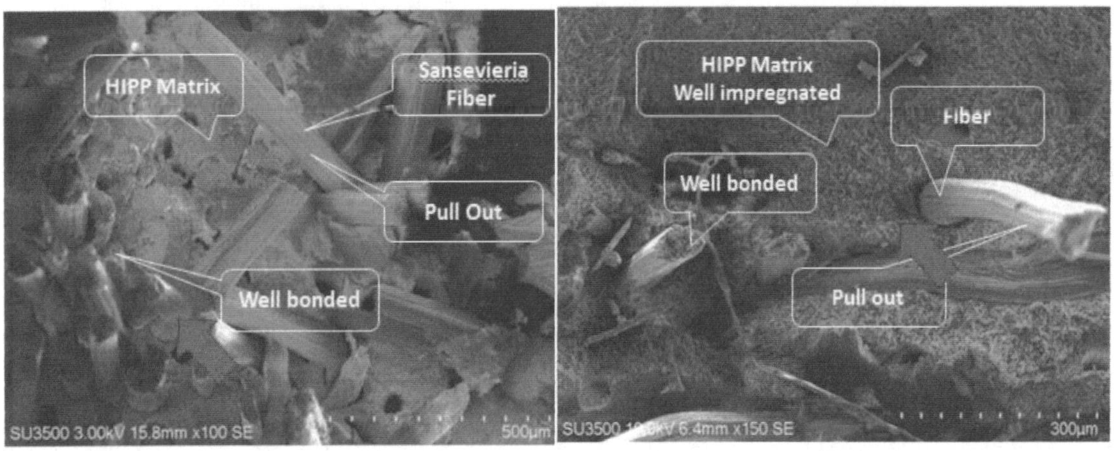

(a) (b)

Fig. 6 SEM images of fracture surface of Unidirectional STF/HIPP with 2% DBP

4. Conclusions

In this study, the effect of dibutyl phthalate (DBP) addition as plasticizer on the mechanical properties of *Sansevieria trifasciata*/high impact polypropylene is investigated. In this research, it is found that tensile strength, modulus of elasticity and elongations of biocomposite *Sansevieria trifasciata*/high impact polypropylene with plasticizer was higher than without added plasticizer. The fracture analysis was investigated by scanning electron microscope (SEM). The highest tensile strength modulus of elasticity and elongation of biocomposites were 65 MPa, 2.31 GPa and 0.42% respectively, obtained from biocomposites with volume fraction of unidirectional alkaline treated fiber 35% and added 2% dibutyl phthalate as plasticizer. Morphological feature based on scanning electron microscope (SEM) observation shows that by added plasticizer had better interface bond between matrix and fibers, so that better mechanical properties and processing can be obtained.

5. Acknowledgements

The Authors are thankful to PT. Chandra Asri Petrochemical for providing high impact polypropylene high impact during this research

6. References

1. A. Elmarakbi, Advanced Composite Material for Automotive Applications: Structural Integrity and Crashworthiness, Chichester: John Wiley & Sons Ltd., 2014.
2. A. K. Mohanty, M. Misra dan T. L. Drzal, Natural Fibers, Biopolyemers, and Biocomposites, New York: Taylor and Francis Group, LLC, 2005.
3. D. D. Stokke, Q. Wu dan G. Han, Introduction to Wood and Natural Fiber Composites, Chichester: John Wiley & Sons, Ltd., 2014.
4. Holbery, James. And Houston, Dan. 2016 Natural-fiber-reinforced Polymer Composite in Automotive Application.. JOM.
5. Xie et.al, 2014 "Effect of Bio-based plasticizer on Mechanical and Thermal Properties of PVC /Wood Flour Composite," Bioresources 9 (4), 7389-7402.
6. Mardiyati , Steven, Raden Reza Rizkiansyah, A.Senoaji , Dan R. Suratman, 2015 "Effects Of Alkali Treatment On The Mechanical And Thermal," *Aip Conference Proceding,* P. 1725.
7. A. D. Shieddiequeet. al. The Effect of Alkaline Treatment and Fiber Orientation on Impact Resistant of Bio-composites Sansevieriaatrifasciata Fiber/Polypropylene as Automotive Components Material". AIP Conference Proceeding, 1945, 0021141 (2018); https://doi.org/10.1063/1.5030263.
8. G. Wypych, 2012, Handbook of Platicizer, Chemtech Publishing, 2nd Edition, Canada.

Study to Enhance Tantalum and Niobium Contents in Bangka Tin Slag by NaOH and $HClO_4$ Leaching

Shiva F Vincia[1], Sulaksana Permana[1], Ahmad Maksum[1,2], Johny W Soedarsono[1], Kurnia S Widana[3], Kurnia Trinopiawan[3], and Badrul Munir [1,*]

[1]Center of Mineral Processing and Corrosion Research, Department of Metallurgy and Materials Engineering, Universitas Indonesia, 16424 Depok, Indonesia
[2]Mechanical Engineering Department, Politeknik Negeri Jakarta, Depok 16425, Indonesia
[3]Center for Nuclear Minerals Technology, National Nuclear Energy Agency Indonesia, Pasar Jum'at, Jakarta 12440, Indonesia

*E-mail: bmunir@ui.ac.id

Abstract. Indonesia is rich in tin ores. Slags from processing of tin ores naturally contain substantial amount of valuable metals, in this case tiobium and tantalum. This research explores how niobium and tantalum contents in local Bangka tin slag can be enhanced using NaOH and $HClO_4$ leaching reagents. First, Bangka tin slag was NaOH-leached and, then, the slag underwent $HClO_4$ leaching. Before undergoing any processes, the contents of niobium and tantalum in Bangka tin slag were 0.47% and 0.23%, respectively. After passing two leaching processes, niobium and tantalum contents reached 1.28% and 0.79%, respectively.

Keywords : Leaching, Niobium, Tantalum, Bangka tin slag, NaOH, $HClO_4$

1. Introduction

Valuables metals such as iron, copper, lead, zinc, tantalum and niobium can be recovered from slags [1]. Information on tantalum supply deficits in 2015 and 2016 [2] challenges researchers to maximize the supply of tantalum from tin slag. Tin slags mainly spread across Nigeria, Australia, South Africa, Brazil, Zaire, Thailand, Malaysia and Indonesia [3-8]. Researchers found that acidic and alkaline leaching in tin slag increases tantalum and niobium contents [3,5,9-14]. HCl, HF, and NaOH leaching, and acid-basic as well as basic-acid leaching of tin slag could recover tantalum up to 60-93% and niobium up to 29-78% [4]. Tin slag with 2.7% $(Ta,Nb)_2O_7$ was dissolved in HF or HF plus H_2SO_4, and the processes effectively recovered both metals up to 85% [11]. HF & H_2SO_4 used to leach tantalum and niobium oxide in tin slag proved to succcessfully optimize the contents of Ta_2O_5 and Nb_2O_5 up to 10.4% and 10.6%, respectively [14].

The latest research successfully upgraded Nb_2O_5 and Ta_2O_5 contents up to 28.4% and 59.1% by HCl leaching followed by NaOH within a particle size of <150 mm [3]. Other investigations utilized 8% HF leaching and HCl leaching followed by NaOH to enrich tantalum and niobium contents up to 0.664% and 1.339%, respectively [8]. In addition, researchers could enhance 0.33% Ta_2O_5 and 0.64% Nb_2O_5 in Bangka tin slag through double leaching so that both metals contents reached 1.56% and 1.11%, respectively [15].

This study aims to enhance tantalum and niobium contents in Bangka Tin Slag by roasting, water-quenching, dewatering, sieving and leaching with NaOH followed by $HClO_4$.

2. Methodology

2.1. Materials and Apparatus
Unprocessed Bangka Island tin slag was used and abbreviated as BTS (Bangka Tin Slag) while the slag which undergone roasting 900^{0}C, water-quenching, dewatering, and sieving is abbreviated as BTS-RQDS. The leaching solutions contain 8M sodium hydroxide, 0.1M, 0.4M, and 0.8M perchloric acid (pa) respectively. The investigation equipment includes a ball mill (Toptek Topvert G1), a Barnstead Thermolyne Furnace, a sieve, a ceramic container, and a magnetic stirrer (Thermo Scientific CIMAREC). X-Ray Flourence (XRF) Bruker handheld analyzer was used for chemical composition characterization of filtrate, in addition to Atomic Absorption Spectroscopy (AAS) PerkinElmer Analyst 400 and an Inductively Coupled Plasma-Optical Emission Spectrometry (ICP-OES) Perkin-Elmer Optima 8300.

2.2. Research Procedure
After XRF characterization, BTS were roasted, water-quenched, dewatered, and +100, -100 + 150, -150 + 200, -200 + 250, and -250 mesh-sieved. Next, -200 + 250 mesh samples were leached with 8M NaOH for 2 hours at room temperature of 25 °C. The volume ratio for NaOH leaching solution and the BTS-RQDS was 1:10. Based on that measure, the researchers used 90 mL tin slag (176 g) and 900 mL of 8M NaOH solution. To separate residues from filtrates in the products of NaOH leaching, a filtering paper was used. The residues were washed with aquadest to completely remove NaOH solution and dried in an oven at 100°C for 2 hours.

The residues from 8M NaOH leaching was then leached by 0.1 M, 0.4 M, and 0.8 M $HClO_4$. The volume ratio between the $HClO_4$ leaching solution and the residues from 8M NaOH leaching was 1:10. Based on this measure, 30 mL (50 grams) residues and 300 ml $HClO_4$ solution were provided. The leaching process was carried out for 2 hours at room temperature of 25 °C. The research procedure is given in Fig.1.

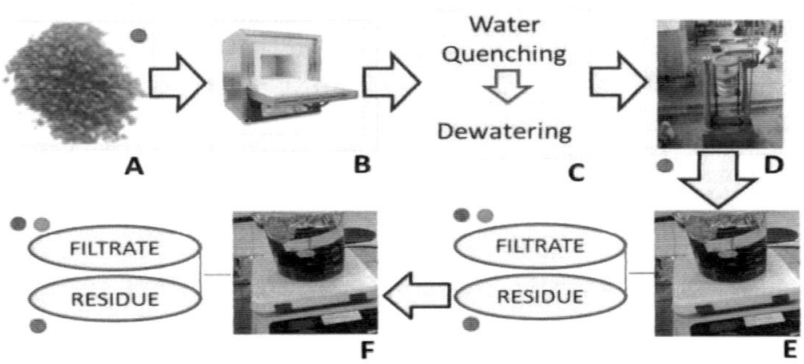

Fig. 1: Research Procedure. (**A**) Bangka Tin Slag. (**B**) Roasting in A Furnace. (**C**) Water-Quenching and Dewatering. (**D**) Sieving. (**E**) NaOH Leaching. (**F**) $HClO_4$ Leaching. (●) XRF Characterization. (●) AAS Characterization. (●) ICP-OES Characterization.

3. Results and Discussions
The stages of BTS processing include XRF characterization, roasting, water-quenching, dewatering, sieving, and NaOH leaching followed by $HClO_4$. The elements in BTS are grouped into 3. The first group contains two valuable oxides, tantalum and niobium. The second one is called the major other oxides (MOO) which are comprised of SiO_2, CaO, TiO_2,

Al_2O_3, Fe_2O_3, Sn, and Zr. Group three are elements and minor other oxides (EMOs) which encompass oxides and chemical elements not categorized into tantalum, niobium, and MOO.

3.1. XRF Characterization of BTS

The BTS characterization results are listed in Table 1, where Ta and Nb contents are 0.23% and 0.47%, respectively. Meanwhile, the total contents of MOO, including SiO_2, TiO_2, Fe_2O_3, CaO, Al_2O_3, Zr and Sn, are 16.15% and those of EMO amount to 83.15%.

Table 1. The results of XRF characterization of Bangka Tin Slag.

Element	Nb	Ta	SiO_2	TiO_2	Fe_2O_3	CaO	Al_2O_3	Zr	Sn	EMO
Content (%)	0.47	0.23	6.56	2.38	2.36	1.22	0.52	1.33	1.78	83.15

The contents of Ta and Nb, 0.23% and 0.47%, in BTS in this investigation are lower than those in the previous research, 0.33% Ta_2O_5 and 0.64% Nb_2O_5 [9,17].

3.2. Roasting, Water-Quenching, Dewatering, and Sieving of BTS

On Fig. 1, the largest tantalum and niobium contents are in the particle size of -100 + 150 mesh, the highest tin contents in that of -200 + 250 mesh, and the biggest SiO_2 contents in that of +100, -100 + 150, -200 +250, and -250 mesh. This implies that the optimal contents of certain oxides are present in certain grain sizes. The grain sizes with the highest tantalum and niobium contents would be the samples that were going to undergo further processes to increase the contents of both metals. On the other hand, previous researches show that +100 mesh BTS grains which have been roasted and water-quenched have the highest weight fraction [17] while in this exploration, the highest contents of oxides are in -100 + 150 mesh.

3.3. NaOH Leaching of BTS

The comparison of the contents of chemical composition between BTS and the residues from NaOH leaching shows a decrease in EMO contents, an increase in MOO contents, and a fall in tantalum contents.

Table 2. The results of XRF characterization of the residues from 8M NaOH leaching.

Element	Nb	Ta	SiO_2	TiO_2	Fe_2O_3	CaO	Al_2O_3	Zr	Sn	EMO
Content (%)	0.72	0.16	11.73	3.23	3.99	4.36	2.61	4.07	6.49	62.64

On Fig. 2, SiO_2 dominates almost all of the grain sizes, except -150 + 200 mesh. This phenomenon indicates that the dominance of SiO_2 dissolution has an effect on increasing tantalum and niobium oxide contents. Meanwhile, a previous research on the same topic also shows the dominance of SiO_2 in all grain sizes [8].

The reaction equations, from 1 to 8, below show the reactions of tantalum, niobium, and all MOO compounds. On the one hand, the reaction equations of (3), (4), (6), (7), and (8) that show the reactions of Si, Al, Ti, Fe and Zr present a small ΔG value. On the other hand, the reaction equation of (5) that shows the Ca reaction presents a positive ΔG value. The small ΔG value indicates that the smaller the ΔG is, the more spontaneous the reaction is while the positive ΔG value infers that the reaction is not working. However, Sn's reaction equation is not discussed because its reaction is not found in the database.

$$Nb_2O_5 + 2NaOH\ (\ell) = 2NaNbO_3 + H_2O \quad \Delta G_{25} = -30.400\ kcal \tag{1}$$

$$Ta_2O_5 + 2NaOH\ (\ell) = 2NaTaO_3 + H_2O \quad \Delta G_{25} = -45.235\ kcal \tag{2}$$
$$SiO_2 + 2NaOH\ (\ell) = Na_2SiO_3 + H_2O \quad \Delta G_{25} = -21.228\ kcal \tag{3}$$
$$Al_2O_5 + 2NaOH\ (\ell) = 2NaAlO_2 + H_2O \quad \Delta G_{25} = -8.165\ kcal \tag{4}$$
$$CaO + 2NaOH\ (\ell) = Ca(OH)_2 + Na_2O \quad \Delta G_{25} = 21.047\ kcal \tag{5}$$
$$6TiO_2 + 2NaOH\ (\ell) = Na_2Ti_6O_{13} + H_2O \quad \Delta G_{25} = -23.582\ kcal \tag{6}$$
$$Fe_2O_3 + 2NaOH\ (\ell) = NaFeO_2 + H_2O \quad \Delta G_{25} = -4.262\ kcal \tag{7}$$
$$ZrO_2 + NaOH\ (\ell) = Na_2ZrO_3 + H_2O \quad \Delta G_{25} = -5.954\ kcal \tag{8}$$

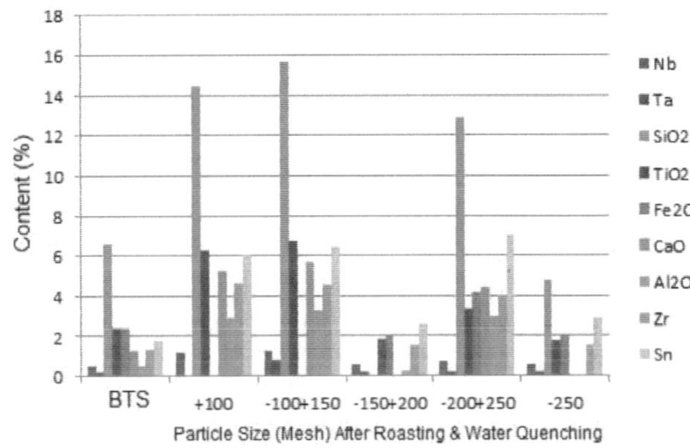

Fig. 2. The contents of oxides and other elements in BTS-RQDS according to the particle sizes.
The characterization results of filtrates from NaOH leaching are listed in Table 3. The tabel shows the dissolution of niobium in filtrates and the not-detected tantalum.

Table 3. The results of ICP-OES characterization of filtrates from 8M NaOH leaching.

Element	Nb	Ta	Ca	Fe
Content (ppm)	0.205	not detected	not detected	2

The Eh-pH diagram maps electrochemical stability and exposes the differences of the redox reaction to an element as a pH function, whereby the presence of metal ions at certain pH and potential conditions can be revealed [16].

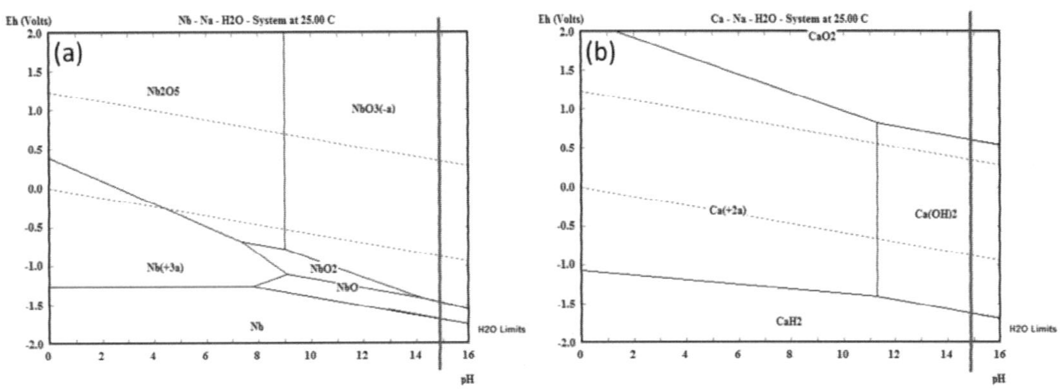

Fig. 3. Eh-pH diagrams of (a) niobium and (b) calcium.

NaOH solution is an alkaline solution, and its pH calculation is carried out by using the equation 9 below.

$$pOH = - \log [OH-] \text{ and } pH = 14 - pOH \quad (9)$$

Previous researches show that niobium, silica, calcium, and titanium dissolve through NaOH leaching under pH condition of 10-16 [16]. Not far from the finding of the previous investigation, the pH of 8M NaOH solution in this study is 14.903. Fig. 3 (a) shows the Pourbaix diagram of Nb-Na- H_2O where at pH 14.903, the niobium element transforms into NbO^{3-} ions.

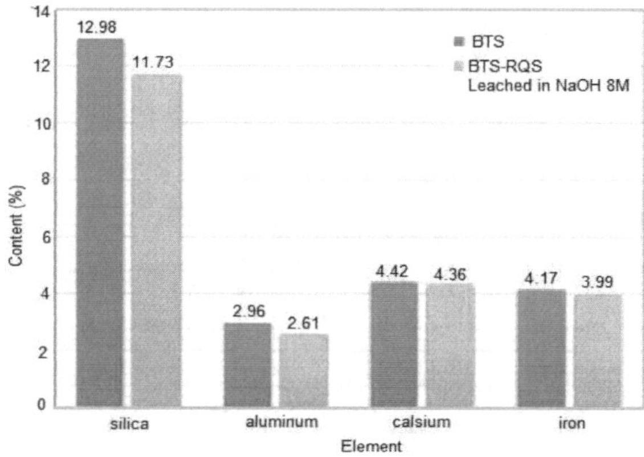

Fig. 4. The graph of silica, aluminum, calcium, and iron contents in BTS and BTS-RQS that was 8M NaOH-leached.

Fig. 4 shows the decrease in silica, aluminum, calcium, and iron contents in the residues from NaOH leaching. The reactions to NaOH and ΔG values of silica, aluminum, calcium, and iron are provided in (3), (4), (5), and (7). Based on the value of ΔG obtained, it is implied that the reaction runs spontaneously and, therefore, selective dissolution can be conducted in silica, aluminum, and iron.

The decrease in calcium contents is the lowest compared with the fall in the contents of aluminum, silica, and iron. This finding corresponds to that of C. Subramanian, et al., that it is not NaOH leaching that dissolves calcium, but acid leaching [14].

The positive ΔG value in the calcium indicates that the reactions do not work or are not spontaneous. On the other hand, pH 14.903 (Fig. 4 (b)) shows that the calcium in NaOH solution is stable in a $Ca(OH)_2$ compound. Table 3 presents the results of AAS characterization of filtrates from 8M NaOH leaching.

3.4. $HClO_4$ Leaching of BTS

The rise process in tantalum contents is from 0.70% at 0.1M $HClO_4$, 0.71% at 0.4M $HClO_4$, and 0.79% at 0.8M $HClO_4$. Meanwhile, niobium contents, 1.22%, at 0.1M $HClO_4$ rises to 1.27% at 0.4M $HClO_4$, and rises again to1.28% at 0.8M $HClO_4$. The results of $HClO_4$ leaching in all concentrations (Fig. 5(a)) show that the increase in niobium and tantalum contents closely relate to that in $HClO_4$ concentration. Fig. 5 (b) shows that the higher $HClO_4$ concentration is, the more the contents of SiO_2, Fe_2O_3, and CaO decrease. The declined contents of TiO_2, Zr, and Al_2O_3 are small while Sn contents are relatively stable. This indicates that as the acid solution concentration gets bigger, the dissolution of certain elements gets easier. The high concentration of acidic solution leads to the presence of H^+ which, then breaks the bonds between Ca, Al, Fe, and other oxides.

In the Pourbaix diagram, the different pH conditions, due to the differences of $HClO_4$ concentrations, show the difference in the form of elements, realized either in compound or ion. Therefore, this requires the calculation of $HClO_4$ acidity (pH), as shown in equation 9.

The results of filtrates, namely Nb, Ta, Ca, and Fe, characterization from $HClO_4$ leaching are provided in Table 4. In that table, calcium and iron contents become higher as $HClO_4$ concentration increases. This phenomenon indicates that the greater the concentration of $HClO_4$ is, the more Ca and Fe contents dissolve in the filtrates. This corresponds to the pH conditions of $HClO_4$ (0.096, 0.4, and 1), as shown on Fig. 6 (b) and (c) where calcium and iron contents are dominant when they transform into ions.

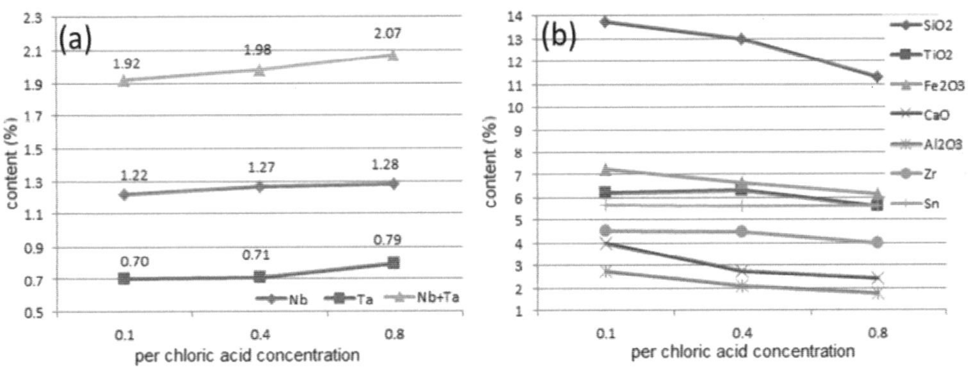

Fig. 5. The results of XRF characterization of BTS after being $HClO_4$-leached. (a) tantalum and niobium. (b) major other oxides (MOO).

Table 4. The results of ICP-OES characterization of filtrates from $HClO_4$ leaching.

Per Chloric Acid Concentration (M)	Element (Content in ppm)			
	Nb	Ta	Ca	Fe
0.1	0,208	not detected	3,466.25	82.33
0.4	0,207	not detected	3,874.75	626.9
0.8	0,206	not detected	4,672.75	1,281.6

Fig. 6 (b) shows calcium ions in $HClO_4$ at pH 1; 0.4 and 0.096, and the stability of Ca^{2+} and $CaCl^+$ ions. This indicates that at pH 1; 0.4 and 0.096, Ca is more soluble. An earlier investigation applying the leaching of acidic solutions also shows the dissolution of Ca and other elements such as Al and Fe, except Si [5]. In addition to that, a study by Pavel Raschman, et al. shows that, various concentrations of HCl can dissolves Ca and Fe as the concentration of the solution grows higher and the longer the dissolution process then greater the solubility of calcium and iron contents is [17].

IOP Conf. Series: Materials Science and Engineering **547** (2019) 012049 doi:10.1088/1757-899X/547/1/012049

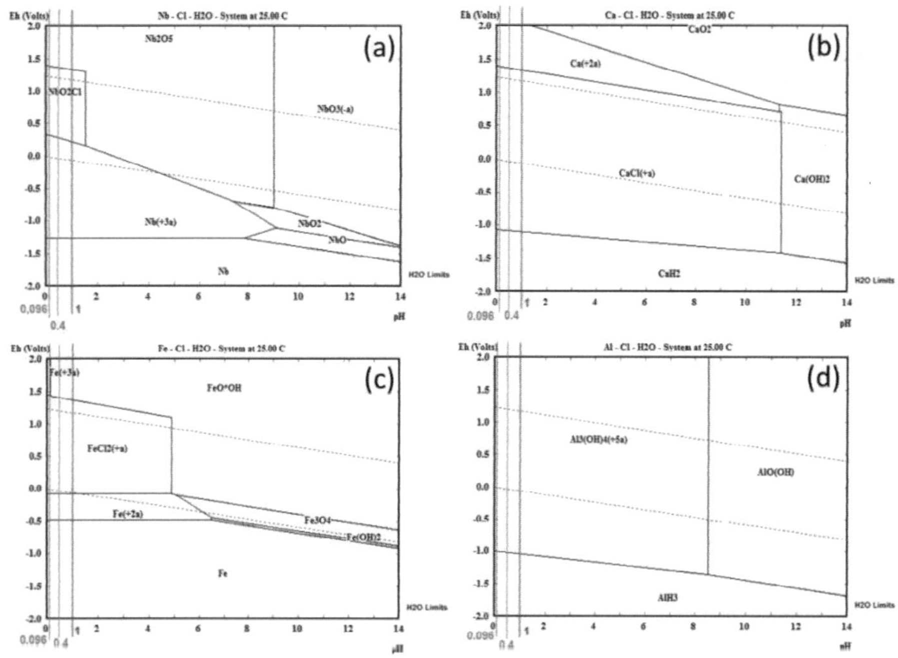

Fig. 6. Pourbaix diagrams of (a) niobium, (b) calcium, (c) iron and (d) alumunium in $HClO_4$ solution.

The important point of all stages of the processes above is the more the concentration of $HClO_4$ rises, the more the contents of Nb+Ta and EMO increase. Fig. 7 shows that the raised concentration of $HClO_4$ results in the increase in the contents of Nb+Ta and EMO.

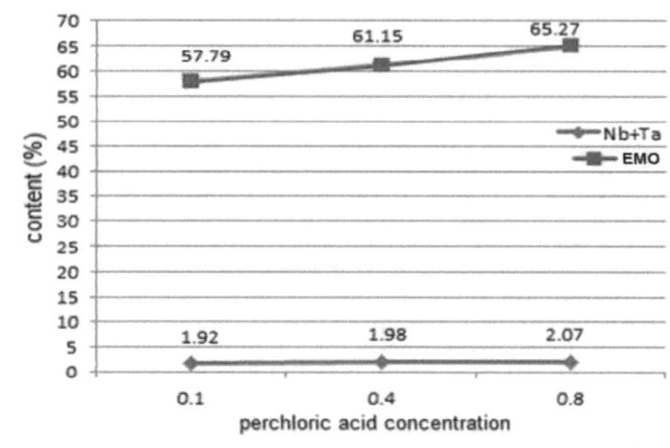

Fig.7: The contents of niobium+tantalum and EMO based on $HClO_4$ concentrations.

4. Conclusions

XRF characterization of pre-processing BTS shows that tantalum and niobium contents in BTS amount to 0.23% and 0.47%, respectively. After roasting, water-quenching, dewatering, and sieving of BTS, the highest contents of certain oxides are not present in all, but certain particle sizes. As the highest Ta and Nb contents are in -200 + 250 mesh, those BTS particles become the samples of the research to be further processed in order to enrich the contents of both metals.

Leaching with NaOH can dissolve of SiO_2 affects the upgrading of Ta and Nb contents and the filtrates from NaOH leaching indicate the solubility of niobium in them and on the contrary, tantalum is undetectable. After BTS is $HClO_4$-leached, the researchers found several results. First, the rise in the contents of Nb and Ta relates to the increase in $HClO_4$ concentration. Secondly, when the concentration of $HClO_4$ rises, the contents of Ca and Fe dissolved in the filtrates also increase. Third, pH 1; 0.4 and 0.096 causes calcium to dissolve easily. In summary, $HClO_4$ leaching effectively upgrades Ta, Nb, and element and minor other oxides (EMO) contents, but decreases the contents of major other oxides (MOO).

Acknowledgements
This work is financially supported by the Directorate of Research and Community Engagement, Universitas Indonesia under the 2018 PITTA program.

References :

[1] Piatak N M 2018 *Environmental Characteristics and Utilization Potential of Metallurgical Slag* (Elsevier B.V.)
[2] Stratton P 2013 Outlook for The Global Tantalum Market *2nd International Tin & Tantalum Seminar, New York, December 11th 2013* (New York: Roskill International Services) p 27
[3] Odo J U, Okafor W C, Ekpe S O and Nwogbu C C 2014 Extraction of Niobium from Tin Slag *Int. J. Sci. Res. Publ.* **4** 1–7
[4] Köck W and Paschen P 1989 Tantalum-processing, properties and applications *Jom* **41** 33–39
[5] Gaballah I, Allain E, Meyer-Joly M-C and Malau K 1992 A Possible Method for the Characterization of Amorphous Slags Recovery of Refractory Metal Oxides from Tin Slags *Metall. Trans.* **23** 249–259
[6] Bunnakkha C and Jarupisitthorn C 2012 Extraction of Tin from Hardhead by Oxidation and Fusion with Sodium Hydroxide *J. Met. Mater. Miner.* **22** 1–6
[7] Sulaiman M Y 1993 Simultaneous determination of thorium and uranium in tin slag *Sci. Total Environ.* **131** 187–195
[8] Permana S, Soedarsono J W, Rustandi A and et.al 2016 Other Oxides Pre-removed from Bangka Tin Slag to Produce a High Grade Tantalum and Niobium Oxides Concentrate *IOP Conference Series : Materials Science and Engineering* vol 12006 p 10
[9] Cenerazzo J A, Mosheim C E, Zionsville and Corrodo E. Marvisa 1971 US Patent 3.585.024-Up Grading The Tantalum and Columbium Contents of Tin Slag 1–4
[10] Gaballah I and Allain E 1994 Recycling of Strategic Metals from Industrial Slag by a Hydro-and Pyrometallurgical Process *Resour. Conserv. Recycl.* **10** 75–85
[11] Allain E, Djona M and Gaballah I 1997 Kinetics of Chlorination and Carbochlorination of Pure Tantalum and Niobium Pentoxides *Metall. Mater. Trans. B* **28** 223–33
[12] M.Charavorthy, S.Srikantth, E.A.Brocchi, And and P.K.Jena 1997 Kinetics of Chlorination of Tantalum Pentoxide with Carbon Tetrachlorida
[13] Gaballah I, Allain E and Djona M 1997 Extraction of Tantalum and Niobium from Tin Slags by Chlorination and Carbochlorination *Metall. Mater. Trans.* **28**
[14] Subramanian C and Suri A K 1998 Recovery of Niobium and Tantalum from Low Grade Tin Slag - A Hydrometallurgical Approach *Environ. Waste Manag.* 100–107
[15] Soedarsono J W, Permana S, Hutauruk J K, Adhyputra R, Rustandi A, Maksum A, Widana K S, Trinopiawan K and Anggraini M 2018 Upgrading tantalum and niobium oxides content in Bangka tin slag with double leaching *IOP Conf. Series : Materials*

IOP Conf. Series: Materials Science and Engineering **547** (2019) 012049 doi:10.1088/1757-899X/547/1/012049

Science and Engineering p 10
[16] Majid R A, Rustandi A and Permana S 2018 Simulation of Tantalum and Niobium Pentoxides Extraction from Bangka Tin Slag Waste *Adv. Sci. Lett.* **24** 767–772
[17] Raschman P, Špáková M and Fedoročková A 2010 Effect of hydrochloric acid concentration on the selectivity of leaching of high-calcium dead-burned magnesite *Acta Montan. Slovaca* **15** 232–237

Enrichment of Tantalum and Niobium Contents in Bangka Tin Slag by NaOH and H_3PO_4 Leaching

Hendri F Akli[1], Sulaksana Permana[1], Ahmad Maksum[1,2], Johny W Soedarsono[1], Kurnia S Widana[3], Mutia Anggraini[3], and Badrul Munir[1,*]

[1]Center of Mineral Processing and Corrosion Research, Department of Metallurgy and Materials Engineering, Universitas Indonesia, 16424 Depok, Jawa Barat, Indonesia
[2]Mechanical Engineering Department, Politeknik Negeri Jakarta, Depok 16425, Indonesia
[3]Center for Nuclear Minerals Technology, National Nuclear Energy Agency Indonesia, Pasar Jum'at, Jakarta 12440, Indonesia

*E-mail: bmunir@ui.ac.id

Abstract. This study explores how 0.23% tantalum and 0.47% niobium in tin slag from the Bangka-Belitung archipelago, Indonesia, are upgraded. To enrich the contents of both metals, this research involves 8M NaOH leaching followed by 0.5M, 1M, and 1.5M phosphoric acid. First, tin slag was roasted at 900^{0}C for 2 hours. Then, the sample was NaOH-leached while H_3PO_4 was exploited to leach its residues. Next, each of the residues from the leaching processes was characterized by XRF while the filtrates characterization used AAS and ICP-OES. On the one hand, NaOH leaching has proven to decrease the contents of iron, calcium, titanium, and aluminum. On the other hand, 0.5 M H_3PO_4 leaching optimizes tantalum contents up to 0.85% while 1.0 M H_3PO_4 leaching maximizes niobium contents up to 1.45%.

Keywords : Leaching, Niobium, Tantalum, Bangka tin slag, NaOH, phosphoric acid

1. Introduction

Investigations on environmental aspects and efforts to maximize the slag potential through metallurgical processes were carried out by previous researchers [1,2]. The results of tin slag characterization in other prior researches [3,4] show that tin slag contains uranium and thorium. Another study mentions that fluoride acid dissolution in tin slag from the Bangka-Belitung archipelago, Indonesia, causes tantalum and niobium contents to increase [5]. The exploration on tin slag from Indonesia also informs that Indonesian tin slag contains 2.7% $(Ta,Nb)_2O_5$ [6]. The above information indicates that, first, the adverse impact of slag processing on the environment becomes a serious problem and, therefore, the involvement of strong acids in slag processing should be minimized. Secondly, tin slag has the potential as a secondary source of tantalum and niobium oxide.

Tantalum and niobium are among 14 critical materials [7]. Having been calculated, the availability of tantalum only remains until 2020 and, more worryingly, the scarcity level of this element reaches mid term [8]. Supply and demand for these metals are unbalanced as the forecast of Ta_2O_5 in 2015 shows that the demand for this metal exceeds 79,000 lbs and in 2016, exceeds 212,000 lbs of total supply [9].

Several previous studies were conducted to upgrade tantalum and niobium contents in tin slag [6,10–13]. Meanwhile, US Patent 3,129,896 did the separation of heavy metal constituents of tin slag, including tantalum and niobium, through electrostatic separation and roasting at 600°F [10]. By collaborating on several techniques, previous researchers extracted tantalum and columbium (niobium) carbide from tin slag through HCl and HF leaching, mixing it with carbide, and calcining it at 800°C [11]. On the other hand, HCl, HF and NaOH leaching in tin slag and acid leaching followed by alkaline and vice-versa (alkaline-acid) were able to recover tantalum from 60 to 93% and niobium from 29 to 78% and eliminate SiO_2 from 40 to 78% [6]. Still involving HF, previous studies resulted in a recovery rate of 85% tin slag with $(Nb+Ta)_2O_5$ contents above 25% through the dissolution of HF or $HF+H_2SO_4$ [12]. In addition to these methods, chlorination was applied to upgrade tantalum and niobium contents in tin slag [14–16]. Other maneuvers such as caustic solution leaching, alkali pugging, alkaline fusion, and acid leaching have also proven effective to optimize the contents of Ta_2O_5 dan Nb_2O_5 to 10.4% and 10.6%, respectively [13]. Among prior researches, there is an investigation applying formation of tin slag into a metal matrix composite (MMC) to effectively increase tantalum and niobium carbide contents up to 12.49% and 31.81%, respectively [17].

Not only the above studies, but some of the following explorations also focus on enriching tantalum and niobium contents. Dissolving hydrochloric acid in Bangka tin slag to upgrade tantalum and niobium contents was able to maximize the two-metals ontents up to 0.664% and 1.339%, respectively [5]. Research on raising rare earth, including tantalum, contained in Bangka tin slag exploited thermodynamic software to predict chemical reactions, and Pourbaix diagrams [18].

This study aims at enriching the contents of both tantalum and niobium in Bangka tin slag through roasting at 900°C, water-quenching, and NaOH followed by H_3PO_4 leaching. These leachings will seperate valuable oxides from the non-valuable ones so that the extraction of tantalum and niobium can be maximized.

2. Methodology

2.1. Materials and Apparatus

The sample, tin slag, was taken from a tin smelter in the Bangka-Belitung archipelago, Indonesia. The leaching solutions consist of 8M sodium hydroxide (technical) and 0.5M, 1M, and 1.5M phosphoric acid (p.a). The research equipment includes a ball mill (Toptek Topvert G1), a sieve, a Barnstead Thermolyne Furnace, a ceramic container, a magnetic stirrer (Thermo Scientific CIMAREC), and characterization equipment consisting of an X-Ray Flourence (XRF), a Bruker handheld XRF analyzer, an Atomic Absorption Spectroscopy (AAS) PerkinElmer Analyst 400, and an Inductively Coupled Plasma-Optical Emission Spectrometry (ICP-OES) PerkinElmer Optima 8300.

2.2. Research Scheme

BTS was roasted at 900°C, water-quenched, dewatered, and sieved with grain sizes of +100, -100+ 150, -150+200, -200+250, and -250 mesh. -200+250 mesh samples were leached with 8M NaOH while the residues of NaOH leaching underwent 0.5M, 1M, and 1.5M H_3PO_4 leaching. The residues from all the leachings were water-quenched to remove chemical solutions and, subsequently, dewatered. The XRF characterization in this study was performed on the residues from both NaOH leaching and H_3PO_4 leaching. Meanwhile, the filtrates were characterized by AAS and ICP-OES.

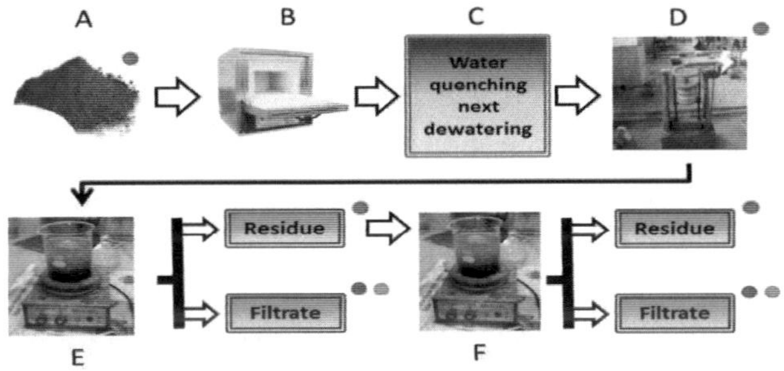

Fig.1: Research Scheme. (A) Bangka Tin Slag. (B) Furnace. (C) Water-Quenching and Dewatering Process. (D) Sieving Machine. (E) NaOH Leaching. (F) H_3PO_4 Leaching.
(●) XRF. (●) AAS. (●) ICP-OES.

3. Results and Discussions

The sample was first homogenized and reduced by grain size to make the particles smoother. Then, tin slag was characterized by X-Ray Fluorescence (XRF) to determine its chemical composition. Next, the sample was roasted, and water-quenched. Last, it was leached with NaOH followed by H_3PO_4.

High-grade oxides such as Sn, SiO_2, CaO, TiO_2, Al_2O_3, Fe_2O_3, and Zr are categorized into major other oxides (MOO) amounting to 16.15%. Elements, other than valuable oxides and MOO, with a total of 83.15% are classified into elements and minor other oxides (EMO).

3.1. XRF Characterization

As shown in Table 1, the tantalum and niobium contents obtained are 0.23% and 0.47%, respectively.

Table 1: XRF characterization results of Bangka tin slag.

Ta	Nb	Sn	SiO_2	CaO	TiO_2	Al_2O_3	Fe_2O_3	Zr	EMO
0.23	0.47	1.78	6.56	1.22	2.38	0.52	2.36	1.33	83.15

0.23% tantalum and 0.47% niobium in Bangka tin slag implies that Bangka tin slag can serve as a secondary source of tantalum and niobium oxides.

3.2. Roasting at 900°C, Water-Quenching, and Sieving

Roasting tin slag at 900°C used a furnace with 1 atm pressure for 2 hours. After being roasted, the sample was water-quenched and dewatered to remove water. Next, the sample was sieved with particle sizes of +100, -100+150, -150+200, -200+250, and -250 mesh. The products of roasting, (water-) quenching, dewatering, and sieving were termed BTS-RQDS.

After being roasted, water-quenched, and sieved, the highest percentage of each-element contents in tin slag is present at different grain sizes. As for tantalum, the highest grade, 0.29%, is in -100+150 mesh while the highest contents of niobium, 0.75%, is in -200+250 mesh. These results determines that the -100+150 mesh tantalum and -200+250 mesh niobium are the right-grain size sample for the enrichment of two metals contents.

Roasting is a process that allows the thermal decomposition reactions to the tin slag pseudo-structure. Reactions that occured in the thermal decomposition process, accompanied by thermodynamic analysis data show that all of Gibbs energy from the oxides roasting reactio n in tin slag is positive. This indicates that thermal decomposition in the roasting process at 900°C did not happen in both tantalum and niobium oxide or other oxides. However, metal oxide compounds having a pseudo-structure, after undergoing both roasting and water-quenching, caused micro cracks which resulted in stretching in tin slag particles as can be seen in Fig. 2.

Fig. 2: SEM microstructure of tin slag in 100x magnification (a) before roasting (b) after roasting at 900°C

The figure above shows that the roasted ting slag have the porosity structure that next could increased the width of wetting contact areas, thus increase the process of leaching. Roasting and water-quenching were also applied to research on saprolite ores where increased roasting temperatures lead to more cracks and porosity so that the grain size became smaller [19].

3.3. NaOH Leaching

In 8M NaOH leaching, the researchers used BTS-RQDS measuring -200+250 mesh where Ta+Nb contents were 0.92% and EMO contents 67.26%. This leaching was performed with a 1:10 solid-liquid ratio and, therefore, the researchers prepared 176 grams of tin slag equivalent to 90 mL, and 900 mL of 8M NaOH solution. The leaching process lasted 120 minutes at a room temperature. Table 2 shows the results of XRF characterization of the residues from 8M NaOH leaching. The table shows a decrease in Si, Ti, Al, Fe, Ta, Nb, and Zr, and an increase in Ca.

Table 2: XRF characterization results of 8M NaOH leaching residues.

Element	Content (%)	
	BTS-RQDS -200+250 mesh	After 8M NaOH Leaching
Ta	0.17	0.18
Nb	0.75	0.76
Si	12.89	12.79
Ca	4.41	4.42

IOP Conf. Series: Materials Science and Engineering **547** (2019) 012050 doi:10.1088/1757-899X/547/1/012050

Ti	3.32	3.23
Al	2.96	2.88
Fe	4.17	3.78
Zr	4.05	4.18

In accordance with the above table, after 8 M NaOH leaching, tantalum contents in the tin slag residues increase to 0.18% and so do niobium's to 0.76%. This suggests that NaOH leaching significantly dissolves other oxides and, therefore, increases the contents of tantalum and niobium. In line with these findings, previous study shows the enhanced contents of tantalum and niobium triggered by the low solubility level of both precious metals in alkaline solutions [13].

$$Ta_2O_5 + NaOH \rightarrow NaTaO_3 + H_2O \qquad -45.235 \text{ kcal} \qquad (1)$$

$$Nb_2O_5 + NaOH \rightarrow NaNbO_3 + H_2O \qquad -30.400 \text{ kcal} \qquad (2)$$

Equation above shows the negative value of Gibss reactions and free energy towards tantalum and niobium oxides, -45.235 kcal and -30.400 kcal. This implies that the reaction is spontaneous in triggering the formation of $NaTaO_3$ and $NaNbO_3$ compounds and, furthermore, contributes to raising tantalum and niobium contents. The formation of these two compounds were observed through tantalum and niobium Pourbaix diagrams on Fig. 2.

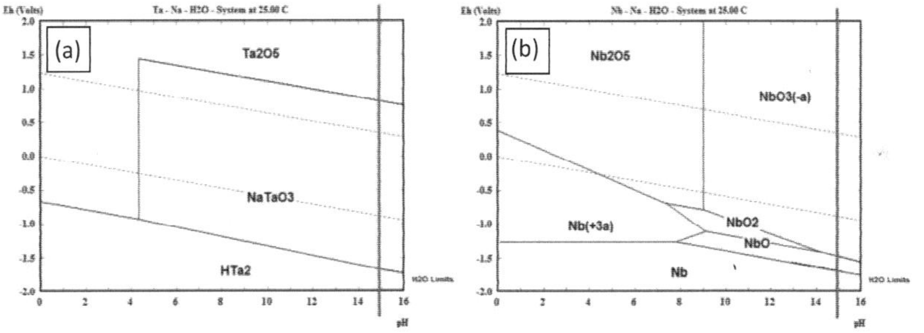

Fig.3: (a) Tantalum and (b) Niobium Pourbaix diagrams in 8M NaOH (pH=14.9).

Fig. 3 (a) shows that in 8 M NaOH (pH = 14.9) leaching at 25°C, a $NaTaO_3$ compound forms. Under the same condition, NbO^{3-} ion also forms, as shown on Fig. 3 (b). This phenomenon indicates the partial dissolution of niobium in 8M NaOH. Meanwhile, the results of ICP-OES analysis in Table 3 shows niobium contents, 0.206 ppm.

Table 3: ICP-OES characterization results of 8M NaOH leaching residues.

Element	Content	
	ppm	%
Tantalum	nd*	nd*
Niobium	0.206	0.00002

*not detected

Of the 0.206 ppm of niobium, the researchers concluded that the 8M NaOH leaching of Bangka tin slag could dissolve some niobium oxide contents.

3.4. H_3PO_4 Leaching

After being NaOH-leached, the residues were again leached with H_3PO_4. The residues from NaOH leaching were divided into three parts, each of which were 40 grams (equivalent to 25 mL). Then, to obtain a 1:10 solid-liquid ratio, the sample was dissolved with 250 mL H_3PO_4 with concentration variations of 0.5 M, 1.0 M, and 1.5 M for 2 hours. The XRF characterization results of H_3PO_4 leaching residues are provided in Table 4.

Table 4: XRF characterization results of H_3PO_4 leaching residues.

Element	Content (%)				
	BTS-RQDS	After Leaching			
	-200+250	NaOH 8M	H_3PO_4 0,5 M	H_3PO_4 1,0 M	H_3PO_4 1.5 M
Ta	0.17	0.18	0.68	0.79	0.85
Nb	0.75	0.76	1.26	1.45	1.41
Si	12.89	12.79	10.75	13.60	13.80
Ca	4.42	4.42	2.60	3.02	2.80
Ti	3.32	3.23	5.34	6.28	6.06
Al	2.96	2.88	1.74	2.54	2.30
Fe	4.17	3.78	5.42	6.02	5.57
Zr	4.05	4.18	4.03	4.83	4.62

The data in Table 4 implies (1) that the increased concentration of H_3PO_4 raises tantalum and niobium contents and, at the same time, (2) the optimal solubility of some other oxides (Fe, Ca, and Al).

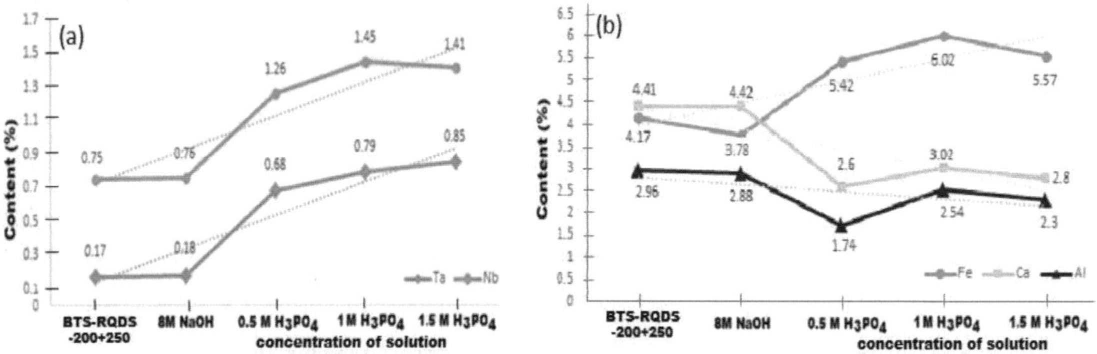

Fig. 4: The results of H_3PO_4 leaching with its concentration variations. (a) tantalum and niobium contents. (b) iron, calcium, and aluminium contents.

On the one hand, Fig. 4 (a) reveals a rise in tantalum and niobium contents in H_3PO_4 leaching residues. This phenomenon implies that tantalum contents rises as the concentration of H_3PO_4 is raised. 1.5 M H_3PO_4 leaching boosted tantalum contents four times compared to those undergoing 8M NaOH leaching.

On the other hand, Fig.4 (a) shows an increase in niobium contents in H_3PO_4 leaching residues. 1.0 M H_3PO_4 leaching optimizes niobium cointents up to 1.45%. This rise is due to the niobium structure that relates directly to that of tantalum. In conclusion, if there is an increase in the contents of tantalum, thereis a rise in those of niobium.

Furthermore, the researchers present the results of Gibss reactions and free energy in H_3PO_4 leaching in the equation below. Detailed description and discussion is provided in the following paragraph.

$$3NaTaO_3 + H_3PO_4 \rightarrow 1.5Ta_2O_5 + Na_3PO_4 + 1.5H_2O \quad 10.0598 \text{ kcal} \quad (3)$$

$$3NaNbO_3 + H_3PO_4 \rightarrow 1.5\, Nb_2O_5 + Na_3PO_4 + 1.5\, H_2O \quad -12.192 \text{ kcal} \quad (4)$$

In equation above, tantalum has a positive ΔG of 10.0598 kcal. This causes no spontaneous reaction during H_3PO_4 leaching even though the Pourbaix diagram (Fig. 5 (a)) shows the reformation of Ta_2O_5. The substance in this case is that the increase in tantalum contents occurs because the contents of the impurities are reduced. As a result, much tantalum and niobium can be extracted since no tantalum and niobium are still attached to the impurities at all. Meanwhile, Fig. 5 presents the results of Pourbaix diagram analysis of tantalum and niobium in H_3PO_4 solution.

In addition, compared to the percentage of raised tantalum contents, that of increased niobium contents is lower. This is due to the dissolution of some niobium in the filtrates. This phenomenon can be seen in Table 8 above where the thermodynamic reaction of niobium with H_3PO_4 produces Gibbs-negative energy, -12.192 kcal.

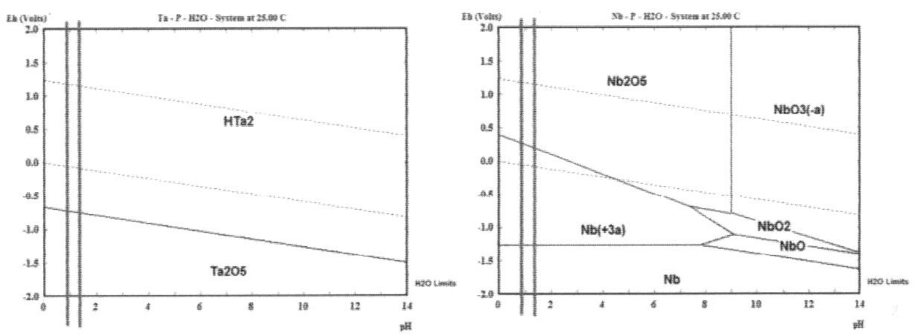

Fig. 5: (a) Tantalum and (b) niobium Pourbaix diagrams in H_3PO_4 solution.

Fig. 5 (b) shows an analysis using the thermodynamic principle of Pourbaix diagram system Nb-Na-O_s-P-H_2O at 25°C. There, precisely at concentrations of 0.5 to 1.5 M H_3PO_4 (pH 0.97 to 1.21), happens the dissociation of niobium which later becomes Nb^{3+} ions. This indicates that the dissociation of Nb to Nb^{+3} occurs at pH 0.97 to 1.21 and -1.25 V to 0.35 V.

Table 5 shows that niobium contents dissolved in H_3PO_4 reaches an optimal value of 0.0008% at 1.5 M. The increase in tantalum and niobium contents is also due to the impurities dissolution.

Table 5: ICP-OES characterization results of tantalum and niobium in H_3PO_4 leaching filtrates

Concentration of H_3PO_4	Tantalum Content		Niobium Content	
	Ppm	%	Ppm	%
0.5 M	nd*	nd*	0.247	0.00002
1.0 M	0.400	0.00004	2.568	0.0003
1.5 M	0.937	0.0001	8.413	0.0008

*not detected

Table above indicates the dissolved tantalum contents rise as the concentration of H_3PO_4 is raised. This is due to the rise in phosphate ions in the leaching solution. Increasing the concentration of the leaching solution also leads to more impurities dissolution. Because of the large amount of dissolved impurities, much tantalum in residues are successfully pulled or extracted from them. Through H_3PO_4 leaching, the optimal tantalum contents reach 0.85% at concentration 1.5 M. In addition, ICP-OES in Table 9 also indicates the dissolution of tantalum in H_3PO_4 solution.

The dissolution of H_3PO_4 in tin slag makes the contents of calcium and aluminum in the slag lessen compared to these two elements'contents in BTS-RQDS and BTS-RQDS that was NaOH-leached (see Fig. 4 (b)). The reduced contents of calcium and aluminum ease the tantalum and niobium extraction so that the two metals contents become higher than their contents to which impurities still attach. This means that calcium and aluminum dissolution raises tantalum and niobium contents.

In line with the upgrading of tantalum and niobium contents in the above paragraph, in mineral ores containing both valuable oxides and non-valuable oxides, the reduction of the latter through comminution processes, in this study 'leaching', will maximize the extraction of the former with the result that these valuable oxides can be wholly taken (nothing still attaches to the impurities) [20]. In other words, the precious oxides which cannot be seperated from the impurities will not be extracted, or thrown away along with the non-valuable oxides.

4. Consclusions

XRF Characterization of Bangka tin slag indicates that this slag can serve as a secondary source of tantalum and niobium oxide as this research reveals, that Bangka tin slag contains 0.23% tantalum and 0.47% niobium.

Through roasting at 900°C, water-quenching, and sieving, only tantalum measuring -100+150 mesh and niobium measuring -200+250 mesh are feasible for the enrichment sample of those two metals in Bangka tin slag and that in tantalum, niobium, and other oxides, no thermal decomposition in the roasting process with a temperature of 900°C happens.

After being NaOH-leached, other oxides contents are massively dissolved and, therefore, tantalum and niobium contents rise. In addition, the negative value of Gibss reactions and free energy towards tantalum and niobium oxides, -45.235 kcal and -30.400 kcal, are spontaneous in triggering the formation of $NaTaO_3$ and $NaNbO_3$ compounds which also play an active role in enriching tantalum and niobium contents.

In the process of H_3PO_4 leaching, the increasing concentration of H_3PO_4 leads to the rise in tantalum and niobium contents. In addition, H_3PO_4 leaching also makes optimum solubility in some other oxides (Fe, Ca, and Al). Another point from this leaching is the increase in tantalum contents due to the decrease in the impurities.

Acknowledgements

This work is financially supported by the Directorate of Research and Community Engagement, Universitas Indonesia under the 2018 PITTA program.

References :

[1] Piatak N M, Parsons M B and Seal R R 2015 *Characteristics and environmental aspects of slag: A review* vol 57(Elsevier Ltd)
[2] Piatak N M 2018 *Environmental Characteristics and Utilization Potential of Metallurgical Slag* (Elsevier B.V.)
[3] Meor Yusoff M.S. and Muslimin M 2007 Uranium and Thorium Analysis In Tin Slag by EDXRF *J. Nucl. Relat. Technol.* **4** 177–80
[4] M.S M Y and Muslimin M 2010 Quantitative Analysis of Uranium and Thorium in Local Zircon and Tin Slag by The EDXRF Technique *J. Nucl. Relat. Technol.* **7** 21–8
[5] Permana S, Soedarsono J W, Rustandi A and Maksum A 2016 Other Oxides Pre-removed from Bangka Tin Slag to Produce a High Grade Tantalum and Niobium Oxides Concentrate *IOP Conf. Ser. Mater. Sci. Eng.* **131**
[6] Gaballah I, Allain E, Meyer-Joly M-C and Malau K 1992 A Possible Method for the Characterization of Amorphous Slags Recovery of Refractory Metal Oxides from Tin Slags *Metall. Trans.* **23** 249–59
[7] Wellmer F and Dalheimer M 2012 The feedback control cycle as regulator of past and future mineral supply *Min. Depos.* **47** 713–29
[8] Laura Talens Peiro, Gara Villalba Mendez R U A 2011 *Rare and Critical Metals as By-products and The Implications for Future Supply*
[9] Stratton P 2013 Outlook for The Global Tantalum Market *2nd International Tin & Tantalum Seminar, New York, December 11th 2013* (New York: Roskill International Services) p 27
[10] Sidney M. Heins 1964 US Patent 3.129.896-Separation of Heavy Metal Contituents of Tin Slag 0–1
[11] Cenerazzo J A, Mosheim C E, Zionsville and Corrodo E. Marvisa 1971 US Patent 3.585.024-Up Grading The Tantalum and Columbium Contents of Tin Slag 1–4
[12] Gaballah I and Allain E 1994 Recycling of Strategic Metals from Industrial Slag by a Hydro-and Pyrometallurgical Process *Resour. Conserv. Recycl.* **10** 75–85
[13] Subramanian C and Suri A K 1998 Recovery of Niobium and Tantalum from Low Grade Tin Slag - A Hydrometallurgical Approach *Environ. Waste Manag.* 100–7
[14] Allain E, Djona M and Gaballah I 1997 Kinetics of Chlorination and Carbochlorination of Pure Tantalum and Niobium Pentoxides *Metall. Mater. Trans. B* **28** 223–33
[15] Gaballah I, Allain E and Djona M 1997 Extraction of Tantalum and Niobium from Tin Slags by Chlorination and Carbochlorination *Metall. Mater. Trans.* **28**
[16] Brocchi E A and Moura F J 2008 Chlorination Methods Applied to Recover Refractory Metals from Tin Slags *Miner. Eng.* **21** 150–6
[17] Mudzanapabwe N T, Chinyamakobvu O S and Simbi D J 2004 In Situ Carbothermic Reduction of a Ferro-Columbite Concentrate in The Recovery of Nb and Ta as Metal Matrix Composite from Tin Smelting Slag Waste Dump *Mater. Des.* **25** 297–302
[18] Permana S, Rustandi A and Majid R A 2017 Thermodynamic Analysis With Software : A Case Study of Ugrading Rare Earth Elements Content in Bangka Tin Slag *Far East J. Electron. Commun.* **17** 1211–20
[19] Chen J, Jak E and Hayes P . 2013 the Recovery of Ferro-Nickel From Saprolite Ores Using Reduction Roasting *Other Ferro Alloy. Fundam.* 713–22
[20] Wills B and Napier-munn 2006 *Mineral Processing Technology: An Introduction to the Practical Aspects of Ore Treatment and Mineral Recovey*

Effects of Austempering Process to Mechanical Properties of Thin Wall Ductile Iron Connecting Rod.

R.D. Sulamet-Ariobimo[1]* M. Fadlan[1], S. Kamili[1], T. Sukarnoto[1], Y. Mujalis[1] and Y. Oktaviano[1]
[1]Mechanical Engineering Department, Faculty of Industrial Technology, Universitas Trisakti, Jakarta, Indonesia

*Corresponding author (E-mail): riantiariobimo@trisakti.ac.id

Abstract. Austempering process was carried out on thin wall ductile iron (TWDI) connecting rod produced by the design built in previous research. Austempering process was applied to improve the mechanical properties of the component, so it will be equal to forged steel. The aim of this work is to study the effects of austempering holding time to microstructure and mechanical properties of TWDI connecting rod. The process was applied to all connecting rods resulting from the vertical casting. The austenitising temperature was 960^{0}C with 35 minutes holding time and the austempering temperature was 350^{0}C with varied austempering holding time, that were 15, 30, 45 and 60 minutes. The process was held in fluidized bed furnace. Metallographic examination, tensile and compression stress-strain test were conducted directly to the connecting rod. The results showed that austempering process improved the mechanical properties of TWDI connecting rod and the result of the compression stress-strain test showed that the austempered TWDI connecting rod met its requirements.

Keywords: TWDI, TWADI, connecting rod, vertical casting, holding time.

1. Introduction

The needs of lighter automotive components for fuel saving have enhanced the application of thin wall casting technology on ductile iron which is known as thin wall ductile iron (TWDI). Casting design to produce TWDI connecting rod was established [1,2] and the connecting rod was produced [3]. Preliminary examination on the connecting rod showed that the casting design was able to produce connecting rod without any major defects such as porosity or shrinkages [1] as presented in Figure 1. The connecting rod was built based on the casting design which was patented with patent number IDP000039503 and IDP000040306 [1-10]. It uses vertical casting system with counter gravity flow to adjust the filling speed. This casting design is different from the design made by Martinez et al [11].

Properties of produced connecting rod should be improved to fulfill the requirement. The improvement will be made through austempering process. Austempering process on ductile iron was able to improve its properties to be similar or even higher than forged steel [11,12] and known as Austempered Ductile Iron (ADI).

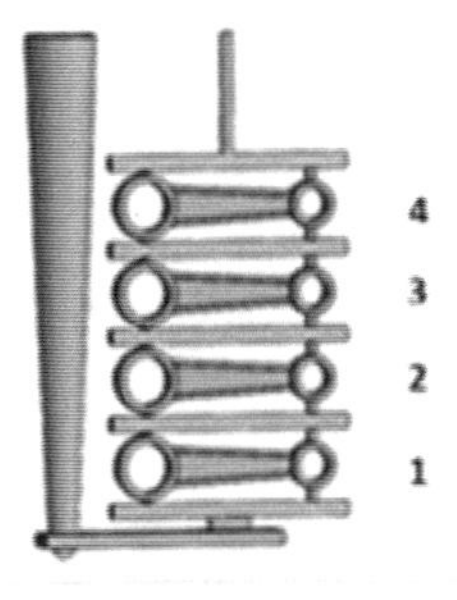

(a) Established design [1,2,3]

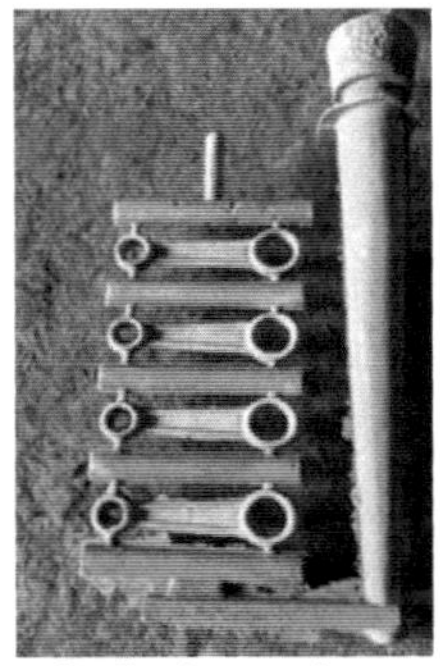

(b) Casting products.

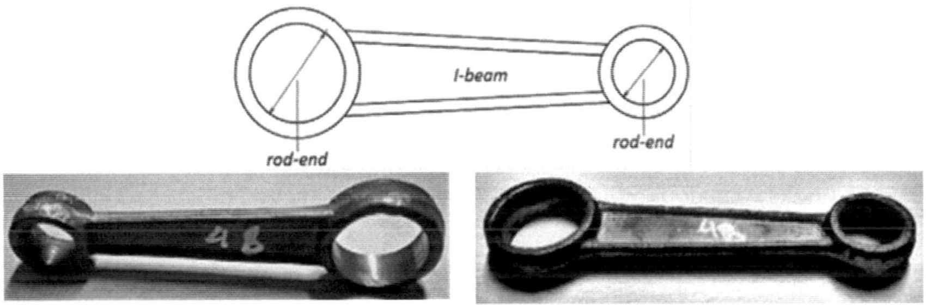

(c) TWDI connecting rod [3]

Figure 1. Casting design and products.

Austempering process will cover heating to austenitization temperature, holding in austenization temperature and following by austempering process which is isothermal quenching to a media with temperature of 250 to 400^{0}C [13] for several times before cooling in air to reach room temperature. Adjustments should be made if austempering process is applied to TWDI since the thickness of TWDI is only 5 mm or below. Soedarsono et al applied austenitization temperature of 960°C for 30 minutes and austempering temperature of 350^{0}C for 10 minutes in their research to make thin wall austempered ductile iron (TWADI) plate which increased the UTS between 44 to 309% and elongation between 50 to 100% [12]. The objective of this study is to analyze the effect of isothermal quenching duration of thin wall ductile iron connecting rod components.

2. Materials and Methods

TWDI Connecting rods were produced in industrial foundry using vertical casting design. The chemical composition was presented in Table 1. Each mold of the casting design produced 4 connecting rods and 4 molds were produced in one pouring. This study used 2 pouring. The first pouring was directly characterized and the second one was austempered.

Austenitization temperature for the austempering process was 960^{0}C with 35 minutes holding time. Then, the connecting rods were directly moved to fluidized bed furnace for isothermal cooling. Fluidized bed furnace which is more environmental friendly was used as isothermal media replacing salt-bath. The durations of the isothermal cooling were varied to 15, 30, 45 and 60 minutes. The durations were applied to every mold. The codes for all connecting rods were presented in Table 2.

IOP Conf. Series: Materials Science and Engineering **547** (2019) 012051 doi:10.1088/1757-899X/547/1/012051

Table 1. Chemical Composition of TWDI

Element		ASTM Standard (% weight)	Testing Result (% weight)	
			Pouring – 1	Pouring – 2
Carbon	(C)	3,60 – 3,80	3.67	3.60
Silicon	(Si)	1,80 – 2,80	2.78	2.80
Manganese	(Mn)	0,15 – 1,00	0.35	0.40
Magnesium	(Mg)	0,03 – 0,06	0.04	0.04
Phosphor	(P)	< 0,30	0.01	0.01
Sulfur	(S)	< 0,02	0.01	0.01
Copper	(Cu)	0,015 – 1,00	0.22	0.20
Chromium	(Cr)	0,03 – 0,07	0.06	0.06
Nickel	(Ni)	0,05 – 2,00	0.04	0.05
Molybdenum	(Mo)	0,01 – 0,10	0.01	0.01
CE – 1	%C + 0,31%Si		4,53	4,47
CE – 2	%C + 0,31%Si +0,55%P-0,027%Mn+0,4%S		4,53	4,47

Table 2. Samples Coding

	Pouring 1 – TWDI		Pouring 2 – TWADI	
	Code	Condition	Code	Condition
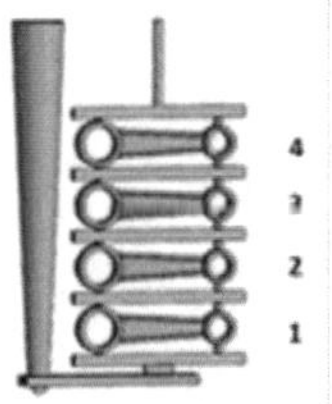	A	Extended Pouring Cup Normal Coating	A	Normal Pouring Condition Isothermal Quenching for 15 min.
	B	Normal Pouring Cup Normal Coating	B	Normal Pouring Condition Isothermal Quenching for 30 min.
	C	Normal Pouring Cup No Coating	C	Normal Pouring Condition Isothermal Quenching for 45 min.
	D	Extended Pouring Cup Special Coating	D	Normal Pouring Condition Isothermal Quenching for 60 min.

Figure 2. Special holder for tensile and compression test

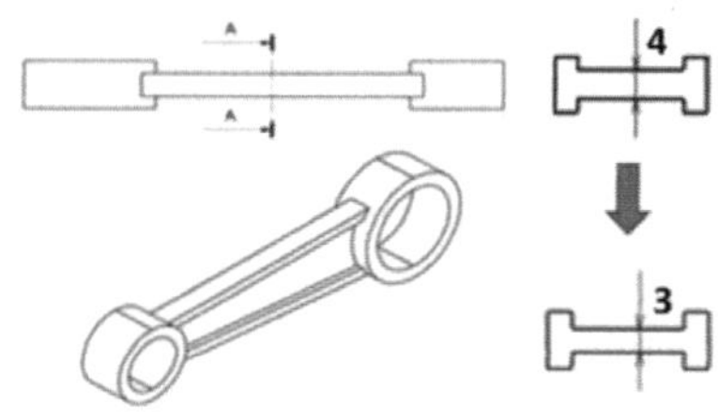

Figure 3. Thickness reduction in I-beam area [1-3].

After austempering process, all the connecting rods were subjected to mechanical testing which cover tensile and compression test. The test was run directly to the connecting rod. Special holder, as presented in Figure 2, was used during the test to hold the connecting rod. Universal Testing Machine (UTM) Shimadzu EHP-EB20186838 was used for tension testing. The maximum load was 20 ton and testing speed was 0.05 mm/mm/minute with level of confidence 95%. Gotech AI – 7000 LA10 Servo Control Computer System Universal Tensile Machine with the maximum capacity of 1000 kg was used for compression test. Calculation for compression strength was based on the area of the I-beam since modification was applied only to that area as shown by Figure 3. On the other hand, tensile strengths were not calculated because the break point happened in rod area as presented by Figure 4. The I-beam area calculation were presented in Table 3.

Samples for metallography examination were taken from the cross section of the connecting rods which have been tested in tensile and compression. The metallography samples were taken from I-beam and end rod area as presented in Figure 5.

Table 3. Area of I-beam

Rod Position	I-beam Area – in mm²							
	Pouring 1 – TWDI				Pouring 2 - TWADI			
	A	B	C	D	A	B	C	D
1	42.07	37.02	39.05	49.15	49.51	46.44	50.40	51.81
2	44.24	40.76	41.08	43.01	49.59	51.04	49.78	52.61
3	41.10	39.24	42.11	40.00	49.04	50.48	50.08	50.41
4	-	-	-	-	50.00	48.82	51.30	53.07

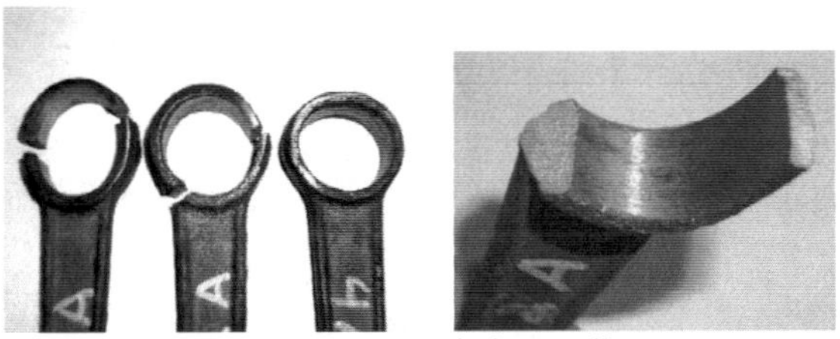
Figure 4. Break point during tensile test.

Figure 5. Metallography samples

3. Results and Discussion

Chemical composition test results showed that all elements are beyond the limit of ASTM standard for both pouring except for nickel in Pouring 1. Nickel content in the first pouring is 20% below the minimum level. Usually, nickel is added to ductile iron to improve its tensile strength and hardenability. Since connecting rods from Pouring 1 were not austempered then the lack of nickel will not give any significant effect.

Carbon equivalent (CE) calculation results presented by Table 1 show similar number for 2 equations in all pouring. There are 1.3% differences between CE values from Pouring 1 to Pouring 2 but this difference is not significant. CE calculation for both pouring are in the limit for TWDI [11,14]. Based on the CE values which are all below 4.67 the matrix of the microstructure will not be 100% ferrite [4,9].

Figure 6 shows the un-etched microstructure of TWDI and TWADI. The size of graphite formed in rod-end area of TWDI tends to be bigger compared to the one formed in I-beam [SENAMM]. Primary graphite is also formed more in rod-end than I-beam. Graphite distribution in rod-end is more even than in I-beam area. The un-etched microstructure of TWADI showed more even graphite distribution and graphite count in both I-beam and rod-end area than TWDI, but as well as TWDI, the size of graphite in I-beam is finer than rod-end. Primary graphite only formed in the end-rod although it is not as much as in TWDI. This condition will result in higher mechanical properties.

The etched microstructure of TWDI as presented in Figure 7 shows that the matrix is not 100% ferrite. This is supported by CE values which is below 4.67. It has the matrix of ferrite-perlite. Matrix formation in I-beam is not similar with rod-end. Austempering process tends to overcome this problem. As seen in Figure 7, after austempering process the matrix in I-beam is similar to rod-end.

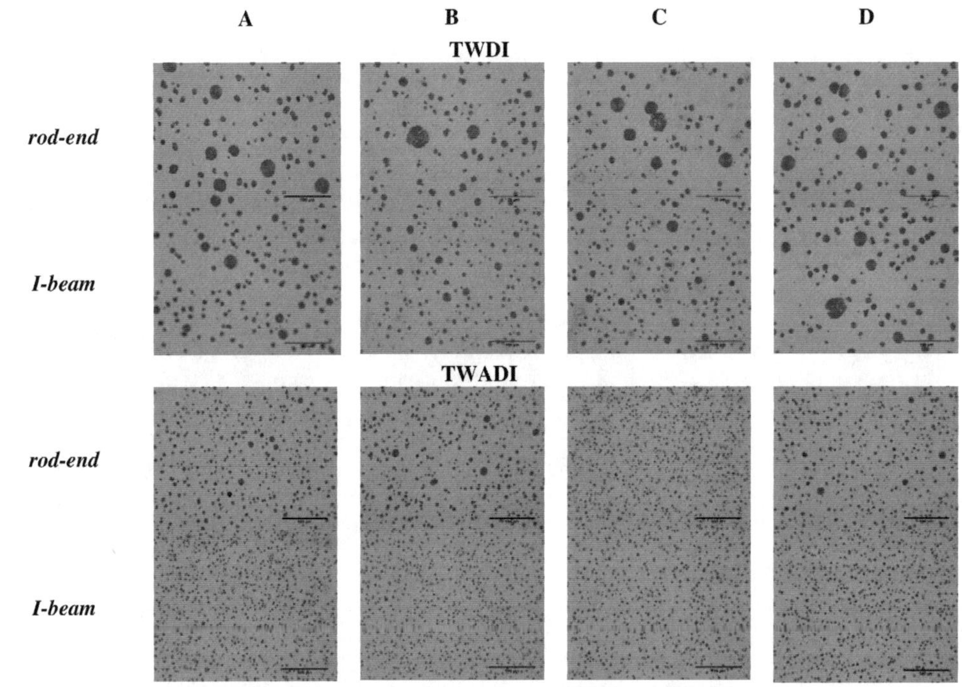

Figure 6. Un etched microstructures of TWDI and TWADI

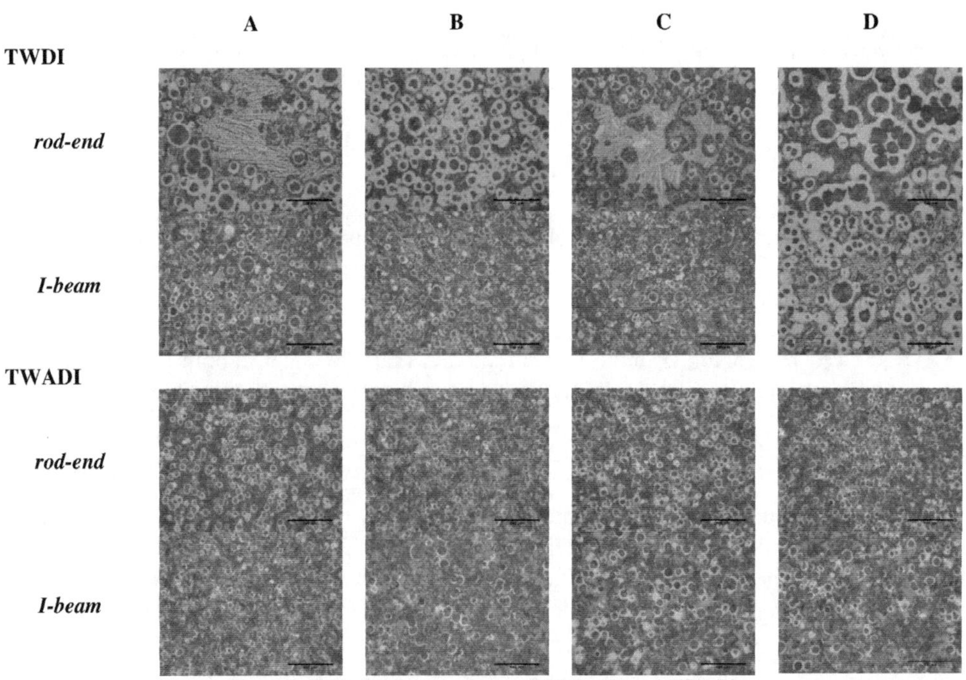

Figure 7. Etched microstructures of TWDI [3] and TWADI

As mention previously, the tension and compression testing were applied directly to the connecting rod. Tensile strength is not calculated since the broken point was not happening in the I-beam. But this condition shows that reducing the thickness of I-beam is not detrimental to the connecting-rod. Evaluation of tensile testing is based on the load used to break the connecting rod. Figure 8 shows that all variations made to the pouring process of TWDI do not give any significant improvement. Even the normal pouring condition gives the highest values for all casting position. Austempering process improves the tensile load as presented in Figure 9. The tensile load of TWADI increases between 115 to 147%. The highest tensile load is found in casting position 3 for all process conditions. As shown by TWDI, TWADI tensile load look similar for all casting position. There are no significant differences between the tensile load. Tensile load tends to decrease as duration of austempering process increases.

Compression load of TWDI, as shown in Figure 8, shows the same tendency as the tensile load, which shows no significant differences between each pouring condition. But not like the tensile load the highest compression load is found in the use of extended pouring cup together with use of special coating treatment. Compression load is 46% higher than tensile load. Compression strengths of TWDI also do not show any significant differences. TWADI compression load increase between 107 to 130% and compression strength increase between 63 to 93%. Compression load and strength tends to increase as the duration of austempering process increases. Closer examination on connecting rod 1 TWADI D reveals that lower compression load and strength shows in Figure 9 is caused by casting defect form in the connecting rod.

In contrast with tensile strength, compression strength should be calculated to determine the ability of TWADI connecting rod replacing the previous one. Compression strength of the previous connecting rod is 15 kg/mm2, while the compression strengths of TWADI connecting rod calculated based on I-beam area are between 67 to 79 kg/mm2. Based on the compression strength the TWADI connecting rod can be used to substitute the previous connecting rod.

4. Conclusions

The result of tensile and compression from TWDI show that variation applied in the pouring process do not give significant effect. The highest tensile load gain is 1280 kg and the lowest is 890 kg. While the highest compression load is 1980 kg and the lowest is 1500 kg. The highest compression strength is 44 kg/mm^2 and the lowest is 40 kg/mm^2.

Austempering TWDI will produce similarity of microstructure formation in I-beam and end-rod which will enhance the mechanical properties. In this study, austempering process improves the tensile load by 115 to 147%. The highest tensile load of TWADI is 2800 kg and the lowest is 2075 kg. While in compression test, austempering process improves the compression load by 107 to 130% and compression strength by 63 to 93%. The highest compression load is 4097 kg and the lowest is 3449 kg. The highest compression strength is 79 kg/mm^2 and the lowest is 67 kg/mm^2. Tensile load tends to decrease as the duration of austempering process increase. Compression load tends to increase as the duration of austempering process increase.

Based on the calculation of compression strength using I-beam area, the TWADI connecting rod can be used as substitute to the Vespa New PX 150.

5. Acknowledgements

The authors wish to show their gratitude to Faculty of Industrial Engineering Universitas Trisakti, PT. Geteka Founindo, PT. ASSAB Indonesia, Department of Metallurgy and Material Faculty of Engineering Universitas Indonesia and Dr. Endang Juwana for their support.

IOP Conf. Series: Materials Science and Engineering **547** (2019) 012051 doi:10.1088/1757-899X/547/1/012051

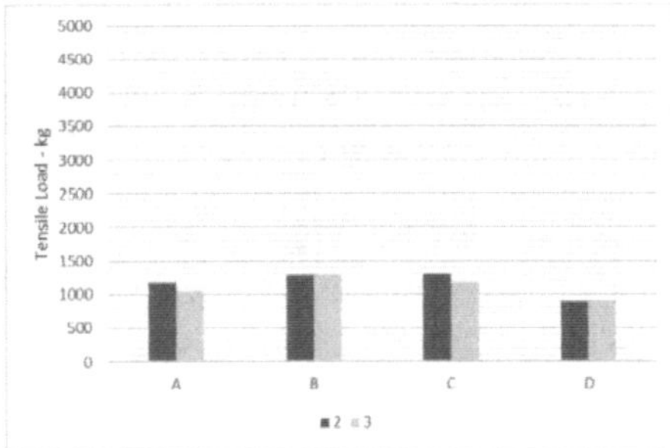

(a)

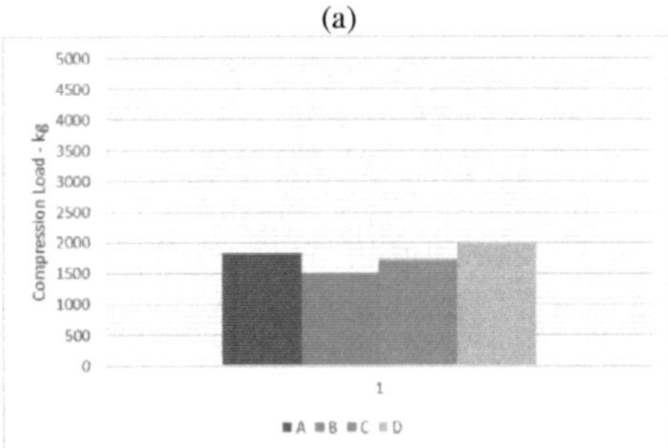

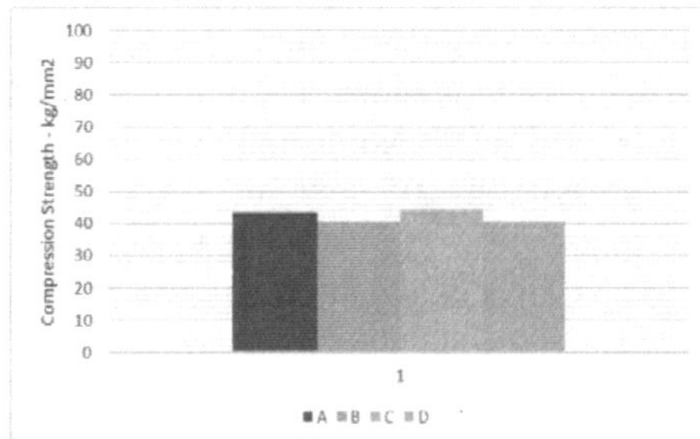

(b) (c)

Figure 8. Result of mechanical testing of TWDI

(a) Tensile Testing : Tensile Load in kg

(b) & (c) Compression Testing : (b) Compression Load in kg & (c) Compression Strength in kg/mm^2

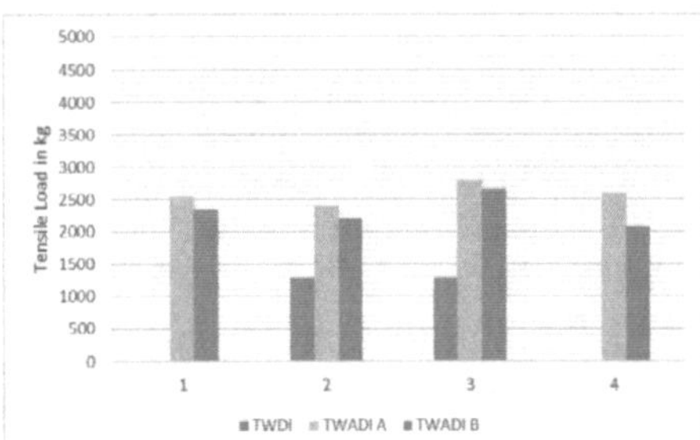

(a)

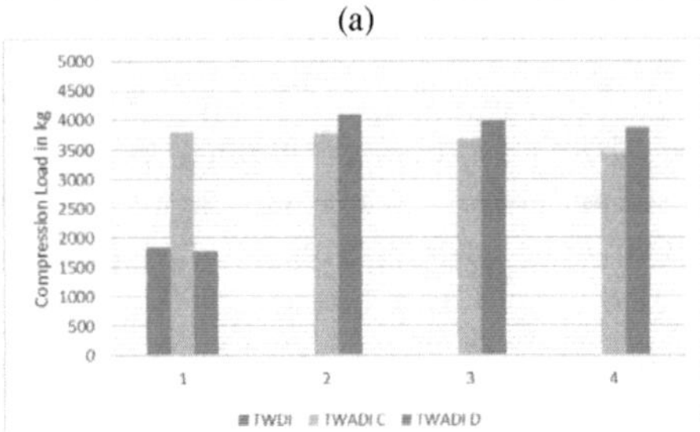

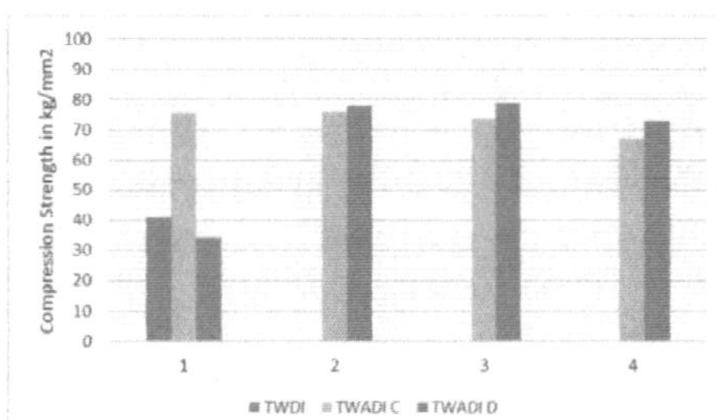

(b) (c)

Figure 9. Result of mechanical testing of TWDI & TWADI

(a) Tensile Testing : Tensile Load in kg

(b) & (c) Compression Testing : (b) Compression Load in kg & (c) Compression Strength in kg/mm^2

6. References

[1] Sulamet-Ariobimo RD, Yudha G, Sukarnoto T, Mujalis Y, Oktaviano Y. Thin wall austempered ductile iron connecting rod for lighter automotive component – Production of thin wall ductile iron connecting rod. JSAE 20179125/SAE 2017-32-0125. 2017.

[2] Sulamet-Ariobimo RD, Yudha G, Oktaviano Y, Mujalis Y, Sukarnoto T. *Penerapan teknologi pengecoran dinding tipis pada batang piston*. Proc. of SENAMM IX. 2016:

[3] Sulamet-Ariobimo RD, Fadlan M, Sukarnoto T, Oktaviano Y, Mujalis Y. *Efek modifikasi cetakan terhadap pembuatan batang piston dinding tipis*. Proc. of SENAMM X. 2017:667-75.

[4] Soedarsono JW, Suharsono B, Sulamet-Ariobimo RD. Effect of casting design to microstructure and mechanical properties of 3 mm TWDI plate. Adv Mat Res. 2012; 415-417:831-37.

[5] Soedarsono JW, Sulamet-Ariobimo RD. Effect of casting design to microstructure and mechanical properties of 1 mm TWDI plate. AMM. 2012; 110-116:3301-7.

[6] Suharno B, Soedarsono JW, Soemardi TP, Sulamet-Ariobimo RD. The effect of plates position in vertical casting producing thin wall ductile iron. Adv Mat Res. 2011; 277:66-75.

[7] Soedarsono JW, Suharno B, Sulamet-Ariobimo RD. Effect of casting design to microstructure and mechanical properties of 5 mm TWDI plate. AMM. 2012; 152-154:1607-11.

[8] Sulamet-Ariobimo RD, Soedarsono JW, Suharno B. Effect of casting design to microstructure and mechanical properties of 2 mm TWDI plate. Adv Mat Res. 2013; 652-654:2404-8.

[9] Sulamet-Ariobimo RD, Soedarsono JW, Suharno B. Effect of casting design to microstructure and mechanical properties of 4 mm TWDI plate. Adv Mat Res. 2013; 702:269 74.

[10] Sulamet-Ariobimo RD, Soedarsono JW, Suharno B. Cooling rate analysis of thin wall ductile iron using microstructure examination and simulation. AMM. 2015; 752-753:845-50.

[11] Martinez RA, Boeri RE, Sikora JA. Application of ADI in high strength thin wall automotive parts. Proc. of 2002 World Conference on ADI. 2002:143-9

[12] Soedarsono JW, Soemardi TP, Suharno B, Sulamet-Ariobimo RD, Damanik EZ, Haryono D. Effect of the austempering process on thin wall ductile iron. JMSE. 2011; 1(2);236-42

[13] Kaczorowski M, Krzyňska A. Mechanical properties and structure of austempered ductile iron – ADI. Arch Foundry Eng. 2007; 7(1):161-6.

[14] Stefanescu DM, Dix LP, Ruxanda RE, Corbitt-Coburn C, Piwonka TS. Production of thin wall low nodularity ductile iron through the delayed in mold process. 5AFS Trans. 2002; 110:1149-58.

Preparation Of Superhydrophobic Zinc Oxide Nanorods Coating On Stainless Steel Via Chemical Bath Deposition

Ahmad Nuruddin[1,3*], Brian Yuliarto[1], Santi Kurniasih[1], Henry Setiyanto[2], Aditianto Ramelan[3]
[1]Engineering Physics Research Group, Faculty of Industrial Technology, Institut Teknologi Bandung, Bandung 40132, Indonesia
[2]Analytical Chemistry Research Group, Faculty of Mathematics and Sciences, Institut Teknologi Bandung, Bandung 40132, Indonesia
[3] Materials Engineering Study Program, Faculty of Mechanical and Aerospace Engineering, Institut Teknologi Bandung, Jalan Ganesha 10 Bandung 40132, Indonesia

*E-mail: nuruddin@tf.itb.ac.id

Abstract. The stearic acid modified zinc oxide nanorod surface was coated with a stainless steel substrate to improve the contact angle and the sliding angle. Facilitated through ZnO seeding on the substrate, the ZnO nanorod was deposited via chemical bath deposition at different precursor concentrations. The surface hydrophobicity was modified by varying concentrations of stearic acid at and submmersion time. The contact angle was investigated in relation to the surface microstructure and modification. Results indicate that the density of ZnO nanorods on the substrate surface and stearic acid coverage are prerequisites for improving surface hydrophobicity. Superhydrophobic properties with a contact angle of 162° was obtained for ZnO deposited with 35 mM zinc nitrate and submerged in 8 mM stearic acid for 36 h, which allowed 5 μL water droplets to slide at 3.8°.

Keywords: stainless steel, chemical bath deposition, stearic acid, contact angle.

1. Introduction

Learning from natural properties of the surface of the lotus leaf, any material surface essentially can be modified into a superhydrophobic. Rigorous observations of physical properties of the lotus leaf surface show that there are two important aspects governing its hydrophobicity, which are (1) specific patterning of the surface microstructure and (2) low surface energy against droplets of water. The surface microstructure should provide air trapping under the water droplet and minimize contact area between droplets and surface, whilst the contact surface with droplet should have low energy [1]. By mimicking lotus leaf properties, numerous works have been conducted that modify various solid surfaces to obtain superhydrophobic behavior [2-5]. Material with rough surface structure is suitable for providing air trapping capability with a less contact area to water droplet. Covering the contact surface area with thin film of low surface energy will enhance water repelling.

Stainless steel (SS) has been widely utilized for construction related works and technological products due to its superior mechanical and physical properties [6]. Stainless steel is an iron based metal containing ~ 18 wt-% chromium and ~ 8 wt-% nickel, and sometimes mixed with small amounts of molybdenum. High chromium content and nickel makes the steel resistant to hazard and chemical environments, and molybdenum improves its resistance to corrosion. There has been a growing interest in utilizing stainless steel in relation with its surface corrosion resistance and hydrophobic property. However, the stainless steel surface is naturally hydrophilic, which required surface modification for particular utilization. Imitating the lotus leaf properties, the stainless steel surface must be microscopically patterned and covered by material with low surface energy.

Up to now, patterning the stainless steel surface was prepared either by engraving its surface or by depositing its micro-nano-texture thin film. Engraving the steel surface includes sand blasting [7], laser ablation [8,9], and wet etching using strong acid solutions ferric chloric acid [10], hydrosulfate acid [11], hydrochloric acid [12]. Depositing thin film with micro-nano-texture surface on steel substrate has been performed with electroless plating of nickel in a bath solution of nickel sulfate and hydrazine [13], electrodeposition of nickel in an aqueous mixed solution of nickel chloride, boric acid and lauryl sodium sulfate [14], spin coating of boehmite thin film from gel containing aluminum tri-sec-butoxide, isopropyl alcohol and ethyl acetoacetate [15], electrodeposition of silica in solution consists of potassium nitrate, tetraethyl orthosilicate, and dodecyltrimethoxysilane [16], electrodeposition of ZnO in mixed solution of zinc chloride and zinc nitrate [17], and spray coating of fluorinated polysiloxane/modified ZnO [18]. Recently, there have been attempts to prepare micro-nano-texture stainless steel surface by combining steel surface patterning and depositing thin film. Motlagh et al. fabricated superamphiphobic surface of stainless steel by surface sandblasted and thin film deposition of silane-encapsulated silica nanoparticles [19]. Meanwhile, Boinovich et al. developed icephobic stainless steel by subsequent process of chemical etch and coat with silica nanoparticles containing hydrophobic agent, leading to the formation of multistructure surface with capability of repelling water even at low temperature [20].

ZnO nanorod was an excellent thin film material with a surface structure that can easily be modified during or post deposition. This material can be deposited on various substrates with different methods [21]. In this study, ZnO nanorod was coated on ZnO seeded chemically etched stainless steel substrate by chemical bath deposition method. Etched steel surface guided to unaligned orientation of ZnO nanorods growth, which lead to a hierarchical surface roughness. Deposition precursor concentration was used to fine-tune the surface microstructure and coverage of ZnO nanorods on the substrate. The surface energy of ZnO nanorods was modified using stearic acid owing its inexpensive and wide technological applications.

2. Materials and Methods

2.1 Materials

Depositions of ZnO nanorods were performed on commercial stainless steel pipe with dimensions of 101.6 mm in diameter and 1.65 mm in thickness. The chemical composition of steel was examined by an Optical Emission Spectrometer (Thermo Scientific, ARL3460). The stainless steel contains wt-%: Fe 69.97, Cr 16.66, Ni 9.67, Mo 2.001, Mn 0.89, Si 0.37, Cu

0.229, V 0.097, W 0.034, P 0.033, Nb 0.016, C 0.014, S 0.007, Sn 0.007, Ti 0.001, and Al 0.001, which is a typical chemical composition of the 316 stainless steel [22]. Zinc acetate dehydrate, zinc nitrate tetrahydrate, diethanolamine (DEA), hexamethylenetetramine (HMT) were obtained from Sigma-Aldrich. Stearic acid (SA), acetone, ethanol and ethylene glycol were purchased from Merck. All chemical substances were used without further purification.

2.2 Preparation of ZnO seed and deposition of ZnO nanorods

The preparation of hydrophobic samples appears in Figure 1. The inner surface of stainless steel pipe was etched with strong acid solution forming a rough surface to facilitate heterogeneous ZnO growth. ZnO nanorods growth on stainless steel substrate was guided by ZnO seeds deposited on an etched stainless surface [23]. The Chemical Bath Deposition (CBD) method was used to deposit ZnO nanorods, and the surface of ZnO nanorods were modified by stearic acid.

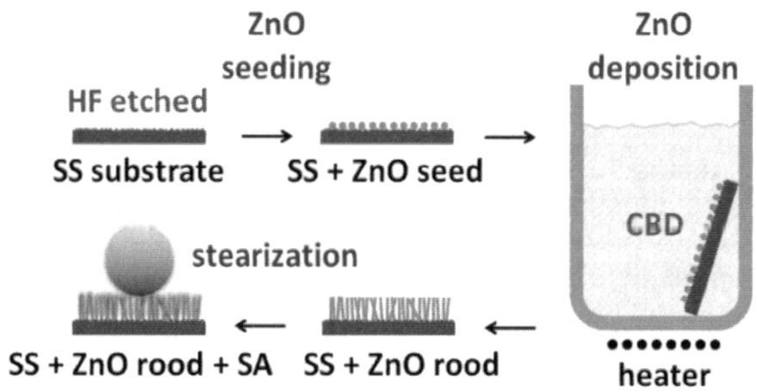

Fig. 1: Preparation of hydrophobic ZnO on stainless steel substrate

A stainless steel pipe was cut into pieces of 10 mm × 10 mm and used as a substrate. The substrate was cleaned under ultrasonic agitation with acetone, ethanol, and distilled water at room temperature for 15 min, and dried by an N_2 blow. The substrate was textured by etching in hydrofluoric acid (HF) 50% for 5 – 30 min at room temperature, and then immediately soaked in distilled water at room temperature for 15 min and dried naturally in air. Zinc oxide seed was deposited on etched stainless steel substrate by using dip coating method. Precursor ZnO sol was prepared from 0.2 M zinc acetate in ethylene glycol. Dropwise of 0.24 M diethanolamine was added to the sol under continuous stirring at 60 °C for 60 min. The sol was aged for 24 h prior the seed deposition. Dipping and withdrawal speed of the substrate was set at rates of 5 cm/min, followed by drying in oven at 135 °C for 30 min. This seeding process was repeated thrice, and then the sample was heat treated at 300 °C for 60 min to crystallize the seeds.

ZnO nanorods were deposited on ZnO seeded stainless steel substrate by using chemical bath deposition (CBD) method. The precursor solution containing hexamethylenetetramine was prepared for various concentration zinc nitrate (5–35 mM) in ethanol-distilled water with volume ratio HMT:zinc nitrate of 1:3. The ZnO seeded stainless steel substrate was immersed vertically in the precursor solution at 80 °C for 6 h. After immersion, the ZnO coated substrate was rinsed with ethanol and distilled water, dried naturally in air, and then annealed at 400 °C for 60 min. To produce hydrophobic surface the ZnO coated substrate was immersed in stearic acid at various molarities (5-35 mM) and time (6-36 h).

2.3 Characterizations

Crystalline structure and surface morphology of the ZnO deposit was observed by X-Ray diffractometer (Philips PW-1830) using Cu K_α radiation (λ = 1.5405 Å). The surface microstructure of the samples was obtained with a scanning electron microscope (JEOL-JSM-6510LV), provided with energy dispersive X-ray (EDX) spectroscopy (JSM-6510LA). Water droplets of 5 μL on samples were captured by an optical charge-couple device (CCD, CNB-GP300CGG1), and the contact angle was analyzed by using ImageJ plug-in DropSnake [24]. Meanwhile, the reported values of contact angle were the average values of measured contact angle at four different locations of the sample surface. Sliding angle was obtained by measuring the tilted angle of the sample holder against the bench surface when the water droplet started to move. The composition of stearic acid and stearic acid modified ZnO nanorod was measured by Fourier transform infrared spectrometer (Shimadzu Prestige 21). All measurements were conducted at room temperature.

3. Results and Discussion

3.1 Surface structure of the substrate and ZnO coating

Figure 2 shows the surface morphology and contact angle of stainless steel substrate before and after etching in an HF solution. The surface morphology of the barestainless steel substrate hassmooth texture in micron scales (Fig. 2a). EDX analysis showed that this bare substrate surface contained Fe, Cr and Ni elements with the atomic percentages of 69.9%, 16.6%, and 9.6%, respectively. This substrate surface has hydrophilic properties with a contact angle of 69.3°, which is typical of chemically untreated SS substrate [25]. Etching the substrate with hydrofluoric acid 50% for 5 min produced a porous surface with an irregular rough surface structure (Fig. 2b). The surface analysis of etched stainless steel indicated that the atomic percentages of Fe, Cr, and Ni on the surface decreased to 67.3%, 16.2%, and 9.6%, respectively, whereas O (oxygen) increased from 2.4% to 25.4%. The decrease of elemental metal and increase of oxygen were due to the formation of their oxides. Using XPS scanner, Lie et al. detected that the etched surface of the 304 stainless steel contains Fe_2O_3, FeF_3, Cr_2O_3, and CrF_2 on the surface of stainless steel substrate [26]. Naturaly, oxide surface has hydrophilic properties. However, rough surface oxide provided air trapping underneath water droplets and increased the water droplet contact angle. As the etching time increased, the surface became rougher and reduced the contact angle (Fig. 2c). The decrease in contact angle is attributed to the high surface roughness in which the surface is physically unable to trap the water droplet anymore.

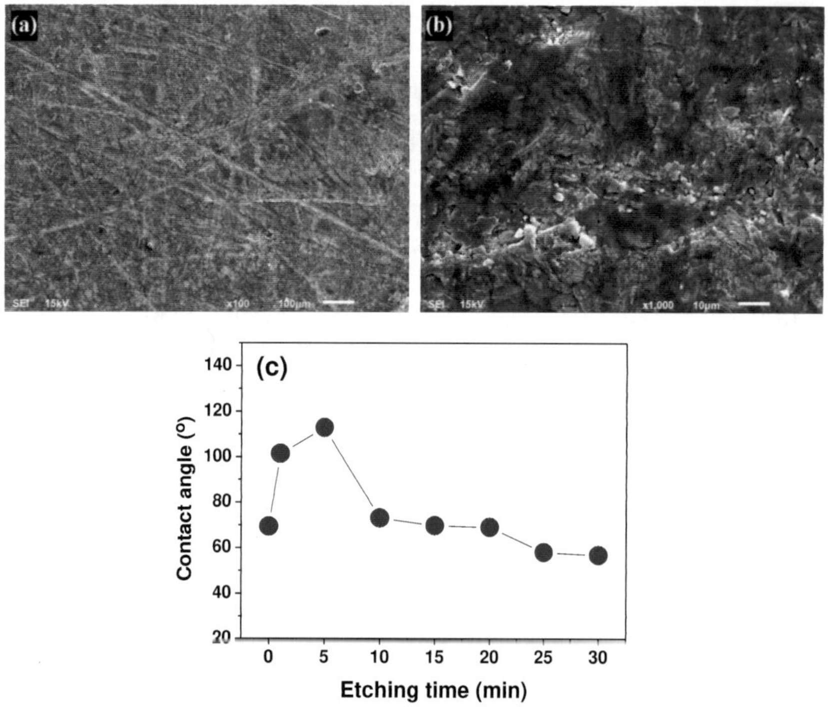

Fig. 2: SEM images of stainless steel substrates surface (a) before and (b) after after 5 min etching in 50% HF, and (c) contact angle of etched stainless steel substrates surface

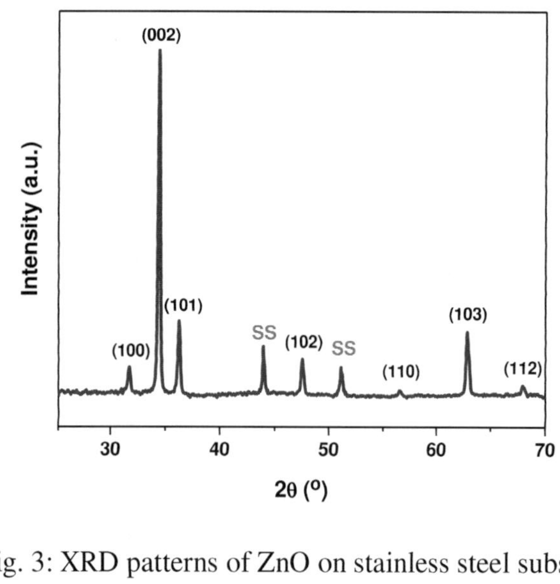

Fig. 3: XRD patterns of ZnO on stainless steel substrate

Figure 3 shows the typical X-ray diffraction (XRD) pattern of ZnO deposited on etched stainless steel substrates. XRD pattern of stainless steel substrate is included to index the diffraction peaks of deposited ZnO. It is clearly shown that diffraction patterns of the deposited ZnO on ZnO seeds etched stainless steel substrate attributed to ZnO polycrystalline, and corresponding a wurzite hexagonal ZnO (JCPDS No. 36-1451). No other peaks corresponding to impurity from zinc precursors was observed, confirming the pureness of the deposited ZnO. Evidently, the relative intensity ratio between (002) and (101) diffraction peak is quiet large indicating that the (002) peak is the preferential orientation in the c-axis alignment [27]. These results implied that ZnO seed on the substrate acted as nucleation center and provided preferred orientation for ZnO growth [28].

3.2 Surface morphology and wettability

Figure 4a-c shows top-view of SEM images of ZnO nanorods deposited on ZnO seeded stainless steel substrate for different ZnO precursor concentrations, under deposition temperature of 80°C for 6 h, immersed in 5 mM stearic acid for 6 h. It can be seen from Figure 4a that the ZnO nanorods grown with 5 mM precursor concentration have random direction staggered over the entire substrate surface. The implication is that the ZnO nanorods have a hexagonal shaped crystal structure with c-axis crystal orientation. ZnO nanorods at their final diameter and length were approximately 47-75 nm and 200-500 nm, respectively. However, as shown in Figure 4a, small portions of the substrate surface at different locations remained exposed. This exposure indicates that the deposition conditions with such low precursor concentrations at relatively low temperature (<90°C) was insufficient to provide the necessary amount of active precursors (Zn^{2+} ions) for growing the ZnO nanorods at all nucleation sites. On the other hand, for higher deposition temperature, i.e. 110°C or higher, Lee at al. succeeded in growing a relatively dense ZnO nanorods on ZnO seeded silicon substrate even when using very low (1 mM) precursor concentrations [29]. These results demonstrated that deposition temperature plays an important role in preparing active deposition precursors whilst the ZnO seed provides nucleation center for growing ZnO nanorods. When the ZnO precursor concentration was increased to 15 mM, denser ZnO nanorods were produced and fully covered the substrate surface in random directions. Increasing the concentration of ZnO precursor supplied larger amount of Zn^{2+} ions to produce denser ZnO nanorods. As seen in Figure 4b, the diameter and length of ZnO nanorods was about 72-98 nm and 500 nm, respectively. Further increasing the concentration of ZnO precursor to 35 mM resulted in much denser but irregular shapes and sizes of ZnO nanorods (Fig. 4c). Higher ZnO precursor concentration meant more Zn^{2+} available in the growth solution, and with availability of sufficient amount of HMTA to supply NH_3 and OH^- through hydrolysis the growth rate of ZnO nanorods increased significantly [30]. There is a competition growth on all available nucleation sites on the substrate surface, as consequence of surface roughness and surface heterogeneity which provides more defects, steps, kinks, and vacancies. In addition, visual comparison of surface roughness of ZnO nanorods portrayed in Figure 4 clearly shows that ZnO nanorod surface produced by higher concentration of ZnO precursor seems to have smaller and more even surface roughness compared to that produced by smaller ZnO concentration, which may provide better air entrapment underneath the water droplet. It is also important to note that all ZnO nanorods deposited on ZnO seeded stainless steel always produces vertically unaligned ZnO nanorods. The explanation of underlying unaligned ZnO nanorods growth can be attributed mainly due to vertically unaligned orientation of ZnO seed deposited on rough stainless steel surface [31]. The proposed model for deposition of ZnO on ZnO seeded stainless steel is presented in Figure 4d.

The effect of concentration of zinc nitrate precursors and molarity of stearic acid on water droplet contact angle was shown in Figure 5. As shown in Figure 5a, the contact angle increases almost linearly with the increase of the concentration of zinc nitrate. In correlation to deposited morphology shown in Figure 4, the more diluted zinc nitrate solution used in CBD, the denser ZnO nanorods was produced. Higher density of ZnO nanorods highlight the possibility that nanoscopic surface roughness not only improved the conformal coverage of stearic acid on nanorods, but also enhanced the surface sites for attaching stearic acid. Both effects simultaneously enhanced hydrophobicity of water droplet. Increasing the molarity of stearic acid from 5 mM to 8 mM improved the water droplet contact angle to no more than 10%, and only produced a maximum contact angle of 145°C. However, increasing the molarity of stearic acid by more than 8 mM decreased the water droplet contact angle, as shown in Figure 5b. This suggests that 8 mM stearic acid provides full coverage of ZnO surface without modifying the surface morphology. Therefore, it seems that increasing the contact angle with stearic acid concentration is due to improvements in the coverage of stearic acid on the ZnO surface. It is commonly known that ZnO deposited by wet chemical contains –OH attached on ZnO surface has hydrophilic properties. Stearic acid molecules on ZnO surface are readily attached to these active –OH through on ZnO surface captured by the carboxyl group of stearic acid leading to lower surface energy. Increasing the concentrations of stearic acid supplied excess of carboxyl group of stearic acid on ZnO surface, which modified both the surface free energy and surface roughness. Excess stearic acid on the ZnO surface reduces the surface roughness and may provide extra hydrophilic stearic acid head, which in turn reduces the water droplet contact angle [32].

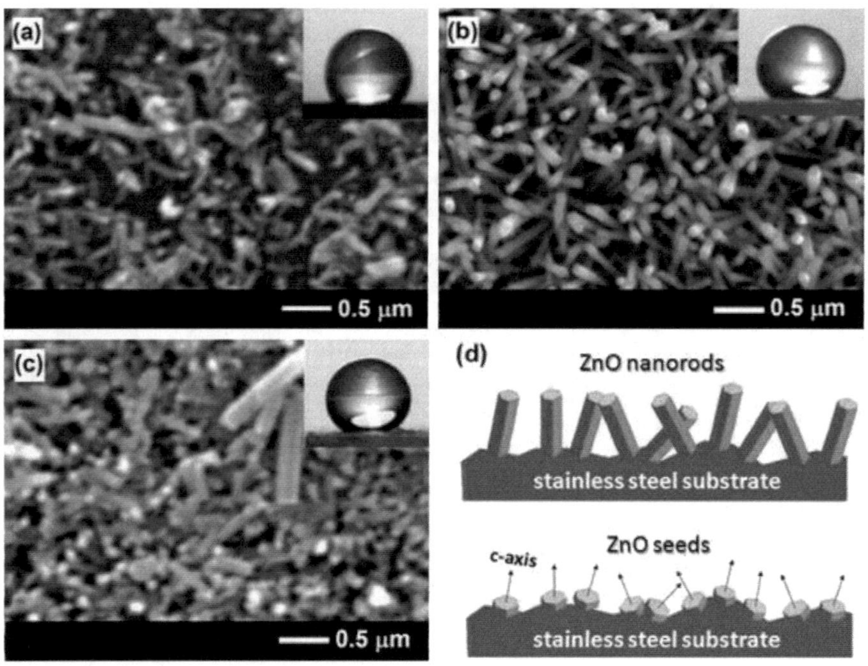

Fig. 4: SEM images of ZnO nanorods on ZnO seeded stainless steel deposited with concentration of zinc-nitrate of (a) 5 mM, (b) 15 mM, and (c) 35 mM, immersed in 5 mM stearic acid for 6 h, and (d) schematic representation of ZnO nanorods growth. Insets show the images of the water droplet on the ZnO/SA surface.

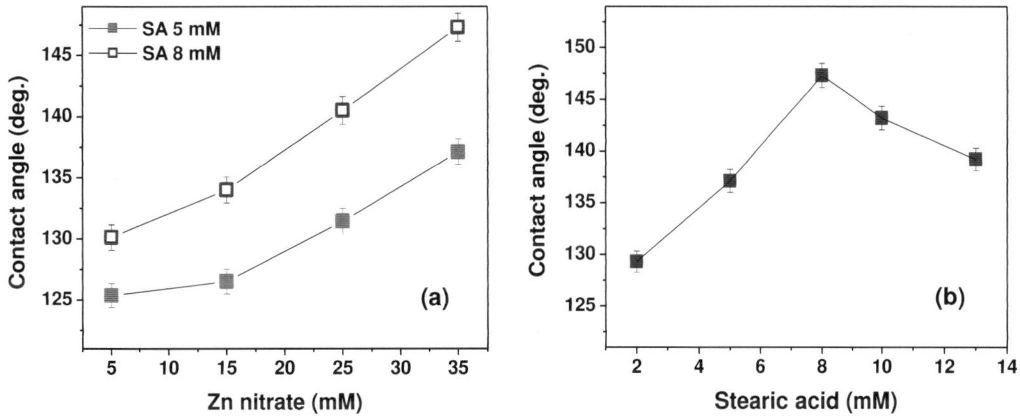

Fig. 5: Contact angle of water droplet on ZnO nanorods coated stainless steel fabricated with, (a) various concentrations of zinc nitrates, immersed in stearic acid for 6 h, (b). 35 mM zinc nitrates, immersed in different concentration of stearic acid for 6 h

The variation of immersion time in stearic acid (8 mM) for a sample deposited with 35 mM zinc nitrate affects the hydrophobicity, as shown in Figure 6 (including the corresponding images of water droplets on ZnO nanorods coated stainless steel). Figure 6a clearly shows that the contact angle increases with immersion time, with the maximum contact angle of 162° obtained at 36 h of immersion. It seems that the carboxyl group of stearic acid rearranges themselves to attach the active –OH on the ZnO nanorod surface, which occurs in a very slow process, and the immersion time of 36 h may not be the final stage in formation of a monolayer stearic acid on ZnO nanorod surface. A full coverage a monolayer stearic acid on ZnO nanorod surface is one requirement for producing maximum water repellent. Spalenka et al. required about 2 h immersion to cover a monolayer stearic acid molecule on a spin coating of ZnO/SiO_2. The use of tetrahydrofuran as a solvent, which has a smaller dissociation constant for carboxylic acid, slows down the acid part of stearic acid to directly attack the ZnO surface and allows stearic acid to react with the ZnO surface. On the other hand, using stearic acid in ethanol with the same concentration completely dissolves the ZnO thin film within 72 h [33]. In this study, ethanol was used as a solvent for preparing 10 mM stearic acid solution. Ethanol dissociated hydrogen ion in stearic acid five orders of magnitude higher than tetrahydrofuran, implying that etching ZnO has taken place during immersion the sample in stearic acid solution. During functionalization, the ZnO surface in stearic acid entered dynamic competition between etching ZnO surface by acid and depositing stearic acid on ZnO surface. The balance reactions may slow down the attachment process of stearic acid to the active hydroxyl on ZnO surface. Immersing the sample in stearic acid ethanol solution of 10 mM for 36 h may not be sufficient to fully cover ZnO surface with stearic acid. For comparison, to produce a SAM of full coverage adsorption of stearic acid on sapphire requires 165.5 h using 1.5 mM stearic acid n-hexadecane solution at room temperature [34].

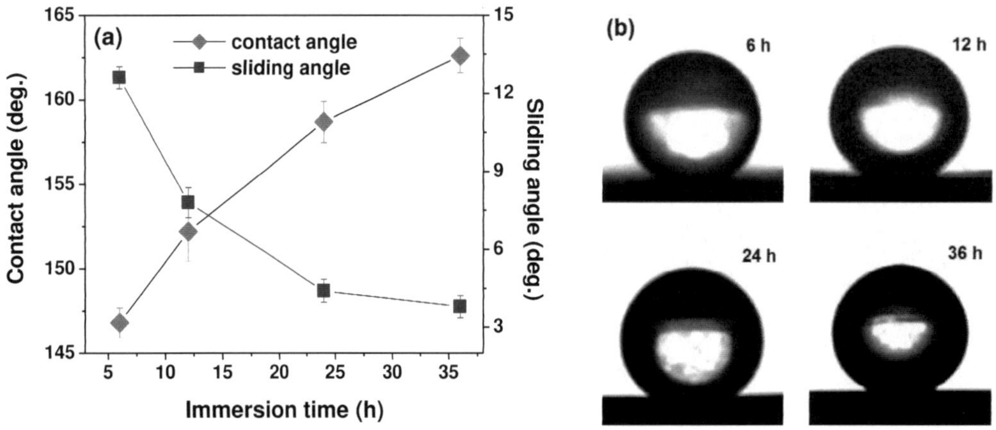

Fig. 6: Relation,(a) contact angle and sliding angle, and (b) corresponding images of water dropet on ZnO nanorodscoated stainless steel, synthesized with 35 mM zinc nitrates, modified in8 mM stearic acid for different immersion time

Thus, the movement of water droplet on ZnO nanorod surface was determined by the friction force of ZnO surface against water droplet movement, which largely depends on sliding angle and weight of water droplet. The sliding angle is affected by the mass of the water droplet and gravitational force. Gravitational force promotes the water droplet to move effortlessly on the ZnO nanorod surface. As the friction force of the ZnO nanorod surface becomes smaller, the water droplet easily moves. Guo et al. formulated the maximum friction force as f_{max} = mg sin θ, in which f_{max} is the maximum friction force, m is the mass of water droplet and g is the earth gravitational constant [35]. Using the aforementioned equation, the friction force of 5 μL water droplet on ZnO nanorod surface treated with 8 mM stearic acid at different immersion time can be calculated. The maximum friction force of ZnO nanorod surface immersed 6 h in stearic acid which allows the water droplet begin to move is calculated to be 10.66 μN, at sample tilting angle of 12.6°. The friction force value drops significantly to 3.24 μN when the ZnO nanorod surface is treated in stearic acid for 36 h, corresponding to the sliding angle of 3.8°.

3.3 Surface composition of ZnO nanorods treated by stearic acid

ZnO coated stainless steel surface covered with hydroxyl groups in water is a hydrophilic property. When the active –OH on ZnO surface is in contact with carboxyl group of stearic acid during the immersion of the stearic acid, heads of hydroxyl groups react with hydroxyl groups and release water, which modifies the ZnO surface into the hydrophobic state. To understand the surface modification of ZnO nanorods with stearic acid, Fourier transform infrared spectroscopy was carried out to obtain infrared transmittance spectra of pure stearic acid and stearic acid modified ZnO surface. Comparison of those spectra provides signature of bonding interaction between carboxylate head of stearic acid, i.e. -OH (C-O-H), C-O and C=O, with hydroxyl on the ZnO nanorod surface.

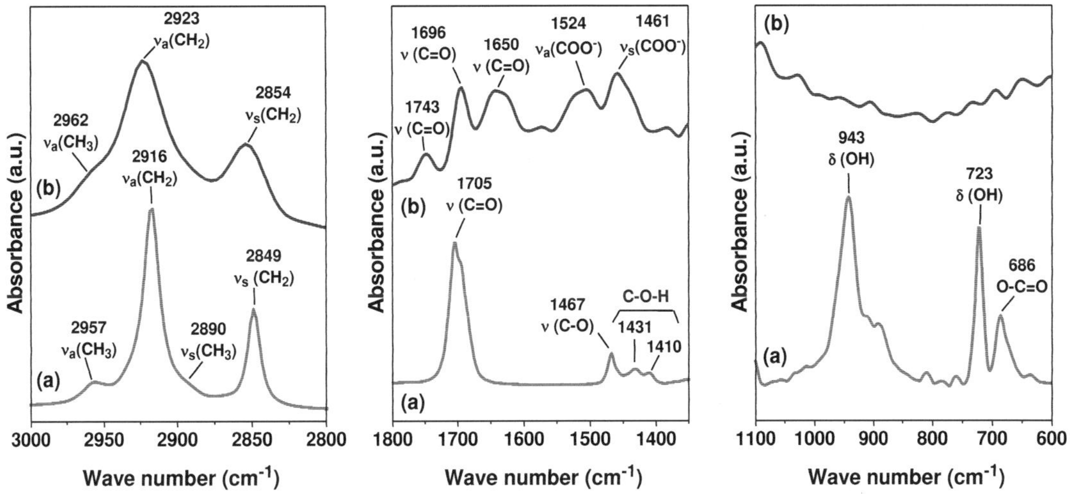

Fig. 7: Infrared absorbance of (a) pure stearic acid, and (b) stearic acid treated ZnO nanorods

Figure 7a shows FTIR spectrum of pure stearic acid prior absorbing onto ZnO surface, with featuring group of infrared absorption peaks at frequency region of the C-H stretching vibration (3000-2800 cm^{-1}), the C-O stretching vibration (1800-1350 cm^{-1}), and the O-H bending and swinging vibrations (1100-600 cm^{-1}) [36]. Two peaks at 2916 cm^{-1} and 2849 cm^{-1} corresponded to asymmetric and symmetric CH_2 vibrations, respectively, of long chain stearic acid. The peak at 2957 cm^{-1} and small shoulder peak at 2890 cm^{-1} are attributed to asymmetric and symmetric CH_3 vibrations, respectively [37]. The absorbance intensity of CH_2 is larger than that of CH_3 because of the long hydrocarbon chain of stearic acid. Two peaks at 1705 cm^{-1} and 1467 cm^{-1} indicate the vibration mode of C=O and C-O, respectively [38]. Peaks 1431 cm^{-1} and 1410 cm^{-1} are assigned to group vibration of C-O-H [38]. The vibration band between 1300 and 1400 cm^{-1} contains a wagging vibration of CH_2 group at 1350 cm^{-1} for double gauche defect [39], and peaks at 943 cm^{-1} and 723 cm^{-1} attributed to the out-of-plane bending and the in-plane swinging vibration of -OH, respectively [35]. Peaks at 1016 cm^{-1}, 810-762 cm^{-1} and 890 cm^{-1} represented CH_2 and CH_3 rocking modes, and peak at 686 cm^{-1} attributed to O-C=O in plane angle deformation [40].

After stearic acid modification of ZnO nanorod surface, the infrared spectra (Fig. 7b) depicts that the asymmetric and symmetric CH_2 and CH_3 vibrations are remain observable, but they shifts to the higher frequencies. This process suggests that the stearic chains are attached to the ZnO surface in random orientations on ZnO nanorod surface. The peak of symmetric CH_3 vibrations is hardly discernible. The peak intensities of the CH_2 vibrations is much larger compared to those of the CH_3 vibrations, implying that the long hydrocarbon chain of stearic acid is slightly tilts away from the ZnO surface normal [41,42]. The vibration peak of C=O remained present in the spectra but shifted and split into three peaks of C=O vibrations at 1743 cm^{-1}, 1696 cm^{-1} and 1650 cm^{-1}. The carboxyl group vibrations at 1743 cm^{-1} is configurative of a unidentate bonding of stearic acid to ZnO surface. The peaks at 1696 cm^{-1} and 1650 cm^{-1} are asymmetric C=O stretching with double bonds attached to the surface as a bidentate modes. The C-O stretching group vibration at 1467 cm^{-1}, 1431 cm^{-1} and 1410 cm^{-1} as well as the group vibration of –OH peaks at 943 cm^{-1} and 743 cm^{-1} disappear, suggesting that the reaction head of stearic acid with –OH on the ZnO surface has taken place. New peaks that appeared at 1524

cm^{-1} and 1461 cm^{-1} are considered stearate characteristic absorption with a symmetric (bidentate bridging) bonding mode of carboxylate [34]. The wagging vibration of CH_2 group remained present in the stearic acid treated ZnO infrared spectra that vibrated at a slightly higher frequency, meaning that the chemical state of CH_2 wagging group was not altered during the chemisorption process. The infrared vibration spectra between 1050 cm^{-1} and 600 cm^{-1} have no peak, indicating that the –OH bending and swinging vibrations of stearic acid has completely reacted with hidroxyls on ZnO nanorod surface.

4. Conclusions

ZnO nanorods coated in stainless steel were successfully prepared by chemical bath deposition of ZnO seeded 316 stainless steel substrate. ZnO seed facilitated the formation of nanorod structures, and variation of precursor concentrations provided means to control the surface coverage and thin film conformity. Higher concentration of zinc-nitrate precursor produced denser ZnO nanorods. Growth of denser ZnO nanorods on unaligned orientation of ZnO seed results in the desired surface roughness structure required for the formation of superhydrophobic surfaces. Tuning the concentration of stearic acid and immersion time for a sample plays an important role in creating the low surface energy of ZnO nanorods.

5. Acknowledgements

This research is supported by the Directorate General of Higher Education, the Ministry of Higher Education, Research and Technology, the Republic of Indonesia, through the Decentralization Research Program.

6. References

[1] Barthlott W, Neinhuis C 1997 *Planta* **202** 1
[2] Roach P, Shirtcliffe N J, Newton M I 2008 *Soft Matter.* **4** 224
[3] Yan Y Y, Gao N, Barthlott W 2011 *Adv. Colloid Interface Sci.* **169** 80
[4] Latthe S S, Gurav A B, Maruti C S, Vhatkar R S 2012 *J. Surf. Eng. Mater. Adv. Technol.* **2** 76
[5] Latthe S S, Terashima C, Nakata K, Fujishima A 2014 *Molecules* **4** 4256
[6] Cobb H M 2010 *The History of Stainless Steel* ASM International Materials Park
[7] Beckford S, Zou M 2011 *Thin Solid Films* **520** 1520
[8] Luo B H, Shum P W, Zhou Z F, Li K Y 2010 *Surf. Coat. Technol.* **204** 1180
[9] Rukosuyev M V, Lee J, Cho S J, Lim G, Jun M B G 2014 *Appl. Surf. Sci.***313** 411
[10] Liu Y, Bai Y, Jin J, Tian L, Han Z, Ren L 2015 *Appl. Surf. Sci.* **355** 1238
[11] Wang N, Xiong D, Li M, Deng Y, Shi Y, Wang K 2015 *Appl. Surf. Sci.* **355** 226
[12] Hai-feng Z, Xiao-wei L, Zhao min Dongbo W, Chang-zhi S 2013 *Key Eng.Mater.* **562-565** pp 33–38
[13] Chen L J, Chen M, Zhou H D, Chen J M 2008 *Appl. Surf. Sci.* **254** 5129
[14] Liang J, Li D, Wang D, Liu K, Chen L 2014 *Appl. Surf. Sci.* **293** 265
[15] Zhang X, Honkanen M, Järn M, Peltonen J, Pore V, Levänen E, Mäntylä T 2008 *Appl. Surf. Sci.* **254** 5129
[16] Wu L-K, Zhang X-F, Hu J-M 2014 *Corros. Sci.* **85** 482
[17] Bassard J D, Sarkar D K, Perron J, Audibert-Hayet A, Melot D 2015 *J. Colloid Interf. Sci.* **447** 240

[18] Qing Y, Yang C, Hu C, Zheng Y, Liu C 2015 *Appl. Surf. Sci.* **326** 48
[19] Motlagh N V, Birjandi F Ch. Sargolzaeia J, Shahtahmassebi N 2013 *Appl. Surf. Sci.* **283** 636
[20] Boinovich L B, Amelyanenko A M, Ivanov V K, Pashinin A S 2013 *ACS Appl. Mater. Inter.* **5** 2549
[21] Shi Y, Yang W, Feng X, Wang Y, Yue G 2015 *Mater. Lett.* **151** 24
[22] Hyde C J, Sun W, Lee S B 2010 *Int. J. Pres. Ves. Pip.* **87** 365
[23] Rodriguez J, Onna D, Sanchez L, Marchi M C, Candal R, Ponce S, Bilmes S A 2013 *Appl. Surf. Sci.* **279** 197
[24] Stalder A, Kulik G, Sage D, Barbieri L, Hoffmann P 2006 *Colloids Surf. A* **286** 92
[25] Kam D H, Bhattacharya S, Mazumder J 2012 *J. Micromech. Microeng.* **22** 1
[26] Li, L, Breedveld V, Hess D W 2012 *ACS Appl. Mater. Inter.* **4** 4549
[27] Nayeri F D, Narimani K, Kolahdouz M, Asl-Soleimani E, Salehi F 2014 *Thin Solid Films* **571** 154
[28] Badre C, Pauporte T, Turmine M, Lincot D 2007 *Nanotechnology* **18** 365
[29] Lee T-H, Ryu H-K, Lee W-J 2014 *J. Alloys Comp.* **597** 85
[30] Pomar F S, Martinez E, Malendrez M F, Tijerina E P 2011 *Nanoscale Res. Lett.* **6**: 524 1
[31] Cha S N, Song B G, Jang J E, Jung J E, Han I T, Ha J H, Hong JP, Kang DJ, Kim JM 2008 *Nanotechnology* **19** 235601
[32] Gurav A B, Latthe S S, Vhatkar R S, Lee J-G, Kim D-Y, Park J-J, Yoon S S 2014 *Ceram. Int.* **40** 7151
[33] Spalenka J W, Gopalan P, Katz H E, Evans PG 2013 *Appl. Phys. Lett.* **102** 041602
[34] Lim M S, Feng K, Chen X, Wu N, Raman A, Nightingale J, Gawalt E S, Korakakis D, Hornak L A, Timperman AT 2007 *Langmuir* **23** 2444
[35] Guo Z-G, Zhou F, Hao J-C, Liang Y-M, Liu W-M, Huck WTS 2006 *Appl. Phys. Lett.* **89** 081911
[36] Latibari S T, Mehrali Mo, Mehrali Me, Afifi A B M, Mahlia T M I, Akhiani A R, Metselaar H S C 2015 *Energy* **85** 635
[37] Ye S, Noda H, Nishida T, Morita S, Osawa M 2004 *Langmuir* **20** 357
[38] Gao J, Li Yp, Li Y, Liu H, Yang W 2012 *Cent. Eur. J. Chem.* **10** (6) 1766
[39] Lee S J, Kim K 1998 *Vibrational Spectroscopy* **18** 187
[40] Pudney P D A, Mutch KJ, Zhu S 2009 *Phys. Chem. Chem. Phys.* **11** 5010
[41] Arndt T, Bubeck C 1988 *Thin Solid Films* **159** 443
[42] Suetaka W 1995 *Surface infrared and Raman Spectroscopy, Methods and Applications* (New York Plenum Press)

Advancement In Forward Osmosis (FO) Membrane For Concentration Of Liquid Foods

L.A. Handojo[1], K. Khoiruddin[1], A.K. Wardani[1], A.N. Hakim[1], I G. Wenten[1,2,*]

[1] Chemical Engineering Department, Institut Teknologi Bandung, Jl. Ganesha 10, Bandung, 40132, Indonesia
[2] Research Center for Nanosciences and Nanotechnology, Institut Teknologi Bandung, Jl. Ganesha 10, Bandung, 40132, Indonesia

*Corresponding author (E-mail): igw@che.itb.ac.id

Abstract. In food processing, concentration of liquid food is one of the important steps required for several purposes. Concentration of liquid food while preserving sensorial and nutritional components is quite challenging, especially for thermal-based concentrating processes. This is due to the significant loss of those components which are heat sensitive. Therefore, considerable efforts have been devoted to develop new concentrating processes which can solve this problem. Among the developed processes, forward osmosis (FO) has been considered as an interesting alternative since it can be operated at low operating pressure and temperature and obtain a concentrated solution with high solid contents. However, there are several challenges in FO operation e.g. fouling phenomena, concentration polarization, and reverse diffusion of solution from draw solution. To address these issues, several developments have been made to prepare membrane which has high hydrophilicity, low fouling tendency, reduced concentration polarization, and low solute diffusion. The desired membrane has been obtained, for example, by modifying selective and support layers of the membrane. This paper reviews advances in FO membrane, including membrane preparation and modification. Principle and important parameters of FO in concentrating liquid foods are overviewed. In addition, challenges and strategies in FO membrane preparation are discussed.

Keywords: food processing; liquid concentration; membrane; osmosis;

1. Introduction

Liquid food concentration is one of the important steps in food processing. The aim is to maintain product stability by reducing water content. The liquid food is generally concentrated by thermal-based processes. However, it is hard to keep the nutritional value of the food in such processes. Most nutritional components are heat sensitive thus exposing these components to elevated temperature results in degradation of bioactive compounds [1,2]. On the other hand, there is an increasing demand on healthy food or functional food due to an improved paradigm aiming for healthy lifestyle. This trend has attracted researcher from both academia and industries to find the alternative process for liquid food concentration which could preserve the nutritional value [3].

Forward osmosis (FO) is considered as an attractive process which can address the drawbacks of the conventional concentration process. As shown in Figure 1, there is a significant increase of publications related to liquid concentration with FO in the last decade. The increasing attention on FO studies was driven by the lower energy consumption and enhanced performance [4]. FO is a membrane-based process which utilizes osmotic pressure as the driving force and semi-permeable membrane as a selective barrier. Osmotic pressure originated from a different solute concentration drives the permeation of pure water from a low salinity solution (feed solution, FS) to high salinity solution (draw solution, DS). The osmotic pressure difference is created by employing a DS which has a higher solute concentration than the FS. FO offers several interesting features including a low operating pressure, ability to treat a high solid content feed, and preserve nutritional compounds due to mild operating condition [5–7]. FO exhibits several advantages compared to pressure-driven membrane processes (reverse osmosis, nanofiltration, etc.). The advantages of FO include no or low hydraulic pressure, high water recovery, lower fouling tendency, and higher energy efficiency [8]. Therefore, FO has gained increasing attention from researchers to explore those advantages indicated by the increasing reported studies in this field (Figure 1).

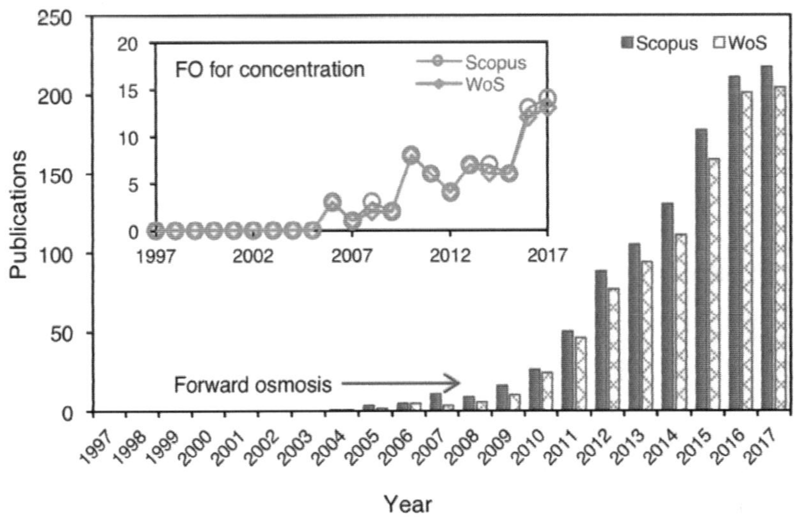

Figure 1. Number of publications related to forward osmosis indexed by Scopus and Web of Science, WoS (query: TITLE(terms); terms: forward osmosis, forward osmosis concentration)

Despite those advantages, FO still has some shortcomings that need to be solved. The first one is the internal concentration polarization (ICP) [9–13]. ICP reduces the osmotic pressure difference between the FS and DS which results in a decrease of water permeation rate or flux. ICP occurs in membrane support as the effect of unstirred boundary layer formed at the interface between the active and support layer. Reverse salt diffusion is another shortcoming of FO operation which also reduces the effectiveness of the process [14–16]. This phenomenon occurs due to imperfect membrane selectivity and the high salt concentration of DS. Reverse salt diffusion may reduce water flux due to DS dilution. Diffusion of salt from DS to FS can contaminate the product. Liquid foods contain various compounds that may have potentials to fouling occurrence. Fouling is the major drawback of membrane operation [17,18] which also occurs in FO operation [19,20]. Fouling reduces membrane productivity and increases membrane operating and maintenance costs. Those shortcomings are mainly related to the membrane characteristics. Therefore, there are considerable efforts to develop membranes which could reduce those drawbacks.

Liquid food concentration with FO has been reviewed in several papers [5–7]. Therefore, this paper will be a complement and an update of the previous published papers. Our main focus in this review is on some advancements in FO membrane and FO process in liquid food concentration including performances and challenges.

2. FO for Concentrating Liquid Foods

Table 1 summarizes several reported studies on liquid food concentration using FO process. DS is one of FO important parameters which determines the effectiveness of FO process. DS should have high osmotic pressure, non-toxic, and easy to regeneration [21]. Petrotos et al. [22] employed FO for concentrating tomato juice and investigated several parameters affecting the FO performance. The performance was analyzed by observing water flux. They compared six DSs, including NaCl, $CaCl_2$, $Ca(NO_3)_2$, sucrose, glucose, and polyethylene glycol 400. Among the solutions, NaCl was found to be the best option due to its low viscosity. In general, the water flux was increased by solute concentration of DS since higher concentration led to a greater osmotic pressure. Results of their study also suggested using a thinner membrane because it resulted in a higher flux. Moreover, the water flux tended to increase exponentially with membrane thickness. Garcia-Castello and McCutcheon [23] compared two different DS concentration, i.e. 2 M and 4 M, for concentrating orange peel press liquor in FO. They found that the liquor could be concentrated up to 3.7 concentration factor when 4 M NaCl was used.

In FO, the membrane can be operated at two modes according to membrane orientation [24]. The first mode is called AL-FS mode in which the membrane active layer facing the feed solution (FS). Meanwhile, the second mode uses a membrane with the active layer facing the draw solution (DS) which is also known as AL-DS mode. Nayak et al. [25] evaluated effects of membrane orientation in FO and molecular weight compounds on FO performance, in term of water flux. According to their results, AL-FS could result in a higher flux rather than AL-DS mode. It was due to a significant external concentration polarization occurred in AL-DS mode. In another study, Petrotos et al. [26] tried to determine the effect of pre-treatment step on FO performance during tomato juice concentration. They proposed a pre-treatment step for clarifying the juice prior to FO. The result showed that pre-treatment step could effectively improve water flux above 100% compared to those unclarified juices. The highest flux was obtained when 25 kDa ultrafiltration membrane was used in the pre-treatment step. A pilot scale study reported by Dova et al. [27] showed membrane characteristics, solute concentrations of feed solution (FS) and DS, and flow rates of FS and DS affected FO performance significantly. Garcia-Castello and McCutheon [23] compared DS with two different concentrations, i.e. 2 M and 4 M, for concentrating orange peel press liquor in FO. The 4 M NaCl DS could achieve concentration factor up to 3.7.

Table 1. Performances of FO in liquid food concentrations

Liquid foods	Membrane	Draw solution	Operating conditions	Results	Ref.
Tomato juice	Polyamide (AFC99, TFC)	22.2% wt. NaCl	Initial TSS = 4.3 °Brix;	Flux = 3.1 kg m^{-2} h^{-1};	[22]
Tomato juice	Polyamide (AFC99, TFC)	23% wt. NaCl	Initial TSS = 4.3 °Brix; Pretreatment = UF 100 kDa	Flux = 7.28 kg m^{-2} h^{-1};	[26]
Red Radish Juice	Osmotek Inc., Corvallis, OR	60 °Brix fructose corn syrup	Initial TSS = 1.2 °Brix;	Final TSS = 5.1 °Brix (single process), 8.2 °Brix (double process).	[28]
Pineapple juice	Cellulose acetate (Osmotek, Inc.)	Mixed 40% sucrose – 12% NaCl	Initial TSS = 12.4 °Brix;	Final TSS = 60°Brix;	[29]
Glucose solution	Polyamide membrane (DS-3-SG); TFC	23% NaCl	Initial glucose conc. = 0.17 molality;	Flux = 1.7 kg m^{-2} h^{-1};	[27]
Sucrose solution	Cellulosic membrane; Albany, OR;	4 M NaCl	Initial TSS = 10 °Brix;	Flux = 5.84 L m^{-2} h^{-1};	[30]
Anthocyanin extract	Cellulose acetate (Osmotek, Inc.)	6 M NaCl	Initial TSS = 2 °Brix; Anthocyanin conc. = 49.6 mg/L;	Final TSS = 52 °Brix; Anthocyanin conc. = 2.7 g/L;	[31]
Beetroot juice	Cellulose acetate (Osmotek, Inc.)	6 M NaCl	Initial TSS = 2.3 °Brix; Betalains initial conc. = 50.92 mg/L;	Final TSS = 52 °Brix; Betalains final conc. = 2.91 g/L; Initial flux = 11-12 L m^{-2} h^{-1};	[25]
Grape juice	Cellulose acetate (Osmotek, Inc.)	6M NaCl	Initial TSS = 8 °Brix; Initial Anthocyanin content = 50.9 mg/L;	Final TSS = 54.6 °Brix; Final Anthocyanin content = 2.9 g/L; Initial flux = ~8 L m^{-2} h^{-1};	[25]
Tuna cooking juice	Cellulosic membrane; Albany, OR;	2M NaCl	Initial protein conc. = 5.5 % (w/v)	Final protein conc. = 9 % (w/v); Flux = 2.54 L m^{-2} h^{-1};	[32]
Jaboticaba (*Myrciaria jaboticaba*) juice	Cellulosic membrane; Albany, OR;	6M NaCl	Initial TSS = 12 °Brix;	Flux = 3.2-3.59 L m^{-2} h^{-1};	[33]
Sweetlime juice	Cellulose acetate (Osmotek, Inc.)	6M NaCl	Initial TSS = 11 °Brix;	Initial TSS = 50 °Brix; Initial flux = 9-10 L m^{-2} h^{-1};	[34]
Raspberries juice	Osmotek's direct-osmotic concentration (DOC) units	69.7 °Brix fructose corn syrup	Initial TSS = 10 °Brix; T = 26°C; t = 5.8 h;	Final TSS = 44.8 °Brix; Flux constant = 1.4 L m^{-2} h^{-1} Δ°$Brix^{-1}$;	[35]

TFC – thin film composite; TSS – total soluble solids;

Concentrating bioactive-containing liquid while preserving the active function is highly needed in various industries. This ability has been performed by using FO in several studies. For example, Yang et al. [36] used FO for concentrating pharmaceutical products by employing a home-made dual-layer membrane. Lysozome product was effectively enriched by FO with a high purity and without properties change. Rodriguex-Saona et al. [28] investigated the performance of FO for the concentration of Red radish extract by employing corn syrup solution (60 °Brix) as the DS. The FO process could increase the total soluble solids (TSS) of Red radish extract from 1.2 °Brix to 5.1 °Brix in a single concentration process and to 8.2 °Brix in a twice concentration process. Furthermore, they also demonstrated that the concentration of TSS could be increased by combining FO with the conventional evaporator. In addition, the concentrated juice also showed an increase in Anthocyanin concentration. The final Anthocyanin concentration increased to 0.899 mg/mL from 0.258 mg/mL initial concentration after concentrated twice. Concentration of Anhocyanin extract was also reported by Nayak and Rastogi [31] using FO process. The FO produced Anthocyanin extract with a final concentration of 2.7 g/L (from 0.05 g/L) in 18 h operation. However, they observed salt back diffusion during the process. Nayak et al. [25] also successfully demonstrated Anthocyanin concentration in Beetroot and grape juices with final concentrations of up to 2.91 g/L (57 fold) and 715.6 mg/L (6.8 fold). Not only able to obtain a high concentration, the antioxidant properties of Anthocyanin in *Jaboticaba juice* were preserved [33]. Wrolstad et al. [35] studied the performance of FO during Raspberries concentration. They analyzed the concentrated juice and compared with the commercial one. Results of aroma and flavor analysis indicated no significant change in concentrated juice. Khongnakorn and Youravong [32] used FO for recovering protein from Tuna cooking juice. The FO could produced Tuna cooking juice with up to 9% protein content and 2.54 LMH water flux. These studies confirmed that FO can concentrate liquid food and is capable to preserve the nutraceuticals content.

Babu et al. [29] investigated Pineapple juice concentration by using mixed sucrose NaCl as the DS. The pineapple juice was successfully concentrated up to 60 °Brix from its initial TSS of 12.4 °Brix when the DS contained 40% sucrose-12% NaCl. The mixed DS solution was used to reduce diffusion of salt to the juice thereby avoiding a salty taste. A study conducted by Garcia-Castello et al. [30] showed that FO could achive up to 5.7 concentration factor which was higher than obtained by RO (up to 2.5). However, FO showed a lower flux than pressure-driven membrane based-processes. Therefore, FO membane needs more improvement for achieving higher water flux and lower salt reverse diffusion.

3. Challenges in FO Operation

Despite the excellent performance of FO in dewatering liquid foods, most FO processes still show lower water flux. Even though the process uses DS with a high solute concentration, the flux is still low, e.g. below 10 LMH [37]. The low water flux will require a large membrane area which not satisfy the practical application. Furthermore, some operational problems should be overcome. FO membrane needs substantial improvement to solve reverse solute diffusion, concentration polarization, and low mechanical strength [8].

3.1. Internal concentration polarization (ICP)

The first major problem encountered in FO process in internal concentration polarization (ICP). ICP occurs due to the permeation of water from FS to DS which reduces the solute concentration of DS at the support-active layer interface. Consequently, this results in osmotic pressure reduction and thereby decreases water flux [38]. It is also accompanied by the low reverse salt diffusion from the bulk DS to the interface [39].

Developing membrane with a low ICP is needed to improve FO performance [40]. Modifying the membrane support layer should be attempted since ICP cannot be controlled by tuning FO operating condition [40]. As ICP is determined by support layer structure and hydrophilicity, extensive efforts have been devoted to improve those parameters [41–44]. Generally, structural parameter of S is used to determine the ICP tendency of the membrane. S is defined as: S = thickness × porosity/tortuosity. The effect of S on water flux is shown in Figure 2. For instance, minimizing ICP can be done by fabricating a very thin support layer

[45]. However, another problem arises, that is a weak mechanical strength.

Another factor affecting ICP and membrane flux is FO operation mode. More severe ICP was found when FO was operated at AL-FS rather than AL-DS [9]. However, AL-FS experienced a higher flux stability than AL-DS [9].

Chanukya and Rastogi [34] proposed ultrasound assisted FO for reducing concentration polarization in concentration of sucrose and pectin containing solutions. Ultrasound assisted FO exhibited a higher water flux and a lower concentration polarization than the conventional FO. However, it makes the equipment more complex.

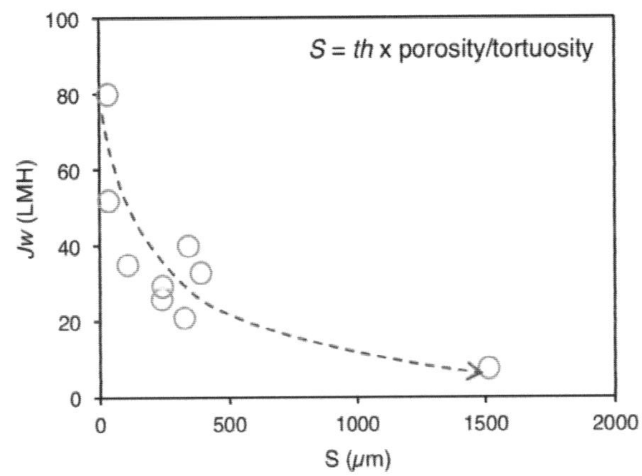

Figure 2. Effect of structural parameter S on water flux (membrane orientation: AL-FS; DS = 2 M NaCl; data from refs. [46–53])

3.2. Reverse solute diffusion from DS to FS

Another problem that should be taken into consideration is salt flux or reverse salt flux [37]. Reverse solute diffusion arises due to the high concentration different originated by a high solute concentration of DS and a lower solute concentration of FS. This problem, for example, was observed by Nayak and Rastogi during the concentration of anthocyanin extract. The diffusion of NaCl was found to be 0.21 moles m^{-2} s^{-1} [31]. Reverse salt diffusion will reduce product quality, especially a salty taste of fruit juice.

Water and salt fluxes of several membranes are depicted in Figure 2. Generally, a trade-off between selectivity and permeability is found in membrane [54,55]. A membrane which has high permeability and selectivity is preferable. Such membrane is needed to achieve a high productivity and to obtain a high-quality product. Therefore, the membrane should be designed to the direction of lower-right side of J_w vs J_s curve (Figure 2).

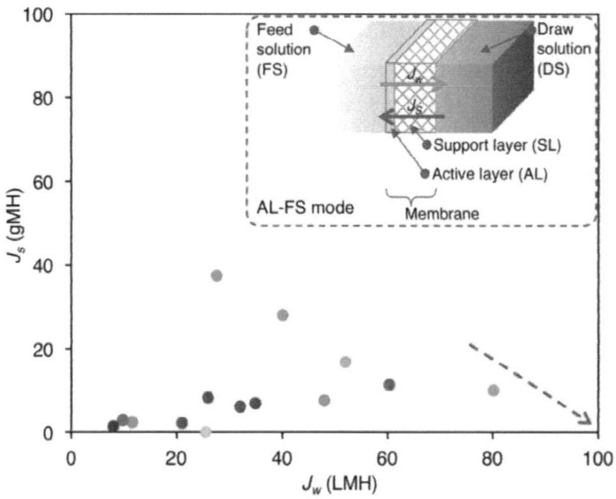

Figure 3. Water and reverse solute fluxes (*Jw*, *Js*) of various FO membranes (FS = deionized water; DS = 2.0 mol/L NaCl; data from refs. [50,51,56–63]).

3.3. Fouling

FO offers several advantages compared to other pressure-driven membrane processes, including the low pressure requirement and the ability to treat a high solid content solution [37]. The low fouling tendency of FO is due to the low hydraulic pressure and low permeating flux resulting in a non-compacted and reversible fouling [21]. Even though the fouling tendency is lower than other pressure-driven membrane processes, fouling is still the major drawbacks in FO process [8]. For instance, during the orange peel press liquor concentration, Garcia-Castello and McCutcheon [23] found that fouling phenomena occurred in FO process. The fouling reduced water permeate flux up to 50%. Pectin was suspected as the dominant fouling contributor. Fouling phenomena in FO process was also found by Khongnakorn and Youravong. They found that protein was the suspected foulant in Tuna cooking juice [32].

Zou et al. reported that [64] more sever fouling was observed when a high concentration of DS was used. This is because a high concentration DS leads to a high permeating flux. Consequently, it will cause more rapid foulant deposition on the membrane surface. Even though FO is also hampered by fouling, the fouling structure was found to be different than those encountered in RO [19]. The fouling layer had looser structure and less compacted thus increasing the efficacy of membrane cleaning [19]. To address this issues, many novels membranes have been developed.

4. FO Membrane Modifications

Generally, two types of membrane are used for FO, i.e. asymmetric cellulose acetate and thin-film composite (TFC) polyamide membranes [65]. The second one shows a higher water flux rather than the first but more susceptible to fouling. TFC membrane is more popular than other FO membranes as it has higher permselectivity and provides more design options, in which active and support layer can be designed separately [66]. TFC membrane exhibits superior characteristics compared to other membranes, in terms of water permeability, salt rejection, mechanical strength, and chemical stability [53,54,67,68]. Most TFC membrane suffers from ICP phenomena due to reverse salt diffusion and support structure thus results in a dramatic water flux reduction. TFC comprises of active layer with a high salt rejection and support layer with high mechanical strength. A support layer with a low tortuosity and a high hydrophilicity is preferable to reduce ICP [54].

Ideally, FO membrane should have high water permeability, high solute rejection, and high mechanical strength. To achieve those properties, membrane active and support layer need to be tuned in order to minimize mass transfer limitation in the support layer and maximize the selectivity of membrane active layer [39]. The active layer is responsible for water permeating flux and reverse solute flux. Accordingly, the active layer should be selective toward water. Meanwhile, the membrane support layer should have high mechanical strength for supporting the active layer. As previously mentioned, it should be noted that the support layer also affects the water flux due to ICP phenomenon. Therefore, the support layer properties including thickness, porosity, and tortuosity should be taken into consideration in the fabrication [69].

Various support optimization strategies were reported, such as additive blending [56], chemical etching [70], in-situ mineralization [46], surface modification [71], and template assisted technique [72]. The aim was to increase membrane porosity and hydrophilicity as well as to reduce membrane tortuosity [53]. Water (J_w) and solute fluxes (J_s) of various support types are shown in Figure 4.

It was demonstrated by numerous studies that the use of nanoparticles in membrane fabrication could improve the membrane performance [73–75]. Incorporating inorganic nanoparticles into membrane support promises an effective ICP reduction. For example, Zhang et al. [53] synthesize polyvinylidene fluoride (PVDF) membrane support by introducing SiO_2@MWNTs as additive. The membrane showed a higher water flux and an improved selectivity. The improved selectivity was observed from a reduction of Js/Jv values from 1.10 to 0.19 g/L. Liu et al. [46] used hydrophilic mineral coating for improving hydrophilicity of polyethersulfone (PES)-based support. The pores surface of the support was coated by CaCO3. The hydrophilic nature of $CaCO_3$ increased PES-based support hydrophilicity leading to the reduced ICP. The prepared TFC membrane had a water flux of 52 LMH when operated at AL-FS mode (FS = deionized water, DS = 2 M NaCl). Even though nanoparticles addition resulted in an improved membrane performance, another problem is emerged that is aggregation of nanoparticles in high concentration due to reducing miscibility [76].

Polymer blending is another facile method for improving support hydrophilicity [77,78]. Chen et al. [40] reported polymer blending method for preparing polysulfone (PSf)/ sulfonated poly (ether ether ketone) (SPEEK) support. PSf/SPPEK showed a high hydrophilicity and open pore structure. The blended polymer resulted in a higher water flux (28.3 LMH). The same method was applied by Zhang et al. [79] in fabrication of PSf support. Disulfonated poly (arylene ether sulfone) hydrophilic-hydrophobic multiblock copolymer (BPSH100-BPS0) and PSf were blended. TFC membrane with a support containing 25 wt% BPSH100-BPS0 could achieve 40.5 LMH water flux when deionized water and 2 M NaCl were used as FS and DS, respectively. The membrane also had a high water flux of 18.6 LMH during seawater desalination.

Several studies also reported that electro-spun membrane support demonstrated a low ICP with a high water flux [66]. Pan et al. [66] fabricated polyamide (PA)/polyacrylonitrile (PAN) TFC membrane. The PAN support was prepared by electrospinning method. Then, the PAN support was laminated by paper laminator. The laminated PAN support showed a high hydrophilicity with 32.3 ± 1.3° water contact angle. Moreover, the membrane could achieve water flux of 57 LMH when 2 M NaCl was used as DS. The scaffold-like structure of electrospun PAN support with interconnected pore was found to increase the support porosity thus reducing ICP phenomenon. However, electro-spun membrane is usually weak.

Chemical modification is another promising strategy for improving FO membrane performance. Chemical modifications can endow a membrane with a high hydrophilicity, anti-fouling property, and reduced ICP [8]. Guan and Wang [80] prepared cellulose triacetate membrane for FO with modified nonwoven fabric for the support. The support surface was modified by polyvinyl alcohol (PVA) and glutaraldehyde via crosslinking process. The modification successfully increased the support hydrophilicity indicated by the reduction of water contact angle (WCA) from 116.9° to 38.5°. Meanwhile, the hydrophilicity of modified cellulose triacetate membrane was increased shown by WCA reduction from 114.1° to 55.8°. The modified membrane had water flux of 55 LMH and structural parameter of 93.65 μm. However, chemical modification still result in a trade-off between membrane performance and mechanical strength [8].

Modifying membrane surface or surface modification is one the most used method in fouling control strategies [81]. For instance, Zhang et al. [65] modified a polyamide membrane by in-situ surface grafting of amine-terminated sulfonated poly(arylene ether sulfone) (NH_2-BPSH100). The modified membrane showed an improved antifouling property during a filtration of soybean oil/water emulsion (40 g/L). Membrane flux could be easily recovered by a simple hydraulic flushing resulting in a higher flux recovery factor (69.8%) rather than the unmodified membrane (11%). Park et al. [82] introduced a simple method for increasing the hydrophilicity of PVDF nanofiber support, that is by dip coating of the membrane support into polyvinyl alcohol (PVA) solution. The modified TFC membrane showed an excellent water flux of 34.2 LMH with 1 M NaCl and deionized water as DS and FS, respectively.

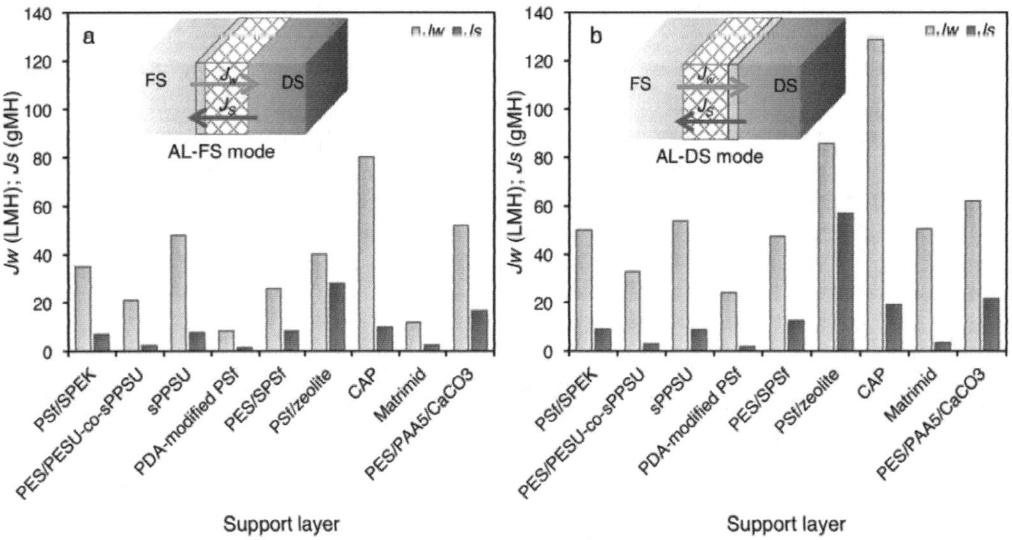

Figure 4. *Jw* and *Js* of FO membranes with various types of support layers. (a) AL-FS mode and (b) AL-DS mode. FS = deionized water; DS = 2.0 mol/L NaCl; data from refs. [46,49–52,56,58,83]; CAP = cellulose acetate propionate; PAA = polyacrylic acid; PDA = polydopamine; PESU = polyethersulfone; PSf = polysulfone; SPEK = sulfonated poly(ether ketone); sPPSU = sulfonated polyphenylenesulfone;

Kang et al. [84] prepared a regenerable FO membrane with multiple polyethylenimine (PEI) and poly(acrylic acid) (PAA) bilayers as the active layer. The fouled membrane can be regenerated by disambling the PEI-PAA bilyers using acid solution followed by coating the new PEI-PAA bilayers on a polydopamine-functionalized polysulfone support. This approach may reduce membrane replacement cost since it only requires to replace the active layer.

To solve the issue of ICP as well as a weak mechanical strength, Li et al. [38] developed a self-standing FO membrane without support layer. The membrane was synthesized from COOH derived polyoxadiazole copolymer. The membrane has a few micron thickness (5-15 μm). The membrane showed a high Na_2SO_4 rejection and an acceptable water permeability

(8.37×10^{-7} LMH bar^{-1}). This result shows the potential of a support-free membrane for eliminating ICP.

Graphene oxide (GO) membranes are considered as a potential next-generation membrane as they have ultra-thin thickness, tailorable pore size, and superior water permeability [85]. Those advantages, for example, have been utilized by Yang et el. [85] to prepare FO membrane. They fabricated polydopamine coated reduced GO membrane. They found that the membrane exhibited a high water flux 36.6 LMH with relatively low reverse solute flux 0.042 mol L^{-1} m^{-2} and high salt rejection 92.0% [85].

5. Conclusions and Future Perspectives

Maintaining nutritional value is quite challenging in liquid food concentration, especially for thermally based processes, since the nutraceutical compounds are heat sensitive. Among the alternative processes, FO offers several attractive features allowing to address the drawbacks of the thermally based processes. In addition to its ability in producing a final product with a high solid concentration, FO can retain the bioactivity of nutraceuticals due to its mild operating condition. Despite the advantages, FO faces several problems that hamper the application, namely concentration polarization, fouling, and reverse salt diffusion [5]. Membrane properties play the significant role in controlling those problems. Therefore, many efforts focus on developing membrane which can alleviate those issues. Despite the advancement in FO membrane preparation, most of the prepared membranes are characterized by deionized water as FS [24]. For practical application, the developed membrane should be tested in a real solution.

6. References

[1] Jiao B, Cassano A and Drioli E 2004 *J. Food Eng.* **63** 303
[2] Morris C, Brody AL and Wicker L 2007 *Packag. Technol. Sci.* **20** 275
[3] de Toledo Guimarães J, Silva EK, de Freitas MQ, de Almeida Meireles MA and da Cruz AG 2018 *Curr. Opin. Food Sci.* **22** 62
[4] Sreedhar I 2017 *A Tryst with Process and Engineering Trends in Forward Osmosis* vol 4
[5] Rastogi NK 2016 *Crit. Rev. Food Sci. Nutr.* **56** 266
[6] Sant'Anna V, Marczak LDF and Tessaro IC 2012 *J. Food Eng.* **111** 483
[7] Terefe NS, Janakievski F, Glagovskaia O, De Silva K, Horne M and Stockmann R 2016 *Innovative Food Processing Technologies: Extraction, Separation, Component Modification and Process Intensification* pp 177–205
[8] Xu W, Chen Q and Ge Q 2017 *Desalination* **419** 101
[9] Tang CY, She Q, Lay WCL, Wang R and Fane AG 2010 *J. Memb. Sci.* **354** 123
[10] Puguan JMC, Kim H-S, Lee K-J and Kim H 2014 *Desalination* **336** 24
[11] Gray GT, McCutcheon JR and Elimelech M 2006 *Desalination* **197** 1
[12] Choi Y, Hwang T-M, Jeong S and Lee S 2018 *Ultrason. Sonochem.* **41** 475
[13] Wang KY, Ong RC and Chung T-S 2010 *Ind. Eng. Chem. Res.* **49** 4824
[14] Touati K and Tadeo F 2016 *Desalination* **389** 171
[15] Heo J, Chu KH, Her N, Im J, Park Y-G, Cho J, Sarp S, Jang A, Jang M and Yoon Y 2016 *Desalination* **389** 162
[16] Nguyen HT, Nguyen NC, Chen S-S, Ngo HH, Guo W and Li C-W 2015 *Sci. Total Environ.* **538** 129
[17] Sianipar M, Kim SH, Khoiruddin, Iskandar F and Wenten IG 2017 *RSC Adv.* **7** 51175
[18] Himma NF, Prasetya N, Anisah S and Wenten IG 2018 *Rev. Chem. Eng.*
[19] Siddiqui FA, She Q, Fane AG and Field RW 2018 *J. Memb. Sci.* **565** 241

[20] Mi B and Elimelech M 2010 *J. Memb. Sci.* **348** 337
[21] Chun Y, Mulcahy D, Zou L and Kim IS 2017 *Membranes (Basel).* **7**
[22] Petrotos KB, Quantick PC and Petropakis H 1998 *J. Memb. Sci.* **150** 99
[23] Garcia-Castello EM and McCutcheon JR 2011 *J. Memb. Sci.* **372** 97
[24] Li D, Yan Y and Wang H 2016 *Prog. Polym. Sci.* **61** 104
[25] Nayak CA, Valluri SS and Rastogi NK 2011 *J. Food Eng.* **106** 48
[26] Petrotos KB, Quantick PC and Petropakis H 1999 *J. Memb. Sci.* **160** 171
[27] Dova MI, Petrotos KB and Lazarides HN 2007 *J. Food Eng.* **78** 422
[28] Rodriguex-Saona LE, Giusti MM, Durst RW and Wrolstad RE 2007 *J. Food Process. Preserv.* **25** 165
[29] Babu BR, Rastogi NK and Raghavarao KSMS 2006 *J. Memb. Sci.* **280** 185
[30] Garcia-Castello EM, McCutcheon JR and Elimelech M 2009 *J. Memb. Sci.* **338** 61
[31] Nayak CA and Rastogi NK 2010 *Desalin. Water Treat.* **16** 134
[32] Khongnakorn W and Youravong W 2016 *J. Eng. Sci. Technol.* **11** 962
[33] Sant'Anna V, Gurak PD, Vargas NSD, da Silva MK, Marczak LDF and Tessaro IC 2016 *Sep. Sci. Technol.* **51** 1708
[34] Chanukya BS and Rastogi NK 2017 *Ultrason. Sonochem.* **34** 426
[35] Wrolstad RE, McDaniel MR, Durst RW, Micheals N, Lampi KA and Beaudry EG 2018 *J. Food Sci.* **58** 633
[36] Yang Q, Wang KY and Chung T-S 2009 *Sep. Purif. Technol.* **69** 269
[37] Wang Y-N, Wang R, Li W and Tang CY 2017 *J. Memb. Sci.* **533** 179
[38] Li M, Karanikola V, Zhang X, Wang L and Elimelech M 2018 *Environ. Sci. Technol. Lett.* **5** 266
[39] Shaffer DL, Werber JR, Jaramillo H, Lin S and Elimelech M 2015 *Desalination* **356** 271
[40] Chen G, Liu R, Shon HK, Wang Y, Song J, Li X-M and He T 2017 *Desalination* **405** 76
[41] Ghanbari M, Emadzadeh D, Lau WJ, Riazi H, Almasi D and Ismail AF 2016 *Desalination* **377** 152
[42] McCutcheon JR and Elimelech M 2008 *J. Memb. Sci.* **318** 458
[43] Liu X and Ng HY 2015 *J. Memb. Sci.* **481** 148
[44] Zhao Y, Wang X, Ren Y and Pei D 2018 *ACS Appl. Mater. Interfaces* **10** 2918
[45] Gai J-G and Gong X-L 2014 *J. Mater. Chem. A* **2** 425
[46] Liu Q, Li J, Zhou Z, Xie J and Lee JY 2016 *Sci. Rep.* **6**
[47] Ong RC, Chung T-S, de Wit J and Helmer BJ 2014 *J. Memb. Sci.* **473** 63
[48] Emadzadeh D, Lau WJ, Matsuura T, Ismail AF and Rahbari-Sisakht M 2014 *J. Memb. Sci.* **449** 74
[49] Ma N, Wei J, Qi S, Zhao Y, Gao Y and Tang CY 2013 *J. Memb. Sci.* **441** 54
[50] Han G, Chung T-S, Toriida M and Tamai S 2012 *J. Memb. Sci.* **423–424** 543
[51] Widjojo N, Chung T-S, Weber M, Maletzko C and Warzelhan V 2011 *J. Memb. Sci.* **383** 214
[52] Han G, Zhang S, Li X, Widjojo N and Chung T-S 2012 *Chem. Eng. Sci.* **80** 219
[53] Zhang X, Shen L, Guan C-Y, Liu C-X, Lang W-Z and Wang Y 2018 *J. Memb. Sci.* **564** 328
[54] Xu G-R, Xu J-M, Feng H-J, Zhao H-L and Wu S-B 2017 *Desalination* **417** 19
[55] Chen X, Xu J, Lu J, Shan B and Gao C 2017 *Desalination* **405** 68
[56] Widjojo N, Chung T-S, Weber M, Maletzko C and Warzelhan V 2013 *Chem. Eng. J.* **220** 15
[57] Pardeshi P and Mungray A 2014 *Int. J. Polym. Mater. Polym. Biomater.* **63** 595
[58] Luo L, Wang P, Zhang S, Han G and Chung T-S 2014 *J. Memb. Sci.* **461** 28

[59] Sun Y, Xue L, Zhang Y, Zhao X, Huang Y and Du X 2014 *Desalination* **336** 72
[60] Pardeshi P and Mungray AA 2014 *J. Memb. Sci.* **453** 202
[61] You S, Tang C, Yu C, Wang X, Zhang J, Han J, Gan Y and Ren N 2013 *Environ. Sci. Technol.* **47** 8733
[62] Sukitpaneenit P and Chung T-S 2012 *Environ. Sci. Technol.* **46** 7358
[63] Zhang S, Wang KY, Chung T-S, Jean YC and Chen H 2011 *Chem. Eng. Sci.* **66** 2008
[64] Zou S, Gu Y, Xiao D and Tang CY 2011 *J. Memb. Sci.* **366** 356
[65] Zhang X, Tian J, Gao S, Zhang Z, Cui F and Tang CY 2017 *J. Memb. Sci.* **527** 26
[66] Pan S-F, Dong Y, Zheng Y-M, Zhong L-B and Yuan Z-H 2017 *J. Memb. Sci.* **523** 205
[67] Yasukawa M, Mishima S, Tanaka Y, Takahashi T and Matsuyama H 2017 *Desalination* **402** 1
[68] Zhao W, Liu H, Meng N, Jian M, Wang H and Zhang X 2018 *J. Memb. Sci.* **565** 380
[69] Kim B, Gwak G and Hong S 2017 *Desalination* **422** 5
[70] Zhao X, Li J and Liu C 2017 *Desalination* **408** 102
[71] Huang L, Arena JT and McCutcheon JR 2016 *J. Memb. Sci.* **499** 352
[72] Li J, Liu Q, Li X, Liu Y and Xie J 2016 *Ind. Eng. Chem. Res.* **55** 5327
[73] Wang Y-N, Goh K, Li X, Setiawan L and Wang R 2018 *Desalination* **434** 81
[74] Akli K, Khoiruddin and Wenten IG 2016 *J. Membr. Sci. Res.* **2** 141
[75] Khoiruddin and Wenten IG 2016 *J. Eng. Technol. Sci.* **48** 1
[76] Suwaileh WA, Johnson DJ, Sarp S and Hilal N 2018 *Desalination* **436** 176
[77] Aryanti PTP, Yustiana R, Purnama RED and Wenten IG 2015 *Membr. Water Treat.* **6** 379
[78] Khoiruddin K, Ariono D, Subagjo S and Wenten IG 2018 *J. Appl. Polym. Sci.*
[79] Zhang X, Tian J, Ren Z, Shi W, Zhang Z, Xu Y, Gao S and Cui F 2016 *J. Memb. Sci.* **520** 529
[80] Guan P and Wang D 2016 *Desalin. Water Treat.* **57** 27505
[81] Khoiruddin, Ariono D, Subagjo and Wenten IG 2017 *J. Appl. Polym. Sci.* **134** 45540
[82] Park MJ, Gonzales RR, Abdel-Wahab A, Phuntsho S and Shon HK 2018 *Desalination* **426** 50
[83] Wang KY, Chung T-S and Amy G 2012 *AIChE J.* **58** 770
[84] Kang Y, Zheng S, Finnerty C, Lee MJ and Mi B 2017 *Environ. Sci. Technol.* **51** 3242
[85] Yang E, Kim C-M, Song J-H, Ki H, Ham M-H and Kim IS 2017 *Carbon N. Y.* **117** 293

Hydrophilic Modification of Polymeric Membrane using Graft Polymerization Method: A Mini Review

J.A. Pinem[1], A.K. Wardani[2], P.T.P. Aryanti[3], K. Khoiruddin[2] and I G. Wenten[2,4,*]
[1]Chemical Engineering Department, Riau University, Pekanbaru 28293, Indonesia
[2]Chemical Engineering Department, Institut Teknologi Bandung, Jl. Ganesa 10, Bandung 40132, Indonesia
[3]Chemical Engineering Department, Universitas Jenderal Achmad Yani, Jl. Terusan Jendral Sudirman, Cimahi 40285, Indonesia
[4]Research Center for Nanosciences and Nanotechnology, Institut Teknologi Bandung, Jl. Ganesha 10, Bandung – 40132, Indonesia

*Corresponding author (E-mail): igw@che.itb.ac.id

Abstract. The separation performance of polymeric membrane is strongly influenced by its surface characteristics. In water treatment applications, it is generally accepted that a hydrophilic membrane surface is favorable to enhance water permeate flux and mitigate membrane fouling. Numerous methods, both chemical and physical modification, have been proposed to improve the hydrophilicity of polymeric membranes. Graft polymerization is one of the promising methods to modify polymer membranes by tethering hydrophilic polymer chains onto membrane surface. The existence of hydrophilic chains on the membrane surface facilitates the reduction of interfacial tension with water, thus higher water flux can be obtained. In addition, the increase of membrane hydrophilicity also improves the antifouling resistance. This review provides a concise summary and discussion of the modification of polymeric membranes using graft polymerization method. The effects of graft polymerization on the membrane hydrophilicity as well as the antifouling property are discussed.

Keywords: antifouling, graft polymerization, hydrophilic, polymeric membrane

1. Introduction

The interest in technologies for water treatment applications is now growing amidst the concern over drinking water security issues and environmental impact. Wide range of water treatment technologies have been developed or are currently developed to overtake traditional technologies, i.e. chemical clarification, granular media filtration, and chlorination. One of the most interesting processes for water treatment is polymeric membrane. It offers some advantages, such as strict solid-liquid separation, high quality of product, and simple operation [1, 2].

IOP Conf. Series: Materials Science and Engineering **547** (2019) 012054 doi:10.1088/1757-899X/547/1/012054

The separation performance of polymeric membrane is strongly influenced by its surface characteristics. In water treatment applications, it is generally accepted that a hydrophilic membrane surface is favorable to enhance water permeate flux. Besides, the impurities components in surface water are mostly hydrophobic (i.e. microbes, particles, humic acids, etc.), thus their interaction with hydrophobic membranes causes the membranes to be more prone to the fouling than hydrophilic membranes [3-8]. It has been proved in the studies by Laine et al. [9] and Malgorzata et al. [10] where better permeability occurred for more hydrophilic membranes (cellulose) than hydrophobic ones (acrylic, polyaramide, or polysulfone).

Fouling not only leads to productivity decline, but also increases the energy requirement to keep the membrane performance constant [11-15]. Therefore, it is necessary to improve the antifouling resistance by increasing membrane surface hydrophilicity. Various methods to increase hydrophilicity of polymeric membranes have been developed either by chemical or physical modification. Hydrophilization by chemical modification can be done by blending polymers with more suitable properties or by addition of polymer layers on the active surface of membrane through ozone treatment, sulfonating treatment, or grafting of strong hydrophilic monomers [16-19]. In the most cases, chemical modifications lead to destruction or a change in the porous structure of the membrane, and typically require expensive instruments and/or complicated process [20]. Meanwhile, physical modification such as wetting or dip-coating is a mild technique and relatively inexpensive. However, the hydrophilicity from physical modification is not stable, thus the water flux decrease over time.

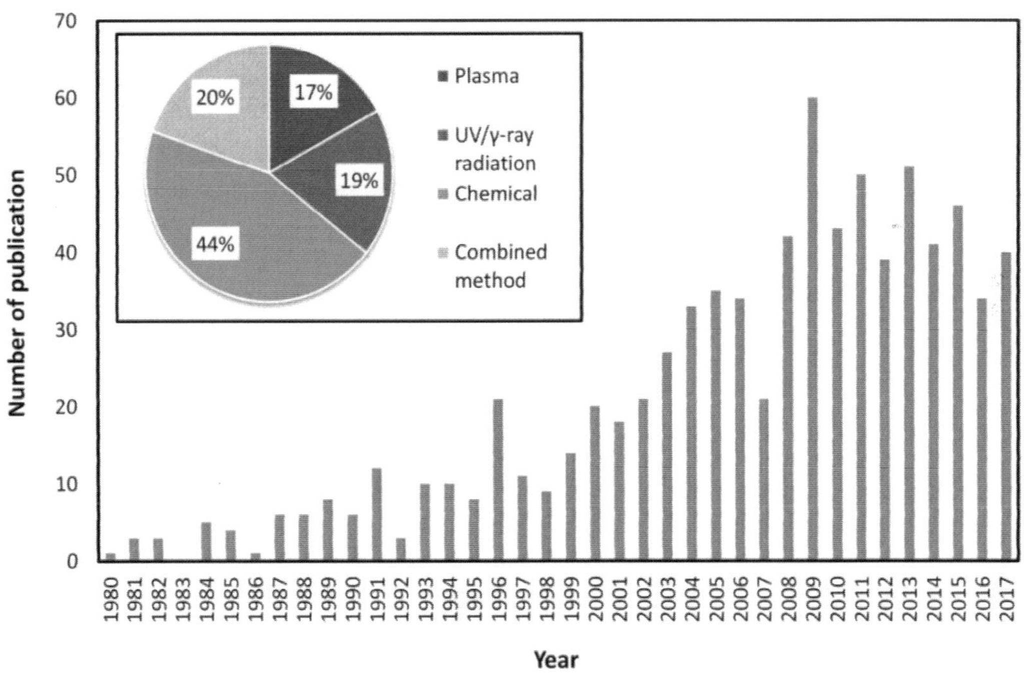

Figure 1 The number of published papers related to the search term of graft polymerization on polymeric membrane. Indexed by Scopus (TITTLE-ABS-KEY (graft polymerization AND polymer membrane); August 2018)

Among the developed hydrophilic modifications of polymeric membrane, graft polymerization is the most promising method. As shown in Figure 1, the number of publications for graft polymerization on polymeric membranes has been increased over year and the chemical-induced graft polymerization has the highest number of publications. Various studies showed that the hydrophilicity and water flux of grafted membranes were stable after several times of operation. To provide an insight for further development of hydrophilic polymeric membrane, this paper gives a review of surface hydrophilization of polymeric membrane using graft polymerization. The effects of graft polymerization on the membrane hydrophilicity as well as the antifouling property are discussed.

2. Methods of Graft Polymerization

Graft polymerization is the promising alternative to increase polymeric membrane hydrophilicity and improves the antifouling resistance by tethering hydrophilic polymer chains onto membrane surface [21, 22]. This method offers advantages of stable hydrophilicity and high water flux. The comparison of graft polymerization to other modification methods is shown in Table 1.

Table 1 Advantages and disadvantages of hydrophilic modification methods [23, 24]

Modification method	Water flux	Simplicity	Hydrophilic stability	Environmental aspect	Cost effectiveness
Coating	Low	High	Low	High	High
Blending	High	High	High	High	High
Composite	High	High	High	High	High
Plasma treatment	High	Low	Low	High	High
Graft polymerization					
• Chemical-induced	High	High	High	Low	High
• Plasma-induced	High	Low	High	High	Low
• Irradiation induced	High	Low	High	Low	Low

In general, graft polymerization can be divided into three main procedures, i.e. chemical-, plasma-, and irradiation- induced graft polymerization [25-28]. For membrane material with reactive groups, it can be easily grafted through the free radical graft polymerization or the chemical reaction. Meanwhile, membrane material without reactive groups needs plasma, γ-ray, or UV to induce graft polymerizations [29].

2.1 Chemical-Induced Graft Polymerization

In chemical-induced graft polymerization, a radical for the polymerization process is formed on the membrane surface through reaction with chemical compounds. One of the most attractive techniques for oxidizing a substrate surface is ozonation [27, 30]. It can be performed either in the gaseous phase or aqueous phase. Ozone acts as a precursor providing radical elements that will initiate the polymerization process [31, 32].

Polypropylene (PP) membranes have been successfully modified using ozone or oxidizing agents such as benzoyl peroxide (BPO) and benzophenonyl bromoisobutyrate (BPBB) as precursor for grafting of hydroxyethyl methacrylate (HEMA) [31, 33], N,N-dimethylaminoethyl methacrylate (DMAEMA) [34] and acrylic acid [35]. Meanwhile, for polyethersulfone (PES) membranes, peroxydisulfate and metabisulfite oxidizing agents have been used to initiate free radical polymerization grafting of methacrylic acid, polyethyleneglycol-methacrylate, and sulfopropylmethacrylate in aqueous solution at ambient temperature [36, 37]. Chemical-induced graft polymerization also has been used to increase hydrophilicity of polysulfone (PSf) membranes [38-40], polyacrylonitrile (PAN) membranes [41], and Poly-vinylidene fluoride (PVDF) membranes [42, 43], as shown in Table 2. In general, polymeric membranes that have been modified by chemical-induced graft polymerization to be less sensitive to fouling due to the presence of the hydrophilic monomer grafted.

2.2 Plasma-Induced Graft Polymerization

Plasma-induced graft polymerization can be achieved through two different procedures, i.e. peroxy radical (P) method and alkyl radical (R) method [44]. In the (P) method, the membrane is introduced to plasma reactor without working gas, then air is added to make peroxy radical that can initiate the grafting process. Meanwhile, the (R) method can be achieved in the absence of working gas or using noble gases such as argon, helium, etc. [44, 45]. This method is operated under vacuum condition where the alkyl radical will initiate the grafting process.

Table 2 Graft polymerization of polymeric membranes

Membrane material	Grafting method	Polymer/additive	Induction	Ref.
PP	Chemical-induced	Acrylic acid, Polyvinyl alcohol, Polyethylene amine, PEGMA, DMAEMA, PNVP, HEMA, TiO_2	Ozone, BPO, BPBB	[28, 30, 31, 46-52]
	Plasma-induced	Acrylic acid, Alkyl glycoside, PNVP, PNIPAM, PSLG, TiO_2	Plasma: Air, N_2, NH_3, Argon	[53-60]
	Irradiation-induced	Acrylic acid, Alkyl glycoside, Acrylamide, PEGMA, HEMA, GMA, PNIPAM, GAMA, SBMA, MPDSAH, TiO_2	UV, γ-ray	[25, 29, 61-73]
PES	Chemical-induced	TiO_2, PEGMA	Peroxydisulfate, Metabisulfite	[36, 37, 74, 75]
	Plasma-induced	Acrylic acid, Acrylamide, PNVP	Plasma: Argon, Helium	[76, 77]
	Irradiation-induced	Acrylic acid, Ethylene diamine, HEMA, PNVP	UV	[78, 79]
PSf	Chemical-induced	PES, PEG, Acrylamide	-	[38-40]
PAN	Chemical-induced	PEO	-	[41]
	Irradiation-induced	Acrylic acid, PEGMA	UV	[80]
PAS	Irradiation-induced	Methacrylic acid, GMA, HEMA	UV	[81]
PVDF	Chemical-induced	PEGMA, PVP	-	[42, 43]
	Irradiation-induced	PES	UV	[82]

Abbreviation: BPBB = benzophenonyl bromoisobutyrate, BPO = benzoyl peroxide, DMAEMA = N-dimethylaminoethyl methacrylate, GAMA = Gluconamidoethyl methacrylate, GMA = Glycidylmethacrylate, HEMA = Hydroxyethyl methacrylate, MPDSAH = [3-(methacryloylamino)propyl]-dimethyl(3-sulfopropyl) ammonium hydroxide, PAN = Polyacrylonitrile, PAS = Poyarylsulfone, PE = Polyethylene, PEG = Poly(ethylene glycol), PEGMA = Poly(ethylene glycol) methacrylate, PEO = Polyethylene oxide, PES = Polyethersulfone, PNIPAM = Poly(N-isopropylacrylamide), PNVP = Poly(N-vinyl-2-pyrrolidone), PP = Polypropylene, PSf = Polysulfone, PSLG = Poly(c-stearyl-L-glutamate), PVDF = Poly(vinylidene fluoride), PVP = Poly(vinyl pyrrolidone), SBMA = Sulfobetaine methacrylate

Several studies of plasma-induced graft polymerization in order to increase the hydrophilicity of polymeric membrane have been conducted using gases such as ammonia [55, 83], oxygen [84], nitrogen [53, 54, 85, 86], hydrogen [87], water [76, 88, 89], argon [58, 77, 90, 91], or air [48, 49, 57]. Meanwhile, the grafting substances consisted of N,N-dimethyl acrylamide (NDMAm) [44], N-isopropyl acrylamide (NIPAm) [53], polyethylene glycol (PEG) [87], acrylamide [92], acrylic acid [58, 90], sulfonated glycidyl methacrylate [91], and tetraethoxylane (TEOS) [93], as shown in Table 2.

While using plasma-induced graft polymerization, the operating condition should be maintained at an appropriate level. As temperature increased, there will be more active sites on the membrane surface, and monomer diffusion is also enhanced due to an increase of grafting degree. Conversely, a high-concentration monomer can decrease the degree of grafting due to increased viscosity that hinders the monomer diffusion to the membrane surface [85] or prevents the plasma treating to reach the surface of the PP membrane, and then no active sites are produced [94].

2.3 Irradiation-Induced Graft Polymerization

Irradiation-induced graft polymerization on PP membrane involves monomer and PP membrane simultaneously to react through radical reaction, with ultraviolet (UV) or γ-ray as the initiator in irradiation induced. UV irradiation is able to cross-link polymer chains and cleave polymer bonds, thus forming functional groups such as hydroxyls, carbonyls, or carboxylic acids on the surface [23].

γ -ray has been used for grafting acrylic acid [69], HEMA [62, 95], N-vinyl-2-pyrrolidone (NVP) [29], and NIPAm/acrylic acid [96] onto PP membrane surfaces. These substances have been successfully grafted onto PP membranes using γ-ray irradiation, resulting hydrophilic surfaces with WCA $< 65°$. Meanwhile, by UV irradiation, several substances have been grafted onto PP membrane surface, such as acrylic acid [66, 67, 97-99], poly-(ethylene glycol)-200-monomethacrylate (PEG200MA) [97], glycidylmethacrylate (GMA) [63], 3-hydroxy-2,2-bis(hydroxymethyl) propylacrylate (PEA) [100], methoxyl polyethylene glycol (MPEG) [101], acrylamide [67], D-gluconamidoethyl methacrylate (GAMA) [68, 102], sulfobetaine methacrylate (SBMA) [71], and α-allyl glucoside [61, 103].

UV-irradiation has also been used to graft several hydrophilic monomers onto PES membranes, i.e. Nvinyl-2-pyrrolidinone, N-vinylcaprolactam, and N-vinylformamide. The results showed there were 25% increase in PES membranes hydrophilicity and 50% decrease in BSA fouling [104-106]. The same method also successfully increased the hydrophilicity of PAN [80], poyarylsulfone (PAS) [81], and PVDF membranes [82].

3. The Influence of Graft Polymerization

Graft polymerization is the promising method to increase hydrophilicity of polymeric membrane. The hydrophilicity of polymeric membrane is mainly measured by water contact angle (WCA). Membrane with WCA less than 90° is classified as hydrophilic membrane while hydrophobic membrane has WCA greater than 90°. Figure 2 shows the effect of graft polymerization on WCA of polymeric membranes from various studies. It can be seen that graft polymerization is able to produce hydrophilic membrane with low WCA using various procedures and chemicals.

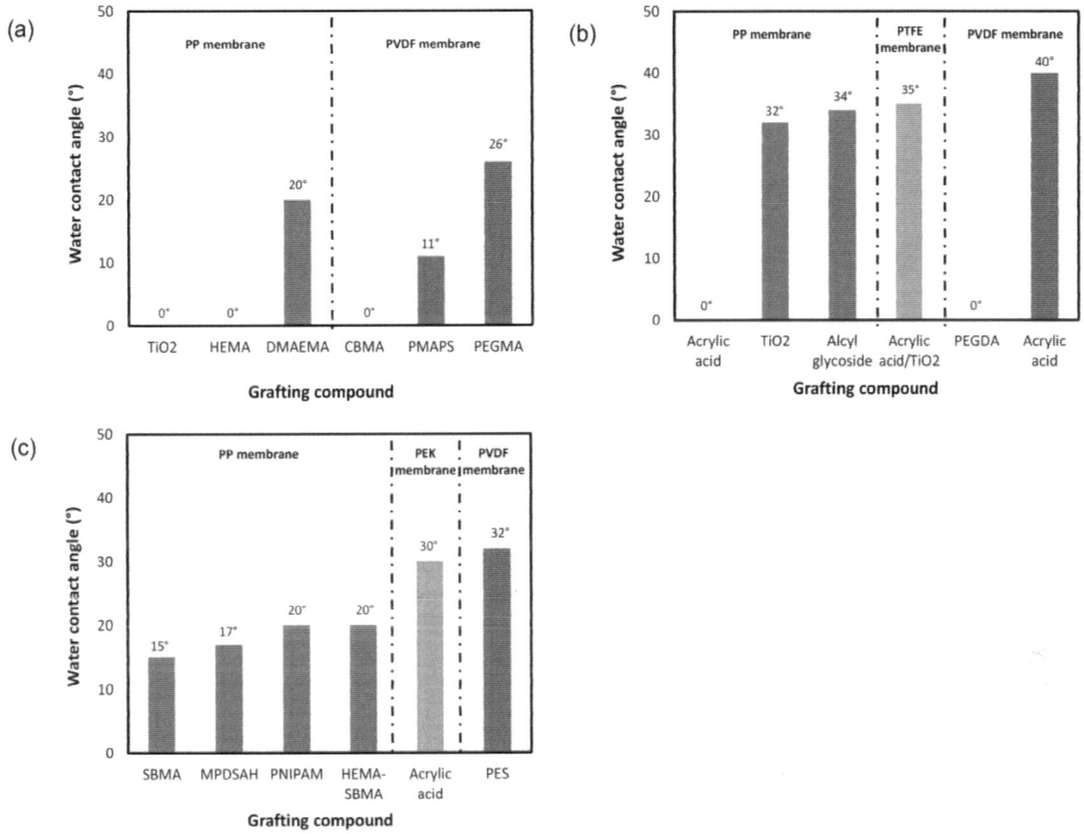

Figure 2 Effect of (a) chemical induced [48, 50, 52, 107-109], (b) plasma induced [54, 59, 60, 110-112], and (c) irradiation induced [70-73, 113, 114] graft polymerization on WCA

Furthermore, hydrophilic modification of polymeric membranes not only improves the water flux, but also endows the membranes with fouling-resistant properties. It makes the modified membranes exhibit higher performance when applied in water and wastewater treatment. Hydrophilic modifications also leads to the increase of biocompatibility of polymeric membranes, thus widen their applications in biotechnology and biomedical applications. The results summary of hydrophilic modification of polymeric membranes via graft polymerization is presented in Table 3. It can be seen that UV irradiation was the most used induction for grafting polymers or additives onto polymeric membrane surface. It is due to the ability of UV to induce cross-linking in the polymer matrix, thus the membrane selectivity was increased [115].

In general, the characteristics of modified membrane mainly depend on the grafting degree. The higher degree of graft polymerization provides a strong driving force to reduce the interfacial free energy by migrating and reorienting of the hydrophilic monomers [51]. As the result, higher grafting degree with a large coverage surface decreases the water contact angle. The study of Yu et al. [67] showed that the water contact angle of PP membrane was decreased from 135° to 42° by increasing the grafting degree to 120%. It is in agreement with the results from other studies [62, 116, 117]. The water flux on modified membrane remained almost unchanged when the grafting degree was low, however the pore-covering soon become the major problem with the increase of the grafting degree, causing the water flux dropping severely. Therefore, the precise design of a graft layer and well-controlled graft architecture is a very important to minimize the membrane fouling.

Table 3 Hydrophilic modification of polymer membranes via graft polymerization

Membrane material	Grafting induction	Polymer/ additive	Influence of graft polymerization	Ref.
PP	Chemical	Polyvinyl alcohol/4-vinylpyridine	Water fluxes were increased up to 500.4 L/m^2.h at operating pressure 40 kPa.	[47]
	Nitrogen plasma	α-allyl glucoside	Water fluxes were increased up to 3826±80 kg/m^2.h at operating pressure 80 kPa.	[54]
	UV	Acrylic acid, Acrylamide	The modified membranes showed better filtration performances in the submerged membrane bioreactor (SMBR) than the unmodified one. Acrylic acid-grafted membrane presented the best antifouling characteristics.	[92]
		GAMA	After continuous operation in the MBR for about 70 h, reduction of water flux was 90.7% for the initial membrane and 80.8-87.2% for the modified membranes, increasing with an increase of a length of the grafted chains.	[68]
		HEMA	The modified membrane showed better protein resistance as well as hemo-compatibility due to the enhancement of surface hydrophilicity. The static water contact angle of the membrane surface decreased from 145° to 42°, with the extent of grafting increasing from 0 to 35.67 wt.%.	[65]
		HEMA, SBMA	Fouling was significantly reduced and most of the fouling on the modified PP membrane surface was reversible.	[72]
PES	Argon plasma	Polyacrylamide	The grafting yield for polyacrylamide on the membrane surfaces increased nearly linearly with the Argon-plasma pretreatment time, with grafting yields (GY) higher than 100 μg/cm^2. The membranes obtained a permanent hydrophilicity (almost no change in contact angle after 1 year), and BSA adsorption was reduced to less than half that of the control membrane (306 to 148 μg/cm^2).	[77]
	UV	Polyvinyl alcohol/4-vinylpyridine	Membranes modified with N-vinyl-2-pyrrolidinone (25% increase in hydrophilicity) exhibited the best combination of low fouling (50% decrease in BSA fouling) and high flux, although membrane permeability was significantly decreased because the grafted polymer chains blocked the membrane pores (over 25% reduction in flux due to modification).	[105]

Table 3 (continue)

Membrane material	Grafting induction	Polymer/ additive	Influence of graft polymerization	Ref.
PES	UV	PNVP	The fouling tendency was reduced for membranes irradiated for 60 s comparing with those irradiated for 180 s. Enlargement of pore structure was observed under long irradiation times facilitating the membrane fouling with a high molecular weight fraction of natural organic matters.	[104]
		PEGMA	Modified membranes showed more resistance to fouling and higher rejection than unmodified membranes.	[118]
PES and PVDF	UV	DMAEMA	The modified membranes were more resistant to biofouling.	[119, 120]
PES/PVDF blend	UV	PNVP	Bovine serum albumin (BSA) adsorption on the membranes decreased from 159±2 to 13±2 μg/cm^2 after 10 min of the grafting. The fouling extent of the blend membrane with 7 min grafting was reduced by 66% and the flux recovery after chemical cleaning increased by 32%.	[82]
PVDF	UV	HEMA, Polydopamine, Acrylic acid	Antifouling properties, e.g., flux recovery, irreversible flux loss, total flux loss, and fouling resistance of modified membranes were enhanced due to an increase in membrane hydrophilicity.	[121]
PSf	UV	MPDSAH	Hydrophilicity and antifouling properties of the modified membranes were enhanced with the increase of the extent of grafting.	[122]
		PNVP, HEMA, Acrylic acid	Modified membranes showed high protein retention, high solution flux, and low irreversible fouling.	[123]
PAN	UV	HEMA, PEGMA, Acrylic acid	Protein-polymer surface interactions were diminished by the membrane modification. Adsorption and fouling were reduced both for the modified membranes.	[80]

4. Conclusions

The application of polymeric membrane is mostly restricted by fouling formation. Hydrophilic modification through graft polymerization is one of the promising methods to enhance membrane hydrophilicity and antifouling properties. This method can be conducted through chemical-induced, plasma-induced, and irradiation-induced using UV or γ-ray. Graft polymerization was able to decrease water contact angle and BSA adsorption on polymeric membranes. In addition, this method offers advantage of stable hydrophilicity and high water flux compared to other hydrophilic modifications.

Reference

[1] Kimura K, Hane Y, Watanabe Y, Amy G and Ohkuma N 2004 *Water Research* **38** 3431-41
[2] Yu C-H, Wu C-H, Lin C-H, Hsiao C-H and Lin C-F 2008 *Separation and Purification Technology* **64** 206-12
[3] Schafer A I, Mastrup M and Jensen R L 2002 *Desalination* **147** 243-50
[4] Ridgway H F, Rigby M G and Argo D G 1985 *Journal-American Water Works Association* **77** 97-106
[5] Fane A, Beatson P and Li H 2000 *Water science and technology* **41** 303-8
[6] Choo K-H and Lee C-H 2000 *Journal of Colloid and Interface Science* **226** 367-70
[7] Ma H, Hakim L F, Bowman C N and Davis R H 2001 *Journal of Membrane Science* **189** 255-70
[8] Mousa H A 2007 *Desalination* **217** 38-51
[9] Laine J-M, Hagstrom J P, Clark M M and Mallevialle J 1989 *Journal (American Water Works Association)* 61-7
[10] Kabsch-Korbutowicz M, Majewska-Nowak K and Winnicki T 1999 *Desalination* **126** 179-85
[11] Zularisam A W, Ismail A F and Salim R 2006 *Desalination* **194** 211-31
[12] Choi H, Zhang K, Dionysiou D D, Oerther D B and Sorial G A 2005 *Separation and Purification Technology* **45** 68-78
[13] Costa A R, de Pinho M N and Elimelech M 2006 *Journal of Membrane Science* **281** 716-25
[14] Du J R, Peldszus S, Huck P M and Feng X 2009 *Water Research* **43** 4559-68
[15] Ariono D, Wardani A K, Widodo S, Putu T P A and Wenten I G 2018 *Mater. Res. Express* **5** 034009
[16] Nunes S P and Peinemann K V 1992 *Journal of Membrane Science* **73** 25-35
[17] Uragami T, Fujimoto M and Sugihara M 1980 *Desalination* **34** 311-23
[18] Sivik B, Wahlgren M and Miezis Y 1990 *Desalination* **77** 181-93
[19] Brink L E S, Elbers S J G, Robbertsen T and Both P 1993 *Journal of Membrane Science* **76** 281-91
[20] Chen H, Kong L and Wang Y 2015 *J. Membr. Sci.* **487** 109-16
[21] Ariono D and Wardani A K 2017 *Adv. Sci. Letters* **23** 5713-5
[22] Ariono D and Wardani A K 2017 *IOP Conf. Series: Mater Sci. Eng.* **214** 012014
[23] Nady N, Franssen M C R, Zuilhof H, Eldin M S M, Boom R and Schroën K 2011 *Desalination* **275** 1-9
[24] Himma N F, Anisah S, Prasetya N and Wenten I G 2016 *J. Polym. Eng.* **36** 329-62
[25] Meng J, Li J, Zhang Y and Ma S 2014 *J. Membr. Sci.* **455** 405-14
[26] Kato K, Uchida E, Kang E-T, Uyama Y and Ikada Y 2003 *Prog. Colloid Polym. Sci.* **28** 209-59
[27] Uyama Y, Kato K and Ikada Y 1998 *Grafting/Characterization Techniques/Kinetic Modeling,* ed H Galina, *et al.* (Springer Berlin Heidelberg) pp 1-39
[28] Zhang C-H, Yang F-l, Wang W-J and Chen B 2008 *Sep. Purif. Technol.* **61** 276-86
[29] Liu Z-M, Xu Z-K, Wang J-Q, Wu J and Fu J-J 2004 *Eur. Polym. J.* **40** 2077-87
[30] Wang C, Feng R and Yang F 2011 *J. Colloid Interface Sci.* **357** 273-9
[31] Wang Y, Kim J-H, Choo K-H, Lee Y-S and Lee C-H 2000 *J. Membr. Sci.* **169** 269-76
[32] Gu H, Wu J, Chan P, Turcotte G and Ye T 2012 *Chem. Eng. Res. Des.* **90** 229-37
[33] Sainbayar A, Kim J S, Jung W J, Lee Y S and Lee C H 2001 *Environ. Technol.* **22** 1035-42
[34] Dai Q-w, Denga H-t, Liua Z-m and Setac P 2004 *Chin. J. Polym. Sci.* **22** 369-77
[35] Duann Y-F, Chen Y-C, Shen J-T and Lin Y-H 2004 *Polymer* **45** 6839-43

[36] Belfer S, Fainchtain R, Purinson Y and Kedem O 2000 *J. Membr. Sci.* **172** 113-24
[37] Reddy A V R, Trivedi J J, Devmurari C V, Mohan D J, Singh P, Rao A P, Joshi S V and Ghosh P K 2005 *Desalination* **183** 301-6
[38] Park J Y, Acar M H, Akthakul A, Kuhlman W and Mayes A M 2006 *Biomaterials* **27** 856-65
[39] Qiu J, Zhang Y, Shen Y, Zhang Y, Zhang H and Liu J 2010 *Applied Surface Science* **256** 3274-80
[40] Yi Z, Zhu L-P, Zhao Y-F, Zhu B-K and Xu Y-Y 2012 *Journal of Membrane Science* **390–391** 48-57
[41] Asatekin A, Kang S, Elimelech M and Mayes A M 2007 *Journal of Membrane Science* **298** 136-46
[42] Hashim N A, Liu F and Li K 2009 *Journal of Membrane Science* **345** 134-41
[43] Bi Q, Li Q, Tian Y, Lin Y and Wang X 2013 *Journal of Applied Polymer Science* **127** 394-401
[44] Masuoka T, Hirasa O, Suda Y and Ohnishi M 1989 *Int. J. Radiat. Appl. Instrum.* **33** 421-7
[45] Kramer P W, Yeh Y S and Yasuda H 1989 *J. Membr. Sci.* **46** 1-28
[46] Xu Z-K, Dai Q-W, Wu J, Huang X-J and Yang Q 2004 *Langmuir* **20** 1481-8
[47] Wang C-c, Yang F-l, Liu L-F, Fu Z-m and Xue Y 2009 *J. Membr. Sci.* **345** 223-32
[48] Yang Y-F, Wan L-S and Xu Z-K 2009 *J. Membr. Sci.* **326** 372-81
[49] Yang Y-F, Wan L-S and Xu Z-K 2009 *J. Membr. Sci.* **337** 70-80
[50] Feng R, Wang C, Xu X, Yang F, Xu G and Jiang T 2011 *J. Membr. Sci.* **369** 233-42
[51] Jin J, Zhang C, Jiang W, Luan S, Yang H, Yin J and Stagnaro P 2012 *Colloids Surf., A* **407** 141-9
[52] Pi J-K, Yang H-C, Wan L-S, Wu J and Xu Z-K 2016 *J. Membr. Sci.* **500** 8-15
[53] Liang L, Shi M, Viswanathan V V, Peurrung L M and Young J S 2000 *J. Membr. Sci.* **177** 97-108
[54] Kou R-Q, Xu Z-K, Deng H-T, Liu Z-M, Seta P and Xu Y 2003 *Langmuir* **19** 6869-75
[55] Liu Z-M, Xu Z-K, Wang J-Q, Yang Q, Wu J and Seta P 2003 *Eur. Polym. J.* **39** 2291-9
[56] Deng H-T, Xu Z-K, Wu J, Ye P, Liu Z-M and Seta P 2004 *J. Mol. Catal. B* **28** 95-100
[57] Liu Z-M, Xu Z-K, Wan L-S, Wu J and Ulbricht M 2005 *J. Membr. Sci.* **249** 21-31
[58] Ciszewski A, Gancarz I, Kunicki J and Bryjak M 2006 *Surf. Coat. Technol.* **201** 3676-84
[59] Zhang S, Cheng C, Lan Y and Meng Y 2009 *Plasma Sci. Technol.* **11** 576
[60] Xu Q, Yang J, Dai J, Yang Y, Chen X and Wang Y 2013 *J. Membr. Sci.* **448** 215-22
[61] Liu Z-M, Tingry S, Innocent C, Durand J, Xu Z-K and Seta P 2006 *Enzyme Microb. Technol.* **39** 868-76
[62] Shim J K, Na H S, Lee Y M, Huh H and Nho Y C 2001 *J. Membr. Sci.* **190** 215-26
[63] Borcherding H, Hicke H-G, Jorcke D and Ulbricht M 2002 *Desalination* **149** 297-302
[64] Yang Q, Xu Z-K, Dai Z-W, Wang J-L and Ulbricht M 2005 *Chem. Mater.* **17** 3050-8
[65] Hu M-X, Yang Q and Xu Z-K 2006 *J. Membr. Sci.* **285** 196-205
[66] Yu H-Y, Xu Z-K, Yang Q, Hu M-X and Wang S-Y 2006 *J. Membr. Sci.* **281** 658-65
[67] Yu H-Y, Xu Z-K, Lei H, Hu M-X and Yang Q 2007 *Sep. Purif. Technol.* **53** 119-25
[68] Gu J-S, Yu H-Y, Huang L, Tang Z-Q, Li W, Zhou J, Yan M-G and Wei X-W 2009 *J. Membr. Sci.* **326** 145-52
[69] Hou X, Zhang T and Cao A 2013 *Polym. Adv. Technol.* **24** 660-7
[70] Wan L-S, Yang Y-F, Tian J, Hu M-X and Xu Z-K 2009 *J. Membr. Sci.* **327** 174-81
[71] Yang Y-F, Li Y, Li Q-L, Wan L-S and Xu Z-K 2010 *J. Membr. Sci.* **362** 255-64
[72] Zhao Y-H, Wee K-H and Bai R 2010 *J. Membr. Sci.* **362** 326-33

[73] Zhao J, Shi Q, Luan S, Song L, Yang H, Shi H, Jin J, Li X, Yin J and Stagnaro P 2011 *J. Membr. Sci.* **369** 5-12
[74] Luo M-L, Zhao J-Q, Tang W and Pu C-S 2005 *Applied Surface Science* **249** 76-84
[75] Razmjou A, Mansouri J and Chen V 2011 *Journal of Membrane Science* **378** 73-84
[76] Wavhal D S and Fisher E R 2002 *J. Membr. Sci.* **209** 255-69
[77] Wavhal D S and Fisher E R 2003 *Langmuir* **19** 79-85
[78] Pieracci J, Crivello J V and Belfort G 1999 *Journal of Membrane Science* **156** 223-40
[79] Rahimpour A 2011 *Desalination* **265** 93-101
[80] Ulbricht M, Matuschewski H, Oechel A and Hicke H-G 1996 *J. Membr. Sci.* **115** 31-47
[81] Yamagishi H, Crivello J V and Belfort G 1995 *Journal of Membrane Science* **105** 237-47
[82] Zhang M, Nguyen Q T and Ping Z 2009 *J. Membr. Sci.* **327** 78-86
[83] Wolff J, Steinhauser H and Ellinghorst G 1988 *J. Membr. Sci.* **36** 207-14
[84] T. V, A. V, U. C, M. K and M. M 2007 *Surf. Interface Anal.* **39** 476-81
[85] Tang Z-Q, Li W, Zhou J, Yu H-Y, Huang L, Yan M-G, Gu J-S and Wei X-W 2009 *Sep. Purif. Technol.* **64** 332-6
[86] Kull K R, Steen M L and Fisher E R 2005 *J. Membr. Sci.* **246** 203-15
[87] Abednejad A S, Amoabediny G and Ghaee A 2014 *Mater. Sci. Eng. C* **42** 443-50
[88] Steen M L, Hymas L, Havey E D, Capps N E, Castner D G and Fisher E R 2001 *J. Membr. Sci.* **188** 97-114
[89] Steen M L, Jordan A C and Fisher E R 2002 *J. Membr. Sci.* **204** 341-57
[90] Ciszewski A, Kunicki J and Gancarz I 2007 *Electrochimica Acta* **52** 5207-12
[91] Choi E-Y, Strathmann H, Park J-M and Moon S-H 2006 *J. Membr. Sci.* **268** 165-74
[92] Yu H-Y, Liu L-Q, Tang Z-Q, Yan M-G, Gu J-S and Wei X-W 2008 *J. Membr. Sci.* **310** 409-17
[93] Prachar J, Novak I, Kleinova A, Gemeiner P, Omastova M, Chmela S and Borsig E 2016 *Vacuum* **127** 38-44
[94] Xu Z, Wang J, Shen L, Men D and Xu Y 2002 *J. Membr. Sci.* **196** 221-9
[95] Yuee F and Tianyi S 1988 *J. Membr. Sci.* **39** 1-9
[96] Kumari M, Gupta B and Ikram S 2012 *Radiat. Phys. Chem.* **81** 1729-35
[97] Ma H, Bowman C N and Davis R H 2000 *J. Membr. Sci.* **173** 191-200
[98] Mohd Yusof A H and Ulbricht M 2008 *J. Membr. Sci.* **311** 294-305
[99] Ulbricht M and Yang H 2005 *Chem. Mater.* **17** 2622-31
[100] Wang L, Wei J, Zhao K and Wu B 2015 *Mater. Letters* **159** 189-92
[101] Wang L-L, Wu J-J, Zhang Z-B, Zhou J, He X-C, Yu H-Y and Gu J-S 2016 *Sep. Purif. Technol.* **164** 81-8
[102] Yang Q, Tian J, Hu M-X and Xu Z-K 2007 *Langmuir* **23** 6684-90
[103] Yang Q, Hu M-X, Dai Z-W, Tian J and Xu Z-K 2006 *Langmuir* **22** 9345-9
[104] Kilduff J E, Mattaraj S, Pieracci J P and Belfort G 2000 *Desalination* **132** 133-42
[105] Pieracci J, Crivello J V and Belfort G 1999 *J. Membr. Sci.* **156** 223-40
[106] Taniguchi M, Pieracci J, Samsonoff W A and Belfort G 2003 *Chem. Mater.* **15** 3805-12
[107] Zhao X, Zhang J, Chen S, Wang X and Wang J 2013 *Colloid Polym. Sci.* **291** 1573-80
[108] Zhou Q, Lei X-P, Li J-H, Yan B-F and Zhang Q-Q 2014 *Desalination* **337** 6-15
[109] Zhu Y, Zhang F, Wang D, Pei X F, Zhang W and Jin J 2013 *Journal of Materials Chemistry A* **1** 5758-65
[110] Qian Y, Chi L, Zhou W, Yu Z, Zhang Z, Zhang Z and Jiang Z 2016 *Appl. Surf. Sci.* **360** 749-57
[111] Huang F, Wang Q, Wei Q, Gao W, Shou H and Jiang S 2010 *Express Polym. Letters* **4**

[112] Ju J, Wang T and Wang Q 2015 *Colloids Surf., A* **481** 151-7
[113] Qiu C, Xu F, Nguyen Q T and Ping Z 2005 *J. Membr. Sci.* **255** 107-15
[114] Zhang M, Nguyen Q T and Ping Z 2009 *J. Membr. Sci.* **327** 78-86
[115] Danu A and Anita Kusuma W 2017 *IOP Conf. Series: Mater Sci. Eng.* **214** 012014
[116] Guo H, Geng C, qin Z and Chen C 2013 *Desal. Water Treat.* **51** 3810-3
[117] Lee S-D, Hsiue G-H, Kao C-Y and Chang P C-T 1996 *Biomaterials* **17** 587-95
[118] Susanto H, Balakrishnan M and Ulbricht M 2007 *J. Membr. Sci.* **288** 157-67
[119] Hilal N, Al-Khatib L, Atkin B P, Kochkodan V and Potapchenko N 2003 *Desalination* **158** 65-72
[120] Hilal N, Kochkodan V, Al-Khatib L and Levadna T 2004 *Desalination* **167** 293-300
[121] Rahimpour A, Madaeni S S, Zereshki S and Mansourpanah Y 2009 *Appl. Surf. Sci.* **255** 7455-61
[122] Yu H, Cao Y, Kang G, Liu J, Li M and Yuan Q 2009 *J. Membr. Sci.* **342** 6-13
[123] Taniguchi M and Belfort G 2004 *J. Membr. Sci.* **231** 147-57

The Role Of Annealing On Biodegradation Behavior Of Mg-Zn-Ca Alloy Prepared From Commercial Mg Sacrificial Anode

Asep Ridwan Setiawan[1], **Arif Basuki**[2]*
[1]Materials Engineering Department, Faculty of Mechanical and Aerospace Engineering, Bandung Institute of Technology, Bandung 40132, Indonesia

*Corresponding author (E-mail): asep.ridwans@material.itb.ac.id

Abstract. Magnesium and its alloys offers great potential application for biodegradable implant due to its low corrosion resistance, high biocompatibility, favorable mechanical properties, and young modulus similar to bone. However, there are still many challenges to be solved. In this work, Mg-Zn-Ca alloy were prepared from commercial Mg sacrificial anode, Mg-Ca master alloy, and pure Zn (99%) through casting process. Melting and alloying operations were conducted in a graphite crucible, using a home made melting facility. The influence of annealing treatment on the alloys performance were studied. Microstructure observations of the alloys were conducted using optical and scanning electron microscope equipped with energy-dispersive X-ray spectroscopy. The microhardness of the alloys were evaluated using Brinell hardness indentation. In vitro biodegradation behaviors of Mg-Zn-Ca alloy were evaluated through immersion test in Ringer Laktat solution. The results show that Mg-Zn-Ca alloy fabricated in our work exhibit minimum corrosion rates. Annealing treatment at 300 °C (8 hour, air cooling) alter the microstructure of Mg-Zn-Ca alloy. The microhardness of Mg-Zn-Ca alloys slightly increased from 43 (as cast) to 53 BHN (as annealed). The corrosion rate of the alloys was increased from 8 mm/year (as cast) to 17 mm/year (as annealed). This is suspected due to grain refinement and extensive formation of needle structures causing anodic couples with the matrix.

Keywords: Biodegradable, Casting, Corrosion, Implant, Magnesium, Mg-Zn-Ca alloy.

1. Introduction

Indonesia is the world's fourth most populated country with 250 million populations [1]. By 2025, the elderly population is expected to grow by more than 40%, making Indonesia one of Asia's fastest aging countries (Fletcher, 2011). This will have a huge impact on an economical and societal level, especially in the healthcare sector. The concentrated population in main Indonesian cities has created chronic crowd traffics dominated by motorcyles that in turn contributes to high traffic accidents causing a high incidence of traumatic bone fractures [3]. As a results, the amount of patients requiring bone surgery in Indonesia is growing, and the need of implanted orthopedic devices for surgeries is huge. According to BMI Research 2017[4], the Indonesian medical device market is projected to rise by a compound annual growth rate (CAGR) of 9.6 %, which should bring it from USD 850 million in 2016 to USD 1,350 million in 2021. Around 1-2% of Indonesian medical devices market are shared by orthopedic/prosthetic implants. Based on those fact, research on orthopedic implant are very challenging in the future.

In bone grafting and healing of broken bones, permanent implant material are widely used, for examples stainless steels, titanium and its alloy [5]. If the bone is declared to have recovered, second removal surgery must be conducted again to remove the residual implant. This additional surgery can itself damage the bone and is an obvious inconvenience to the patient. Therefore it would be desirable to make an implanting a material that would degrade or be resorbed in the body.

The concept of non permanent implant materials departs from the phenomenon of bone grafting as follows: (i) In the early stages, bone connections should be reinforced by implantable material. (ii) Along with the healing time, the bone connection will be stronger so that the strengthening of the implant material can be reduced. (iii) In the final stages, bone connections are no more require reinforcement from the implants. This is only possible, when the implant material dissolve or be absorbed in the body, a characteristic which is called biodegradable [6]. Thus, the rate of bone strengthening and absorption rate of the implant material by the body is a parameter that must be determined precisely, research on the topic is widely open and challenging in the future [7].

Magnesium are excellent candidates for biodegradable implant materials. Magnesium is required in human body for metabolism and biological mechanism. The human body can absorb magnesium in relatively high amounts, 400 mg daily for men and 310 mg per day for women [8]. Magnesium based alloy, has been proposed as better biodegradable materials for load-bearing applications due to their superior combination of strength and ductility over polymers. However, concern about their rapid corrosion, corresponding hydrogen evolution, and subsequent loss of mechanical strength impedes the development of Mg alloys in clinical applications [9]. Therefore, maintaining the appropriate corrosion rate is very critical for biodegradable implants made from magnesium.

Corrosion behavior of magnesium in general depends upon its composition. Alloying magnesium with others element such as zinc, calcium, aluminium, manganese, lithium, silver, etc., enhances the corrosion resistance [10]. Zinc and Calcium have been alloyed with magnesium to increase the corrosion resistance, and have given good results [11]. However, further information is essential to optimize the biocompatibility of Mg-Zn-Ca alloy in body. Heat treatment process have been widely reported able to improve the corrosion resistance of various Mg alloys [12-13]. Heat treatment may homogenize and refine the microstructure typically present in magnesium alloys. Despite many studies have been done on mechanical and corrosion properties of Mg-Zn-Ca alloys, there are still problems regarding the optimal composition, alloy source/purity and processing method to achieve optimum mechanical and corrosion properties. In this works, for the first time, the fabrication of Mg-Zn-Ca alloys from commercial Mg sacrificial anode are demonstrated. The influence of heat treatment on corrosion resistance of Mg-Zn-Ca alloys in Ringer solution are reported.

2. Materials and Methods

Pure Mg and Mg-Zn-Ca alloy were prepared from commercial Mg anode, Mg-Ca master alloy, and pure Zn (99%). Melting and alloying operations were conducted in a graphite crucible, using a home made melting facility, as shown in Figure 1. A mixture of HFC 134a (5% volume ratio) and CO_2 gas (balance) was used as a protection gas to prevent Mg from oxidation/burning. The melt were kept for 10 minutes to ensure that all the requiring alloying elements were dissolved in the melt alloy, and then poured into a steels mould to make Mg or Mg-4%Zn-0.2 %Ca ingots. Chemical composition of Mg and Mg-4Zn-0.2Ca were analyzed using atomic absorption spectroscopy (AAS). Annealing heat treatment at 300 °C for 8 hour was carried out followed by air cooling.

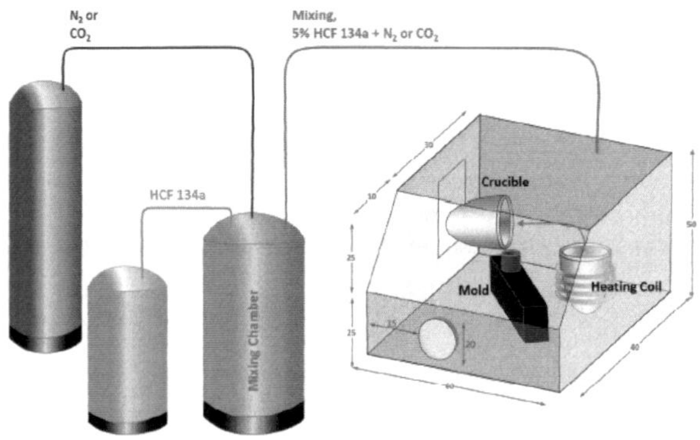

Figure 1. Inhouse Magnesium Melting Facilities developed in ITB

For microstructure observation, the alloys were ground with SiC abrasive papers with successive grades from 200 to 2000 and polished up to mirror finish with 0.5 micron diamond suspension. Polished Mg-Zn-Ca alloy were etched by picric solution (5 mL acetic acid, 10 mL water, 6 gr picric acid and 100 mL ethanol). As polished Mg alloy were etched by oxalic acid solution (2 gr oxalic acid, 100 mL water). After etching, the alloys were cleaned with distillated water and alcohol, and then dried by hot air. Microstructure observations of the alloys were conducted using Nikon optical microscope. Scanning electron microscope (SEM, Hitachi) equipped with energy-dispersive X-ray spectroscopy (EDX) were used to reveal the microstructure at higher magnification. EDX point analysis were conducted to determine the phase available within the microstructure. Phase identification also was carried out by X-ray diffractometry (XRD). The microhardness of the alloy were evaluated using Brinell hardness indentation. In vitro biodegradation behaviors of Mg and Mg-Zn-Ca alloy were evaluated through immersion test in Ringer Laktat solution. The immersion test procedure was carried out according to the ASTM: G1-03 [14].

3. Results and Discussion

Microstructural analysis

The chemical composition of as-cast Mg alloys obtained from the atomic absorption spectroscopy (AAS) were listed in Table 1. The amount of impurities element in Mg and Mg-Zn-Ca alloys were relatively low.

Table 1. Chemical composition of as cast Mg and Mg-Zn-Ca alloys

Alloy	Zn	Ca	Al	Mn	Fe	Ni	Cu	Si	Mg
Pure Mg	0.006	0.005	0.58	0.007	0.028	0.0001	0.002	0.04	Balance
Mg-Zn-Ca	2.65	0.05	0.02	0.008	0.007	0.000	0.001	0.02	Balance

IOP Conf. Series: Materials Science and Engineering **547** (2019) 012055 doi:10.1088/1757-899X/547/1/012055

For Mg-Zn-Ca alloys, the concentration of Zn and Ca element are relatively low compared to the initial targeted design. This might be due to evaporation during melting operation.

The optical microstructure of as-cast Mg and Mg-Zn-Ca alloys with different heat treatment are summarized in Figure 2. The as-cast Mg alloy consisted of primary α-Mg. On the other hands, the as-cast Mg-4Zn-0.2Ca alloy consisted of primary α-Mg and secondary phases with needle-like appearance. After annealing heat treatment, the microstructures of Mg-4Zn-0.2Ca alloys were slightly changes. The grain size was smaller compared to that as-cast specimen. Moreover, extensive formation of needle like secondary phase were found to increases at the annealed Mg-Zn-Ca alloys microstructure. Figure 3 shows SEM micrographs of as-cast Mg, as-cast Mg-Zn-Ca and as-annealed Mg-Zn-Ca alloy. The corresponding EDS analysis suggested that the Mg alloys consist only of α-Mg solid solution phase.

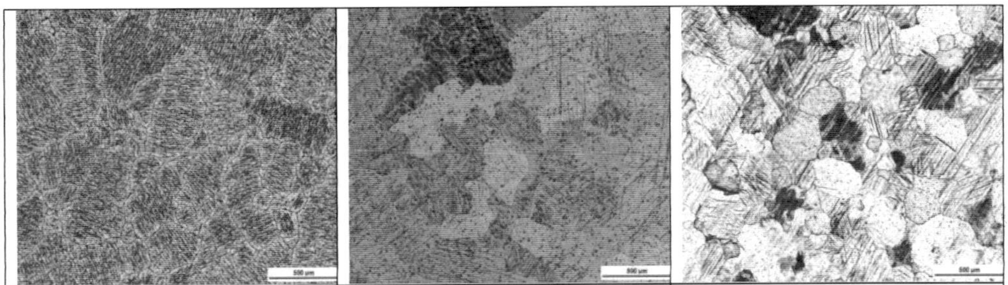

a.Mg, as cast	b. Mg-Zn-Ca, as cast	c. Mg-Zn-Ca, as annealed

Figure 2. Optical microstructure image of (a) Mg as-cast, (b) Mg-4Zn-0.2Ca, as-cast (c) Mg-4Zn-0.2Ca, as annealed.

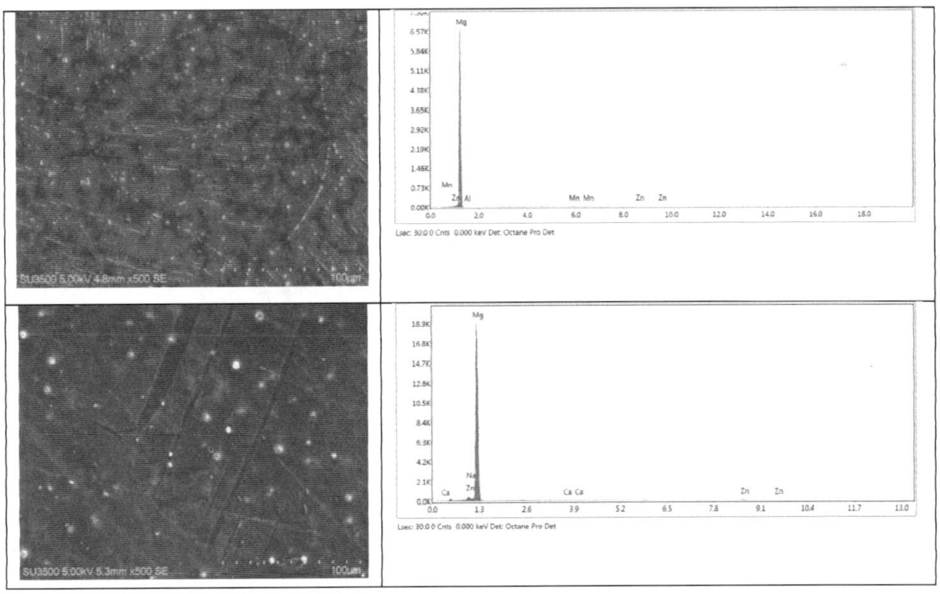

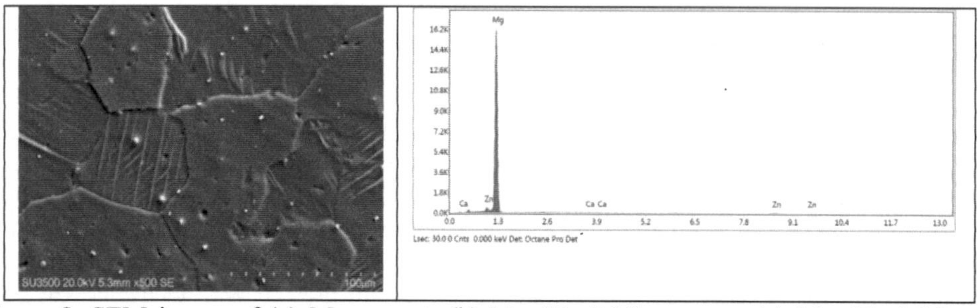

Figure 3. SEM image of (a) Mg as-cast, (b) Mg-4Zn-0.2Ca, as-cast (c) Mg-4Zn-0.2Ca, as annealed

The XRD pattern of the as-cast Mg alloy, as-cast Mg-Zn-Ca alloy and as annealed Mg-Zn-Ca alloys is shown in Figure 4. The XRD patterns indicates that only peak corresponding to α-Mg matrix have been observed in Mg and Mg-Zn-Ca alloys. The formation of second phase particles in Mg-Zn-Ca alloys were not observed both for as cast and as annealed condition. Compared to that of pure Mg alloys, the intensity of (101) peak increases, while the intensity of (002) peak decreases in Mg-Zn-Ca alloys, both for as cast and annealed condition. This indicates that the addition of Zn and Ca element into Mg matrix might influence the preferred crystal growth orientation. Annealing treatment conducted on Mg-Zn-Ca alloys do not change the phase constituent in the alloys.

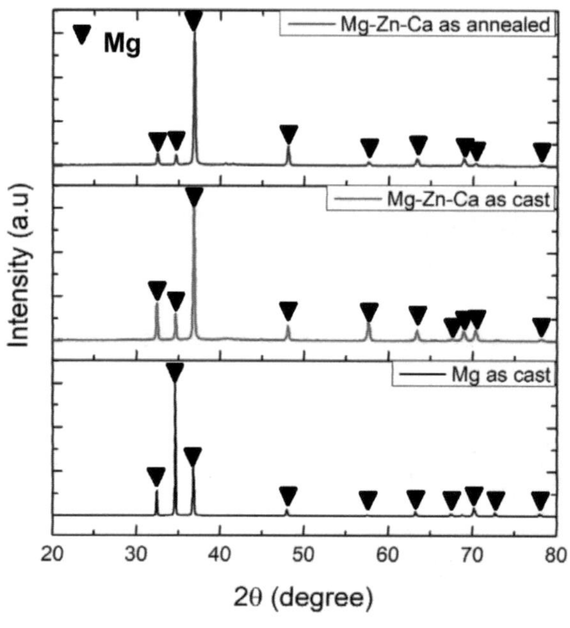

Figure 4. X-ray diffraction patterns of Mg and Mg-Zn-Ca alloy at different heat treatment condition.

Hardness test results

Figure 5 shows the average hardness of Mg alloys after casting and annealing. There is not a large difference in final hardness between different alloys. The hardness of Mg in as cast condition is 38 BHN. The hardness of Mg-Zn-Ca alloy in as cast condition is 43 BHN, while that for annealed condition is 53 BHN. The addition of Zn and Ca in Mg-Zn-Ca alloys does not change the hardness of Mg alloys, as both element was diluted in α-Mg solid solution. Annealing treatment conducted on Mg-Zn-Ca alloys slightly modify the final hardness. This is related to the extensive formation of needle like structures in the alloys and grain size refinement. This indicates that the effect of Zn and Ca addition is more significant on the recrystallization behavior and final grain size after annealing.

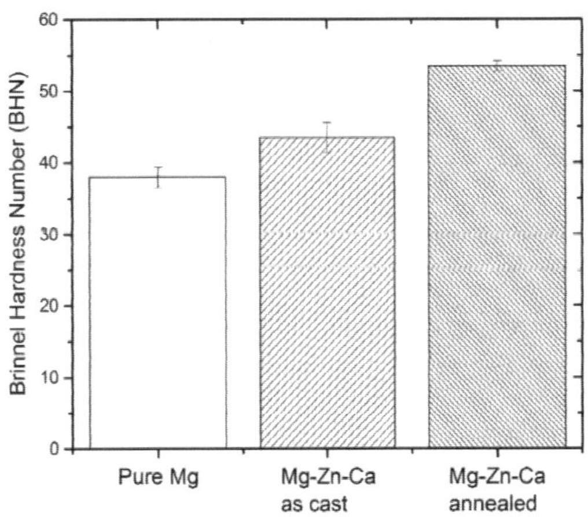

Figure 5. Hardness of Mg alloys with various condition

Immersion test results

Figure 6 shows the corrosion rate obtained from weight loss of Mg, Mg-Zn-Ca as cast and annealed Mg-Zn-Ca alloys, after immersion in ringer solution for 168 hour. The average corrosion rate for Mg, Mg-Zn-Ca and annealed Mg-Zn-Ca alloys were 65.9, 8.33 and 17.59 mm/year, respectively. For comparison, the average corrosion rate of biodegradable Mg-4Zn-0.2Ca alloys (as-cast) in Hank's solution are 2.05 mm/years [11]. The addition of Zn and Ca element into Mg matrix improves the corrosion resistance of the alloys significantly. After annealing, the corrosion rates of Mg-Zn-Ca alloys slightly increases compared to that as cast condition. This was related to the extensive formation of needle like structures after annealing and grain size reduction. Needle like structures in Mg matrix may act as a micro-cathode, resulted in the formation of galvanic couples with the Mg matrix. The increases of needle like structures in Mg-Zn-Ca alloys may increases the formation of micro-cathode, which finally increase the galvanic corrosion rates in the alloys. Furthermore, the reduction in grain size after heat treatment on Mg-Zn-Ca alloys increases the corrosion rate of the alloys. The finer the grains, the larger the effective grain boundaries which lead to the higher risk of corrosion. Grain boundaries act as a weak or critical section for various type of corrosions, mainly intergranular type.

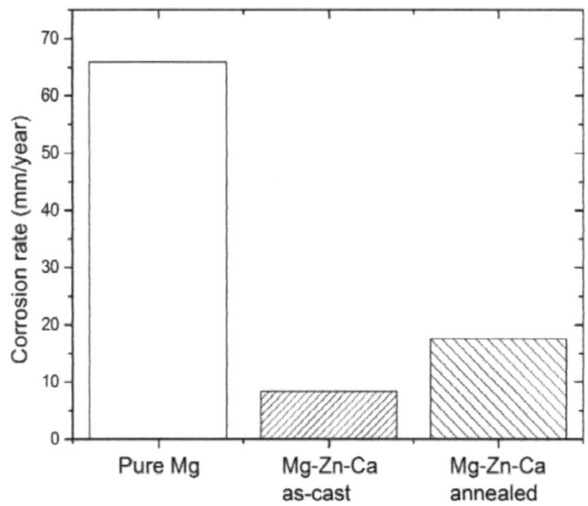

Figure 6. Corrosion rate of Mg alloys obtained from weight loss of Mg, Mg-Zn-Ca as cast and annealed Mg-Zn-Ca alloys, after immersion in ringer solution for 168 hour

4. Conclusions

This study shows that the performance of Mg-Zn-Ca alloys prepared from the commercial Mg anode is promising for implant application in the future. Our result shows that Mg-Zn-Ca alloy fabricated in our work exhibit minimum corrosion rates in ringer solution. Annealing treatment at 300 ^{o}C (8 hour, air cooling) alter the microstructure of Mg-Zn-Ca alloy, by introducing grain size reduction and high density of needle like structures. The microhardness of Mg-Zn-Ca alloys slightly increased from 43 (as cast) to 53 BHN (as annealed). The corrosion rate of the Mg-Zn-Ca alloys was increased from 8 mm/year (as cast) to 17 mm/year (as annealed). This is suspected due to grain refinement and extensive formation of needle structures causing anodic couples with the matrix.

5. Acknowledgements

The Indonesian Ministry of Research, Technology and Higher education for providing research fund through PUPT 2018 programs are gratefully acknowledged.

6. References

1. United Nations 2017 *World Population Prospects 2017 – Data Booklet* (ST/ESA/SER.A/401)
2. Fletcher R 2011 *Challenging opportunity for aging population*, (the jakarta post).
3. Santosa SP, Mahyuddin AI and Sunoto FG 2017 *J. Eng. Technol. Sci.* **49**(3).
4. Fitchsolutions 2017 *Executive summary of Indonesia Medical Devices Report*
5. Staiger MP, Pietak AM, Huadmai J and Dias G 2006 *Biomaterials*. 2006 **27**(9) 1728.
6. Witte F 2010 *Acta Biomater.* **6**(5) 1680.
7. Kirkland NT 2012 *Corr. Eng. Sci. Tech.* **47**(5) 322.
8. Yates AA, Schlicker SA and Suitor CW 1998 *J. Am. Diet. Assoc.* **98**(6) 699.
9. Zberg B, Uggowitzer PJ and Löffler JF 2009 *Nature Materials.* **8** 887.
10. Song GL and Atrens A 2000 *Adv. Eng. Mat.* **1**(1) 11.
11. B.P. Zhang 2011 *Research on Mg-Zn-Ca Alloy as Degradable Biomaterial.*, ed Y. Wang (IntechOpen) chapter. 9

12. Zhou W, Shen T and Aung NN 2010 *Corr. Sci.* **52**(3):1035
13. Zhang X, Zhang K, Li X, Deng X, Li Y and Ma M 2012 *J. Rare Earths.* **30**(11) 1158
14. ASTM G1-03 2017 *Standard Practice for Preparing, Cleaning, and Evaluating Corrosion Test Specimens* (ASTM International)

IOP Conf. Series: Materials Science and Engineering **547** (2019) 012056 doi:10.1088/1757-899X/547/1/012056

Effect Of Silicon On Corrosion Behaviour Of Al-Zn As A Low Voltage Sacrificial Anode For Marine Environment

Yudha Pratesa*[1], Ahmad A. Utama[1], Ardi Erianto[1], and Deni Ferdian[1]

[1]Department of Metallurgy and Materials, Faculty of Engineering, Universitas Indonesia, 16424, Indonesia

*Corresponding author (E-mail): yudha@metal.ui.ac.id

Abstract. A sacrificial anode is a method for material protection in the sea and coastal area. This method is widely applied in marine, subsea and ship structure. Al-Zn-In, a commercial grade of the sacrificial anode had a weakness in the possibility of stress corrosion cracking due to overprotection. This study tried to substitute indium with silicon as the alloying element with composition 0.5% and 1% wt. However, commercial grade Al ingot contained trace elements such as iron in the significant amount, which could make a β-$Al_9Fe_2Si_2$ as an iron-rich intermetallic phase. Therefore, this study will also examine the effect of β-$Al_9Fe_2Si_2$ on alloys corrosion behavior. The alloy was produced through casting process with resistance furnace. Corrosion behavior was characterized by cyclic polarization in 3.5% NaCl solution. Microstructure characterization was performed using Scanning Electron Microscope, Energy Dispersive Spectroscopy, and optical metallography. The result showed that the silicon increases the corrosion rate of Aluminum Zinc (Al-Zn) and reduced the potential protection of the alloy in high strength steel. Our research showed the corrosion took place along the interdendritic area where the β-$Al_9Fe_2Si_2$ intermetallic phase formed in a micro galvanic manner

Keywords: Aluminum, Marine Corrosion, Low Voltage, Sacrificial Anode, Ship Maintenance

1. Introduction

Various ways are already applied to inhibit corrosion attacks, such as inhibitors, coating, and application of cathodic protection. Cathodic protection is considered a cheaper option than the application of corrosion-resistant alloys such as stainless steel or a nickel-based alloy. Furthermore, there are two methods of cathodic protection, impressed current and sacrificial anode. The latter methods will be discussed further in this research.

Nowadays, the sacrificial anode Al-Zn-In has been used extensively as an aluminum sacrificial anode. Salinas et al. found interdendritic zona is a susceptible area to be attacked by corrosion. [1] Moreover, this phenomena will produce self-corrosion that was the principal cause of efficiency loss. Generally, indium (In) element is used to prevent the passivation of Aluminum. However, in some cases have shown that Aluminum – Zinc – Indium (Al-Zn-In) sacrificial anode could trigger hydrogen induced cracking in high-strength steel.[2,3] Apparently, it happened due to overprotection which made the steel potential reaches the hydrogen evolution potential area.

The sacrificial aluminum anode can be referred as "low voltage" if protection potential is between -0.80 V to -0.85 V vs. SCE.[4] Hence, a sacrificial anode with lower voltage potential was developed with the additions of silicon or copper. Some other alloying element has been proposed for replacement of indium such as mercury and tin. [5,6] Nevertheless, the problem with mercury is an environmental issue while Tin did not give sufficient efficiency for the sacrificial anode. Pautasso found Al-0.1% gallium alloy could be used as the low voltage sacrificial anode.[7]

Other researchers stated that the addition of bismuth could improve the efficiency of sacrificial anode alloys Al-5Zn-0.25Sn. [8] Bismuth acted by widening the aluminum (Al) matrix and make tin (Sn) dissolved into the aluminum matrix. Murai et al. developed a new alloy using bismuth ranging between 0.005-1.0% with improved efficiency between 65 and 71%.[9] Furthermore, the efficiency has been increased more than 83% by the addition of 2% Mg. Muazu et al. found protection potential of carbon steel is less than or equal to -0.5 V SHE after a couple with the low voltage sacrificial anode. It is sufficient to change the steel behavior from active to immune region based on pourbaix diagram. [10]

Effect of Silicon on precipitate will vary depending on the bonded metal in the precipitate. Zeng Feng et al. found the preferential dissolution of metal in precipitate Mg_2Si. The first step was Mg dissolution. After a significant amount of Mg dissolve, Mg_2Si potential changed to a positive direction.[11]

Schnatterer et al. found another effect of silicon in an aluminum alloy. They found precipitate silicon formation will affect the adjacent area of precipitate due to the reduction of silicon content in this area. [12]

Most of the research is used more than one alloy to increase the efficiency of sacrificial anode. However, they are considered more expensive compared with Al-Zn-In. Nevertheless, previous work by Pratesa et al. used copper to substitute indium and found a significant effect in the reduction of Open Circuit Potential (OCP) the aluminum alloys. [13,14] This study used silicon as a substitution for indium in the aluminum sacrificial anode. Therefore, this alloy could be more easy to produce and cheaper than the other proposed alloying element

2. Materials and Methods

Fabrication of sacrificial anode was performed using casting methods with resistance melting furnace. Commercially high purity aluminum ingot (99.98%), Zinc ingot (99.7%) and Silicon (99.5%) were mixed to achieve the target composition (Al-5Zn, Al-5Zn-0.5Si, and Al-5Zn-1Si). Melting process was conducted at temperature 800°C without any degassing treatment in a graphite crucible. The casting product was characterized by the chemical composition test using Optical emission spectroscopy (OES).

Corrosion behavior of the alloy was analyzed using cyclic potentiodynamic polarization in 3.5% NaCl solution with scanning rate 0,01 V/s at room temperature. Three electrode systems were used for cyclic potentiodynamic polarization with graphite as a counter electrode and Ag/AgCl as a reference electrode. The reversed of scan direction happened when the electrode potential reached the anodic potential of 0 V. Polarization testing was conducted to determine the value of pitting potential (Epitt), re-passivation potential (Eprot) and corrosion potential (Ecorr). Electrochemical impedance spectroscopy (EIS) was conducted in 3.5% NaCl water with PGSTAT302N and Nova 1.8 for the analysis. It used 100kHz – 10mHz and amplitude 10mV after preconditioned in the same solution for 30 minutes.

Immersion test was conducted for three days for corrosion morphology examination. Protection potential of low carbon steel pipe was measured using galvanic coupling methods and measured using saturated calomel electrode (SCE) reference electrode. After the immersion test, the sample was viewed under a microscope directly without the etching process. For phase identification, Scanning Electron Microscope (SEM), Energy Dispersive Spectroscopy (EDS), and Differential scanning calorimetry (DSC) with scanning rate 10°K/min were done.

3. Results and Discussion

3.1 Chemical Composition

The chemical composition of the alloy was shown in Table 1. The result was achieved with some minor deviation from the target in zinc composition. This condition is a result of the melting temperature differences between zinc and aluminum, which make zinc easier to fade out during the casting process.

Table 1 **Chemical composition of product**[14]

	Element (%)				
Alloy	**Al**	**Si**	**Ni**	**Zn**	**Fe**
Al-5%Zn	94.9	0.120	<0.005	4.90	0.04
Al-5%Zn-0.5%Si	94.8	0.597	0.006	4.60	0.07
Al-5%Zn-1%Si	93.6	1.2	0.03	4.85	0.035

3.2 Cyclic polarization

Cyclic polarization test was conducted to analyze the pitting behavior of the alloys such as breakdown potential (Epit), repassivation potential (Erep) and corrosion potential (Ecorr). Breakdown potential (Epit) of the alloy was reduced as the silicon content increases in the alloy. This condition could be a contribution of the intermetallic phase that formed in the interdendritic region. However, the corrosion potential (Ecorr) escalated with the percentage of silicon added to the alloy. The reason is due to the silicon that has higher potential than aluminum and some of the silicon atoms dissolved in an aluminum matrix. Aluminum has a maximum solubility limit for silicon at 1.25% atomic percent, while this alloy contains only 1% of silicon. It means silicon dissolved in the matrix α Al and caused increasing of corrosion potential of aluminum.

Figure 1 showed the effect of silicon in Al-5Zn alloy. Al-Zn-Si corrosion potential is -0.9 mV while Al-Zn-In corrosion potential is about -1.3 mV[15]. It is an indication that Al-Zn-Si is a potential candidate for the low voltage sacrificial anode. Size of the hysteresis loop (gap between Epit and Erep) reduced after the addition of silicon. The larger hysterical loop will vulnerable to the pitting corrosion. This is caused by harder reformation of the passive layer in the alloys. The largest area of the loop is Al-5Zn, and it reduced after the addition of silicon. Silicon has improved the potential repassivation (ERep) of the alloy and reduced the size of the hysterical loop.

This result is different with the condition on Al-Mn/Al-Mg alloy. Addition of Mn or Mg to the pure aluminum alloy with a composition between 0.5% wt and 1% wt did not cause any difference on pitting behavior.[16] Silicon acted differently compared with those metals. It could be an effect of the intermetallic phase that more cathodic behavior than Mn or Mg precipitate.

Based on figure 1 a and b, the Ecorr is always lower than the E repp. It showed that all of the alloys would suffer by pitting corrosion mechanism and no crevice corrosion will form.

3.3 Galvanic Coupling

Galvanic coupling was measured between Al-Zn-Si and low carbon steel pipe. The coupled potential of Al-5Zn is about -0.901 V, however, the addition of 1% silicon can increase the steel potential to -0.860 V vs. SCE in Al-5Zn. Figure 2 showed the result of galvanic coupling between Al-Zn-Si alloy with the other study for the comparison. Al-Zn-Si gave the lowest potential compared with Al-Zn-In and existing "low voltage sacrificial anode (Al-Ga) "[5]. Based on this result, Al-Zn-Si alloy is a promising alloy for the low voltage sacrificial anode

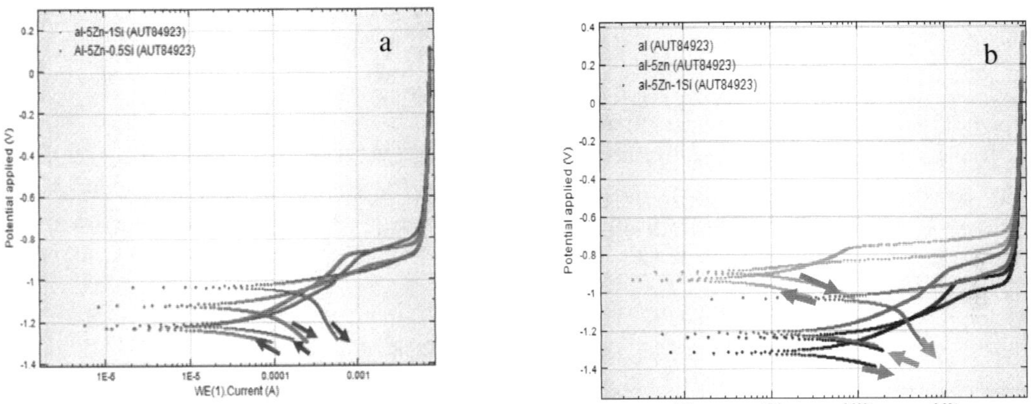

Figure 1 (a) Effect of Silicon in Al-5Zn (b) Cyclic polarization after addition of Zn and Silicon

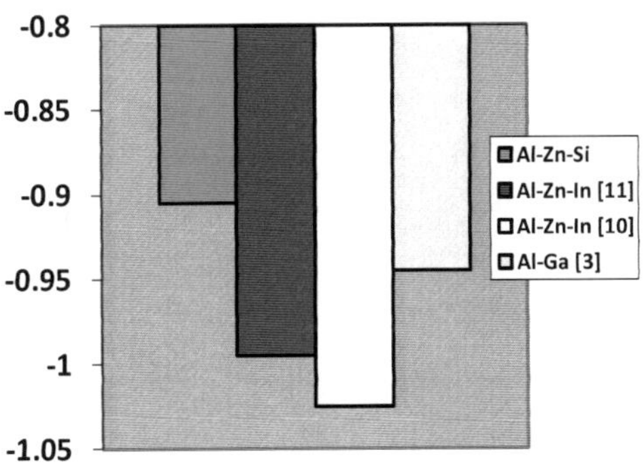

Figure 2. Carbon Steel potential protection with different aluminum anode alloy [5,15]

3.4 Phase Analysis

Phase analysis was conducted using metallography method and EDS analysis. Figure 3 showed the presence of intermetallic and eutectic silicon formed at the interdendritic area. The intermetallic phase had a needle-like shape which is typical for the Al-Fe-Si intermetallic compound. Iron (Fe) came from the impurities in aluminum and silicon ingot. The addition of silicon reduced dendrite arm spacing of aluminum, which contributes to the escalating of the intermetallic formation Energy dispersive spectroscopy (EDS) analysis confirm the formation of an iron-rich intermetallic precipitate as shown in figure 4. The alloy had a composition ratio of Fe/Si which less than 2, which tend to form as a β-$Al_9Fe_2Si_2$ phase.[17]

IOP Conf. Series: Materials Science and Engineering **547** (2019) 012056 doi:10.1088/1757-899X/547/1/012056

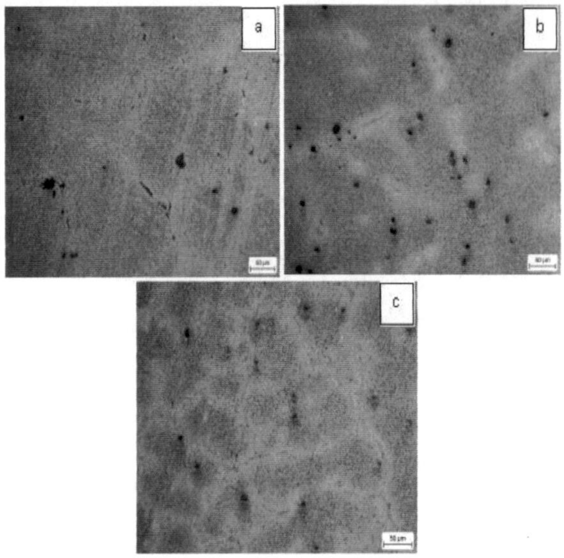

Figure 3 Microstructure of (a) Al5Zn, (b) Al5Zn0.5Si and (c) Al5Zn1Si

EDS also showed that zinc composition is higher in the area around the intermetallic precipitate (table 2). This condition is a result of low zinc solubility in the intermetallic precipitate.

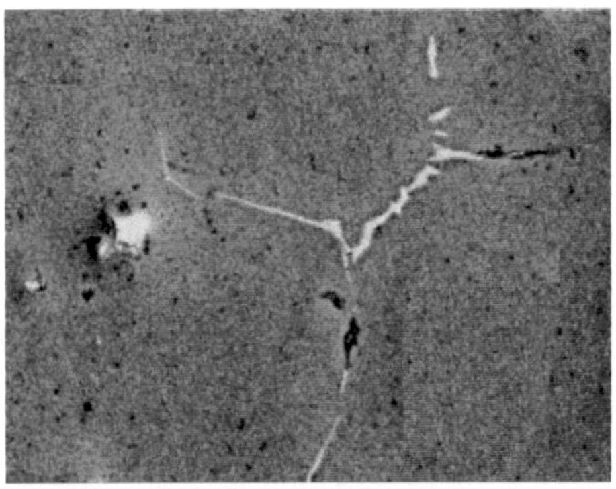

Figure 4 location of EDS testing

Table 2 Composition of each point EDS analysis

Location	Element	Wt%	Possibility of phase
1	Al	61.7	β-$Al_9Fe_2Si_2$
	Zn	4.41	
	Si	8.56	
	Fe	12.46	
2	Al	80.2	α-Al
	Zn	0.69	
	Si	6.48	
	Fe	0.05	
3	Al	83.7	α-Al
	Zn	2.68	
	Si	7.54	
	Fe	2.95	

Differential scanning calorimetry showed the formation of each phase by their solidification temperature. Table 3 showed AlFeSi was formed at a temperature around 560°C. [17] The first solidification was α-Al matrix phase at temperature 640 °C.

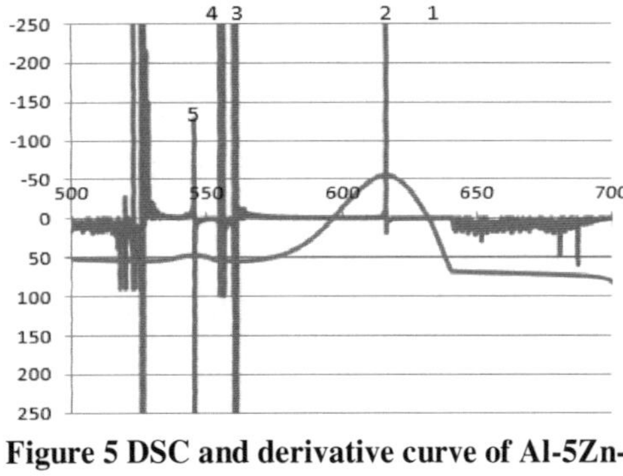

Figure 5 DSC and derivative curve of Al-5Zn-1Si

Table 3 Phase formation during solidification on Al-5Zn-1Si

No	Fasa	Suhu (°C)
1	Start α-Al	640
2	Peak α-Al	616
3	β-AlFeSi	560
4	Start Eutektik Si	555
5	Peak Eutektik Si	545

3.5 Electrochemical impedance spectroscopy (EIS)

EIS was conducted for the surface electrochemical behavior of the alloy. The results were analyzed by Randall's circuit. Figure 6 showed the Nyquist plot that silicon will make a semi-circular diagram. Semi-circular Nyquist diagram showed silicon made double layer phenomena in the material surface which also indication of passive layer formation. The corrosion process is charge transfer controlled. Pure aluminum has the largest loop. However, the addition of another alloying element, i.e zinc and silicon, reduce the size of the loop. This reduction indicate the charge transfer resistance reduce, which also indicated silicon and zinc reduce the uniformity of the passive layer of aluminum as showed in figure 6b. Table 4 shows the result of EIS analysis with simple randals cell.

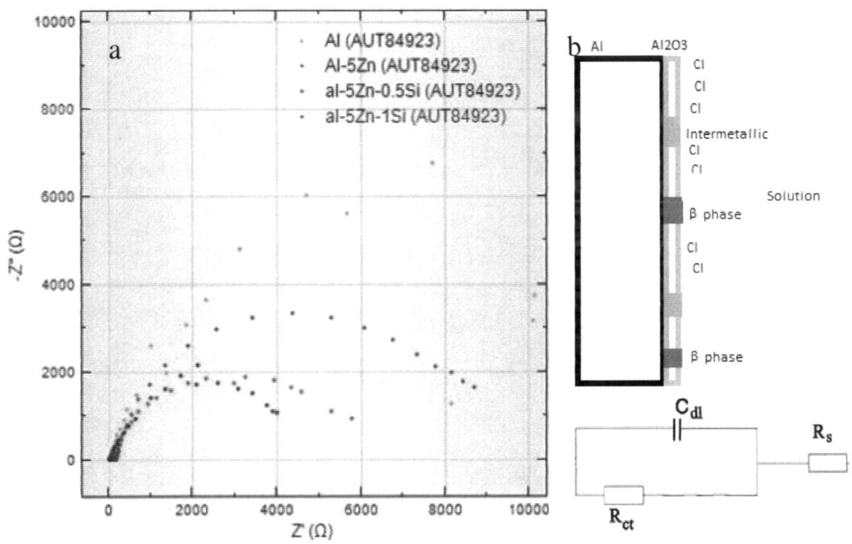

Figure 6 Nyquist diagram from Al, AlZn, and AlZnSi (a) and the electrochemical circuit of each alloy (b)

3.6 Immersion test

Immersion test was conducted to analyze the effect of each phase in this alloy in 3.5% NaCl solution. For 48 hours, Bruzzone et al. found the effect of the intermetallic phase to the sacrificial anode efficiency. [18] They found self-corrosion at any phases in aluminum alloy would be the reason for the sacrificial anode efficiency reduction. Based on figure 7, the intermetallic and matrix creating an electrochemical cell. The intermetallic phase β-$Al_9Fe_2Si_2$ acted as an anode while the matrix acted as a cathode. The corrosion that attacked the intermetallic could be caused a breakdown on a passive layer of aluminum by chloride ion

Table 4 EIS result form the EIS test

Alloy	Rct	N	CPE
Al	13,2kΩ	0,798	19,4
Al-5Zn	8,72kΩ	0,77	64,8
Al-5Zn-0.5Si	5,81kΩ	0,736	107
Al-5Zn-1Si	4,54kΩ	0,827	44,8

This study showed different result compare with the previous study by Yoshikawa et al. that stated intermetallic acted as an active cathode while the matrix is the anode.[19] The reason for this condition is due to zinc percentage on the intermetallic phase that higher compared with the matrix as shown in table 2. Higher zinc composition contributes to reducing the corrosion potential of the intermetallic phase, hence acted as the anode.

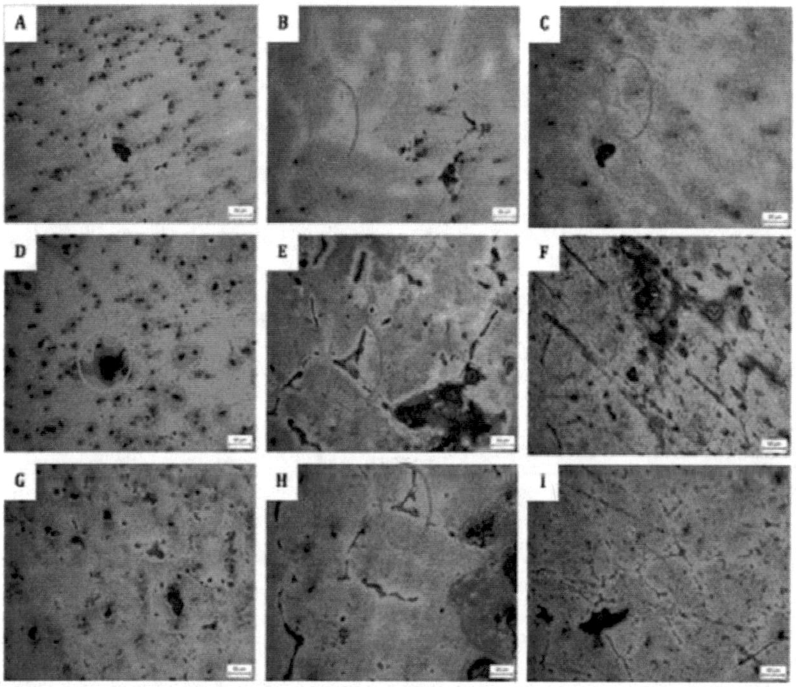

Figure 7 Pit formation before immersion (a,b,c), after immersion without polishing (d,e,f) and after polishing (g,h,i). Blue circle showed porosity condition before and after immersion while red circle showed corrosion propagated from interdendritic phase

4. Conclusions

Silicon in Al-5Zn alloy will form a eutectic Si phase and in the presence of iron as an impurity in commercial aluminum ingot will create intermetallic β-$Al_9Fe_2Si_2$ phase. β-$Al_9Fe_2Si_2$ phase and eutectic Si make a passive layer that did not distribute uniformly in aluminum surface. Silicon improved the Potential Repassivation (E_{rep}) of the alloy and reduced the size of the hysterical loop.

Immersion test showed that β-$Al_9Fe_2Si_2$ phase acted as anode area while the aluminum matrix served as cathode area. The amount of dissolved zinc changed the intermetallic β-$Al_9Fe_2Si_2$ phase behavior to be more anodic than the matrix. The result showed that Al-Zn-Si alloy increased coupling potential of low carbon steel and suitable for low voltage sacrificial anode candidate.

5. Acknowledgments

The authors are grateful to PT INALUM Indonesia for providing the aluminum ingot and DRPM-UI through PITTA grant 2017 for the financial support.

6. References

1. Salinas, D. R., Bessone, J. B. & Garcia, S. G. Influence of alloying elements and microstructure on aluminum sacrificial anode. *J. Appl. Electrochem.* **29,** 1063–1071 (1999).
2. Ma, L., Li, K., Yan, Y. & Hou, B. Low Driving Voltage Aluminum Alloy Anode for Cathodic Protection of High Strength Steel. *Multi-Functional Mater. Struct. Ii, Pts 1 2* **79–82,** 1047–1050 (2009).
3. Batt, C., Dodson, J. & Robinson, M. J. Hydrogen embrittlement of cathodically protected high strength steel in sea water and seabed sediment. *Br. Corros. J.* **37,** 194–198 (2002).
4. Monzel, William (Jacob); Druschitz, Alan P.; Maxfield, M. Development of New, Low-Voltage, Aluminum, Sacrificial Anode Chemistries. in *Corrosion* 4284 (2014).
5. Lemieux, E. J., Hogan, E. A., Lucas, K. E. & Grolleau, A. M. Performance Evaluation of Low Voltage Anodes for Cathodic Protection. *CORROSION 2002* 11 (2002).
6. Ma L, Li K, Yan Y, et al. Development of Low Driving Voltage Sacrificial Anode. in *NACE East Asia & Pacific Rim Area Conference & Expo 2012* Paper No.12060 (NACE International, 2012).
7. Pautasso, J.-P., Le Guyader, H. & Debout, V. Low Voltage Cathodic Protection for High Strength Steels: Part 1- Definition of a New Aluminum Galvanic Anode Material. *CORROSION 98* 27 (1998).
8. Gurrappa, I. Cathodic protection of cooling water systems and selection of appropriate materials. *J. Mater. Process. Technol.* **166,** 256–267 (2005).
9. Murai Tosuke, Miura Shutoshi; Tamura Yuichi. Study on aluminum galvanic anode (1st report) page 191–195 (1974).
10. Muazu, A. & Yaro, S. . Effects of Zinc Addition on the Performance of Aluminium as Sacrificial Anode in Seawater. *J. Miner. Mater. Charact. Eng.* **10,** 185–198 (2011).
11. Zeng, F. L. *et al.* Corrosion mechanism associated with Mg2Si and Si particles in Al-Mg-Si alloys. *Trans. Nonferrous Met. Soc. China (English Ed.* **21,** 2559–2567 (2011).
12. Schnatterer, C. & Zander, D. Influence of the grain boundary chemistry on the intergranular corrosion mechanisms of a high-strength Al-Mg-Si alloy. *Surf. Interface Anal.* **48,** 750–754 (2016).
13. Ferdian, D., Pratesa, Y., Togina, I. & Adelia, I. Development of Al-Zn-Cu Alloy for Low Voltage Aluminum Sacrificial Anode. *Procedia Eng.* **184,** 418–422 (2017).
14. Pratesa, Y., Ferdian, D. & Togina, I. Effect of Copper and Silicon on Al-5%Zn Alloy as a Candidate Low Voltage Sacrificial Anode. in *IOP Conference Series: Materials Science and Engineering* **204,** (2017).
15. Muñoz, A. G., Saidman, S. B. & Bessone, J. B. Corrosion of an Al–Zn–In alloy in chloride media. *Corros. Sci.* **44,** 2171–2182 (2002).
16. Aziz, P. M. & Godard, H. P. Pitting Corrosion Characteristics of Aluminum - Influence of Magnesium and Manganese. *Ind. Eng. Chem.* **44,** 1791–1795 (1952).
17. Ferdian, D. Effet de la vitesse de refroidissement sur la taille des grains, la modification eutectique et la précipitation d'intermétalliques riches en fer dans des alliages Al-Si hypoeutectiques. (2014).
18. Barbucci, A., Cerisola, G., Bruzzone, G. & Saccone, A. Activation of aluminium anodes by the presence of intermetallic compounds. *Electrochim. Acta* **42,** 2369–2380 (1997).

19. Yoshikawa, D. S., Terada, M., Assis, S. L., Costa, I. & Padilha, A. F. Correlation between microstructure and corrosion behavior of two Al–Fe–Si alloys. *Mater. Corros.* **67,** 286–296 (2015).

Effect of 5-15% Carbon Black Reinforcement on Mechanical Properties and Resistivity of Polyurethane Composites

Dwi Rahmalina[1]*, Jefri Irwanto, Hendri Sukma, Susanto, Indra C. Setiawan, Amin Suhadi
[1]Department of Mechanical Engineering, Faculty of Engineering, Universitas Pancasila
[2]National Laboratory for Structural Strength Technology, BPPT

*Corresponding author (E-mail): drahmalina@univpancasila.ac.id

Abstract. This research investigated the effects of carbon black-reinforced thermoplastic polyurethane matrix composite with variation of 5, 10, and 15 vol.% on mechanical properties and electrical resistance. The composites were produced using injection molding process with temperature of 193 °C, pressure of 30 bar and injection speed of 176 mm/sec. Characterization of composites were conducted by hardness and tensile testing, electrical resistivity measurement and SEM examinations. The results showed that the highest tensile strength of composite of 20.44 MPa is obtained with carbon black 10 vol.% reinforcement. The hardness improved with the increase of carbon black reinforcement, which the highest hardness was produced on carbon black 15 vol.% composite by 89.5 Shore A. The lowest resistivity was achieved in polyurethane composite reinforced with 15 vol.% carbon black of 8.20×10^5 Ω.

Keywords: Polyurethane composite, carbon black, mechanical properties, resistivity.

1. Introduction

Polymer composites are improved nowadays for many component, for instance as castor wheel of hospital bed. The polyurethane materials have insulated properties with resistivity values of 1.8×10^{12} Ω, which must be reduced to avoid electrostatic discharge (ESD) especially when the wheel moved on the carpets [1]. ESD could be very dangerous because it can harm electronic components related to human health [2]. Several studies have been conducted to decrease resistivity of the polymer materials, especially adding conductive reinforcement as carbon basis materials. Reinforcing 25 wt.% of carbon basis material on polymer improved antistatic properties, whereas decrease mechanical strength and process ability [3,4]. Islam, et al. [5] studied role of 30 wt.% carbon black reinforcement on PVC composite which resulting resistivity value of 9.92×10^3 Ω. Carbon black is one of the commercially produced Nano-particles which have three main characteristics of particle size, structure and chemical surface. Carbon black basically has element of carbon in the form of very fine particles that improve mechanical properties [6-8]. Carbon-filled fillers have recently emerged as one of the promising substitute materials because of their high aspect ratio, low cost, easy production of low resistance values, low density and high conductive values [9-10]. Ibrahim et al. [11] conducted a research by adding carbon black to the polyurethane matrix to improve mechanical properties. Therefore, this research studied carbon black-reinforced polyurethane composite, which is expected to decrease electrical resistance value and improve mechanical properties.

2. Materials and Methods

The matrix materials used in this study are thermoplastic polyurethanes; consisting of Isothane and Desmopan with a mixture ratio of 2:1. The reinforcement material is carbon black with grade of N330, with variation of 5,10 and 15 vol.%. The composites were produced using injection-molding process, with temperature of 193 C, pressure of 30 bar and injection speed of 176 mm/sec. The instruments used in this research are a 215 kg mixing machine; a 16.6 kW typical hopper-dryer machine; and injection molding machine with capacity of 150 Ton. Composites characterization were conducted by hardness testing using shore durometer A based on standard of ASTM D2240-15[12]; tensile testing according to ASTM D 638-14 [13] with force of 50 kN; SEM observation refer to ASTM E 986-97[14] to determine surface morphology. Electrical resistivity of the composites was measured based on ASTM D257 [15] through frequency flow of 5000-100000 Hz.

3. Results and Discussion

The result of mechanical properties of the thermoplastic polyurethane composites reinforced with 5, 10, 15 vol.% carbon black is presented in Figure 1 (a) and (b) and Table 1. It shows that the highest tensile strength is discovered in a 10 vol.% composition of carbon black with a value of 20.44 MPa and the lowest value tensile strength is presented in the composite with carbon black 5 vol.% of reinforcement. The increasing tensile property is significant to the use of carbon black, which has relatively good adhesion that occurs between matrix and reinforcement materials. Composite reinforced with 15 vol.% of carbon black obtained 18.9 MPa, which is lower than 10 vol.% of carbon black. The decrease of tensile strength is could be caused by the weakening of bond between the interface of polyurethane and carbon black due to cavities caused by increasing reinforcement material. Figure 1 (b) describes strain value of the composites, which the highest strain is 501.7 % on composite reinforced with 10 vol.% of carbon black. From the figure, it can be concluded that the highest mechanical properties are produced on the composition 90 vol.% polyurethane composites reinforced with 10 vol.% carbon black.

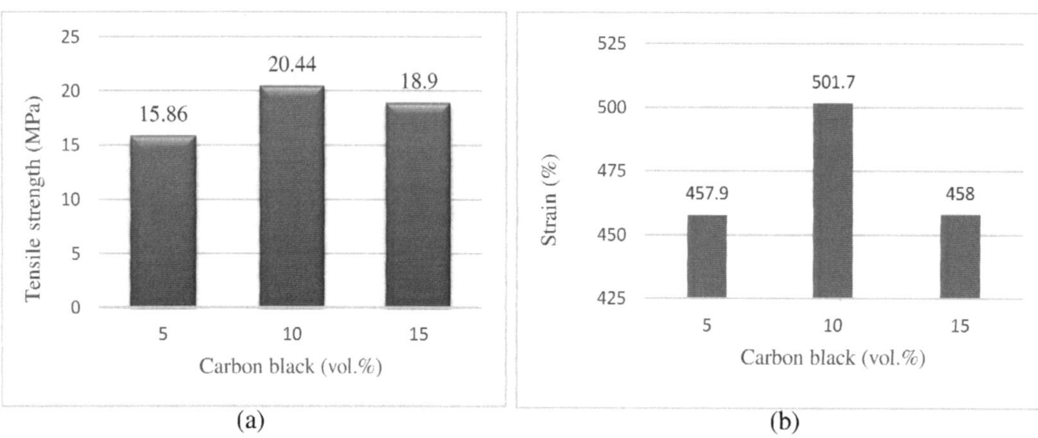

Figure 1. Mechanical properties of polyurethane composite with variation of carbon black of 5, 10, 15 vol.%: (a) tensile strength and (b) strain

Table 1 Mechanical properties and resistivity of polyurethane resistance with 5, 10, 15 vol.% CB

Sample	Characterization				Void
	Tension test	Strain strength	Hardness	Resistance	
Carbon black 5 vol.%	15.86	457.9	86.5	1.20 x 10^6	✓
Carbon black 10 vol.%	20.44	501.7	88.5	1.41 x 10^6	✓
Carbon black 15 vol.%	18.9	458	89.5	8.20 x 10^6	✓

With the significant addition of carbon black of 5 and 10 vol.%, the value of tensile strength and strain will increase until the optimal composition of 10 vol.%. These results are supported by previous studies by Broza et al. [3], that having a decrease of tensile strength is due to the embrittlement caused by the addition of carbon black. Also, reinforcing 15 vol.% of carbon black will decrease tensile strength because of the difficulty in distributing reinforcement particles and the tendency of the particles to agglomerate.

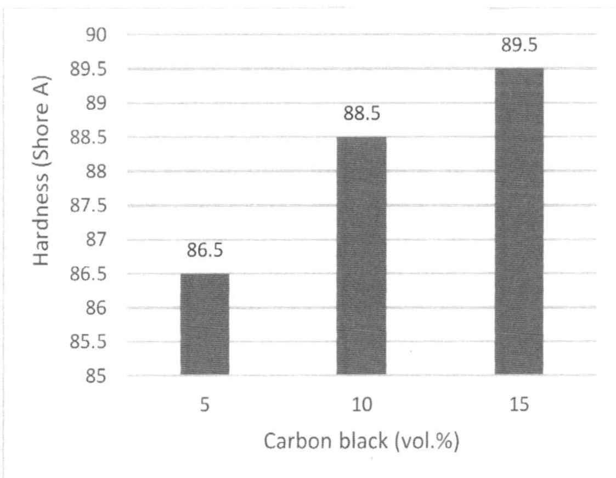

Figure 2. Hardness of polyurethane composite with variation of carbon black of 5, 10, 15 vol.%

Figure 2 shows the highest hardness is achieved by the reinforcement composition of 15 vol.% carbon black of 89.5 shore A, while the lowest hardness of 86.5 shore A is obtained by the composite reinforced with carbon black 5 vol.%. It can be said that increasing vol.% of carbon black could improve hardness of the composites.

Figure 3 illustrates a noticeable change in the structure shown by the red arrow. This change is caused by the increasing of carbon black, which allows the formation of high structures thus increasing contact between the particles. Carbon black with high structure has good disperse-ability properties and is able to fill voids in composites. The high particle interaction will simultaneously increase the value of electrical conductivity. Figure 3 (a) and (b) shows cavities, which are marked by yellow arrows, and visible blobs, marked by red rectangles. These swellings are caused by in-homogeneous distribution of reinforcements. The cause of the cavities in the composite could be affected by high moisture of composite in certain area

IOP Conf. Series: Materials Science and Engineering **547** (2019) 012057 doi:10.1088/1757-899X/547/1/012057

so that it would absorb air and produce an uneven density. This is in accordance with the results of SEM examination, which is seen as voids.

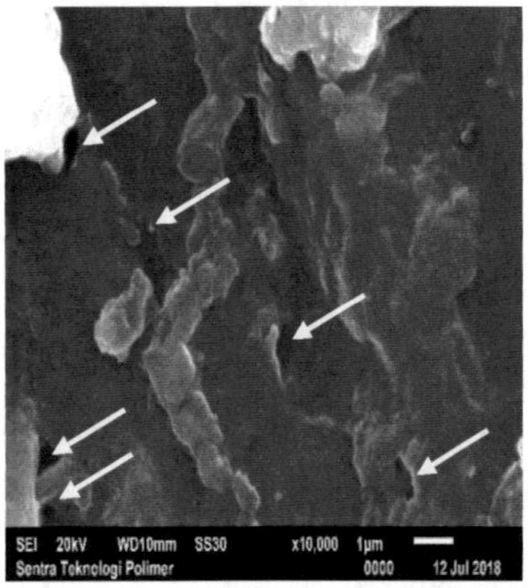

(a)

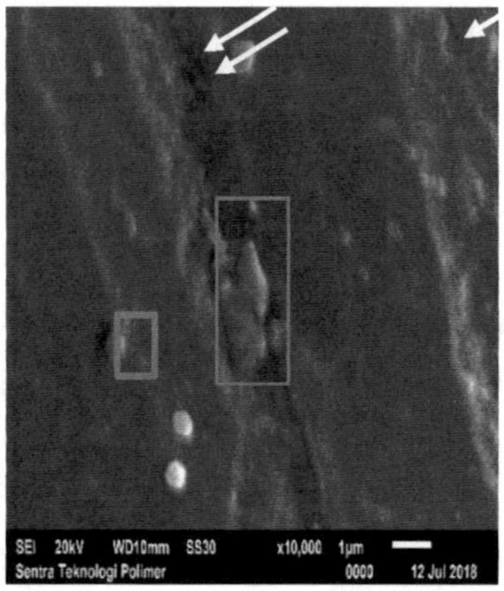

(b)

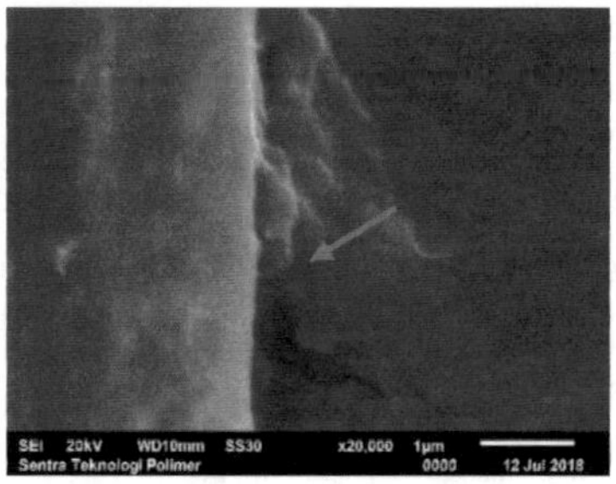

(c)

Figure 3 SEM examination of polyurethane composite-carbon black with variation: (a) 5 vol.%, (b) 10 vol.%, (c) 15 vol.%

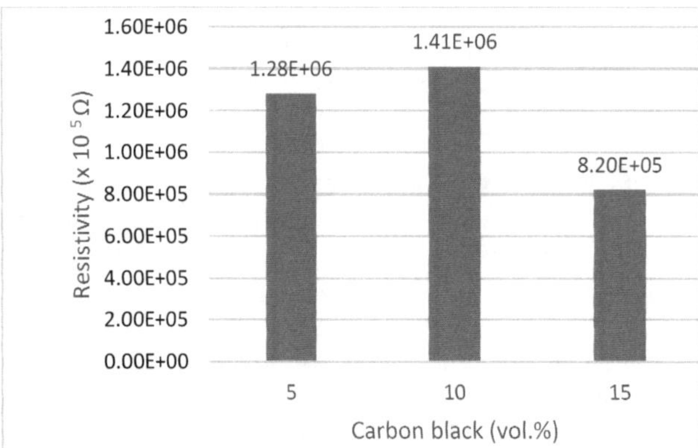

Figure 4. Resistance of polyurethane composite with variation of carbon black of 5, 10, 15 vol.%

Figure 4 explains the variation of surface electrical resistance. It can be seen that the composite with carbon black of 10 vol.% obtained the highest resistance value of 1.41 x 10^6 Ω. The smallest resistance value is shown on carbon black of 15 vol.%, with resistance value of 8.20 x 10^5 Ω. Based on previous research, the addition of carbon black can reduce the value of electrical resistance efficiently by minimum addition, because the particles have a rounded and branched round structure, high surface area and small particle size.

4. Conclusions

Addition of carbon black to the thermoplastic polyurethane improved the following properties of the composite. The maximum tensile strength and strain was achieved at carbon black 10 vol.%. Increasing carbon black from 5 to 15 vol.% will improve hardness of the polyurethane composites, with the highest value of 89.5 shore A. As for resistivity, carbon black of 15 vol.% has the least resistance value of 8.20 x 10^5 Ω. SEM examination revealed that voids appeared on all composite and increasing carbon black reinforcement generated the particles tend to agglomerate.

5. Acknowledgements

The authors are grateful to Universitas Pancasila, Jakarta, Indonesia by funding this work under the In-house Research Grant 2017.

6. References

[1] The European Standard: EN-British Standard (12531:1999) for castors and wheels-hospital bed castor.
[2] ESD-Book. Protection from electrostatic dischargers in the electronic environments. BJZ Industriedienst and Vertieb.
[3] Broza G, Piszczek K, Schulte K, Sterzynski T. 2007. Nanocomposites of Poly(Vinyl Chloride) With Carbon Nanotubes (CNT), Composite Science and Technology, 67:5, 890-894.
[4] Li J, Wang Q, Li M, Feng J, Jia Z, Su Y. 2014. An Facile High-Density Polyethylene – Exfoliated Graphite – Aluminium Hydroxide Composite: Manufacture, Morphology, Structure, Antistatic and Fireproof Properties, Materials Science, 20:3, 289-294.
[5] Islam I, Sultana S, Ray SK. 2018. Electrical and tensile Properties of Carbon Black Reinforced Polyvinyl Chloride Conductive Composite. Journal of Carbon Research. 4:15.
[6] Wei T, Song L. 2010. The Synergy of A three Filler Combination in the Conductivity of Epoxy Composite. Materials Letters. 64:21, 2376-2379
[7] Hamza MS, Oleiwi JK and Nassir NA. 2011. The Study Effect of Carbon Black Powder on The Physical Properties of SBR/NR Blends Used In Passenger Tine Treads. Eng Tech. J., 29:5. 856-864.
[8] Bruchell TD. 1999. Carbon Materials for Advanced Technology. Pergamon. 77-89.
[9] Komameni S. 1992. Nanocomposites. J Mat Chem. 2:12.
[10] Peng, YK. 2007. The Effect Of Carbon Black And Silica Fillers On Cure Characteristics And Mechanical Properties Of Breaker Compounds. Thesis. University Science Malaysia.
[11] Ibrahim, Ismail M., 2017. Mechanical And Physical Properties Of Polyurethane Composite Reinforced With Carbon Black N990 Particles. International Journal Of Scientific And Technologi Research. 6:08.

IOP Conf. Series: Materials Science and Engineering **547** (2019) 012057 doi:10.1088/1757-899X/547/1/012057

[12] Standar test method for rubber property durometer hardness. 2007. Designation : D 2240-05, Annual Book of ASTM Standar, Volume 08.01 Plastics (I) : D 256- D3159
[13] Standar Test Methode for tensile properties of Plastics D638M-87b. 1988. Annual book of ASTM Standard, Volume 08.01 Plastics (I) : D 256 – D3159
[14] Standar partice for scanning electron microsope beam size characteristization ASTM E986-97.
[15] Standar test methods for DC resistance or conductance of insulating materials ASTM D257-07.2007.annual Book of ASTM Standar, Volume 08.01 Plastics (I) : D 256- D315

Thermal Behaviour Studies on Building Walls based on Type and Composition of the Materials

Heri Andoni[1], Aldissain Jurizat[1], Steven[2], Dixon Thomas.[3], Rizky Amalia Achsani[1], Inge Magdalena Sutjahja[3], Mardiyati[2], Surjamanto Wonorahardjo[1]*
[1]Building Technology Research Group, SAPPK, Institut Teknologi Bandung, Jl. Ganesha No. 10, Bandung 40132, Indonesia
[2] Materials Engineering Department, FTMD, Jl. Ganesha No. 10, Bandung 40132, Indonesia
[3] Physics Department, FMIPA, Institut Teknologi Bandung, Jl. Ganesha No. 10, Bandung 40132, Indonesia
*Corresponding author (E-mail): titus@ar.itb.ac.id

Abstract. Controlling the thermal behaviour of buildings is very important because it affects the energy use in the buildings and impacts the thermal environment of the area. Thermal behaviour of the building is determined by the type and composition of the building material. This paper shows the thermal behaviour of several types and composition of commonly used building wall materials, i.e. brick, aerated concrete, laminated wood with glass wool insulation, and gypsum with glass wool insulation. For the thermal behaviour measurements, we used wall models exposed to halogen lamps of 2000 watts for 4 hours heating and 4 hours for cooling with a measurement interval of 5-15 minutes. Data retrieval using 8 thermocouples mounted on the surface and inserted inside the wall. The observed thermal behaviour phenomena are the heat flow on the walls, the time delay, and the heat capacity. The results show different maps of the thermal behaviour on the walls and indicating the need for additional thermal insulation material to control the heat release to the outside and inside of the building.

Keywords: thermal behaviour, building material, thermal insulation

1. Introduction

Many experts claim that the building has a negative effect on energy use and environmental damage such as the urban heat island (UHI) phenomenon. The negative effects of the building are mainly due to the material properties used, especially the role of the building envelope [1]. Currently, in major cities in the world, there is a phenomenon of increased air temperature due to heat dissipation effects of buildings.

The material properties that play a major role in this phenomenon are the ability to save or Specific heat and release heat or IR emissivity. The use of brick, concrete in the building and asphalt on the road body reported giving the greatest warming effect on city air [2]. The building materials have a large specific heat so that when exposed to sunlight throughout the day will store large amounts of heat. The intensity of sunlight received is influenced by a number of factors namely the orientation of buildings, seasons, weather and location of a latitude of the location.

This phenomenon is difficult to control because the materials attached to the building cannot be replaced easily and cheaply. Some experts claim that the effect of shadowing can reduce the rate of heating the building so that the effects of caloric release are also reduced. Differently, other experts claim that the replacement of glass material can reduce the effects of energy use in buildings. But both ways can not optimally control the recovery and release of heat. One of the more effective ways is to add thermal insulation materials on the walls of the building, but the addition of insulation requires a thermal behavioural review of commonly used building materials.

This paper presents the results of an experimental study on the commonly used thermal properties of building materials. The result of this research is to formulate requirement insulation for improvement of thermal behaviour of building envelope wall [3-13].

2. Materials and Methods

The study of the thermal behaviour of buildings due to sun exposure is carried out in the laboratory or indirectly. Exposure to sunlight in buildings is modelled by exposure to light from two halogen lamps @ 1000 Watt. The exposure process is carried out for 4 hours then followed by a cooling process for 4 hours. The duration of exposure is similar to the sun irradiation period at 08.00-12.00 on the wall facing east then cooling when the position of the sun is inclined westward at 12.00 to 16.00.

The thermal behaviour of the buildings is represented by the east and west walls of the buildings which are made from various possibilities of popular building wall materials namely; 1) multiplex and glass-wool, 2) gypsum and glass-wool 3) aerated concrete, 4) brick. Comparison of the four wall materials of the building is carried out in a 1x1 m model. This study method was adopted from Aversa et al. [14].

Each wall model is exposed to a halogen lamp with a distance of 65 cm. The temperature measurement is carried out every 15 minutes using a Type K thermocouple with an accuracy of 0.2% + 1 ° C and an infrared (IR) camera from the outer and inner surfaces (Figure 1).

IOP Conf. Series: Materials Science and Engineering **547** (2019) 012058 doi:10.1088/1757-899X/547/1/012058

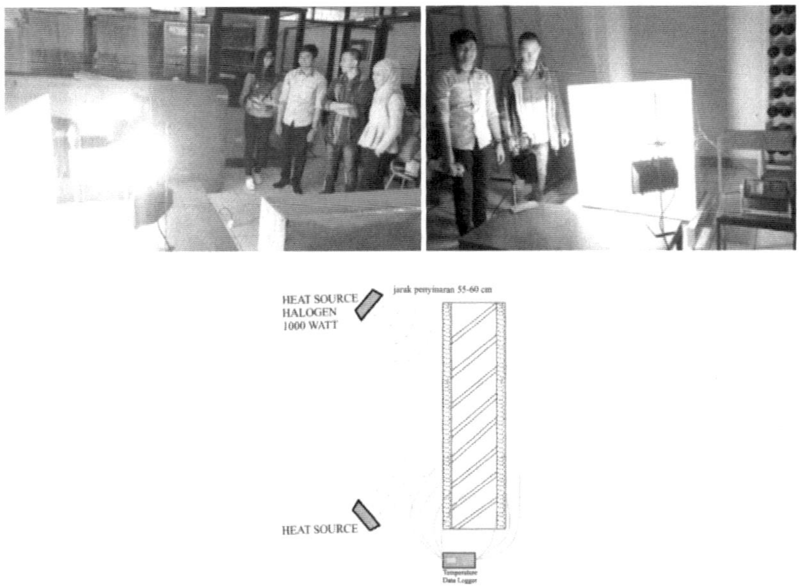

Figure 1: Experimental Setup

The irradiation is directed to the center of the wall, where the thermocouple is placed on the outer side (Tso), the inside of the outer layer (Tm1), the center of the wall (Tm2), the inside of the indoor layer (Tm3) and the indoor surface (Tsi) (Figure 2).

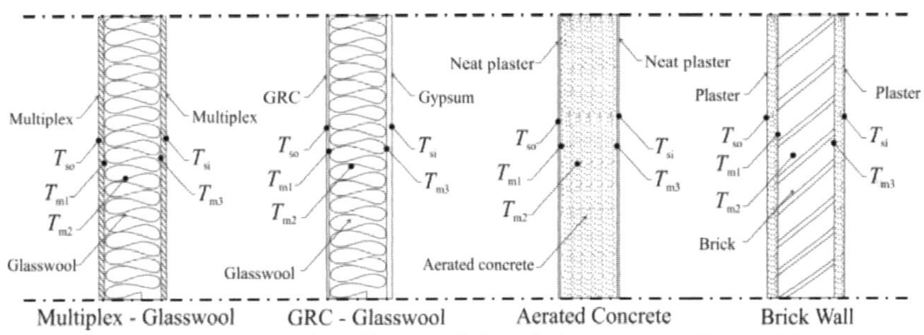

Figure 2: Type of wall material and thermocouple position

The multiplex and GRC walls represent the type of sandwich while the walls of aerated concrete and brick represent the block type. On the sandwich wall the type of material and composition in general are:

- 15 mm multiplex on the outside and inside with 4-inch glass wool insulation (density 16 kg/m^3) in the centre. On the outside is given a layer of waterproofing and weather shield paint, while the inner multiplex is given a layer of ordinary wall paint.
- 4 mm GRC on the outside and 9 mm gypsum on the inner side with 4-inch glass wool insulation in the middle. The outer and inner layers are the same as multiplex walls.

IOP Conf. Series: Materials Science and Engineering **547** (2019) 012058 doi:10.1088/1757-899X/547/1/012058

While the type of block walls, types of materials and compositions that are commonly used are:

- 12cm thick aerated concrete wall. The outer and inner layers are the same as multiplex walls.
- Brick wall with 1.5cm cement mortar plaster on the outer and inner surfaces. Finishing on the outer and inner layers is the same as the multiplex wall.

3. Results and Discussion

The measurement results show the heating and cooling profiles on the four walls (Figure 3). On the sandwich wall, the outer surface temperature is relatively stable after 90 minutes of heating, while in the block wall type the wall temperature continues to increase during the heating period. While the surface temperature in the sandwich wall tends to be stable, while in the block wall type the temperature continues to increase with a linear gradient. The difference in thermal behaviour in the two types shows the role of glass-wool which significantly inhibits the heat flow from the outside. In the block system, aerated concrete and brick materials tend to continue heat from outside to inside. In the block system, the difference between outside and inside temperatures continue to increase during the heating period while the on the sandwich system the different temperature is relatively stable.

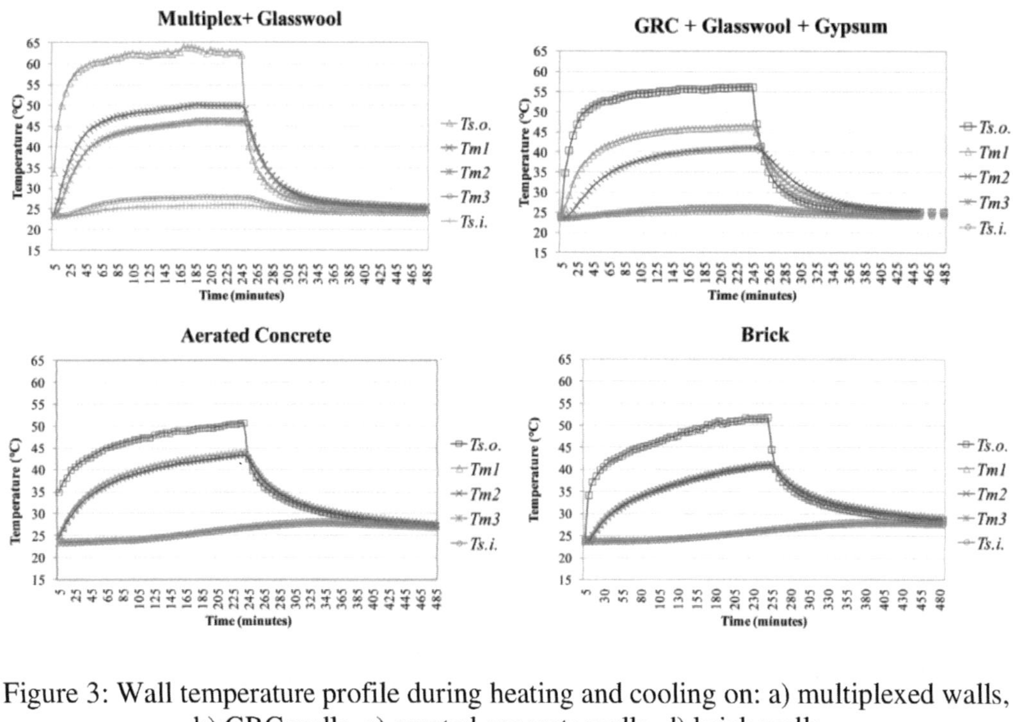

Figure 3: Wall temperature profile during heating and cooling on: a) multiplexed walls, b) GRC walls, c) aerated concrete walls, d) brick walls

In the cooling process, the outer surface of the sandwich wall quickly cools to lower than the middle temperature of the wall. On the block wall, cooling runs slower even though the surface temperature is still rising in the cooling period. This shows the block wall works as a thermal mass.

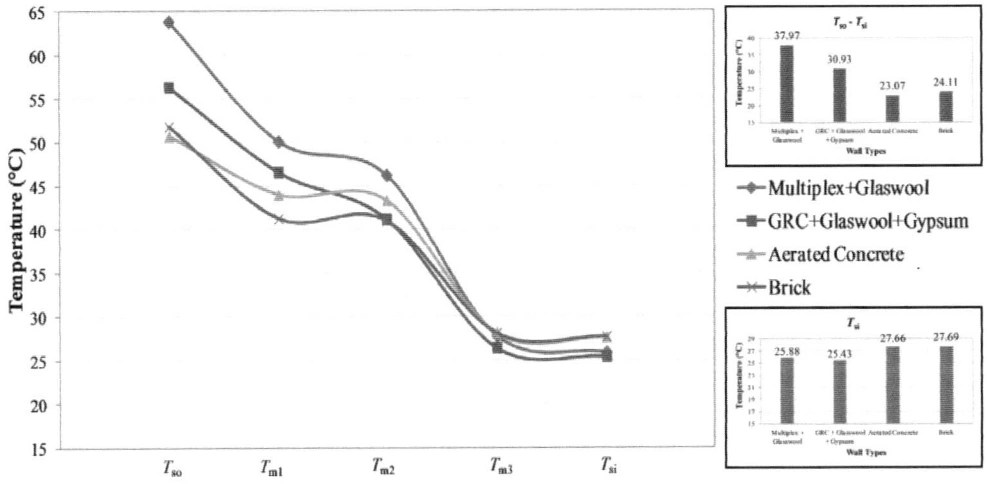

Figure 4: Wall temperature profile in maximum heating

In Figure 4, the wall temperature profile shows that the sandwich wall with glass wool insulation can hold the heat flow well enough so that the difference between the outer and inner surface temperatures is greatest. But the temperature of the outer surface of the sandwich wall increases rapidly to a temperature of 65 ° C, adversely affecting the thermal outdoor environment. The role of surface properties needs to be considered in the use of this sandwich wall.

Unlike the block wall, the surface temperature does not soar when exposed to sunlight, because it requires a lot of heat energy to increase the temperature. Block walls have a large thermal capacity compared to sandwich walls.

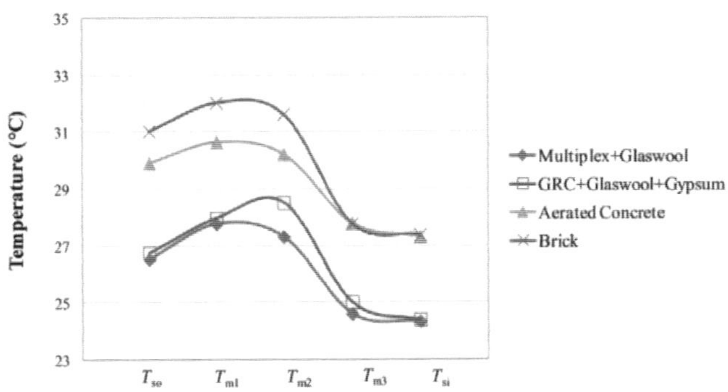

Figure 5: Wall temperature profile after cooling

In figure 5, it can be seen that the cooling process starts from the surface, both outside and inside surfaces. In the above conditions, it can be seen that the middle wall temperature is higher than the surface temperature.

Sandwich walls are lower than the wall block after 4 hours of cooling. Brick walls store the most heat compared to other walls, this shows the ability to store the largest brick wall heat. Negative effects of heat storage are seen at higher external and inner surface temperatures than sandwich walls. The temperature of the wall surface which is higher than the air temperature indicates the release of heat from the wall to the air which is felt as heating both indoors and outdoors. The phenomenon of warming outdoor space by buildings is known as the urban heat island, where accumulatively the buildings in the region release heat into the air so that the temperature of the region's air rises significantly.

Thus it has become clear that the use of heavy materials such as bricks is not suitable for future use. Likewise, aerated concrete material recommended as a substitute for bricks also shows thermal behaviour that is not good compared to sandwich walls.

This phenomenon requires the study of building materials in the future, where lightweight bricks and concrete that have large heat capacity must be repaired in several ways, namely; 1) reduce the heat capacity, 2) reduce the thermal conductivity.

Currently, mitigation technology is needed to improve the thermal behaviour of buildings [2] especially those using brick and concrete materials. New insulation material technology applied to the outer surface of the wall so that heat gain from the sun can be avoided.

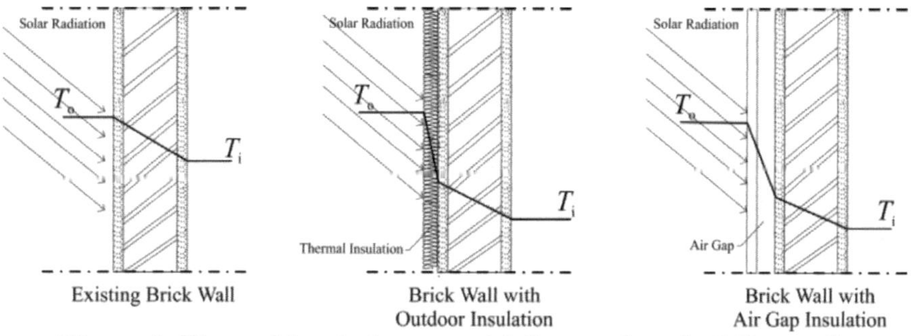

Figure 6: Thermal insulation material for envelope building wall

The study of the thermal behaviour of building materials shows the need for many studies of thermal insulation material technology suitable for the outer walls. This compatibility can be seen from the UHI effect parameters [2] and the material's weather resistance effect. Insulation material for application outside the wall requires water-resistant properties and deformation due to possible loads [6]. The use of natural and sustainable materials such as cellulose has enormous potential in Indonesia [7, 8, 9, 10, 15].

The behaviour of the thermal wall of the building affects indoor temperature and space occupants [16]. In the next study, it is expected that the role of the insulation of the outer walls in the outdoor and indoor environment can be well mapped.

4. Conclusions

Thermal behaviour of buildings is strongly influenced by its material properties. Physical environment parameters both outdoor and indoor can be a direction for the study of insulation material technology. Based on the study above, it was proven that aerated brick and concrete materials showed poor thermal behaviour for outdoor and indoor environments. To improve the thermal behaviour of the building, it is necessary to have appropriate insulation material on the outside of the wall.

It is necessary to study the thermal properties of the thermal insulation of these walls which are in accordance with the needs of the tropical environment where UHI parameters and thermal comfort are applied.

Sustainability parameters must also be considered, where the use of natural insulation such as cellulose is very potential in a tropical environment. But the cellulose material needs an in-depth study related to fire and water resistance.

5. Acknowledgements

This research and publication is financed by the Multidisciplinary Research Program, ITB 2018 with Number: 074a / I1.C10 / KU / 2018.

We would like to express our deepest gratitude to Suwardi Tedja who has prepared the sample so that this research can work well.

6. References

[1] Wonorahardjo, S.; Sutjahja, I.M.; Kurnia, D.; Fahmi, Z.; Putri, W.A., 2018, Potential of Thermal Energy Storage Using Coconut Oil for Air Temperature Control. Buildings 2018, 8, 95.
[2] Wonorahardjo, S., 2011, New concepts in districts planning, based on heat island investigation. Procedia Soc. Behav. Sci. 2012, 36, 235–242
[3] Diamant, R. M. E., Thermal and Acoustic Insulation, 1st Edition, eBook ISBN: 9781483279954
[4] Ojanena T., Seppäa I. P., Nykänena E., 2015, Thermal insulation products and applications - Future road maps, , Energy Procedia 78 (2015) 309 – 314.
[5] Deshmukh G., Birwal P., Datir R. and Patel S., 2017, Thermal Insulation Materials: A Tool for Energy Conservation, Journal of Food J Processing & Technology 2017, 8:4 DOI: 10.4172/2157-7110.1000670.
[6] Fibres, Zach J., Brožovský J., Hroudová J., May 2010, Research and Development of Thermal-Insulating Materials Based On Natural, Modern Building Materials, Structures and Techniques.
[7] Dhivar P.S., Patil A.R., Thermal Insulation Using Agricultural By-Product: A Review, IOSR Journal of Mechanical and Civil Engineering (IOSR-JMCE) e-ISSN: 2278-1684,p-ISSN: 2320-334X PP. 53-61.
[8] Paivaa A., Pereiraa S., Sáa A., Cruz D., Varumc H., Pintoa J., 2012, A contribution to the thermal insulation performance characterization of corn cob particleboards, , Energy and Buildings 45 (2012) 274–279.
[9] Pinto J., Vieira B., Pereira H., Jacinto C., Vilela P., Paiva A., Pereira S., Cunha V.M.C.F., Varum H,, Corn cob lightweight concrete for non-structural applications, Construction and Building Materials 34 (2012) 346–351
[10] Zacha J., Hroudováb J., Brožovskýc J., Krejzad Z., Gailiuse A., 2013, Development of Thermal Insulating Materials on Natural Base for Thermal Insulation Systems, , Procedia Engineering 57 (2013) 1288 – 1294.
[11] Asdrubali F., D'Alessandro F., Schiavoni S., 2015, A review of unconventional sustainable building insulation materials, Sustainable Materials and Technologies 4 (2015) 1–17.
[12] Terry N., Palmer J., Cooper I., 15 June 2012, State-of-the-Art Review: Insulation and Thermal Storage Materials Pre-Publication Draft, Cambridge Architectural Research Limited.

[13] Azimi E. A., Bakri Abdullah M. M.A., Ming L.Y., Yong H. C., Hussin K.and Aziz I.H., 2016, Processing and Properties of Geopolymers as Thermal Insulating Materials: a Review, , Rev. Adv. Mater. Sci. 44 (2016) 273-285.

[14] Aversa P., Palumbo D., Donatelli A., Tamborrino R., Ancona F., Galietti U., Luprano V. A.M., 2017, 2017, Infrared thermography for the investigation of dynamic thermalbehaviour of opaque building elements: Comparison between emptyand filled with hemp fibres prototype walls, Energy and Buildings 152 (2017) 264–272

[15] Steven ., Mardiyati ., R. Suratman, 2014, Pembuatan mikrokristalin selulosa rotan manau (calamus manan sp.) serta karakterisasinya, Journal Selulosa Vol 4, No 02 (2014) , DOI: http://dx.doi.org/10.25269/jsel.v4i02.84

[16] Damiati, S.A.; Zaki, S.A.; Rijal, H.B.; Wonorahardjo, S., 2016, Field study on adaptive thermal comfort in office buildings in Malaysia, Indonesia, Singapore, and Japan during hot and humid season. Build. Environ. 2016, 109, 208–223.

IOP Conf. Series: Materials Science and Engineering **547** (2019) 012059 doi:10.1088/1757-899X/547/1/012059

Effect of heat treatment on the stress corrosion cracking (scc) susceptibiliy of the 13Cr martensitic stainless steel for steam turbine blade

Rusnaldy[1]* **Efendi Mabruri**[2] **Luky W. Nugroho** [1]
[1]Department of Mechanical Engineering, Diponegoro University, Semarang, Indonesia
[2]Research Center for Metallurgy & Material, Indonesia Institute of Science

*Corresponding author (E-mail): rusnaldy@undip.ac.id

Abstract. Two different constant loads have been used to evaluate the effect of heat treatment on susceptibility to stress corrosion cracking (SCC) of 13 Cr martensitic stainless steel in 3.5% NaCl. The steel specimens were austenitized at 1050°C for 1 hour followed by oil quench to obtain martensite structure. Then, specimens were tempered at 400, 500, 600, 650 and 700°C for 1 hour. To investigate the SCC susceptibility, SCC tests were carried out at two different constant loads of 40% and 80% of ultimate tensile strength (UTS) in accordance with the method developed by Nishimura et.al. After the SCC tests, scanning electron microscope (SEM) observations of the fracture surface on the fractured specimens were conducted to investigate fracture mode. The experimental results showed that the 13Cr martensitic stainless steel was highly susceptible to SCC in a solution 3.5% NaCl at high constant load (80% of UTS) in almost tempered conditions. The least susceptible steel was the tempered one at 700°C. At lower constant load (40% of UTS), the highly susceptible steels were the tempered ones at 400 and 500°C. Steels which were tempered at high tempering temperature were not failed for 240 hours of SCC testing. The mechanism responsible for this susceptibility is thought to be secondary phenomenon where carbide precipitation of M_7C or $M_{23}C_6$ occurs. The fracture surfaces exhibited intergranular cracks and dimple rupture fracture for steels tempered at 400, 600 and 650°C and transgranular cracks and cleavage fracture for steels tempered at 500°C.
Keywords: heat treatment, martensitic stainless steel, stress corrosion cracking, turbine blades

1. Introduction

Steam turbine blades are the critical component in power plant which convert the linier motion of high temperature and high pressure steam flowing down a pressure gradient into a rotary motion of the turbine shaft [1]. The blades are susceptible to failure due to stresses arising primarily from centrifugal loads and bending forces related to the steam mass flow [2]. And also during operation, the blades operate in a wet steam environment and their failure modes are typically either stress corrosion cracking or corrosion fatigue [3].

In blade design, attention is being focused on using the higher strength alloys and improving resistance to corrosion and environment induced cracking [4]. Among the various blade materials, the most popular is 12%Cr martensitic stainless steel (MSS) which has an excellent combination of strength, toughness and corrosion resistance as well as high inherent damping characteristics [5]. The other martensitic stainless steel that has been used for turbine blades due to its mechanical properties and moderate corrosion resistance is MSS with 13 wt% Cr [6,7].

The corrosion resistance and mechanical properties of martensitic stainless steel can be optimized by chemical composition modification and heat treatment process. Heat treatments have a significant effect on microstructures and properties of steels. Mabruri et.al tried to modify 13Cr MSS by adding Mo and Ni [8]. They investigated the influence of Mo and Ni in tempered condition on their tensile properties. They found that the addition of 1% and 3% Mo will increase both the tensile strength and the elongation of the steels, but the addition of 3% Ni will decrease both the tensile strength and the elongation of the steels. The effect of heat treatment on hardness and microstructure of the modified 13Cr steam turbine blade steel was also investigated by Prifiharni et.al [9]. The steels were austenitized at 1050°C and tempered at 300°C until 700°C. They showed that increasing tempering temperature until 500°C can improve hardness due to formation of precipitation carbide $M_{23}C_6$ into microstructure. Increasing tempering temperature from 500 to 650°C can decrease Cr content on carbide and increase Mo content on carbide.

In this work, the effect of tempering temperature of 13Cr martensitic stainless steels was investigated to evaluate the susceptibility of 13Cr martensitic stainless steel to SCC in a solution of 3.5% NaCl and the fracture modes caused by SCC.

2. Materials and Methods

The material used in this study was 13Cr martensitic stainless steel (AISI 410). The chemical compositions are listed in Table 1. For SCC test, tensile specimen configuration was used in this study and it shown in Fig. 1. The specimens were austenitized at 1050°C for 1 hour followed by oil quench to obtain martensite structure. Then, specimens were tempered at 400, 500, 600, 650 and 700°C for 1 hour.

Tensile tests were also carried out in this study. Heat-treated plate specimens were machined to make the tensile specimens with dimension complied with ASTM E8 [10]. The tensile testings were carried out by using universal testing machine until the specimens broke.

To investigate the SCC susceptibility, SCC tests were carried out at two different constant loads of 40% and 80% of ultimate tensile strength (UTS) for each heat-treated specimen in a solution of 3.5% NaCl, in accordance with the method developed by Nishimura et.al [11]. After the SCC tests, scanning electron microscope (SEM) observations of the fracture surface on the fractured specimens were conducted to investigate fracture mode.

Table 1. Chemical composition (wt %) of the 13Cr martensitic stainless steel (AISI 410)

C	S	P	Mn	Si	Cr	Mo	Ni	Fe
0,034	0	0,0079	0,310	0,861	13,535	0,00474	0,0842	Bal.

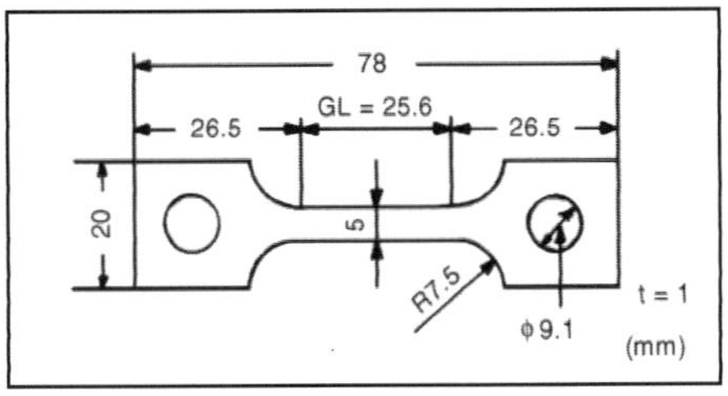

Fig. 1. Specimen's configuration for SCC test

3. Results and Discussion

Table 2 shows the effect of tempering temperature of the 13Cr martensitic stainless steel on the mechanical properties. It can be seen that the yield strength and ultimate tensile strength decrease as tempering temperature increases. At tempering temperature of 500°C, embrittlement occurs. According to Bezuidenhout and van Rooyen [12], sensitization happens when martensitic stainless steel is held at a temperature of 450-550°C causing the steel to be susceptible to intergranular stress corrosion cracking. But when the tempering temperature increases from 500 to 600°C, the yield strength and ultimate tensile strength increase. The increasing of strength from 500 to 600°C is maybe related to secondary hardening phenomenon due to formation of M_7C_3 carbides within the martensite lath [9]. On the other side, decreasing strength when tempering temperature increases from 600 to 700°C occurs due to the M_7C_3 carbide started to coarsen and partially transform to $M_{23}C_6$ carbides [13]. The highest strength was achieved when specimen was tempered at temperature of 400°C.

Figs. 2 and 3 show the results of the SCC test. The effect of tempering temperatures on time to failure at constant load of 40% and 80% of ultimate tensile strength (UTS) of each heat-treated 13Cr martensitic stainless steel was investigated. The tests were carried out for 240 hours. At constant load of 40% of UTS (see Fig. 2), it can be seen that specimens which were tempered at 400 and 500°C were failed and were tempered at 600, 650 and 700°C were not failed. At tempering temperature of 400 and 500°C, steels were failed after 28.32 and 17.36 hours of SCC testing respectively. Meanwhile, at tempering temperature of 650 and 700°C, steels have the lowest increase of length during testing, i.e 0.12 and 0.10 mm respectively. When the constant load increases to 80% of UTS (see Fig. 3), among the heat-treated steels in this work, only steel which was tempered at 700°C was not failed for 240 hours of testing. It can also be seen that the lower tempering temperature more susceptible the steel to SCC. Carbide transformation from M_7C_3 to $M_{23}C_6$ at high tempering temperature has significant role to increase corrosion resistance and decrease pitting potential [14]. It could be related to the high susceptibility of steel which was tempered at 700°C.

The fracture surfaces for tempered specimens at constant load of 80% of UTS were shown in Fig. 4. The fracture surface for tempered specimens at 400, 600 and 650°C were similar (see Figs 4a, 4c and 4d). The fracture surfaces exhibited intergranular cracks and dimple rupture fracture and these fracture surfaces correspond to ductile failure. At tempering temperature of 500°C, steel has transgranular cracks and cleavage fracture on its fracture surface (see Fig. 4b). This kind of fracture could be caused by embrittlement and secondary hardening where carbide precipitation of M7C or M23C6 occurs [9]

IOP Conf. Series: Materials Science and Engineering **547** (2019) 012059 doi:10.1088/1757-899X/547/1/012059

Table 2. Mechanical properties of heat-treated 13Cr martensitic stainless steel (AISI 410)

Tempering Temperature (°C)	Yield Strength (MPa)	Ultimate Tensile Strength (MPa)	Elongation (%)
400	673.70	778.95	8.30
500	648.32	686.57	9.10
600	658.34	865.73	6.50
650	405.86	598.33	9.90
700	358.10	518.57	11.10

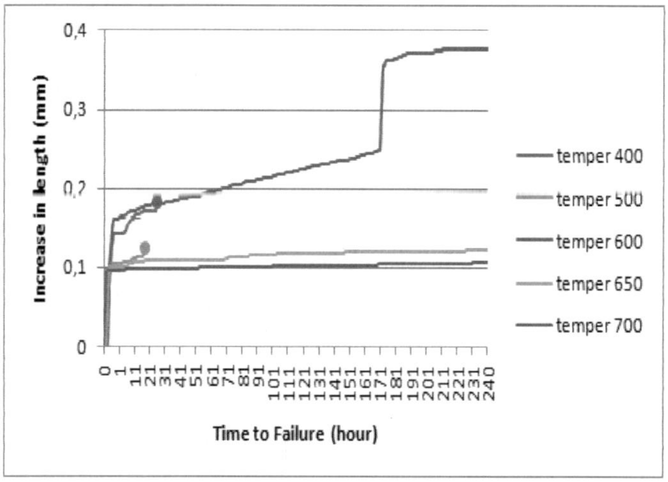

Fig. 2. The effect of tempering temperature on time to failure of 13Cr martensitic stainless steel at constant load of 40% of ultimate tensile strength

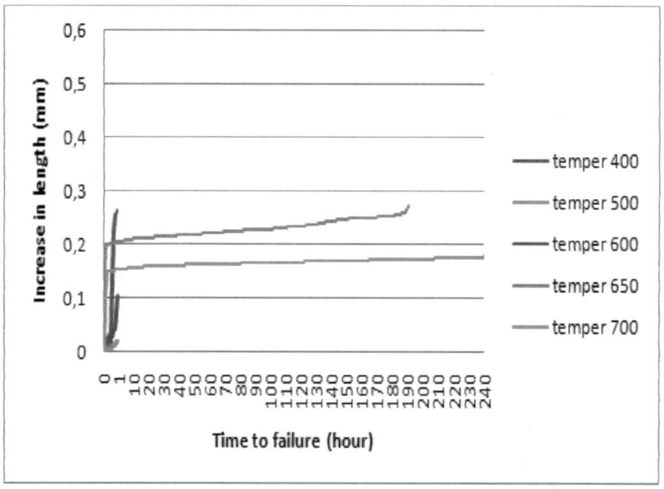

Fig. 3. The effect of tempering temperature on time to failure of 13Cr martensitic stainless steel at constant load of 80% of ultimate tensile strength

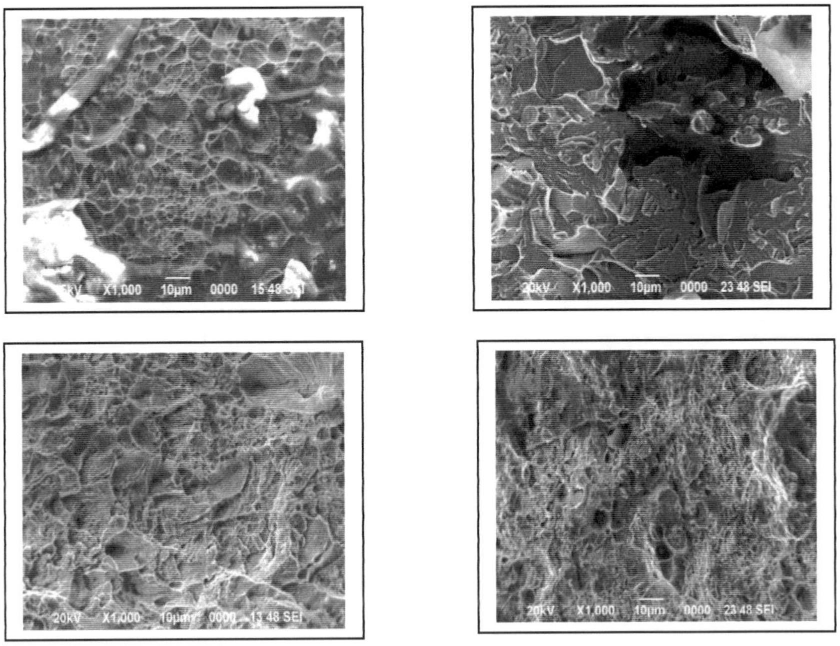

Fig. 4. Fracture surfaces for tempered specimens at constant load of 80% of UTS at tempering temperature of (a) 400°C, (b) 500°C, (c) 600°C and (d) 650°C

4. Conclusions

The susceptible of 13Cr martensitic stainless steel (AISI 410) at various tempered conditions was investigated. The 13Cr martensitic stainless steel was highly susceptible to SCC in a solution 3.5% NaCl at high constant load (80% of UTS) in almost tempered conditions. The least susceptible steel was the tempered one at 700°C. At lower constant load (40% of UTS), the highly susceptible steels were the tempered ones at 400 and 500°C. Steels which were tempered at high tempering temperature were not failed for 240 hours of SCC testing. The mechanism responsible for this susceptibility is thought to be secondary phenomenon where carbide precipitation of M_7C or $M_{23}C_6$ occurs. The fracture surfaces exhibited intergranular cracks and dimple rupture fracture for steels tempered at 400, 600 and 650°C and transgranular cracks and cleavage fracture for steels tempered at 500°C.

5. References

[1] S. Saxena, J.P. Pandey, R.S. Solanki, G.K. Gupta, O.P. Modi, 2015, "Coupled mechanical, metallurgical and FEM based failure investigation of steam turbine blade" Engineering Failure Analysis 52 pp. 35-44.

[2] J. Baran, 2016, "Redesign of steam turbine rotor blades and rotor packages – Environmental analysis within systematic eco-design approach" Energy Conversion and Management 116 pp. 18-31.

[3] M.N James, M. Newby, D.G. Hattingh, A. Steuwer, 2010, " Shot-peening of steam turbine blades: Residual stresses and their modification by fatigue cycling", Procedia Engineering 2 pp 441-451.

[4] A. Turnbull, S. Zhou, 2010, "Comparative evaluation of environment induced cracking of conventional and advanced steam turbine blade steels. Part1: Stress Corrosion Cracking", Corrosion Science 52 pp. 2936-2944.

IOP Conf. Series: Materials Science and Engineering **547** (2019) 012059 doi:10.1088/1757-899X/547/1/012059

[5] D. Ziegler, M. Puccinelli, B. Bergallo, A. Picasso, 2013, "Investigation of turbine blade failure in a thermal power plant" Case Studies in Engineering Failure Analysis 1 pp. 192-199.
[6] L. Yu-rong, Y.E. Dong, Y. Qi-long, S.U. Jie, Z. Kun-yu, J. Wen, 2011, "Effect of heat treatment on microstructure and property of Cr13 Super Martensitic Stainless Steel" Journal of Iron and Steel Research, 18 (11) pp. 60-66.
[7] B.S. Kumar, V. Kain, M. Singh, B. Vishwanadh, 2017, "Influence of hydrogen on mechanical properties and fracture and fracture of tempered 13 wt% Cr martensitic stainless steel" Materials Science & Engineering A 700, pp. 140-151.
[8] E. Mabruri, M.S. Anwar, S. Prifiharni, T.B. Romijarso, B. Adjiantoro, 2015, "Tensile properties of the modified 13 Cr martensitic stainless steels" AIP Conf. Proce. 1725, 020039-1 – 020039-5.
[9] S. Prifiharni, H. Perdana, T.B. Romijarso, B. Adjiantoro, A. Juniarsih, E. Mabruri, 2017, "The hardness and microstructure of the modified 13Cr steam turbine blade steel in tempered conditions", The International Journal of Engineering and Technology Vol. 8 No. 6 pp.2672-2675, Dec 2016-Jan 2017.
[10] ASTM E8: Standard Test Methods for Tension Testing of Metallic Materials
[11] R. Nishimura, I. Katim, Y. Maeda, 2001, "Stress corrosion cracking of sensitized type 304 stainless steel in hydrochloric acid solution predicting time-to-failure and effectc of sensitizing temperature", Corrosion Vol 57 No. 10.
[12] M. Bezuidenhout and G.T. van Rooyen, 2004, "Heat Treatment Bench Mark Testing of FV566 Blade Material", TSI, Eskom, Germiston, South Africa
[13] A.N. Isfahany, H. Saghafian, and G. Borhani, 2011, "The effect of heat treatment on mechanical properties and corrosion behavior of AISI420 martensitic stainless steel," J. Alloys Compd., vol. 509, no. 9, pp. 3931–3936.
[14] I. Taji, M.H. Moayed, M. Mirjalili, 2015, "Correlation between sensitisation and pitting corrosion of AISI 403 martensitic stainless steel", Corrosion Science, Vo. 92 pp. 301-308.

IOP Conf. Series: Materials Science and Engineering **547** (2019) 012060 doi:10.1088/1757-899X/547/1/012060

Effect of Annealing Cycle on Drawability and Microstructure of Cold Rolled Steel Sheet

Iskandar Muda[1] **and Druce P Dunne**[2]
[1] University of Jenderal Achmad Yani, Bandung, Metallurgical Engineering, Indonesia
[2] University of Wollongong, School of Mechanical, Materials & Mechatronic Engineering, Australia

E-mail: dr.ir.iskandarmuda@gmail.com

Abstract. The effect of the annealing cycle on microstructure, mechanical properties and draw ability of aluminium killed steel sheets at three different heating rates and holding times were investigated. Tensile test, hardness test, optical microscope and transmission electron microscope (TEM) techniques were utilized. The results of samples for different times in laboratory experiments were compared with the production samples. The results for these experiments showed that lower heating rates (12 °C/h) occurred at a lower annealing temperature, increase draw ability. This effect is attributable to the longer time available to initiate nucleation at the lowest heating rate. However, for all three heating rates (12 °C/h, 24 °C/h and 36 °C/h), the aluminium and nitrogen combine to form atmospheres or pre-precipitation clusters at polygonised sub-grains and as rolled boundaries, modifying the development of the recrystallized structure. Different amounts of AlN were precipitated prior to recrystallization for each heating rate resulting in different final recrystallized grain sizes, mechanical properties and drawabilities. The heating rate strongly influences recrystallization and drawability with a lower heating rate reducing both the temperature of the start of recrystallization and the recrystallized grain size. It is inferred that the lower rate promotes nucleation of recrystallization overgrowth of recrystallized. The effect of holding time on mechanical properties was to: (i) decrease tensile strength, yield strength and hardness and (ii) increase drawability, elongation and grain size. Tensile strength and hardness results from laboratory experiments were lower than those for production samples, but the trends were similar.

Keywords: Aluminium killed steel, mechanical properties, annealing, recrystallization, drawability.

1. Introduction

Low carbon aluminium killed steel is one of the most important products of the steel industry today. It had become important because it has many capabilities that no other commercial material can offer at low cost, properties such as relatively high drawability the ability to be readily formed at high production rates into various shapes, attractive surface finish, and good weldability. An important requirement for many applications involving sheet steel is, of course, good formability and the most common sheet forming method is press forming. The development of formability of low carbon aluminium killed steels depends on the alloying elements, impurities and inclusions in steel making, slab reheating furnace, finishing and coiling temperature in hot strip mill, total reduction, lubrication, heating rate, temperature and time annealing in cold rolling mill. All the process parameters above must be controlled to produce excellent deep drawing properties.

There are five major steps in the manufacturing of cold rolled and annealed sheet steels for extra deep drawing [1-9], as measured by the normal anisotropy (r-value). This property strongly controls the ability of the material to be deep-drawn into a flat-bottomed cup, and the strain hardening exponent (n-value) determines the ability of the material to be stretched. The first step is steel melting and casting, which includes control of elements in charging, such as copper, chromium, nickel, tungsten and eliminating as for as possible harmful elements such as sulphur. An increase in carbon content will decrease the r-value and excessive levels of impurities such as N_2, O_2, and inclusions including Al_2O_3 and SiO_2 will also decrease the r-value. Therefore, close composition control is required. The second step is hot rolling practice control. A high finishing temperature will increase the r-value for Al-killed steel because a high finishing temperature produces a more favourable annealed texture of cube-on corner rather more than a cube-on face. The most important step for aluminium killed steels is coiling practice, a lower coiling temperature leads to higher r-value because ALN is retained in solution in ferrite and can precipitate out on annealing after cold working with an increase in {111} grain orientation at the expense of {100} grain [10-15]. Therefore, the r-value becomes higher if the coiling temperature is low (<600°C). In the third step of cold rolling, the major variable is the reduction ratio, with the optimum r-value being obtained on annealing following a reduction ratio of 60% - 70%. The fourth step of batch annealing includes the rate of heating to the annealing temperature, the temperature and time of the hold. In aluminium killed steels the response to annealing varies with the aluminium content the r-value decreases monotonically with the heating rate in the case of low carbon rimmed steel, but as the Al content becomes higher, a peak r-value appears. The final step is an extension in the skin pass mill, which is important for stretching in the forming of cold rolled steel [16-20].

2. Experimental

The investigation was divided into two stages, the first stage consisted of a "mass production" investigation, including variations in reduction ratio in the cold rolling mill and the holding time during batch annealing. The second stage was the "laboratory scale simulation of industrial batch annealing" which allowed a closer study of the structure and properties of cold rolled and annealed samples, including effects of heating time, holding time and heating rate. These investigations in totality were used to obtain more detailed information about the mechanisms underlying the changes in the n and r-values with annealing conditions.

ID	C (%)	Si (%)	Mn (%)	P (%)	S (%)	Al (%)	N (%)	Cr (%)	Ni (%)	Cu (%)	Nb (%)	V (%)	Mo (%)	Al/N
A	0.046	0.004	0.250	0.008	0.007	0.040	0.006	0.012	0.02	0.016	-	0.005	0.003	6.7
B	0.034	0.019	0.270	0.013	0.007	0.050	0.006	0.007	0.017	0.014	0.001	0.006	0.004	8.3
C	0.046	0.003	0.252	0.010	0.006	0.040	0.006	0.008	0.015	0.023	0.001	0.003	0.002	10
D	0.035	0.010	0.290	0.007	0.009	0.052	0.005	0.006	0.012	0.010	0.001	0.005	0.002	10.4

Table 1. Chemical compositions of steels used.

The normal anisotropy (r) of each annealed steel sheet was determined using a tensile specimen with a gauge length (l) and width (w) 50 and 25 mm, respectively. The r value was determined (similar to JIS No. 5) by stretching the sample by 15% and then using measurements of the width in ten locations before (wo) and after elongation (w15%) in Equation 1.

$$r = \frac{\varepsilon_w}{\varepsilon_h} = \frac{\ln\left(\frac{w_o}{w_{15\%}}\right)}{\ln\left(\frac{lw_o}{lw_{15\%}}\right)} \quad (1)$$

Tensile samples were taken 0 (L)° 45° (D) and 90° (T) relative to the rolling direction: An average value of r was determined from Equation 2.

$$r = \frac{(r_L + 2 \cdot r_D + r_T)}{4} \quad (2)$$

The strain hardening exponent, n, of each steel was calculated using the measured stress at 10% strain and thc ultimate strength according to the Nelson-Winlock Table. The samples were examined using optical microscopy and transmission electron microscopy to detect precipitation of aluminium nitride and observation of the structures associated with recovery, recrystallization and grain growth. The mean grain size was measured using quantitative metallographic analysis according to ASTM E 112.

3. Results and Discussions

Effect of Annealing Cycle on Drawability. The results of mechanical tests for various holding times at 680 °C are shown in Figs.1 and 7. Figure 1 shows the effect of holding time on r-value and n-value. For holding at 680 °C from zero to 8 hours the r value increased, after which it decreased on holding time for up to 16 hours. The peak r-value occurred in 6 hours, for a heating rate of 24°C/h and the peak r-value was obtained in 4 hours for heating rate 36°C/h. From these data, it is evident that slow heating results in a higher r-value occurring at a longer holding time than for rapid heating. Figure 2 shows the effect holding time on n-value for the three heating rates. The trend in n value is similar to the trend in r-value above. For heating rates of 12, 24 and 36 °C/h, peaks in n-value occurred after 8 hours, 6 hours and 4 hours, respectively. The influence of grain size on r-value is significant during heating, with increasing grain size being associated with an increase in r and n value is shown in Figure 3. The grain size increases from 20 μm at 640 °C after 8 hours, to 35 μm after 20 hours. However, for holding time at 680 °C, the influence of grain size on r and n value is not significant.

The results of tensile tests for various holding times at 680 °C are shown in Figures 4 and 7. These curves show that the hardness, tensile and yield strength decrease with increasing holding time. It can be seen that an increase in the holding time increases the elongation, as a result of the increasing grain size The ratio of tensile strength to yield strength gradually increased with increased holding time up to 20 hours. The effect of a high ratio of tensile to yield strength is to increase the exponent for strain hardening. While the strength properties decreased with increasing time for the laboratory samples studied, the drawability of the experimental aluminium killed steel improved with increasing temperature or heating time to an optimal value corresponding to a particular holding time.

Comparing the r-values obtained from samples treated by the three different laboratory annealing cycles with the production scale. The curve of yield strength from the production samples was sharper than for laboratory samples, but both showed a decrease in the yield strength with time. Holding times that produced the coarsest grain sizes also resulted in lowest yield strength, tensile strength and hardness. Increasing the holding time, increased the r and n values, but the curves were not linear, with peaks being obtained at 4, 6 and 8 hours for heating rates of 12 °C/h, 24 °C/h and 36 °C/h, respectively. Figures 1 - 2 show there was no essential effect of holding on r and n-value from the actual production samples. The difference in the laboratory and production results is caused by differences in equipment and other practical factors. For example, the production scale treatment in a batch annealing furnace has many variables: coil stacking practice, furnace temperature, maximum permitted top outer edge temperature and annealing equipment (direct fired vs radiant-tube furnaces, fan size, convector plates). The annealing cycles and practices are compromise, in that the furnace temperature, firing time and permitted top outer temperatures are specified to sufficiently anneal the most difficult part of the coil to heat, the cold spot, which in current practice is approximately one third of the way into the bottom or middle coil from the eye of the coil.

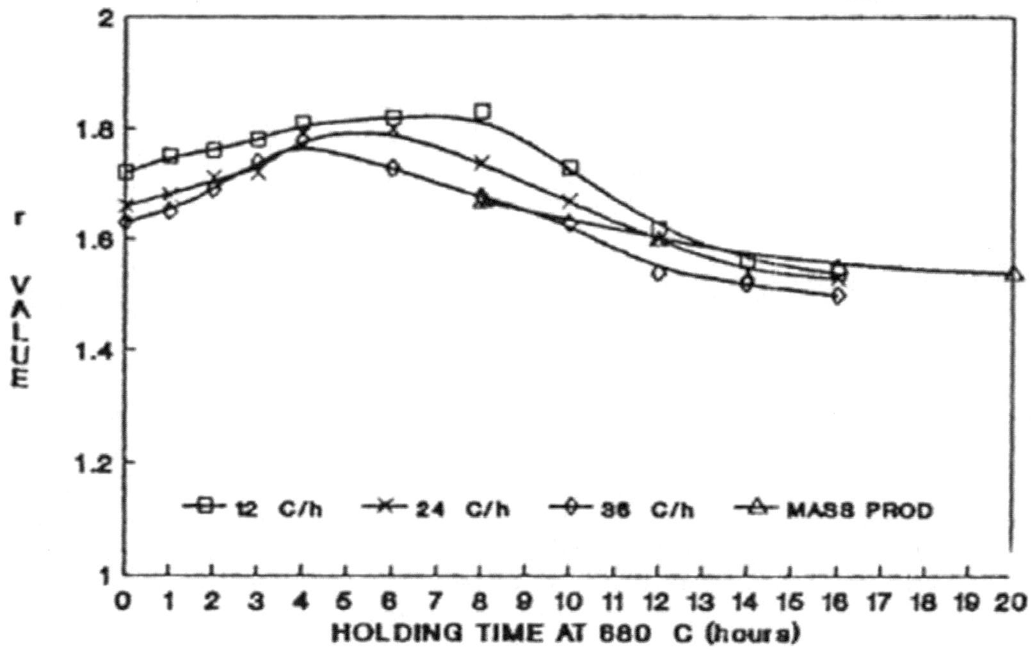

Figure 1. Effect of holding time on r value.

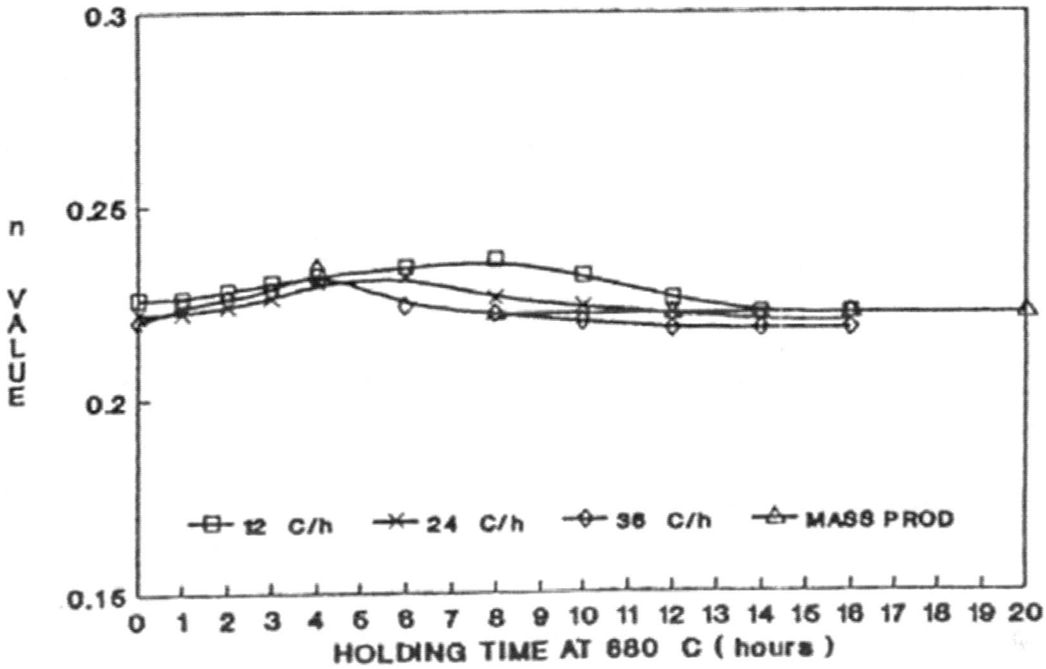

Figure 2. Effect of holding time on n value

Time at temperature during box annealing is considerably less important than maximum temperature attained. The difference in the laboratory and production results is also caused by extension in temper pass mill. Production scale is utilised extension in temper pass mill, but not used for laboratory scale. Temper pass is the term applied to the final processing of the cold rolled strip to impart the correct roughness of the surface and to strain sufficiently to suppress the yield point for the elimination of stretcher strains during forming treatment of the customer, so that the materials continue to stretch evenly up to the yield strength. An investigation by Shimizu et al [21] indicated that increasing the holding time and holding temperature resulted in a decrease in tensile strength of cold rolled rimmed steel sheet.

Pakkala et al [22] reported that very little improvement in properties is obtained by holding at subcritical temperatures for periods longer than 10 hours or in many instances beyond 5 hours. Furthermore, when annealing at 732 °C, no improvement is obtained from using time in excess of one hour. A rule of thumb that is frequently followed maintains that an increase in the temperature of 100 °C is roughly equivalent to a doubling of the time of annealing. Pakkala et al [22] studied the effect of annealing temperature and time on two nominally 0.040 inch thick rimmed steels of similar composition and processing (nominal 60 % cold reduction), except for coiling temperature. The results were generally representative of the trends expected for rimmed steel, i.e. the as-annealed hardness, yield, tensile strength and yield point elongation generally decreased as annealing temperature and annealing time are increased. Also, as would be expected, grain size increased with increases these variables. Elongation values unexpectedly showed little change with a variation of only 3 % and no consistent trend with the variables studied. The inconsistent relationship between elongation and the annealing variables may be related to local variations in composition and local variations in micro cleanliness.

Selected data from a study by Schwer et al [23] of the effects of annealing temperature on the as-annealed properties of low carbon rimmed steels (nominally containing 0.07 % C, 0.44 % Mn, 0.008 % P, 0.025 % S, 0.008 % Si, 0.02 % Cu, 0.01 % Ni and 0.004 % N.). indicated similar trends to those of Pakkala et al [22]. In addition, Schwer et al [23], showed an increase in the r- value and the n-value, associated with increasing annealing temperature. It should be noted that significant undesirable grain coarsening resulted from annealing these steels at 690 °C and higher, however, inasmuch as the annealing was conducted in stagnant air in a welded box, this grain coarsening could have been due in part to partial decarburisation. In this study, slow heating and cooling rates were used.

Espey [24] reported the results of his work with an 88 % cold reduction steel containing 0.09 % C, 0.45 % Mn, 0.007 %P, 0.026 %S, 0.003 %Si, 0.098 % Cu and 0.02 %Mo and related the mechanical properties to the effective temperature and location within a coil. Effective temperature is defined by Espey [24] as the temperature with slight upward variations at which the steel is held at for 3 hours. However, these result probably represent the most complete time/temperature work done on box annealing and are useful for predicting trends.

Trishman et al [25] conducted an extensive investigation of the effects of annealing temperature and time on the hardness and grain size of four low carbon steels. Their annealing practice was to heat at 482 °C/h to 538 °C, then at 1 °C/h from 482 to the annealing temperature, and furnace cooling after a specified hold at the annealing temperature at a rate that reportedly approximates plant cooling rates. These results indicate that increasing the holding time and holding temperature resulted in a decrease in hardness and an increase the grain size.

Thesima and Shimizu [26] reported on the effect of increasing the annealing temperature of rimmed steel. They concluded that whereas a temperature above Ac-1 may be good for drawability, on the basis of conical-cup values, inter-critical annealing may be deleterious to stretchability. Also, during annealing above the Ac1, cementite segregates to grain boundaries, and during cooling the precipitation of carbon is retarded, resulting in increased aging tendency due to carbon in solution.

Matsuda and Nakaoka [27] investigated the effect of increasing annealing temperature and cold reduction on the r-value and grain size of rimmed steel. Their data indicated that r-value is increased by increasing the annealing temperature from 700 °C to 800 °C. They further reported an almost linear relationship between ferrite grain growth and increase in r-value, which they expressed as:

$$r = ro + k^{-1/2} + do^{-1/2}$$

where:
k = cold reduction kept almost constant at 50-80 % (k<0)
r = final average r-value
ro= average r-value after completion of of recrystallization.
d = average diameter (mm) of final ferrite grains
do= average diameter of grains upon completion of recrystallization.

Matsuda and Nakaoka [27] suggested that this relationship is also applicable to low carbon Al-killed steel and extra low carbon Ti steels. Matsuda and Nakaoka [27] in the previously cited review paper reported the results of Takechi on the effects of increasing annealing temperature on the r-value of rimmed and killed steels. Takechi concluded that r-values of both rimmed and Al-killed steel increased as the annealing temperature is increased from about 600 °C-850 °C then decreased significantly when the annealing temperature was further increased. However, the increase in r-value for Al-killed steel appeared to be slight from about 732 °C-850 °C. Schwer et al [23] reported data for four heats of Al-killed steel properly processed on the hot mill and cold reduced 45 to 50 % before being annealed for 16 hours at temperatures ranging from 650 °C-746 °C. These results, which are the averages of data obtained for the four heats, indicated the average change in mechanical properties that can be expected for Al-killed steel sheet for an increasing annealing temperature. As expected, hardness, yield and tensile strengths decreased and r, n value, grain size and elongation increased with an increase in annealing temperature.

A study conducted at the USS Research laboratory [28] was concentrated primarily on the effects of cold reduction, but two annealing practices were included: 6 hours at 677 °C and 16 hours at 705 °C. The results of this study of sheet cold reduced 50, 60 and 70 %, showed a modest improvement in hardness, yield and tensile strength elongation, associated with the higher temperature, longer-time anneal. However, identical r-values and essentially identical r-values resulted from both annealing practices. These results also showed the expected increase in r-value and decrease in r-value with increasing cold reduction from 50 to 70 %. They did not show any deleterious effect of increasing cold reduction from 50 to 70 % on hardness, yield strength or tensile strength. Although the elongation values showed a decrease with increasing cold reduction, calculations indicate that this decrease is essentially due to the reduced thickness resulting from increased cold reduction, and is not an effect of cold reduction. Thesima and Shimizu [26] studied the effect of annealing temperature (690 °C-750 °C) and holding time (2 - 18 hours) on Erichsen, conical cup (CCV), and r-value. Their results, which are somewhat different for the two steels studied, generally showed a slight improvement in r-value with an increase in annealing temperature. For one steel, the Erichsen value (stretchability) decreased with increasing annealing temperature, whilst for the other, there was no consistent effect. Likewise, conical-cup values (drawability) improved with increasing temperature for one steel, but increased annealing temperature showed no consistent effect for other steel. Holding time increases from 2 to 18 hours showed no consistent effect on the properties.

Sardana [29] studied the effect of annealing at 700 °C (20 and 30 hours) vs annealing at 740 °C (23 and 30 hours). The latter temperature was in the two-phase or inter-critical region. Sardana's data suggest that whereas yield strength is generally reduced by increasing annealing temperature and holding time, increasing temperature and is always associated with improvements in elongation and cup test performance. In general, the references discussion tend to support the result obtained in the present work.

IOP Conf. Series: Materials Science and Engineering **547** (2019) 012060 doi:10.1088/1757-899X/547/1/012060

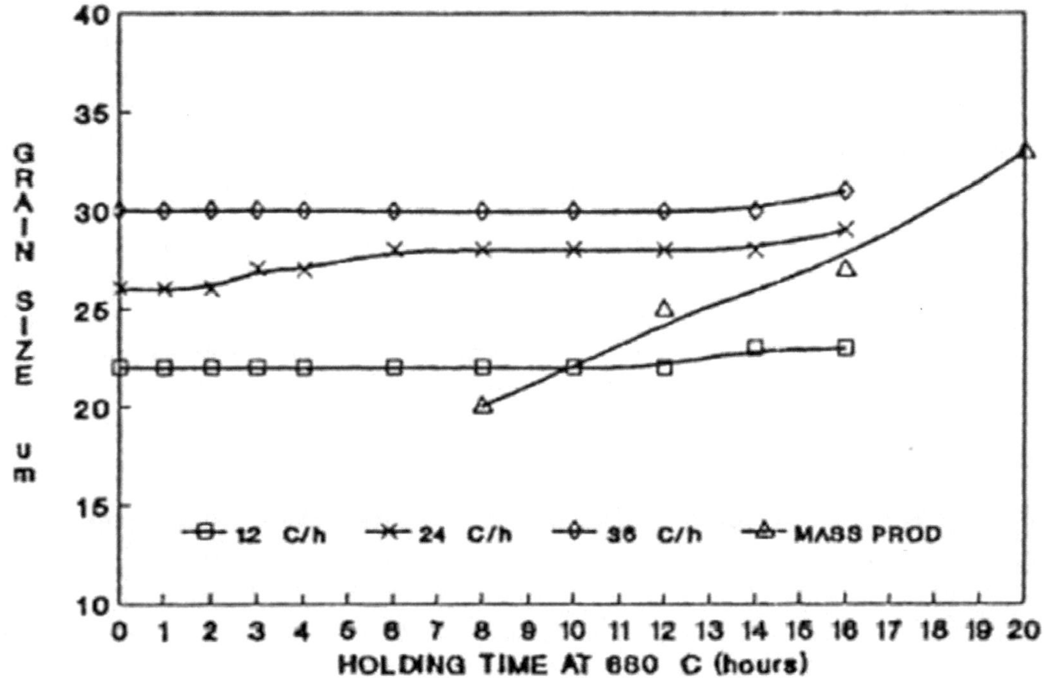

Figure 3. Effect of holding time on grain size

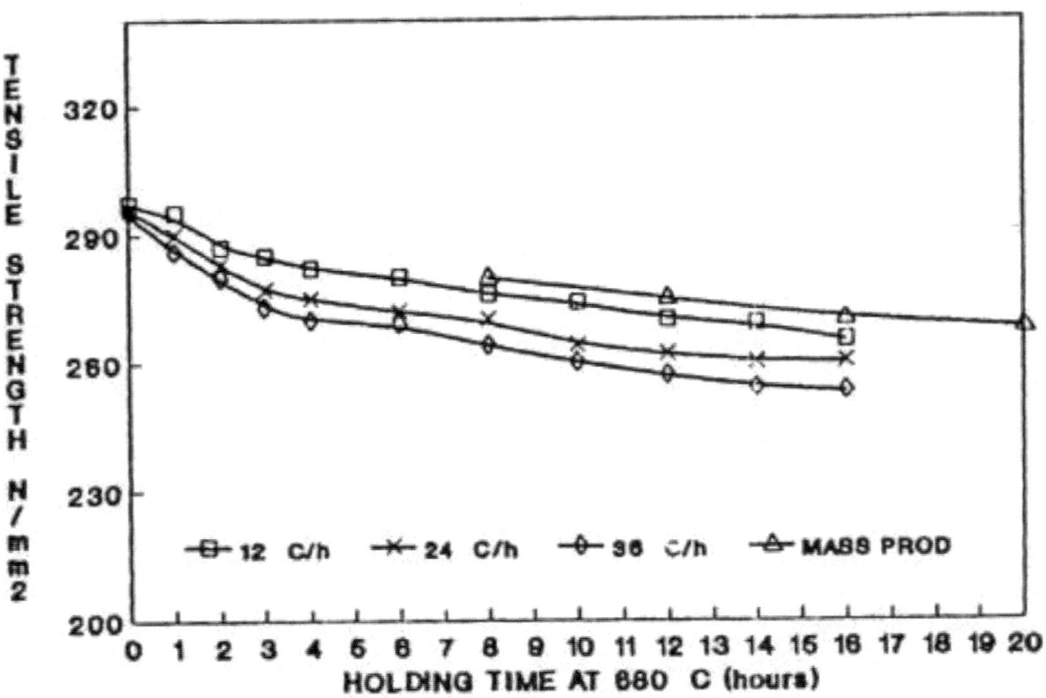

Figure 4. Effect of holding time on tensile strength.

IOP Conf. Series: Materials Science and Engineering **547** (2019) 012060 doi:10.1088/1757-899X/547/1/012060

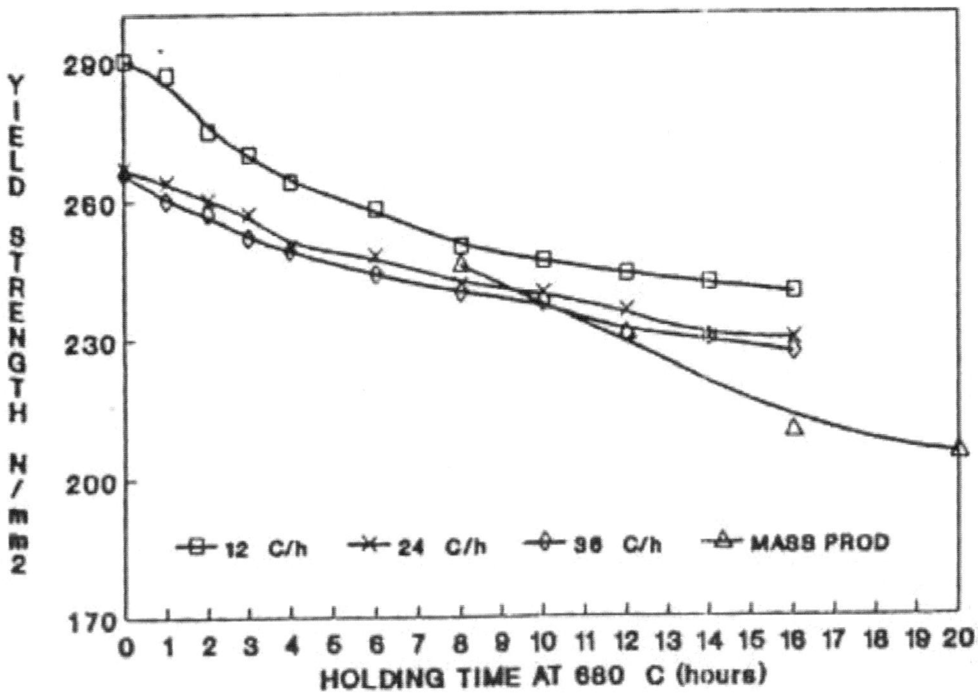

Figure 5. Effect of holding time on yield strength.

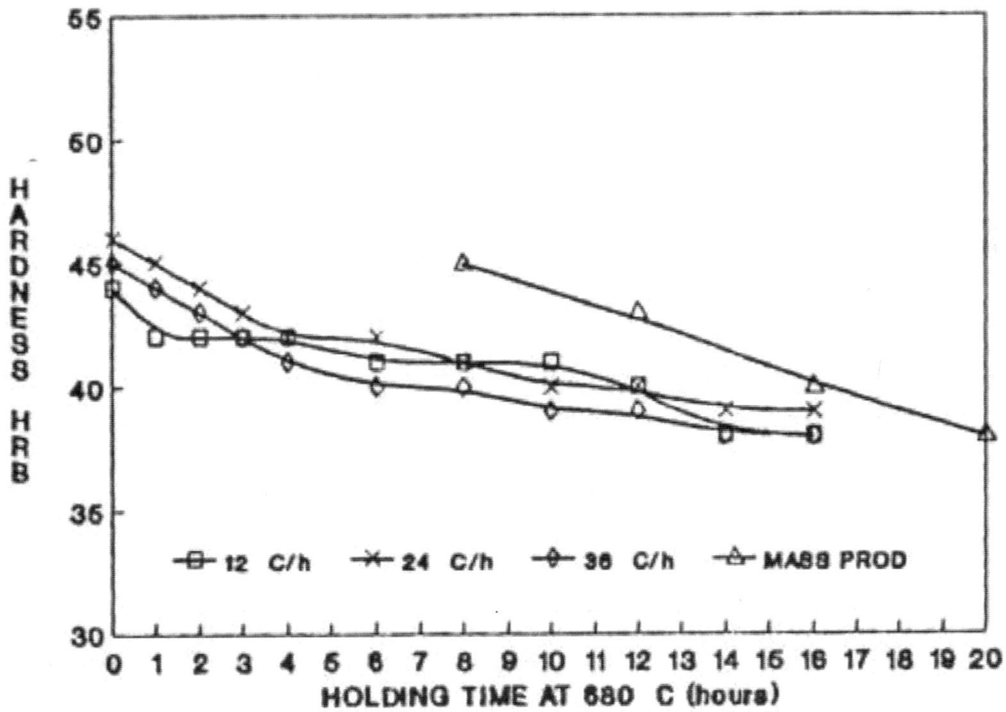

Figure 6. Effect of holding time on hardness.

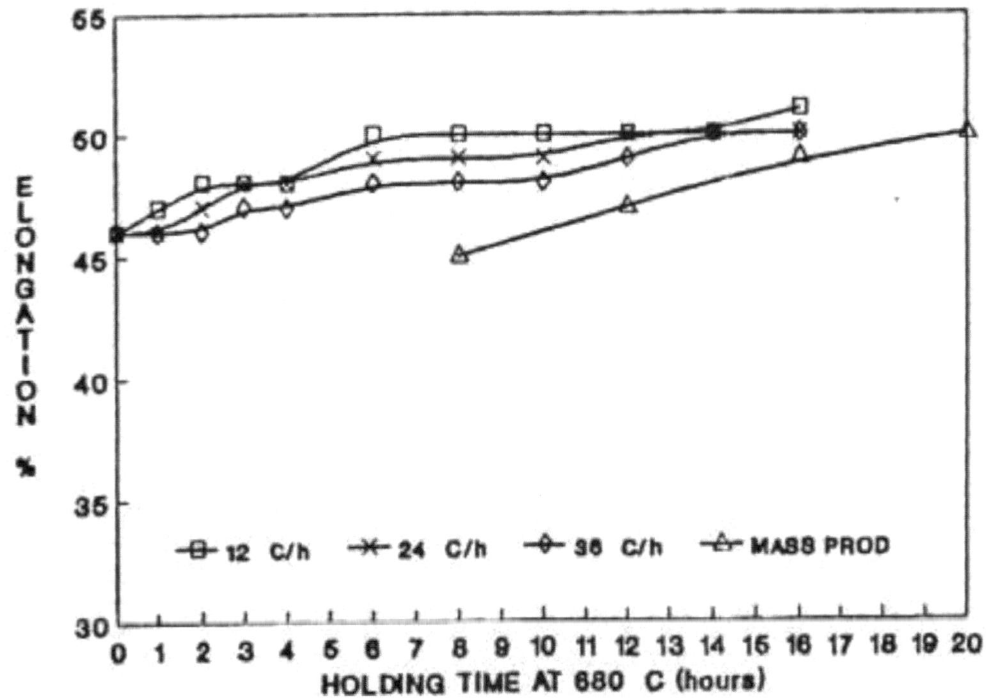

Figure 7. Effect of holding time on elongation

Effect of Annealing Cycle on Microstructure. Figure 8 shows the microstructure of mass-produced batch annealed steel. These photomicrographs show that the width of the grains is much smaller than the length of the grains. In other words, the structure is elongated or has a "pancake" shape. The effect of a holding temperature of 680 °C for times up to 16 hours is that grain size increases gradually, starting from a fully recrystallized grain size of approximately 20 μm and finishing at a size of about 35 μm after 20 hours. The grain dimensions after a holding time of 20 hours still produced a pancake shape.

The results obtained by metallographic point counting are shown as plots of volume fraction recrystallized versus time in Fig 9. The curves for different heating rates show the classic sigmoidal shape associated with recrystallization. There is an incubation stage, which depends on heating rate, followed by rapid recrystallization and a final stage during which recrystallization rate slows down as grain impingement occurs. A lower recrystallization temperature was associated with a lower heating rate. The recrystallized grain size of the steel used was determined for heating rates of 12 °C/h, 24 °C/h and 36 °C/h by the circular intercept procedure. The results are shown in Figs 3 and 10 The starting grain size as rolled grain condition was approximately 10 to 11 um. For a heating rate of 12 °C/h recrystallization started at a temperature of 508 °C, when the grain size was approximately 13 μm. For a heating rate of 24 °C/h recrystallization started at 544 °C and for a heating rate of 36 °C/h the recrystallization started at temperature was 556 °C. The grain size at the start of recrystallization was similar in the three cases. During recrystallization, the grain growth at a heating rate of 36 °C/h was more pronounced than for a heating rate of 24 °C/h and also for 36 °C/h. The grain size averaged approximately 22 μm for a heating rate 12 °C/h, 26 μm for

IOP Conf. Series: Materials Science and Engineering **547** (2019) 012060 doi:10.1088/1757-899X/547/1/012060

rate strongly influences recrystallization. The smaller the heating rate the lower the temperature of the start of recrystallization and the smaller the recrystallized grain size. The implication heating rate 24 °C/h and 30 μm for 36 °C/h. From the grain size versus temperature curves, it is evident that the heating is that the lower rate promotes nucleation of recrystallization overgrowth of recrystallized grains. The effect of holding time on the recrystallized grain size at 680 °C is also presented in Fig 3. This plot shows that grain size mostly changed a little on holding and was lower at the lower heating rate. The grain size for heating rate 12 °C/h was approximately constant at 22 - 23 μm, for 24 °C/h, 26 to 29 um, and for 36 °C/h 30 to 31 μm.

Optical microscopy also identified features associated with the mode of nucleation of recrystallization. From observations of the grain boundaries during the course of recrystallization, it was evident that there was a large degree of grain boundary migration occurs as a precursor to recrystallization. Fig 11 an extraction replica of this study shows that the AlN has precipitated at the former as rolled grain and sub-grain boundaries by the time recrystallization are substantially complete. It presented that the recrystallized grains are elongated. Inspection with TEM of the areas within these grains reveals traces of aluminium nitride precipitates at the former as rolled grain boundaries, the results for all three heating rates indicated that the grain size is relatively constant with increasing time up to 16 hours at a holding temperature of 680 °C, because of retardation of grain motion by layer of precipitates of AlN at prior cold worked boundaries. Dunne and Dunlea [16] concluded that in aluminium killed steel, nucleation by sub-grain growth is selectively inhibited by clusters or precipitates at sub-grain boundaries and growth of recrystallized grains is anisotropically retarded by precipitates at prior cold worked boundaries. They concluded that texture development is primarily dependent on nucleation events and that grain shape and growth are determined by other physical conditions. Elongated grain is frequently associated with a favourable texture, but in some cases, the link between texture and elongated grains can be broken. By increasing the time and temperature of pre-recrystallization anneal, and hence the amount of AlN precipitated.

Goodenow, Michalak and Leslie [9-11] observed that elongation of the recrystallized grains increased. Precipitation of AlN occurred throughout the annealing process and the extent of precipitation was determined by standard analytical techniques, which are said to account only for precipitate particles greater than about 200 A.

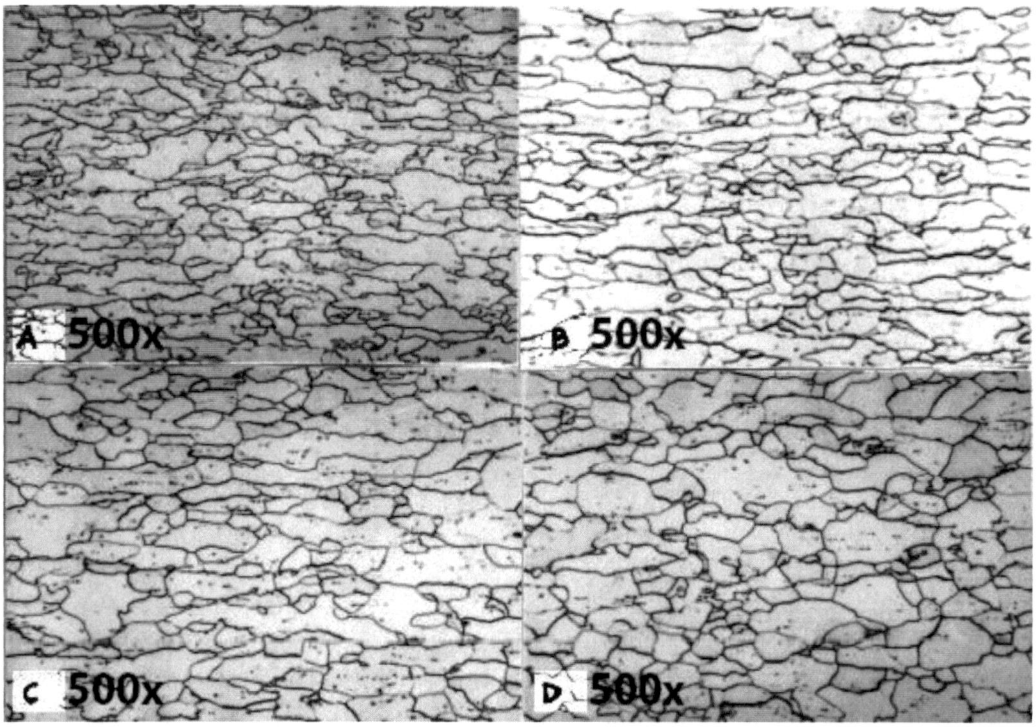

Figure 8. Photomicrographs (500 X) of Al-killed steel fully recrystallized after holding at (a) 640 °C for 8 hours (b) 680 °C for 12 hours (c) 680 °C for 16 hours (d) 680 °C for 20 hours.

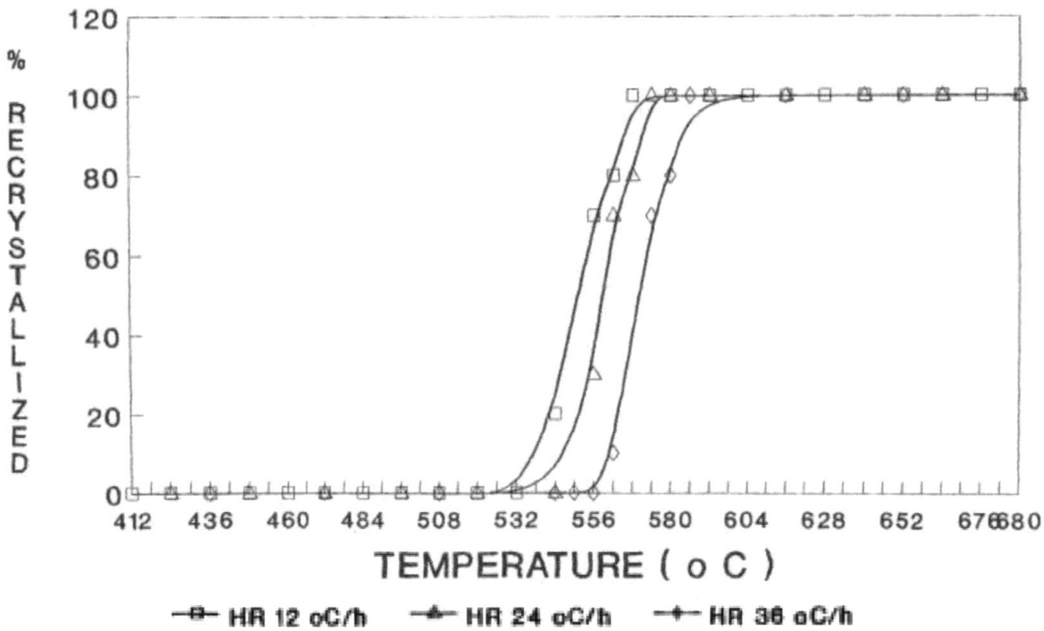

Figure 9. The volume fraction of recrystallized grain as a function of temperature.

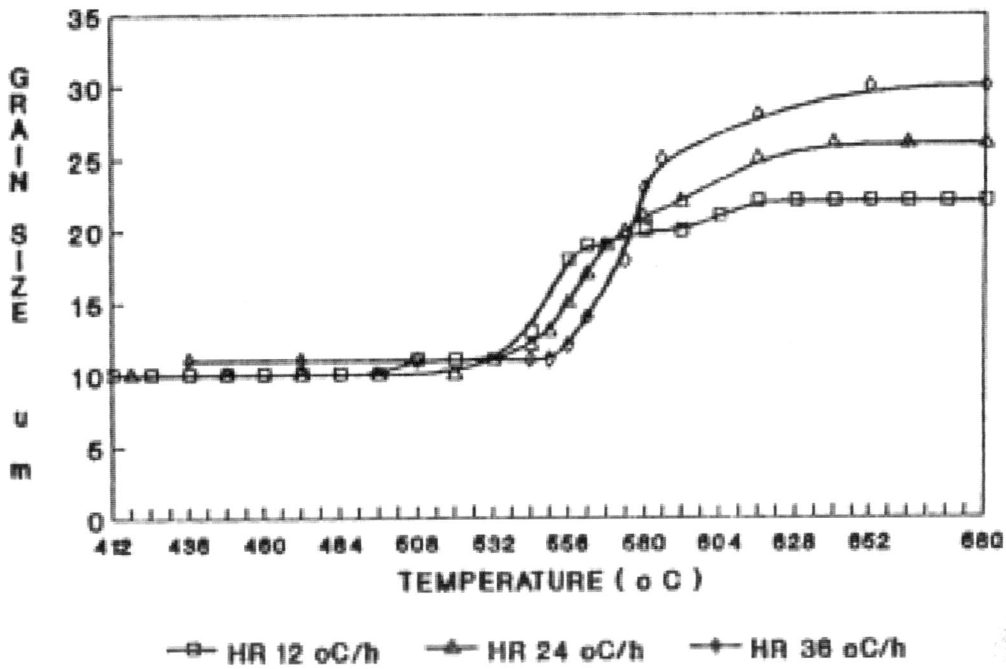

Figure 10. Effect of temperature on grain size

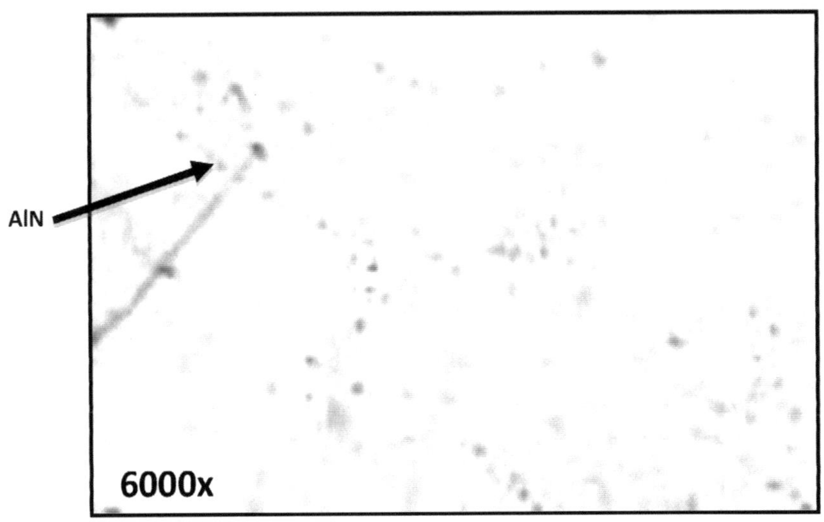

Figure 11. Carbon extraction replica by transmission electron microscope (TEM) of aluminium killed steel annealed by holding time at 680 °C for 16 hours, with the heating rate of 24 °C/ h: Aluminium nitride precipitate delineates cold worked grain and sub-grain boundaries.

4. Conclusions

The low carbon aluminium killed steels studied in this work have the r-value increased gradually as the temperature increased for the heating rate of 12°C/h, and for isothermal holding at 680°C, the r-value increased for up to 8 hours, after which it is decreased. Likewise, for the heating rate of 24°C/h the r-value increased to a peak after 6 hours, and for the heating rate of 36°C/h the peak occurred after 4 hours. It is caused by the aluminium and nitrogen combine to form atmospheres or pre-precipitation cluster at polygonized sub-grain and as rolled boundaries, modifying the development of the recrystallized structure to elongated, nucleation by sub-grain growth is selectively inhibited by clusters or precipitates at sub-grain boundaries. However, after holding time of 8, 6, 4 hours for the heating rate of 12, 24, 36 °C/h, AlN precipitate cannot lock the elongated structure anymore.

The effect of holding time on mechanical properties was: (i) decrease in tensile strength, yield strength and hardness; (ii) increase in elongation and grain size. Tensile strength and hardness results from laboratory experiments were lower than those for production samples, but the trends of curves were similar. Holding times that produced the coarsest grain size also resulted in the lowest yield strength, tensile and hardness, for all three heating rates.

5. Acknowledgments

The authors gratefully acknowledge PT Krakatau Steel (Persero) Tbk Cilegon Banten for providing material, plant and research facilities.

IOP Conf. Series: Materials Science and Engineering **547** (2019) 012060 doi:10.1088/1757-899X/547/1/012060

References

[1] W. T. Lankford, S. C. Snyder and C. A Bausher, ASM Trans 1950, 42 1197.
[2] I. S. Brammar, T. R. Thomson and R. M. Hobbs, The Journal of Australian Institute of Metals, Vol 17, No 3. 1972, 147, 148.
[3] S. R. Goodman and Hsun Hu, Metallurgical Transaction, vol 1, 1970, 1629
[4] R. L. Whiteley and D. E. Wise, Flat Rolled Product III, Interscience, NY, 1962, 47-65
[5] Sanak Mishra and C. Darman, International Metals Reviews, vol 27 no 6, 1982, 308.
[6] W. Jolley, Journal of The Iron Steel Institute, March, 1967, 322.
[7] R. L. Solter and C. W. Beattie, Journal of Metals Transaction AIME, Sep, 1951.
[8] D. J. Blickwede, Transactions of the ASM, vol 61, 1968, 662-663.
[9] R. H. Goodenow, Transaction of the ASM, vol 59, 1966, 810.
[10] J. T. Michalak and R. D. Schoone, Transaction of Metallurgical Society of AIME, vol 242, 1954, 1968, 1151.
[11] W. C. Leslie, R. L. Rickett, C. L. Dotson and C. S. Walton, Transaction of the ASM vol 46, 1954, 1470-1499.
[12] P. N. Richards, Journal Australian Institute of Metals, vol 12, 1967, 2-9.
[13] W. C. Leslie, J. T. Michalak and F. W. Aul, "*The annealing of cold worked iron, Iron and Its Dilute Solid Solutions*", John Wiley & Sons, New York 1963, 119.
[14] H. Hu, Trans AIME, 224, 1962, 75.
[15] J. T. Michalak and H. W. Paxton, Trans AIME, 221, 1961, 850.
[16] D. P. Dunne and R. L. Dunlea, Metallurgical Forum 12, vol 2, no 3, September 1978, 160, 161.
[17] I. L. Dillamore, C. J. Smith and T. W. Watson, Met Sc Journal I, 1967, 49.
[18] I. L. Dillamore and S. F, H. Fletcher "*Recrystallization, Grain Growth and Texture*" pp 448-53. ASM, Metals Park 1966.
[19] W. B. Hutchinson, International Metals Reviews, 1984, vol 29 no1, p 30-31.
[20] W. B. Hutchinson, International Conference on Texture of Metals, 1993, Toronto.
[21] M. Shimizu, K Matsakura, Takahasi and Y Shinagawa, Japan Iron Steel Institute, Japan, 1964, 50.
[22] M. H. Pakkala, R. L Rickett, J. T. Mackenzie, Jr. and N. A Matthews "*Experimental Annealing of Cold Reduced Low Carbon Steel*", USS Research Report 398, Jan 11, 1940.
[23] J. W. Schwer, E. Umene, J. W. Stewart and H. C. Schneider, "*Effect of variations in Annealing cycle on Tensile Properties and Plastic-flow characteristic of sheet steel* "USS Research Report, Oct 1958.
[24] G. B. Espey "*Metallurgical Control of Coil Annealing in Radiant Convention Furnace* " Iron and Steel Engineer, December, 1949, p 99.
[25] L. E. Trishman, D. H. Rowlan, A. G Thayer and J. H. Steven, "*Experiments Annealing of Low Carbon Reduced Sheet Steel*" USS Research, February 26, 1941.
[26] S. Thesima and M. Shimizu "Recrystallization Behaviour of cold rolled mild steel" Sixth AIME Mechanical Working and Steel Processing Confe, Chicago, January 31. 1964.
[27] Kazua Matsuda and Kazuhide Nakaoka, "*Effect of Processing Conditions on the Deep Drawability of Cold Rolled Steel Sheet*" Source Book on Forming of Steel Sheet, ASM Engineering Bookshelf, Editor: John R Newby, 1978, p 197-200

[28] H. M. Alworth and T. E. Johnson "*Effects of Cold Reduction on Mechanical Properties*" USS Research Report.
[29] *V. P. Sardana, "Metallurgical Factors Influencing Sheet Metal Formability*" Supplement to Transactions of the Iron and Steel Institute of Japan, vol 11, 1971, 911.

IOP Conf. Series: Materials Science and Engineering **547** (2019) 012061 doi:10.1088/1757-899X/547/1/012061

Failure Analysis of Blade on Coconut Shell Crusher Machine

Surasno[1]*, Apriardi Ihlas[2]
Center for Material and Technical Product
Jl. Sangkuriang No. 14 Bandung 40135, Indonesia
*E-mail: surasno2005@yahoo.co.id

Abstract. Blade mounted on coconut shell crusher machine was frequently having damaged while being operated in making of coconut shell powder at small and medium industries. This took place repeatedly during the operation, therefore blade repairing and replacement has to be done for every 16 hours at the wear rate of 24.28%. A research was carried out to find out the cause of damage to the metal blade material on the coconut shell crusher machine. The scope of this research included the observation on visual examination, dimensions examination, chemical composition, hardness, and metallography. The results of observations and tests showed that blade was broken due to abrasion and deformation during operation. The metal material used for the blade was low carbon steel. Chemical composition of the metal content of the blade were: C (0.103 up to 0.144 %) and Mn (0.657 up to 1.400 %). Hardness value of blade metal materials were 146 up to 148 Hv. Microstructure consisted of ferrite and pearlite. This low carbon steel material is not suitable to be used as a blade on coconut shell crusher machines for producing of coconut shell. Repairing and maintenance of welding coating methods on blades using low carbon electrodes does not give increased significant value of hardness. Blade replacement in coconut shell crusher machines must comply with the requirements of tool steel with suitable heat treatment or the use of low carbon steel with the carburization process.

Keywords: blade, coconut shell machine, low carbon steel, failure analysis

1. Introduction

Coconut shell powder at small and medium industries as seen in Figure 1 was an organic fiber to be used in the manufacture of mosquito repellent. This material is flammable, but burnt out slowly. The process of making coconut shell powder uses a coconut shell crusher machine with 150 HP and a 3.100 rpm engine speed. In this machine, 16 blades were used to rotate and grind the coconut shells into fine crusher. Each coconut shell crusher machine has a production capacity of 1 ton/day of coconut shell powder and every day operates 16 hours. (See Figure 2).

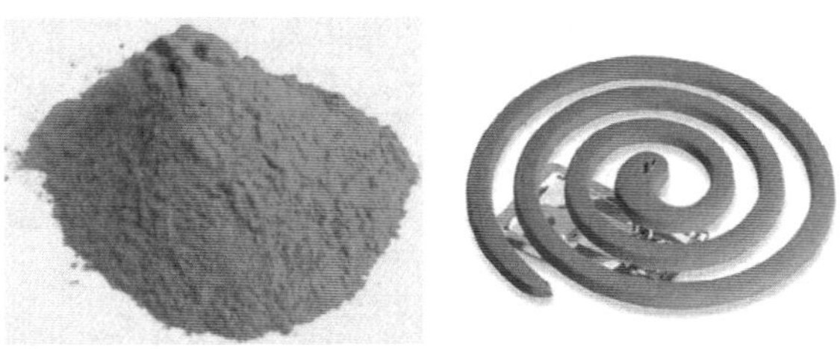

Figure 1. Coconut shell powder (left) [1] and Mosquito repellent (right) [2]

Figure 2. Mosquito coil factory

The use of blades on coconut shell crusher machines is often a failure, resulting in a decrease in lifetime of blade. Thus the steel material is not suitable to be used as a tool steel material, because the blades must be replaced and repaired frequently. This is an obstacle for machine operators and company expenses. Therefore, the inappropriate use of metal blade material is a case of failure that must be analyzed.

Investigation of failure analysis is a research process to determine the causes of failure of a production equipment through: investigation of steel materials used for blades, investigation of the operation of coconut shell crusher equipment operated by factory operators, investigation of blade designs installed on machine tools coconut shell crusher, and operating environment investigations on the durability of steel materials used [3].

This research will be focused on steel blade materials on coconut shell crusher machines that have been damaged and are still new (See Figure 3). Broken blade can occur due to the use of metal as an unsuitable blade raw material. This is indicated by the appearance of defects on the blade surface as an impact of improper abrasion process resulting in a shorter lifetime using this blade [4].

Abrasion is the initial phenomenon of damage to the blade steel surface by removal and release of friction material [5]. Abrasion occurs on the surface of the blade material when particles of organic material, stone, or sand rub against the surface of a blade steel material that rotates in a coconut shell crusher machine. The process of abrasion and collision of blade surface due to the rotation of the coconut crusher machine affect the blade weight and dimensional reduction progressively [3, 10].

Figure 3. Blade mounting position

Changes in the dimensions of the blade due to the abrasive process cause unstable rotation and the occurrence of collisions/impacts of two different materials, so that there will be deformation on the surface of the blade surface in contact with different levels of deformation. Normally, wear problems are different from direct damage. Wear of hard metal parts produces a smooth surface, glides or rolls with pressure on the surface [6].

The surface of the metal material with low hardness value is easier to undergo an abrasion [3]. The chemical composition of carbon steel with low carbon and manganese will not raise significant hardness even though the heat treatment process is carried out [7]. Blade repair by coating the surface using the SMAW welding process using filler metal electrode RB-26 (Kobe Steel) as a metal surface welding surfacing hardener, the blade hardness value remains low [8]. The abrasion process of the blade surface is generally uneven will affect the rotation of the engine imbalance and will accelerate surface deformation by the process of pounding and collision [9]. From the problems faced by the manufacturer of coconut shell `crusher`, it is necessary to conduct research to improve the performance of coconut shell `crusher` machines.

2. Materials and Methods

The blade material used for this research is a broken blade material and a new blade material equivalent to low carbon steel AISI 1010.

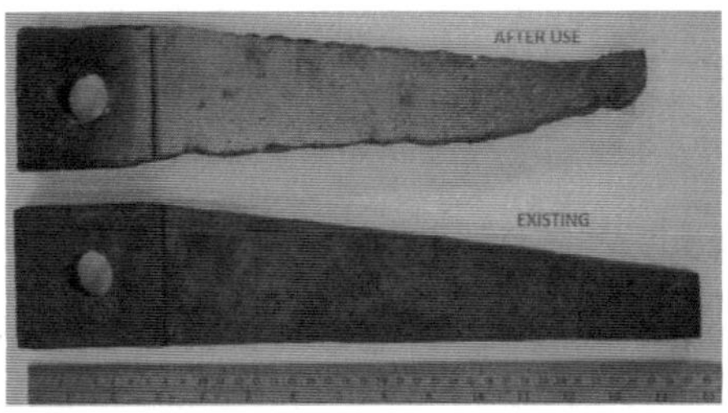

Figure 4. Difference between used blades and new blades

The research was carried out through visual observation on the surface of the damaged blade (existing) and visual observation of the blade before being used as a comparison. The purpose of this observation was to obtain an overview of broken blade from observing the cross-section surface of the broken section.

The purpose of a macrostructure examination is to observe the cross section of a blade, compare a new blade and a broken blade, as well as the welding treatment of a broken blade.

Chemical composition examination using spectrometer was objected to determine the element of the basic metal composition of the blade, the composition of the chemical elements of the blade material as a basic for determining the type of parent metal material used, whether it is suitable to be used as a coconut shell crusher. Chemical composition testing was carried out on both coconut shell crusher machine blades which were new and damaged.

Testing the hardness of the cross section of blade was carried out by means of the Vickers Hardness test. Hardness value were used as a consideration for the performance of coconut shell cutting tools, compared it with the hardness of tool steel. Hardness test were mainly carried out on welded surfaces. In addition, hardness test was also carried out on the two parent metals to estimate the correlation between hardness values and the value of chemical composition, especially carbon and manganese content. The test was carried out using the Vickers method with a load of 10 kgf.

Metallographic observation by means of an optical microscope was conducted to determine the type of microstructure of metal materials for new blades and material of broken metal blade.

3. Results and Discussion

Visual examination

After 16 hours of use, all coconut shell crusher blades were damaged in the form of wear, abrasion and deformation (See Figure 5). Therefore, every blade usage always uses the new one or that has been repaired. Visual examination on the surface of the blade of coconut shell crusher indicated wear defects of the surface (surface abrasion) and deformation. Due to the wear abrasion, surface erosion has occured (See Figure 4). Deformation defects and surface wear on the damaged blade, in the form of surface digestion, estimated by metal blade material with low hardness.

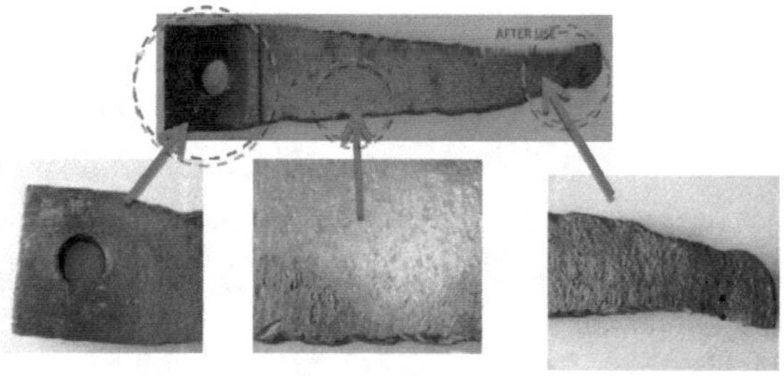

Figure 5. Texture of blade after the abrasion process

Coating weld of the back side and the sharp side of the blade by SMAW welding process with electrode of RB 26 kobe steel (See Figure 6). Welding patches on sharp parts are intended for direct collision with coconut shell product, appear to be damaged due to abrasion and deformation by strong friction. The abrasion resistance and strength of the metallic material of the weld layer on the blade is insignificant.

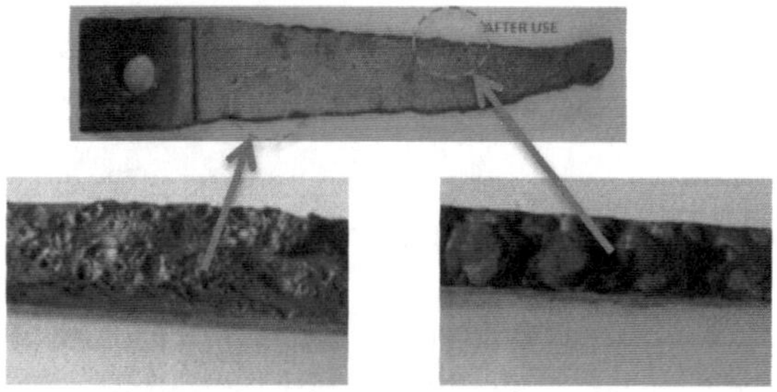

Figure 6 Welding layer on the blade

The bolt hole for tightening is deformed which can occur by the force of pounding from the imbalanced engine speed. This phenomenon may be due to the uneven blade surface abrasion resulting in changes and different sizes. The unstable rotation generate the pounding which resulting in deformation. Uneven blade abrasion resulting in unstable engine speed, so that deformation arises in the bolt hole of the blade.

IOP Conf. Series: Materials Science and Engineering **547** (2019) 012061 doi:10.1088/1757-899X/547/1/012061

Dimension reduction

The initial dimensions of blade on the coconut shell crusher machine was trapezoidal shape as seen in Figure 5. After the coconut shell crusher machine was operated for 16 hours there was a reduction in the size of the blade thickness of 7 mm, the size of the blade thickness was originally 14 to 8 mm thick, while the reduction towards the blade width is 20 mm, from original size of 70 to 50 mm.

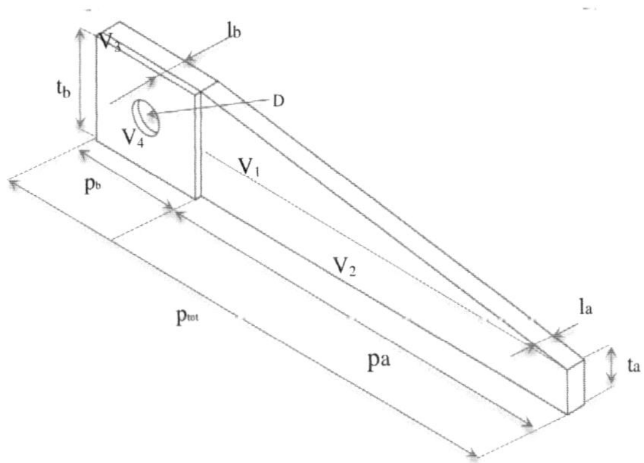

Figure 7. The dimension of crusher blade

Weight reduction

From figure 4 and 7, the new coconut shell crusher blade (1) and broken coconut shell crusher blade (2) were obtained:

- ptot1 = 389 mm, (pa1 = 309 mm, pb1 = 80 mm),
  la1 = 14 mm, lb1 = 19 mm,
  ta1 = 30 mm, tb1 = 70 mm.
- Hole diameter, D = 22 mm, r = 11 mm.

V = V1 + V2 + V3- V4
= (½ (pa x (tb-ta)) x la) + (pa x la x ta) + (pb x tb x lb) - (pi x radius^2 x lb).

Volume is obtained for the new blade, $V_{new} = 328.850\ cm^3 = 0.00032885\ m^3$.

In the mass measurement for the new blade, (M_{new}) = 2.9 kg.
Then the density of the blade material, $\rho_{new} = 8{,}818.61\ kg/m^3$.
Since the new and damaged blade is the same material, ($\rho_{broken} = \rho_{new} = 8{,}818.61\ kg/m^3$)

To measure the volume of an irregular surface,
calculate mass of the broken blade i.e, M_{broken} = 2.2 kg., then:

$V_{broken} = M_{broken}/\rho_{broken} = 0.000249\ m^3$.

These the dimension reduction on the broken blade is,
$\Delta V = ((V_{\_new}\text{-}V_{\_broken}) / V_{\_new})$ x100%, ΔV = 24.28%.

IOP Conf. Series: Materials Science and Engineering **547** (2019) 012061 doi:10.1088/1757-899X/547/1/012061

Loss of weight and dimensions resulting in damage to all blades as many as 16 pieces mounted on the coconut shell crusher machine after the machine is operated. Similarly, the high of coconut shell hardness affect the deformation and abrasion of the surface of the blade metal due to impact and friction with low carbon steel surfaces. The hard coconut shell and its non-uniform dimensions are the obstacles to the operation process of making coconut shell crusher which affect the unstable collisions so that the deformed blade bolt hide occured (Figure 5).

Macrostructure

Macrostructure testing results showed that the cross section of the new blade was rectangular (Figure 8a), while that of broken blade has decreased in thickness and width after 16 hours of operation (Figure 8b).

Broken blade material has been welded and subjected to abrasion during use and deformation of the bolt hole, and no indication of heat treatment. It was known that improper repairing has been applied.

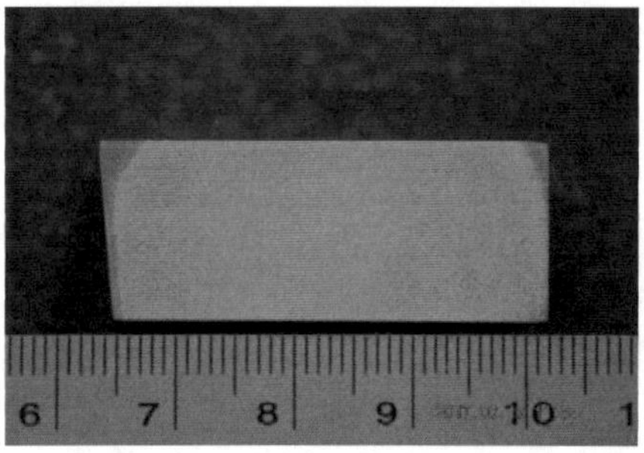

Figure 8a. Cross section of the New Blade

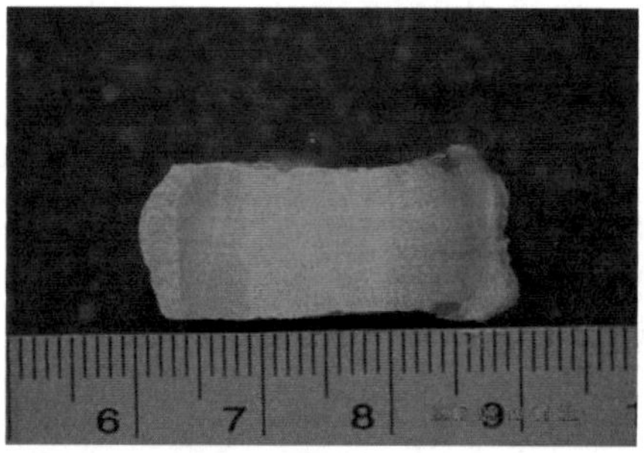

Figure 8b. Cross section of the Broken Blade

Chemical Composition

The chemical composition of blade is shown in Table 1. New blades and broken blades include low carbon steel groups (AISI/SAE 1010 standard). The broken blade contain of higher carbon compared to new blade.

Table 1. Chemical composition of blade

No.	Elements	New blade, %	Broken blade, %	AISI 1010, %
1	**Carbon**	**0.103**	**0.144**	**0.08-0.13**
2	Silicon	0.233	0.359	0.15-0.35
3	**Manganese**	**0.657**	**1.40**	**0.30-0.60**
4	Phosphor	0.0156	0.0132	≤ 0.040
5	Sulfur	0.0056	0.0034	≤ 0.050
6	Chrom	0.0094	0.0158	≤ 0.20
7	Molibdenum	0.00049	0.00057	-
8	Nickel	0.0134	0.0158	-
9	Aluminium	0.0343	0.0325	-
10	Cupper	0.0142	0.0118	≤ 0.25
11	Vanadium	< 0.0001	<0.0020	-
12	Wolfram	< 0.0002	<0.00020	-

Hardness

Hardness of the new and broken blades are presented in Table 2. Hardness of new blade cross section was in the range of 117-123 Hv, while the hardness of the broken blade cross section included the base metal (BM) zone (146-148 Hv), HAZ (152-154 Hv) and the weld metal (WM) zone (171-179 Hv). The welding process resulting in the difference of blade hardness. Hardness of the blade with BM is lower than that of WM and HAZ. The highest hardness was found in the WM zone. High increase in hardness was occured due to an increasing in heating welding process. The low value of hardness of new blade cross section compared to that of broken one were due to the higher content of carbon in broken blade.

Table 2. New and broken blades

No.	**New blade**	**Broken blade**
1.	118 (BM)	179 (WM))
2.	117 (BM)	171 (WM)
3.	123 (BM)	152 (HAZ)
4.		154 (HAZ)
5.		146 (BM)
6.		148 (BM)

Microstructure

The microstructure testing results showed the presence of ferrite and pearlite with the grain size of 8 (See Figure 9). Pearlite compounds tend to be much more in the broken blade; this comply with its higher carbon content. The change in microstructure for welded broken blade was due to the heat effect in the parent metal zone by welding process.

The microstructure of ferritic was found in the WM zone with the grain size of 9 which were finer than its parent metal. This will contribute to the higher hardness of broken blade.

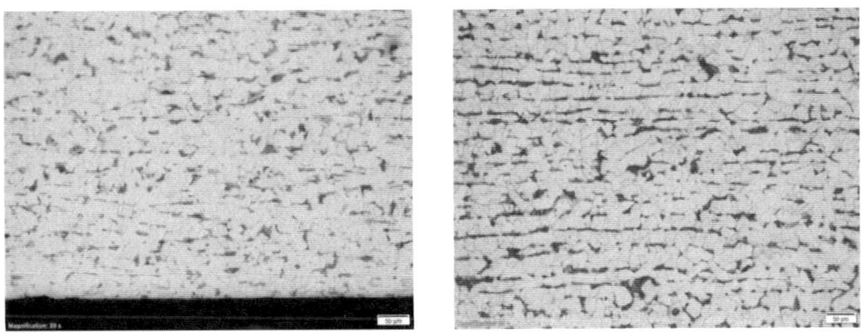

Figure 9. Microstructure of new blade (left) and broken blade (right)

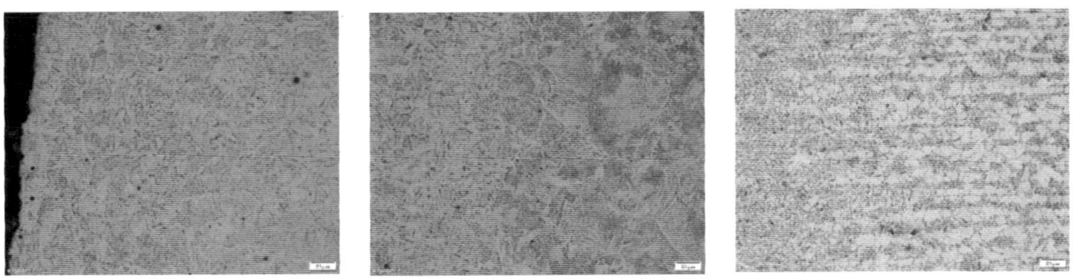

Figure 10. Photograph of the broken blade microstructure WM (left), MW-dilution-HAZ (middle), HAZ-BM (right)

Blade repair and maintenance by welding SMAW using low carbon steel electrodes does not improve its performance.

4. Conclusions

1. Blade failure in coconut shell machine were occured due to abrasive wear and deformation for every 16 hours operation at the wear rate of 24.28%.
2. The use of low carbon steel material AISI 1010, is not suitable for coconut shell machine blade. The low of hardness as well as its microstructure of ferrite pearlite will not be resistant to abrasion and impact.
3. Heat treatment of blades from low carbon steel materials will not increase the value of hardness.
4. In order to improve the materials blade for coconut shell crusher machines, it is recommended to use steel materials suitable for heat treatment tool steel (M3, M4, or M10) for increased abrasion resistance or using low carbon steel with carburizing hardening heat treatment [13, 14].

5. Acknowledgements

The authors would like to express gratitude to the Center for Material and Technical Product (B4T) and PT. Indratma Sahitaguna which have supported this research. We would like to thank our colleagues and technicians who have assisted conducting this research. We would also like to thank Mr. Kosasih as Head of Technical Inspection B4T for the assistance in discussion and analysis as well as Mrs. Wieke Pratiwi as senior researcher for comments to improve the manuscript.

6. References

[1] Coconut shell powder, 12 March 2014, http://aliranalam.blogspot.com/2014/03/tepung-batok-kelapa.html.

[2] Mosquito repellent, 20 August 2018, http://anti-nyamuk.blogspot.com/p/penangkal-nyamuk.html.

[3] A. Srithar, K. Palanikumar, and B. Durgaprasad, "Experimental Investigation and Surface roughness Analysis on Hard turning of AISI D2 Steel using Coated Carbide Insert", *Procedia Eng.*, vol. 97, pp. 72–77, 2014.

[4] V. Ratia, I. Miettunen, and V. Kuokkala, "Surface deformation of steels in impact-abrasion : The effect of sample angle and test duration", *Wear*, vol. 301, no. 1–2, pp. 94–101, 2013.

[5] S. Schuldt, G. Arnold, J. Roschy, Y. Schneider, and H. Rohm, "Defined abrasion procedures for cutting blades and comparative mechanical and geometrical wear characterization", *Wear*, vol. 300, no. 1–2, pp. 38–43, 2013.

[6] M. Stability and C. Resistance, "Mechanical Stability, Corrosion Resistance of Superhydrophobic", *J. Colloid Interface Sci.*, 2017.

[7] C. Study, W. Mechanisms, E. Abbasi, Q. Luo, and D. Owens, "Author ' s Accepted Manuscript Case Study : Wear Mechanisms of NiCrVMo-steel and CrB- steel Scrap Shear Blades", *Wear*, 2017.

[8] T. Bakša, T. Kroupa, P. Hanzl, and M. Zetek, "Durability of Cutting Tools during Machining of Very Hard and Solid Materials", *Procedia Eng.*, vol. 100, pp. 1414–1423, 2015.

[9] T. Rostek and W. Homberg, "ScienceDirect Locally Graded Steel Materials for Self-Sharpening Cutting Blades", *Procedia Eng.*, vol. 207, pp. 2185–2190, 2017.

[10] T. Materials, ASM Handbook Vol 11. "*Failure Analysis of Prevention*", Printed in the United States of America, pp. 1916-2009, 2001.
[11] V. J. Colangelo and F. A. Heiser, "Analysis of Metallurgical Failures", New York: A Wiley-Interscience Publication, pp. 206-223, 1974.
[12] T. Materials, ASM Handbook Vol 6. "Welding Brazing and Soldering", Printed in the United States of America, pp. 1916-2009, 2001.
[13] T. Materials, ASM Handbook Vol 1. "Properties and Selection: Irons Steels and High Performance Alloys", Printed in the United States of America, pp. 1763-1840, 2001.
[14] T. Materials, ASM Handbook Vol 4. "Heat Treating", Printed in the United States of America, pp. 749-755, 2001.

Fabrication of Chitosan based-Scaffold as Potential Cornea Implant

Satria Rusdiputra[1], Arie Wibowo[1,2], Lia Amelia Tresna Wulan Asri[1], Bambang Sunendar Purwasasmita[3,*]

[1]Department of Materials Engineering, Faculty of Mechanical and Aerospace Engineering, Institut Teknologi Bandung

[2] Research Center for Nanoscience and Nanotechnology, Institut Teknologi Bandung.

[3]Department of Engineering Physics, Faculty of Industrial Technology, Institut Teknologi Bandung

*Corresponding author: purwa@tf.itb.ac.id

Abstract. Corneal damage is the second cause of blindness in the world due to limitation of high quality corneal from donor. This condition promote rapid development of biodegradable and biocompatible scaffold for corneal tissue because it can reduce the patient's dependence on the cornea donor. The purpose of this research is to synthesize chitosan-based hydrogel scaffold with glutaraldehyde (GA) as crosslink agent and observe the influence of concentration chitosan and GA to transparency and mechanical properties of the scaffold. Scaffolds were prepared by solution casting combined with salt leaching method. Among all of chitosan-based scaffolds that have been prepared, sample with 6.7% of chitosan and 0.0% of GA possess the highest transparency (85%), while sample with 6.7% of chitosan and 0.3% of GA gave the highest Young's modulus (3.09 MPa). However, sample with 6.7% of chitosan and 0.1% of GA might possess the highest potency as corneal tissue implant because of its transparency fulfill the minimum value of corneal tissue (80%) and its Young's modulus (1.41 MPa) is higher than any chitosan-based scaffold with minimum transparency 80%.

Keywords: Chitosan, Cornea, Hydrogel, Scaffold, Tissue engineering

1. Introduction

Corneal is a transparent front part of the eyes with role as first protection of eyes and optical element that contribute to most of the eyes's focusing power [1]. Corneal damage is the second cause of blindness in the world and current acceptable treatment to treat it is corneal transplantation from donor [1-4]. Nowadays, around 60,000 of corneal transplantation were performed every year globally, which is still too low compare to 10 million of blindness cases worldwide [2,3]. This problem occurred mainly because of shortage stock of corneal donor with high quality [4]. Therefore, development of synthetic scaffold biomaterials for corneal tissue application is needed.

For fruitfulness of corneal tissue engineering, scaffold for cornea tissue application should be non-toxic, possess good transparency, good biocompatibility, good permeability, suitable mechanical properties with corneal tissue, and promote cell growth [1-3]. Hydrogels-based scaffold is one of attractive materials for corneal tissue engineering application due to similarity of their properties with extracellular matrix (ECM) of cornea [3]. Hydrogels can be prepared by both of synthetic and natural based-polymers. One of natural based-polymers that widely explored is chitosan due to its advantages such as non-toxic, biodegradable, biocompatible, high antibacterial activities and high chelating abilities [5-8]. Nevertheless, chitosan-based scaffold suffer with their weakness such as low mechanicals properties ($\sigma_{u\ compressive}$ (2% w/v, dry) = 0.967 MPa) and low stability to maintain their shape in long term condition [9-12]. To overcome this challenge, cross linking agent such as glutaraldehyde are commonly used to increase mechanical properties and stability of scaffolds [13]. In this experiment, the influence of glutaraldehyde was used as cross linking agent in fabrication of chitosan-based scaffold and studying their optical and mechanical properties.

2. Materials and Methods

Materials: Chemicals that were used in this study are pro analyst (p.a) grade and purchased from Sigma Aldrich. Chitosan was purchased from CV. Biochitosan Indonesia.

Preparation of Chitosan-based Scaffold: Chitosan based-hydrogels was prepared by solution casting and particulate leaching method [14]. Briefly, chitosan and NaCl with various concentration were dissolved in acetic acid solution and mixed for 2 hours. Then, hydrogels was molded by pouring chitosan solution in glass substrate, followed by immersion in 2M NaOH solution for about three minutes. Obtained hydrogels was immersed in demineralized water for about 30 minutes to remove remaining NaOH and let the NaCl leach from scaffold. Excess amount of water in hydrogels was dried with tissue. Preparation parameter that were used in this experiment are concentration of chitosan, scaffold thickness and concentration of Glutaraldehyde (GA). Thus, samples code can be seen in table 1.

Table 1. Molecular weight from dilution viscometry method

Sample code	[chitosan] (%)	Sample thickness (mm)	[GA] (%)
Ch15-1L-GA0.0%	6.7	0.1	0.0
Ch20-1L-GA0.0%	5.0	0.1	0.0
Ch15-2L-GA0.0%	6.7	0.2	0.0
Ch20-2L-GA0.0%	5.0	0.2	0.0
Ch25-2L-GA0.0%	4.0	0.2	0.0
Ch15-1L-GA0.1%	6.7	0.1	0.1
Ch15-1L-GA0.2%	6.7	0.1	0.2
Ch15-1L-GA0.3%	6.7	0.1	0.3

Characterization: Functional group of chitosan were identified determined by Fourier Transform Infra-red (FTIR; Prestige 21 Shimadzu) method and their degree of deacetylation (DD) were determined by baseline method following Eq. 2.1 [15,16]:

$$DD[\%] = 100 - (\frac{A_{1655}}{A_{3450}} x\ 115) \qquad (2.1)$$

Transparency of scaffolds were studied by UV-VIS Spectrophotometer (Agilent 8453 G1103A) method and mechanical properties of scaffolds were determined by Tensilon

IOP Conf. Series: Materials Science and Engineering **547** (2019) 012062 doi:10.1088/1757-899X/547/1/012062

3. Results and Discussion

3.1. FTIR result

Before used, it is important to ensure that DD of purchased chitosan are above 70%, to be classified as chitosan [17]. Thus, purchased chitosan was characterized using FTIR method and imaginary line on FTIR spectra of sample for calculation of DD of chitosan can be seen in Fig. 1. FTIR spectra showed that typical peaks of chitosan are observed in sample (~3400 cm^{-1} for O-H and/or N-H functional groups; 1656 cm^{-1} for amide I functional groups; 1600 cm^{-1} for amide II functional groups) [15,16]. Based on baseline method, calculated DD of sample was 76%. Since DD of sample is above 70%, it can be claimed that purchased chitosan is categorized as chitosan and can be used in this experiment.

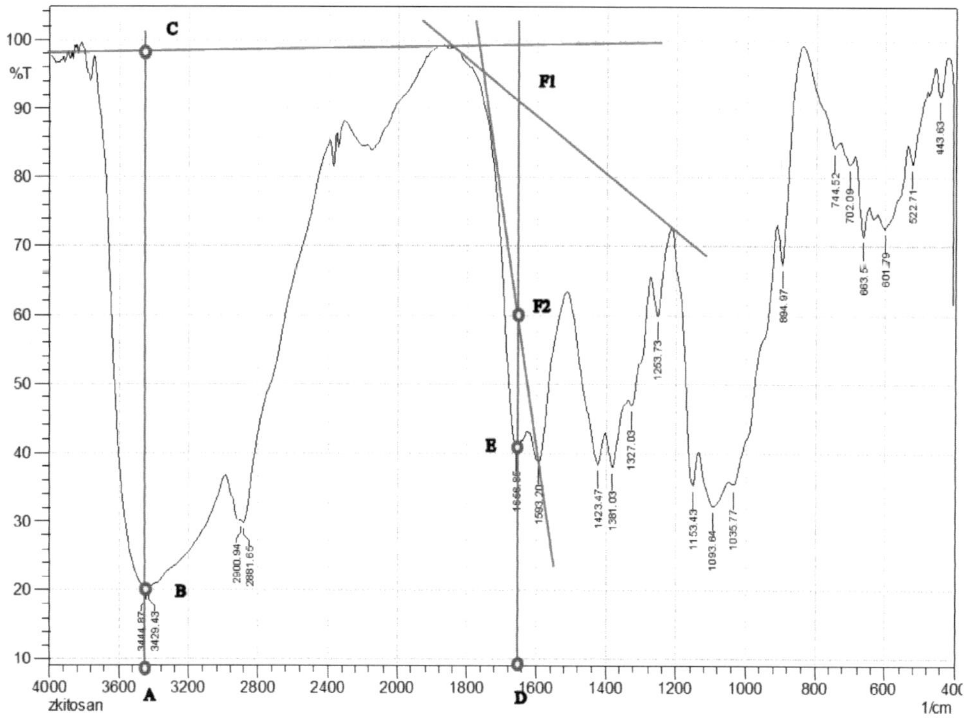

Figure 1. FTIR spectra of purchased chitosan

3.2. UV-Vis results

Since transparency is important properties for synthetic corneal tissue, transparency of all of obtained chitosan scaffolds have been evaluated using UV-Vis spectroscopy method. UV-Vis results of chitosan scaffold can be seen in Fig. 2. UV-Vis results of samples with various chitosan concentration and without GA (Fig. 2a) revealed that obtained sample with 6.7% of chitosan and without GA showed the highest transparency, ± 85%, which is higher than minimum value for corneal tissue (80%) [3]. Thus, samples with chitosan 6.7% was cross-linked by GA as a function of concentration of GA (0.1; 0.2; and 0.3%) and their transparency were evaluated by UV-Vis spectroscopy method. UV-Vis results (Fig. 2b) showed that sample with cross-linked agent have lower transparency than sample without cross linking agent.

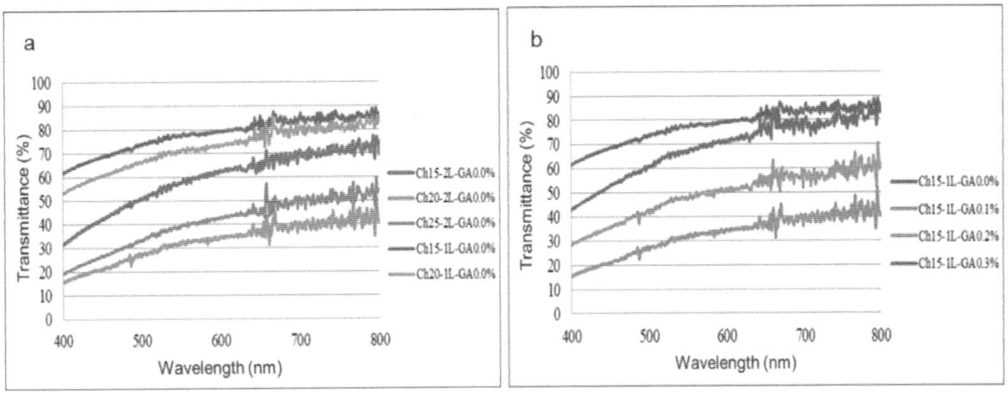

Figure 2. UV-Vis spectra of scaffold samples a) with various chitosan concentration and without GA, and b) with chitosan 6.7% and with various GA concentration.

Rolf Klein mentioned that deep of light penetration (denoted as a) and thickness of sample (z) can be used to predict interaction between materials with incident light, and can be calculated by equation:

$$a = \frac{-z}{lnT} \qquad \text{(eq. 2.2)}$$

If a >> z : Incident light will not be absorbed by materials

a ≅ z : Incident light will be absorbed by bulk of materials

a << z : Incident light will be absorbed by surface of materials

Summary of deep of penetration light (a), thickness of sample (z) and materials characteristic is presented in table 1. As can be seen in Table 1, this rough estimation can be used to explain why sample with chitosan 6.7% without GA have the highest transparency while sample with chitosan 6.7% and GA 0.3% showed the lowest transparency.

Table 2. Summary of deep of penetration light (a), thickness of sample (z) and materials characteristic of all of samples

No	Sample	a (mm)	z (mm)	Characteristic	Transparency (%)
1	Ch15-1L-GA0.0%	0,519	0,116	a >> z	85
2	Ch20-1L-GA0.0%	0,394	0,116	a >> z	80
3	Ch15-2L-GA0.0%	0,403	0,194	a >> z	70
4	Ch20-2L-GA0.0%	0.188	0.206	a ≅ z	40
5	Ch25-2L-GA0.0%	0.229	0,198	a ≅ z	50
6	Ch15-1L-GA0.1%	0,137	0,132	a ≅ z	80
7	Ch15-1L-GA0.2%	0,168	0,114	a ≅ z	60
8	Ch15-1L-GA0.3%	0,355	0,114	a >> z	40

3.3. Mechanical Properties of Samples

IOP Conf. Series: Materials Science and Engineering **547** (2019) 012062 doi:10.1088/1757-899X/547/1/012062

Considering that mechanical properties is important properties for corneal tissue, mechanical properties of chitosan scaffold as function of GA concentration were performed and their results are presented in Fig. 3. Measurement of tensile strength (Fig. 3a) and elastic modulus/Young's modulus (Fig. 3b) of samples showed that their strength and Young's modulus are increased as concentration of GA increased. Meanwhile, measurement of elongation of samples (Fig. 3c) showed that elongation of samples decreased as concentration of GA increased. These results are reasonable since GA is cross linking agent that provide additional bond between polymer chain and 3D network in scaffold. Thus, scaffold become stronger and stiffer by addition of GA. Since corneal Young's modulus that measured in vitro are varies from 0.1 to 57 MPa [17], all of chitosan-based scaffold that prepared with chitosan 6.7% (Young's modulus were varied from 0.67 – 3.09 MPa) could be considered as acceptable materials for corneal tissue application. However, scaffold with higher Young's modulus are more favorable for corneal tissue implant.

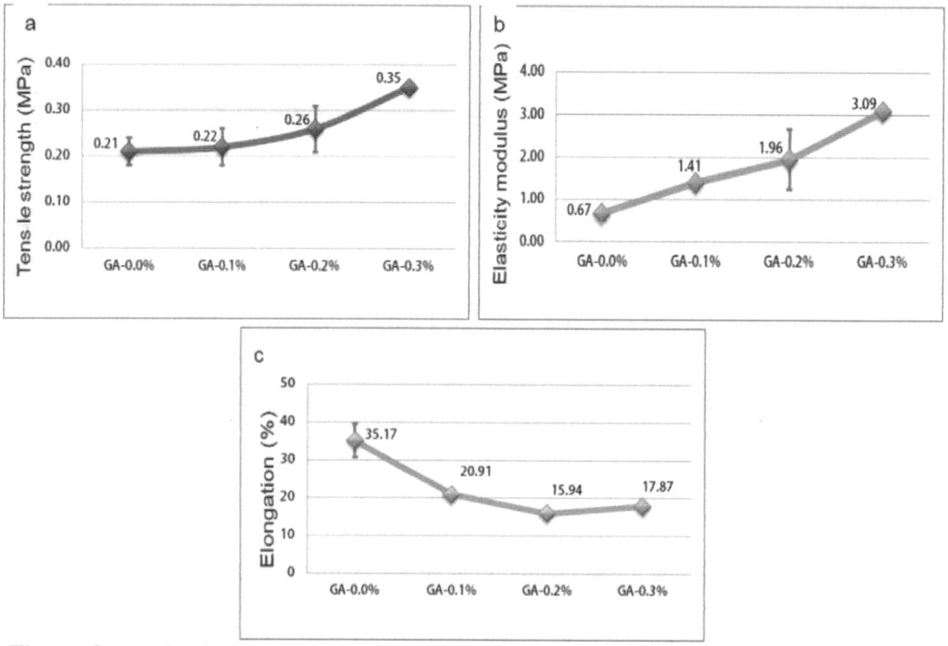

Figure 3. Mechanical properties of samples with chitosan 6.7% as a function of GA concentration: a) tensile strength of samples, b) elasticity modulus of samples, and c) elongation of samples

4. Conclusions

Chitosan based scaffold have been successfully prepared by SCPL method. Sample with chitosan 6.67% and GA 0.0% showed the highest transparency (± 85%). As concentration of GA increased, tensile strength and Young's modulus of scaffold increased but elongation of scaffold tend to decreased. Considering that transparency of corneal tissue is higher than 80% and Young's modulus that measured in vitro are varies from 0.1 to 57 MPa, sample that prepared with chitosan 6.67% and GA 0.1% showed the highest potency for corneal tissue implant with transparency 80% and its Young's modulus (1.41 MPa) is higher than any chitosan-based scaffold with minimum transparency 80%.

5. Acknowledgements

This research was supported by Applied Excellent Research from Indonesian Ministry for research, Technology and Higher Education 2018.

6. References

1. Zhao, X.; Liu, Y.; Li, W.; Long, K.; Wang, L.; Liu, S.; Wang, Y.; ren, L.; 2015. *Materials Science and Engineering C*. 55 (2015) 201-208.
2. Ionescu, A.-M.; Alaminos, M.; Cardona, J. C.; Duran, J. D. G.-L.; Gonzales-Andrades, M.; Ghinea, R.; Campos, R.; Hita, E.; Perez, M. M.; *Journal of the Mechanical Behavior of Biomedical Materials*. 4 (2011) 1963-1973
3. Tonsomboon, K.; Oyen, M. L.; *Journal of the Mechanical Behavior of Biomedical Materials*. 21 (2013) 185-194.
4. Hashimoto, Y.; Funamoto, S.; Sasaki, S.; Honda, T.; Hattori, S.; Nam, K.; Kimura, T.; Mochizuki, M.; Fujisato, T.; Kobayashi, H.; Kishida, A.; *Biomaterials*. 31 (2010) 3941-3948.
5. Rabea, E. I.; Badawy, M. E.-T.; Stevens, C. V.; Smagghe, G.; Steurbaut, W.; *Biomacromolecules*, **2003**, *4(6)*, 1457 – 1465
6. Zargar, V.; Asghari, M.; Dashti, A.; *ChemBioEng Reviews*, **2015**, *2(3)*, 204-226
7. Kumar, M. N. V. R.; *Reactive and Functional Polymers*, **2000**, *46(1)*, 1-27
8. Asri, L. A. T. W.; Nuruddin, A.; Hidayatullah, S.; Simatupang, M.; Wibowo, A.; Septawendar, R.; Purwasasmita, B. S.; *Journal of Australian Ceramics Society*, **2017**, (DOI 10.1007/s41779-017-0103-1)
9. Levengood, S. L., Zhang, M., *Journal of Materials Chemistry B*, 21 (2014) 3161–3184.
10. Li, Z., Ramay, H. R., Hauch, K. D., Xiao, D., Zhang, M., *Biomaterials*, 26 (2005) 3919-3928.
11. Venkatesan, J., Kim, S.-K., *Marine drugs*, 8 (2010) 2252-2266.
12. Han, J., Zhou, Z., Yin, R., Yang, D., Nie, J., *International Journal of Biological Macromolecules*, 46 (2010) 199-205.
13. Niknejad, H., Mahmoudzadeh, R., *Iranian Journal of Pharmaceuticals Research*, 14 (2015) 385 – 394.
14. Liao, C.-J.; Chen, C.-F.; Chen, J.-H.; Chiang, S.-F.; Lin, Y.-J.; Chang, K.-Y.; *Journal of Biomedical Materials Research,* 59 (2001) 676 – 681.
15. Baxter, A., Dillon, M., Taylor, K. D. A., Roberts, G. A. F., *International Journal of Biological Macromolecules,* 14 (1992) 166 – 169
16. Kasaai, M. R., *Carbohydrate Polymers,* 71 (2008) 497 – 508.
17. Rodríguez-Vázquez, M., Vega-Ruiz, B., Ramos-Zuniga, R., Saldana-Koppel, D. A., Quinones-Olvera, L. F., *BioMed Research International*, Volume 2015, Article ID 821279, 2015

The Preparation of Electrical Double Layer Capacitor (EDLC) from Boron-doped Reduced-Graphene Oxide (B-rGO) Material

Diah Susanti[1]* Yurian Ariandi Andrameda[1] Haniffudin Nurdiansah[1] Hariyati Purwaningsih[1] Yusuf Pradesar[1] Hosta Ardhyananta[1]
[1]Materials and Metallurgical Engineering Department, Institut Teknologi Sepuluh Nopember (ITS), Surabaya, INDONESIA

*Corresponding author (E-mail): santiche@mat-eng.its.ac.id

Abstract. The increase of human needs to electronic instruments requires sufficient energy storage devices. One of the electrochemistry-based energy storage devices is electrochemical capacitor (EC). The challenge in EC manufacturing is to prepare ECs with high energy and power densities. It depends very much on the ECs material. In this research, graphene, theoretically a single layer of carbon material, was used as the material for electrical double layer capacitor (EDLC), a type of ECs. To enhance the electrical conductivity, graphene was doped with boron using 1 M H_3BO_3 (boric acid) with volume variations of 1, 2, and 3 ml. X-Ray Diffractometer (XRD) and Scanning Electron Microscope (SEM) equipped with Energy Dispersive X-Ray Spectroscopy (EDXS) were used to characterize the structure, morphology and element content of the material. Four Point Probe (FPP) method was used to measure the materials electrical conductivity, while Cyclic Voltammetry (CV) measurement was used to inspect the capacitive properties of the material in 1 M Na_2SO_4 electrolyte solution. The best result, 192.5 F/g, was found from B-doped graphene with 2 ml volume of boric acid (equivalent to 4.37 %B). From this experiment, boron doping significantly improved the capacitive properties of graphene-based EDLC.

Keywords: Graphene, electrochemical capacitor, EDLC, Doping, Boron, Capacitance

1. Introduction

Recently energy storage device is needed to support human daily life. Electrochemical capacitor (EC) is one of potential energy storage device based on electrochemistry principle. It could be used for primary power, back-up power and alternating power sources. Referring to Ragone's plot ECs has higher energy density than conventional capacitor and higher power density than battery and fuel cell. Therefore, EC is often hybridized with other energy storage devices such as battery and fuel cell. The applications are including for toys, cell phones, laptops, PCs, and hybrid or electric vehicles for civilians and military applications [1,2].

ECs are categorized into two main types, pseudocapacitor and electric double layer capacitor (EDLC). Materials for pseudocapacitor usually are metal oxide and polymer, while for EDLC are carbon-based material with high active surface area. Graphene is a good candidate for EDLC material, because it was reported to have high electrical conductivity ~$1.6x10^5$ Scm^{-1} [3] and active surface area ~2600 m^2g^{-1} [4].

The challenge is to improve the characteristics of graphene in order to increase the capacitive properties of graphene-based EDLC. One of the methods is by doping graphene with other elements, especially the neighbor elements having similar atomic radius to carbon, such as nitrogen and boron. Nitrogen of 5 electron valence will provide more free electrons to carbon while boron of 3 electron valence will provide more holes. Both free electron and hole are charge carriers which are important for facilely transporting the charges in EDLC. Some researchers such as Yeom et.al [5] and Zhuo et.al [6] have reported B-doped EDLC. The highest specific capacitances reported by Yeom et.al and Zuo et.al were 448 F/g and 281 F/g, respectively.

In this research we analyzed the effects of doping content of boron into graphene towards the structure, morphology, and the capacitive properties of graphene-based EDLC. Briefly, graphene was synthesized via a modified Hummer method followed by reduction and hydrothermal processes. Boric acid (H_3BO_3) was used as the source of boron for doping. All materials were characterized by Scanning Electron Microscope (SEM)-Energy Dispersive Spectrometer (EDXS), and X-Ray Diffractometer (XRD). The electrical conductivity was measured by Four Point Probe (FPP) and the capacitive properties were measured by cyclic voltammetry (CV) electrochemical measuring instrument.

2. Experimental Methods

Graphite powder (150 μm, 99% purity) was dissolved by H_2SO_4 and oxidized into graphite oxide (GO) by $KMnO_4$ and $NaNO_3$ in an ice bath according to the modified Hummer's method and stirred for 20 hours. H_2O_2 was then added to complete the oxidation process. The solution was centrifuged to separate the precipitate and the liquid. The precipitate was washed with 5% HCl and dried to get the GO. During the oxidation process, the lattice spacings between carbon-carbon bonds were enlarged and the oxygen and hydroxyl functional groups were bonded to the outer carbon atoms. This process was intended to chemically exfoliate the carbon arrangement of graphite into layers.

The GO was dissolved into DI water and ultrasonicated at 20 kHz frequency for 2 hours to exfoliate the layers so that graphite oxide becoming graphene oxide because the graphite oxide layers became thinner. Graphene oxide was then hydrothermally treated at 200 °C for 4 hours to reduce the graphene oxide into reduced-graphene oxide (rGO) by eliminating the oxygen and hydroxyl bonds. In this experiment, the resulted graphene was near to the structure of rGO, since chemical process hardly produced high purity graphene. For short, the term graphene was used in this report to name the impure graphene.

Boron was doped into the graphene structure by adding 1 M H_3BO_3 solution with variation of volume 1, 2, and 3 ml after ultrasonication process of GO. The solution was washed with warm DI water for removing boric acid. Then the solution was hydrothermally treated at 200 °C for 4 hours to result in B-rGO materials.

To prepare the electrodes for CV measurement, r-GO or each B-rGO were dissolved in DI water and deposited into nickel foam with a dimension of 5x1 cm^2 by ultrasonic process. The active materials of rGO or B-rGO were weighed prior to being electrochemically measured using CV. The rGO or B-rGO deposited Ni foam was then used as the EDLC working electrode.

The structure and phase of the synthesized materials were identified using X-ray diffractometer (XRD, X'Pert Pro PAN Analytical) with CuKα radiation (α = 1.54060°A). The morphology, topography, and composition investigation were carried out with scanning electron microscopy equipped with energy dispersive X-ray spectroscopy (SEM and SEM-EDX, Jeol JSM-7001F, FEI Inspect S50, USA).

The electrical properties were investigated by using four-point probe (FPP) method was used to measure the electrical conductivity. Cyclic voltammetry (CV) was conducted to measure the specific capacitance of the EDLC electrodes. The CV measurement was carried out in 1 M Na_2SO_4 with Pt as counter electrode and 0 – 0.8 V as the potential window by using AUTOLAB PG STAT 302 METHROM (Netherlands) at scan rates of 5, 10, 50 and 100 mV/s.

3. Results and Discussion

Figure 1 shows comparison pattern between rGO and B-rGO with addition 1 M H_3BO_3 of 1, 2, and 3 ml. XRD result shows with the increasing addition volume of 1 M H_3BO_3 of 1, 2 and 3 ml, the peaks of rGO was shifted to lower angles, 25.06°, 24.72° and 24.54°, respectively. The shifted peak showed that boron has successfully doped in rGO.

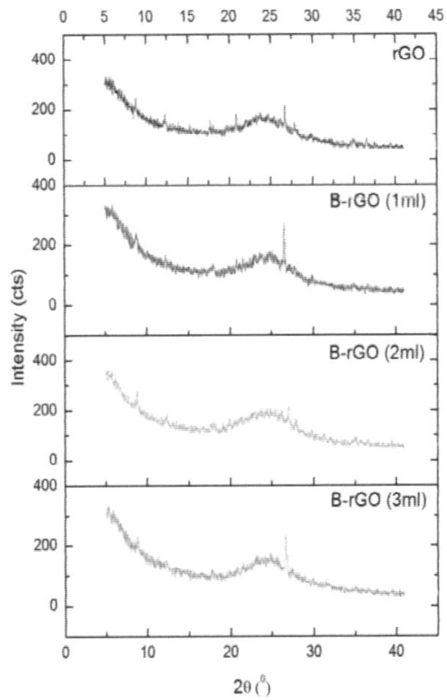

Figure 1. XRD pattern of rGO and B-rGO

To understand the morphology and particle size of materials, SEM was conducted with magnification 500x. Figure shows the morphology of graphite, graphite oxide and graphene. As shown in Figure 2 (a), graphite had flakes structure. The particle of graphite was small and irregular. After the oxidation to graphite oxide, the morphology became thick stacking layers as shown in Figure (b). The thickness was shown by the color gradation, as thicker the material the color become strong white. Figure (c) shows thin stacking layers of rGO which formed by mechanical exfoliation of graphite oxide.

IOP Conf. Series: Materials Science and Engineering **547** (2019) 012063 doi:10.1088/1757-899X/547/1/012063

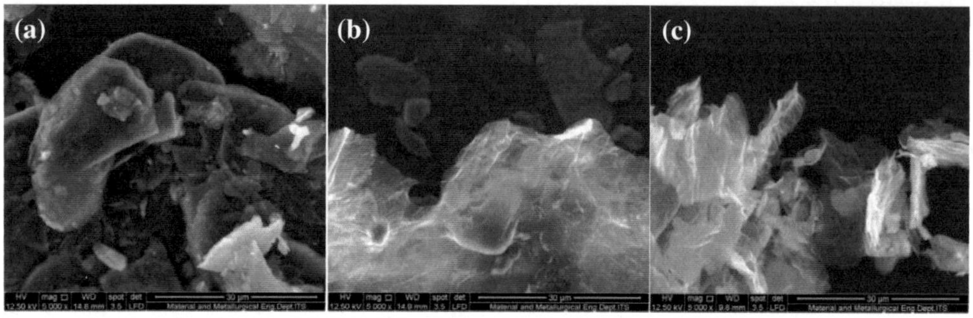

Figure 2. SEM images of (a) graphite, (b) graphite oxide and (c) rGO

Table was a representative data of elemental analysis of B-rGO based on EDX results. The addition 1, 2 and 3 mL of 1 M H_3BO_3 has atomic percentage of boron 3.59%, 4.37% and 5.17%, respectively.

Table 1. Elements composition of rGO and B rGO with addition 1, 2 and 3 mL of 1M H_3BO_3

Materials	Element	At%
B-rGO with 1 mL of H_3BO_3	B	3.59
B-rGO with 2 mL of H_3BO_3	B	4.37
B-rGO with 2 mL of H_3BO_3	B	5.17

Table shows the results of FPP test and calculation. The conductivity of rGO and B-rGO with addition 1, 2 and 3 mL of 1 M H_3BO_3 was 195.84, 209.72, 216.34 and 218.05, respectively. As the boron content increase, the conductivity also increasing. The doping of boron certainly increased the electrical conductivity. As stated by Marconcini et al [6], the increasing of electrical conductivity on B-rGO was due to the addition of hole impurities in the band gap of rGO, so the electrons easily move through the band gap from the valence band to the conduction band.

Table 2. FPP test and calculation result of rGO and B-rGO

No	Materials	%B	Conductivity (S/m)
1	rGO	0	195.84
2	B-rGO (1 mL 1 M H_3BO_3)	3.59	209.72
3	B-rGO (2 mL 1 M H_3BO_3)	4.37	216.34
4	B-rGO (3 mL 1 M H_3BO_3)	5.17	218.05

The CV test was conducted to investigate the number of specific capacitances of rGO and B-rGO with addition 1, 2 and 3 mL of H_3BO_3. The CV curve with current vs potential (I vs V) was showed in Figure . By using Na_2SO_4 neutral electrolyte at working potential of 0 – 0.8 V, there was no hump produced. A hump was indicated an existence of redox reaction. The CV data was converted into specific capacitance as shown in Table 3.

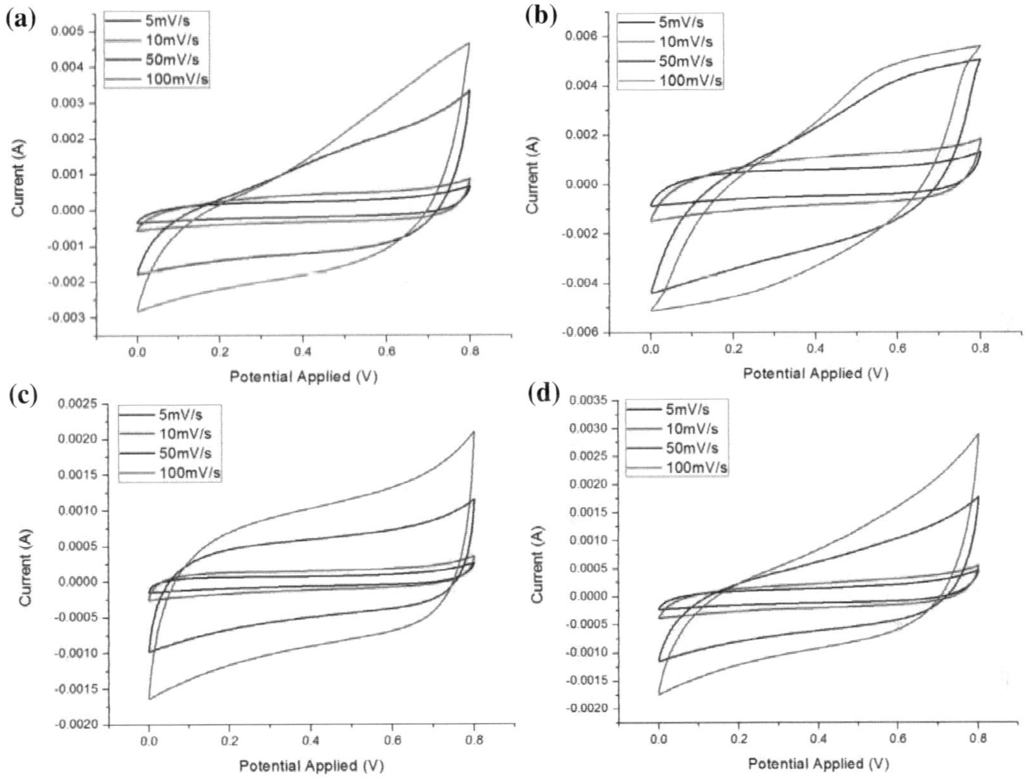

Figure 3. I vs E CV curve of (a) Ni-rGO and Ni-B-rGO with addition (b) 1, (c) 2 and (d) 3 mL of H_3BO_3

Table 3. Relationship Between Scan Rate and Specific Capacitance of rGO and B-rGO at scan rate of 10 mV/s

No	Materials	%B	Specific Capacitance (F/g)
1	rGO	0	135.53
2	B-rGO (1 mL 1 M H_3BO_3)	3.59	156.16
3	B-rGO (2 mL 1 M H_3BO_3)	4.37	150.97
4	B-rGO (3 mL 1 M H_3BO_3)	5.17	169.58

As shown in **Figure**

Table 3 the highest specific capacitance of 169.58 F/g was obtained from the B-rGO after doped with 5.17% Boron. The results were in-line with the conductivity results as shown in Table 2. Hence the specific capacitance was greatly influenced by the materials conductivity. These results were lower than those reported by Yeom, et.al [5] and Zhuo, et.al [6]. However, this EC could still be used as an energy storage device for small electronic equipment.

4. Conclusions

The rGO and B-rGO were successfully synthesized via Hummer's method, reduction method and the addition of H_3BO_3 solution as indicated by the XRD results. The increase of the boron content slightly shifted the XRD peaks as the indication of the shifting in the atomic structure. The FPP result shows that the B-rGO has better conductivity value than rGO, because the boron doped was p-type dopant which increase the hole in structure. The CV curves show that all materials (rGO and B-rGO) had almost rectangular shapes which indicate the ELDC properties. At scan rate 10 mV/s, B-rGO from 5 ml of H_3BO_3 had the highest capacitance. The trend in the capacitances were in line with the electrical conductivity trend. As the electrical conductivity increased, the specific capacitance also increased.

5. Acknowledgements

Authors greatly acknowledge ITS for the financial support to this research through Laboratory Based Education (LBE) Research Grant 2018.

6. References

[1] Kötz R, Carlen M, Principles and applications of electrochemical capacitors, Electrochimica Acta, 2000, 45: 2483-2498.

[2] Conway B E, Electrochemical Supercapacitors, Scientific Fundamentals and Technological Applications, Springer Science, New York, 1999

[3] Singh V, Joung D, Zhai L, Das S, Khondaker S I, Seal S, Graphene based materials: past, present and future, Progress in Materials Science, 2011, 56(8): 1178-1271.

[4] Pei S and Cheng H M, The reduction of graphene oxide, Carbon, 2012, 50(9): 3210-3228.

[5] Yeom DY, Jeon W, Tu ND, Yeo SY, Lee SS, Sung BJ, Chang H, Lim JA, Kim H. High-concentration boron doping of graphene nanoplatelets by simple thermal annealing and their supercapacitive properties. Scientific reports. 2015, 5 srep09817.

[6] Z. Zuo, Z. Jiang, and A. Manthiram, "Porous B-doped graphene inspired by Fried-Ice for supercapacitors and metal-free catalysts," Journal of Materials Chemistry A, 2013, 1(43), 13476–13483.

[7] Marconcini P, Cresti A, Triozon F, Fiori G, Biel B, Niquet YM, Macucci M, Roche S. Atomistic boron-doped graphene field-effect transistors: a route toward unipolar characteristics, ACS nano. 2012, 6(9):7942-7.

Effect of Titanium Dioxide Particle Size in Water-based Micro/Nanofluid as Quench Medium in Heat Treatment Process

I.F Prayogo[1], F. Muhammad[1], D. Rakhman[1], W.N. Putra[1,*], S. Harjanto[1]

[1] Department of Metallurgy & Materials Engineering,
Faculty of Engineering, Universitas Indonesia, Depok, Indonesia

*Email: wahyuaji@metal.ui.ac.id

Abstract. Recently, nanofluid is used to improve the thermal conductivity of the quench medium in the heat treatment industry. In this research, ball-milled micro-sized TiO_2 powder and nano-sized TiO_2 particle were used and compared for their cooling characteristic in a micro/nanofluid. The micro/nanofluids were produced by mixing 0.1%, 0.3%, and 0.5% volume of both micro- and nano-sized particle into 100 ml of distilled water. The planetary ball mill was used at 500 rpm for 15 hours to reduce the dimension of micron-sized TiO_2. Composition characterization by Energy Dispersive Spectroscopy (EDS) showed that the powder used were free from impurities. Nanofluids were then used to quench S45C carbon steel samples, which heated at 1000°C for 1 hour. The hardness test result showed that the sample quenched with 0.5%addition of the nano-sized particle in nanofluid had the highest number up to 691 HV, almost 100HV increment from a water-quenched sample where the hardness was 598 HV, showing that the cooling rate in the nanofluid was much higher. The addition of micro-size particle in fluid generally had a lower cooling rate than the addition of nano-size particle.

Keywords: Nanofluid, TiO_2, Quenching

1. Introduction

Heat treatment is a standard process used in the metal industry to improve the properties of materials, especially steels. In this process, steel properties could be adjusted to have higher hardness or more ductility, depend on the application. Quenching is a process in heat treatment, where steel parts are heated to austenizing temperature and rapid-cooled by immersing into the quenchant or quenching medium [1]. Water, brine, polymers, oils, salts, and gas are commonly used as a quenching medium which has their cooling characteristics and results. Improper quenchant can cause the material to become too brittle, suffers some geometric distortion, and undesirable residual stress.

Microstructure and mechanical properties after quenching depend on several factors such as surface temperature, thermo-physical properties of saturated liquid, size and shape of the material, the roughness of the material, and cooling rate of quench process. Typically, after the quenching process, martensite is formed in steel, and the hardness is increased. To ensure the

martensite formation, cooling rate during rapid-cooling must be equal or higher than the critical cooling rate for martensite forming [2].
Studies on quenching medium that influenced cooling rate has been done by many researchers [3-5]. One of the popular quenching medium is nanofluid. Nanofluid refers to a fluid containing a small amount dispersed nano-sized (< 100 nm) solid nanoparticles. It has been proved that the nanoparticles will enhance thermal conductivity and improve the thermal performance of heat transfer of the base fluid [6]. Commonly used nanoparticles include Al_2O_3, CuO, TiO_2, and Carbon. Nanoparticles used in nanofluid are considered by their chemical stability, thermophysical properties, toxicity, and cost [7]. The thermal conductivity of TiO_2 nanofluid can increase a maximum of 38.7% than distilled water [8].
TiO_2 micro- and nanoparticle, and aqueous-distilled water were used for micro/nanofluid in this research. In this study, the effect of the particle size of TiO_2 in the quench medium after heat treatment will be compared and discussed.

2. Materials and Methods

2.1 Preparation of TiO_2 Micro/Nanofluids

TiO_2 micro/nanofluids are obtained by a two-step process, particle synthesis, and particle dispersion. The two-step method was more economical compared than the one-step method to produce micro/nanofluids commercially, and also considered the best method in dispersing the particles in a base fluid [9]. The TiO_2 powder was purchased from Sigma-Aldrich in micro-size (44 μm) and nano-size (21nm) form. This study used three particle-size variations. The mentioned size variations were micro-sized laboratory grade TiO_2, reduced micro-sized laboratory grade TiO_2, and nano-meter sized laboratory grade TiO_2. Reduced micro-sized laboratory grade milled was obtained using a high energy ball mill for 15 hours at 500 rpm. The second step was the dispersing TiO_2 in distilled water with ultrasonic agitation. Variations of TiO_2 particle addition in the fluid were 0.1%, 0.3%, 0.5%.

2.2 Material for Quenching Experiments

The material used in this experiment was S45C carbon steel bar. S45C steel is classified as a medium carbon steel material that is often used in manufacturing industries.

Table 1. Chemical composition of S45C Carbon Steel

Fe	C	Si	Mn	P	S	Cr	Mo
98.3	0.47	0.287	0.718	0.0261	0.005	0.028	0.005
Ni	**Al**	**Co**	**Cu**	**Nb**	**Ti**	**V**	**W**
0.005	0.02	0.003	0.018	<0.002	0.035	<0.002	<0.01
Pb	**Sn**	**B**	**Ca**	**Zr**	**As**	**Bi**	
0.025	<0.002	0.0013	0.0005	<0.002	0.008	<0.03	

Table 1 showed the composition of the S45C steel sample used in this research. Medium carbon steel was chosen because its microstructure is sensitive to the cooling rate during quenching.

2.3 Experimental Methodology

The TiO_2 particle was characterized by Field Emission Scanning Electron Microscope (FE-SEM) Inspect F50 from FEI, coupled with Energy Dispersive Spectroscopy (EDS) from EDAX for shape and purity checking. Cooling rate observations of the quench medium were carried out by using microstructure and hardness test of the steel sample.

The S45C steel was prepared by cutting the sample with a size of 15mm x 10mm x 10mm with a saw machine. Samples were then preheated in 600^0C for 15 minutes and then heated at 1000^0C for one hour. After that, samples were quenched in water-based nanofluid with varying particle sizes. The microstructure was obtained by metallographic observation. Structure of specimens was made visible by using 3% nital etching. An optical microscope was then used to record the microstructure observation. Hardness measurement was obtained by a Vickers Hardness Test with a 300gf load.

3. Results and Discussion

3.1 TiO_2 Particle Characterization and S45C Microstructure Observation

FE-SEM was used to observe the morphology of TiO_2 particles, while EDS was used to check any impurities present in the powder. FE-SEM and EDS examination were done at CMPFA University of Indonesia.

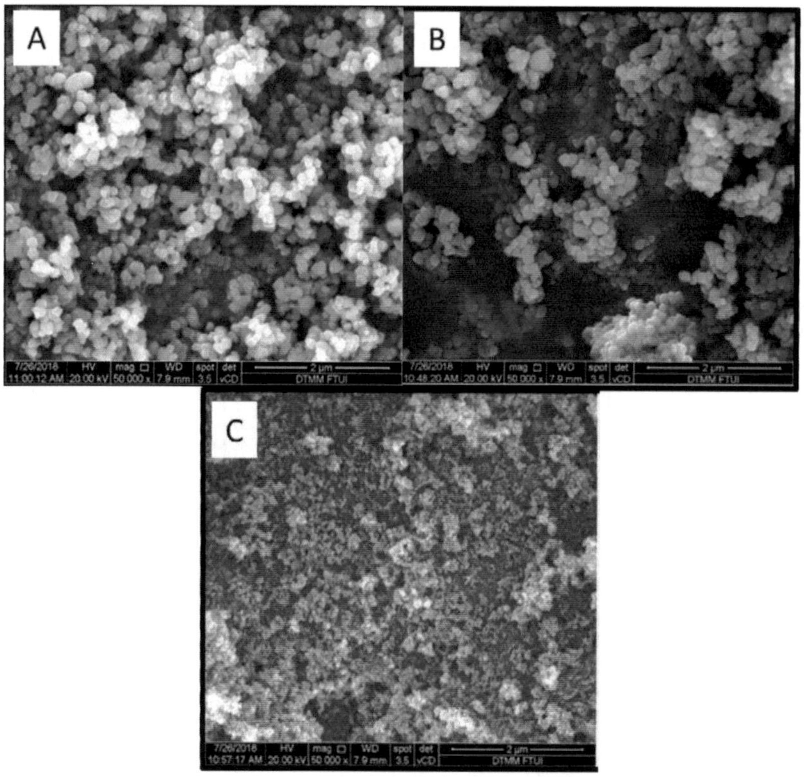

Figure 1. FE-SEM image of laboratory grade TiO_2particle. (A. micro-size, B. reduced micro-size, C. nano-size). Magnification: 50.000X

IOP Conf. Series: Materials Science and Engineering **547** (2019) 012064 doi:10.1088/1757-899X/547/1/012064

Figures 1 showed the particle shape of micro-size, reduced micro-size, and nano-size TiO_2 respectively. Manual measurement of reduced micro-size showed an average of 29μm particle size.

Table 2 showed EDS result for micro-size TiO_2 particles. From the result, it can be seen that no significant metal impurities present on the powder. There was a tiny amount of Phosphorous and Chloride elements, probably due to contamination during EDS sample preparation. The main element on the particle was Titanium as expected.

Table 2. Chemical composition of TiO_2 powder by EDS

Element	Weight %
P	0.32
Cl	0.32
Ti	99.36

Figure 2. Microstructures of S45C after quenching in distilled water. Mag: 200X

Figure 2 showed the microstructure of S45C steel quenched by water without nanoparticle taken by an optical microscope. Figure 3 showed the microstructure of S45C steel quench with varies addition of TiO_2 particle. All of the mentioned figures showed typical quenched steel microstructure where martensite phase present in the material. However, it was challenging to differentiate the martensite percentage from the sample quenched with water and nanofluid. Hardness test was then conducted to observe the difference of hardness from these samples. The difference of hardness number will show the martensite percentage on each sample.

IOP Conf. Series: Materials Science and Engineering **547** (2019) 012064 doi:10.1088/1757-899X/547/1/012064

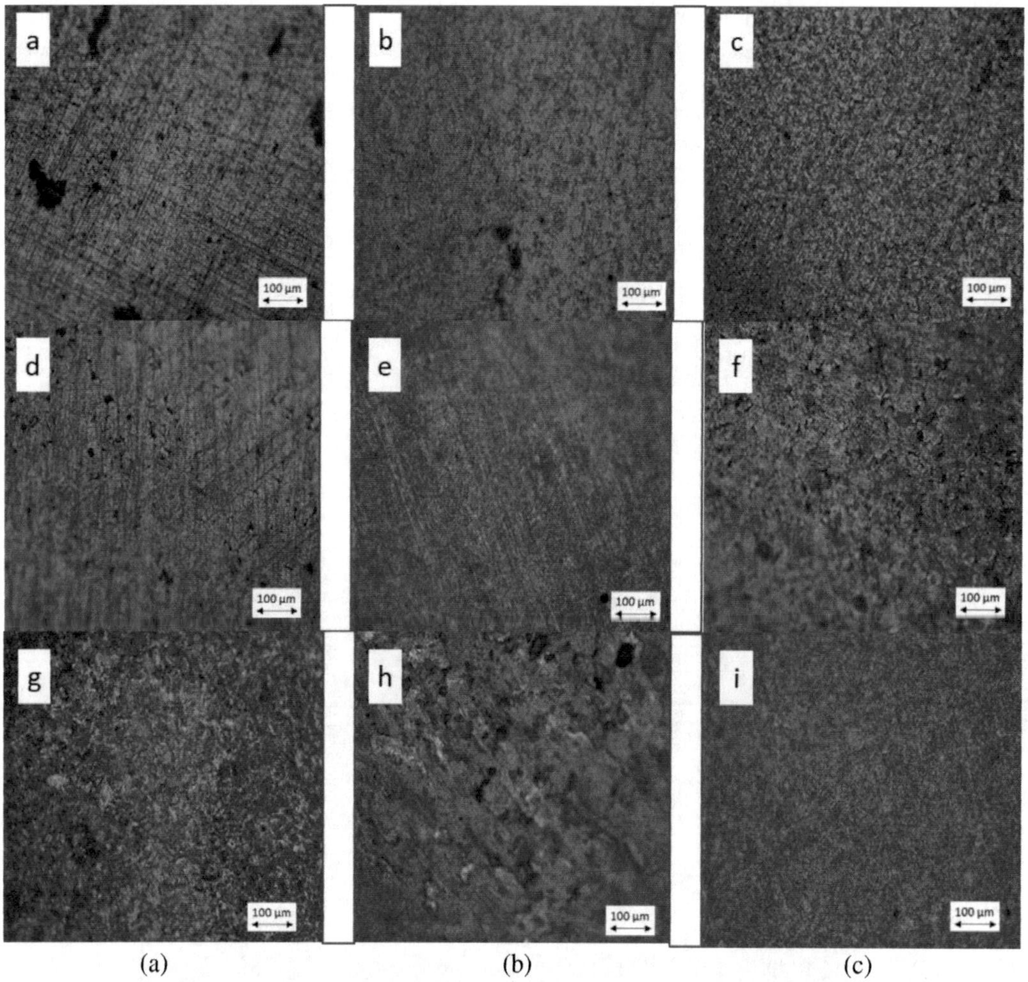

Figure 3. Microstructures of S45C after quenching in micro-size (a) 0.1 % (b) 0.3% (c) 0.5%, reduced micro-size (d) 0.1 % (e) 0.3% (f) 0.5%, and nano-size (g) 0.1 % (h) 0.3% (i) 0.5% TiO_2 particle. Magnification: 200X

3.2 Hardness Testing

Vickers hardness testing was done to support the microstructure data. In this research, higher hardness means higher cooling rate which resulted in more martensite phase formed in the steel. The hardness of the steel sample quenched in the normal water was 598 HV. Addition of micro-size TiO_2 particle into fluid resulted in higher hardness number i.e. 579 HV, 686 HV, and 628 HV for particle concentration of 0.1%, 0.3%, and 0.5% respectively. Reduced micro-size particle quench medium appeared to have generally lower hardness number, i.e. 605 HV, 608 HV, and 639 HV for the same particle concentration. This phenomenon was not expected as smaller particle size should have more surface area, thus increasing the cooling rate. However, there was no surfactant used in this research. Therefore, agglomeration may take place during particle mixing into the fluid.

Nano-size TiO_2 particle addition in the quench medium showed higher cooling rate, as seen in the higher hardness number at 611 HV, 644 HV, and 691 HV for a particle concentration of

0.1%, 0.3%, and 0.5% respectively. Increasing the particle concentration may increase the cooling rate of the quench medium as well. Figure 4 showed the Vickers hardness number trend for each sample with different quenching medium variations.

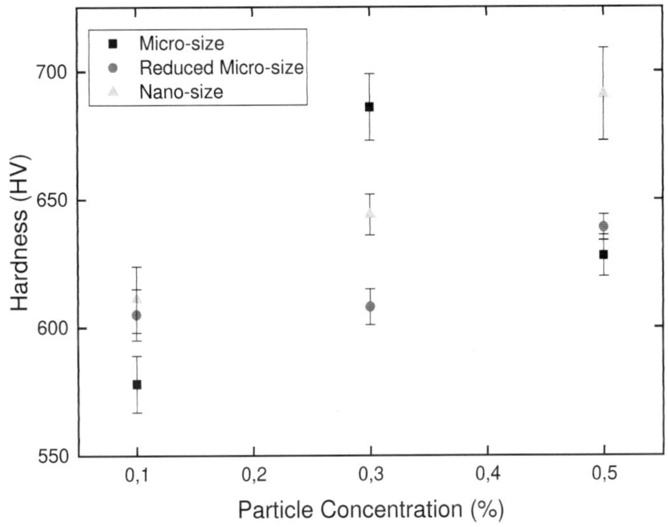

Figure 4. The hardness value of steel quenched in a different medium

This result was in accordance with the initial hypothesis where the smaller the size of the fluid particles, the heat transfer will be evenly distributed and increasing the cooling rate. The higher cooling rate will result in more martensite phase formed in the steel.

4. Conclusions

The cooling rate of the quench medium with several variations of particle size was conducted. Micro-size (44 µm), reduced micro-size (29µm), and nano-size (21nm) were used as size variation and then dispersed in distilled water as the fluid base. Vickers hardness test showed that the highest steel hardness was archived by quenching in 0.5% nano-size TiO_2 particle water-based quench medium. In this condition, the steel hardness could reach up to 691 HV. This increment was quite significant. It was almost 100HV increment from steel quenched in normal water-based quench which only had 598 HV. The hardness improvement may occur because of the TiO_2 nanoparticle, which has higher thermal conductivity, provide better heat transfer in the medium during quenching.

In other words, nanoparticle in the quench medium produces a faster cooling rate. Thus, more martensite phase was formed. Hence, the higher the hardness number. The smaller size and a higher percentage of the particle may affect the cooling rate as well.

5. Acknowledgments

The authors would like to thank Ms. Ghiska Ramahdita for the fruitful discussion, and gratefully acknowledged the financial support from Hibah PITTA 2018 funded by DRPM Universitas Indonesia No. 2537/UN2.R3.1/HKP.05.00/2018.

6. References

[1] Babu, K., Arularasan, R., & Srinath Ramkumar, S. (2017). *Quenching performance of AISI 1010 in CNT nanofluids*. Materials Today: Proceedings, *4*(10), 11044–11049. http://doi.org/10.1016/j.matpr.2017.08.065

[2] S.S. Yahya, S. Harjanto, W.N. Putra, G. Ramahdita, Kresnodrianto, and E. P. Mahiswara. *Characterization and observation of water-based nanofluids quench medium with carbon particle content variation,* AIP Conference. Proceedings 1964, 020006(2018);https://doi.org/10.1063/1.5038288

[3] J Župan, T Filetin & D Landek (2012) *The effect of TiO_2 nanoparticles on fluid quenching characteristics*, International Heat Treatment and Surface Engineering, 6:2, 56-60, DOI: 10.1179/1749514812Z.00000000014

[4] Ramesh, G. and Narayan K. Prabhu., "*Effect of Addition of Aluminum Nanoparticles on Cooling Performance and Quench Severity of Water during Immersion Quenching*." In STP1567-EB Nanofluids, edited by K. Prabhu, (pp. 122-135). West Conshohocken, PA: ASTM International, 2012. doi:10.1520/STP156720120008

[5] Ramesh, G. & Prabhu, N.K. *Review of thermo-physical properties, wetting and heat transfer characteristics of nanofluids and their applicability in industrial quench heat treatment.* Nanoscale Res Lett (2011) 6: 334. https://doi.org/10.1186/1556-276X-6-334

[6] Dhinesh K D and Valan A A. 2016. *A review on preparation, characterization, properties and applications of nanofluids*. (India: Elsevier) 21-40.

[7] Yu, W., France, D., Choi, S., & Routbort, J *Review and Assessment of Nanofluid Technology for Transportation and Other Application.*. 2007, Argonne: Energy Systems Division, Argonne National Laboratory.

[8] Akhgar, A., & Toghraie, D. (2018). An experimental study on the stability and thermal conductivity of water-ethylene glycol/TiO_2-MWCNTs hybrid nanofluid: Developing a new correlation. *Powder Technology*, *338*, 806–818. https://doi.org/10.1016/j.powtec.2018.07.086 Developing a new correlation. *Powder Technology*, *338*,806–818. https://doi.org/10.1016/j.powtec.2018.07.086

[9] A.M. Rashidi, M.M. Akbarnejad, A.A. Khodadadi, Y. Mortazavi, A. Ahmadpour, Single-wall carbon nanotubes synthesized using organic additives to Co-Mo catalysts supported on nanoporous MgO, Journal of Nanotechnology 18 (31) (2007). No.315605

Institute of Physics
Dirac House, Temple Back
Bristol BS1 6BE UK

ISSN: 1757-8981
ISBN 978-1-5108-9382-5

CURRAN ASSOCIATES INC.
proceedings.com